AN

ILLUSTRATED FLORA

OF THE

NORTHERN UNITED STATES

AND CANADA

FROM NEWFOUNDLAND TO THE PARALLEL OF THE SOUTHERN BOUNDARY OF
VIRGINIA, AND FROM THE ATLANTIC OCEAN WESTWARD
TO THE 102D MERIDIAN

BY

NATHANIEL LORD BRITTON, Ph.D., Sc.D., LL.D.

DIRECTOR-IN-CHIEF OF THE NEW YORK BOTANICAL GARDEN; PROFESSOR IN COLUMBIA UNIVERSITY

AND

HON. ADDISON BROWN, A.B., LL.D.

PRESIDENT OF THE NEW YORK BOTANICAL GARDEN

THE DESCRIPTIVE TEXT

CHIEFLY PREPARED BY PROFESSOR BRITTON, WITH THE ASSISTANCE OF SPECIALISTS IN
SEVERAL GROUPS; THE FIGURES ALSO DRAWN UNDER HIS SUPERVISION

SECOND EDITION—REVISED AND ENLARGED

IN THREE VOLUMES

VOL. III.

GENTIANACEAE TO COMPOSITAE

GENTIAN TO THISTLE

DOVER PUBLICATIONS, INC., NEW YORK

CONTENTS OF VOLUME III.

Gamopetalae (continued)   1–560

ENGLISH FAMILY NAMES

Petals wholly or partly united, rarely separate or wanting (continued)   1–560

Published in Canada by General Publishing Company, Ltd., 30 Lesmill Road, Don Mills, Toronto, Ontario.

Published in the United Kingdom by Constable and Company, Ltd., 10 Orange Street, London WC 2.

This Dover edition, first published in 1970, is an unabridged and unaltered republication of the second revised and enlarged edition as published by Charles Scribner's Sons in 1913 under the title *An Illustrated Flora of the Northern United States, Canada and the British Possessions*.

International Standard Book Number: 0-486-22644-1
Library of Congress Catalog Card Number: 76-116827

Manufactured in the United States of America
Dover Publications, Inc.
180 Varick Street
New York, N.Y. 10014

° is used after figures to indicate feet.
′ is used after figures to indicate inches.
″ is used after figures to indicate lines, or twelfths of an inch.
˘ over syllables indicates the accent, and the *short* English sound of the vowel.
ˋ over syllables indicates the accent, and the long, broad, open or close English sound of the vowel.

IN THE METRIC SYSTEM.
The metre = 39.37 inches, or 3 feet 3.37 inches.
The decimetre = 3.94 inches.
The centimetre = ⅖ of an inch, or 4¾ lines.   ⎫ very nearly
The millimetre = ₂₅ of an inch, or ½ a line.   ⎬
2⅕ millimeters = 1 line.   ⎭

Family 15. GENTIANÀCEAE Dumort. Anal. Fam. 20. 1829.

GENTIAN FAMILY

Bitter mostly quite glabrous herbs, with opposite (rarely verticillate) exstipulate entire leaves, reduced to scales in *Bartonia,* and regular perfect flowers in terminal or axillary clusters, or solitary at the ends of the stem or branches. Calyx inferior, persistent, 4–12-lobed, -toothed or -divided (of 2 sepals in *Obolaria*), the lobes imbricated or not meeting in the bud. Corolla gamopetalous, funnelform, campanulate, club-shaped or rotate, often marcescent, 4–12-lobed or -parted, the lobes convolute or imbricated in the bud. Stamens as many as the lobes of the corolla, alternate with them, inserted on the tube or throat; anthers 2-celled, longitudinally dehiscent; filaments filiform, or dilated at the base. Disk none, or inconspicuous. Ovary superior in our genera, 1-celled or partly 2-celled; ovules numerous, anatropous or amphitropous; style simple, or none; stigma entire, or 2-lobed, or 2-cleft. Capsule mostly dehiscent by 2 valves. Seeds globose, angular or compressed; endosperm fleshy, copious; embryo small, terete or conic.

About 70 genera and 700 species, widely distributed, most abundant in temperate regions.
Leaves normal; corolla-lobes convolute in the bud.
  Style filiform; anthers usually twisting or recurving when old.
    Corolla salverform.   1. *Centaurium.*
    Corolla rotate.   2. *Sabbatia.*
    Corolla campanulate-funnelform.   3. *Eustoma.*
  Style short, stout or none; anthers remaining straight.
    Corolla without nectariferous pits, glands or scales.
      Corolla funnelform, campanulate or clavate.
        Corolla without plaits in the sinuses; calyx without an interior membrane.
                                                                          4. *Gentiana.*
        Corolla with plaits in the sinuses; calyx with an interior membrane. 5. *Dasystephana*
      Corolla rotate.   6. *Pleurogyna.*
    1–2 nectariferous pits, glands or scales at the base of each corolla-lobe.
      Corolla rotate, a fringed gland at each lobe.   7. *Frasera.*
      Corolla campanulate, spurred at the base.   8. *Halenia.*
Leaves, at least those of the stem, reduced to scales; corolla-lobes imbricated in the bud.
  Calyx of 2 foliaceous spatulate sepals; upper leaves normal.   9. *Obolaria.*
  Calyx of 4 lanceolate sepals; leaves all reduced to scales.   10. *Bartonia.*

1. CENTAÙRIUM Hill. Brit. Herb. 62. 1756.

[ERYTHRAEA Neck. Elem. 2: 10. 1790.]

Herbs, mostly annual or biennial, with sessile or amplexicaul leaves, and small or middle sized, commonly numerous, pink, white or yellow flowers in cymes or spikes. Calyx tubular, 5–4-lobed or -divided, the lobes or segments narrow, keeled. Corolla salverform, 5–4-lobed, the tube long or short, the lobes spreading, contorted, convolute in the bud. Stamens 5 or 4, inserted on the corolla-tube; filaments short-filiform; anthers linear or oblong, becoming spirally twisted. Ovary 1-celled, the placentae sometimes intruded; style filiform; stigma 2-lobed. Capsule oblong-ovoid or fusiform, 2-valved. Seed-coat reticulated. [Latin, 100 gold pieces, with reference to its supposed medicinal value.]

About 25 species, natives of the Old World, western North and South America, and in the West Indies. Besides the following, about 8 others occur in the western and southwestern parts of the United States. Type species: *Gentiana Centaurium* L.
Flowers spicate-racemose.   1. *C. spicatum.*
Flowers cymose or cymose-paniculate.   2. *C. Centaurium.*
  Basal leaves tufted.
  No tuft of basal leaves.
    Corolla-lobes 1½″–2½″ long.

1

Flowers short-pedicelled; naturalized species.   3. *C. pulchellum.*
Flowers slender-pedicelled; native western species.
   Leaves oblong to linear-oblong.   4. *C. exaltatum.*
   Upper leaves mere subulate bracts.   5. *C. texense.*
Corolla-lobes 3½"–5" long.   6. *C. calycosum.*

1. Centaurium spicàtum (L.) Fernald. Spiked Centaury. Fig. 3330.

Gentiana spicata L. Sp. Pl. 230. 1753.

Erythraea spicata Pers. Syn. 1: 283. 1805.

Centaurium spicatum Fernald, Rhodora 10: 54. 1908.

Annual, glabrous, erect, strict, usually branched, 6′–18′ high. Leaves oblong or lanceolate-oblong, sessile, obtusish at the apex, clasping at the base, ½′–1½′ long, 2″–7″ wide; flowers pink, sessile, distant and spicate-racemose on the mostly simple and leafless branches, about 8″ long; tube of the corolla somewhat longer than the subulate calyx-segments, 2–3 times as long as the linear-oblong lobes; capsule 4″–5″ long.

Coast of Nantucket, and at Portsmouth, Va. Naturalized from Europe. May–Sept.

2. Centaurium Centaùrium (L.) W. F. Wight. Lesser or European Centaury. Bitter-herb. Bloodwort. Fig. 3331.

Gentiana Centaurium L. Sp. Pl. 229. 1753.
Erythraea Centaurium Pers. Syn. 1: 283. 1805.
C. Centaurium W. F. Wight, Contr. Nat. Herb. 11: 449. 1906.

Annual, glabrous, erect, usually branched, 6′–15′ high. Leaves oblong, apex obtuse, the base narrowed; the lower forming a basal tuft, 1′–2½′ long, 3″–6″ wide; stem leaves smaller, distant, rounded at the sessile or slightly clasping base; flowers numerous, 6″–8″ long, nearly sessile, in compound terminal mostly dense bracteolate cymes; corolla-lobes obtuse, 2½″–3″ long, about as long as the calyx-segments and one-third to one-half as long as the corolla-tube; stigmas oval.

In waste places, Nova Scotia and Quebec to Massachusetts, Illinois and Michigan. Naturalized from Europe. Earth-gall. Sanctuary. June–Sept.

3. Centaurium pulchéllum (Sw.) Druce. Branching Centaury. Fig. 3332.

Gentiana pulchella Sw. Act. Holm. 1783: 84. f. 8, 9. 1783.

Gentiana ramosissima Vill. Hist. Pl. Dauph. 2: 530. 1787.

Erythraea ramosissima Pers. Syn. 1: 283. 1805.

Erythraea pulchella Fries, Novit. 74. 1828.

C. pulchellum Druce, Fl. Oxf. 342. 1897.

Annual, glabrous, much branched, 3′–8′ high. Leaves oval, ovate or lanceolate, the lower mostly obtuse, 3″–8″ long, the upper usually acutish or acute and smaller; no basal tuft of leaves; flowers pink, cymose-paniculate, all or nearly all of them short-pedicelled, 5″–6″ long; tube of the corolla 1½–2 times longer than the calyx-segments, its lobes oblong, obtuse, 1½″–2″ long; stigma oval; anthers oblong.

In fields and waste places, southern New York to Pennsylvania, Illinois and Maryland; also in the West Indies. Naturalized from Europe. June–Sept.

4. Centaurium exaltàtum (Griseb.) W. F. Wight. Tall or Western Centaury. Fig. 3333.

Cicendia exaltata Griseb. in Hook. Fl. Bor. Am. **2**: 69. *pl. 157.* 1834.

Erythraea Douglasii A. Gray, Bot. Cal. **1**: 480. 1876.

Erythraea exaltata Coville, Contr. Nat. Herb. **4**: 150. 1893.

C. exaltatum W. F. Wight, Contr. Nat. Herb. **11**: 449. 1906.

Annual, erect, glabrous, branched, 6′–18′ high, the branches few, erect, slender. Leaves oblong or linear-oblong, sessile, mostly acute at both ends, 5″–10″ long, distant, the basal not tufted; flowers few, terminal and axillary, all slender-pedicelled, 6″–8″ long; tube of the corolla about one-third longer than the calyx-segments, the lobes oblong, obtuse, 1½″–2″ long.

In sandy soil, western Nebraska to Wyoming, Washington, Arizona and California. May–Sept.

5. Centaurium texénse (Griseb.) Fernald. Texan Centaury. Fig. 3334.

Erythraea texensis Griseb; Hook. Fl. Bor. Am. **2**: 58. 1838.

C. texense Fernald, Rhodora **10**: 54. 1908.

Annual, corymbosely branched above, slender, 2′–8′ high. Stem-leaves linear or linear-lanceolate, 8″ long or less, acute, sessile, the upper ones reduced to subulate bracts; pedicels slender, as long as the calyx or longer; calyx 4″–5″ long, its lobes subulate; corolla light rose color, its tube longer than the calyx, its acute oblong to oblong-lanceolate lobes about half as long as the tube; capsule longer than the calyx.

In rocky soil, Missouri to Texas. May–Sept.

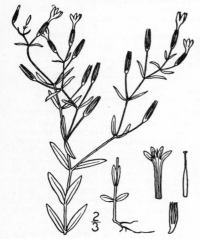

6. Centaurium calycòsum (Buckley) Fernald. Buckley's Centaury. Fig. 3335.

Erythraea calycosa Buckley, Proc. Acad. Phil. **1862**: 7. 1863.

Centaurium calycosum Fernald, Rhodora **10**: 54. 1908.

Annual, corymbosely branched or sometimes simple, 2° high or less, the branches ascending or spreading. Leaves oblong to spatulate or linear, ½′–1½′ long, acute, sessile; pedicels as long as the calyx or longer; calyx 4″–5″ long, its lobes narrowly linear; corolla pink, its tube a little longer than the calyx, its oval or oblong obtuse lobes nearly as long as the tube.

In wet or moist soil, Missouri to Texas, Mexico and New Mexico. April–June.

2. SABBÀTIA Adans. Fam. Pl. **2**: 503. 1763.

Annual or biennial erect usually branched glabrous herbs, with opposite or sometimes verticillate sessile or rarely petioled or clasping leaves, and rather large terminal and solitary or cymose pink rose or white flowers. Calyx 4–12-parted or -divided, the tube campanulate, sometimes very short, the lobes or segments usually narrow. Corolla rotate, deeply 4–12-parted. Stamens 4–12, inserted on the short tube of the corolla; filaments filiform, short;

anthers linear or oblong, curved, revolute or coiled in anthesis. Ovary 1-celled, the placentae intruded; style 2-cleft or 2-parted, its lobes filiform, stigmatic along their inner sides. Capsule ovoid or globose, 2-valved, many-seeded. Seeds small, reticulated. [In honor of L. and C. Sabbati, Italian botanists, according to Salisbury, Parad. Lond. *pl. 32,* therefore *Sabbatia,* though Adanson's spelling was *Sabatia.*]

About 18 species, natives of eastern North America, the West Indies and Mexico. Besides the following, some 6 others occur in the southern United States. Type species: *Chironia dodecandra* L.

Flowers normally 4–5-parted, sometimes 6–7-parted.
  Branches opposite.
    Style 2-parted to below the middle or nearly to the base; flowers white.
      Leaves lanceolate or ovate, acute; flowers 8″–12″ broad.   1. *S. lanceolata.*
      Leaves linear-oblong or lanceolate, obtuse; flowers 6″–9″ broad.   2. *S. paniculata.*
    Style 2-cleft to about the middle; flowers normally pink.
      Leaves linear-lanceolate, sessile; stem slightly 4-angled.   3. *S. brachiata.*
      Leaves ovate, cordate-clasping; stem strongly 4-angled.   4. *S. angularis.*
  Branches alternate, the lower sometimes opposite in nos. 5 and 6.
    Calyx-segments foliaceous, longer than the corolla.   5. *S. calycina.*
    Calyx-segments linear or lanceolate, not longer than the corolla.
      Calyx-tube 5-ribbed; flowers 1′–2′ broad; leaves ovate to oblong.   6. *S. campestris.*
      Calyx-tube scarcely ribbed; flowers 1′–1½′ broad; leaves linear to lanceolate.
        Calyx shorter than the corolla; style 2-parted.
          Flower pink with a yellow eye (rarely white); corolla-segments obovate.
                                                   7. *S. stellaris.*
          Flowers white, very numerous; corolla-segments spatulate or oblanceolate.
                                                    8. *S. Elliottii.*
      Calyx-segments filiform, as long as the corolla; style 2-cleft.   9. *S. campanulata.*
Flowers normally 8–12-parted, 1½′–2½′ broad.   10. *S. dodecandra.*

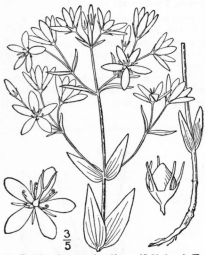

$\frac{3}{5}$

2. Sabbatia paniculàta (Michx.) Pursh. Branching Sabbatia. Fig. 3337.

C. paniculata Michx. Fl. Bor. Am. 1: 146. 1803.
S. paniculata Pursh, Fl. Am. Sept. 138. 1814.

Stem usually freely branching, 4-angled, 1°–2½° high, the branches all opposite. Leaves linear, linear-oblong, or lanceolate, obtuse, ½′–1½′ long, the lower commonly shorter and broader, the uppermost small and bract-like; flowers white, 6″–9″ broad, usually very numerous in corymbed cymes; pedicels mostly short, the central flowers of the cymes often nearly sessile; calyx-lobes linear, not more than one-half the length of the corolla; corolla-segments spatulate-oblong; anthers recurved or coiled; style 2-parted; capsule oblong, about 3″ high.

In dry or moist soil, Virginia to Florida. May–Sept.

1. Sabbatia lanceolàta (Walt.) T. & G. Lance-leaved Sabbatia. Fig. 3336.

Chironia lanceolata Walt. Fl. Car. 95. 1788.
S. lanceolata T. & G.; A. Gray, Man. 356. 1848.

Stem branched above, or simple, slender, somewhat 4-angled, or terete below, 1°–3° high, the branches all opposite. Leaves lanceolate to ovate, acute, or the lower sometimes obtuse, 3–5-nerved, 1′–2′ long, or the lowest shorter, the uppermost reduced to narrow bracts; flowers white, fading yellowish, 8″–12″ broad, usually numerous in bracteolate corymbed cymes; pedicels slender, 2″–7″ long; calyx-lobes filiform-linear, much shorter than the corolla; corolla-segments oblong or slightly obovate; anthers recurved; style 2-parted; capsule ovoid, about 3″ high.

In pine-barren swamps, New Jersey to Florida. May–Sept.

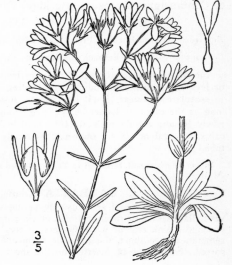

$\frac{3}{5}$

3. Sabbatia brachiàta Ell. Narrow-leaved Sabbatia. Fig. 3338.

Chironia angularis var. *angustifolia* Michx. Fl.
Bor. Am. 1: 146. 1803.
S. brachiata Ell. Bot. S. C. & Ga. 1: 284. 1817.
S. angustifolia Britton, Mem. Torr. Club 5: 259.
1894.

Stem slender, branched above, slightly 4-angled, 1°–2° high, the branches all opposite. Leaves linear, linear-oblong, or linear-lanceolate, obtuse, or the upper acute, sessile, 1′–2′ long, the uppermost small and bract-like; flowers pink with a yellowish or greenish eye, few in the racemed or short-corymbed cymes, or solitary at the ends of the branches, about 1′ broad; calyx-lobes linear, usually more than one-half the length of the corolla; corolla-segments obovate-oblong; style 2-cleft to about the middle; capsule oblong, 3″–4″ high.

In dry or moist soil, Indiana to Louisiana, east to North Carolina and Florida. May–Sept.

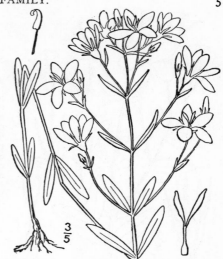

4. Sabbatia angulàris (L.) Pursh. Bitter-bloom. Rose-Pink. Square-stemmed Sabbatia. Fig. 3339.

Chironia angularis L. Sp. Pl. 190. 1753.
S. angularis Pursh, Fl. Am. Sept. 137. 1814.

Stem usually rather stout and much branched, sharply 4-angled, 2°–3° high, the branches all opposite or the lowest rarely alternate. Leaves ovate, acute at the apex, cordate-clasping at the base, 9″–18″ long, or the lower oblong and obtuse, those of the branches smaller; flowers rose-pink, with a central greenish star, occasionally white, 1′–1½′ broad, usually solitary at the ends of the branches; calyx-lobes linear, one-half the length of the corolla, or less; corolla-segments obovate; style 2-cleft; capsule oblong, about 3″ high.

In rich soil, often in thickets, New York and Pennsylvania to western Ontario, Michigan, Florida, Arkansas, Oklahoma and Louisiana. Bitter clover. Pink-bloom. American centaury. July–Aug. Flowers fragrant.

5. Sabbatia calycìna (Lam.) Heller. Coast Sabbatia. Fig. 3340.

Gentiana calycina Lam. Encycl. 2: 638. 1786.
C. calycosa Michx. Fl. Bor. Am. 1: 146. 1803.
Sabbatia calycosa Pursh, Fl. Am. Sept. 138. 1814.
S. calycina Heller, Bull. Torr. Club 21: 24. 1894.

Stem somewhat 4-angled, freely branched, 6′–12′ high, the branches alternate, or the lowest sometimes opposite. Leaves oblong or some of them slightly obovate, obtuse or acute, 3-nerved, 1′–2′ long, narrowed to the sessile base or the lower into petioles; flowers usually few, solitary at the ends of the branches or peduncles, 1′–1½′ broad; calyx-lobes linear or spatulate, leaf-like, longer than the spatulate segments of the pink rose-purple or whitish corolla; style 2-parted, capsule ovoid-oblong, 3″–4″ high.

In moist soil, Virginia to Florida, near the coast. Cuba; Santo Domingo. June–Aug.

6. Sabbatia campéstris Nutt. Prairie Sabbatia. Fig. 3341.

Sabbatia campestris Nutt. Trans. Amer. Phil. Soc. (II.) **5**: 197, 1833–37.

Stem 4-angled, branched, 6′–15′ high, the branches alternate or dichotomous, or the lowest opposite. Leaves ovate, oblong or lanceolate, mostly obtuse at the apex, sessile or slightly clasping and subcordate at the base, ½′–1′ long; flowers solitary at the ends of the branches and peduncles, 1′–2′ broad; peduncles, 1′–2′ long; calyx 5-ribbed or almost 5-winged, its lobes lanceolate, acute, 6″–12″ long, about as long as the obovate lilac corolla-segments, becoming rather rigid in fruit; style 2-cleft; capsule oblong, 3″–4″ high.

On prairies, Missouri and Kansas to Texas.

7. Sabbatia stellàris Pursh. Sea or Marsh Pink. Fig. 3342.

Sabbatia stellaris Pursh, Fl. Am. Sept. **1**: 137. 1814.
Chironia stellata Muhl. Cat. Ed. 2, 23. 1818.

Stem nearly terete, or slightly 4-angled, usually much branched, 6′–2° high, the branches alternate. Leaves lanceolate-oblong to linear, sessile, ½′–2′ long, 2″–5″ wide, or the lowest smaller, somewhat obovate and obtuse, the uppermost narrowly linear and bract-like; flowers 9″–18″ broad, usually numerous, solitary at the ends of the branches or slender peduncles; calyx not ribbed, its lobes narrowly linear, shorter than or nearly equalling the oblong or obovate corolla-segments; corolla pink to white with a yellowish starry eye bordered with red; style 2-cleft to below the middle; capsule about 2½″ high.

In salt-meadows, coast of Massachusetts to Florida. Recorded from Maine. Called also Rose-of-Plymouth. July–Sept.

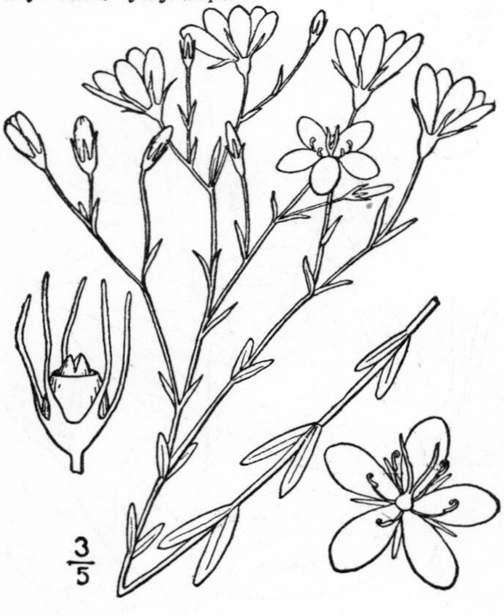

8. Sabbatia Ellióttii Steud. Elliott's Sabbatia. Fig. 3343.

Swertia difformis L. Sp. Pl. 226. 1753?

Sabbatia paniculata Ell. Bot. S. C. & Ga. **1**: 282. 1817. Not Pursh, 1814.

S. Elliottii Steud. Nomencl. Ed. 2, **2**: 489. 1841.

Stem paniculately branched, terete or slightly ridged, 1°–2° high, the branches alternate. Lower leaves obovate or lanceolate-oblong, obtuse, 6″–9″ long, the upper and those of the branches narrowly linear or subulate; flowers very numerous, white, solitary at the ends of the branches and short peduncles, 1′ broad or less; calyx not ribbed, its lobes subulate-linear, one-half as long as the spatulate or oblanceolate segments of the corolla, or less; style 2-parted; capsule about 2½″ high.

In pine-barrens, Virginia and North Carolina to Florida. Quinine-flower. July–Sept.

9. Sabbatia campanulàta (L.) Torr.   Slender Marsh Pink.   Fig. 3344.

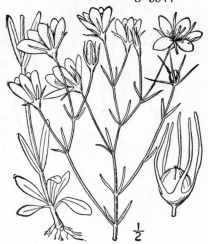

Chironia campanulata L. Sp. Pl. 190. 1753.
Chironia gracilis Michx. Fl. Bor. Am. 1: 146. 1803.
Sabbatia gracilis Salisb. Parad. Lond. *pl. 32.* 1806.
Sabbatia campanulata Torr. Fl. U. S. 1: 217. 1824.

Similar to the preceding species. Stem usually very slender and much branched, 1°–2° high, the branches alternate. Leaves linear, or linear-lanceolate, 1′–1½′ long, sessile, acute, or the lowest much shorter, obtuse, oblong or oblanceolate, sometimes narrowed into short petioles, the uppermost almost filiform; flowers pink with a yellow eye, about 1′ broad, solitary at the ends of the branches and peduncles, mostly 5-parted; calyx-lobes filiform-linear, equalling the oblong-obovate corolla-segments, or somewhat shorter; style 2-cleft to about the middle; capsule obovoid, about 2½″ high.

In salt marshes and along brackish rivers, rarely in fresh-water swamps, Nantucket to Florida and Louisiana. Also on the summits of the southern Alleghanies. Bahamas; Cuba. May–Aug.

10. Sabbatia dodecándra (L.) B.S.P.   Large Marsh Pink.   Fig. 3345.

Chironia dodecandra L. Sp. Pl. 190. 1753.
Chironia chloroides Michx. Fl. Bor. Am. 1: 147. 1803.
Sabbatia chloroides Pursh, Fl. Am. Sept. 138. 1814.
Sabbatia dodecandra B.S.P. Prel. Cat. N. Y. 36. 1888.

Stem 1°–2° high, little branched or simple, terete or nearly so, the branches alternate. Basal leaves spatulate, obtuse, 1½′–3′ long; stem leaves lanceolate or oblong-lanceolate, acute, the uppermost usually narrowly linear; flowers few, pink, sometimes white, solitary at the ends of the branches or peduncles, 1½′–2½′ broad; calyx-lobes narrowly linear, about one-half as long as the 8–12 spatulate-obovate corolla-segments; anthers coiled; style deeply 2-cleft, its divisions clavate; capsule globose-oval, 3″ high.

In sandy borders of ponds and along salt marshes, Massachusetts to North Carolina, near the coast. July–Sept. Plants of the Gulf States, previously referred to this species, prove to be distinct.

3. EUSTÒMA Salisb. Parad. Lond. *pl. 34.* 1806.

Erect usually branched glaucous annual herbs, with opposite sessile or clasping entire leaves. Flowers large, blue, purple or white, long-peduncled, axillary and terminal, solitary or paniculate. Calyx deeply 5–6-cleft, the lobes lanceolate, acuminate, keeled. Corolla broadly campanulate, deeply 5–6-lobed, the lobes oblong or obovate, usually erose-denticulate, convolute in the bud. Stamens 5–6, inserted on the throat of the corolla; filaments filiform; anthers oblong, versatile, at length recurved, or remaining nearly straight. Ovary 1-celled; style filiform; stigma 2-lamellate. Capsule oblong or ovoid, 2-valved. Seeds small, numerous, foveolate. [Greek, open-mouth, referring to the corolla.]

Four species, natives of the southern United States, New Mexico and the West Indies. Type species: *Eustoma silenifolium* Salisb.

1. Eustoma Russelliànum (Hook.) Griseb.
Russell's Eustoma.   Fig. 3346.

Lisianthus Russellianus Hook. Bot. Mag. *pl. 3626.* 1839.

Lisianthus glaucifolius Nutt. Trans. Am. Phil. Soc. (II.) 5: 197. 1833–37. Not. Jacq. 1786.

E. Russellianum Griseb. in DC. Prodr. 9: 51. 1845.

Stem terete, rather stout, 1°–2½° high. Leaves oblong or ovate-oblong, 3–5-nerved, the upper usually acute at the apex, cordate-clasping at the base, 1½′–3′ long, the lower usually obtuse at the apex and narrowed to a sessile or slightly auricled base; peduncles 1′–4′ long, stout, bracted at the base, the bracts lanceolate-subulate, small; flowers 2′–3′ broad; calyx-lobes long-acuminate, shorter than the purple corolla; corolla-lobes obovate, about 4 times as long as the tube; style slender; anthers remaining nearly straight; capsule oblong, pointed.

On prairies, Nebraska to Louisiana, Colorado, Texas, New Mexico and Mexico. Canada pest. May–Aug.

4. GENTIÀNA [Tourn.] L. Sp. Pl. 227. 1753.

Erect mostly glabrous herbs, with opposite or rarely verticillate, entire sessile or short-petioled leaves. Flowers blue, purple, yellow or white, solitary or clustered, terminal or axillary. Calyx tubular, 4–7- (usually 5-) cleft. Corolla tubular, clavate, campanulate, salverform or funnelform (rotate in some exotic species), 4–7-lobed, often gland-bearing within, the lobes entire or fimbriate. Stamens as many as the lobes of the corolla and inserted in its tube, included; anthers connate into a tube, or separate, not recurved or coiled. Ovary 1-celled; ovules very numerous; style short or none; stigma cleft into 2 lamellae. Capsule sessile or stipitate, 2-valved. Seeds numerous, sometimes covering the whole inner wall of the capsule, wingless or winged. [Named for King Gentius of Illyria.]

About 150 species, mostly natives of the north temperate and arctic zones and the Andes of South America. Besides the following, some 20 others occur in the western parts of North America. Type species: *Gentiana lutea* L.

Corolla-lobes fringed or serrate; flowers 1′–3′ long.
  Corolla enclosed in the swollen wing-angled calyx.   1. *G. ventricosa.*
  Corolla conspicuously longer than the wingless calyx.
    Corolla-lobes fringed all around their summits; leaves lanceolate.   2. *G. crinita.*
    Corolla-lobes fringed mainly on the margins; leaves linear.   3. *G. procera.*
Corolla-lobes with entire or rarely denticulate margins; flowers 6″–12″ long.
  Corolla-lobes fimbriate-crested at the base, acute.   4. *G. acuta.*
  Corolla-lobes naked, subulate-acuminate.
    Calyx-lobes unequal; flowers mostly 4-parted.   5. *G. propinqua.*
    Calyx-lobes equal; flowers 5-parted.   6. *G. quinquefolia.*

1. Gentiana ventricòsa Griseb.   Swollen Gentian.   Fig. 3347.

Gentiana ventricosa Griseb. in Hook. Fl. Bor. Am. 2: 65. *pl. 152.* 1834.

Annual; stem strict, terete, branched above, about 1° high. Basal leaves obovate, small, those of the stem ovate-oblong, obtuse or acute at the apex, rounded or subcordate at the base, 1′ long, 4″–6″ wide; flowers few, solitary at the ends of stout peduncles, 4-parted; calyx inflated, ovoid, about 9″ high, 4″–5″ thick, wing-angled, enclosing the corolla; corolla-lobes oblong, obtuse, lacerate-serrate; ovary elliptic-oblong.

Grand Rapids of Saskatchewan, between Cumberland House and Hudson Bay (Drummond). Not recently collected.

2. Gentiana crinìta Froel. Fringed Gentian. Fig. 3348.

Gentiana crinita Froel. Gen. 112. 1796.

Annual or biennial; stem leafy, usually branched, 1°–3¼° high, the branches erect, somewhat 4-angled. Basal and lower leaves obovate, obtuse, the upper lanceolate or oblong-lanceolate, 1′–2′ long, acute or acuminate at the apex, sessile by a rounded or subcordate base; flowers mostly 4-parted, several or numerous, about 2′ high, solitary at the ends of the usually elongated peduncles; calyx-lobes lanceolate, acuminate, unequal, their midribs decurrent on the angles of the tube; corolla bright blue, rarely white, narrowly campanulate, its lobes obovate, rounded, conspicuously fringed all around their summits, scarcely fringed on the sides, spreading when mature; capsule spindle-shaped, stipitate; seeds scaly-hispid.

In moist woods and meadows, Quebec to Ontario, Minnesota, south to Georgia and Iowa. Sept.–Oct.

3. Gentiana prócera Holm. Smaller Fringed Gentian. Fig. 3349.

Gentiana procera Holm, Ottawa Nat. **15** : 11. 1901.

Annual; similar to the preceding species but smaller; stem simple, or little branched, 3′–18′ high. Basal and lower leaves spatulate, obtuse, the upper linear or linear-lanceolate, 1′–2½′ long, 2″–4″ wide; flowers 1–6, solitary at the ends of elongated erect peduncles, mostly 4-parted, about 1½′ high; calyx-lobes lanceolate, acuminate, their midribs decurrent on the tube; corolla narrowly campanulate, bright blue, its lobes spatulate-oblong, strongly fringed on both sides, entire or somewhat fimbriate or toothed around the apex; capsule short-stipitate; seeds scaly-hispid.

In wet places, New York and Ontario to Minnesota, South Dakota and Manitoba. July–Sept. Previously confused with *G. detonsa* Rottb. and with *G. serrata* Gunner, Old World species. Northwestern and Rocky Mountain plants formerly referred to this species prove to be distinct from it.

Gentiana nesóphila Holm, of Anticosti, differs in merely denticulate corolla-lobes.

4. Gentiana acùta Michx. Northern Gentian. Fig. 3350.

Gentiana acuta Michx. Fl. Bor. Am. **1** : 177. 1803.
Gentiana Amarella var. *acuta* Herder, Act. Hort. Petrop. **1** : 428. 1872.

Annual; stem leafy, slightly wing-angled, simple or branched, 6′–20′ high. Basal and lower leaves spatulate or obovate, obtuse, the upper lanceolate, acuminate or acute at the apex, rounded or subcordate at the base, sessile, or somewhat clasping, ½′–2′ long; flowers numerous, racemose-spicate, 5″–8″ high, the pedicels 2″–6″ long, leafy-bracted at the base; calyx deeply 5-parted (rarely 4-parted), its lobes lanceolate; corolla tubular-campanulate, 5-lobed (rarely 4-lobed), blue, its lobes lanceolate, acute, each with a fimbriate crown at the base; capsule sessile.

In moist or wet places, Labrador to Alaska, Maine, Minnesota, south in the Rocky Mountains to Arizona and Mexico. Also in Europe and Asia. Closely resembles the Old World *G. Amarella.* Felwort. Bastard-gentian. Baldmoney. Summer.

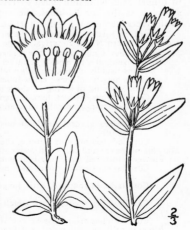

5. Gentiana propínqua Richards.   Four-parted
Gentian.   Fig. 3351.

Gentiana propinqua Richards. Frank. Journ. 734. 1823.

Stems slender, usually branched from the base and
sometimes also above, slightly wing-angled, 2′–7′ high.
Basal leaves spatulate, obtuse, the upper distant, oblong
or lanceolate, ½′–1½′ long, acute or obtusish at the apex,
rounded at the slightly clasping base, obscurely 3-nerved;
flowers few or several, solitary on slender peduncles,
mostly 4-parted, 8″–10″ high; calyx 4-cleft, 2 of the
lobes oblong, the others linear-lanceolate; corolla blue
or bluish, glandular within at the base, its lobes ovate-
lanceolate, entire or denticulate; capsule linear, at length
a little longer than the corolla.

Labrador to Alaska and British Columbia.   Summer.

6. Gentiana quinquefòlia L.   Stiff Gentian.
Ague-weed.   Fig. 3352.

Gentiana quinquefolia L. Sp. Pl. 230. 1753.
Gentiana quinqueflora Lam. Encycl. 2: 643. 1786.

Annual; stem ridged, usually branched, 2′–2½°
high, quadrangular. Basal leaves spatulate, obtuse,
those of the stem ovate, acute at the apex, clasping
and subcordate or rounded at the base, 3–7-nerved,
½′–2′ long; flowers in clusters of 1–7 at the ends
of the branches, or also axillary; pedicels slender,
2″–7″ long; calyx one-fourth to one-third the length
of the corolla, its lobes narrow, or sometimes foliaceous, acute; corolla tubular-funnelform, 5″–10″
long, blue or yellowish, glandular within at the base,
its lobes triangular, very acute, entire; capsule stipitate; seeds globose, wingless.

In dry or moist soil, Maine and Ontario to Michigan,
south to Florida and Missouri. Ascends to 6300 ft. in North Carolina. Consists of several or
many races, differing in size of plant, size of corolla; the calyx-lobes narrow or broad. Five-
flowered gentian. Gall-of-the-Earth. Gall-weed. Aug.–Oct.

5. DASYSTÉPHANA [Reneal.] Adans. Fam. Pl. **2**: 502. 1763.

Mostly perennial herbs with opposite entire leaves, and large sessile or short-stalked
flowers in terminal and axillary clusters, or sometimes solitary, usually 2-bracted under the
calyx. Calyx mostly 5-lobed, with an interior membrane. Corolla mostly 5-lobed, clavate
to funnelform, with thin lobed or toothed plaits in the sinuses, but without glands. Stamens
as many as the corolla-lobes, the anthers cohering in a ring or separate. Ovary 1-celled;
ovules very numerous; style short; capsule stipitate. [Greek, rough garland.]

About 75 species, natives of the north temperate zone. Besides the following, several others
occur in western North America. Type species: *Gentiana asclepiadea* L.

Margins of leaves and calyx-lobes scabrous or ciliate; seeds winged.
 Anthers separate, or merely connivent.
  Stems usually clustered; calyx-lobes unequal; corolla narrowly funnelform.   1. *D. affinis.*
  Stems mostly solitary; calyx-lobes equal; corolla campanulate-funnelform.   2. *D. puberula.*
 Anthers cohering in a ring or short tube.
  Corolla-lobes distinct, longer than or equalling the plaits.
   Flowers 1–4; corolla campanulate-funnelform, its lobes 2–3 times as long as the plaits.
                                                          3. *D. parvifolia.*
   Flowers several or numerous; corolla club-shaped, its lobes not much longer than the
    plaits.   4. *D. Saponaria.*
  Corolla-lobes none or minute, the plaits very broad.   5. *D. Andrewsii.*
Margins of leaves and calyx-lobes smooth or nearly so.
 Flowers clustered, sessile, 2-bracteolate under the calyx.
  Corolla-lobes ovate, twice as long as the plaits; leaves broad, acuminate; seeds winged.
                                                          6. *D. flavida.*
  Corolla-lobes rounded, little longer than the plaits; leaves narrow; seeds winged.
                                                          7. *D. linearis.*
  Corolla-lobes ovate, acute, much longer than the broad plaits; leaves broad; seeds winged.
                                                          8. *D. Grayi.*
  Corolla-lobes triangular-lanceolate; leaves obovate; seeds wingless.   9. *D. villosa.*
 Flowers solitary, pedunculed, not bracteolate; leaves linear.   10. *D. Porphyrio.*

1. Dasystephana affinis (Griseb.) Rydb. Oblong-leaved Gentian. Fig. 3353.

Gentiana affinis Griseb. in Hook. Fl. Bor. Am. **2**: 56. 1834.
D. affinis Rydb. Bull. Torr. Club **33**: 149. 1906.

Perennial; stems clustered from deep roots, minutely puberulent, simple, 6′–18′ high. Leaves linear-oblong to lanceolate-oblong, obtuse or acutish, rounded or narrowed at the base, firm, roughish-margined, indistinctly nerved, ½′–1½′ long, the floral smaller; flowers few, numerous, or rarely solitary, 5-parted, sessile and solitary or clustered in the axils of the upper leaves, about 1′ high, not bracted under the calyx; calyx-lobes linear or subulate, unequal, the longer about equalling the tube, the smaller sometimes minute; corolla narrowly funnelform, blue, its lobes ovate, acute or mucronate, entire, spreading, with laciniate appendages in the sinuses; anthers separate; seeds broadly winged.

In moist soil, Minnesota to British Columbia, south in the Rocky Mountains to New Mexico. Aug.–Oct.

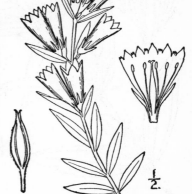

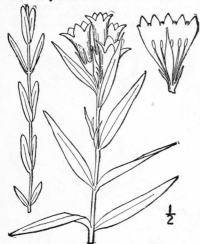

2. Dasystephana pubérula (Michx.) Small. Downy Gentian. Fig. 3354.

Gentiana puberula Michx. Fl. Bor. Am. **1**: 176. 1803.
D. puberula Small, Fl. SE. U. S. 930. 1903.

Perennial; stems usually solitary, leafy, 8′–18′ high, minutely puberulent or glabrous, nearly terete. Leaves firm, lanceolate, or the lower oblong, roughish-margined, indistinctly nerved, pale beneath, narrowed or rounded at the base, 1′–3′ long; flowers sessile or nearly so in the upper axils, rarely solitary and terminal, 2-bracteolate under the calyx, 1½′–2′ high; calyx-lobes linear-lanceolate, equal, about as long as the tube, rough; corolla campanulate-funnelform, 2–3 times as long as the calyx, its lobes ovate, entire, spreading, much longer than the cleft or laciniate appendages; anthers at first connivent, soon separate; seeds oblong, broadly winged.

On prairies, Maryland to Ohio, Minnesota, South Dakota, Georgia and Kansas. Aug.–Oct.

3. Dasystephana parvifòlia (Chapm.) Small. Elliott's Gentian. Fig. 3355.

?*Gentiana rigida* Raf. Med. Fl. **1**: 212. 1832.
Gentiana Elliottii Chapm. Fl. S. States, 356. 1860. Not Raf.
?*Gentiana scaberrima* Kusnezow, Act. Hort. Petrop. **13**: 59. 1893.
D. parvifolia Small, Fl. SE. U. S. 930. 1903.
D. latifolia Small, Fl. SE. U. S. 930. 1903.

Perennial; roots a cluster of thick fibres; stems slender, leafy, terete, minutely rough-puberulent, simple, 8′–2° high. Leaves ovate or lanceolate, acute or acutish at the apex, rounded or narrowed at the base, thin, roughish-margined, 1′–2′ long, 3-nerved, or the lower much smaller and obtuse; flowers 1–4, terminal, or rarely 1 or 2 in the upper axils, about 2′ long, sessile, 2-bracteolate under the glabrous calyx; calyx-lobes oblong or lanceolate, foliaceous, longer than the tube; corolla campanulate-funleform, blue, its lobes ovate, obtuse, sometimes mucronate, entire, 3″–4″ long, about twice as long as the fimbriate or toothed appendages; anthers cohering in a tube; seeds oblong, broadly winged.

In moist soil, Virginia to Florida. Sampson snake-root. Sept.–Oct.

Gentiana decòra Pollard, of the same range, differs in having the calyx-tube pubescent.

4. **Dasystephana Saponària** (L.) Small.
Soapwort or Blue-Gentian.   Fig. 3356.

Gentiana Saponaria L. Sp. Pl. 228. 1753.
Gentiana Catesbaei Walt. Fl. Car. 109. 1788.
D. Saponaria Small, Fl. SE. U. S. 930. 1903.

Perennial; stem erect or ascending, terete, slender, simple or with short erect axillary branches, glabrous, or minutely rough-puberulent above, $1°-2\frac{1}{2}°$ high, leafy. Leaves lanceolate, ovate-lanceolate or oblong, usually acute at both ends, 3–5-nerved, roughish-margined, $2'-4'$ long, the lowest obovate and smaller; flowers $1'-2'$ high, in sessile terminal and usually also axillary clusters of 1–5, 2-bracteolate under the calyx; calyx-lobes oblong or spatulate, ciliolate; corolla blue, club-shaped, its lobes erect, obtuse, equalling or longer than the cleft or lacerate appendages; anthers cohering in a tube; capsule stipitate; seeds broadly winged.

In wet soil, Ontario to Minnesota, Connecticut, Florida and Louisiana. Calathian violet. Harvestbells. Rough or marsh-gentian. Sampson snakeroot. Aug.–Oct.

5. **Dasystephana Andréwsii** (Griseb.) Small.
Closed Blue or Blind Gentian.   Fig. 3357.

?Gentiana alba Muhl. Cat. Ed. 2, 29. 1818.
?Gentiana clausa Raf. Med. Fl. 1 : 210. 1832.
G. Andrewsii Griseb. in Hook. Fl. Bor. Am. 2 : 55. 1834.
D. Andrewsii Small, Fl. SE. U. S. 930. 1903.

Perennial; stout, glabrous, $1°-2°$ high, simple, leafy. Leaves ovate to lanceolate, 3–7-nerved, acuminate at the apex, narrowed or sometimes rounded at the base, $2'-4'$ long, rough-margined, the lowest oblong or obovate, smaller; flowers $1'-1\frac{1}{2}'$ high in a terminal sessile cluster and commonly 1 or 2 in the upper axils, 2-bracteolate under the calyx; calyx-lobes lanceolate or ovate, ciliolate, usually spreading; corolla oblong, club-shaped, blue, or occasionally white, nearly or quite closed, its lobes obsolete, the intervening appendages very broad, light colored, opposite the stamens; anthers cohering in a tube; capsule stipitate; seeds oblong, winged.

In moist soil, Quebec to Manitoba, Georgia and Nebraska. Cloistered-heart. Bottle- or barrel-gentian. Aug.–Oct.

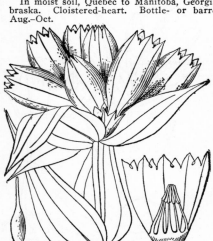

6. **Dasystephana flávida** (A. Gray) Britton.   Yellowish Gentian.   Fig. 3358.

G. flavida A. Gray, Am. Journ. Sci. (II.) 1 : 80. 1846.
G. alba A. Gray, Man. 360. 1848. Not Muhl. 1818.

Perennial; stem stout, terete, glabrous, simple, erect or ascending, $1°-3°$ high. Leaves ovate-lanceolate or oblong-lanceolate, smooth-margined, acuminate at the apex, subcordate, clasping or rounded at the base, $2'-5'$ long, $1'-2'$ wide; flowers several or numerous in a sessile terminal cluster and sometimes 1 or 2 in the upper axils, $1\frac{1}{2}'-2'$ high, 2-bracteolate under the calyx; calyx-lobes ovate, acute, smooth or minutely rough-margined, shorter than the tube, at length spreading; corolla campanulate-funnelform, open, greenish or yellowish white, its lobes ovate, acute or obtuse, twice as long as the erose-denticulate appendages; anthers cohering in a tube; capsule stipitate; seeds winged.

In moist soil, Ontario to Minnesota, Missouri, Virginia and Kentucky. Aug.–Oct.

7. Dasystephana lineàris (Froel.) Britton. Narrow-leaved Gentian. Fig. 3359.

Gentiana linearis Froel. Gent. 37. 1796.
Gentiana Saponaria var. *linearis* Griseb. in Hook. Fl. Bor. Am. **2**: 55. 1834.
Gentiana rubricaulis Schwein. in Keating's Narr. Long's Exp. **2**: 384. 1824.
Gentiana linearis var. *lanceolata* A. Gray, Syn. Fl. **2**: Part 1, 123. 1878.

Perennial; glabrous throughout; stem slender, terete, simple, 6′–2° high, leafy. Leaves linear or linear-lanceolate, smooth-margined, acute at both ends, 3-nerved, 1½′–3′ long, 2″–5″ wide; flowers 1′–1¾′ high in a terminal cluster of 1–5 and sometimes also in the upper axils; calyx-lobes linear or lanceolate, shorter than the tube; corolla erect, blue, oblong-funnelform, open, its lobes erect, rounded, 1″–2″ long, slightly longer than the entire or 1–2-toothed appendages; anthers coherent in a tube, or at length distinct; capsule stipitate; seeds winged.

In bogs and on mountains, New Brunswick and Ontario to Maryland and Minnesota. Ascends to 5000 ft. in the Adirondacks. Aug.–Sept.

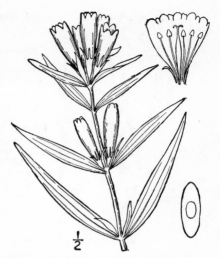

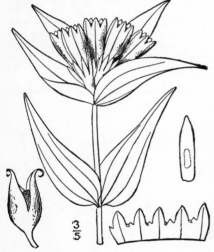

8. Dasystephana Gràyi (Kusnezow) Britton. Gray's Gentian. Fig. 3360.

Gentiana linearis var. *latifolia* A. Gray, Proc. Am. Acad. **22**: 309. 1887.
Gentiana Grayi Kusnezow, Act. Hort. Petrop. **13**: 59. 1893.

Perennial; glabrous; stem terete, 1°–2° high. Leaves rather distant, lanceolate or ovate-lanceolate, acuminate, rounded at the base, smooth-margined, 2′–3′ long, the lower narrower and obtuse; flowers about 1½′ high in a sessile terminal cluster, 2-bracteolate under the calyx; calyx-lobes unequal, the larger about as long as the tube; corolla greenish blue or bright blue, club-shaped, its lobes ovate, acute or acutish, much longer than the broad 1–2-toothed appendages; anthers coherent, or at length distinct; seeds oblong, winged.

In wet soil, New Brunswick to western Ontario, Michigan and Minnesota. Referred in our first edition to *Gentiana rubricaulis* Schwein., which proves to be a synonym of the preceding species. Recorded from central New York. Aug.–Sept.

9. Dasystephana villòsa (L.) Small. Striped Gentian. Fig. 3361.

Gentiana villosa L. sp. Pl. 228. 1753.
Gentiana ochroleuca Froel. Gent. 35. 1796.
D. villosa Small, Fl. SE. U. S. 931. 1903.

Perennial, glabrous or nearly so; stem simple, slender, terete, 6′–18′ high. Leaves obovate, obtuse or the upper acute, narrowed at the base, faintly 5-nerved, 1′–3′ long, the lower much smaller; flowers several in a terminal sessile cluster and sometimes also in the upper axils, nearly 2′ long, 2-bracteolate under the calyx; calyx-lobes unequal, linear, longer than the tube; corolla greenish white, striped within, oblong-funnelform, open, its lobes triangular-ovate or ovate-lanceolate, erect, much longer than the oblique entire or 1–2-toothed appendages; seeds oval, wingless.

In shaded places, southern New Jersey and Pennsylvania to Florida and Louisiana. Marsh- or straw-colored gentian. Sampson snake-root. Sept.–Nov.

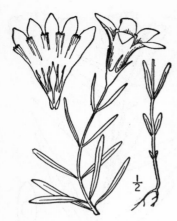

10. Dasystephana Porphýrio (J. F. Gmel.) Small.
One-flowered Gentian.　Fig. 3362.

Gentiana purpurea Walt. Fl. Car. 109.　1788.　Not. L. 1753.
Gentiana Porphyrio J. F. Gmel. Syst. 2 : 462.　1791.
Gentiana angustifolia Michx. Fl. Bor. Am. 1 : 177.　1803.
D. Porphyrio Small, Fl. SE. U. S. 931.　1903.

Perennial, glabrous; stem erect or ascending, simple or branched, 6′–18′ high.　Leaves linear, firm, acute or blunt at the apex, 1′–2′ long, 1″–2″ wide, or the upper and lower shorter; flowers solitary at the ends of the stem or branches, short-peduncled, about 2′ high, not bracteolate under the calyx; calyx-lobes linear, longer than the tube; corolla funnelform, bright blue, sometimes brown-dotted within, its lobes ovate, acutish, spreading, 5″–7″ long, three times as long as the conspicuously laciniate appendages, or more; seeds oblong, wingless.

　　In moist pine barrens, New Jersey to Florida.　Aug.–Oct.

6. PLEURÓGYNA Eschol. Linnaea 1 : 187.　1826.

　　Slender usually branched annual glabrous herbs, with opposite leaves, and rather large blue or white flowers in terminal narrow racemes or panicles, or solitary at the ends of the slender peduncles.　Calyx deeply 4–5-parted; segments narrow, often unequal.　Corolla rotate, 4–5-parted; lobes　vate or lanceolate, convolute, acute, with a pair of narrow appendages at the base.　Stamens 4 or 5, inserted on the corolla-tube near its base; filaments slender or filiform; anthers ovate, sagittate, straight.　Ovary 1-celled; ovules numerous; style none; stigma decurrent along the sutures of the ovary.　Capsule 2-valved.　Seeds small and numerous.　[Greek, referring to the lateral stigmatic surfaces.]

　　About 7 species, of the colder parts of the northern hemisphere, only the following typical one in North America.

1. Pleurogyna rotàta (L.) Griseb.　Marsh Felwort.
Fig. 3363.

Swertia rotata L. Sp. Pl. 226.　1753.

Pleurogyne rotata Griseb. Gent. 309.　1839.

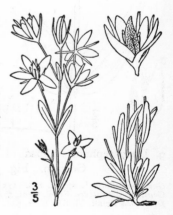

　　Stem erect, usually 6′–15′ high, sometimes lower, simple, or with nearly erect branches.　Leaves linear to lanceolate, ½′–2′ long, 1″–2″ wide, or the basal spatulate or oblong, shorter and sometimes broader; sepals narrowly linear to lanceolate, about the length of the lanceolate to oblong-lanceolate corolla-segments which are 4″–7″ long; capsule narrowly oblong.

　　Quebec, Greenland and Labrador to Alaska, south in the Rocky Mountains to Colorado.　Recorded from the White Mountains of New Hampshire.　Also in Europe and Asia.　Summer.

7. FRÀSERA Walt. Fl. Car. 87.　1788.

　　Perennial or biennial erect glabrous mostly tall herbs, with opposite or verticillate leaves, and rather large white yellowish or bluish flowers, in terminal cymose panicles or thyrses.　Calyx 4-parted, the segments narrow.　Corolla rotate, 4-parted, the lobes convolute in the bud, each bearing 1 or 2 fimbriate or fringed glands within.　Stamens 4, inserted on the base of the short corolla-tube; filaments subulate or filiform, sometimes united at the base; anthers oblong, versatile.　Ovary ovoid, 1-celled; style slender or short, but distinct; stigma 2-lobed or nearly entire.　Capsule ovoid, coriaceous, somewhat compressed, 2-valved, few-seeded.　Seeds flattened, smooth, margined or narrowly winged.　[In honor of John Fraser, a botanical collector.]

　　About 15 species, natives of North America, all but the following typical one far western.

1. Frasera carolinénsis Walt. American Columbo. Fig. 3364.

Frasera carolinensis Walt. Fl. Car. 87. 1788.

Stem 3°–7° high, stout. Leaves mostly verticillate in 4's, those of the stem lanceolate or oblong, acute or acutish, 3'–6' long, the basal ones spatulate or oblanceolate, usually much larger, the uppermost small and bract-like; inflorescence thyrsoid-paniculate, large, often 2° long; flowers slender-pedicelled, about 1' broad; corolla yellowish white with brown-purple dots, its lobes broadly oblong, bearing a large circular long-fringed gland at or below the middle; style 2''–3'' long; stigma 2-lobed; capsule much compressed, 6''–8'' high, longer than the calyx.

In dry soil, western New York and Ontario to Wisconsin, south to Georgia and Tennessee. Yellow gentian. Pyramid-flower or -plant. June–Aug.

8. HALÈNIA Borck. in Roem. Archiv 1: 25. 1796.
[Tetragonanthus S. G. Gmel. Fl. Sib. 4: 114. Hyponym. 1769.]

Annual or perennial usually tufted glabrous herbs, with opposite leaves, and middle-sized white yellowish purple or blue flowers in terminal and axillary often panicled cymes. Calyx deeply 4-cleft or 4-parted, the segments lanceolate or oblong. Corolla campanulate, 4–5-cleft, the lobes convolute in the bud, each with a hollow spur or projection below, which is glandular at the bottom within, or sometimes spurless. Stamens 4 or 5, inserted near the base of the corolla; filaments filiform or subulate; anthers oblong, versatile. Ovary 1-celled, the placentae more or less intruded; ovules numerous; style very short, sometimes none; stigma 2-lobed. Capsule ovoid or oblong, 2-valved. Seeds globose-ovoid to oblong, compressed, smooth. [In honor of Jonas Halen, 1727–1810, a pupil of Linnaeus.]

About 30 species, natives of mountainous regions of North America, South America and Asia. Besides the following, another occurs in the southwestern United States. Type species: *Halenia sibírica* Borck.

1. Halenia defléxa (J. E. Smith) Griseb. Spurred Gentian. Fig. 3365.

Swertia deflexa J..E. Smith in Rees' Cyclop. no. 8. 1816.
Halenia deflexa Griseb. in Hook. Fl. Bor. Am. 2: 67. *pl.* 155. 1834.
H. Brentoniana Griseb. in Hook. Fl. Bor. Am. 2: 68. 1834.
H. heterantha Griseb. loc. cit. 1834.
Tetragonanthus deflexus Kuntze, Rev. Gen. Pl. 431. 1891.

Annual or biennial; stem simple or branched, slender, erect, 6'–20' high, usually with long internodes. Basal leaves obovate or spatulate, obtuse, narrowed into petioles; stem-leaves ovate or lanceolate, acute, sessile, 3–5-nerved, 1'–2' long, the uppermost much smaller; calyx-segments lanceolate or spatulate, acute or acuminate; corolla purplish or white, about 4'' high, its lobes ovate, acute, the spurs deflexed or descending, one-fourth to one-half the length of the corolla or none; capsule narrowly oblong, 6''–7'' long, about twice as long as the calyx.

In moist woods and thickets, Newfoundland and Labrador to Massachusetts, New York, Saskatchewan, Montana, Michigan and South Dakota. Races differ in size of the plant and of the flowers and in the development of the corolla-spurs. Recorded from the "Indian Territory" (Oklahoma), apparently erroneously. July–Aug.

9. OBOLÀRIA L. Sp. Pl. 632. 1753.

A low glabrous perennial herb, the stem simple or branchèd, the lower leaves reduced to opposite scales, the upper foliaceous, subtending the racemose-spicate or thyrsoid white or purplish flowers. Calyx of 2 spatulate sepals. Corolla oblong-campanulate, 4-cleft, the lobes imbricated, at least in the bud. Stamens 4, inserted in the sinuses of the corolla; filaments slightly longer than the ovate sagittate anthers. Ovary 1-celled, with 4 internal placental projections; ovules numerous; style distinct; stigma 2-lamellate. Capsule ovoid, 2-valved or irregularly bursting. Seeds minute, covering the whole interior of the capsule. [Greek, obolus, a coin, alluding to the thick round leaves.]

A monotypic genus of eastern North America.

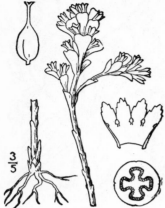

1. **Obolaria virgínica** L.   Pennywort.   Fig. 3366.

Obolaria virginica L. Sp. Pl. 632. 1753.

Stems 3'–6' high from a perennial base with thick fibrous roots, bearing 2–6 pairs of thick small obtuse scales in place of leaves. Floral leaves broadly obovate-cuneate, obtuse, purplish, 4"–6" long; flowers sessile or nearly so in the axils, in clusters of 1–4 (usually 3), and terminal; corolla about 5" long, cleft to about the middle, the lobes oblong, obtuse, entire, or denticulate; stamens included; capsule 2½" long.

In rich woods and thickets, New Jersey to Georgia, west to Illinois and Texas. Ascends to 2600 ft. in Virginia. April–May.

10. **BARTÒNIA** Muhl.; Willd. Neue Schrift. Ges. Nat. Fr. Berlin **3** : 444.   1801.

Slender or filiform erect glabrous annual or biennial herbs, the leaves reduced to minute opposite subulate scales, or some of them alternate, and white purplish or yellowish racemose or paniculate flowers, or these rarely solitary and terminal. Calyx deeply 4-parted, the segments lanceolate, acuminate, keeled. Corolla campanulate, deeply 4-cleft, the lobes imbricated in the bud. Stamens 4, inserted at the sinuses of corolla; filaments slender, longer than the ovate sagittate anthers. Ovary 1-celled, the placentae intruded; ovules numerous; style very short or none; stigma 2-lobed. Capsule ovoid-oblong, compressed, acute, 2-valved. Seeds minute, covering the whole inner surface of the capsule. [In honor of Professor Benjamin Smith Barton, 1766–1815, of Philadelphia.]

Four species, natives of eastern North America.   Type species: *Bartonia tenella* Willd.

Corolla yellowish, 1½"–2" long.
    Corolla-lobes oblong, abruptly tipped, erose.   1. *B. virginica.*
    Corolla-lobes lanceolate, acute or acuminate, entire.   2. *B. paniculata.*
Corolla white, 3"–4" long, its lobes spatulate.   3. *B. verna.*

1. **Bartonia virgínica** (L.) B.S.P.   Yellow Bartonia. Fig. 3367.

Sagina virginica L. Sp. Pl. 128. 1753.
B. tenella Willd. Neue Schrift. Ges. Nat. Fr. Berlin **3** : 445. 1801.
Bartonia virginica B.S.P. Prel. Cat. N. Y. 36. 1888.

Stem rather stiff, almost filiform, 4'–15' high, simple, or with few erect branches above, 5-angled, yellowish green, sometimes twisted. Subulate scales 1"–2" long, appressed, mostly opposite, the basal pairs close together, the upper distant; flowers mostly opposite; pedicels ascending or erect, 2"–6" long; corolla greenish yellow or whitish, 1½"–2" long, its lobes oblong, obtuse, denticulate or erose, somewhat exceeding the calyx; stamens included; ovary 4-sided; stigma about ½" long; capsule about 1½" long.

In moist soil, Nova Scotia to Florida, Michigan, Minnesota and Louisiana. Screw-stem. July–Sept.

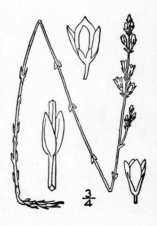

2. Bartonia paniculàta (Michx.) Robinson. Branched Bartonia. Fig. 3368.

Centaurella paniculata Michx. Fl. Bor. Am. 1: 98. 1803.
?*C. Moseri* Steud. & Hochst.; Griseb. Gent. 308. 1839
B. paniculata Robinson, Rhodora 10: 35. 1908.
B. lanceolata Small, Fl. SE. U. S. 932. 1903.

Slender, 8′–16′ high, mostly paniculately branched above, the branches and scales mainly alternate, the slender pedicels spreading or ascending. Corolla yellowish-white or greenish, about twice as long as the calyx, its lanceolate acute or acuminate lobes 1″–1½″ long; anthers yellow.

In wet soil, Massachusetts to Florida, Arkansas and Louisiana. Aug.–Oct.

Bartonia iodándra Robinson, of bogs in Newfoundland and Nova Scotia, has larger purplish flowers with broader ovate-lanceolate corolla-lobes and purple-brown anthers.

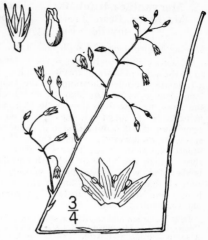

3. Bartonia vérna (Michx.) Muhl. White Bartonia. Fig. 3369.

Centaurella verna Michx. Fl. Bor. Am. 1: 97. *pl. 12. f. 1.* 1803.
Bartonia verna Muhl. Cat. 16. 1813.

Stem thicker and less rigid than that of the two preceding species, usually simple, 2′–15′ high, racemosely or corymbosely 1-several-flowered. Bracts distant, appressed, some of them commonly alternate; flowers solitary at the ends of the elongated erect peduncles, 3″–4″ long; peduncles sometimes 2′ long; corolla white, its lobes spatulate or obovate, obtuse, denticulate or entire, about 3 times the length of the calyx; ovary compressed; capsule about 2½″ high.

In moist sand, southern Virginia to Florida and Louisiana. March–May.

Family 16. MENYANTHÀCEAE G. Don, Gard. Dict. 4: 167. 1837.

BUCKBEAN FAMILY.

Perennial aquatic or marsh herbs, with basal or alternate entire crenate or 3-foliolate leaves, and clustered regular perfect flowers. Calyx inferior, deeply 5-parted, persistent. Corolla funnelform to rotate, 5-lobed or 5-cleft, the lobes induplicate-valvate, at least in the bud. Stamens 5, borne on the corolla, and alternate with its lobes; filaments mostly short; anther-sacs longitudinally dehiscent; pollen-grains 3-angled. Ovary 1-celled, the 2 placentae sometimes intruded; style long, short or none. Fruit a capsule, or indehiscent.

Five genera and about 35 species, widely distributed.

Leaves 3-foliolate; swamp plant.   1. *Menyanthes.*
Leaves simple, entire, cordate, floating.   2. *Nymphoides.*

1. MENYÁNTHES [Tourn.] L. Sp. Pl. 145. 1753.

A perennial glabrous marsh herb, with creeping rootstocks, alternate long-petioled 3-foliolate basal leaves, and white or purplish flowers, racemose or panicled on long lateral scapes or peduncles. Calyx 5-parted, the segments oblong or lanceolate. Corolla short-funnelform, 5-cleft, its lobes induplicate-valvate, fimbriate or bearded within, spreading. Stamens 5, inserted on the tube of the corolla; filaments filiform; anthers sagittate. Disk of 5 hypogynous glands. Ovary 1-celled; style subulate, persistent; stigma 2-lamellate. Capsule oval, indehiscent or finally rupturing. Seeds compressed-globose, shining. [Greek, perhaps month-flower, name used by Theophrastus.]

A monotypic genus of the cooler parts of the northern hemisphere.

1. Menyanthes trifoliàta L. Buckbean. Marsh or Bean Trefoil. Bog-bean or -myrtle. Fig. 3370.

Menyanthes trifoliata L. Sp. Pl. 145. 1753.

Rootstock thick, scaly, sometimes 1° long, marked by the scars of bases of former petioles. Leaves 3-foliolate; petioles sheathing at the base, 2′–10′ long; leaflets oblong or obovate, entire, obtuse at the apex, narrowed to the sessile base, pinnately veined, 1½′–3′ long; raceme borne on a long scape-like naked peduncle, arising from the rootstock, 10–20-flowered; pedicels stout, 3″–12″ long, bracteolate at the base; flowers 5″–6″ long; calyx shorter than the white or purplish corolla, which is bearded with white hairs within; stamens shorter than the corolla and style exserted, or longer and style nearly included; capsule ovoid, obtuse, about 4″ long.

In bogs, Greenland to Alaska, south to Long Island, Pennsylvania, Minnesota, Nebraska and California. Also in Europe and Asia. Water- or bitter trefoil. Water-shamrock. Moon-flower. Marsh-clover. Bitter worm. Bog-nut. Brook-bean. May–July.

2. NYMPHOÌDES Hill, Brit. Herb. 77. 1756.

[LIMNANTHEMUM S. G. Gmelin, Nov. Act. Acad. Petrop. **14**: 527. *pl. 17. f. 2.* 1769.]

Aquatic perennial herbs, with slender rootstocks. Leaves floating, petioled, ovate or orbicular, deeply cordate, entire or repand, or the primary ones different; flowers yellow or white, polygamous, umbellate at the summit of filiform stems at the bases of the petioles, or axillary, often accompanied by a cluster of thick elongated root-like tubers. Calyx 5-parted. Corolla nearly rotate, deeply 5-cleft, the lobes induplicate-valvate in the bud, sometimes fimbriate on the margins, glandular at the base. Stamens 5, inserted on the base of the corolla; filaments short; anthers sagittate, versatile. Ovary 1-celled; style short or none; stigma 2-lamellate. Capsule ovoid or oblong, indehiscent or irregularly bursting. Seeds numerous or few, smooth or rough. [Greek, like *Nymphaea*.]

About 20 species, widely distributed in temperate and tropical regions. The following are the only ones known to occur in North America. Type species: *Nymphoides flava* Hill.

Flowers white, accompanied by tufts of root-like tubers; native species.
  Floating leaves 1′–2′ long; flowers 3″–6″ broad; seeds smooth.   1. *N. lacunosum.*
  Floating leaves 2′–6′ long; flowers 6″–10″ broad; seeds rough.   2. *N. aquaticum.*
Flowers not accompanied by tufts of tubers; corolla bright yellow, 1′ broad or more; introduced species.   3. *N. nymphaeoides.*

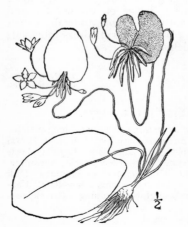

1. Nymphoides lacunòsum (Vent.) Kuntze. Floating Heart. Fig. 3371.

Villarsia lacunosa Vent. Choix des Plantes, 9. 1803.
Limnanthemum lacunosum Griseb. Gent. 347. 1839.
Nymphoides lacunosum Kuntze, Rev. Gen. Pl. 429. 1891.

Rootstock buried in the mud, the roots long and fibrous. Primary leaves membranous, submerged, short-petioled; stems (stolons) filiform, greatly elongated, sometimes 10° long, bearing a short-petioled floating leaf, an umbel of flowers and a cluster of narrow tubers at its summit, or some of the floating leaves on petioles as long as the stems; floating leaves ovate-orbicular, purplish beneath, 1′–2′ long, the basal sinus narrow or broad; pedicels slender; tubers linear-conic, ½′–1′ long; flowers nearly white, 3″–6″ broad; style none; capsule ovoid, covered by the connivent calyx-segments; seeds numerous, smooth.

In ponds, Nova Scotia to Florida, west to Ontario, Minnesota and Louisiana. July–Aug.

2. Nymphoides aquáticum (Walt.)
Kuntze.   Larger Floating Heart.
Fig. 3372.

Anonymos aquatica Walt. Fl. Car. 109. 1788.
Villarsia aquatica Gmel. Syst. 1: 447. 1791.
Menyanthes trachysperma Michx. Fl. Bor. Am. 1:
126. 1803.
Limnanthemum trachyspermum A. Gray, Man. Ed.
5, 390. 1867.
Limnanthemum aquaticum Britton, Trans. N. Y.
Acad. Sci. 9: 12. 1889.
Nymphoides aquaticum Kuntze, Rev. Gen. Pl.
429. 1891.

Similar to the preceding species but stouter
and larger. Floating leaves cordate-orbicular,
thick, entire or repand, 2′–6′ long, spongy, and
with the petioles and stolons densely covered
with minute pits; primary leaves spatulate;
pedicels slender, 1′–3′ long; tubers linear-
oblong, thicker; corolla white, 6″–10″ broad;
style none; seeds rough; capsule longer than
the calyx.

In ponds, southern New Jersey and Delaware
to Florida and Texas. May–Aug.

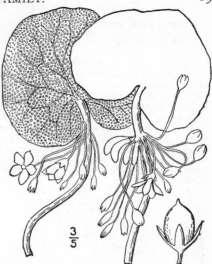

$\dfrac{3}{5}$

3. Nymphoides nymphaeoìdes (L.)
· Britton.   Water Lily.   Floating
Heart.   Fig. 3373·

Menyanthes nymphaeoides L. Sp. Pl. 145. 1753·

Limnanthemum nymphaeoides Hoffm. & Link,
Fl. Port. 1: 344. 1809.

Stems long, stout, creeping or buried in
the mud, ascending to the surface of the
water, branched, the upper nodes bearing
a petioled leaf and a cluster of flowers, or
the upper internodes very short and their
leaves apparently opposite. Petioles stout,
mostly longer than the ovate-orbicular firm
blades, which are 2′–4′ broad; pedicels
stout, becoming 3′–6′ long, not accompanied
by tufts of tubers; flowers bright yellow,
1′ in diameter, or more; corolla segments
short-fringed; seeds with fimbriate margins.

Naturalized in ponds, District of Columbia.
Native of Europe and Asia. May–July.

$\dfrac{3}{5}$

Family 17.   **APOCYNÀCEAE** Lindl. Nat. Syst. Ed. 2, 299. 1836.
DOGBANE FAMILY.

Perennial herbs, shrubs, vines, or some tropical genera trees, mostly with an
acrid milky juice, with simple opposite alternate or verticillate exstipulate leaves,
and perfect regular 5-parted cymose solitary or paniculate flowers. Calyx inferior, persistent, the lobes imbricated in the bud. Corolla gamopetalous, its lobes
convolute in the bud and often twisted. Stamens as many as the lobes of the
corolla, alternate with them, inserted on the tube or throat; anthers linear-oblong,
or sagittate, 2-celled; pollen-grains simple, often glutinous. Ovary superior, or
its base adherent to the calyx, of 2 distinct carpels, or 1-celled, with 2 parietal
placentae, or 2-celled; ovules few or numerous, anatropous or amphitropous; style
simple, or 2-divided; stigma simple. Fruit usually of 2 follicles or drupes. Seeds
often appendaged by a coma; endosperm fleshy, not copious; embryo straight;
cotyledons flat or concave; radicle terete, usually shorter than the cotyledons.

About 130 genera and 1100 species, very widely distributed, mostly in tropical regions.
Leaves alternate; erect herbs.   1. *Amsonia.*
Leaves opposite; vines or herbs.
    Flowers large, axillary, solitary.   2. *Vinca.*

Flowers small, cymose.
   Erect or diffuse herbs; corolla campanulate.   3. *Apocynum.*
   High-climbing vines; corolla funnelform.   4. *Trachelospermum.*

1. AMSÒNIA Walt. Fl. Car. 98. 1788.

Perennial herbs, with alternate membranous leaves, and rather large blue or bluish flowers, in terminal thyrsoid or corymbose cymes. Calyx 5-parted, the segments narrow, acuminate. Corolla mostly salverform, the tube cylindric, but somewhat dilated at the summit, villous within. Stamens inserted on the throat of the corolla, included; anthers ovate or oblong. Disk none. Ovary of 2 carpels, connected at the top by the filiform style; ovules in 2 rows in each cavity, numerous; stigma appendaged by a reflexed membrane. Fruit of 2 cylindric several-seeded follicles. Seeds cylindric or oblong, obliquely truncate at each end, not appendaged. [Named for Charles Amson of South Carolina.]

About 8 species, natives of North America and eastern Asia. Besides the following, 5 others occur in the southern and southwestern United States. Type species: *Amsonia Tabernaemontana* Walt.

1. Amsonia Amsònia (L.) Britton. Amsonia. Fig. 3374.

Tabernaemontana Amsonia L. Sp. Pl. Ed. 2, 308. 1762.
Amsonia Tabernacmontana Walt. Fl. Car. 98. 1788.
A. salicifolia Pursh, Fl. Am. Sept. 184. 1814.
A. Amsonia Britton, Mem. Torr. Club 5: 262. 1894.

Glabrous or nearly so, simple, or branched above, 2°–4° high. Leaves ovate, ovate-lanceolate or lanceolate, entire, acuminate at the apex, narrowed at the base, sometimes pubescent beneath, 2′–4′ long, ½′–2′ wide; petioles 2″–4″ long; flowers numerous; pedicels bracteolate at the base; calyx about 1″ long, its segments subulate; corolla 6″–9″ long, beaked by the convolute limb in the bud, its lobes linear and about as long as the tube; follicles 2′–4′ long, about 2″ thick, attenuate at the apex, glabrous, divergent or ascending; seeds papillose.

In moist soil, New Jersey to Illinois, Kentucky, Missouri, Florida and Texas. Consists of several races, differing in leaf-form. April–July.

2. VÍNCA L. Sp. Pl. 209. 1753.

Erect or trailing herbs, some species slightly woody, with opposite leaves, and large solitary blue pink or white axillary flowers. Calyx 5-parted, the segments narrow, acuminate. Corolla salverform, the tube cylindric, or expanded above, pubescent within, the lobes convolute, at least in the bud, oblique. Stamens included. Disk of 2 glands, alternate with the 2 carpels. Ovules several in each carpel; style filiform; stigma annular, its apex penicillate. Follicles 2, erect or spreading, cylindric, several-seeded. Seeds oblong-cylindric, truncate at each end, not appendaged. [The Latin name.]

About 12 species, natives of the Old World. Typ⟨ species: *Vinca major* L.

1. Vinca mìnor L. Periwinkle. Myrtle. Fig. 3375.

Vinca minor L. Sp. Pl. 209. 1753.

Perennial, trailing, glabrous; stems 6′–2° long. Leaves oblong to ovate, entire, firm, green both sides, obtuse or acutish at the apex, narrowed at the base, short-petioled, 1′–2½′ long, ½′–1′ wide; flowers not numerous, solitary in some of the axils, blue, 9″–15″ broad; peduncles slender, ½′–1½′ long; calyx very deeply parted, the segments subulate-lanceolate, glabrous, about 1½′ long; corolla-tube expanded above, as long as or slightly longer than the obovate, nearly truncate lobes; anther-sacs with a broad connective; follicles few-seeded.

Escaped from gardens to roadsides and woods, Ontario to Connecticut, southern New York and Georgia. Native of Europe. Leaves shining. Also called running myrtle or small periwinkle. Feb.–May.

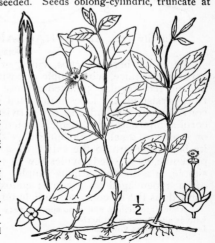

3. APÓCYNUM L. Sp. Pl. 213.   1753.

Perennial branching herbs, with opposite entire leaves, and small white or pink flowers in terminal and sometimes axillary corymbed cymes. Calyx 5-parted, the segments acute. Corolla urceolate to campanulate, the tube bearing within 5 small appendages alternate with the stamens, the limb 5-lobed. Stamens inserted on the base of the corolla; anthers sagittate, connivent around the stigma and slightly adherent to it. Disk 5-lobed. Ovary of 2 carpels; ovules numerous in each carpel; stigma ovoid, obtuse, obscurely 2-lobed. Follicles slender, elongated, terete. Seeds numerous, small, tipped with a long coma. [Greek, dogbane.]

About 11 species, of the north temperate zone, most abundant in North America. Type species: *Apocynum androsaemifolium* L.

Corolla campanulate, not angled, 4″–4½″ long, pink, its lobes widely spreading or recurved.
1. A. androsaemifolium.

Corolla urceolate to short-tubular, or narrowly campanulate, angled, 1½″–3½″ long, greenish, white, or faintly pink, its lobes a little spreading, or erect.

Corolla narrowly campanulate, 2½″–3½″ long, lobes nearly equal the tube.   2. *A. medium.*
Corolla urceolate to short-tubular, 1½″–3″ long, lobes much shorter than the tube.

Calyx-lobes ovate, much shorter than the corolla tube.   3. *A. Milleri.*
Calyx-lobes lanceolate, about as long as the corolla-tube.

Leaves and cymes glabrous, or sparingly pubescent.
Leaves all narrowed at the base and distinctly petioled; flowers greenish; plant rather bright green.   4. *A. cannabinum.*
At least the lower leaves sessile, and mostly rounded or subcordate at the base; plant pale glaucous green; flowers white.   5. *A. sibiricum.*
Leaves, cymes, and often the whole plant densely pubescent.   6. *A. pubescens.*

1. Apocynum androsaemifòlium L.
Spreading Dogbane.   Fig. 3376.

Apocynum androsaemifolium L. Sp. Pl. 213.   1753. Syst. Ed. 10, 946.   1759.

A. divergens Greene, Leaflets 1: 56.   1904.

Rootstock horizontal; stem 1°–4° high; branches broadly spreading, mostly glabrous. Leaves ovate or oval, acute or obtuse and mucronate at the apex, rounded or narrowed at the base, glabrous above, pale and usually more or less pubescent beneath, 2′–4′ long, 1′–2½′ wide; petioles 2″–4″ long; cymes loose; pedicels 2″–3″ long, subulate-bracted at the base; flowers about 4″ broad; calyx-segments shorter than the tube of the pinkish corolla; corolla-lobes revolute; follicles about 4′ long, narrowed at the apex.

In fields and thickets, Anticosti to British Columbia, south to Georgia, Missouri, Nebraska and Arizona. Ascends to 3500 ft. in Virginia. Honey-bloom. Bitter-root or -dogbane. Rheumatism-wood. Wild ipecac. Wandering milk-weed. Western wall-flower. Fly-trap. June–July. Linnaeus inadvertently failed to affix a binominal specific name in the first edition of his "Species Plantarum," but corrected this six years later.

2. Apocynum mèdium Greene.   Intermediate Dogbane.
Fig. 3377.

Apocynum medium Greene, Pittonia 3: 29.   1897.

Apocynum speciosum G. S. Miller, Proc. Biol. Soc. Wash. 13: 83. 1899.

Rootstock horizontal; stem stout, 4° high or less, the branches ascending, glabrous. Leaves oblong to oval or elliptic, mucronulate, ascending, 2½′–4′ long, somewhat pubescent beneath; petioles 2″–4″ long; cymes terminal, compact; pedicels 1½″–2½″ long; flowers erect; calyx-segments ovate, about half as long as the 5-angled corolla-tube; corolla white or pinkish, 2½″–3½″ long, urceolate-campanulate, its lobes acutish, somewhat spreading, nearly as long as the tube; follicles 3′–4′ long.

Fields and hillsides, Quebec to the District of Columbia, west to Iowa.   June–Aug.

½

3. Apocynum Mílleri Britton. Miller's Dogbane. Fig. 3378.

Apocynum Milleri Britton, Manual 739. 1901.

Stem slender, 3° high or less, the branches spreading. Leaves oblong to ovate-lanceolate, 2½′–3½′ long, pubescent beneath, the pubescent petioles 1½″–3″ long; cymes small, terminal or also in the upper axils, the pedicels 1″–1½″ long; flowers nearly erect; corolla pinkish, 2½″–3″ long, its rounded segments spreading, much shorter than the tube, which is longer than the ovate calyx-segments; follicles about 4′ long.

Dry soil, New York to Maryland and the District of Columbia. June–July.

4. Apocynum cannábinum L. Indian Hemp. Amy-root. Fig. 3379.

Apocynum cannabinum L. Sp. Pl. 213. 1753.
A. cannabinum glaberrimum DC. Prodr. 8: 439. 1844.
A. nemorale G. S. Miller, Proc. Biol. Soc. Wash. 13: 87. 1899.
Apocynum urceolifer G. S. Miller, loc. cit.

Root deep, vertical, soon branching. Stem extensively branched, the branches erect or ascending, glabrous or nearly so, more or less glaucous. Leaves oblong, lanceolate-oblong or ovate-oblong, acute or obtuse and mucronate at the apex, narrowed or rounded at the base, glabrous above, sometimes pubescent beneath, 2′–6′ long, ½′–3′ wide; petioles 1″–6″ long, or sometimes none; cymes dense; pedicels short, bracteolate at the base; calyx-segments about as long as the tube of the greenish-white corolla; corolla-lobes nearly erect; follicles similar to those of the preceding species.

In fields and thickets, Connecticut to Wisconsin, Alabama, Tennessee, Missouri and Kansas, perhaps extending farther north. Rheumatism-root. Wild cotton.

½

5. Apocynum sibíricum Jacq. Clasping-leaved Dogbane. Fig. 3380.

A. sibiricum Jacq. Hort. Vind. 3: 37. *pl. 66.* 1776.
A. hypericifolium Ait. Hort. Kew. 1: 304. 1789.
A. cannabinum var. *hypericifolium* A. Gray, Man. 365. 1848.
Apocynum album Greene, Pittonia 3: 230. 1897.

Glabrous, pale green, often glaucous; stem 1°–2° high, the branches ascending. Leaves oblong, oblong-lanceolate to oval, 1′–3′ long, ½′–1½′ wide, obtuse or acutish at the apex, cordate-clasping, rounded, truncate, or most of the upper narrowed at the base, short-petioled, or sessile, the primary venation forming broad angles with the midvein; cymes many-flowered, dense to loose; pedicels mostly not longer than the flowers, bracteolate; calyx-segments about as long as the corolla-tube, lanceolate, acute; corolla-lobes nearly erect; follicles 2′–3½′ long.

Mostly along streams, Quebec to British Columbia, Long Island, Ohio, Kansas and New Mexico. St. John's-dogbane. June–Aug.

6. Apocynum pubéscens R. Br. Velvet Dogbane. Fig. 3381.

A. pubescens R. Br. Mem. Wern. Soc. 1 : 68. 1811.
Apocynum cannabinum var. *pubescens* A. DC. Prodr. 8 :
440. 1844.

Whole plant, including the pedicels and calyx,
densely velvety-pubescent, or the stem sometimes
glabrate. Branches ascending; leaves oval to elliptic,
obtuse or acute at the apex, strongly mucronate,
obtuse or obtusish at the base, the veins impressed
in the pubescence of the lower surface; petioles 1″–2″
long; cymes dense; calyx-segments about as long as
the tube of the corolla, lanceolate, acute; corolla-
lobes erect; follicles about 4′ long.

In dry sandy soil, Ontario to Rhode Island, Maryland,
Alabama, Iowa and Kansas. April–Aug. Perhaps a pubescent race of *A. cannabinum* L.

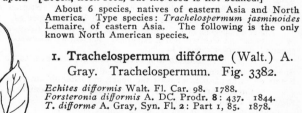
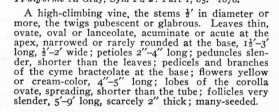

$\frac{1}{2}$

4. TRACHELOSPÉRMUM Lemaire, Jard. Fleur. 1 : *pl. 61.* 1851.

Twining woody vines (some exotic species nearly erect shrubs), with opposite entire
deciduous leaves, and small yellow greenish or white flowers in terminal and axillary compound cymes. Calyx small, deeply 5-parted, glandular within, the segments narrow. Corolla
funnelform or salverform, the tube nearly cylindric, expanded above, the lobes convolute,
more or less twisted. Stamens included, or short exserted; anthers sagittate, acuminate,
connivent around the stigma and slightly adherent to it. Disk of 5 glandular lobes. Ovary
of 2 carpels; ovules numerous in each carpel; style slender, its apex thickened below the
narrow ring of the ovoid stigma. Follicles much elongated, slender. Seeds linear or oblong,
not beaked, long-comose at the apex. [Greek, neck-seed, but the seed is not beaked.]

About 6 species, natives of eastern Asia and North
America. Type species: *Trachelospermum jasminoides*
Lemaire, of eastern Asia. The following is the only
known North American species.

1. Trachelospermum difförme (Walt.) A. Gray. Trachelospermum. Fig. 3382.

Echites difformis Walt. Fl. Car. 98. 1788.
Forsteronia difformis A. DC. Prodr. 8 : 437. 1844.
T. difforme A. Gray, Syn. Fl. 2 : Part 1, 85. 1878.

A high-climbing vine, the stems $\frac{1}{2}$′ in diameter or
more, the twigs pubescent or glabrous. Leaves thin,
ovate, oval or lanceolate, acuminate or acute at the
apex, narrowed or rarely rounded at the base, 1$\frac{1}{2}$′–3′
long, $\frac{1}{2}$′–2′ wide; petioles 2″–4″ long; peduncles slender, shorter than the leaves; pedicels and branches
of the cyme bracteolate at the base; flowers yellow
or cream-color, 4″–5″ long; lobes of the corolla
ovate, spreading, shorter than the tube; follicles very
slender, 5′–9′ long, scarcely 2″ thick; many-seeded.

In moist woods and along streams, Delaware to Florida, Texas and Mexico, mostly near the coast, north to
Arkansas and Missouri. June–Aug.

Family 18. ASCLEPIADÀCEAE Lindl. Nat. Syst. Ed. 2, 302. 1836.*

MILKWEED FAMILY.

Perennial herbs, vines or shrubs, mostly with milky juice, with opposite alternate or verticillate exstipulate leaves, and mostly umbellate perfect regular flowers.
Calyx inferior, its tube very short, or none, its segments imbricated or separate
in the bud. Corolla campanulate, urceolate, rotate or funnelform, 5-lobed or
5-cleft, the segments commonly reflexed, mostly valvate in the bud. A 5-lobed
or 5-parted crown (corona) between the corolla and the stamens and adnate to
one or the other. Stamens 5, inserted on the corolla, usually near its base; filaments short, stout, mostly monadelphous, or distinct; anthers attached by their

* Text revised for our first edition by Miss ANNA MURRAY VAIL, here somewhat modified.

bases to the filaments, introrsely 2-celled, connivent around the stigma, or more or less united with each other ; anther-sacs tipped with an inflexed or erect scarious membrane, or unappendaged at the top, sometimes appendaged at the base ; pollen coherent into waxy or granular masses, one or rarely two such masses in each sac, connected with the stigma in pairs or fours, by 5 glandular corpuscles alternate with the anthers.   Disk none.   Ovary of 2 carpels ; styles 2, short, connected at the summit by the peltate discoid stigma ; ovules numerous in each carpel, mostly anatropous, pendulous.   Fruit of 2 several–many-seeded follicles.   Seeds compressed, usually appendaged by a long coma ; endosperm cartilaginous, mostly thin ; embryo nearly as long as the seed ; cotyledons flat.

About 220 genera and 2000 species, of very wide geographic distribution, most abundant in tropical or warm-temperate regions.

Erect or decumbent herbs.
    Corona-hoods each with an incurved horn within ; eaves mostly opposite.   1. *Asclepias.*
    Corona-hoods prominently crested within ; leaves alternate.   2. *Asclepiodora.*
    Corona-hoods unappendaged or with a thickened crest-like keel.   3. *Acerates.*
Twining vines.
    Corolla-lobes erect ; corona-lobes 1–2-awned.   4. *Gonolobus.*
    Corolla rotate.
        Anthers tipped with a scarious membrane ; pollen-masses pendulous.   5. *Cynanchum.*
        Anthers merely tipped ; pollen-masses horizontal.   6. *Vincetoxicum.*

1. ASCLEPIAS L. Sp. Pl. 214. 1753.

Perennial erect or decumbent herbs, with opposite verticillate or rarely alternate entire leaves, and middle-sized or small flowers in terminal or axillary umbels. Calyx 5-parted or 5-divided, usually small, the segments or sepals acute, often glandular within. Corolla deeply 5-parted, the segments mostly valvate, reflexed in anthesis. Corona-column generally present. Corona of 5 concave erect or spreading hoods, each bearing within a slender or subulate incurved horn, either included or exserted. Filaments connate into a tube ; anthers tipped with an inflexed membrane, winged, the wings broadened below the middle ; pollen-masses solitary in each sac, pendulous on their caudicles. Stigma nearly flat, 5-angled or 5-lobed. Follicles acuminate. Seeds comose in all but one species. [Dedicated to Aesculapius.]

About 95 species, mostly natives of the New World ; besides the following, some 25 others occur in southern and western North America.   Known as Milkweed, Silkweed, or Swallow-wort.   Type species: *Asclepias syriaca* L.

    * **Corolla and corona orange; leaves alternate or opposite.**
Stem erect or ascending ; leaves nearly all alternate.   1. *A. tuberosa.*
Stems reclining ; leaves, at least the upper, opposite, oblong or oval.   2. *A. decumbens.*
    ** **Corolla bright red or purple (sometimes greenish in A. rubra) leaves opposite.**
Flowers 4″–6″ broad ; corona-hoods 2″–3″ high.
    Leaves lanceolate or linear ; hoods oblong, obtuse.   3. *A. lanceolata.*
    Leaves ovate or ovate-lanceolate ; hoods lanceolate.   4. *A. rubra.*
    Leaves oblong, ovate or ovate-oblong ; hoods oblong, acutish.   5. *A. purpurascens.*
Flowers 2″–3″ broad ; corona-hoods 1″–1½″ high.
    Plant nearly or quite glabrous ; leaves lanceolate or oblong-lanceolate.   6. *A. incarnata.*
    Plant pubescent ; leaves oblong.   7. *A. pulchra.*
    *** **Corolla greenish, purplish, yellowish or white; leaves opposite or verticillate.**
Leaves ovate, oblong, ovate-lanceolate, obovate or orbicular.
    Plants glabrous throughout, or minutely pubescent above.
        Leaves sessile, clasping or very short-petioled.
            Peduncles of the solitary or several umbels short.
                Leaves ovate-oblong ; hoods 2-auriculate at the base.   8. *A. Sullivantii.*
                Leaves nearly orbicular ; hoods truncate.   9. *A. latifolia.*
            Peduncle of the usually solitary umbel elongated.
                Leaves wavy-margined.
                    Leaves cordate-clasping.   10. *A. amplexicaulis.*
                    Leaves rounded at the base, short-petioled.   11. *A. intermedia.*
                Leaves sessile, flat ; horn not exceeding the hood.   12. *A. Meadii.*
        Leaves manifestly petioled.
            Corolla greenish ; umbels loose, the pedicels drooping.   13. *A. exaltata.*
            Corolla white ; umbels dense.   14. *A. variegata.*
            Corolla pink ; some of the leaves verticillate in 4's.   15. *A. quadrifolia.*
    Plants, at least the lower surfaces of the leaves, canescent or tomentose.
        Follicles tomentose, covered with soft spinose processes.
            Corona-hoods obtuse, short.   16. *A. syriaca.*
            Corona-hoods elongated, lanceolate.   17. *A. speciosa.*
        Follicles with no spinose processes, glabrous or pubescent.
            Leaves wavy-margined ; corolla-segments 4″–5″ long.   18. *A. arenaria.*
            Leaves flat ; corolla-segments 2″–3″ long.   19. *A. ovalifolia.*
Leaves lanceolate, oblong-lanceolate or linear.
    Leaves opposite, lanceolate or oblong-lanceolate.
        Leaves thick, short-petioled ; inflorescence woolly.   20. *A. brachystephana.*
        Leaves thin, slender-petioled ; inflorescence downy.   21. *A. perennis.*

Leaves mostly verticillate in 3's–6's, narrowly linear.
  Hoods entire.
  Hoods dorsally hastate-sagittate.
Leaves scattered, densely crowded, filiform-linear.

                22. *A. verticillata.*
                23. *A. galioides.*
                24. *A. pumila.*

1. Asclepias tuberòsa L.   Butterfly-weed or -flower. Pleurisy-root. Fig. 3383.

Asclepias tuberosa L. Sp. Pl. 217. 1753.

Hirsute-pubescent; stems stout, simple, or branched near the summit, ascending or erect, very leafy, 1°–2° high, the milky sap scanty. Leaves alternate, lanceolate or oblong, acute or sometimes obtuse at the apex, narrowed, rounded or cordate at the base, sessile or short-petioled, 2′–6′ long, 2″–12″ wide; umbels cymose, terminal, many-flowered; peduncles shorter than the leaves; pedicels ½′–1′ long; corolla-segments about 3″ long, greenish orange; corona-column about ½″ long; hoods erect, oblong, bright orange, or yellow, 2–3 times as long as the stamens, longer than the filiform horns; fruiting pedicels decurved; follicles nearly erect, finely pubescent, 4′–5′ long.

In dry fields, Maine and Ontario to Minnesota, Florida, Texas, Chihuahua and Arizona. Consists of numerous races, differing in shape and size of the leaves and color of the flowers. June–Sept. Wind- or orange-root. Canada-, flux-, tuber- or white-root. Orange swallow-wort. Yellow milkweed. Indian-posy.

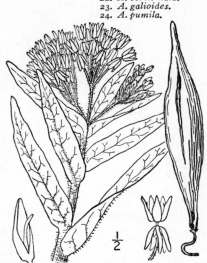

2. Asclepias decúmbens L.   Decumbent Butterfly-weed. Fig. 3384.

Asclepias decumbens L. Sp. Pl. 216. 1753.

Hirsute-pubescent; stems decumbent, 2°–3° long, the ends ascending or erect. Leaves sessile or short-petioled, oblong or elliptic, obtuse at the apex, narrowed and often inequilateral at the base, 1′–3′ long, ½′–2½′ wide, the upper opposite, the lower commonly alternate, the uppermost very small; umbels several or numerous, many-flowered, racemose along the branches, one usually in each of the upper axils; peduncles stout, short; pedicels slender, pubescent, about ½′ long; corolla-segments oblong, acutish, dark orange, about 3″ long; column about ½″ high, the hoods erect, oblong, orange, longer than the subulate horn; follicles slender.

In dry fields, Illinois and Ohio to North Carolina and Florida. Creeping milkweed. June–Aug.

3. Asclepias lanceolàta Walt.   Few-flowered Milkweed. Fig. 3385.

Asclepias lanceolata Walt. Fl. Car. 105. 1788.
A. paupercula Michx. Fl. Bor. Am. 1: 118. 1803.

Glabrous; stem slender, usually simple, naked above, 2°–4° high. Leaves opposite, distant, linear or narrowly lanceolate, elongated, acuminate, narrowed at the base, short-petioled, 4′–10′ long, 2″–7″ wide, roughish on the margins, the primary nerves widely spreading; umbels few-flowered, solitary or 2–4 at the summit; peduncles about equalling the slender puberulent pedicels; corolla-segments oblong, 4″–5″ long, deep red; column thick, about 1″ high; hoods obovate or oblong, obtuse, orange, 2-toothed near the base, nearly twice the length of the anthers and longer than the subulate incurved horn; fruiting pedicels decurved; follicles erect, minutely puberulent, fusiform, about 4′ long.

In swamps, southern New Jersey to Florida and Texas, mostly near the coast. June–Aug.

4. Asclepias rùbra L. Red Milkweed.
Fig. 3386.

Asclepias rubra L. Sp. Pl. 217. 1753.

Nearly glabrous throughout; stem usually simple, 1°-4° high. Leaves opposite, rather distant, short-petioled, ovate, lanceolate or the lower sometimes oblong, rounded or subcordate at the base, gradually acuminate, rather firm, 3′-8′ long, 1′-2½′ wide, the primary nerves wide-spreading; umbels 1-4, many-flowered; peduncles shorter than or equalling the upper leaves; pedicels slender, downy, ½′-1′ long; corolla-segments and hoods lanceolate-oblong, purplish red, or the hoods orange-red, 3″-4″ long, or flowers sometimes greenish; horns of the hoods very slender, nearly straight; fruiting pedicels deflexed, the follicles erect, spindle-shaped, glabrous, about 4′ long.

In moist soil, New Jersey and Pennsylvania to Florida, Missouri, Louisiana and Texas. June–July.

5. Asclepias purpuráscens L. Purple Milkweed. Fig. 3387.

Asclepias purpurascens L. Sp. Pl. 214. 1753.
?*Asclepias amoena* L. Sp. Pl. 214. 1753.

Stem stout, puberulent or glabrous, usually simple, 2°-4° high, leafy to the top. Leaves ovate, elliptic or oblong, short-petioled, acute or obtuse and mucronulate at the apex, narrowed or rounded at the base, nearly glabrous above, finely tomentose beneath, 3′-8′ long, 1½′-3′ wide, the primary nerves very wide-spreading; umbels many-flowered, borne in several of the upper axils, or sometimes solitary; peduncles stout; pedicels slender, puberulent, 1′-1½′ long; corolla deep purple, its segments oblong to oblong-lanceolate, about 3″ long; column very short and thick; hoods oblong or ovate, nearly twice as long as the anthers, pale red or purple, the horns broad at the insertion, short-subulate and incurved at the apex; fruiting pedicels deflexed, the downy follicles nearly erect, 4′-5′ long.

In dry fields and thickets, New Hampshire to North Carolina, west to southern Ontario, Minnesota and Arkansas. Ascends to 2000 ft. in the Catskills. June–Aug.

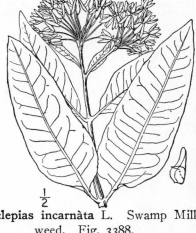

6. Asclepias incarnàta L. Swamp Milkweed. Fig. 3388.

Asclepias incarnata L. Sp. Pl. 215. 1753.

Stem slender, glabrous, or puberulent in 2 lines above, branched or rarely simple, 2°-4° high, leafy to the top. Leaves lanceolate or oblong-lanceolate, acuminate at the apex, narrowed, obtuse or sometimes subcordate at the base, 3′-6′ long, ½′-1½′ wide, the primary nerves not wide-spreading; petioles 3″-6″ long; umbels usually numerous, corymbed, many-flowered; pedicels pubescent, 5″-10″ long; corolla red or rose-purple, rarely white, its lobes oblong, about 2″ long; column more than one-half as long as the obtuse pink or purplish hoods; horns slender, incurved, longer than the hoods; anther-wings entire, or obscurely notched at the base; fruiting pedicels erect or incurved; follicles erect, 2′-3½′ long, sparingly puberulent.

In swamps, New Brunswick to Ontario, Saskatchewan, Tennessee, Louisiana and Colorado. Ascends to 3000 ft. in West Virginia. Rose- or swamp-silk-

weed. Water nerve-root. White Indian-hemp. July–Sept.

7. Asclepias púlchra Ehrh. Hairy Milkweed. Fig. 3389.

Asclepias pulchra Ehrh.; Willd. Sp. Pl. 1: 1267. 1798.
A. incarnata var. *pulchra* Pers. Syn. 1: 276. 1805.

Similar to the preceding species and perhaps hybridizing with it where the two grow together; stem stout, tomentose-pubescent, usually branched, 2°–3½° high, leafy to the top. Leaves broadly lanceolate, acute, acuminate or some of them obtusish at the apex, subcordate, rounded, or the upper narrowed at the base, puberulent or glabrous above, pubescent, at least on the veins beneath, 3′–5′ long, ½–2′ wide; petioles usually stout and short; flowers similar to those of *A. incarnata*, but the corolla commonly lighter red or pink, rarely white; peduncles and pedicels tomentose; fruiting pedicels erect or incurved; follicles erect, densely pubescent, 2′–3′ long.

In moist fields and swamps, Nova Scotia to Minnesota, south to Georgia. White Indian-hemp. July–Sept.

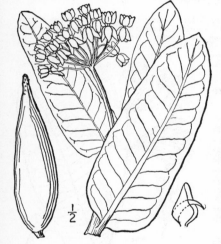

8. Asclepias Sullivántii Engelm. Sullivant's Milkweed. Fig. 3390.

A. Sullivantii Engelm.; A. Gray, Man. 366. 1848.

Glabrous throughout; stem stout, simple or sometimes branched above, 2°–4° high, leafy to the top. Leaves thick, sessile, or on petioles less than 1″ long, oblong or ovate-oblong, usually obtuse and mucronulate at the apex, subcordate, rounded or slightly clasping at the base, 4′–6′ long, 1¾–3′ wide, the primary nerves very wide-spreading; umbels terminal and sometimes also in the upper axils, many-flowered; peduncles shorter than the leaves; corolla-segments oval-oblong, 5″–6″ long, purplish; column very short and thick; hoods oval, obtuse or truncate, gibbous at each side near the base, longer than the anthers and the subulate incurved horn; follicles erect, glabrous, 3′–4′ long, usually with blunt processes near the apex.

In moist soil, southern Ontario to Ohio, Minnesota, Nebraska, Missouri and Kansas. July–Sept.

9. Asclepias latifòlia (Torr.) Raf. Broad-leaved Milkweed. Fig. 3391.

Asclepias obtusifolia var. *latifolia* Torr. Ann. Lyc. N. Y. 2: 117. 1826.

Asclepias latifolia Raf. Atl. Journ. 146. 1832–33.

A. Jamesii Torr. Bot. Mex. Bound. Surv. 162. 1859.

Minutely puberulent when young, glabrous when old; stem stout, usually simple, 1°–2½° high, very leafy. Leaves very thick, oval or orbicular, sessile or nearly so, commonly broadly emarginate and mucronulate at the apex and cordate or subcordate at the base, 4′–6′ long and nearly as wide, primary nerves very wide-spreading; umbels 2–4, many-flowered, short-peduncled in the upper axils or rarely terminal; pedicels slender, canescent, nearly 1′ long; corolla-segments ovate, acute, 4″–6″ long, greenish; column short and thick; hoods truncate, about equalling the anthers, the horn projecting from a short crest over the edge of the stigma; follicles erect on deflexed pedicels, ovoid, acutish, 2′–3′ long, about 1′ thick.

On dry plains, Nebraska to Colorado, Texas and Arizona. July–Sept.

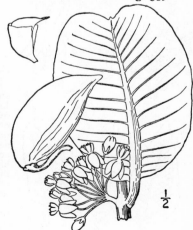

10. **Asclepias amplexicaùlis** J. E. Smith. Blunt-leaved Milkweed. Fig. 3392.

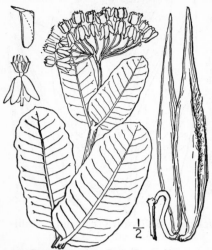

A. amplexicaulis J. E. Smith, Georgia Insects **1**: 13. *pl. 7.* 1797.

A. obtusifolia Michx. Fl. Bor. Am. **1**: 115. 1803.

Nearly glabrous, pale green, somewhat glaucous; stem stout, erect or ascending, 2°–3° high. Leaves sessile, oblong or ovate-oblong, obtuse and mucronulate at the apex, cordate-clasping at the base, 3′–5′ long, 1′–1½′ wide, margins wavy-crisped; umbels many-flowered, usually solitary on the long terminal peduncle, rarely with a second shorter-peduncled one at its base; pedicels slender, downy, about 1′ long; corolla-segments oblong, greenish-purple, about 4″ long; column thick; hoods pink, nearly truncate and toothed at the summit, shorter than the subulate incurved horn, longer than the anthers; follicles erect on the stout decurved fruiting pedicels, downy, 4′–6′ long.

In dry fields, mostly in sandy soil, New Hampshire to Florida, northern New York, Minnesota, Nebraska and Texas. Ascends to 3000 ft. in Virginia. May–Aug.

A. Bicknéllii Vail, Bull. Torr. Club **31**: 458, is apparently a hybrid between *A. amplexicaulis* J. E. Smith and *A. exaltata* (L.) Muhl.

11. **Asclepias intermedia** Vail. Intermediate Milkweed. Fig. 3393.

A. intermedia Vail, Bull. Torr. Club **31**: 459. 1904.

Stem erect, glabrous, purplish, not glaucous, about 1° high. Leaves oblong-elliptic, glabrous above, minutely pubescent beneath, 6′ long or less, obtuse at both ends or the upper subcordate at the base, the petioles very short; umbels 2 or more, terminal, peduncled, the peduncles and slender pedicels pubescent; corolla green-purple, its oblong-lanceolate segments 3″–4″ long, the erect hoods ovate-lanceolate, obtuse, about 3″ long, pink-purple, with a darker stripe on the back, the margins with an erect tooth above the middle, the horn slender.

Lawrence, Long Island. Possibly a hybrid between *A. syriaca* and *A. amplexicaulis.*

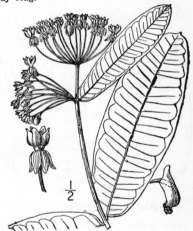

12. **Asclepias Mèadii** Torr. Mead's Milkweed. Fig. 3394.

A. Meadii Torr.; A. Gray, Man. Ed. 2, Add. 704. 1856.

Nearly glabrous throughout, pale green or glaucous; stem simple, or rarely branched above, 1°–2° high. Leaves opposite, sessile, flat, mostly distant, ovate, ovate-lanceolate, lanceolate or the lower oblong, acute or sometimes obtuse at the apex, the margins scabrous; umbel solitary, terminal, several-flowered, borne on a peduncle 3′–6′ long; corolla-segments greenish, ovate, acute, 3″–4″ long; column very short, thicker than high; hoods ovate, purplish, nearly twice as long as the anthers, rounded and truncate at the summit, longer than the subulate inflexed horn, with a small tooth at each side on the inner infolded margin; follicles erect on decurved pedicels, minutely puberulent, narrow, 4′–5′ long.

In dry soil, Illinois to Iowa and Wisconsin. June–Aug.

13. Asclepias exaltàta (L.) Muhl. Poke or Tall Milkweed. Fig. 3395.

A. Syriaca var. *exaltata* L. Sp. Pl. Ed. 2, 313. 1762.
Asclepias exaltata Muhl. Cat. 28. 1813.
A. phytolaccoides Pursh, Fl. Am. Sept. 180. 1814.

Nearly glabrous throughout, with two opposite lines of pubescence on the usually simple stem, 3°–6° high. Leaves opposite, thin or membranous, oval, ovate or oblong, acuminate at both ends, 4′–9′ long, 1½′–4′ wide, the lower sometimes obovate, obtuse, shorter; petioles ¼′–1′ long; peduncles 1′–3′ long; umbels usually several; pedicels slender, drooping or spreading, 1′–2′ long, puberulent; corolla green-purple, the segments ovate or oblong, obtusish, 3″–4″ long; column short; hoods white or pink, slightly shorter than the anthers, much shorter than the subulate horn, at the summit truncate and entire or erose, with 1 or 2 slender teeth on each of the inner margins; follicles erect on the deflexed pedicels, downy, long-acuminate, 4′–6′ long.

In thickets and woods, Maine to Minnesota, Georgia, Missouri and Arkansas. Ascends to 5500 ft. in North Carolina. June–Aug.

14. Asclepias variegàta L. White Milk-weed. Fig. 3396.

Asclepias variegata L. Sp. Pl. 217. 1753.

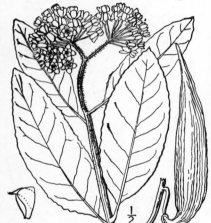

Stem glabrous below, pubescent above when young, simple, 1°–3° high. Leaves opposite, thick, oval, ovate, oblong or the lower somewhat obovate, obtuse and cuspidate or acutish at the apex, narrowed or rounded at the base, dark green above, pale beneath, 3′–6′ long, 1′–3′ wide, the middle ones sometimes verticillate in 4's; petioles 3″–12″ long; umbels 1–4, terminal, or rarely 1 or 2 in the upper axils, densely many-flowered; peduncles 1′–2′ long; pedicels ½′–1½′ long, erect or ascending, usually densely puberulent; corolla-segments ovate or oval, about 3″ long, white or purple near the base; column very short and thick, purplish; hoods globose-obovoid, obtuse, spreading, longer than the anthers, about equalling the semi-lunate horizontally pointed horn; follicles downy, erect on the deflexed fruiting pedicels, 4′–5′ long.

In dry woods or thickets, Connecticut (?), southern New York to Illinois, Arkansas, Florida and Texas. June–July.

15. Asclepias quadrifòlia Jacq. Four-leaved Milkweed. Fig. 3397.

A. quadrifolia Jacq. Obs. Part 2, 8. *pl. 33.* 1767.

Stem slender, simple, 1°–2° high, usually leafless below. Leaves thin, sparingly pubescent on the veins beneath, ovate to lanceolate, 2′–6′ long, ½′–2½′ wide, acute or acuminate, narrowed or rounded at the base, or the lowest pair much smaller, obovate and obtuse, the upper and lower opposite, the middle ones usually verticillate in 4's; umbels 1–4, terminal, or rarely in the upper axils; peduncles slender, ½′–2½′ long; pedicels about 1′ long; corolla pink or nearly white, its lobes lanceolate-oblong, 2″–3″ long; column short; hoods white, obtuse at the apex, broadly 2-toothed above the base, twice as long as the anthers and the short incurved horn; follicles erect on the erect fruiting pedicels, 3′–5′ long, glabrous.

Woods and thickets, Maine and Ontario to Minnesota, Alabama and Arkansas. May–July.

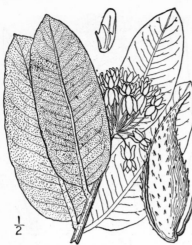

16. Asclepias syrìaca L. Common Milkweed. Silkweed. Fig. 3398.

Asclepias syriaca L. Sp. Pl. 214. 1753.
Asclepias Cornuti Dec. in DC. Prodr. 8: 564. 1844.

Stem stout, usually simple, 3°–5° high, finely pubescent at least above. Leaves oblong, oval or ovate, finely and densely pubescent beneath, soon glabrous above, acute or obtuse and cuspidate at the apex, obtuse, narrowed or subcordate at the base, 4′–9′ long, 2′–4½′ wide, the primary nerves wide-spreading; petioles stout, 3″–8″ long; umbels several or numerous; peduncles pubescent or tomentose, 1½′–3½′ long; pedicels 1′–2′ long; corolla green-purple or greenish-white, its segments oblong-lanceolate, 3″–4″ long; column short and thick, the hoods ovate-lanceolate with a tooth on each side, longer than the anthers and the incurved horn; follicles 3′–5′ long, erect on recurved pedicels, tomentose and covered with short soft processes.

In fields and waste places, New Brunswick to Saskatchewan, North Carolina and Kansas. Leaves rarely lanceolate. Silky swallow-wort. Virginia silk. Wild cotton. June–Aug.

Asclepias kansàna Vail, of Kansas, differs by erect-spreading hoods of the corolla and more densely tomentose follicles.

17. Asclepias speciòsa Torr. Showy Milkweed. Fig. 3399.

Asclepias speciosa Torr. Ann. Lyc. N. Y. 2: 218. 1826.
A. Douglasii Hook. Fl. Bor. Am. 2: 53. *pl. 152.* 1834.

White-tomentose or canescent all over, or glabrate below, pale; stem simple, stout, 1°–2½° high. Leaves thick, broadly ovate or oval, obtuse and cuspidate or acute at the apex, subcordate, rounded or narrowed at the base, petioled, 3′–8′ long, 2′–4′ wide; peduncles 1′–3′ long; umbels several or rarely solitary, many-flowered; pedicels stout, 9″–18″ long; corolla purple-green, its segments oblong or ovate-oblong, 4″–6″ long, tomentose on the outer face; column very short or none; hoods lanceolate, 5″–7″ long, obtusish, expanded and with 2 blunt teeth below, the apex ligulate, 5–7 times as long as the anthers; horn short, inflexed; follicles erect or spreading on the recurved fruiting pedicels, 3′–4′ long, densely woolly and covered with soft spinose processes.

In moist soil, Minnesota to British Columbia, south to Kansas, Utah and California. May–July.

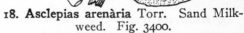

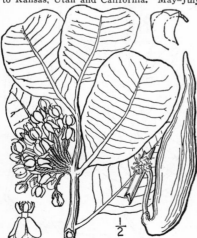

18. Asclepias arenària Torr. Sand Milkweed. Fig. 3400.

A. arenaria Torr. Bot. Mex. Bound. Surv. 162. 1859.

Densely tomentose-canescent all over, stems simple, ascending or erect, stout, 1°–2° high. Leaves obovate or oval, wavy-margined, thick, obtuse or retuse and cuspidate at the apex, truncate, subcordate, obtuse, or rarely some of them narrowed at the base, 2′–4′ long, 1½′–3′ wide, the angle of the primary nervation broad; umbels densely many-flowered, short-peduncled or sessile; corolla greenish-white, its segments oval-oblong, 4″–5″ long; column 1″–2″ high; hoods oblong, truncate at each end, oblique at the apex, longer than the anthers, with a broad tooth on each side within; horn semi-lunate with an abruptly incurved subulate apex; follicles puberulent, 4′–5′ long, erect on the decurved fruiting pedicels.

On sand-bars and hills along rivers, Nebraska and Colorado to Mexico and New Mexico. June–Sept.

19. Asclepias ovalifòlia Dec. Oval-leaved or Dwarf Milkweed. Fig. 3401.

Asclepias ovalifolia Dec. in DC. Prodr. **8** : 567. 1844.

Finely tomentose all over; stem simple, usually slender, erect, 10′–2° high. Leaves oval, ovate, oblong or ovate-lanceolate, acute or obtuse and mucronulate at the apex, rounded or narrowed at the base, 2′–3′ long, ½′–1½′ wide, short-petioled, the upper surfaces becoming glabrate at maturity; umbels solitary or few, several–many-flowered; peduncles short; corolla greenish-white or purplish, its segments ovate-oblong, obtuse, 2″–3″ long; column very short; hoods oval-oblong, yellowish, nearly twice as long as the anthers, bearing a large acute tooth on each of the inner margins; horn subulate, incurved over the stigma; follicles ascending on the reflexed fruiting pedicels, pubescent.

In woods and on prairies, Illinois to North Dakota, Wisconsin, Manitoba and Athabasca. June–July.

20. Asclepias brachystéphana Engelm. Short-crowned Milkweed. Fig. 3402.

Asclepias brachystephana Engelm.; Torr. Bot. Mex. Bound. Surv. 163. 1859.

Puberulent when young, soon glabrate; stems clustered, often branched, spreading or ascending, 6′–12′ long. Leaves mostly opposite, lanceolate or linear-lanceolate, thick, long-acuminate at the apex, rounded, subcordate or narrowed at the base, 2′–5′ long, 2″–6″ wide, or the lowest shorter; petioles 1″–4″ long; umbels several, terminal and axillary, few-flowered; peduncles short; pedicels densely woolly, equalling or longer than the peduncles; corolla greenish-purple, its segments about 2″ long; column very short or none; hoods ovate, obtuse, shorter than the anthers, the short erect-incurved horn slightly exserted; follicles erect on the spreading or decurved fruiting pedicels, downy or hoary, acuminate, 2′–3½′ long.

In dry soil, Kansas (according to B. B. Smyth), Wyoming to Texas, Arizona and Mexico. June–Aug.

21. Asclepias perénnis Walt. Thin-leaved Milkweed. Fig. 3403.

Asclepias perennis Walt. Fl. Car. 107. 1788.

Puberulent above, glabrous below; stem slender, simple or branched, erect, 1°–3° high. Leaves thin, opposite, lanceolate, oblong or ovate-lanceolate, slender-petioled, acuminate or acute at both ends, 2′–6′ long, ¼′–1′ wide, nearly glabrous; umbels solitary or several and corymbose; peduncles 1′–2′ long; pedicels slender, ½′–1′ long; flowers small, white; corolla-segments oblong, 1″–2″ long; column about ½″ high; hoods oval, erect, entire, about as long as the anthers, shorter than the subulate-filiform incurved horn; follicles glabrous, erect on the nearly erect fruiting pedicels; seeds 5″–6″ long, 3½″–4½″ wide, thin, usually without coma.

On river-shores and in wet places, North Carolina to Indiana, Illinois and Missouri, south to Florida and Texas. May–Aug.

22. Asclepias verticillàta L. Whorled Milk-
weed. Fig. 3404.

Asclepias verticillata L. Sp. Pl. 217. 1753.

Roots slender, fascicled; stem slender, simple or
branched, pubescent in lines above, leafy, 1°–2½°
high. Leaves narrowly linear, sessile, verticillate in
3's–7's or some of them alternate, glabrous or very
nearly so, their margins narrowly revolute; umbels
numerous, many-flowered; peduncles slender, ½'–1½'
long; pedicels almost filiform, shorter than the peduncles; corolla greenish white, its segments oblong,
1½''–2'' long; column about ½'' high; hoods white,
oval, entire, about equalling the anthers, much
shorter than the subulate incurved horn; follicles
erect on the erect fruiting pedicels, narrowly spindle-
shaped, glabrous, 2'–3' long.

In dry fields and on hills, Maine and southern Ontario to Saskatchewan, south to Florida, Mexico and
New Mexico. July–Sept.

23. Asclepias galioìdes H.B.K. Bedstraw Milk-
weed. Fig. 3405.

Asclepias galioides H.B.K. Nov. Gen. **3**: 188. 1818.

Glabrous, except the minutely pubescent stems and pedicels. Stems erect, 1° high or more, from a horizontal rootstock; leaves erect or spreading, in whorls of 2–6, narrowly
linear, 2'–3' long, the margins revolute; peduncles longer
than the pedicels and shorter than the leaves; umbels
9''–13'' in diameter; flowers greenish-white; corolla-segments 2'' long; hoods as high as the anthers, broadly
rounded at the summit, dorsally hastate-sagittate, the ventral margins slightly involute, entire; horn arising from
the base of the hood, long-exserted over the anthers;
anther-wings minutely notched at the base; follicles erect
on erect fruiting pedicels, attenuated, 2'–2¾' long, glabrous
or minutely puberulent.

Kansas to Colorado, Arizona and Mexico. May–July.

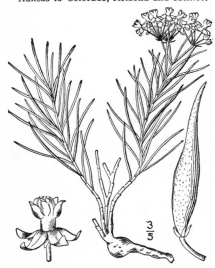

24. Asclepias pùmila (A. Gray) Vail.
Low Milkweed. Fig. 3406.

Asclepias pumila Vail; Britton & Brown, Ill. Fl. **3**:
12. 1898.

Asclepias verticillata var. *pumila* A. Gray, Proc.
Am. Acad. **12**: 71. 1876.

Stems 4'–10' high, tufted from a woody root.
Leaves very numerous, crowded, sometimes obscurely whorled, filiform-linear, 1'–2' long,
smooth or minutely roughened, the margins revolute; umbels 2–several, short-peduncled, few-
flowered; pedicels filiform, puberulent, 3''–4''
long; corolla greenish white, its segments oblong, 1½''–2'' long; column short; hoods white,
erect, oblong, entire, equalling the anthers,
shorter than the slender incurved horn; follicles
erect on erect fruiting pedicels, narrowly spindle-
shaped, 1½'–3' long, finely puberulent.

Dry plains, Iowa to South Dakota, Arkansas,
Texas, Wyoming, Colorado and New Mexico.

2. ASCLEPIODÒRA A. Gray, Proc. Am. Acad. 12 : 66. 1876.

Erect or decumbent perennial herbs, similar to *Asclepias,* with alternate or opposite entire leaves, and rather large flowers in terminal solitary or corymbed umbels. Sepals lanceolate. Corolla rotate, its segments spreading. Hoods oblong, inserted over the whole of the very short corona-column, curved upward, obtuse, crested within, at least in the upper part, slightly longer than the anther; at the sinuses between the hoods a small lobe or appendage, alternate with the anther-wings, simulating an inner crown. Anthers tipped with a scarious membrane, their wings horny, narrowed below, sometimes angled above the middle. Pollen-masses pendulous, pyriform, longer than their caudicles. Follicles ovoid or oblong, acuminate, with or without soft spinose processes, erect or ascending on the decurved or twice bent fruiting pedicels. Seeds comose. [Greek, gift of Aesculapius.]

Five or six species, natives of the southern United States and Mexico. Type species : *Asclepiodora viridis* (Walt.) A. Gray.

Glabrous or nearly so ; leaves oblong to ovate-lanceolate ; umbels corymbose.　　1. *A. viridis.*
Stem rough-puberulent ; leaves lanceolate or linear, acuminate ; umbels solitary.　2. *A. decumbens.*

1. **Asclepiodora víridis** (Walt.) A. Gray.
Oblong-leaved Milkweed. Fig. 3407.

Asclepias viridis Walt. Fl. Car. 107. 1788.

Asclepiodora viridis A. Gray, Proc. Am. Acad. 12 : 66. 1876.

Stem erect, puberulent above, simple, 1°–2° high. Leaves oblong to ovate-lanceolate, rather thin, obtuse and mucronulate or acute at the apex, narrowed or rounded at the base, short-petioled, 2½′–5′ long, ½′–1½′ wide; umbels 2–4, or sometimes solitary; peduncles 1½′–2′ long; pedicels slender, about ¾′ long; corolla greenish, its segments, when expanded, oblong, obtuse or acute, 4″–6″ long, 2–3 times as long as the purplish or violet entire-margined hoods; anther-wings narrow, scarcely angled above; fruiting pedicels twice bent; follicles ascending, puberulent, 2′–3′ long, sometimes with soft spinose projections.

In dry soil, Illinois to Kansas, Texas, South Carolina and Florida. May–July.

2. **Asclepiodora decúmbens** (Nutt.) A. Gray. Decumbent Milkweed. Fig. 3408.

Ananthrix decumbens Nutt. Trans. Am. Phil. Soc. (II.) 5 : 202. 1833–37.
Asclepiodora decumbens A. Gray, Proc. Am. Acad. 12 : 66. 1876.

Stems decumbent or ascending, rough-puberulent, 10′–2° long. Leaves firm, linear to lanceolate, glabrous above, puberulent on the veins beneath, acuminate at the apex, narrowed at the base, 3′–7′ long, 2″–8″ wide; umbel solitary, many-flowered; peduncle 1′–5′ long; pedicels stout, ½′–1′ long; corolla depressed-globose in the bud, greenish, its segments, when expanded, ovate or broadly oval, longer than the hoods; hoods purple, obtusely 3-lobed on the ventral margins, about 3″ long, their tips incurved; anther-wings broad, angled above; follicles nearly erect on the recurved fruited pedicels, 3′–4′ long, puberulent, at least when young, with or without soft projections.

In dry soil, Kansas to Texas and Mexico, west to Utah and Arizona. April–June.

3. ACERÀTES Ell. Bot. S. C. & Ga. 1 : 316. 1817.

Perennial herbs, similar to *Asclepias,* with alternate or opposite thick leaves, and green or purplish flowers in terminal or axillary and short-peduncled or sessile umbels. Calyx 5-parted or 5-divided, the segments acute, glandular within. Corolla deeply 5-cleft, the segments valvate, reflexed in anthesis. Corona-column very short. Corona of 5 involute-concave or somewhat pitcher-shaped hoods, neither horned nor crested within or in one species having a small interior crest and usually a few small processes at the base of the anther-wings, forming an obscure inner crown. Pollen-masses solitary in each sac, oblong, pendulous. Stigma 5-lobed. [Greek, without horn, referring to the crown.]

About 7 species, natives of North America. Type species : *Acerates longifolia* (Michx.) Ell.

Umbels sessile, or very nearly so, mostly axillary.
    Leaves oval to linear; hoods entire at the apex.   1. *A. viridiflora.*
    Leaves narrowly linear; hoods 3-toothed.   2. *A. angustifolia.*
Umbels, at least the lower, distinctly peduncled.
    Plants glabrous, or nearly so; umbels usually several; leaves narrow.
       Hoods obtuse, entire; column ½″ long; stem roughish puberulent.   3. *A. floridana.*
       Hoods emarginate; column very short; stem glabrous.   4. *A. auriculata.*
    Plant hirsute; umbel solitary, terminal; leaves ovate to oblong.   5. *A. lanuginosa.*

1. Acerates viridiflòra (Raf.) Eaton. Green Milkweed. Fig. 3409.

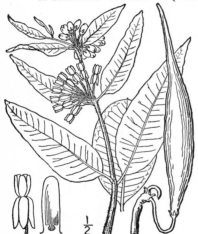

Asclepias viridiflora Raf. Med. Rep. (II.) **5** : 360. 1808.

Acerates viridiflora Eaton, Man. Ed. 5, 90. 1829.

Puberulent or tomentulose, at least when young; stems simple, reclined or ascending, rather stiff, 1°–3° high. Leaves slightly rough, alternate or opposite, thick, oval, oblong or ovate to lanceolate or linear, 1′–5′ long, ¼′–2′ wide, short-petioled, the margins usually undulate; umbels several, or rarely solitary, axillary, densely many-flowered, sessile or very nearly so; pedicels very slender, tomentose, 4″–8″ long; flowers green; corolla-segments narrowly oblong, 2″–3″ long; column very short or none; hoods lanceolate-oblong, obtuse, minutely 2-auricled at the base; mass of anthers longer than thick; anther-wings tapering below, semi-rhomboid above; follicles puberulent, 2′–4′ long.

In dry, sandy or rocky soil, Massachusetts to southern Ontario, Saskatchewan, Florida and Texas. Consists of several races, differing mainly in leaf-form. June–Sept.

2. Acerates angustifòlia (Nutt.) Dec. Narrow-leaved Milkweed. Fig. 3410.

Polyotus angustifolius Nutt. Trans. Am. Phil. Soc. (II.) **5** : 201. 1833–37.

A. angustifolia Dec. in DC. Prodr. **8** : 522. 1844.

Asclepias stenophylla A. Gray, Proc. Am. Acad. **12** : 72. 1876.

Stems mostly several together, erect, straight, 1°–2° high, puberulent above, glabrate below. Leaves opposite, or some of the lower alternate, sessile, narrowly linear, 2′–5′ long, glabrous, the revolute margins and the thick midvein rough beneath; umbels 10–15-flowered, short-peduncled or subsessile, axillary, usually numerous; pedicels puberulent; corolla-segments oblong, greenish; hoods white, not exceeding the anthers, 3-toothed at the apex, the acute middle tooth merely a prolongation of the thickened crest-like midvein, shorter than the obtuse lateral ones; anther-wings notched at about the middle; follicles slender, erect, about 3′ long or more.

On dry plains, Missouri, Nebraska and Colorado to Texas.

3. Acerates floridàna (Lam.) A. S. Hitchc. Florida Milkweed. Fig. 3411.

Asclepias floridana Lam. Encycl. 1: 284. 1783.
Acerates longifolia Ell. Bot. S. C. & Ga. 1: 317. 1817.
Acerates floridana A. S. Hitchc. Trans. St. Louis Acad. 5: 508. 1891.

Rough-puberulent; stems slender, simple or little branched, erect or ascending, 1°–3° high. Leaves mostly alternate, linear or rarely linear-lanceolate, acute or acuminate, short-petioled, 2′–8′ long, 1½″–6″ wide, commonly rough-ciliolate on the margins and midrib; umbels several or solitary, peduncled, usually many-flowered; peduncles 3″–15″ long; pedicels slender, hirsute, ½′–1′ long; corolla greenish white, its segments narrowly oblong, about 2″ long; column short but distinct; hoods oblong, obtuse, entire, shorter than the anthers; anther-wings narrowed to the base; follicles densely puberulent, 4′–5′ long.

Moist soil, Ohio to southern Ontario and Minnesota, North Carolina, Florida and Texas. June–Sept.

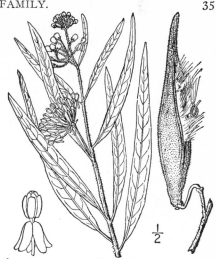

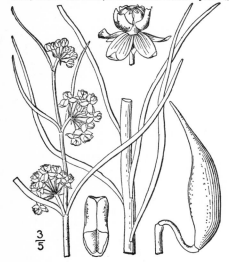

5. Acerates lanuginòsa (Nutt.) Dec. Woolly Milkweed. Fig. 3413.

Asclepias lanuginosa Nutt. Gen. 1: 168. 1818.

Acerates lanuginosa Dec. in DC. Prodr. 8: 523. 1844.

Hirsute all over; stems erect, slender, simple, 6′–18′ high. Leaves oblong, ovate-lanceolate or lanceolate, obtuse at the apex, narrowed or rounded at the base, short-petioled, 1′–4′ long, 4″–15″ wide; umbel solitary, terminal, densely many-flowered, peduncled; peduncle stout, densely hirsute, ½′–1½′ long; pedicels slender, puberulent or hirsute; corolla greenish, its segments oblong, about 2½″ long; column none; hoods purplish, oblong, obtuse, entire, with a flat fold or auricle on the infolded lower ventral margins, shorter than the anthers; anther-wings broadest below the middle.

On prairies, northern Illinois to Minnesota, Nebraska and Wyoming. June–Aug.

4. Acerates auriculàta Engelm. Auricled Milkweed. Fig. 3412.

Acerates auriculata Engelm. Bot. Mex. Bound. Surv. 160. 1859.
Asclepias auriculata Holzinger, Bot. Gaz. 17: 125. 1892.

Stem glabrous, often glaucous, sinuous above, rarely branched below, usually stout, 1°–3° high. Leaves mostly alternate, narrowly linear, glabrous, short-petioled, 3′–8′ long, ½″–2½″ wide, becoming leathery, the rough margins not revolute; umbels commonly several, densely many-flowered, peduncled; peduncles 2″–1′ long, pubescent; pedicels slender, pubescent; flowers greenish white tinged with dull purple; corolla-segments oblong, 2″–2½″ long; column short, but distinct; hoods yellow, often with a purplish keel, entire, or emarginately truncate at the apex, not exceeding the anthers, the involute margins spreading at the base into auricles; follicles 2′–3′ long, curved.

In dry soil, Nebraska and Colorado to Texas and New Mexico. June–Sept.

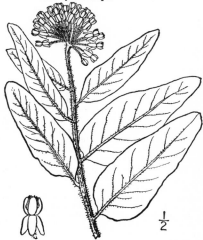

4. GONÓLOBUS Michx. Fl. Bor. Am. 1: 119. 1803.

[Ampelanus Raf.; Britton, Bull. Torr. Club **21**: 314. 1894.]

[Enslenia Nutt. Gen. **1**: 164. 1818. Not Raf. 1817.]

Perennial twining herbaceous vines, with petioled opposite cordate thin leaves, and small whitish flowers in axillary peduncled clusters. Calyx 5-parted, minutely glandular within, the segments lanceolate. Corolla campanulate, deeply 5-cleft, the lobes slightly contorted, nearly erect. Crown nearly sessile, of 5 membranous truncate lobes, each appendaged by a simple or 2-cleft awn. Stamens inserted at the base of the corolla, the filaments connate into a short tube; anthers terminated by an inflexed membrane; pollen-masses solitary in each sac, ellipsoid, pendulous. Stigma conic, slightly 2-lobed. Follicles thick, acuminate. Seeds comose. [Greek, referring to the somewhat angled pod.]

Three species, natives of America. Type species: *Gonolobus laevis* Michx.

1. Gonolobus laèvis Michx. Sand Vine. Enslen's-vine. Fig. 3414.

Gonolobus laevis Michx. Fl. Bor. Am. 1: 119. 1803.
Enslenia albida Nutt. Gen. 1: 164. 1818.
A. albidus Britton, Bull. Torr. Club **21**: 314. 1894.

Stem sparingly puberulent, at least above, high-climbing, slender. Leaves slender-petioled, ovate, gradually acuminate, deeply cordate, palmately veined, glabrous or very nearly so, entire, 3′–7′ long, 1½′–5′ wide; petioles 1′–4′ long; cymes numerous, densely flowered; peduncles stout, 3′′–2′ long; flowers 2′′–3′′ long; corolla-segments lanceolate, acute, twice as long as those of the calyx and exceeding the 2-cleft awns of the corona-lobes; follicles on ascending fruiting pedicels, 4′–6′ long, somewhat angled, glabrous when mature.

Along river-banks and in thickets, Pennsylvania to Illinois, Kansas, Florida and Texas. June–Aug.

5. CYNÁNCHUM L. Sp. Pl. 212. 1753.

[Vincetoxicum Moench, Meth. 717. 1794.]

Perennial twining herbaceous or slightly woody vines (some species erect herbs), with opposite or rarely verticillate or alternate leaves, and small yellowish green or purplish flowers in axillary cymes. Calyx 5-parted, minutely glandular within. Corolla rotate, deeply 5-cleft, the segments spreading, somewhat twisted. Crown flat or cup-like, entire, 5-lobed or 5-parted, the lobes not appendaged. Stamens attached to the base of the corolla, their filaments connate into a tube; anthers appendaged by an inflexed membrane. Pollen-masses solitary in each sac, pendulous. Stigma flat or conic. Follicles acuminate, glabrous. Seeds comose. [Greek, dog-strangling.]

About 100 species, natives of both the Old World and the New. Besides the following, 2 native species occur in the southeastern United States. Type species: *Cynanchum acutum* L.

1. Cynanchum nìgrum (L.) Pers. Black Swallow-wort. Fig. 3415.

Asclepias nigra L. Sp. Pl. 216. 1753.
Vincetoxicum nigrum Moench, Meth. 317. 1794.
Cynanchum nigrum Pers. Syn. 1: 274. 1805.

Twining, or at first erect, puberulent, slender, 2°–5° high. Leaves ovate or ovate-lanceolate, entire, thin, acuminate at the apex, rounded at the base, pinnately veined, petioled, 2′–5′ long, ½′–2½′ wide; petioles 2′′–12′′ long; pedicels 1½′′–3′′ long; flowers dark purple, about 2½′′ broad; corolla-segments pubescent within; crown fleshy, 5-lobed; follicles on nearly straight fruiting pedicels, about 2′ long, glabrous.

In waste places, escaped from gardens, Massachusetts to Pennsylvania and Ohio and in British Columbia. Introduced from Europe. June–Sept.

Cynanchum Vincetóxicum (L.) Pers., with greenish-white glabrous corollas, another Old World species, is recorded as escaped from cultivation in southern Ontario.

6. VINCETÓXICUM Walt. Fl. Car. 104. 1788.

Twining or trailing perennial vines, with opposite usually cordate leaves, and rather large purple, brown, white or greenish flowers in axillary cyme-like umbels or fascicles. Calyx 5-parted or deeply 5-cleft, mostly 5-glandular within. Corolla rotate, very deeply 5-parted, the tube very short, the segments convolute in the bud. Corona (crown) annular or cup-shaped, entire, lobed or divided, adnate to the corolla. Stamens inserted on the base of the corolla, the filaments connate into a tube; anthers not appendaged, merely tipped, borne along or just under the margin of the flat-topped stigma, the sacs more or less transversely dehiscent. Pollen-masses solitary in each sac, horizontal or nearly so. Follicles thick, acuminate, smooth, angled or tuberculate. Seeds comose. [Greek, subduing poison.]

About 75 species, natives of America. Besides the following, some 10 others occur in the southern and southwestern United States. Type species: *Vincetoxicum gonocarpos* Walt.

Crown annular, 10-crenate; follicles angled, not warty.
  Corolla about twice as long as the calyx.   1. *V. suberosum.*
  Corolla 3-4 times as long as the calyx.   2. *V. gonocarpos.*
Crown cup-shaped, about as high as the anthers; follicles warty.
  Flowers purple to dull yellow.
    Corolla-segments oblong, 3″-4″ long; crown crenate.   3. *V. hirsutum.*
    Corolla-segments linear or linear-oblong, 5″-7″ long.
      Crown merely crenate.   4. *V. obliquum.*
      Crown toothed or lobed.
        Crown 5-lobed, with a subulate 2-cleft tooth in each sinus.   5. *V. carolinense.*
        Crown 10-toothed, the alternate teeth thinner and longer.   6. *V. Shortii.*
  Flowers white; crown deeply cleft.   7. *V. Baldwinianum.*

1. Vincetoxicum suberòsum (L.) Britton. Coast Vincetoxicum. Fig. 3416.

Cynanchum suberosum L. Sp. Pl. 212. 1753.
G. suberosus R. Br. in Ait. Hort. Kew. Ed. 2: 82. 1811.
V. suberosum Britton, Mem. Torr. Club 5: 266. 1894.

Stem pubescent or glabrous, slender, twining. Leaves thin, 2′-5′ long, 1′-3′ wide, ovate or ovate-oval, acute or abruptly acuminate at the apex, cordate at the base; petioles ½-2′ long; umbels commonly few-flowered; peduncles ¼′-1′ long; pedicels ½′-1′ long, fleshy, nearly glabrous; corolla brown-purple, broadly conic in the twisted bud, its segments lanceolate or ovate-lanceolate, acute, pubescent or granulose within, 3″-4″ long, about twice as long as the calyx; crown an annular fleshy undulately 10-crenate disk; follicles glabrous, 3-5-angled, when young fleshy, when mature dry and spongy, 4′-6′ long, 1′ in diameter or more.

In thickets, Virginia to Florida, mainly near the coast. May-July.

2. Vincetoxicum gonocàrpos Walt. Large-leaved Angle-pod. Fig. 3417.

Vincetoxicum gonocarpos Walt. Fl. Car. 104. 1788.
G. macrophyllus Michx. Fl. Bor. Am. 1: 119. 1803.
Gonolobus laevis var. *macrophyllus* A. Gray, Syn. Fl. 2: Part 1, 103. 1878.

Glabrous or pubescent, stems slender, climbing high. Leaves broadly ovate, thin, 3′-8′ long, 2′-6′ wide, acuminate at the apex, deeply cordate at the base, the sinus narrow or the rounded auricles overlapping; petioles 1′-4′ long; umbels few-flowered; peduncles 1′-3′ long; pedicels rather stout, glabrous or nearly so; corolla conic in the bud, not twisted, its segments lanceolate, glabrous, 4″-5″ long, 3-4 times as long as the calyx; crown a low obtusely undulate disk; follicles glabrous, similar to those of the preceding species but usually shorter.

Along rivers and in moist thickets, Virginia to South Carolina, Georgia, Indiana Missouri and Texas.

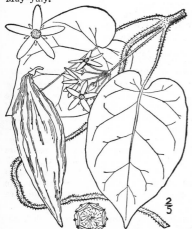

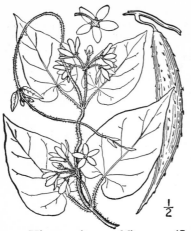

3. Vincetoxicum hirsùtum (Michx.) Britton. Hairy Vincetoxicum. Fig. 3418.

Gonolobus hirsutus Michx. Fl. Bor. Am. 1 : 119. 1803.
V. hirsutum Britton, Mem. Torr. Club 5 : 266. 1894.

Stem downy, slender. Leaves ovate, acuminate at
the apex, deeply cordate at the base, downy, 2′–4′
long, 1′–2½′ wide, the sinus narrow, or the lobes
sometimes overlapping; petioles slender, pubescent,
½′–2′ long; peduncles usually about equalling the
petioles, sometimes longer; umbels few–several-
flowered; corolla brown-purple to greenish yellow,
ovoid in the bud, its segments oblong, very obtuse,
3″–4″ long, minutely puberulent without, about 4
times as long as the densely pubescent calyx; crown
cup-shaped, fleshy, about as high as the anthers, the
margin 10-crenate; follicles lanceolate, 3′–5′ long,
muricate, puberulent; seeds entire.

In thickets, Maryland to Florida, west to Tennessee.
Running milkweed. Negro-vine. July–Aug.

4. Vincetoxicum oblìquum (Jacq.) Britton. Large-flowered Vincetoxicum. Fig. 3419.

Cynanchum hirtum L. Sp. Pl. 212. 1753?
Cynanchum obliquum Jacq. Coll. 1 : 148. 1786.
G. obliquus R. Br.; R. & S. Syst. 6 : 64. 1820.
V. obliquum Britton, Mem. Torr. Club 5 : 266. 1894.

Stem puberulent or hirsute. Leaves
pubescent, broadly ovate, short-acuminate at the
apex, deeply cordate at the base with an open or
closed sinus, 2′–8′ long, 1½′–6′ wide; petioles
rather stout, pubescent, 1′–4′ long; umbels slender-peduncled, few–several-flowered; pedicels very
slender, 1′–2′ long; corolla narrowly conic in the
bud, red-purple within, greenish and minutely
pubescent without, its segments elongated-linear,
obtuse, 6–7 times as long as the hirsute calyx;
crown cup-shaped, as high as the anthers, fleshy,
its margin 10-crenulate, the intermediate crenulations sometimes 2-dentate; follicles ovoid-lan-
ceolate, 2′–3′ long, muricate.

In thickets, Pennsylvania to Ohio, Missouri, Virginia and Kentucky. July–Aug.

5. Vincetoxicum carolinénse (Jacq.) Britton. Carolina Vincetoxicum. Fig. 3420.

Cynanchum carolinense Jacq. Coll. 2 : 228. 1788.
G. carolinensis R. Br.; R. & S. Syst. 6 : 62. 1820.
V. carolinense Britton, Mem. Torr. Club 5 : 265. 1894.

Stem hirsute. Leaves broadly ovate, acute or
short-acuminate at the apex, deeply cordate at the
base with a narrow or closed sinus, 3′–7′ long, 2′–5½′
wide, pubescent, at least beneath; petioles hirsute,
1½′–4′ long; peduncles 2′–4′ long; pedicels very slender, 1′ long or more; corolla brown-purple, oblong-
conic in the bud, puberulent without, its segments
linear-oblong or linear-lanceolate, obtusish, 5″–6″
long, 5–6 times longer than the hirsute calyx; crown
cup-shaped, scarcely fleshy, 5-lobed, with a subulate
longer 2-cleft erect tooth in each sinus; follicles
muricate.

In thickets, Virginia to Missouri, south to South Carolina and Louisiana. May–July.

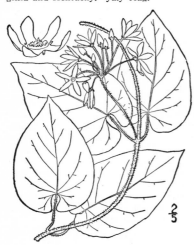

6. Vinceoxicum Snórtii (A. Gray) Britton.   Short's Vincetoxicum.   Fig. 3421.

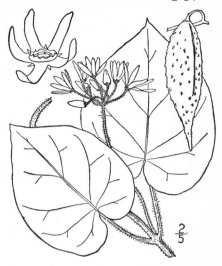

Gonolobus obliquus var. *Shcrtii* A. Gray, Syn. Fl. 2:
  Part 1, 104. 1878.
G. Shortii A. Gray, loc. cit. Ed. 2, 404. 1886.
V. Shortii Britton, Mem. Torr. Club 5: 266. 1894.

Stem pubescent, or hirsute with spreading
hairs. Leaves downy, broadly ovate, acute or
short-acuminate at the apex, deeply cordate at
the base and when old with a narrow or closed
sinus, 4'–7' long, 1½'–5½' wide; petioles stout,
pubescent, 1½'–3' long; peduncles usually longer
than the petioles; umbels several-flowered; pedicels 1' long or more; corolla oblong-conic in
the bud, dark crimson-purple, its lobes linear,
5"–7" long, 5–7 times as long as the hirsute
calyx; crown cup-shaped, fleshy, as high as the
anthers, its margin about 10-toothed, the alternate teeth thinner and longer, emarginate or
2-parted, the others broader, thicker, with an
obscure internal crest or ridge below the summit; follicles warty.

In thickets, Pennsylvania to eastern Kentucky
and Georgia. Flowers with the odor of the straw-
berry-shrub. June–Aug.

7. Vincetoxicum Baldwiniànum (Sweet) Britton.   Baldwin's Vincetoxicum. Fig. 3422.

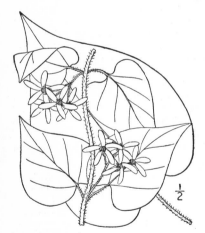

Gonolobus Baldwinianus Sweet; A. Gray, Syn. Fl. 2:
  Part 1, 104. 1876.
Vincetoxicum Baldwinianum Britton, Mem. Torr. Club
  5: 265. 1894.

Stem pubescent and hirsute. Leaves downy, broadly
ovate, acute, or short-acuminate at the apex, deeply
cordate at the base, 3'–6' long, or more; petioles hirsute, 1'–2' long; peduncles 6"–12" long, usually longer
than the pedicels; umbels several–many-flowered;
corolla white or cream-color, the lobes thin, oblong,
or becoming spatulate, 4"–5" long; crown thin, the 5
broader lobes quadrate, emarginate, or obscurely
toothed; in their sinuses a pair of very slender
linear-subulate teeth of more than double their
length, much surpassing the stigma.

Missouri and Arkansas to Georgia. May–June.

Periploca graèca L., silk-vine, a handsome woody
climber, with glabrous ovate-oblong leaves and brownish flowers in umbels, the obtuse corolla-segments villous on the inner side has been collected as an escape
from cultivation.

Family 19.   DICHONDRACEÀE Dumort. Anal. Fam. 20, 25.   1829.

DICHONDRA FAMILY.

Consists only of the following genus:

1. DICHÓNDRA Forst. Char. Gen. Pl. 39. *pl. 40.*   1776.

Prostrate or creeping slender annual (sometimes perennial?) silky-pubescent or glabrous
herbs, with nearly orbicular cordate or reniform petioled entire leaves, and very small solitary axillary peduncled flowers. Sepals nearly equal, oblong or spatulate. Corolla open-
campanulate, deeply 5-parted, the lobes induplicate in the bud. Stamens shorter than the
corolla; filaments filiform. Ovary villous, deeply 2-parted, each lobe 2-celled; styles 2, simple,
arising from the bases of the ovary-lobes; stigmas capitate. Fruit of 2 pubescent 2-valved
or indehiscent 1–2-seeded capsules. [Greek, two-grained, referring to the capsules.]

About 5 species, natives of warm and tropical regions. Besides the following, another occurs
in the southwest. Type species: *Dichondra repens* Forst.

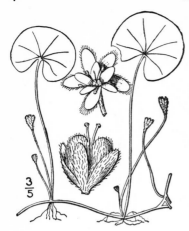

1. Dichondra carolinénsis Michx. Dichondra.
Fig. 2423.

Dichondra carolinensis Michx. Fl. Bor. Am. 1 : 136. 1803.

Somewhat pubescent, or glabrous; stems almost filiform, creeping, rooting at the nodes, 6'–2° long. Leaves
orbicular to reniform, deeply cordate, ¼'–1½' in diameter,
palmately veined; petiole often much longer than the
blade; flowers 1"–2" broad; peduncles filiform; sepals
obtuse, spatulate or obovate; corolla yellow to white,
shorter than the sepals, its lobes ovate to oblong; capsule 1" high or less.

In moist or wet places, Virginia to Texas and Mexico,
near the coast. Widely distributed in Central and South
America. Has been regarded as referable to the Old
World *D. repens* Forst.; the specific name *evolvulacea* was
used for it in our first edition, in error.

Family 20. CONVOLVULÀCEAE Vent. Tabl. 2 : 394. 1799.
MORNING-GLORY FAMILY.

Herbs, some tropical species shrubs or trees, the stems twining, ascending,
trailing or erect, with alternate exstipulate entire dentate lobed or dissected leaves,
and regular perfect axillary cymose or solitary flowers. Calyx inferior, 5-parted
or 5-divided, usually persistent, the segments or sepals imbricated. Corolla gamopetalous, funnelform, salverform, campanulate, tubular or rarely subrotate, the
limb 5-angled, 5-lobed or entire. Stamens 5, inserted low down on the tube of
the corolla and alternate with its lobes, all anther-bearing, the filaments filiform,
or dilated at the base, equal or unequal; anthers 2-celled, the sacs longitudinally
dehiscent. Disk annular or none. Ovary superior, sessile, 2–3-celled, with 2
ovules in each cavity, or falsely 4–6-celled with a single ovule in each cavity,
entire or 2–4-divided; styles 1–3, terminal, or arising from between the ovary-
divisions; ovules anatropous. Fruit a 2–4-valved capsule or of 2–4 distinct carpels, in our species. Seeds erect, the testa villous, pubescent or glabrous; embryo
plaited or crumpled; cotyledons foliaceous; endosperm fleshy or cartilaginous,
usually scanty.

About 45 genera and probably 1000 species, of wide geographic distribution, most abundant
in the tropics.

Style 2-cleft or 2-divided.	
Style 2-cleft or 2-parted.	1. *Stylisma.*
Style 2-divided to the ovary, each division 2-cleft.	2. *Evolvulus.*
Style entire up to the stigma.	
Stigma or stigmas capitate or globose.	
Corolla salverform; stamens and style exserted.	3. *Quamoclit.*
Corolla funnelform or campanulate; stamens and style included.	4. *Ipomoea.*
Stigmas 2, filiform to oblong.	5. *Convolvulus.*

1. STYLÍSMA Raf. Neog. 2. 1825.

Herbs, mostly perennial and procumbent, with entire short-petioled or sessile leaves, and
1–5-flowered axillary peduncles; flowers white, purple, pink, or yellow. Sepals acute or
obtuse. Corolla campanulate or funnelform-campanulate; limb plaited, 5-angled or slightly
5-lobed. Stamens included; filaments filiform, or dilated at the base. Ovary 2-celled; style
2-cleft or 2-parted; stigmas capitate. Capsule globose to ovoid, 2-celled, 2–4-valved. Style
rarely 3-parted and ovary 3-celled. Seeds 1–4, glabrous or pubescent. [Greek, referring to
the 2-parted style.]

Seven known species, of the southeastern United States and Mexico. In our first edition this
genus was referred to the Australian *Breweria* R. Br. Type species: *Convolvulus aquaticus* Walt.

Sepals acute or acuminate; leaves oblong, elliptic or linear.	
Corolla white; filaments pubescent; plant pubescent or puberulent.	1. *S. humistrata.*
Corolla purple; filaments glabrous; plants silky-tomentose.	2. *S. aquatica.*
Sepals obtuse; leaves narrowly linear.	3. *S. Pickeringii.*

1. Stylisma humistràta (Walt.) Chapm.
Southern Breweria. Fig. 3424.

Convolvulus humistratus Walt. Fl. Car. 94. 1788.
Stylisma humistrata Chapm. Fl. S. States, 346. 1860.
Bonamia humistrata A. Gray, Man. Ed. 5, 376. 1867.
Breweria humistrata A. Gray, Syn. Fl. **2** : Part 1, 217. 1878.

Pubescent or puberulent; stems slender, 1°–2° long, simple, or with a few long branches. Leaves elliptic, oblong-elliptic, or ovate-oblong, obtuse and mucronulate or some of them emarginate at the apex, subcordate, rounded or narrowed at the base, ½′–1′ wide, 1′–2′ long; petioles 1″–3″ long; peduncles slender, longer than the leaves, 1–7-flowered, minutely bracted at the summit; sepals glabrous or puberulent, oblong, acuminate, 2″–3″ long; corolla white, 6″–8″ long; filaments pubescent; style 2-cleft; capsule ovoid, acute, glabrous, about as long as the calyx.

In dry pine barrens, Virginia to Florida and Louisiana. May–Aug.

2. Stylisma aquática (Walt.) Chapm. Water Breweria. Fig. 3425.

Convolvulus aquaticus Walt. Fl. Car. 94. 1788.
Stylisma aquatica Chapm. Fl. S. States, 346. 1860.
Bonamia aquatica A. Gray, Man. Ed. 5, 376. 1867.
Breweria aquatica A. Gray, Syn. Fl. **2** : Part 1, 217. 1878.

Finely and densely silky-tomentose, branched, the branches long and slender. Leaves oblong, elliptic or oblong-lanceolate, obtuse at both ends, mucronate or emarginate at the apex, sometimes subcordate at the base, ½′–1½′ long, 2″–8″ wide; peduncles 1–3-flowered, longer than the leaves, minutely bracted at the summit; sepals densely silky-tomentose, oblong, acute or acuminate, about 2″ long; corolla purple or pink, 5″–7″ long; filaments glabrous; style 2-parted nearly to the base.

In wet soil, especially in pine barrens, Missouri to Texas, east to North Carolina and Florida. May–Aug.

3. Stylisma Pickeríngii (M. A. Curtis) A. Gray. Pickering's Breweria. Fig. 3426.

Convolvulus Pickeringii M. A. Curtis, Bost. Journ. Nat. Hist. **1** : 129. 1837.

Stylisma Pickeringii A. Gray, Man. Ed. 2, 335. 1856.

Bonamia Pickeringii A. Gray, Man. Ed. 5, 376. 1867.

Breweria Pickeringii A. Gray, Syn. Fl. **2** : Part 1, 217. 1878.

Stem pubescent or puberulent, very slender, simple or branched, 1°–2° long. Leaves puberulent or glabrous, narrowly linear, obtuse or acutish at the apex, narrowed at the base, 1′–2½′ long, ½″–2″ wide, the lowest sometimes narrowly spatulate; petioles very short; peduncles slender, about as long as the leaves, with 1 or 2 linear bracts at the summit which are usually longer than the pedicels and calyx; sepals pubescent or hirsute, ovate to oval, obtuse, about 2″ long; corolla white, about 1′ long; filaments nearly glabrous; style 2-cleft, above, exserted; capsule ovoid, acute, pubescent, longer than the calyx.

In dry pine barrens, New Jersey to North Carolina; Illinois to Iowa, Louisiana and Texas. June–Aug.

2. EVÓLVULUS L. Sp. Pl. Ed. 2, 391. 1762.

Erect or diffuse branching, mostly silky-pubescent or pilose, annual or perennial herbs, with small usually entire leaves, and axillary solitary, racemose or paniculate, small blue pink or white flowers. Sepals nearly equal, acute or obtuse. Corolla funnelform, campanulate or rotate, the limb plaited, 5-angled or 5-lobed. Stamens included or exserted; filaments filiform; anthers ovate or oblong. Ovary entire, 2-celled; style 2-divided to the base, or near it, each division deeply 2-cleft; stigmas linear-filiform. Capsule 2-celled, globose to ovoid, 2–4-valved, 1–4-seeded. Seeds glabrous. [Latin, unrolling.].

About 85 species, natives of warm and tropical regions. Besides the following, some 7 others occur in the southern United States. Type species: *Evolvulus nummulàrius* L.

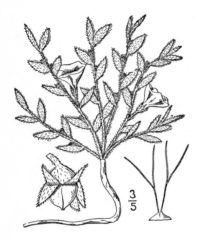

3/5

1. Evolvulus pilòsus Nutt.   Evolvulus.
Fig. 3427.

Evolvulus argenteus Pursh, Fl. Am. Sept. 187. 1814. Not R. Br. 1810.

Evolvulus pilosus Nutt. Gen. 1: 174. 1818.

Perennial, densely silky-pubescent or villous; stems ascending or erect, 3′–9′ high, very leafy. Leaves sessile, oblong, lanceolate or spatulate, 3″–9″ long, 1″–3″ wide, acute or obtuse at the apex, narrowed at the base; flowers solitary and nearly sessile in the axils; peduncles 2-bracted at the base, recurved in fruit, 1″–2″ long; sepals lanceolate, acute or acuminate; corolla funnelform-campanulate, purple or blue, 3″–6″ broad; capsule 1½″–2″ in diameter, about as long as the sepals.

On dry plains, North Dakota to Missouri, Nebraska, Mexico and Arizona. May–July.

3. QUÁMOCLIT [Tourn.] Moench, Meth. 453. 1794.

Twining herbaceous vines, with petioled entire lobed or pinnately parted leaves, and cymose racemose or solitary peduncled axillary flowers. Sepals 5, herbaceous, equal, acuminate, mucronate or appendaged. Corolla salverform (usually scarlet in the following species), the tube narrow, somewhat dilated above, mostly longer than the spreading 5-lobed limb. Stamens and simple style more or less exserted; stigma capitate; ovary 2-celled or falsely 4-celled, 4-ovuled. Fruit usually 4-celled and 4-seeded. [Greek, dwarf kidney-bean.]

About 10 species, of warm and tropical regions, only the following in North America. Type species: *Ipomoea coccinea* L.

Leaves pinnately parted into very narrow segments.   1. *Q. Quamoclit.*
Leaves cordate, acuminate, entire or angulate-lobed.   2. *Q. coccinea.*

1. Quamoclit Quámoclit (L.) Britton.   Cypress Vine.   Indian Pink.   Fig. 3428.

Ipomoea Quamoclit L. Sp. Pl. 159. 1753.
Q. vulgaris Choisy in DC. Prodr. 9: 336. 1845.
Q. Quamoclit Britton, in Britt. & Brown, Ill. Fl. 3: 22. 1898.

Annual, glabrous; stem slender, twining to a height of 10°–20°. Leaves ovate in outline, petioled or nearly sessile, 2′–7′ long, pinnately parted nearly to the midvein into narrowly linear entire segments less than 1″ wide; peduncles slender, commonly much longer than the leaves, 1–6-flowered; pedicels 1′ long or more, thickening in fruit; sepals oblong, obtuse, usually mucronulate, 2″–3″ long; corolla scarlet, rarely white, salverform, 1′–1½′ long, the tube expanded above, the limb nearly flat, the lobes ovate, acutish; stamens and style exserted; ovary 4-celled; ovule 1 in each cell; capsule ovoid, 4-valved, about 5″ high, twice as long as the sepals.

In waste and cultivated ground, Virginia to Florida, Kansas and Texas. Sparingly escaped from gardens farther north. Naturalized from tropical America. July–Oct. American red bell-flower. Sweet-william-of-the-Barbadoes. Cupid's-flower. Red jasmine.

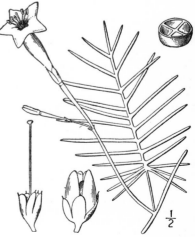

1/2

2. Quamoclit coccínea (L.) Moench.   Small Red Morning-glory.   Fig. 3429.

Ipomoea coccinea L. Sp. Pl. 160. 1753.

I. hederaefolia L. Syst. Ed. 10, 925. 1759.

Quamoclit coccinea Moench, Meth. 453. 1794.

Annual, glabrous or puberulent, stem twining to a height of several feet or trailing. Leaves ovate to orbicular, deeply cordate, long-acuminate, 2′–6′ long, entire or angulate-lobed, slender-petioled; peduncles few–several-flowered, usually not longer than the leaves; sepals oblong, obtuse, about 2″ long, subulate-appendaged; corolla scarlet, salverform, 10″–20″ long, the limb obscurely 5-lobed; stamens and style slightly exserted; ovary 4-celled with 1 ovule in each cell; capsule globose, 4-valved, 3″–4″ in diameter.

Along river-banks and in waste places, Rhode Island to Pennsylvania, Florida, Ohio, Missouri, Texas and Arizona. Naturalized from tropical America, or native in the Southwest. A hybrid of this species with the preceding is sometimes cultivated. American jasmine. July–Oct.

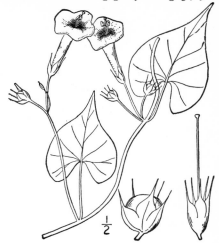

4. IPOMOÈA L. Sp. Pl. 159. 1753.

Twining trailing ascending or rarely erect herbs, annual or perennial, with large showy axillary solitary or cymose flowers. Sepals equal or unequal. Corolla funnelform or campanulate, the limb entire, 5-angled or 5-lobed, the tube more or less plaited. Stamens equal or unequal, included; filaments filiform, or dilated at the base; anthers ovate, oblong, or linear. Ovary entire, globose or ovoid, 2–4-celled, 4–6-ovuled; style filiform, included; stigmas 1 or 2, capitate or globose. Capsule globose or ovoid, usually septifragally 2–4-valved, 2–4-seeded. [Greek, worm-like.]

About 400 species, of wide geographic distribution. Besides the following, some 30 others occur in southern and western North America. Known as Morning-Glory or False Bindweed. Type species: *Ipomoea pes-tigridis* L.

Ovary 2-celled (rarely 4-celled); stigma entire or 2-lobed.
  Leaves cordate; stems trailing or twining.
    Perennial from an enormous root; corolla 2′–3′ long.   1. *I. pandurata.*
    Annual; roots fibrous; corolla 4″–6″ long, white.   2. *I. lacunosa.*
    Annual; corolla 1′–1½′ long, pink or purple.   3. *I. trichocarpa.*
  Leaves linear; stems ascending or erect.   4. *I. leptophylla.*
Ovary 3-celled; stigmas 3; leaves cordate. (Genus Pharbitis.)
    Leaves entire; corolla 2′–2½′ long.   5. *I. purpurea.*
    Leaves deeply 3-lobed, corolla 1′–1½′ long.   6. *I hederacea.*

1. Ipomoea panduràta (L.) Meyer. Wild Potato Vine. Fig. 3430.

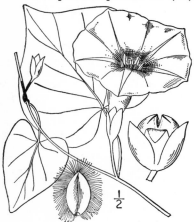

Convolvulus panduratus L. Sp. Pl. 153. 1753.
I. pandurata Meyer, Prim. Fl. Esseq. 100. 1818.

Perennial from an enormous fleshy root, glabrous or puberulent; stems trailing or feebly climbing, 2°–12° long. Leaves broadly ovate, cordate, acuminate at the apex, 2′–6′ long, slender-petioled, entire, sometimes contracted in the middle, or some of the later ones rarely angulate-dentate or 3-lobed; peduncles 1–5-flowered, much elongated in fruit; sepals oblong, obtuse or acutish, 6″–8″ long, glabrous; corolla funnelform, white, or with pinkish purple stripes in the throat, 2′–3′ long, the limb 5-lobed; ovary 2-celled; capsule ovoid, 2-valved, 2–4-seeded, the seeds densely woolly on the margins and pubescent on the sides.

In dry soil, in fields or on hills, Ontario to Connecticut, Florida, Michigan, Kansas and Texas. Occurs rarely with double flowers. Man-of-the-Earth. Mecha-meck (Indian). Wild sweet potato. Man-root. Wild jalap. Scammony. May–Sept.

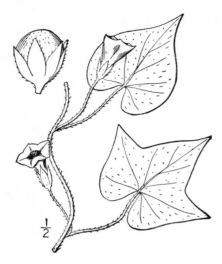

2. Ipomoea lacunòsa L. Small-flowered White Morning-glory. Fig. 3431.

Ipomoea lacunosa L. Sp. Pl. 161. 1753.

Annual, pubescent or hirsute, rarely glabrous; stem twining, 2°–10° long. Leaves slender-petioled, broadly ovate, cordate, acute or acuminate at the apex, entire, angled or 3-lobed, 2′–4′ long, the lobes acute; peduncles 1–3-flowered, shorter than the leaves; pedicels slender; sepals oblong or lanceolate, acute or acuminate, pubescent or ciliate, about 5″ long; corolla funnelform, 6″–10″ long, white, or the limb purple; ovary 2-celled; stigma capitate; capsule globose, 2-valved, shorter than or about equalling the sepals.

In moist soil, Pennsylvania to South Carolina, Illinois, Missouri, Kansas and Texas. In ballast at Atlantic seaports. White star. Morning-glory. July–Sept.

3. Ipomoea trichocàrpa Ell. Small-flowered Pink Morning-glory. Fig. 3432.

Convolvulus carolinus L. Sp. Pl. 154. 1753.

Ipomoea trichocarpa Ell. Bot. S. C. & Ga. 1 : 258. 1817.

Ipomoea commutata R. & S. Syst. 4 : 228. 1819.

Ipomoea carolina Pursh, Fl. Am. Sept. 145. 1814. Not L. 1753.

Similar in habit to the preceding species, but the leaves usually more lobed; peduncles often longer than the leaves, 1–3-flowered; sepals lanceolate or oblong-lanceolate, acuminate, pubescent or ciliate; corolla 1′–1½′ long, pink or purple; capsule glabrous or pubescent.

Kansas to Texas, east to South Carolina and Florida.

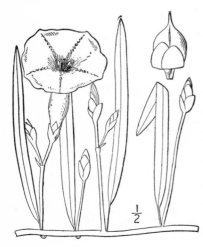

4. Ipomoea leptophýlla Torr. Bush Morning-glory. Fig. 3433.

Ipomoea leptophylla Torr. in Frem. Rep. 95. 1845.

Perennial from an enormous root, which sometimes weighs 25 lbs., glabrous throughout; stems erect, ascending or reclining, rather stout, 2°–4° long, much branched. Leaves narrowly linear, entire, acute, 2′–5′ long, 1″–3″ wide; petioles very short; peduncles stout, nearly erect, usually shorter than the leaves, 1–4-flowered; pedicels shorter than the peduncles; sepals broadly ovate, obtuse, 3″–4″ long, or the outer shorter; corolla funnelform, purple or pink, about 3′ long, the limb scarcely lobed; capsule ovoid, acute, 8″–12″ long, 2-celled, much longer than the sepals; seeds pubescent.

In dry soil, South Dakota to Nebraska, Wyoming, Texas and New Mexico. Man-root. May–July.

5. Ipomoea purpùrea (L.) Lam. Morning-glory. Fig. 3434.

Convolvulus purpureus L. Sp. Pl. Ed. 2, 219. 1762.
Ipomoea purpurea Lam. Tabl. Encycl. 1 : 466. 1791.
Pharbitis purpurea Voigt. Hort. Sub. Calcutta 354.
1845.

Annual, pubescent; stem retrorsely hairy, twining or trailing, 4°–10° long. Leaves broadly ovate,
deeply cordate, acute or acuminate, 2′–4′ wide,
slender-petioled; peduncles slender, 1–5-flowered,
often longer than the petioles; sepals lanceolate
or oblong, acute, pubescent or hirsute near the
base, 6″–8″ long; corolla funnelform, blue, purple, pink, variegated or white, 2′–2½′ long; ovary
3-celled (rarely 2-celled); stigmas 3 (rarely 2);
capsule depressed-globose, about 5″ in diameter,
shorter than the sepals.

In waste places, commonly escaped from gardens,
Nova Scotia to Florida, west to Ontario, Nebraska
and Texas. There is a double-flowered form in cultivation. Adventive or naturalized from tropical
America. Ropewind. July–Oct.

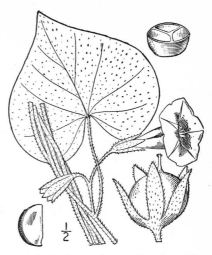

6. Ipomoea hederàcea Jacq. Ivy-leaved Morning-glory. Fig. 3435.

Ipomoea hederacea Jacq. Icon. Rar. *pl. 36.* 1781.
Pharbitis hederacea Choisy, Mem. Soc. Gen. 6: 440.
1833.

Annual, pubescent; stem twining or climbing to
a height of 2°–5°, slender, retrorsely hairy. Leaves
ovate-orbicular in outline, long-petioled, deeply
3-lobed, cordate at the base, 2′–5′ long, the lobes
ovate, acuminate, entire, or the lateral ones sometimes repand or dentate; peduncles 1–3-flowered,
much shorter than the petioles; flowers opening in
early morning, soon closing; sepals lanceolate with
long linear often recurved tips, densely hirsute below, sparingly so above, 8″–12″ long; corolla funnelform, the tube usually nearly white, the limb light
blue or purple, 1′–1½′ long; ovary 3-celled; stigmas
3; capsule depressed-globose, 3-valved, about as
long as the lanceolate portion of the sepals.

In fields and waste places, Maine to Florida, Pennsylvania, Nebraska and Mexico. Naturalized or adventive from tropical America. July–Oct.

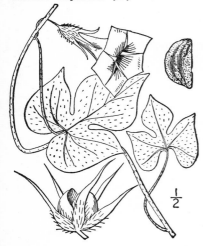

6. CONVÓLVULUS L. Sp. Pl. 153. 1753.

Herbs (the following species perennials with slender roots or rootstocks) with trailing,
twining or erect stems. Leaves entire dentate or lobed, mostly cordate or sagittate and
petioled. Flowers axillary, solitary or clustered, large, pink, purple or white. Sepals nearly
equal or the outer larger, the calyx bractless or with a pair of bracts at its base. Corolla
funnelform or campanulate, the limb plaited, 5-angled, 5-lobed, or entire. Stamens inserted
on the tube of the corolla, included; filaments filiform, or dilated at the base. Ovary 1–2-
celled, 4-ovuled; style filiform; stigmas 2, filiform, oblong, or ovoid. Capsule globose or
nearly so, 1–4-celled, 2–4-valved. Seeds glabrous. [Latin, to roll together, or entwine.]

About 200 species, of wide distribution in tropical and temperate regions. Besides the following, some 30 others occur in the southern and western United States. Type species: *Convolvulus
sepium* L.

Calyx with two large bracts at the base, which enclose it.
  Stems trailing or climbing.
    Peduncles long, much longer than the petioles.
      Stems 3°–10° long; leaves hastate, the auricles often dentate.   1. *C. sepium.*
      Stems 1°–3° long; leaves sagittate, the auricles rounded, entire.   2. *C. repens.*
    Peduncles short, mostly not longer than the petioles.   3. *C. fraterniflorus.*
  Stem erect or ascending; flowers white; bracts not cordate.   4. *C. spithamaeus.*
Calyx not bracted; peduncle bracted at the summit.
  Glabrous or nearly so; leaves entire, auriculate.   5. *C. arvensis.*
  Canescent; leaves with 2–4 basal lobes.   6. *C. incanus.*

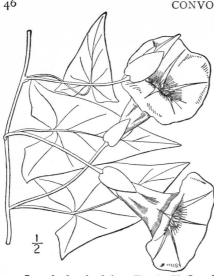

$\frac{1}{2}$

1. Convolvulus sèpium L.　Hedge or Great Bindweed.　Lily-bind.　Fig. 3436.

Convolvulus sepium L. Sp. Pl. 153.　1753.
Convolvulus sepium var. *americanus* Sims, Bot. Mag. *pl. 732*.　1804.
Calystegia sepium R. Br. Prodr. Fl. Nov. Holl. 1: 483.　1810.

Glabrous or sparingly pubescent; stems extensively trailing or high-twining, 3°–10° long. Leaves slender-petioled, triangular in outline, hastate, 2′–5′ long, acute or acuminate at the apex, the basal lobes divergent, usually acute, angulate-dentate or entire; petioles ½′–2′ long; peduncles 1-flowered, longer than the petioles, often 2–3 times as long; flowers pink with white stripes or white throughout, about 2′ long; bracts at the base of the corolla, large, ovate, acute or obtuse, cordate; stigmas oblong.

In fields and thickets, usually in moist soil, Newfoundland to North Carolina, British Columbia, Illinois, Nebraska and New Mexico. Also in Europe and Asia. June–Aug. Bell-bind. Woodbind. Pear- or Devil's-vine. Lady's-nightcap. Hedge- or harvest-lily. Rutland beauty. Woodbine. German scammony. Creepers. Bracted-bindweed.

Convolvulus japónicus Thunb. Fl. Jap. 85.　1784, a species with narrow hastate leaves and smaller pink flowers, cultivated in a double-flowered form, has in this form escaped from cultivation from southeastern New York to the District of Columbia and Missouri.

2. Convolvulus rèpens L.　Trailing or Hedge Bindweed.　Fig. 3437.

Convolvulus repens L. Sp. Pl. 153.　1753.
Convolvulus sepium var. *repens* A. Gray, Syn. Fl. 2: Part 1, 215.　1878.

More or less pubescent or tomentose; stem trailing or twining, 1°–3° long, simple, or sparingly branched.　Leaves ovate or oblong, petioled, 1′–2′ long, obtuse, acute or abruptly acuminate at the apex, sagittate or cordate at the base, entire, the basal lobes rounded, scarcely or not at all divergent; petioles ½′–1′ long; peduncles 1-flowered, equalling or longer than the leaves; flowers white (sometimes pink?) about 2′ long; calyx enclosed by 2 ovate acute or obtusish slightly cordate bracts; stigmas oblong.

In moist and dry soil, Quebec to Florida and Louisiana.　Recorded from the Great Lake region. May–Aug.

C. intèrior House, of the western plains, with broader leaves and smaller corollas, is found in Kansas and Nebraska.

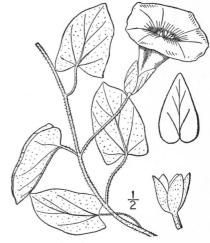

$\frac{1}{2}$

$\frac{2}{5}$

3. Convolvulus fraterniflòrus MacKenzie & Bush.　Short-stalked Bindweed.　Fig. 3438.

C. Sepium fraterniflorus Mack. & Bush, Fl. Jackson Co. 153.　1902.

C. fraterniflorus Mack. & Bush, Rep. Mo. Bot. Gard. 16: 104.　1905.

Sparingly pubescent; stems trailing or twining, much branched, 3°–6° long.　Leaves hastate or hastate-sagittate, short-pubescent on both sides, 4′ long or less, acute at the apex, the basal lobes entire or dentate, spreading; peduncles often 2 in each axil, wing-angled, mostly not longer than the petioles; bracts large, cordate, pubescent, concealing the sepals at flowering time; sepals glabrous, obtusish, 5″–8″ long; corolla white, about 2′ long.

Dry banks and prairies, western Missouri; recorded eastward to the District of Columbia.　July–Sept.

4. Convolvulus spithamaèus L. Upright or Low Bindweed. Fig. 3439.

Convolvulus spithamaeus L. Sp. Pl. 158. 1753.
Calystegia spithamaea Pursh, Fl. Am. Sept. 143. 1814.
Volvulus spithamaeus Kuntze, Rev. Gen. Pl. 447. 1891.
Convolvulus camporum Greene, Pittonia 3: 328. 1898.

Pubescent, or glabrate; stem erect or ascending, straight, or the summit sometimes feebly twining, 6′–12′ high. Leaves oval, short-petioled or the uppermost sessile, usually obtuse at both ends, sometimes acutish at the apex, and subcordate at the base, 1′–2′ long, ½–1¼′ wide; peduncles 1-flowered, longer than the leaves; flowers white, nearly 2′ long; calyx enclosed by 2 large oval acutish bracts which are narrowed at both ends and not cordate at the base; stigmas oblong, thick.

In dry sandy or rocky fields or on banks, Nova Scotia to Ontario, Manitoba, Florida and Kentucky. Dwarf morning-glory. Low or bracted-bindweed. May–Aug.

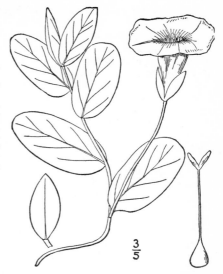

3/5

5. Convolvulus arvénsis L. Small Bindweed. Fig. 3440.

Convolvulus arvensis L. Sp. Pl. 153. 1753.

Glabrous, or nearly so; stems trailing or decumbent, very slender, 1°–2½° long, simple or branched. Leaves slender-petioled, ovate or oblong, entire, obtusish and mucronulate or acutish at the apex, sagittate or somewhat hastate at the base, 1′–2′ long, the basal lobes spreading, acute or obtuse; peduncles 1–4-flowered (commonly 2-flowered), shorter than the leaves, 1–3-bracted at the summit, usually with another bract on one of the pedicels; sepals oblong, obtuse, 1½″ long; corolla pink or nearly white, 8″–12″ broad; calyx not bracted at the base; stigmas linear.

In fields and waste places, Nova Scotia to Ontario, Montana, New Jersey, Pennsylvania, Kansas, New Mexico and California. Naturalized from Europe. Native also of Asia. May–Sept. Hedge-bells. Bearbind. Corn-lily. Withwind. Bellbine. Corn-bind. Lap-love. Sheep-bine.

3/4

6. Convolvulus incànus Vahl. Hoary Bindweed. Fig. 3441.

Convolvulus incanus Vahl, Symb. Bot. 3: 23. 1794.

Finely and densely canescent, pale, or sometimes greener; stems procumbent or trailing, usually branched, 1°–3° long. Leaves rather short-petioled, lanceolate, ovate to linear in outline, usually with 2–4 divergent lobes at the base, or the lower pair of lobes reflexed, otherwise entire or irregularly dentate, obtuse and mucronulate at the apex, 1′–2′ long; peduncles 1–2-flowered, as long as or longer than the leaves, minutely bracted at the summit; pedicels 3″–6″ long; sepals oblong, obtuse or mucronulate, about 3″ long; corolla white to rose-color; stigmas narrowly linear; capsule globose, about as long as the sepals.

In waste places, near Lincoln, Neb. (according to Webber). In dry soil, Kansas and Arkansas to Texas, Arizona and Mexico. Also in southern South America. April–Aug.

2/3

Family 21. **CUSCUTÀCEAE** Dumort, Anal. Fam. 20. 1829.

DODDER FAMILY.

White or yellow slender parasites, dextrorsely twining, the leaves reduced to minute alternate scales, the small white, yellowish or pinkish flowers cymosely clustered. Calyx inferior, 5-lobed or 5-parted (rarely 4-lobed or 4-parted), or of 5 distinct sepals. Corolla campanulate, ovoid, urceolate or cylindric, 5-lobed (rarely 4-lobed), the lobes imbricated in the bud, the tube bearing as many fimbriate or crenulate scales as there are lobes and alternate with them, or these sometimes obsolete. Stamens as many as the corolla-lobes and alternate with them, inserted in the throat or sinuses above the scales, short-exserted or included; filaments short or slender; anthers short, ovate or oval, obtuse, 2-celled, the sacs longitudinally dehiscent. Ovary globose or oblong, 2-celled; ovules 2 in each cavity; styles 2, terminal, separate, or rarely united below; stigmas linear or capitate. Capsule globose or ovoid, circumscissile, irregularly bursting or indehiscent, 1–4-seeded. Seeds glabrous, globose or angular; embryo linear, terete, curved or spiral, its apex bearing 1–4 minute scales, endosperm fleshy; cotyledons none.

1. CÚSCUTA [Tourn.] L. Sp. Pl. 124. 1753.

Characters of the family. The filiform twining stems are parasitic on herbs and shrubs by numerous minute suckers. The seeds germinate in the soil and the plantlet attaches itself to its host, its root and lower portion soon perishing. The subsequent nutrition of the parasite is apparently wholly through its suckers. [Name from the Arabic.]

About 100 species, of wide geographic distribution. Besides the following, some 15 others occur in the southern and western parts of North America. Known as Dodder, or Strangle-weed. Type species: *Cuscuta europaèa* L.

* Corolla-scales crenulate; stigmas slender; capsule circumscissile; introduced species.

Scales crenulate above, not incurved.	1. *C. Epilinum.*
Scales crenulate all around, strongly incurved.	2. *C. Epithymum.*

** Corolla-scales fringed; stigmas capitate; capsule indehiscent; native species.

Sepals united below into a gamosepalous calyx.
  Flowers very nearly sessile; corolla persistent at the base of the capsule.
    Corolla-scales ovate, fringed all around; calyx-lobes obtuse.   3. *C. arvensis.*
    Corolla-scales abortive, or of a few processes; calyx-lobes acutish.   4. *C. Polygonorum.*
  Flowers distinctly pedicelled; corolla enclosing or capping the capsule, or at length deciduous.
    Tips of the corolla-lobes incurved or reflexed.
      Scales ovate, fringed all around; capsule enclosed by the corolla.   5. *C. indecora.*
      Scales abortive, or of a few slender processes; corolla capping the capsule. 6. *C. Coryli.*
    Corolla-lobes spreading or recurved.
      Scales small, irregularly fringed; capsule depressed-globose.   7. *C. Cephalanthi.*
      Scales long, fringed mainly above; capsule pointed.
        Corolla 1½" long; capsule globose, short-pointed.   8. *C. Gronovii.*
        Flowers 2"–3" long; capsule oval, long-pointed.   9. *C. rostrata.*
Sepals separate, subtended by similar bracts.
  Flowers cymose, pedicelled; scales short; bracts entire.   10. *C. cuspidata.*
  Flowers closely sessile in dense clusters; bracts serrulate.
    Bracts few, broad, appressed; styles as long as the ovary.   11. *C. compacta.*
    Bracts numerous, narrow, their tips recurved; styles longer than the ovary. 12. *C. paradoxa.*

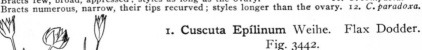

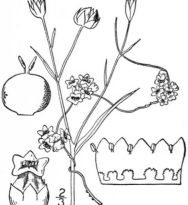

2/3

1. Cuscuta Epílinum Weihe.   Flax Dodder.
Fig. 3442.

Cuscuta Epilinum Weihe, Archiv. Apoth. **8**: 54. 1824.
Cuscuta densiflora Soyer-Willem. Act. Soc. Linn. Paris **4**: 281. 1826.

Stems very slender, yellow or red; flowers sessile in dense clusters, yellowish white, about 1½" long. Calyx hemispheric, 5-lobed, the lobes acute, nearly as long as the corolla-tube; corolla yellowish-white, short, cylindric, becoming urceolate, 5-lobed, the lobes ovate, acutish, spreading, its scales short, erect, less than one-half the length of the tube, 2-cleft or emarginate, crenulate above, the crenulations not extending to the base; stigmas linear-filiform; capsule circumscissile, the withering-persistent corolla borne on its summit.

On flax, Nova Scotia to New Jersey and Pennsylvania. Introduced from Europe. Native also of Asia. July–Aug.

2. Cuscuta Epíthymum Murr. Thyme Dodder. Lesser Lucerne or Clover Dodder. Fig. 3443.

Cuscuta Epithymum Murr. in L. Syst. Ed. 13, 140. 1774.

Cuscuta Trifolii Bab. Phytol. 1: 467. 1843.

Stems filiform, red; flowers sessile in small dense clusters, pinkish, about 1″ long. Calyx variable, 4–5-lobed, more than one-half the length of the cylindric corolla-tube, the lobes acute; corolla 4–5-lobed, the lobes erect, about one-half as long as the tube, acute, its scales strongly incurved, crenulate nearly or quite to the base; stigmas filiform; capsule circumscissile, capped by the withering corolla.

Usually on clover, Maine and Ontario to South Dakota and Pennsylvania. Introduced from Europe, where it occurs on thyme, clover and other low plants. Hailweed. Hairweed. July–Sept.

Cuscuta europaèa L., another Old World species, with obtuse calyx-lobes and shorter styles, is recorded from Maine.

3. Cuscuta arvénsis Beyrich. Field Dodder. Love-vine. Fig. 3444.

Cuscuta arvensis Beyrich; Hook. Fl. Bor. Am. 2: 77. As synonym. 1834.

Plant pale yellow; stems filiform, the flowers nearly sessile in small clusters. Calyx broad, 5-lobed, the lobes broad, obtuse; corolla nearly campanulate, 5-lobed, the lobes acute or acuminate, as long as the tube, their tips reflexed, its scales large, ovate, as long as or longer than the tube, densely fringed all around with short irregular processes; stamens not exserted; style shorter than the ovary; stigmas capitate; capsule depressed-globose, indehiscent, the withering corolla and usually the stamens persistent at its base.

On various herbs and low shrubs, Massachusetts to Manitoba, Florida, Texas, Mexico and California. Also in the West Indies and South America. July–Aug.

4. Cuscuta Polygonòrum Engelm. Smartweed Dodder. Fig. 3445.

Cuscuta Polygonorum Engelm. Am. Journ. Sci. 43: 342. *pl. 6. f. 26–29.* 1842.

C. chlorocarpa Engelm.; A. Gray. Man. 350. 1848.

Plant orange-yellow; stems slender but rather coarse; flowers sessile or nearly so in dense clusters. Calyx short, 4–5-lobed, the lobes ovate-oblong, acute or acutish; lobes of the corolla 4 or 5, triangular-ovate, acute, mostly as long as the tube, the scales usually obsolete, wanting, or consisting of only 2 or 3 slender processes on each side of the attached lower portion of the filament; filaments mostly slender; styles shorter than the ovary; stigmas capitate; capsule globose, the withering corolla persistent at its base.

On *Polygonum* and other herbs, Pennsylvania and Delaware to Minnesota, Wisconsin and Arkansas. July–Sept. Has been referred to the South American *C. obtusiflora* H.B.K.

4

5. Cuscuta indecòra Choisy.   Pretty Dodder.   Fig. 3446.

Cuscuta indecora Choisy, Mem. Soc. Gen. **9**: *278.*
pl. 3. f. 5. 1841.
C. pulcherrima Scheele, Linnaea **21**: 750. 1848.
Cuscuta decora Choisy; Engelm. Trans. St. Louis
  Acad. 1 : 501. 1859.

Stems rather stout; flowers 1½″ long, pedicelled in loose cymes, more or less papillose.
Calyx 5-lobed, the lobes ovate to lanceolate,
acute, mostly shorter than the corolla-tube;
corolla campanulate, 5-lobed, the lobes triangular, minutely crenulate, spreading, nearly as
long as the tube, their tips inflexed; scales
ovate, erect, irregularly fringed with short
processes all around; stamens slightly exserted
or included; stigmas capitate; capsule oblong,
acute, enveloped by the withering corolla.

On various herbs and low shrubs, Illinois to
Nebraska, south to Florida, Texas and Mexico,
in several races.   Also in the West Indies and
South America.   Corolla white; stigmas often
yellow or purple.   June–Aug.

6. Cuscuta Còryli Engelm.   Hazel Dod-der.   Fig. 3447.

Cuscuta Coryli Engelm. Am. Journ. Sci. **43**: 337.
  f. 7–11. 1842.
Cuscuta inflexa Engelm. Trans. St. Louis Acad. **1** :
  502. 1859.

Stems coarse; flowers about 1″ long, pedicelled in loose or rather dense cymes.   Calyx
4–5-lobed, the lobes triangular or triangular-
lanceolate, acutish, about as long as the corolla-
tube; corolla campanulate, 4–5-lobed, the lobes
minutely crenulate, nearly erect, triangular,
acute, about as long as the tube, their tips
inflexed; scales small, oval, obtuse, often with
only a few processes on each side; stamens
scarcely exserted; styles shorter than the
ovary; stigmas capitate; capsule oblong, pointed, enveloped or at length capped by the withering corolla.

On the hazels and other shrubs or tall herbs,
Connecticut to Virginia, South Dakota and Arkansas.   July–Aug.

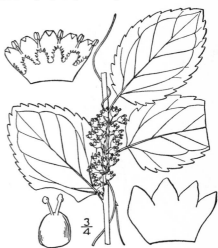

7. Cuscuta Cephalánthi Engelm.   Button-bush Dodder.   Fig. 3448.

Cuscuta Cephalanthi Engelm. Am. Journ. Sci. **43**: 336.
  pl. 6. f. 1–6. 1842.

Cuscuta tenuiflora Engelm.; A. Gray, Man. 350. 1848.

Plant yellow, stems rather coarse; flowers about
1″ long, short-pedicelled, clustered; calyx 5-lobed,
the lobes ovate, obtuse, shorter than the corolla-tube;
corolla cylindric-campanulate, its lobes ovate, obtuse
and rounded, spreading, one-half the length of the
tube or less; scales about as long as the lobes,
fringed mainly toward the apex with irregular processes; stamens included; styles slender, about as
long as the ovary, shorter than the ripe capsule;
stigmas capitate; capsule depressed-globose, 1½″ in
diameter, surrounded or capped by the withering, at
length deciduous corolla.

On shrubs and tall herbs, Pennsylvania to Minnesota,
Texas and Arizona.   July–Aug.

8. Cuscuta Gronòvii Willd. Gronovius' Dodder. Love-vine. Fig. 3449.

C. Gronovii Willd.; R. & S. Syst. 6: 205. 1820.
Cuscuta vulgivaga Engelm. Am. Journ. Sci. 43: 338. *pl. 6. f. 12–16.* 1842.

Stems yellow to orange, slender, high-climbing; flowers usually short-pedicelled, numerous in dense cymes. Calyx not bracted, its lobes ovate, obtuse, shorter than the corolla-tube; corolla campanulate, about 1½″ long, the lobes ovate, obtuse, rounded, spreading, nearly as long as the tube, the scales narrow, equalling or longer than the tube, thickly fringed about the summit and sparingly along the sides with long slender processes; styles slender, not as long as the ovary; stigmas capitate; capsule globose, short-pointed or pointless, 1½″ in diameter, enveloped or capped by the withering deciduous corolla.

On herbs and low shrubs, Nova Scotia to Manitoba, Montana, Florida and Texas. Scald-weed. Devil's-gut July–Aug.

9. Cuscuta rostràta Shuttlw. Beaked Dodder. Fig. 3450.

Cuscuta rostrata Shuttlw.; Engelm. Bost. Journ. Nat. Hist. 5: 225. 1845.

Stems coarse, yellowish white; flowers larger than in any of our other species, loosely cymose, pedicelled. Calyx gamosepalous, 5-lobed, the lobes ovate-triangular, shorter than the corolla-tube; corolla campanulate, 2″–3″ long, white, its lobes broadly ovate, obtuse, about as long as the calyx-lobes; scales narrow, sometimes spatulate, shorter than the tube, heavily fringed at the summit and sparingly along the sides with long slender processes; stamens included; styles slender, about as long as the flask-shaped ovary; stigmas capitate; capsule oval, long-beaked.

On herbs and shrubs, Maryland to South Carolina and Georgia. July–Sept.

10. Cuscuta cuspidàta Engelm. Cuspidate Dodder. Fig. 3451.

Cuscuta cuspidata Engelm. Bost. Journ. Nat. Hist. 5: 224. 1845.

Plant yellowish; stems slender; flowers about 1½″ long in loose panicled cymes. Calyx of 5 distinct entire sepals, shorter than the corolla-tube, with 2–4 similar bracts at its base and often others on the pedicels; sepals orbicular to lanceolate, cuspidate, mucronate or acuminate; corolla nearly salverform, its lobes triangular-lanceolate or oblong, acute or cuspidate, spreading, about one-half the length of the tube; scales narrow, usually less than one-half as long as the tube, fringed all around with short irregular processes; stamens not exserted; styles very selnder, longer than the ovary; stigmas capitate; capsule bearing the withered corolla on its summit.

On coarse herbs, Nebraska to Missouri and Texas. July–Sept.

11. Cuscuta compácta Juss.   Compact Dodder.   Love-vine.   Fig. 3452.

Cuscuta compacta Juss.; Choisy, Mem. Soc. Gen. **9**: 281. *t. 4. f. 2.* 1841.

Plant yellowish white, stems rather stout; flowers about 2″ long, closely sessile in dense clusters. Calyx of 5 (rarely 4) distinct oval crenulate obtuse sepals, subtended by 3–5 similar rhombic-orbicular appressed serrulate bracts; corolla salverform, persistent, the tube cylindric, its 5 (rarely 4) lobes oblong or ovate, obtuse, spreading, much shorter than the tube, the scales narrow, one-half the length of the tube, fringed with numerous long processes; stamens included; styles slender; capsule oblong, enveloped or capped by the withering corolla.

On shrubs, Ontario to Massachusetts, New York and Alabama, west to Kansas and Texas.   July–Sept.

12. Cuscuta paradóxa Raf.   Glomerate or American Dodder.   Fig. 3453.

Cuscuta paradoxa Raf. Ann. Nat. **13.** 1820.
Cuscuta glomerata Choisy, Mem. Soc. Gen. **9**: 184. *pl. 4. f. 1.* 1841.

Plant yellowish white, stems slender; flowers sessile, 1½″ long, exceedingly numerous in dense confluent clusters covering portions of the stem of the host-plant.   Calyx of 5 distinct concave oblong obtuse serrulate sepals, subtended by 8–15 narrower serrulate much imbricated bracts with recurved tips; corolla tube oblong-cylindric, its lobes oblong-lanceolate or triangular-lanceolate, obtuse, spreading or recurved, persistent; scales copiously fringed at the summit and sparingly along the sides with numerous long processes; styles 2–4 times as long as the ovary; capsule capped by the withering corolla.

On tall herbs, mainly Compositae, Ohio to South Dakota, Nebraska and Texas.   July–Sept.

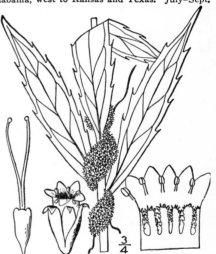

Family 22.   **POLEMONIÀCEAE** DC. Fl. Franc. **3**: 645.   1805.

PHLOX FAMILY.

Herbs, some species slightly woody, with alternate or opposite entire lobed or dissected leaves.   Flowers perfect, corymbose-capitate, cymose or paniculate, regular, or nearly regular.   Calyx inferior, persistent, tubular or campanulate, 5-cleft, the lobes or teeth slightly imbricated.   Corolla gamopetalous, funnelform, saucer-shaped, campanulate or rotate, the limb 5-parted, convolute in the bud.   Stamens 5, inserted on the tube of the corolla and alternate with its lobes; filaments slender or filiform; anthers ovate, oblong or linear, versatile, 2-celled, the sacs longitudinally dehiscent.   Ovary superior, mostly 3-celled; ovules 2–∞ in each cavity, amphitropous; style simple filiform; stigmas 3, linear.   Capsule mostly loculicidally 3-valved.   Seeds various, sometimes winged, sometimes enveloped in mucilage and emitting spiral tubes when wetted; endosperm abundant; embryo straight; cotyledons flat; radicle inferior.

About 20 genera and over 200 species, most abundant in western America.
Calyx distended and at length ruptured by the ripening capsule.
  Calyx scarious between the lobes.
    Corolla salverform; leaves opposite, entire.
      Seeds not mucilaginous when wetted; mostly perennials with large flowers; leaves
        opposite.   1. *Phlox.*
      Seeds mucilaginous when wetted; annuals; floral leaves alternate; flowers small.
                                           2. *Microsteris.*
    Corolla funnelform, tubular, salverform or campanulate; leaves alternate or opposite.
                                                3. *Gilia.*
  Calyx not scarious between the lobes; leaves alternate, deeply cleft.   4. *Leptodactylon.*

Calyx not distended nor ruptured by the capsule; leaves alternate.
   Calyx-teeth herbaceous, not spinulose-tipped.
      Stamens declined; leaves pinnate.
      Stamens straight and leaves entire in our species.
   Calyx-teeth spinulose-tipped; leaves pinnatifid.

5. *Polemonium.*
6. *Collomia.*
7. *Naverretia.*

1. PHLÓX L. Sp. Pl. 151. 1753.

   Perennial or rarely annual, erect or diffuse herbs, with opposite entire leaves, or some of the upper ones alternate, and large blue purple red or white flowers, in terminal cymes or cymose panicles. Calyx tubular or tubular-campanulate, 5-ribbed, 5-cleft, the lobes acute or acuminate, mostly scarious-margined and the sinuses commonly scarious. Corolla salver-form, the tube narrow, the limb 5-lobed; lobes obovate, orbicular or obcordate, spreading. Stamens straight, short, unequally inserted on the corolla-tube, included. Ovary oblong or ovoid, 3-celled; style usually slender; ovules 1-4 in each cavity. Capsule ovoid, 3-valved, at length distending and rupturing the calyx-tube. Seeds usually only 1 in each cavity of the capsule, ovoid, wingless or narrowly winged, not emitting spiral threads when wetted. [Greek, flame.]

   About 40 species, natives of North America and Russian Asia, many of them widely cultivated. Besides the following, some 24 others occur in the southern and western parts of North America. Type species: *Phlox glaberrima* L.

Leaves flat, ovate, oblong, lanceolate or linear.
  Cymes panicled; flowers short-pedicelled or sessile.
    Calyx-teeth subulate.
      Stem glabrous or puberulent; leaves lanceolate to oblong.
      Stem villous, glandular above; leaves ovate to ovate-lanceolate.
    Calyx-teeth lanceolate, acute; leaves lanceolate or ovate, acuminate.
  Cymes corymbose, simple, or flowers scattered.
    Flowering stems erect or ascending, simple.
      Plants glabrous or nearly so.
        Leaves ovate or oblong; calyx-teeth acute.
        Leaves lanceolate or linear; calyx-teeth subulate-lanceolate.
      Plants pubescent, hirsute or villous.
        Stems erect or ascending; no prostrate sterile shoots.
          Leaves linear or lanceolate, acuminate, spreading.
          Leaves linear-oblong, acute or obtuse, nearly erect.
        Stems ascending or reclining; sterile shoots prostrate.
          Lower leaves and those of the sterile shoots oblong or ovate.
          Lower leaves and those of the sterile shoots obovate.
    Stems diffusely branched, usually creeping; leaves narrow.
      Corolla-lobes cleft to or about the middle.
      Corolla-lobes cleft only at the apex.
      Corolla-lobes rounded; western.
Leaves subulate, fascicled or crowded; plants low.
  Stems creeping or ascending; flowers cymose; eastern.
    Corolla-lobes shallowly emarginate; plant not glandular.
    Corolla-lobes deeply emarginate; upper part of plant glandular.
  Densely tufted; flowers mostly solitary; western.
    Leaves densely white-woolly, 1″ long; plant moss-like.
    Leaves less woolly or merely ciliate, 2″-6″ long.
      Corolla-tube shorter than or equalling the calyx.
      Corolla-tube longer than the calyx.

1. *P. paniculata.*
2. *P. amplifolia.*
3. *P. maculata.*

4. *P. ovata.*
5. *P. glaberrima.*

6. *P. pilosa.*
7. *P. amoena.*

8. *P. divaricata.*
9. *P. stolonifera.*

10. *P. bifida.*
11. *P. Stellaria.*
12. *P. Kelseyi.*

13. *P. subulata.*
14. *P. Brittonii.*

15. *P. bryoides.*

16. *P. Hoodii.*
17. *P. Douglasii.*

1. Phlox paniculàta L. Garden Phlox.
Fig. 3454.

Phlox paniculata L. Sp. Pl. 151. 1753.

   Stem erect, stout or slender, simple or branched above, glabrous or puberulent, 2°-6° high. Leaves thin, sessile or short-petioled, oblong to oblong-lanceolate, acute or acuminate at the apex, narrowed at the base, or the uppermost subcordate, 2′-6′ long, ¼-1½′ wide; flowers short-pedicelled in compact paniculate cymules, the inflorescence often 12′ long; calyx-teeth subulate, glabrous, puberulent or glandular, more than one-half as long as the tube; corolla pink, purple or white, its lobes broadly obovate, rounded, entire, shorter than its tube; capsule oval, obtuse, slightly longer than the ruptured calyx-tube.

   In woods and thickets, Pennsylvania to Florida, Illinois, Kansas and Louisiana. Freely escaped from gardens in the north and east. Consists of many races, differing in leaf-form, size and color of flowers, and in pubescence. July–Sept.

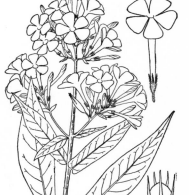

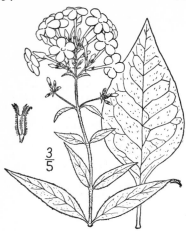

2. Phlox amplifòlia Britton. Large-leaved Phlox. Fig. 3455.

Phlox amplifolia Britton, Man. 757. 1901.

Stem villous or glandular-villous, at least above, 2°–3½° high. Leaves large and broad, 2½′–6′ long, 1½′–2½′ wide, roughish above, the upper sessile, the lower ones, or some of them, narrowed, usually abruptly, into winged petioles which are sometimes one-third as long as the blade; flowers similar to those of *P. paniculata,* the inflorescence often 1° long; calyx glandular-villous; corolla-tube glabrous, the lobes obovate, rounded or retuse; capsules 4″–5″ long.

Woods and thickets, Indiana to Missouri, Kentucky and Tennessee. June–Aug.

3. Phlox maculàta L. Wild Sweet-William. Fig. 3456.

Phlox maculata L. Sp. Pl. 152. 1753.
Phlox suaveolens Ait. Hort. Kew. 1: 206. 1789.

Stem slender, erect, simple or branched above, glabrous or puberulent, usually flecked with purple, 1½°–3° high. Leaves lanceolate or the upper ovate-lanceolate, glabrous, rather firm, long-acuminate, sessile, rounded or subcordate at the base, 2′–5′ long, widest just above the base, the lowest sometimes linear-lanceolate; flowers short-pedicelled, the compact cymules forming an elongated narrow thyrsoid panicle; calyx-teeth triangular-lanceolate, acute, or acuminate, about one-fourth the length of the tube; corolla pink or purple, rarely white, its lobes rounded, shorter than the tube; capsule similar to that of the two preceding species.

In moist woods and along streams, Connecticut to Florida, Ohio, Minnesota and Mississippi. Occasionally escaped from gardens further north. *P. maculata* var. *cándida* Michx. (*P. suavèolens* Ait.) is a race with white flowers and unspotted stem, occurring with the type. June–Aug.

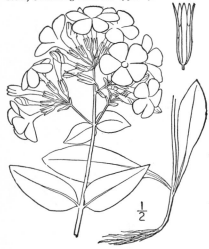

4. Phlox ovàta L. Mountain Phlox. Fig. 3457.

Phlox ovata L. Sp. Pl. 152. 1753.

Phlox carolina L. Sp. Pl. Ed. 2, 216. 1762.

Glabrous or nearly so throughout; stems simple, slender, ascending from a decumbent base, 1°–2° high. Leaves rather firm, the upper ovate or ovate-lanceolate, sessile by a rounded or subcordate base, acute at the apex, 1′–2′ long, the lower and basal ones longer, oblong or ovate-oblong, acute at both ends, narrowed into slender often margined petioles; flowers short-pedicelled in corymbed or sometimes simple cymes; calyx-teeth lanceolate or triangular-lanceolate, acute, or acuminate, one-third to one-half the length of the tube; corolla pink or red, its lobes obovate, rounded, entire.

In woods, Pennsylvania to North Carolina, Georgia and Alabama, mostly in the mountains. May–Aug.

5. Phlox glabérrima L.   Smooth Phlox.
Fig. 3458.

Phlox glaberrima L. Sp. Pl. 152. 1753.

Glabrous or nearly so throughout; stem simple, slender, erect or ascending, 1°–3° high. Leaves lanceolate or linear, rather firm, mostly 1-nerved, acuminate at the apex, narrowed at the base, 1½′–4′ long, 2″–6″ wide, sessile, or the lowest linear or oblong, obtusish, shorter, and short-petioled; flowers short-pedicelled, the cymules corymbed; calyx-teeth subulate-lanceolate, one-third to one-half the length of the tube; corolla commonly pink, its lobes obovate, rounded or obcordate, longer than the tube.

In open woods and on prairies, Virginia to Illinois, Wisconsin, Florida, Mississippi, Missouri and Arkansas. Ascends to 2200 ft. in Virginia.   May–July.

6. Phlox pilòsa L.   Downy or Prairie Phlox.   Fig. 3459.

Phlox pilosa L. Sp. Pl. 152. 1753.

Soft downy or hairy, often glandular; stem erect or ascending, simple or branched, slender, 1°–2° high. Leaves linear or lanceolate, spreading or divaricate, long-acuminate, 1′–4′ long, 1½″–4″ wide, sessile, the base narrowed or rounded; cymules corymbed; flowers short-pedicelled; calyx glandular, viscid, its teeth setaceous-subulate, longer than the tube; corolla pink, purple or white, its lobes obovate, entire, the tube usually pubescent; capsule shorter than the calyx.

In dry soil, Ontario to Manitoba, Connecticut, New Jersey, Florida, Arkansas and Texas. Hairy phlox. Sweet-william. April–June.

Phlox argillàcea Clute & Ferriss is a recently described relative or race of this species, growing on prairies in Indiana and Illinois.

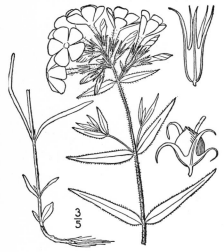

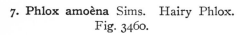

7. Phlox amoèna Sims.   Hairy Phlox.
Fig. 3460.

Phlox amoena Sims, Bot. Mag. *pl. 1308.* 1810.

Usually quite hairy; stems simple, slender, ascending, 6′–18′ high. Leaves linear-oblong to ovate-oblong, sessile, acute or obtuse at the apex, mostly narrowed at the base, nearly erect, ¾′–2′ long, 1½″–2½″ wide, the lowest much shorter; flowers very nearly sessile in a dense terminal simple or somewhat compound cyme, which is subtended by the uppermost pair of leaves; calyx hirsute, its teeth subulate, as long as or shorter than the tube; corolla pink or white, its lobes obovate, entire or rarely emarginate, shorter than the glabrous tube.

In dry soil, Virginia to Kentucky, Tennessee, Florida and Alabama.   April–June.

8. Phlox divaricàta L.   Wild Blue Phlox.   Fig. 3461.

Phlox divaricata L. Sp. Pl. 152. 1753.

Finely viscid-pubescent; stems ascending or diffuse, slender, producing creeping or ascending leafy shoots from the base. Leaves of the sterile shoots oblong or ovate, obtuse, 1′–2′ long, those of the flowering stems lanceolate, ovate-lanceolate, or oblong, mostly acute or acutish; flowers pedicelled in open corymbed cymules, faintly fragrant; calyx-teeth subulate, longer than the tube; corolla bluish, its lobes obcordate, emarginate or entire, not much longer than the tube, sometimes shorter; capsule oblong-globose, about 2″ high.

In moist woods, Quebec to Ontario, Minnesota, Pennsylvania, Florida, Louisiana and Arkansas. Ascends to 3700 ft. in Virginia. Sometimes called wild sweet william. April–June.

9. Phlox stolonífera Sims.   Crawling Phlox.   Fig. 3462.

Phlox stolonifera Sims, Bot. Mag. *pl. 563.* 1802.
Phlox reptans Michx. Fl. Bor. Am. 1 : 145. 1803.

Hirsute or pubescent; stems slender, diffuse, producing sterile creeping leafy shoots from the base. Leaves of the sterile shoots obovate, obtuse at the apex, 1′–3′ long, narrowed at the base into petioles; flowering stems 4′–10′ high, their leaves oblong or lanceolate, acute or obtuse, smaller; flowers in a simple or barely compound cyme, slender-pedicelled; calyx-teeth linear-subulate, as long as the tube or longer; corolla pink, purple or violet, its lobes rounded, mostly entire, about one-half the length of the tube; capsule subglobose, 1½″ high.

In woods, Pennsylvania to Georgia and Kentucky, mainly in the mountains. Ascends to 4500 ft. in Virginia. April–June.

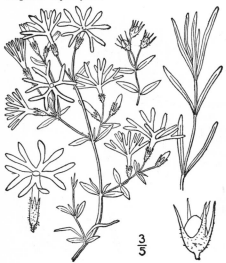

10. Phlox bífida Beck.   Cleft Phlox. Fig. 3463.

P. bifida Beck, Am. Journ. Sci. 11 : 170. 1826.

Puberulent or pubescent; stems diffuse, somewhat woody, much branched, slender, often 1° long, the branches erect or ascending, 4′–8′ high. Leaves of sterile shoots linear, sessile, 1′–2′ long, 1″–2″ wide, acute, those of flowering branches linear-oblong or lanceolate, much shorter; flowers in simple cymes or solitary in the axils, slender-pedicelled; pedicels 3″–12″ long; calyx-teeth lanceolate-subulate, somewhat longer than the tube; corolla pale purple, its lobes as long as the tube or somewhat shorter, cuneate, cleft to about the middle into linear or oblong obtuse diverging segments; capsule oblong-globose, 1″–1½″ high.

In dry places, Indiana to Tennessee, Michigan and Missouri. April–June.

11. Phlox Stellària A. Gray. Chick-weed Phlox. Fig. 3464.

Phlox Stellaria A. Gray, Proc. Am. Acad. **8** : 252. 1870.

Glabrous or puberulent; stems diffuse, some-what woody, much branched, the branches nearly erect, 3′–8′ high. Leaves all linear, or linear-lanceolate, sessile, acute, 1′–2′ long, 1″–1½″ wide; flowers in simple cymes or solitary in the axils, slender-pedicelled; calyx-teeth subulate-lanceolate, shorter than the tube; corolla pale blue or nearly white, its lobes cuneate, 2-lobed at the apex, nearly as long as the tube.

On cliffs, southern Illinois and Kentucky and in Tennessee. April–May.

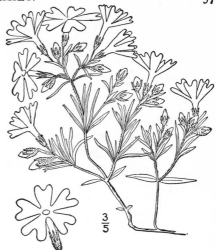

12. Phlox Kélseyi Britton. Kelsey's Phlox. Fig. 3465.

P. Kelseyi Britton, Bull. Torr. Club **19** : 225. 1892.

Many-stemmed from a woody root, the stems spreading, creeping, or ascending, sometimes 8′ long, glabrous, or slightly pubescent above, very leafy. Leaves oblong, or linear-oblong, sessile, glabrous, or nearly so 3″–12″ long, 1″–2″ wide, or the upper longer and narrower, thick, rigid, the apex spinose-mucronate, the revolute margins ciliate; flowers sessile or short-peduncled; peduncles and calyx somewhat glandular-pubescent or glabrous; calyx-teeth subulate, as long as the tube or longer; corolla-tube somewhat exceeding the calyx, the limb about 8″ broad, blue or lilac, the obovate-cuneate lobes rounded or truncate.

North Dakota to Nebraska, Montana and Wyoming. May–June.

13. Phlox subulàta L. Ground or Moss Pink. Fig. 3466.

Phlox subulata L. Sp. Pl. 152. 1753.

Pubescent or becoming glabrate. Stems tufted, forming mats, diffuse, much branched, the branches 2′–6′ long. Leaves persistent, subulate-linear, linear-lanceolate or linear-oblong, acute or acuminate, 4″–10″ long, ½″–1″ wide, spreading, ciliate, rigid, commonly fascicled at the nodes; flowers in simple cymes, slender-pedicelled; calyx-teeth subulate from a broader base, about as long as the tube; corolla pink, purple or white, with a darker eye, its lobes emarginate or entire, shorter than the tube; capsule oblong, nearly 2″ high.

In dry sandy or rocky soil, New York to Florida, west to Michigan and Kentucky. Ascends to 3500 ft. in West Virginia. Wild or mountain-pink. Flowering moss. April–June.

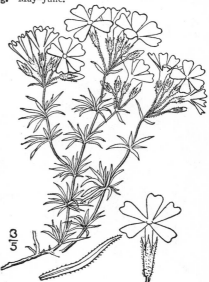

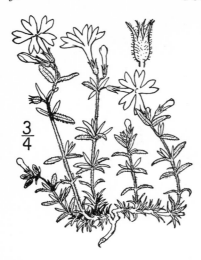

14. **Phlox Brittònii** Small. Britton's Phlox.
Fig. 3467.

Phlox Brittonii Small, Bull. Torr. Club **27**: *279.* 1900.

Glandular-pilose, deep green; stems copiously branched, forming wide mats. Leaves numerous, with small ones often clustered in the axils of the larger, subulate or narrowly linear-subulate, 2½″–5″ long, ciliate, especially near the base; calyx glandular-pubescent like the branches, the lobes subulate, about as long as the tube; corolla mainly white, the limb 12–13 mm. broad, its lobes cuneate, much shorter than the tube, with 2 pale magenta spots at the base, cleft by a V-shaped sinus, a minute tooth in each sinus; capsule oblong, 2″ high.

On dry mountain slopes, Virginia, West Virginia and North Carolina. May.

15. **Phlox bryoìdes** Nutt. Moss Phlox.
Fig. 3468.

Phlox bryoides Nutt. Journ. Acad. Phila. (II.) **1**: **153.** 1848.

Depressed, shrubby, moss-like, densely branched from a deep woody root, forming compact tufts 2′–3′ high. Leaves minute (about 1″ long), closely imbricated in 4 ranks, copiously white-woolly, triangular-lanceolate, pale, acute, the margins infolded; flowers solitary and sessile at the ends of the branches, about 2½″ long; tube of the corolla longer than the calyx, its lobes broadly cuneate, entire.

On dry hills, western Nebraska, Colorado and Wyoming. May–July.

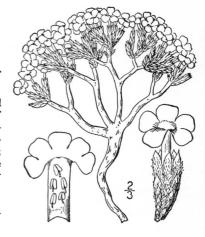

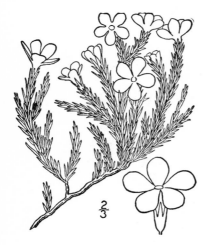

16. **Phlox Hoòdii** Richards. Hood's Phlox.
Fig. 3469.

Phlox Hoodii Richards. App. Frank. Journ. *733. pl. 28.* 1823.

Densely tufted and branched from a woody root, 2′–4′ high. Leaves imbricated, erect, rigid, subulate, mucronate, somewhat woolly or ciliate, becoming glabrate, 2″–6″ long; flowers solitary and sessile at the ends of the branches, about 5″ long; calyx-teeth lanceolate, acuminate, rigid, longer than the tube; tube of the corolla shorter than or equalling the calyx, its lobes obovate, entire.

In dry sandy or rocky soil, North Dakota to Manitoba, Alberta, western Nebraska and Wyoming. May–July.

17. Phlox Douglásii Hook. Douglas' Phlox. Fig. 3470.

P. *Douglasii* Hook. Fl. Bor. Am. **2** : 73. *pl. 158.* 1834.
Phlox Douglasii andicola Britton, Mem. Torr. Club **5** : 269. 1894.
Phlox Douglasii longifolia A. Gray, Proc. Am. Acad. **8** : 254. 1870. Not *P. longifolia* Nutt.

Similar to the preceding species; leaves pubescent or glabrous, less imbricated, sometimes spreading, rigid, usually fascicled at the nodes, 4″–12″ long. Flowers solitary and sessile or short-pedicelled at the ends of the branches, 5″–8″ long; calyx-teeth narrowly lanceolate, acuminate, about equalling the tube; tube of the purple or white corolla longer than the calyx, its lobes obovate, entire.

Dry soil, Nebraska and Montana to Utah, California and British Columbia. May–July.

2. MICROSTÈRIS Greene, Pittonia 3 : 300. 1898.

Much-branched annual herbs, with entire leaves, all but the floral ones opposite, the small flowers solitary or in pairs in the upper axils. Calyx mostly campanulate, 5-cleft, scarious between the lobes. Corolla salverform, with a slender tube and a 5-lobed limb. Stamens short. Ovary 3-celled. Capsule at length distending and rupturing the calyx-tube. Seeds few and large, mucilaginous when wetted, but not emitting spiral tubes. [Greek, small *Steris*.]

About 6 species, of western North America. Type species : *Microsteris grácilis* (Dougl.) Greene.

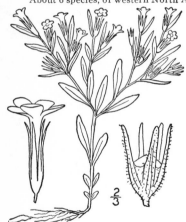

1. Microsteris micràntha (Kellogg) Greene. Small-flowered Microsteris. Fig. 3471.

Collomia micrantha Kellogg, Proc. Cal. Acad. **3** : 18. 1863.
M. micrantha Greene, Pittonia **3** : 303. 1898.

Annual, pubescent, at length corymbosely much branched, 2′–6′ high, the branches ascending. Lower and basal leaves oblong to spatulate, obtuse, commonly opposite and nearly sessile, the upper linear or lanceolate, sessile, ½′–1′ long, 1″–2″ wide, opposite or alternate, entire; cymes 1–5-flowered; calyx-lobes linear-subulate, usually longer than the tube; corolla about 4″ long, the tube yellowish, narrow, equalling or slightly longer than the calyx, the lobes purple or violet, short; ovules 2 or 3 in each cavity; capsule oblong, obtuse, as long as or longer than the calyx-tube.

In dry or moist soil, western Nebraska to Wyoming, Colorado, New Mexico, Arizona and California. In our first edition referred to the northwestern *M. gracilis* (Dougl.) Greene. April–Aug.

3. GÌLIA R. & P. Prodr. Fl. Per. 47. 1798.

Herbs with opposite or alternate, entire pinnatifid palmatifid or dissected leaves. Flowers small or large, solitary, cymose, capitate, thyrsoid, or paniculate. Calyx campanulate or tubular, 5-toothed or 5-cleft, the sinuses scarious. Corolla funnelform, tubular, campanulate, rotate or salverform, 5-lobed, the lobes ovate, oblong, or obovate. Stamens equally or unequally inserted on the corolla, included or exserted. Ovary oblong or ovoid, 3-celled; ovules solitary or several in each cavity. Capsule ovoid or oblong, 3-celled, at length distending and rupturing the calyx. Seed-coat commonly mucilaginous when wetted, in some species emitting thread-like tubes. [Named for Philip Gil, a Spanish botanist.]

About 100 species, natives of America. Besides the following, many others occur in the southern and western parts of North America. Type species : *Gilia laciniata* R. & P.

Corolla funnel-form to salverform; leaves pinnatifid or pinnately divided, the segments linear, not rigid.

  Flowers thyrsoid-paniculate or corymbose-paniculate.

    Corolla 1′–2′ long; plants 1°–4° tall.

Flowers paniculate, white.	1. *G. longiflora.*
Flowers narrowly thyrsoid, red.	2. *G. aggregata.*
Corolla 3″–5″ long, violet or blue.	3. *G. pinnatifida.*
Flowers narrowly thyrsoid-spicate.	4. *G. spicata.*

Flowers in dense or capitate cymes, or heads; flower-clusters leafy-bracted.
  Perennial; corolla-tube not longer than the calyx.   5. *G. iberidifolia.*
  Annual; corolla-tube 2–3 times as long as the calyx.   6. *G. pumila.*
Corolla rotate; leaf-segments acicular.   7. *G. acerosa.*

1. Gilia longiflòra (Torr.) Don. White-flowered Gilia. Fig. 3472.

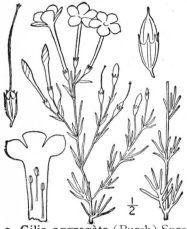

Cantua longiflora Torr. Ann. Lyc. **2**: 221. 1827.
Gilia longiflora Don, Gard. Dict. **4**: 245. 1838.
Collomia longifl. A. Gray, Proc. Am. Acad. **8**: 261. 1870.

Annual, glabrous, paniculately branched, 1°–2°
high. Leaves all alternate, sessile, 1′–2½′ long, pinnately divided into linear-filiform segments or the
uppermost entire; flowers numerous, white, paniculate, about 2′ long; calyx-teeth triangular-lanceolate,
acuminate, shorter than the tube; corolla salverform,
its tube narrow, 4 or 5 times as long as the orbicular
or ovate, rounded or pointed, spreading lobes; stamens unequally inserted; ovules 8–12 in each cell;
capsule narrowly oblong, exceeding or equalling the
calyx; seed-coat mucilaginous and emitting spiral
threads when wetted.

In dry soil, Nebraska and Colorado to Texas and Arizona. May–Sept.

Gilia rùbra (L.) Heller (*Gilia coronopifòlia* Pers.), a
related species with less spreading corolla-lobes, is commonly cultivated, and rarely escapes from gardens to
roadsides and waste grounds.

2. Gilia aggregàta (Pursh) Spreng. Scar-let Gilia. Fig. 3473.

Cantua aggregata Pursh, Fl. Am. Sept. 147. 1814.
Gilia aggregata Spreng, Syst. **1**: 626. 1825.

Biennial, pubescent or puberulent; stem simple
or sparingly branched, 2°–4° high, leafy at least
below. Leaves alternate, the basal often tufted,
mostly petioled, 1′–3′ long, pinnately parted into
narrowly linear segments; inflorescence narrowly
thyrsoid-paniculate, often 12′ long; flowers sessile or very nearly so in small peduncled clusters,
scarlet or red; corolla tubular-funnelform, the
tube 1′–1½′ long, slightly thicker upward, the limb
cleft into ovate or lanceolate acute or acuminate
spreading or recurved lobes; stamens unequally
or about equally inserted in the throat; ovules
numerous; seeds mucilaginous and emitting spiral
threads when wetted.

In dry soil, western Nebraska (according to
Coulter) to Texas and Mexico, west to British
Columbia and California. June–Aug.

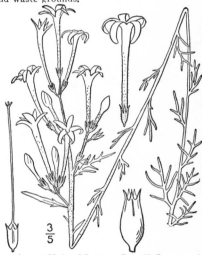

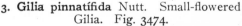

3. Gilia pinnatífida Nutt. Small-flowered Gilia. Fig. 3474.

Gilia pinnatifida Nutt.; A. Gray, Proc. Am. Acad. **8**:
  276. 1870.

Biennial or perennial from a deep root, much
branched, viscid-glandular, 6′–2° high. Leaves
thick, pinnatifid, the basal tufted, 1′–3′ long, the
segments linear-oblong, sometimes toothed, obtuse or acutish, 2″–6″ long, those of the stem
alternate, smaller, the uppermost minute and entire; flowers very numerous, small, paniculate,
some sessile, some petioled; calyx 5-lobed, the
lobes lanceolate to ovate; corolla salverform,
3″–5″ long, the limb violet or blue, its white tube
longer than the calyx and its obovate lobes; stamens exserted; seeds not mucilaginous nor emitting spiral threads when wetted.

In sandy soil, western Nebraska and Wyoming to
New Mexico. Has been mistaken for *G. inconspicua,*
June–Aug.

4. Gilia spicàta Nutt. Spicate Gilia. Fig. 3475.

Gilia spicata Nutt. Journ. Acad. Nat. Sci. Phila. (II.) **1**: 156. 1848.

Perennial, woolly-tomentose; stems erect, rather stout, simple, solitary, or 2–4 from the woody root, 6′–18′ high. Leaves alternate, narrowly linear, pinnately parted into 3–5 linear segments, or some or many of them entire, 1′–2′ long; flowers in an elongated narrow spicate thyrsus, sessile in small clusters, purplish, 4″–6″ long; tube of the corolla somewhat exceeding the calyx, considerably longer than the ovate-oblong lobes; calyx-lobes acuminate; anthers equally inserted in the throat.

In dry soil, western Nebraska to Wyoming and Utah. May–Aug.

5. Gilia iberidifòlia Benth. Round-headed Gilia. Fig. 3476.

Gilia iberidifolia Benth. in Hook. Kew. Journ. Bot. **3**: 290. 1851.

Perennial by a deep root, woolly-tomentose, at least when young, branched from the base or also above, 3′–18′ high. Leaves mostly petioled, ½′–2′ long, pinnately divided into 3–9 narrowly linear sharp-pointed segments, or the uppermost entire; flowers white, densely capitate-clustered, 2″–3″ long, the clusters bracted by the upper leaves, ½′–1′ broad, sometimes corymbed; calyx-lobes awn-like; corolla-tube about the length of the calyx, slightly longer than the oval lobes; filaments equally inserted in or below the sinuses of the corolla; ovules 1–4 in each cavity.

In dry soil, South Dakota to Nebraska, Colorado, California, Montana and Idaho. In our first edition included in the similar *G. congesta* Hook. May–Aug.

6. Gilia pùmila Nutt. Low Gilia. Fig. 3477.

Gilia pumila Nutt. Journ. Acad. Nat. Sci. Phila. (II.) **1**: 156. 1848.

Annual, branched from the base and sometimes also above, woolly at least when young, 3′–8′ high. Leaves alternate, thick, ½′–1′ long, pinnately divided into linear mucronulate sometimes lobed segments, or the uppermost entire; flowers in dense or at length looser simple or compound cymes, sessile; corolla 3″–4″ long, its tube about 3 times the length of the lobes and twice as long as the calyx; calyx-lobes awn-like; stamens inserted in or below the sinuses of the corolla, somewhat exserted; ovules 5 or 6 in each cavity.

In dry soil, western Nebraska to Texas, Idaho, Nevada and New Mexico. April–June.

Gilia trìcolor Benth., of California, admitted into our first edition as recorded escaped from gardens to roadsides at Lincoln, Nebraska, is not known to have become established within our area.

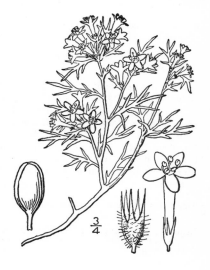

7. Gilia aceròsa (A. Gray) Britton. Needle-leaved Gilia. Fig. 3478.

G. rigidula var. *acerosa* A. Gray, Proc. Am. Acad. **8**: 280. 1870.

Gilia acerosa Britton, Man. 761. 1901.

Perennial, from a woody base, 4'–7' high, glandular-pubescent, bushy-branched, the slender branches erect-ascending. Leaves 1' long or less, pinnately parted into 3–7 acicular entire segments; flowers scattered, on slender pedicels 1' long or less; corolla rotate, about 10" wide, its lobes rounded at the apex, abruptly contracted at the base; filaments filiform; anthers oblong; capsule about as long as the calyx-tube.

Rocky and gravelly soil, Kansas to Texas, Mexico and New Mexico. April–Aug.

4. LEPTODÁCTYLON H. & A. Bot. Beech. Voy. 369. 1841.

Low perennial herbs, somewhat woody, with alternate deeply pinnatifid or palmatifid leaves, their segments subulate and spinescent, the rather large flowers terminal, clustered or solitary. Calyx 4–5-lobed, the lobes spinescent, the sinuses not scarious but membranous. Corolla funnelform, 4–5-lobed, the lobes somewhat spreading. Stamens short. Capsule at length distending the calyx-tube, many-seeded. [Greek, referring to the deeply divided leaves.]

Three or four species, natives of western North America. Type species: *Leptodactylon californicum* H. & A.

1. Leptodactylon caespitòsum Nutt. Tufted Sharp-leaved Gilia. Fig. 3479.

L. caespitosum Nutt. Journ. Acad. Phila. II. **1**: 157. 1847.

Gilia pungens caespitosa A. Gray, Proc. Am. Acad. **8**: 268. 1870.

Gilia caespitosa A. Nelson, Bull. Torr. Club **25**: 546. 1898. Not A. Gray.

Much branched and tufted from a thick buried woody base, 3'–5' high. Leaves densely fascicled and imbricated, 3–5-parted, 4" long or less, the segments subulate, spinulose-tipped, glabrous, or the margins ciliate; bases of the old leaves persistent; calyx about half as long as the corolla-tube, 4-lobed, the lobes subulate; corolla white or yellowish, salverform, the tube about 6" long, the limb 4-lobed; stamens 4.

Dry bluffs, western Nebraska and Wyoming. June–Sept.

Leptodactylon púngens Nutt. [*Gilia púngens* (Torr.) Benth.], of western North America, illustrated in our first edition, where the above species was regarded as a variety of it, is not known to occur within our area.

5. POLEMONIUM [Tourn.] L. Sp. Pl. 162. 1753.

Perennial or rarely annual herbs, with alternate pinnate membranous leaves, and mostly large cymose-paniculate or thyrsoid flowers. Calyx herbaceous, not angled nor ribbed, campanulate, 5-cleft to about the middle, accrescent in fruit, the segments lanceolate or ovate, entire, erect or connivent over the capsule. Corolla tubular-campanulate or funnelform, rarely rotate, blue, white or yellow, the limb 5-lobed. Stamens about equally inserted near the base of the corolla, declined, the filaments slender, often pilose at the base. Ovary ovoid; ovules few or several in each cavity. Capsule ovoid, obtuse, 3-valved. Seeds wingless, or narrowly winged, mucilaginous and emitting spiral threads when wetted. [Name not explained.]

About 15 species, natives of the cooler parts of the north temperate zone. Besides the following, some 10 others occur in the western parts of North America. Type species: *Polemonium coeruleum* L.

Anthers exserted; flowers 8″–10″ broad; stem erect.   1. *P. Van Bruntiae.*
Anthers included; flowers 5″–6″ broad; stem reclining.   2. *P. reptans.*

1. Polemonium Van Brúntiae Britton. American Jacob's Ladder. Fig. 3480.

Polemonium coeruleum A. Gray, Man. Ed. 4, App. 1863. Not L. 1753.
Polemonium Van Bruntiae Britton, Bull. Torr. Club **19**: 224. *pl. 131.* 1892.

Rootstock stout, horizontal, clothed with fibrous roots. Stems erect, glabrous below, somewhat glandular-pubescent above, 1½°–2½° high, leafy to the top; leaflets of the lower leaves short-stalked or sessile, ovate or lanceolate, acute, ½′–1½′ long, those of the upper fewer, the uppermost leaves 3–5-foliolate or simple; cymose clusters panicled or solitary, rather loosely 3–5-flowered; pedicels 2″–4″ long; flowers bluish-purple, 8″–10″ broad; corolla-lobes rounded; calyx 5-lobed to about the middle, much enlarged in fruit, the lobes acute; stamens exserted; ovules 3 or 4 in each cavity; capsule several-seeded.

In swamps and along streams, Vermont and northern New York to Maryland. Differs from the Old World *P. coeruleum* L. in its stout rootstocks, more leafy stem, exserted stamens, and rounded corolla-lobes. May–July.

2. Polemonium réptans L. Greek Valerian. Blue-bell. Fig. 3481.

Polemonium reptans L. Syst. Ed. 10, no. 1. 1759.

Glabrous or very nearly so throughout, usually not more than 1° high; stems weak, slender, at length reclining or diffuse, the rootstock short. Leaflets oblong, ovate-oblong, or lanceolate-oblong, ½′–1½′ long, the uppermost leaves 3–5-foliolate or simple; flowers blue, 5″–8″ broad; calyx 5-lobed, its lobes obtuse or acute; stamens not exserted; ovules 3 or 4 in each cavity; seeds about 3 in each capsule.

In woods, New York to Minnesota, south to Georgia and Kansas. Ascends to 2200 ft. in Virginia. April–May. American abscess-root. Sweat-root. American-or creeping great-valerian.

6. COLLÒMIA Nutt. Gen. 1: 126. 1818.

Annual or rarely perennial herbs, with alternate mostly entire leaves, and purple white or reddish capitate or cymose flowers. Calyx obpyramidal or cup-shaped, 5-cleft, scarious in the sinuses, accrescent in fruit, not distended by nor ruptured by the ripening capsule, its lobes lanceolate or triangular, entire, erect, the sinuses often at length enlarged into a revolute lobe. Corolla tubular-funnelform or salverform, the limb 5-lobed, spreading, the lobes obtuse. Stamens unequally inserted on the tube of the corolla, mostly straight, the filaments unequal. Ovules 1 or few in each cavity. Capsule oval to obovoid. Seeds of most species mucilaginous and emitting spiral threads when wetted. [Greek, gluten, referring to the glutinous seeds when wetted.]

About 15 species, natives of western America. Besides the following typical one, some 10 others occur in the western United States and British Columbia.

1. Collomia lineàris Nutt. Narrow-leaved Collomia. Fig. 3482.

Collomia linearis Nutt. Gen. 1: 126. 1818.
Gilia linearis A. Gray, Proc. Am. Acad. 17: 223. 1882.

Annual, viscid-puberulent; stem erect, leafy, simple or branched, slender, 3'-18' high. Leaves linear-oblong, lanceolate or linear-lanceolate, entire, acuminate at the apex, narrowed at the base, sessile, or the lower short-petioled, 1'-2½' long, 1½"-6" wide; flowers 5"-7" long, numerous in terminal capitate leafy-bracted clusters; calyx-lobes triangular-lanceolate, acute; corolla light purple or nearly white, the tube very slender, longer than the calyx, the lobes 1"-2" long; capsule at maturity about as long as the calyx.

In dry soil, Manitoba to Minnesota and Nebraska, west to British Columbia, Arizona and California. Also in Quebec and New Brunswick. May–Aug.

7. NAVARRÈTIA R. & P. Fl. Per. 2: 8. 1799.

Annual glabrous or viscid-pubescent herbs, with alternate spinose-pinnatifid leaves, or the lowest entire, and numerous small flowers in dense terminal bracted clusters. Calyx prismatic or obpyramidal, the tube 5-angled, 5-cleft, not accrescent in fruit, not distended by nor ruptured by the ripening capsule, the sinuses scarious, the lobes mostly unequal, erect or spreading, spiny-tipped, entire, or often toothed. Corolla tubular-funnelform or salverform, 5-lobed, the lobes oval or oblong. Stamens straight or declined, equally inserted in or below the throat of the corolla. Ovary 2-3-celled; ovules solitary, few or several in each cavity. Capsule 1-3-celled, dehiscent or indehiscent. Seeds mostly mucilaginous and emitting spiral threads when wetted. [In honor of Navarrete, a Spanish physician.]

About 24 species, natives of western America. Besides the following, some 20 others occur in the western United States. Type species: *Navarretia involucràta* R. & P.

1. Navarretia mínima Nutt. Small Navarretia. Fig. 3483.

Navarretia minima Nutt. Journ. Acad. Nat. Sci. Phila. (II.) 1: 160. 1848.

Gilia minima A. Gray, Proc. Am. Acad. 8: 269. 1870.

Depressed, tufted, somewhat pubescent; stem usually branched, 1'-3' high. Leaves sessile, ½'-1' long, 1-2-pinnatifid into almost filiform rigid acicular segments; flowers about 2" long, white, densely capitate; calyx-lobes awl-shaped, mostly toothed, about as long as the tube and equalling the corolla, the sinuses more or less white-pubescent; calyx-tube about equalling the indehiscent 1-6-seeded capsule.

In dry soil, Nebraska and South Dakota to Washington and Arizona. Summer.

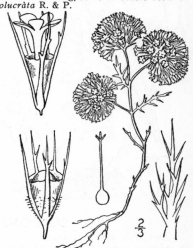

Family 23. HYDROPHYLLÀCEAE Lindl. Nat. Syst. Ed. 2, 271. 1836.

WATER-LEAF FAMILY.

Herbs, mostly hirsute, pubescent or scabrous, with alternate or basal, rarely opposite leaves, and perfect regular 5-parted flowers, in scorpioid cymes, spikes or racemes, or rarely solitary. Calyx inferior, deeply cleft or divided, the sinuses sometimes appendaged. Corolla gamopetalous, funnelform, salverform, campanulate, or rotate. Stamens 5, inserted on the tube or base of the corolla, and alternate with its lobes; filaments filiform; anthers ovate, oblong, or linear, mostly versatile, 2-celled, the sacs longitudinally dehiscent. Disk annular or none. Ovary superior, 2-celled, or 1-celled with 2 placentae; styles 2, separate, or partly united; stigmas small, terminal; ovules few or numerous, anatropous or amphitropous. Capsule 1–2-celled, mostly loculicidally 2-valved, rarely septicidally or irregularly dehiscent. Seeds oblong, globose, or angular, usually pitted, rugose or reticulated; endosperm fleshy or cartilaginous; embryo small; cotyledons half-terete or plano-convex.

About 17 genera and 175 species, mostly natives of western North America.
Styles united below; ovary 1-celled; leaves mostly lobed or dentate.
  Corolla-lobes convolute (rarely imbricated) in the bud; placentae dilated.
    Stamens exserted.
    Stamens not exserted.   1. *Hydrophyllum.*
      Sinuses of the calyx appendaged.
      Calyx much enlarged in fruit, its sinuses not appendaged.   2. *Nemophila.*
  Corolla-lobes imbricated in the bud; placentae narrow.   3. *Nyctelea.*
Styles distinct to the base; ovary 2-celled; leaves entire.   4. *Phacelia.*
                                                 5. *Nama.*

1. HYDROPHÝLLUM [Tourn.] L. Sp. Pl. 146. 1753.

Perennial or biennial herbs, with large lobed pinnatifid or pinnately divided leaves, and rather large, white blue or purple flowers, in terminal or lateral peduncled more or less scorpioid cymes. Calyx deeply 5-parted, the segments lanceolate or subulate, the sinuses naked or appendaged. Corolla tubular-campanulate or campanulate, 5-lobed, the lobes convolute in the bud, each with a linear appendage within, which extends to the base of the corolla and is incurved into a groove. Stamens 5, exserted; filaments pilose below or at the base; anthers linear or oblong, versatile. Ovary 1-celled, hispid-pubescent; placentae fleshy, dilated so as to nearly fill the cavity, free from the ovary-wall except at the top and bottom, each enclosing 2 ovules; styles united nearly to the summit. Capsule 2-valved. Seeds 1–4, globose-obovoid. [Greek, water-leaf, referring to the supposed cavity for water in each leaf.]

About 12 species, natives of North America. Type species: *Hydrophyllum virginianum* L.
Leaves, at least the lower, pinnatifid or pinnately divided.
  Calyx not appendaged in the sinuses or scarcely so.
    Plant sparingly pubescent; leaf-segments acute.   1. *H. virginianum.*
    Plant villous-hirsute; leaf-segments blunt.   2. *H. macrophyllum.*
  Calyx with a reflexed appendage in each sinus.   3. *H. appendiculatum.*
Leaves palmately 5–9-lobed.   4. *H. canadense.*

1. Hydrophyllum virginiànum L. Virginia Water-leaf. Fig. 3484.

H. virginianum L. Sp. Pl. 146. 1753.

Perennial by scaly rootstocks; stems slender, glabrous or nearly so, simple or sparingly branched, ascending or erect, rather weak, 1°–3° long. Lower and basal leaves long-petioled, 6'–10' long, pinnately divided into 5–7 oblong ovate or ovate-lanceolate, acute or acutish, sharply toothed or incised segments 1'–2' long, glabrous or with few scattered hairs; upper leaves similar, short-petioled, smaller, with fewer segments; cymes slender-peduncled, simple or forked, dense or at length open; flowers white or violet, purple, darkest at high altitudes, short-pedicelled; pedicels strigose-pubescent; calyx-segments narrowly linear, hispid, spreading, the sinuses not appendaged; corolla about 4" long, its segments erect; capsule globose, nearly 2" in diameter.

In woods, Quebec to Ontario, South Dakota, South Carolina and Kansas. Ascends to 5000 ft. in North Carolina. Brook-flower. May–Aug.

$\frac{3}{5}$

5

H. pàtens Britton, of Minnesota, differs in having the calyx-segments appressed to the corolla, the corolla-segments with spreading tips.

½

3. Hydrophyllum appendiculàtum Michx.
Appendaged Water-leaf. Fig. 3486.

H. appendiculatum Michx. Fl. Bor. Am. 1 : 134. 1803.

Biennial, rough-hairy all over; stem slender, simple or usually branched, weak, 1°–2° long, somewhat viscid above. Lower and basal leaves long-petioled, pinnatifid or pinnately divided into 5–7 ovate or oval, acute or obtusish, irregularly dentate or incised, membranous segments; upper leaves smaller and shorter-petioled, ovate to orbicular, variously lobed, the lobes acute and dentate; cymes loose, branched, the slender pedicels 4″–10″ long; calyx parted to near the base, enlarging in fruit, the segments triangular-lanceolate, acuminate, spreading, with a short reflexed appendage in each sinus; corolla violet or purple, 6″–7″ long; stamens little exserted; capsule about 1½″ in diameter.

In woods, Ontario to New York, North Carolina, west to Minnesota and Kansas. May–June.

½

2. Hydrophyllum macrophýllum Nutt.
Large-leaved Water-leaf. Fig. 3485.

H. macrophyllum Nutt. Journ. Phila. Acad. 7 : 111. 1834.

Perennial by scaly rootstocks, villous-hirsute all over; stem rather stout, 2°–3° high. Lower leaves long-petioled, 8′–12′ long, deeply pinnatifid or pinnately divided into 7–13 oval or ovate obtuse coarsely dentate segments 1′–3′ long; upper leaves similar, smaller, shorter-petioled and with fewer segments; cymes long-peduncled, simple or forked, very dense; peduncles stout; pedicels short; flowers nearly white; calyx cleft to below the middle, its lobes lanceolate, erect, white-hispid, the sinuses not appendaged; corolla 5″–6″ long; anthers oblong; capsule globose, densely white-hispid, 1½″ in diameter, enclosed by the calyx.

In rich woods, Virginia to Ohio, Illinois, Alabama and Tennessee. Ascends to 4000 ft. in Virginia.

½

4. Hydrophyllum canadénse L. Broad-leaved Water-leaf. Fig. 3487.

H. canadense L. Syst. Ed. 10, 919. 1759.

Perennial by scaly rootstocks; stems rather slender, commonly simple, glabrous or nearly so, 1°–2½° high. Leaves nearly orbicular, cordate, sparingly pubescent, at least above, palmately 5–9-lobed, the lower long-petioled, often 12′ broad, occasionally with 1 or 2 pairs of small segments on the petiole; upper leaves smaller, but usually 4′–7′ broad; lobes ovate, acuminate, dentate; peduncles shorter than the leaves; cymes dense or becoming loose, simple or forked; pedicels short, nearly glabrous; calyx cleft nearly to the base, its segments linear-lanceolate, acute, nearly or quite glabrous, sometimes with a minute tooth in each sinus; corolla campanulate-rotate, white to purplish anthers linear-oblong; capsule 2″ in diameter.

In woods, Vermont to New York, North Carolina, Ontario, Illinois and Kentucky. Ascends to 4000 ft. in Virginia. June–Aug.

2. NEMÓPHILA Nutt. Journ. Phil. Acad. 2: 179. 1822.

Annual diffuse pubescent slender and fragile herbs, with alternate or opposite mostly pinnatifid or lobed leaves. Flowers white, blue or variegated, solitary, peduncled, lateral or terminal. Calyx deeply 5-cleft or 5-parted, with a reflexed or spreading appendage in each sinus. Corolla campanulate or rotate-campanulate, mostly longer than the calyx, usually with 10 small appendages within at the base, the lobes convolute in the bud. Stamens included; anthers ovate or oblong. Ovary 1-celled with placentae similar to those of *Hydrophyllum;* styles partly united; ovules 2-12 on each placenta. Capsule 2-valved. Seeds 1-4. [Greek, grove-loving.]

About 10 species, natives of North America, mostly Californian. Type species: *Nemophila phacelioides* Nutt.

1. Nemophila micrócalyx (Nutt.) F. & M. Small-flowered Nemophila. Fig. 3488.

Ellisia microcalyx Nutt. Trans. Am. Phil. Soc. (II.) **5:** 191. 1833-37.
Nemophila microcalyx F. & M. Sert. Petrop. *pl. 6.* 1846.

Stems very slender, diffuse, branched, 2′-15′ long. Leaves membranous, petioled, 1′-2½′ long, pinnatifid or pinnately divided into 3-5 obovate cuneate or oblique, obtuse 2-3-dentate or -lobed, approximate or confluent segments, the upper all alternate, the lowest opposite; peduncles slender, 4″-12″ long, opposite the leaves, shorter than or equalling the petioles; flowers white or blue, 1½″-2″ long; appendages in the sinuses of the calyx minute; calyx scarcely enlarged in fruit; ovules 2 on each placenta; corolla-appendages obsolete or none; anthers oval; capsule about 1½″ in diameter, much longer than the calyx, 1-2-seeded.

In woods, Virginia to Florida, west to Arkansas and Texas. April-June.

3. NYCTÈLEA Scop. Introd. 183. 1777.

[MACROCALYX Trew, Nov. Act. Nat. Cur. **2:** 330-332. *pl. 7. f. 1.* Hyponym. 1761.]
[ELLISIA L. Sp. Pl. Ed. 2, 1662. 1763. Not Syst. Ed. 10, 1121. 1759.]

Annual hirsute or pubescent branching slender herbs, with opposite or alternate, pinnately divided or 1-3-pinnatifid leaves, and solitary or racemose small white or bluish flowers. Calyx 5-lobed or 5-parted, spreading, much enlarged in fruit, destitute of appendages in the sinuses. Corolla campanulate or nearly cylindric, shorter than or slightly exceeding the calyx, usually with 5 minute appendages on the tube within, its lobes convolute or imbricated in the bud. Stamens included; anthers oval or oblong. Ovary 1-celled; styles united below; ovules 2-4 on each of the placentae, which are similar to those of the two preceding genera. [Name unexplained.]

About 3 species, natives of North America, the following typical.

1. Nyctelea Nyctèlea (L.) Britton. Nyctelea. Fig. 3489.

Ipomoea Nyctelea L. Sp. Pl. 160. 1753.
Polemonium (?) *Nyctelea* L. Sp. Pl. Ed. 2, 231. 1762.
Ellisia Nyctelea L. Sp. Pl. Ed. 2, 1662. 1763.
Macrocalyx Nyctelea Kuntze, Rev. Gen. Pl. 434. 1891.

Sparingly hirsute-pubescent; stem several times forked, 4″-12′ high. Leaves pinnately divided, petioled, 2′-4′ long, ovate-oblong in outline, the upper alternate, the lower opposite, the segments oblong or lanceolate, dentate, entire or lobed; peduncles slender, 1-flowered, opposite the leaves; calyx in flower about 2″ long, about equalling the corolla, enlarging, widely spreading and becoming 8″-14″ broad in fruit, its lobes lanceolate or ovate-lanceolate, acuminate; fruit at length pendulous; capsule globose, 2″-3″ in diameter.

In moist soil, New Jersey to Minnesota, Saskatchewan, Virginia, Nebraska and Kansas. April-July.

4. PHACÈLIA Juss.; J. F. Gmel. Syst. 330. 1791.

Annual, biennial or perennial, mostly hirsute hispid or scabrous herbs, with alternate entire dentate lobed pinnatifid or dissected leaves, the lowest rarely opposite, and blue purple violet or white flowers in terminal scorpioid cymes or racemes. Calyx 5-lobed, somewhat enlarging in fruit; sinuses not appendaged. Corolla campanulate, nearly rotate, tubular or funnelform, the tube sometimes appendaged within, opposite the lobes, the lobes imbricated in the bud. Stamens exserted or included, attached near the base of the corolla; anthers mostly ovate. Ovary 1-celled, the 2 placentae narrow, affixed to the walls; styles united below; ovules 2 or several on each placenta. Capsule 1-celled, or falsely nearly 2-celled by the intrusion of the placentae, 2-valved. Seeds usually reticulated. [Greek, a cluster, referring to the clustered flowers of some species.]

About 90 species, natives of the New World. Besides the following, some 60 others occur in the western parts of North America. Type species: *Phacelia secunda* J. F. Gmel.

Corolla-lobes entire.
  Corolla manifestly appendaged within, between the stamens.
    Leaves entire.   1. *P. leucophylla.*
    Leaves crenate-dentate.   2. *P. integrifolia.*
    Leaves pinnately divided, or pinnatifid, the segments incised.
      Racemes loose; pedicels slender; ovules 2 on each placenta.   3. *P. bipinnatifida.*
      Racemes dense; pedicels short; ovules numerous.   4. *P. Franklinii.*
  Appendages of the corolla inconspicuous or none.
    Filaments pubescent; calyx-lobes oblong.
      Puberulent; flowers 4″–5″ broad.   5. *P. dubia.*
      Hirsute; flowers 6″–7″ broad.   6. *P. hirsuta.*
    Filaments glabrous; calyx-lobes linear.   7. *P. Covillei.*
Corolla nearly rotate, its lobes fimbriate.
  Lobes of the leaves and calyx acute.   8. *P. Purshii.*
  Lobes of the leaves and calyx obtuse.   9. *P. fimbriata.*

1. Phacelia leucophýlla Torr. Silky Phacelia. Fig. 3490.

Phacelia leucophylla Torr. Frem. Rep. 93. 1845.

Perennial by a stout rootstock, pale, densely silky-pubescent, the hairs appressed or ascending. Stem simple or branched, 1°–1½° high; leaves lanceolate to oblong, entire, pinnately veined, 2′–4′ long, 4″–12″ wide, the lower long-petioled, the upper sessile or nearly so; spike-like branches of the scorpioid cymes very dense, nearly straight and 1′–3′ long when expanded; flowers sessile, very numerous, about 4″ high; calyx-lobes hispid, oblong-lanceolate or linear, somewhat shorter than the white or bluish, 5-lobed corolla; corolla-appendages conspicuous, in pairs between the filaments; filaments exserted, glabrous; ovules 2 on each placenta; capsule ovoid.

In dry soil, South Dakota to Idaho, British Columbia, Nebraska and Colorado. May–Aug. The species has been taken for *P. heterophylla* Pursh, of the far west, which has spreading brown hairs, some of the leaves usually pinnatifid, and pilose filaments.

2. Phacelia integrifòlia Torr. Crenate-leaved Phacelia. Fig. 3491.

P. integrifolia Torr. Ann. Lyc. N. Y. 2: 222. *pl. 3.* 1827.

Annual or biennial; stem erect or ascending, rather stout, very leafy, commonly branched above, viscid-hirsute, 6′–2° high. Leaves finely strigose-pubescent, ovate-oblong or oblong-lanceolate, irregularly crenate-dentate, obtuse at the apex, rounded or cordate at the base, 1′–2½′ long, petioled or the uppermost sessile; spike-like branches of the scorpioid cymes dense, 2′–4′ long when expanded; flowers sessile, about 4″ long; calyx-segments oblong, acute; corolla tubular-campanulate, white or blue, its tube longer than the calyx; filaments glabrous, exserted; ovules 2 on each placenta; capsule ovoid, obtuse.

In saline soil, western Kansas (according to B. B. Smyth); Oklahoma to Colorado, Mexico, Utah and Arizona. April–Sept.

3. Phacelia bipinnatífida Michx. Loose-flowered Phacelia. Fig. 3492.

Phacelia bipinnatifida Michx. Fl. Bor. Am. 1 : 134. *pl. 16.* 1803.

Biennial, hirsute-pubescent; stem erect, usually much branched, glandular-viscid above, 1°–2° high. Leaves slender-petioled, 2′–5′ long, pinnately divided or deeply pinnatifid into 3–7 ovate or oblong acute or acutish, dentate or incised segments, or these again pinnatifid; flowers blue or violet, 6″–8″ broad, numerous, slender-pedicelled in loose racemes, the inflorescence only slightly scorpioid; pedicels 4″–10″ long, recurved in fruit; calyx-segments linear; appendages of the rotate-campanulate corolla in pairs between the stamens, conspicuous, villous on the margins, corolla-lobes entire; filaments pilose, exserted; ovules 2 on each placenta; capsule globose.

In moist thickets and along streams, Ohio to Illinois, Missouri, Georgia, Alabama and Tennessee. Ascends to 4000 ft. in North Carolina. April–June.

4. Phacelia Franklínii (R. Br.) A. Gray. Franklin's Phacelia. Fig. 3493.

Eutoca Franklinii R. Br. App. Frank. Journ. 51. *pl. 27.* 1823.
Phacelia Franklinii A. Gray, Man. Ed. 2, 329. 1856.

Annual, villous-pubescent; stem erect, 6′–18′ high, simple, or corymbosely branched at the summit. Leaves 1½′–3′ long, pinnately parted into 7–15 linear or linear-oblong acute entire dentate or incised segments; flowers blue or nearly white, short-pedicelled in dense scorpioid racemes; calyx-segments linear-lanceolate, acute; longer than the tube of the rotate-campanulate corolla; appendages of the corolla free at the apex; anthers scarcely exserted; filaments glabrous, or nearly so; styles united nearly to the summit; ovules numerous on each placenta; capsule ovoid, acute.

Western Ontario, Michigan and Minnesota to British Columbia, Wyoming and Idaho. Summer.

5. Phacelia dùbia (L.) Small. Small-flowered Phacelia. Fig. 3494.

Polemonium dubium L. Sp. Pl. 163. 1753.
Phacelia parviflora Pursh, Fl. Am. Sept. 140. 1814.
Phacelia dubia Small, Bull. Torr. Club 21 : 303. 1894.

Annual, puberulent or glabrate, branched from the base, the branches very slender, erect or ascending, 5′–12′ high. Lower and basal leaves petioled, 1′–2′ long, pinnatifid or pinnately divided into 3–5 oblong obtuse entire or dentate segments, or rarely merely dentate, or even entire; upper leaves much smaller, sessile, less divided; flowers light blue or white, racemose, 4″–5″ broad; racemes 5–15-flowered, elongated in fruit; pedicels 3″–7″ long; calyx-lobes oblong or oblong-lanceolate; corolla rotate-campanulate, the appendages obsolete; filaments pubescent; anthers slightly exserted; ovules 4–8 on each placenta; capsule globose, 1½″ in diameter, 6–12-seeded; fruiting pedicels ascending.

In moist soil, New York and Pennsylvania to Georgia, Missouri, Kansas and Texas. Ascends to 2000 ft. in Virginia. April–June.

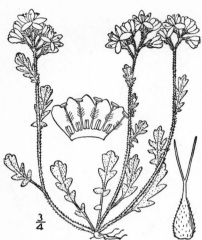

6. Phacelia hirsùta Nutt. Hairy Phacelia.
Fig. 3495.

Phacelia hirsuta Nutt. Trans. Am. Phil. Soc. (II.) 5: 191. 1834–37.
Phacelia parviflora var. *hirsuta* A. Gray, Proc. Am. Acad. 10: 321. 1875.

Similar to the preceding species but usually stouter and larger, hirsute-pubescent. Leaves petioled, pinnatifid or deeply pinnately divided into 5–9 oblong obtuse usually entire segments, or the upper 3–5-lobed or entire, sessile; racemes rather dense, scorpioid when unfolding; flowers blue, 6″–7″ broad; calyx-segments oblong-lanceolate, hirsute; corolla rotate-campanulate, its appendages short; ovules 3–4 on each placenta; stamens scarcely exserted; capsule globose, 4–8-seeded.

In dry soil, Virginia to Georgia, west to Missouri, Kansas and Texas. April–June.

7. Phacelia Covíllei S. Wats. Coville's
Phacelia. Fig. 3496.

Phacelia Covillei S. Wats. in A. Gray, Man. Ed. 6, 360. 1890.

Annual, similar to the two preceding species, branched from the base, pubescent; branches very slender, weak, 6′–12′ long. Leaves deeply pinnatifid or pinnately divided into 3–7 oblong or obovate, obtuse segments; racemes only 1–5-flowered; pedicels filiform, 6″–8″ long; calyx-segments linear, elongating in fruit; corolla tubular-campanulate, about 3″ long and broad when expanded; filaments glabrous; anthers not exserted; appendages of the corolla obsolete; capsule globose, 1½″–2″ in diameter; fruiting pedicels recurved.

Along the Potomac River above Washington, D. C., and in Illinois. April–May.

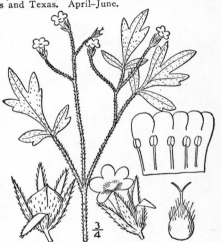

8. Phacelia Púrshii Buckl. Pursh's Phacelia. Fig. 3497.

Phacelia Purshii Buckl. Am. Journ. Sci. 45: 171. 1843.

Annual, pubescent; stem erect, usually much branched, 6′–18′ high. Lower and basal leaves petioled, 1½′–3′ long, pinnately parted or pinnatifid into 9–15 oblong or lanceolate, acute or obtusish, entire or sometimes incised segments; upper leaves sessile, with fewer segments; racemes strongly 1-sided, 10–20-flowered, much elongated in fruit; pedicels 3″–12″ long; calyx-segments lanceolate or linear-lanceolate; corolla 5″–6″ broad, blue or white, nearly rotate, not appendaged within, its lobes fimbriate; filaments slightly exceeding the corolla; ovules 2 on each placenta; capsule globose-ovoid.

In moist woods or thickets, Pennsylvania to Minnesota, south to North Carolina, Alabama and Missouri. Ascends to 2000 ft. in Virginia. April–June.

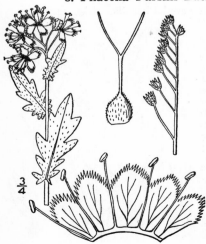

9. Phacelia fimbriàta Michx. Fringed or Mountain Phacelia. Fig. 3498.

Phacelia fimbriata Michx. Fl. Bor. Am. 1: 134. 1803.

Annual, sparingly pubescent; stems simple or branched, ascending or diffuse, 6′-20′ long. Basal and lower leaves slender-petioled, 2′-4′ long, pinnately divided into 5-9 orbicular obovate or oblong obtuse dentate or entire segments; upper leaves sessile, their segments oblong or lanceolate, acute; clusters loose, 3-12-flowered; pedicels 4″-15″ long; flowers 4″-5″ broad, white; calyx-segments linear to spatulate, obtuse; corolla rotate-campanulate, not appendaged within, its lobes strongly fimbriate; filaments pilose, about equalling the corolla; ovules 2 on each placenta; capsule depressed-globose, 2″ in diameter.

In woods, mountains of Virginia to Alabama. May-June.

5. NÀMA L. Sp. Pl. 226. 1753.

[Hydrolea L. Sp. Pl. Ed. 2, 328. 1762.]

Perennial herbs, some tropical species shrubby, with alternate entire leaves, sometimes with spines in their axils, and blue cymose-clustered or racemose flowers. Calyx-segments distinct to the base, ovate or lanceolate. Corolla rotate-campanulate, not appendaged within, 5-cleft, the lobes imbricated in the bud. Stamens inserted on the base of the corolla; filaments filiform, dilated below; anthers sagittate. Ovary 2-celled (rarely 3-celled); ovules numerous in each cavity, on fleshy adherent placentae; styles 2, rarely 3, slender or filiform, distinct to the base; stigma capitellate. Capsule globose or ovoid, septicidally or irregularly dehiscent.

About 15 species, natives of warm and tropical regoins of both the Old World and the New. Besides the following, another occurs in the southern United States. Type species: *Nama zeylanica* L.

Leaves lanceolate; flowers mostly in axillary clusters.
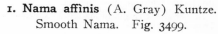
  Glabrous, or very nearly so, throughout; sepals ovate to ovate-lanceolate.   1. *N. affinis.*
  Villous, at least above, and on the calyx; sepals lanceolate.   2. *N. quadrivalvis.*
Leaves ovate; flowers mostly in terminal clusters; sepals villous.   3. *N. ovata.*

1. Nama affìnis (A. Gray) Kuntze. Smooth Nama. Fig. 3499.

Hydrolea affinis A. Gray, Man. Ed. 5, 370. 1867.

Nama affinis Kuntze, Rev. Gen. Pl. 435. 1891.

Glabrous or very nearly so throughout, with or without slender spines in the axils; stems ascending, 1°-2½° high. Leaves oblong-lanceolate, petioled, acute or acuminate at both ends, 2′-5′ long, 4″-8″ wide; flowers 6″-7″ broad, short-pedicelled, in rather dense peduncled leafy-bracted axillary clusters; sepals ovate or ovate-lanceolate, acute or acuminate, about equalling the corolla; capsule 2″ in diameter when mature, somewhat longer than the styles, shorter than the sepals.

In wet places, southern Illinois to Missouri, Louisiana and Texas. June-Aug.

2. Nama quadrivàlvis (Walt.) Kuntze.
Hairy Nama. Fig. 3500.

Hydrolea quadrivalvis Walt. Fl. Car. 110. 1788.
H. caroliniana Michx. Fl. Bor. Am. 1 : 177. 1803.
N. quadrivalvis Kuntze, Rev. Gen. Pl. 435. 1891.

Similar to the preceding species, but pubescent, at least above, and on the calyx, with spreading hairs, usually bearing slender spines in the axils; stem ascending, 1°–2° high. Leaves lanceolate, acute or acuminate, glabrous, or sparingly pubescent, petioled, 2′–5′ long, 3″–8″ wide; lower petioles ½′ long, or more; flowers in axillary clusters; sepals lanceolate or linear-lanceolate, acuminate, about as long as the corolla; capsule 2″–3″ in diameter, longer than the styles, about the length of the sepals.

In wet soil, southeastern Virginia to Florida and Louisiana. June–Aug.

3. Nama ovàta (Nutt.) Britton. Ovate-leaved Nama. Fig. 3501.

Hydrolea ovata Nutt. Trans. Am. Phil. Soc. (II.)
    5 : 196. 1833–37.
N. ovata Britton, Mem. Torr. Club 5 : 272. 1894.

Stem erect or nearly so, 1°–3° high, usually branched near the summit, puberulent, or somewhat hirsute, at least above, usually spine-bearing in most of the axils. Leaves ovate, rarely ovate-lanceolate, puberulent or glabrous, short-petioled, or the upper almost sessile, 1′–2½′ long, ½′–1¼′ wide, acute at the apex, narrowed or rounded at the base; flowers often 1′ broad or more, in terminal clusters; sepals lanceolate, acuminate, very villous, shorter than the corolla, longer than the capsule; styles longer than the sepals.

In wet soil, Georgia to Missouri, Louisiana and Texas. May–Sept.

Family 24. BORAGINÀCEAE Lindl. Nat. Syst. Ed. 2, 274. 1836.

BORAGE FAMILY.

Annual, biennial or perennial herbs, shrubs, or some tropical species trees. Leaves alternate, very rarely opposite or verticillate, exstipulate, mostly entire and hispid, pubescent, scabrous or setose. Flowers perfect, usually regular, mostly blue, in one-sided scorpioid spikes, racemes, cymes, or sometimes scattered. Calyx inferior, mostly 5-lobed, 5-cleft, or 5-parted, usually persistent, its lobes valvate. Corolla gamopetalous, mostly regular and 5-lobed, sometimes crested or appendaged in the throat, rarely irregular, its lobes imbricated, convolute, plicate or induplicate in the bud. Stamens as many as the corolla-lobes and alternate with them, inserted on the tube or throat; filaments slender or short; anthers 2-celled, the sacs longitudinally dehiscent. Disk annular, entire, or 5-lobed, or none, commonly inconspicuous. Ovary superior, of 2 2-ovuled carpels, entire, or the carpels commonly deeply 2-lobed, making it appear as of 4 1-ovuled carpels; style simple, entire or 2-cleft in our genera; ovules anatropous or amphitropous. Fruit mostly of 4 1-seeded nutlets, or of 2 2-seeded carpels. Endosperm none; embryo straight or curved; cotyledons mostly flat or plano-convex; radicle short.

About 85 genera and 1500 species, of wide geographic distribution.

* Ovary entire or 2–4-grooved; style terminal. 1. *Heliotropium.*
** **Ovary 4-divided or deeply 4-lobed, the style arising from the center.**
Flowers regular.
 Nutlets armed with barbed prickles.
  Nutlets spreading or divergent, covered by the prickles. 2. *Cynoglossum.*
  Nutlets erect or incurved, the prickles on their backs or margins. 3. *Lappula.*
 Nutlets unarmed.
  Nutlets attached laterally to the receptacle, sometimes just above their bases.
   Fruiting calyx not greatly enlarged nor membranous.
    Corolla small, usually white; receptacle conic or elongated.
     Annuals; calyx nearly closed in fruit; inflorescence naked or bracteolate.
      Lowest leaves mostly opposite; calyx persistent. 4. *Allocarya.*
      Leaves all alternate; calyx at length deciduous. 5. *Cryptantha.*
     Perennials or biennials; calyx-segments more or less spreading in fruit; inflo-
     rescence leafy. 6. *Oreocarya.*
    Corolla tubular-funnelform; receptacle flat or convex.
     Corolla blue, rarely white; nutlets attached just above their bases.
      Maritime; nutlets fleshy, smooth and shining. 7. *Pneumaria.*
      Not maritime; nutlets wrinkled when mature and dry. 8. *Mertensia.*
     Corolla yellow; nutlets laterally attached. 9. *Amsinckia.*
   Fruiting calyx much enlarged, membranous, veiny. 10. *Asperugo.*
  Nutlets attached to the receptacle by their very bases.
   Scar of attachment small, flat.
    Corolla salverform or funnelform, its lobes rounded, spreading.
     Racemes not bracted; corolla-tube short. 11. *Myosotis.*
     Racemes bracted; corolla-tube cylindric, usually slender. 12. *Lithospermum.*
    Corolla tubular, its lobes erect, acute. 13. *Onosmodium.*
   Scar of attachment large, concave.
    Corolla tubular, 5-toothed. 14. *Symphytum.*
    Corolla rotate; anthers erect in a cone. 15. *Borago.*
Flowers irregular.
 Stamens included; throat of the corolla closed by scales. 16. *Lycopsis.*
 Stamens exserted; throat of the corolla dilated, open. 17. *Echium.*

1. HELIOTRÒPIUM [Tourn.] L. Sp. Pl. 130. 1753.

Herbs or shrubs, with alternate mostly entire and petioled leaves, and small blue or white flowers, in scorpioid spikes, or scattered. Calyx-lobes or -segments lanceolate or linear. Corolla salverform or funnelform, naked in the throat, its tube cylindric, its lobes imbricated, plicate or induplicate in the bud, spreading in flower. Stamens included; filaments short, or none. Style terminal, short or slender; stigma conic or annular. Fruit 2–4-lobed, separating into 4 1-seeded nutlets, or into 2 2-seeded carpels. Ovary entire, or 2–4-grooved. [Greek, sun-turning, *i. e.*, turning to or with the sun.]

 About 125 species, widely distributed in warm-temperate and tropical regions. Besides the following, some 10 others occur in the southern and western parts of North America. The species are called Turnsole. Type species: *Heliotropium europaeum* L.
Fruit 4-lobed, each lobe becoming a 1-seeded nutlet.
 Flowers in scorpioid spikes.
  Plant rough-puberulent; leaves oval. 1. *H. europaeum.*
  Glabrous, fleshy; leaves linear to spatulate.
   Leaves linear to oblanceolate; corolla 2″ broad. 2. *H. curassavicum.*
   Leaves broadly spatulate; corolla 3″–4″ broad. 3. *H. spathulatum.*
 Flowers solitary, terminating short branches. 4. *H. tenellum.*
Fruit 2-lobed, or of 2 carpels.
 Style elongated; flowers large, scattered, white. 5. *Cryptantha.*
 Style very short; flowers blue, in scorpioid spikes. 6. *H. indicum.*

1. Heliotropium europaèum L. European Heliotrope. Fig. 3502.

Heliotropium europaeum L. Sp. Pl. 130. 1753.

 Annual, much branched, rough-puberulent, 6′–18′ high. Leaves oval, 1′–2′ long, obtuse at the apex, narrowed at the base, slender-petioled, pinnately veined; flowers white, 1″–2″ broad, in dense 1-sided scorpioid, bractless spikes; terminal spikes in pairs, the lateral ones commonly solitary, becoming 1′–3′ long in fruit; calyx-segments lanceolate to linear-lanceolate, shorter than the corolla-tube; anthers distinct, obtuse; stigma-tip long-conic; fruit depressed-globose, pubescent, 4-lobed, at length separating into 4 nutlets.

 In waste places, Massachusetts to New York, Pennsylvania and Florida. Adventive or naturalized from Europe. June–Oct.

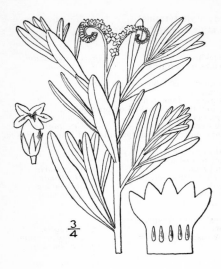

$\frac{3}{4}$

2. Heliotropium curassávicum L. Seaside Heliotrope. Fig. 3503.

Heliotropium curassavicum L. Sp. Pl. 130. 1753.

Annual, fleshy, glabrous throughout, more or less glaucous, branched, diffuse, the branches 6'-18' long. Leaves linear, or linear-oblong, entire, very inconspicuously veined, 1'-2' long, 1½''-3'' wide, obtuse at the apex, narrowed into petioles, or the upper sessile, sometimes with smaller ones fascicled in the axils; scorpioid spikes densely flowered, bractless, mostly in pairs; flowers about 2'' broad; calyx-segments lanceolate, acute; corolla white with a yellow eye or changing to blue; stigma umbrella-shaped; anthers acuminate; fruit globose, at length separating into 4 nutlets.

On sandy seashores, Delaware to Texas and Mexico. Widely distributed in saline and maritime soil in the West Indies, South America and the Old World. In ballast about the northern seaports. May–Sept.

3. Heliotropium spathulàtum Rydb. Spatulate-leaved Heliotrope. Fig. 3504.

H. spathulatum Rydb. Bull. Torr. Club **30**: 262. 1903.

Perennial, glabrous, fleshy, glaucous, branched, 1°–1½° high, the branches ascending. Leaves spatulate, indistinctly veined, 2' long or less, obtuse at the apex; scorpioid spikes 2–5; flowers 3''–4'' broad, white, or bluish; calyx-segments ovate-lanceolate, acute; fruit rather larger than that of *H. curassavicum*.

Prairies, plains and meadows, Iowa to North Dakota, Assiniboia, Chihuahua and California. June–Sept.

$\frac{3}{4}$

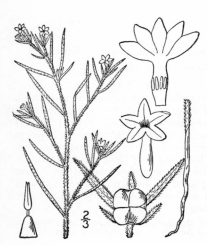

$\frac{2}{3}$

4. Heliotropium tenéllum (Nutt.) Torr. Slender Heliotrope. Fig. 3505.

Lithospermum tenellum Nutt. Trans. Am. Phil. Soc. (II.) **5**: 188. 1833–37.
H. tenellum Torr. in Marcy's Rep. 304. *pl. 14.* 1853.

Annual, strigose-canescent; stem erect, slender, paniculately branched, commonly leafless below, 6'-18' high. Leaves linear, entire, ½'-1½' long, 1''-2'' wide, narrowed at both ends, sessile, or the lower petioled; flowers white, about 2½'' long, sessile at the ends of short lateral branches, bracted by 1 or 2 leaves; calyx-segments unequal, the 2 or 3 larger ones about as long as the corolla; corolla-tube canescent, slightly longer than the limb, its lobes entire; anthers obtuse; stigma subulate-tipped; fruit depressed, 4-lobed, strigose-pubescent, separating into 4 1-seeded nutlets.

In dry soil, Kentucky to Kansas, south to Alabama, Texas and New Mexico. April–Aug.

5. Heliotropium convolvulàceum (Nutt.) A. Gray.   Bindweed Heliotrope. Fig. 3506.

Euploca convolvulacea Nutt. Trans. Am. Phil. Soc. (II.)
  5: 189. 1833–37.
H. convolvulaceum A. Gray, Mem. Am. Acad. 6: 403.
  1857.

Annual, strigose-canescent, usually much branched,
6′–15′ high, the branches ascending. Leaves oblong,
ovate, or lanceolate, entire, short-petioled, obtuse or
acute at the apex, narrowed at the base, ½′–1½′ long;
flowers numerous, fragrant, very short-peduncled,
terminal and lateral, mostly solitary and opposite the
leaves; calyx-segments lanceolate, acuminate, equal;
corolla white, strigose, 8″–10″ long, about 6″ broad,
its tube narrowed at the throat, longer than the calyx
and the angulate-lobed limb; anthers inserted on the
tube of the corolla, slightly cohering by their tips;
style filiform; stigma with a tuft of bristly hairs;
fruit 2-lobed, pubescent, each lobe splitting into 2
1-seeded nutlets.

In dry sandy soil, Nebraska to Texas, Utah and
Mexico. July–Sept.

6. Heliotropium índicum L.   Indian Heliotrope. Fig. 3507.

Heliotropium indicum L. Sp. Pl. 134. 1753.

Annual, more or less hirsute or hispid; stem commonly branched, 1°–3° high. Leaves ovate or oval,
obtuse or acute at the apex, obtuse rounded or subcordate at the base, 2′–6′ long, 1′–3½′ wide, repand or
undulate, borne on margined petioles ½′–2½′ long;
flowers blue, 2″–3″ broad, sessile in terminal dense
bractless usually solitary scorpioid spikes which become 3′–6′ long in fruit; calyx-segments lanceolate,
acute, shorter than the strigose corolla-tube; style
very short, deciduous; fruit deeply 2-lobed, glabrous,
the lobes divergent, each finally splitting into 2
nutlets, each of which is ribbed on the back.

In waste places, North Carolina to Kentucky, Illinois,
Missouri, Florida and Texas. Naturalized from India.
Also in ballast about the northern seaports. Widely
distributed in warm regions as a weed. Turnsole.
May–Nov.

2. CYNOGLÓSSUM [Tourn.] L. Sp. Pl. 134. 1753.

Hirsute or hispid (rarely glabrous) mostly tall herbs, with alternate entire leaves, the
basal long-petioled, and purple blue or white flowers in panicled, more or less scorpioid
racemes. Calyx 5-cleft or 5-parted, enlarged and spreading or reflexed in fruit. Corolla
funnelform or salverform, the tube short, the throat closed by 5 scales opposite the imbricated rounded lobes. Stamens included; filaments short; anthers ovate or oblong. Ovary
deeply 4-lobed, separating into 4 diverging nutlets in fruit; style mostly slender. Nutlets
oblique, flat or convex above, attached laterally to the convex or conic receptacle, covered
with short barbed prickles. [Greek, dog's tongue.]

About 75 species of wide geographic distribution. Besides the following, some 3 others occur
in western North America. Type species: *Cynoglossum officinale* L.

Stem leafy to the top; flowers reddish, purple or white; nutlets flat.	1. *C. officinale.*
Stem leafless above; flowers blue; nutlets convex.	
Flowers about 5″ broad; nutlets about 4″ long.	2. *C. virginianum.*
Flowers about 3½″ broad; nutlets about 2½″ long.	3. *C. boreale.*

1. Cynoglossum officinàle L.   Hound's-tongue.   Gipsy Flower.   Fig. 3508.

Cynoglossum officinale L. Sp. Pl. 134. 1753.

Biennial, pubescent; stem erect, leafy to the top, stout, usually branched, $1\frac{1}{2}°-3°$ high. Basal and lower leaves oblong or oblong-lanceolate, slender-petioled, sometimes obtuse, 6'–12' long, 1'–3' wide; upper leaves lanceolate, acute or acuminate, sessile, or the uppermost clasping; racemes several or numerous, bractless or sparingly bracted, simple or branched, much elongated in fruit; pedicels 3''–6'' long; calyx-segments ovate-lanceolate, acute; corolla reddish-purple or rarely white, about 4'' broad; fruit pyramidal, about 5'' broad, each of the 4 nutlets forming a side of the pyramid, flat on their upper faces, margined, splitting away at maturity, but hanging attached to portions of the subulate style.

In fields and waste places, Quebec and Ontario to Manitoba, South Carolina, Alabama, Kansas and Montana. Often a troublesome weed. Naturalized from Europe. Native also of Asia. Called also dog's-tongue, rose noble.   Canadian or dog-bur.   Sheep-lice.   **Tory-weed.**   Wood-mat.   May–Sept.

2. Cynoglossum virgínianum L.   Wild Comfrey.   Fig. 3509.

Cynoglossum virginianum L. Sp. Pl. 134. 1753.

Perennial, hirsute; stem usually simple, leafless above, stout, $1\frac{1}{2}°-2\frac{1}{2}°$ high. Basal and lower leaves oval or oblong, 4'–12' long, obtuse at the apex, narrowed into petioles; upper leaves oblong, or ovate-lanceolate, sessile and clasping by a cordate base, acute, nearly as large, or the one or two uppermost quite small; racemes 2–6, corymbose, bractless, long-peduncled; flowers blue, about 5'' broad; corolla-lobes obtuse; calyx-segments oblong-lanceolate, obtuse, about 2'' long at flowering time; fruit depressed, 4'' broad, the nutlets convex on the upper face, not margined, separating and falling away at maturity, about 4'' long.

In woods, New Jersey to Kentucky, Missouri, Florida, Louisiana and Kansas. Ascends to 2500 ft. in Virginia. Dog-bur. April–May.

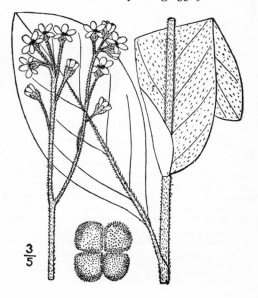

3. Cynoglossum boreàle Fernald. Northern Wild Comfrey. Fig. 3510.

Cynoglossum boreale Fernald, Rhodora **7 : 250.** 1906.

Similar to the preceding species, the stem more slender, villous below, appressed-pubescent above. Upper leaves clasping the stem; lower and basal leaves oblong, acute or acutish, long-petioled; racemes few, the flowers distant; calyx-segments only about 1″ long at flowering time; corolla 3″–4″ broad; nutlets 2″–2½″ long.

Woods and banks, Quebec to Ontario, Connecticut, New York and Minnesota. In our first edition included in the preceding species, of which it may be a northern race. May–June.

3. LÁPPULA [Rivin.] Moench, Meth. 416. 1794.

[Echinospermum Sw.; Lehm. Asperif. 113. 1818.]

Annual or perennial rough-pubescent or canescent erect branching herbs, with alternate narrow entire leaves, and small or minute blue or white flowers, in terminal bracted or bractless racemes. Calyx deeply 5-cleft or 5-parted, the segments narrow. Corolla salverform or funnelform, the tube very short, the throat closed by 5 scales, the lobes obtuse, spreading, imbricated in the bud. Stamens included; filaments very short. Ovary 4-lobed; style short. Nutlets 4, erect or incurved, laterally attached to the receptacle, at length separating, the margins or backs armed with stout often flattened barbed prickles, the sides usually papillose or tuberculate. [Diminutive of the Latin *lappa,* a bur.]

About 40 species, mostly natives of the north temperate zone. Besides the following, several others occur in western North America. Type species: *Lappula Myosòtis* Moench.

Racemes bracted; fruiting pedicels not deflexed.
  Prickles in 2 rows on the margins of the nutlets, distinct.   1. *L. Lappula.*
  Prickles in 1 row on the margins, more or less confluent.   2. *L. texana.*
Racemes bracted only at the base; fruiting pedicels deflexed.
  Stem-leaves ovate-oblong, the basal cordate; fruit globose.   3. *L. virginiana.*
  Leaves oblong, oblong-lanceolate or linear; fruit pyramidal.
    Flowers 3″–5″ broad; fruit about 3″ broad.   4. *L. floribunda.*
    Flowers 1″–2″ broad; fruit about 2″ broad.   5. *L. deflexa.*

1. Lappula Láppula (L.) Karst. European Stickseed. Burseed. Fig. 3511.

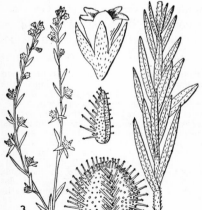

Myosotis Lappula L. Sp. Pl. 131. 1753.
Lappula echinata Gilib. Fl. Lithuan. 1 : 25. 1781.
Lappula Myosotis Moench, Meth. 417. 1794.
Echinospermum Lappula Lehm. Asperif. 121. 1818.
Lappula Lappula Karst. Deutsch. Fl. 979. 1880–83.

Annual, pale, leafy, hispid or appressed-pubescent, branched, 1°–2° high, the branches erect. Leaves linear, linear-oblong or the lowest spatulate, sessile or the lower narrowed into petioles, ascending or erect, obtuse or obtusish at the apex, ½′–1½′ long; racemes leafy-bracted, more or less 1-sided; pedicels very short, stout, not deflexed in fruit; calyx-segments lanceolate, becoming unequal and spreading; corolla blue, about 1″ broad; fruit globose-oval, 1½″ in diameter; the nutlets papillose or also prickly on the back, the margins armed with 2 rows of slender distinct prickles.

In waste places, Nova Scotia to British Columbia, south to New Jersey and Kansas. Naturalized from Europe. Native also of Asia. Stick-tight. Small sheep-bur. May–Sept.

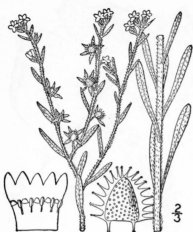

2. Lappula texàna (Scheele) Britton. Hairy Stickseed. Fig. 3512.

Cynoglossum pilosum Nutt. Gen. **1** : 114. 1818. Not R. & P. 1794.
Echinospermum texanum Scheele, Linnaea **25** : 260. 1852.
Echinospermum Redowskii var. *cupulatum* A. Gray in Brewer & Wats. Bot. Cal. **1** : 530. 1876.
Lappula texana Britton, Mem. Torr. Club **5** : 273. 1894.
L. Redowskii occidentalis Rydb. Contr. Nat. Herb. **3** : 170. 1895.

Annual, similar to the preceding species, 6'–2° high, paniculately branched, the branches ascending or erect. Leaves linear or linear-oblong, mostly obtuse, the lower narrowed into petioles; racemes leafy-bracted; pedicels short, not deflexed in fruit; flowers about 1'' broad; nutlets papillose-tuberculate on the back, the margins armed with a single row of flat, usually more or less confluent bristles, or these united into a cup.

In dry soil, Ontario to Manitoba, British Columbia, south to Nebraska, Texas and Arizona. Found also in ballast at Atlantic seaports. Nutlets with nearly distinct bristles and others with bristles united into a cup sometimes occur on the same fruit. April–Aug.

3. Lappula virginiàna (L.) Greene. Virginia Stickseed. Fig. 3513.

Myosotis virginiana L. Sp. Pl. 131. 1753.
Cynoglossum Morisoni DC. Prodr. 10 : 155. 1846.
E. virginicum Lehm. Asperif. 120. 1818.
Lappula virginiana Greene, Pittonia **2** : 182. 1891.

Biennial, pubescent; stem paniculately branched, 2°–4° high, the branches slender, spreading. Basal leaves (seldom present at flowering time) ovate or nearly orbicular, cordate, long-petioled, mostly obtuse; stem leaves ovate-oblong or oval, acute or acuminate at the apex, narrowed to the base, petioled, 3'–8' long, 1'–4' wide, the uppermost smaller, sessile; racemes very slender, divergent, bracted at the base, the bracts similar to the upper leaves; pedicels slender, short, recurved in fruit; corolla nearly white, about 1'' broad; fruit globose, nearly 2'' in diameter; nutlets covered on the margins and usually also on the back by the slender distinct flattened barbed prickles, the backs commonly also more or less papillose.

In dry woods and thickets, New Brunswick to Ontario, Minnesota, Georgia, Louisiana, Nebraska and Kansas. Called beggar's-ticks or -lice. Virginia mouse-ear. Soldiers. Stick-tight. Dysentery-root or -weed. June–Sept.

4. Lappula floribúnda (Lehm.) Greene. Large-flowered Stickseed. Fig. 3514.

Echinospermum floribundum Lehm. in Hook. Fl. Bor. Am. **2** : 84. *pl. 164.* 1834.
Lappula floribunda Greene, Pittonia **2** : 182. 1891.

Biennial or perennial, rough-pubescent; stem stout, paniculately branched, 2°–5° high, the branches nearly erect. Leaves oblong, oblong-lanceolate, or linear-lanceolate, 2'–4' long, 2''–10'' wide, sessile, acute or obtuse at the apex, or the lower narrowed into petioles; racemes numerous, erect or nearly so, very densely flowered, bracted at the base, many of them in pairs; pedicels 2''–4'' long, reflexed in fruit; flowers blue or white, 3''–5'' broad; fruit pyramidal, about 3'' broad; nutlets keeled, papillose-tuberculate on the back, the margins armed with a single row of flat prickles, which are sometimes confluent at the base.

Western Ontario and Minnesota to Saskatchewan, British Columbia, south to New Mexico and California. June–Aug.

5. Lappula defléxa (Wahl.) Garcke. Nodding Stickseed. Fig. 3515.

Echinospermum deflexum var. *americanum* A. Gray,
    Proc. Am. Acad. **17**: 224. 1882.
Lappula deflexa Garcke, Fl. Deutsch. Ed. 6, 275. 1863.
Lappula americana Rydberg, Bull. Torr. Club **24**: 294.
    1897.

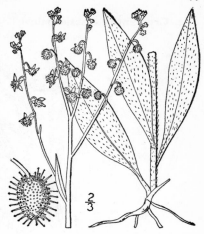

Annual, rough-puberulent; stem slender, erect, paniculately branched, 1°–3° high, the branches spreading or ascending. Leaves oblong or oblong-lanceolate, mostly narrowed at both ends, 2'–4' long, 2½"–6" wide, the lower petioled, the upper sessile; racemes slender, many-flowered; pedicels slender, 2"–4" long, deflexed in fruit; corolla white or bluish, about 1" broad; fruit pyramidal, about 2" broad; nutlets keeled, papillose-tuberculate on the back, rarely with a few prickles on the keel, the margins armed with a single row of flat prickles, these sometimes few.

In thickets, Quebec to Manitoba, British Columbia, Iowa, Nebraska and Wyoming. Also in Europe and Asia. May–Aug.

4. ALLOCÀRYA Greene, Pittonia, 1: 12. 1887.

Mostly annual low herbs, with linear entire leaves, the lowest often opposite, and small flowers in terminal spikes or racemes. Pedicels thickened at the summit, persistent. Calyx 5-divided, persistent, the segments narrow. Corolla salverform, white, yellow in the throat. Stamens included. Ovary 4-divided; style short. Nutlets crustaceous, smooth, or rough, attached at their base or below the middle to the receptacle, the scar of attachment concave or raised. [Greek, different nuts.]

About 25 species, natives of western North America. Type species: *Allocarya lithocarya* (A. Gray) Greene.

1. Allocarya scopulòrum Greene. Mountain Allocarya. Fig. 3516.

Eritrichium californicum var. *subglochidiatum* A. Gray, Bot. Cal. 1: 526. In part. 1876.

Allocarya scopulorum Greene, Pittonia 1: 16. 1887.

Somewhat succulent, pubescent with scattered stiff appressed hairs, branched, the slender spreading branches 1'–8' long. Leaves 6"–18" long, 1"–1½" wide, sessile or very short-petioled; flowers about 1" broad, distant, borne in most of the axils, very short-pedicelled; floral bracts similar to the leaves, but shorter; calyx segments linear-lanceolate; nutlets reticulate on the back, lightly grooved on the ventral side.

Western Nebraska to Montana, Wyoming and Colorado. June–Sept.

5. CRYPTÁNTHA Lehm. Sem. Hort. Hamburg. 1832. F. & M. Ind. Sem. Hort. Petrop. 2: 35. 1836.

[KRYNITZKIA F. & M. Ind. Sem. Hort. Petrop. **7**: 52. 1841.]

Low annual setose or hispid branched herbs, with narrow alternate entire leaves, and small mostly white flowers, in scorpioid bractless or bracteolate spikes. Calyx 5-parted or 5-cleft, at length deciduous from the spike, the lobes or segments erect, mostly connivent in fruit. Corolla small, funnelform, usually with 5 scales closing the throat, the lobes imbricated in the bud. Stamens included; filaments short. Ovary 4-divided; style short; stigma capitellate. Nutlets erect, rounded on the back, not keeled, the margins obtuse, acute or wing-margined, attached laterally to the conic or elongated receptacle, the scar of attachment mostly longer than broad. [Greek, hidden-flowered.]

About 50 species, natives of North and South America, mostly of the western United States.
Type species: *Cryptantha glomerata* Lehm.

Nutlets, at least some of them, with short processes.   1. *C. crassisepala.*
All four nutlets smooth and shining.   2. *C. Fendleri.*

1. Cryptantha crassisépala (T. & G.) Greene.   Thick-sepaled Cryptanthe.
Fig. 3517.

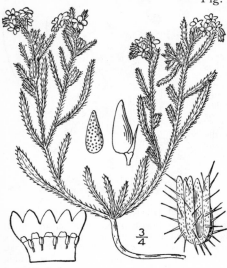

Eritrichium crassisepalum T. & G. Pac. R. R.
  Rep. **2**: 171.   1854.
Krynitzkia crassisepala A. Gray, Proc. Am.
  Acad. **20**: 268.   1885.
Cryptanthe crassisepala Greene, Pittonia **1**:
  112. 1887.

Densely hispid, at length much branched,
3′–6′ high.   Leaves linear, or linear-
spatulate, ½′–1½′ long, sessile, or the lower
narrowed into petioles; spikes very densely
flowered; flowers about 2″ broad, sessile,
bracteolate, the bractlets slightly longer
than the calyx; fruiting calyx 3″ long,
closing over the fruit, its segments linear,
obtusish, their midribs much thickened;
fruit of 3 finely muricate nutlets, and 1
larger smooth and shining nutlet about 1″
long, attached to the receptacle from the
base to near the middle.

In dry soil, Saskatchewan to Nebraska,
Kansas, Texas and New Mexico.   Rosita.
June–Aug.

2. Cryptantha Féndleri (A. Gray) Greene.   Fendler's Cryptanthe.
Fig. 3518.

Krynitzkia Fendleri A. Gray, Proc. Am. Acad.
  20: 268. 1885.
Cryptanthe Fendleri Greene, Pittonia **1**: 120.
  1887.

Erect, hispid; stem slender, paniculately
branched, 6′–15′ high. Leaves linear, or the
lowest linear-spatulate, 1′–2½′ long; spikes
slender, bracteolate only at the base; flowers
sessile, 1″–2″ broad; fruiting calyx nearly
closed, its segments linear, leaf-like, herbaceous, about 2″ long; nutlets 4, all alike,
brown, smooth and shining, less than 1″ long,
attached to the receptacle from the base to
about the middle.

In dry soil, Saskatchewan to Washington, south
to Nebraska and Arizona. June–Aug.

6. OREOCÀRYA Greene, Pittonia, **1**: 57.   1887.

Perennial or biennial hispid or strigose-pubescent herbs, mostly with thick woody roots,
alternate or basal narrow leaves, and small white racemose-paniculate or densely thyrsoid
flowers. Calyx very deeply 5-parted or 5-divided, the segments lanceolate, more or less
spreading or recurved in fruit. Corolla funnelform or salverform, mostly crested in the
throat, 5-lobed. Stamens included. Ovary 4-divided; style mostly short. Nutlets 4, laterally attached to the receptacle, not keeled, their margins acute or winged. [Greek, mountain nut.] Called White Forget-me-not.

About 9 species, natives of western North America and Mexico. Type species: *Oreocarya
suffruticòsa* (Torr.) Greene.

Inflorescence racemose-paniculate; nutlets smooth.	1. *O. suffruticosa.*
Inflorescence thyrsoid or thyrsoid-glomerate; nutlets rough.	
Corolla-tube not longer than the calyx, little longer than the lobes.	
Densely rough-hairy, 6′–18′ high.	2. *O. glomerata.*
Silvery appressed-pubescent, 3′–6′ high.	3. *O. sericea.*
Corolla-tube longer than the calyx, 2 or 3 times as long as the lobes.	4. *O. fulvocanescens.*

1. Oreocarya suffruticòsa (Torr.) Greene.
Shrubby Oreocarya. Fig. 3519.

Myosotis suffruticosa Torr. Ann. Lyc. N. Y. **2** : 225.
1827.
Eritrichium Jamesii Torr. in Marcy's Rep. 294. 1853.
Krynitzkia Jamesii A. Gray, Proc. Am. Acad. **20** : 278.
1885.
Oreocarya suffruticosa Greene, Pittonia **1** : 57. 1887.

Perennial, rather stout, branched from the
base and sometimes also above, strigose-pubescent
or somewhat hirsute, 5′–12′ high. Upper leaves
linear, ½′–1′ long, the lower oblanceolate, somewhat longer, obtuse or acute; racemes slender,
panicled; pedicels about 1″ long; calyx canescent
and somewhat hispid, the segments slightly
spreading, or erect in fruit; bractlets longer than
the fruiting calyx; corolla 1½″–3″ broad; its tube
about equalling the calyx; nutlets smooth, nearly
1″ long, shining, closely fitting together, triangular, acute-margined, nearly as wide as high.

In dry soil, western Nebraska and Wyoming to
Texas and Arizona. May–Aug.

2. Oreocarya glomeràta (Pursh) Greene.
Clustered Oreocarya. Fig. 3520.

Cynoglossum glomeratum Pursh, Fl. Am. Sept. 729.
1814.
Eritrichium glomeratum DC. Prodr. **10** : 131. 1846.
Krynitzkia glomerata A. Gray, Proc. Am. Acad. **20** :
279. 1885.
O. glomerata Greene, Pittonia **1** : 58. 1887.

Perennial or biennial, densely hispid; stem
erect, stout, simple or branched, 6′–18′ high.
Leaves spatulate or the upper linear, obtuse,
1′–2½′ long, the basal commonly tufted; inflorescence of thyrsoid clusters, the short dense lateral
spike-like clusters mostly longer than the subtending bracts; calyx densely bristly; corolla
3″–5″ broad; fruit pyramidal, the nutlets triangular-ovate, acute, acutely margined, papillose
on the back.

In dry soil, Manitoba to Nebraska, New Mexico
and Utah. May–Sept.

Oreocarya thyrsiflòra Greene, a related species
of the Rocky Mountain region, enters our limits in
western Nebraska.

3. Oreocarya serícea (A. Gray) Greene.
Low Oreocarya. Fig. 3521.

Eritrichium glomeratum var. *humile* A. Gray, Proc.
Am. Acad. **10** : 61. 1874. Not *E. humile* DC.
Krynitzkia sericea A. Gray, Proc. Am. Acad. **20** : 279.
1885.
Oreocarya sericea Greene, Pittonia **1** : 58. 1887.

Perennial, low, tufted from the woody root;
stems usually simple, 3′–6′ high, silvery appressed-
pubescent, or hirsute above. Leaves linear-
spatulate, ½′–1′ long, 1″–1½″ wide, obtuse or
acutish, imbricated on the short sterile shoots
and at the bases of the flowering stems; inflorescence thyrsoid or glomerate, usually short; calyx
densely hispid; corolla 2″–3″ broad, its tube not
longer than the calyx; style short; nutlets acutely
margined, acute, papillose on the back.

In dry soil, Northwest Territory to Nebraska and
Utah. May–Sept.

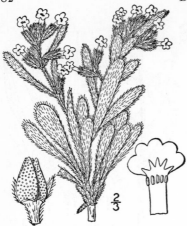

4. Oreocarya fulvocanéscens (A. Gray) Greene. Tawny Oreocarya. Fig. 3522.

Eritrichium fulvocanescens A. Gray, Proc. Am. Acad. 10: 61. 1874.
Eritrichium glomeratum var. (?) *fulvocanescens* S. Wats. Bot. King's Exp. 243. *pl. 23. f. 7.* 1871.
Oreocarya fulvocanescens Greene, Pittonia 1: 58. 1887.

Perennial, tufted, similar to the preceding species but densely strigose or hirsute. Leaves spatulate, or oblanceolate, obtuse, the lower and basal ones 1′–1½′ long; inflorescence of thyrsoid clusters; calyx densely setose with yellowish hairs; corolla about 2″ broad, its tube longer than the calyx, 2 or 3 times the length of the lobes; style filiform; nutlets acutely margined, tuberculate on the back.

In dry soil, western Nebraska (according to Webber), Wyoming to Nevada and New Mexico. May–Aug.

7. PNEUMÀRIA Hill, Veg. Syst. 7: 40. *pl. 37.* 1764.

A perennial fleshy glabrous glaucous diffusely branched herb, with alternate entire leaves, and small blue pinkish or white flowers in loose terminal leafy-bracted racemes. Calyx-lobes triangular-ovate or lanceolate, somewhat enlarging in fruit. Corolla tubular-campanulate, crested in the throat, 5-lobed, the lobes imbricated in the bud, slightly spreading. Filaments slender, scarcely exserted. Ovary 4-divided; style slender. Nutlets erect, fleshy, attached just above their bases to the somewhat elevated receptacle, smooth, shining, acutish-margined, becoming utricle-like when mature.

A monotypic genus of sea-beaches of the north temperate zone.

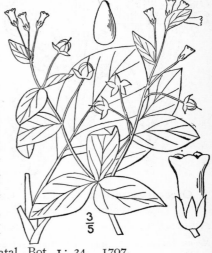

1. Pneumaria marítima (L.) Hill. Sea Lungwort. Sea Bugloss. Oyster Plant. Fig. 3523.

Pulmonaria maritima L. Sp. Pl. 136. 1753.
P. maritima Hill, Veg. Syst. 40. *pl. 37. f. 3.* 1764.
Mertensia maritima S. F. Gray, Nat. Arr. Brit. Pl. 354. 1821.

Pale green, the branches spreading or ascending, 3′–15′ long. Leaves thick, ovate, obovate, or oblong, 1′–4′ long, acute or obtuse at the apex, narrowed at the base, the lower and basal ones contracted into margined petioles, the uppermost smaller; flowers blue or nearly white, about 3″ long, all pedicelled; pedicels very slender, 5″–18″ long; calyx shorter than the corolla-tube; corolla with a crest in the throat opposite each lobe; nutlets about as long as the calyx-lobes when mature.

On sea-beaches, Long Island (?), Massachusetts to Newfoundland and Greenland, Oregon to Alaska. Also on the coasts of Europe and Asia. May–Sept.

8. MERTÉNSIA Roth, Catal. Bot. 1: 34. 1797.

Perennial glabrous or pubescent herbs, with alternate sometimes punctate leaves, and rather large blue purple or white flowers, in panicles, cymes, or racemes. Calyx-lobes lanceolate or linear, little enlarged in fruit. Corolla tubular-funnelform or trumpet-shaped, crested or unappendaged in the throat, its lobes obtuse, imbricated, little spreading. Stamens inserted on the tube of the corolla, included, or scarcely exserted; filaments flattened, or filiform; anthers oblong or linear, obtuse. Ovary 4-divided; style filiform. Nutlets erect, coriaceous, wrinkled when mature, attached above their bases to the convex or nearly flat receptacle. [In honor of Prof. C. F. Mertens, a German botanist.]

About 40 species, natives of the northern hemisphere. Besides the following, many occur in the western part of North America. The species are called Smooth Lungwort. Type species: *Mertensia pulmonarioides* Roth.

Corolla trumpet-shaped, not crested in the throat, the limb barely 5-lobed.   1. *M. virginica.*
Corolla funnelform-campanulate, crested in the throat, the limb manifestly 5-lobed.
    Stem-leaves ovate or ovate-lanceolate, acuminate.   2. *M. paniculata.*
    Stem-leaves oblong or lanceolate, obtuse or acute.   3. *M. lanceolata.*

1. Mertensia virgínica (L.) DC. Virginia Cowslip. Tree Lungwort. Roanoke-bells. Fig. 3524.

Pulmonaria virginica L. Sp. Pl. 135. 1753.
Mertensia virginica DC. Prodr. 10: 88. 1846.

Glabrous; stem erect, or ascending, simple or sometimes branched, 1°–2° high, rather stout. Leaves oblong, oval, or obovate, pinnately veined, obtuse at the apex, 2′–5′ long, the uppermost sessile, the lower narrowed into margined petioles; racemes short, corymb-like; pedicels 2″–6″ long; flowers blue-purple, very showy, about 1′ long; calyx-lobes oblong-lanceolate, obtusish, 1″ long or less; corolla trumpet-shaped or nearly salverform, its tube cylindric, a little expanded above, longer than the 5-lobed plaited limb, pubescent at the base within, not crested in the throat; disk with two opposite linear lobes; filaments filiform, much longer than the anthers; nutlets not shining, rounded.

In low meadows and along streams, southern Ontario to New Jersey and South Carolina, Minnesota, Nebraska and Kansas. Blue bells. March–May.

2. Mertensia paniculàta (Ait.) G. Don. Tall Lungwort. Fig. 3525.

P. paniculata Ait. Hort. Kew. 1: 181. 1789.
M. paniculata G. Don, Gen. Syst. 4: 318. 1838.

Roughish-pubescent, dark green; stem erect, branched above, 1½°–3° high, the branches slender. Leaves thin, pinnately veined, those of the stem ovate or ovate-lanceolate, acuminate at the apex, narrowed at the base, 2′–5′ long, the lower narrowed into slender petioles; basal leaves ovate, rounded or cordate at the base; racemes several-flowered, panicled; pedicels filiform, 4″–10″ long; flowers purple-blue, 6″–7″ long; calyx-lobes lanceolate, acute; corolla tubular-campanulate, crested in the throat, the tube about twice as long as the calyx and exceeding the 5-lobed limb; filaments flattened, slightly longer than the anthers; style filiform, usually somewhat exserted; nutlets rounded.

In woods or thickets, Hudson Bay to Alaska, south to Michigan, Nebraska, Idaho and Washington. July–Aug.

3. Mertensia lanceolàta (Pursh) DC. Lance-leaved Lungwort. Fig. 3526.

P. lanceolata Pursh, Fl. Am. Sept. 729. 1814.
Mertensia lanceolata DC. Prodr. 10: 88. 1846.
M. linearis Greene, Pittonia 3: 197. 1897.

Glabrous or somewhat hirsute; stem simple or branched, slender, 6′–18′ high. Leaves papillose, indistinctly veined, light green, the upper lanceolate, acute, sessile or slightly clasping at the base, the lower oblong or oblanceolate, obtuse, 3′–4′ long, narrowed into margined petioles; racemes few-flowered, usually panicled; flowers blue, 5″–6″ long; pedicels 3″–7″ long; calyx-lobes lanceolate, obtuse, or acutish; corolla tubular-campanulate, the tube longer than the calyx and longer than the 5-lobed limb, hairy at the base within, the throat crested; filaments a little longer than the anthers; style filiform, scarcely exserted.

In thickets, western Nebraska to Manitoba, Wyoming, Idaho and New Mexico. Races differ in amount or absence of pubescence and in width of leaves. June–Aug.

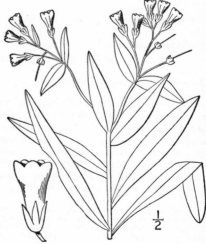

9. AMSÍNCKIA Lehm. Del. Sem. Hort. Hamb. 7. 1831.

Annual hispid or setose herbs, with narrow leaves, the yellow flowers in scorpioid spikes. Calyx 5-parted, the segments linear. Corolla salverform or funnelform, the tube slender, naked or minutely crested in the throat, the 5 lobes spreading. 'Stamens 5, borne on the corolla-tube, included. Ovary deeply 4-lobed. Nutlets ovoid, rough, laterally attached to the receptacle below the middle. [In honor of William Amsinck, a burgomaster of Hamburg and friend of the Hamburg botanical garden.]

About 15 species, natives of western North America and Chile, the following typical.

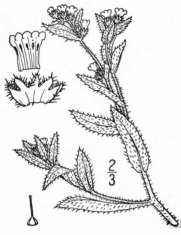

1. Amsinckia lycopsioìdes Lehm. Amsinckia. Fig. 3527.

Lithospermum lycopsioides Lehm. Pug. **2**: 28. 1830.

Amsinckia lycopsioides Lehm.; DC. Prodr. **10**: 117. 1846.

Diffusely branched, loosely hispid with long, bristly hairs, the branches often 1° long, decumbent or ascending. Leaves lanceolate, oblong-lanceolate or ovate-lanceolate, repand-dentate or entire, 3' long or less, sessile; scorpioid spikes short in flower, elongating in fruit, the lower flowers bracteolate, the upper ones commonly bractless; corolla about 4" long, its tube somewhat longer than the calyx; nutlets rugose-reticulate.

Waste grounds, Massachusetts and Connecticut. Adventive from California. May–July.

Amsinckia intermédia F. & M., an erect Californian species, with orange-yellow flowers and linear leaves, has been found in eastern Long Island and Nantucket.

10. ASPERÙGO [Tourn.] L. Sp. Pl. 138. 1753.

An annual rough-hispid procumbent herb, with alternate entire leaves, or the uppermost sometimes opposite, and small blue or nearly white flowers, short-pedicelled and 1–3 together in the upper axils. Calyx campanulate, unequally 5-cleft, much enlarged and folded together in fruit, the lobes incised-dentate. Corolla tubular-campanulate, 5-lobed, the lobes imbricated. Stamens 5, included, inserted on the corolla-tube; filaments very short. Ovary 4-divided; style short; stigma capitate. Nutlets 4, ovoid, erect, granular-tuberculate, keeled, laterally attached above the middle to the elongated-conic receptacle. [Latin, rough, referring to the leaves.]

A monotypic genus of Europe and Asia.

1. Asperugo procúmbens L. German Madwort. Catchweed. Fig. 3528.

Asperugo procumbens L. Sp. Pl. 138. 1753.

Stems slender, branched, diffusely procumbent, 6'–18' long, very rough with stiff bristly hairs. Leaves oblong, lanceolate, or the lower spatulate, obtuse or acutish at the apex, ½'–1½' long, the lower narrowed into margined petioles; flowers very short-pedicelled, about 1" broad, blue, the pedicels recurved in fruit; fruiting calyx dry and membranous, strongly veined, 4"–6" broad; nutlets obliquely ovoid.

In waste places and ballast, Massachusetts to southern New York, New Jersey, District of Columbia and Minnesota. Adventive from Europe. Called also small wild bugloss and great goose-grass. May–Aug.

11. MYOSÒTIS [Dill.] L. Sp. Pl. 131. 1753.

Low annual biennial or perennial, more or less pubescent, branching, diffuse or erect herbs, with alternate entire leaves, and small blue pink or white flowers in many-flowered elongated bractless more or less 1-sided racemes, or these sometimes leafy at the base. Calyx 5-cleft, the lobes narrow, spreading or erect in fruit. Corolla salverform, the limb 5-lobed, the lobes convolute in the bud, rounded, the throat crested. Stamens 5, included, inserted on the corolla-tube; filaments filiform; anthers obtuse. Ovary 4-divided, style filiform. Nutlets erect, glabrous or pilose, attached by their bases to the receptacle, the scar of attachment small, flat. [Greek, mouse-ear.]

About 35 species of wide geographic distribution. Besides the following, 1 or 2 others occur in the southern and western parts of North America. Called forget-me-not and scorpion-grass. Type species: *Myosotis scorpioides* L.

Hairs of the calyx all straight; perennial swamp or brook plants.
　Calyx-lobes shorter than the tube; corolla 3″–4″ broad.　　　　　　　　1. *M. scorpioides.*
　Calyx-lobes as long as the tube; corolla 2″–3″ broad.　　　　　　　　2. *M. laxa.*
Hairs of the calyx, or some of them, with hooked tips; annuals or biennials.
　Fruiting pedicels longer than the calyx.　　　　　　　　　　　　　　3. *M. arvensis.*
　Fruiting pedicels not longer than the calyx.
　　Calyx-lobes equal; corolla blue or yellowish, changing to violet and blue.
　　　Corolla yellowish, changing to violet and blue; style longer than the nutlets.
　　　　　　　　　　　　　　　　　　　　　　　　　　　　　　　4. *M. versicolor.*
　　　Corolla blue; style not longer than the nutlets.　　　　　　　　5. *M. micrantha.*
　　Calyx-lobes unequal; corolla white.　　　　　　　　　　　　　　6. *M. virginica.*

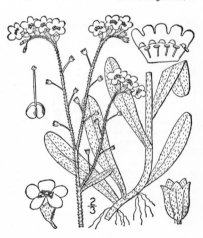

1. Myosotis scorpioìdes L.　Forget-me-not. Mouse-ear Scorpion-grass.　Fig. 3529.

Myosotis scorpioides var. *palustris* L. Sp. Pl. 131.　1753.
Myosotis palustris Lam. Fl. Fr. 2: 283.　1778.

Appressed-pubescent, perennial, with slender root-stocks or stolons; stems slender, decumbent or ascending, rooting at the lower nodes, 6′–18′ long. Leaves oblong, oblanceolate or oblong-lanceolate, obtuse, narrowed at the base, 1′–3′ long, 2″–6″ wide, those of the stem sessile or very nearly so, or the lower petioled; racemes loosely many-flowered; pedicels longer than the calyx; calyx with straight appressed hairs, its lobes equal, triangular-ovate, acute, shorter than the tube, spreading in fruit; corolla blue with a yellow eye, the limb flat, 3″–4″ broad; nutlets angled and keeled on the inner side.

In brooks and marshes, Newfoundland to New York, Pennsylvania and Tennessee. Escaped from cultivation. Native of Europe and Asia. Called also marsh scorpion-grass, snake-grass and love-me. May–July.

2. Myosotis láxa Lehm.　Smaller Forget-me-not.　Fig. 3530.

Myosotis laxa Lehm. Asperif. 83.　1818.

Myosotis palustris var. *laxa* A. Gray, Man. Ed. 5, 365.　1867.

Perennial, appressed-pubescent, similar to the preceding species; stems decumbent, spreading, rooting at the nodes, 6′–20′ long. Leaves oblong, oblong-lanceolate or spatulate, obtuse; racemes very loosely many-flowered; pedicels spreading, much longer than the fruiting calyx; hairs of the calyx straight, appressed, its lobes equal, ovate-lanceolate, acutish, spreading in fruit, quite as long as the tube; corolla blue with a yellow eye, its limb concave, about 2″ broad; nutlets convex on both the inner and outer sides.

In wet muddy places, Newfoundland to Ontario, south to Virginia and Tennessee. Also in Europe. Ascends to 3500 ft. in Virginia. May–July.

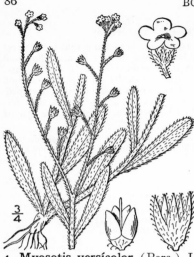

3. Myosotis arvénsis (L.) Hill. Field Scorpion-grass or Mouse-ear. Fig. 3531.

Myosotis scorpioides var. *arvensis* L. Sp. Pl. 131. 1753.
Myosotis arvensis Hill, Veg. Syst. **7**: 55. 1764.

Annual or biennial, hirsute-pubescent; stem erect, branched, 6'–18' high. Basal and lower leaves oblanceolate, obtuse, petioled or sessile; stem leaves mostly oblong or oblong-lanceolate, obtuse or acutish at the apex, narrowed to the sessile base, ½'–1½' long, 2''–4'' wide; racemes loosely flowered; fruiting pedicels longer than the calyx; hairs of the calyx, or some of them, with minutely hooked tips, the lobes equal, erect, or connivent in fruit, triangular-lanceolate, acute, about as long as the tube; corolla blue or white, the limb concave, 1''–1½'' broad; nutlets convex on the outer side, somewhat keeled on the inner.

In fields, Newfoundland to western Ontario and Minnesota, south to West Virginia. Perhaps not indigenous. Also in Europe. June–Aug.

4. Myosotis versícolor (Pers.) J. E. Smith. Yellow and Blue Scorpion-grass. Fig. 3532.

M. arvensis var. (?) *versicolor* Pers. Syn. **1**: 156. 1805.
Myosotis versicolor J. E. Smith, Engl. Bot. *pl. 480.* 1813.

Annual, hirsute-pubescent, with mostly straight hairs, often much branched above; stems slender, erect or ascending, 4'–12' high. Leaves oblong, obtuse or obtusish, sessile, or nearly so, or the lower spatulate and narrowed into margined petioles; racemes slender, mostly naked below; pedicels shorter than the fruiting calyx, appressed-pubescent, erect; calyx equally 5-cleft, the lobes linear-lanceolate, erect or connivent in fruit, longer than or equalling the tube, the hairs, or some of them, with minutely hooked tips; corolla pale yellow changing to violet and blue, its limb about 1'' broad; nutlets convex on the outer, slightly keeled on the inner side, shorter than the style.

In fields and along roadsides, southern New York and Delaware. Naturalized from Europe. May–July.

5. Myosotis micrántha Pall. Blue Scorpion-grass. Fig. 3533.

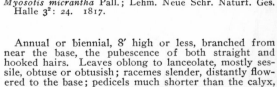

Myosotis micrantha Pall.; Lehm. Neue Schr. Naturf. Ges. Halle **3**²: 24. 1817.

Annual or biennial, 8' high or less, branched from near the base, the pubescence of both straight and hooked hairs. Leaves oblong to lanceolate, mostly sessile, obtuse or obtusish; racemes slender, distantly flowered to the base; pedicels much shorter than the calyx, pubescent, and usually with some hooked hairs; calyx equally 5-cleft, about as long as the corolla-tube; style not longer than the nutlets.

Fields and roadsides, Ontario to Massachusetts and Ohio. May–July. Naturalized from Europe. Has been mistaken in America for *M. collina* Hoffm.

6. Myosotis virgínica (L.) B.S.P. Spring or Early Scorpion-grass. Fig. 3534.

Lycopsis virginica L. Sp. Pl. 139. 1753.
Myosotis verna Nutt. Gen. 2 : Add. 1818.
Myosotis virginica B.S.P. Prel. Cat. N. Y. 37. 1888.

Annual or biennial, hirsute-pubescent or hispid, erect, branched, 3′–15′ high, the branches erect. Leaves oblong or linear-oblong, sessile, 3″–12″ long, obtuse, or the lower spatulate and narrowed into short petioles; racemes usually bracted at the base, strict; pedicels ascending or erect, or slightly spreading at the apex, shorter than the fruiting calyx; calyx somewhat 2-lipped, unequally 5-cleft, the lobes lanceolate, acute, longer than the tube, connivent in fruit, very hispid, the hairs, or most of them, with minutely hooked tips; corolla white, the limb 1½″ broad or less; nutlets convex on the back, slightly keeled and margined on the inner side.

On dry hills and banks, Maine and Ontario to Minnesota, Florida and Texas. Forget-me-not. April–June.

Myosotis macrospèrma Engelm., of the Southern States, with larger flowers and fruit, the ripe calyx nodding or spreading, ranging north to Virginia and Kentucky, appears to be a race of this species.

$\frac{3}{4}$

12. LITHOSPÉRMUM [Tourn.] L. Sp. Pl. 132. 1753.

Annual or perennial, erect branching or rarely simple, pubescent hirsute or hispid herbs, with alternate entire leaves, and small or large, white yellow or blue flowers in leafy-bracted spikes or racemes. Calyx 5-parted or 5-cleft, the segments or lobes narrow. Corolla funnelform or salverform, 5-lobed, naked, pubescent or crested in the throat, the lobes entire or erose-denticulate, the tube sometimes pubescent at the base within. Stamens 5, included, inserted on the throat of the corolla; filaments short. Ovary 4-divided; style slender, or filiform; stigma capitate, or 2-lobed. Nutlets 4, or fewer, erect, white, smooth and shining, or brown and wrinkled, attached by their bases to the nearly flat receptacle, the scar of attachment not concave. [Greek, stone-seed, from the hard nutlets.]

About 40 species, natives of the northern hemisphere, a few in South America and Africa. Besides the following, some 7 others occur in the southern and southwestern parts of the United States. Type species: *Lithospermum officinale* L.

Corolla white or yellowish, its tube shorter than or equalling the calyx; flowers distant.
  Nutlets brown, wrinkled and pitted; annual or biennial.   1. *L. arvense.*
  Nutlets white, smooth and shining; perennials.
    Leaves lanceolate, acute; nutlets ovoid.   2. *L. officinale.*
    Leaves ovate, acuminate; nutlets globose-ovoid.   3. *L. latifolium.*
Corolla dull yellow, its tube longer than the calyx; leaves lanceolate; flowers dense.   4. *L. pilosum.*
Corolla bright yellow, its tube much longer than the calyx; flowers dense; red-rooted perennials.
  Corolla-lobes entire; flowers all complete.
    Hispid-pubescent; corolla-tube bearded at the base within.   5. *L. carolinense.*
    Hirsute, somewhat canescent; corolla-tube not bearded at the base.   6. *L. canescens.*
  Corolla-lobes erose-denticulate; later flowers cleistogamous.   7. *L. linearifolium.*

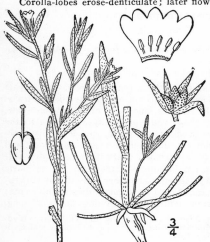

$\frac{3}{4}$

1. Lithospermum arvénse L. Bastard Alkanet. Corn Gromwell. Fig. 3535.

Lithospermum arvense L. Sp. Pl. 132. 1753.

Annual or biennial, appressed-pubescent; stem erect, usually branched, 6′–20′ high. Leaves bright green, lanceolate, linear or linear-oblong, sessile or the lowest short-petioled, mostly appressed, obtuse or acutish at the apex, narrowed at the base, indistinctly veined, ½′–1½′ long, ½″–3″ wide, the uppermost smaller; flowers sessile or very nearly so in the spikes, becoming distant, white, about 3″ long; calyx-segments linear-lanceolate, longer than or equalling the corolla-tube; corolla funnelform, puberulent in the throat but not crested; nutlets brown, wrinkled and pitted, glabrous, about 1″ high, convex on the back, keeled on the inner side, one-third to one-half the length of the calyx-segments.

In waste places and fields, Quebec to Ontario and Michigan, south to Georgia and Kansas. Naturalized from Europe. Native also of Asia. Pearl-plant. Salfern-stoneseed. May–Aug.

2. Lithospermum officinàle L.   Gromwell.
Fig. 3536.

Lithospermum officinale L. Sp. Pl. 132. 1753.

Perennial, finely puberulent; stem usually much branched, 2°–4° high, leafy. Leaves lanceolate or oblong-lanceolate, acute at the apex, narrowed at the base, few-veined, sessile, 1½'–4' long, 3"–12" wide, the upper surface rough, the lower pubescent; flowers yellowish-white, about 2" long, sessile; calyx-segments linear-lanceolate, about equalling the corolla-tube; corolla funnelform, crested in the throat; style about as long as the stamens; nutlets, when mature, white, smooth, shining, about 1½" high, ovoid, obtuse, more than one-half as long as the calyx-segments, seldom all ripening.

In fields and waste places, Quebec to southern New York, New Jersey and Minnesota. Plant grayish. Naturalized from Europe. Native also of Asia. Graymile. Littlewale. Pearl-plant. May–Aug.

3. Lithospermum latifòlium Michx.
American Gromwell. Fig. 3537.

Lithospermum latifolium Michx. Fl. Bor. Am. **1**: 131. 1803.

Perennial, rough-puberulent; stem branched, 2°–3° high, the branches long and slender. Leaves ovate or ovate-lanceolate, acuminate at the apex, pinnately veined, 2'–5' long, 1'–2' wide, or the uppermost smaller, the lowest obtuse; flowers yellowish white or pale yellow, 2"–3" long, few, solitary, distant; calyx-segments linear-lanceolate, about as long as the corolla or a little longer; corolla funnelform, crested in the throat; style shorter than the stamens; nutlets white, shining, globose-ovoid, about 2" long, more than one-half as long as the calyx-segments.

In dry thickets and fields, Quebec to New York, Minnesota, Kansas, Virginia, Tennessee and Arkansas. May.

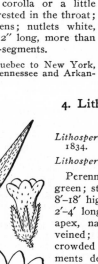

4. Lithospermum pilòsum Nutt.   Woolly Gromwell. Fig. 3538.

Lithospermum pilosum Nutt. Journ. Phil. Acad. **7**: 43. 1834.

Lithospermum Torreyi Nutt. loc. cit. 44.   1834.

Perennial from thick roots, hirsute, rather pale green; stems usually stout and clustered, very leafy, 8'–18' high. Leaves lanceolate or linear-lanceolate, 2'–4' long, 2"–5" wide, gradually acuminate to the apex, narrowed at the base, sessile, indistinctly veined; flowers dull yellow, very numerous and crowded in a terminal leafy thyrsus; calyx-segments densely hirsute, shorter than the cylindric corolla-tube; corolla salverform, the throat puberulent below each lobe; style longer than the filaments; nutlets ovoid, acute, white, shining, about 2" long.

Western Nebraska (according to Williams); Wyoming to Montana, Alberta, British Columbia and California. May–July.

5. Lithospermum carolinènse (Walt.) MacM.
Hairy or Gmelin's Puccoon. Fig. 3539.

Anonymos carolinensis Walt. Fl. Car. 91. 1788.
Batschia carolinensis Gmel. Syst. 2: Part 1, 315. 1791.
Lithospermum carolinianum Lam. Tabl. Encycl. 1: 397. 1791.
Lithospermum hirtum Lehm. Asperif. 305. 1818.
Lithospermum carolinense MacM. Met. Minn. 438. 1892.

Perennial, hispid-pubescent, or scabrous; stems usually clustered, rather stout, simple, or branched above, 1°–2½° high, very leafy. Leaves narrowly lanceolate, sessile, obtuse or acute at the apex, narrowed at the base, 2′–3′ long, the lowest commonly reduced to appressed scales, the uppermost oblong; flowers 6″–8″ long, in dense short terminal leafy racemes, dimorphous; pedicels 1″–3″ long; calyx-segments linear-lanceolate, shorter than the tube of the orange-yellow salverform corolla; corolla-lobes entire, rounded, the throat crested, the tube bearded at the base within by 10 hirsute teeth; nutlets white, shining, about 2″ high, ovoid, very much shorter than the calyx-segments.

In dry woods, western New York to Florida, Minnesota, Montana and New Mexico. April–June.

6. Lithospermum canéscens (Michx.)
Lehm. Hoary Puccoon. Fig. 3540.

Batschia canescens Michx. Fl. Bor. Am. 1: 130. *pl. 14.* 1803.
Lithospermum canescens Lehm. Asperif. 305. 1818.

Perennial, hirsute, somewhat canescent, at least when young; stems solitary or clustered, simple or often branched, 6′–18′ high. Leaves oblong, linear-oblong, or linear, obtuse or acutish at the apex, sessile by a narrowed base, ½′–1½′ long, 2″–5″ wide, the lowest often reduced to appressed scales; flowers about 6″ long, sessile, numerous in dense short leafy racemes, dimorphous; calyx-segments linear-lanceolate, shorter than the tube of the orange-yellow salverform corolla; corolla crested in the throat, its lobes rounded, entire, its tube glandular but not bearded at the base within; nutlets white, smooth, shining, acutish, shorter than the calyx-segments.

In dry soil, Ontario to western New Jersey and Alabama, Saskatchewan, North Dakota and Texas. April–June.

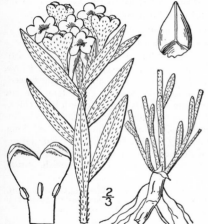

7. Lithospermum linearifòlium Goldie.
Narrow-leaved Puccoon. Fig. 3541.

L. angustifolium Michx. Fl. Bor. Am. 1: 130. 1803. Not Forsk.
L. linearifolium Goldie, Edinb. Phil. Journ. 1822: 322.

Perennial by a deep root, strigose-pubescent and scabrous; stem branched, 6′–2° high, the branches erect or ascending. Leaves linear, sessile, acute or acutish, ½′–2′ long, 1½″–2½″ wide; flowers of two kinds, in terminal leafy racemes; corolla of the earlier ones salverform, about 1′ long, bright yellow, the tube 3–5 times as long as the linear-lanceolate calyx-segments, the lobes erose-denticulate, the throat crested, the base of the tube not bearded within; later flowers (sometimes all of them) much smaller, pale yellow, cleistogamous, abundantly fertile, their pedicels recurved in fruit; nutlets white, smooth, shining, ovoid, 1½″–2″ high, more or less pitted, keeled on the inner side.

In dry soil, especially on prairies, Ontario and Indiana to Illinois, Kansas and Texas, west to British Columbia, Utah and Arizona. Yellow puccoon. April–July.

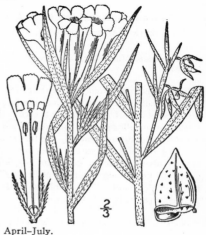

13. ONOSMÒDIUM Michx. Fl. Bor. Am. 1 : 132. 1803.

Perennial stout hispid or hirsute branching herbs, with alternate entire strongly veined leaves, and rather small yellowish or greenish white proterogynous flowers, in terminal leafy-bracted scorpioid spikes or racemes. Calyx deeply 5-parted, the segments narrow. Corolla tubular or tubular-funnelform, 5-lobed, the lobes erect, the throat not appendaged, the sinuses slightly inflexed, the tube with a glandular 10-lobed band within at the base. Stamens 5, inserted on the tube or throat of the corolla, included; filaments short. Ovary 4-parted; style filiform, exserted. Nutlets 4, or commonly only 1 or 2 perfecting, ovoid, sometimes sparingly pitted, shining, smooth, white, attached by the base to the nearly flat receptacle, the scar of attachment small, flat. [Greek, like *onosma,* or ass-smell.]

About 10 species, natives of North America and Mexico. Besides the following, 3 others occur in the southern and southwestern United States. Type species: *Onosmodium hispidum* Michx.

Corolla-lobes 2–3 times as long as wide.　　　　　　　　　　　　　1. *O. virginianum.*
Corolla-lobes scarcely longer than wide.
　　Stem glabrous below.　　　　　　　　　　　　　　　　　　　　2. *O. subsetosum.*
　　Stem hirsute or pubescent to the base.
　　　　Pubescence silky; nutlets distinctly pitted:　　　　　　　　　3. *O. molle.*
　　　　Pubescence hirsute to strigose; nutlets indistinctly pitted.
　　　　　　Nutlets not constricted.　　　　　　　　　　　　　　4. *O. occidentale.*
　　　　　　Nutlets distinctly constricted just above the base.　　　5. *O. hispidissimum.*

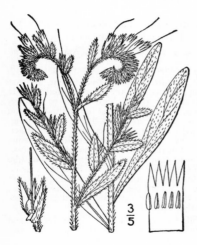

3/5

1. Onosmodium viriniànum (L.) DC.　Virginia False Gromwell.　Fig. 3542.

Lithospermum virginianum L. Sp. Pl. 132. 1753.
Onosmodium virginianum DC. Prodr. 10: 70. 1846.

Densely appressed-hispid or strigose, with stiff hairs; stem rather slender, usually branched above, 1°–2½° high. Leaves oblong, oval, or oblong-lanceolate, obtuse or acutish, sessile, 1′–3½′ long, or the lower oblanceolate and narrowed into petioles; calyx-segments linear-lanceolate, acuminate; corolla cylindric or nearly so, yellowish-white, about 4″ long, the lobes narrowly lanceolate, acuminate, 2 or 3 times as long as wide, nearly as long as the tube, strigose without; nutlets ovoid, obtuse or obtusish, pitted, 1″–1½″ long.

In dry thickets or on hillsides, Massachusetts to Pennsylvania, Florida and Louisiana. Ascends to 3000 ft. in Virginia. Wild job's-tears.　May–July.

2. Onosmodium subsetòsum Mack. & Bush.　Ozark False Gromwell.　Fig. 3543.

O. subsetosum Mack. & Bush; Small, Fl. SE. U. S. 1001. 1903.

Stem erect, glabrous, or with a few scattered appressed hairs above, somewhat branched, 3° high or less, the branches appressed-pubescent. Leaves lanceolate, acute, papillose and appressed-hispid above, whitish appressed-pubescent beneath, the larger about 3½′ long; bracts ½′–1′ long; calyx-lobes oblong, obtuse, 3″ long; corolla about 5″ long, canescent, its lobes triangular, acute, about 1″ long; fruiting pedicels 2″–3″ long; nutlets whitish, ovoid, 1½″ long, obtuse or acutish, not constricted, sparingly pitted.

Barrens, Ozark Mountains, Missouri and Arkansas. June–Aug.

2/3

3. Onosmodium mólle Michx. Soft-hairy False Gromwell. Fig. 3544.

O. molle Michx. Fl. Bor. Am. 1 : 133. 1803.

Stem erect, branched above, about 2° high, hispid-pubescent or strigose, the branches soft-pubescent. Leaves lanceolate to ovate-lanceolate, the larger about 2′ long, densely soft-pubescent on both sides; bracts similar to the leaves, 1′ long or less; calyx-lobes linear-oblong, 3″ long, obtusish; corolla 4″–6″ long, its lobes triangular, acute, 1″–1½″ long, pubescent outside; nutlets about 1″ long, usually distinctly pitted.

Barrens, Kentucky, southern Illinois and Tennessee. May–July.

4. Onosmodium occidentàle Mackenzie. Western False Gromwell. Fig. 3545.

O. occidentale Mackenzie, Bull. Torr. Club 32 : 502. 1905.
O. occidentale sylvestre Mackenzie, loc. cit. 504. 1905.

Stem 1°–3½° high, branched above or also from the base, strigose or hirsute-pubescent. Leaves lanceolate to ovate-lanceolate, acute, appressed-pubescent on both sides, or the hairs somewhat spreading, 2′–3′ long, strongly veined; bracts similar to the leaves but much smaller; calyx-lobes lanceolate, acute to obtuse, 3″–6″ long; corolla 6″–10″ long, canescent all over outside, its lobes 1½″–2″ long, broadly triangular and acute; nutlets ovoid, acutish, about 2″ long, dull, scarcely if at all pitted, not constricted at the base.

On prairies and plains, Illinois to North Dakota, Manitoba, Alberta, Kansas, Texas and New Mexico. Included in *O. molle* Michx., in our first edition, and by previous authors. May–July.

5. Onosmodium hispidíssimum Mackenzie. Shaggy False Gromwell. Fig. 3546.

Onosmodium hispidissimum Mackenzie, Bull. Torr. Club 32 : 500. 1905.

Spreading-hirsute with rough bristly hairs; stem stout, usually much branched, 1°–4° high. Leaves lanceolate, ovate-lanceolate or oblong, acute or acuminate at the apex, narrowed to the sessile base, 5–9-ribbed, 2′–4½′ long, ½′–1½′ wide; flowers very numerous and crowded; pedicels 1″–2″ long in fruit; calyx-segments linear, somewhat shorter than the corolla-tube; corolla yellowish-white, pubescent outside, 5″–9″ long, its lobes triangular-lanceolate, acute, one-third to one-half as long as the tube; nutlets obtuse, about 1½″ long, distinctly constricted at the base, little if at all pitted.

In dry fields or thickets, or on banks, Ontario and western New York to Minnesota, Missouri, Georgia and Texas. Ascends to 2200 ft. in Virginia. Previously referred to *Onosmodium carolinianum* (Lam.) DC. May–July.

14. SÝMPHYTUM [Tourn.] L. Sp. Pl. 136. 1753.

Erect coarse hairy perennial branching herbs, with thick mucilaginous roots, alternate entire leaves, those of the stem mostly clasping, the uppermost tending to be opposite, the lower long-petioled. Flowers yellow, blue, or purple, in terminal simple or forked scorpioid racemes. Calyx deeply 5-cleft. Corolla tubular, slightly dilated above, 5-toothed or 5-lobed, the lobes short, the throat with 5 crests below the lobes. Stamens 5, included, inserted on the corolla-tube; filaments slender. Ovary 4-divided; style filiform. Nutlets 4, obliquely ovoid, slightly incurved, wrinkled, inserted by their bases on the flat receptacle, the scar of the attachment broad, concave, dentate. [Greek, grow-together, from its supposed healing virtues.]

About 15 species, natives of the Old World. Type species: *Symphytum officinale* L.

Leaf-bases decurrent.   1. *S. officinale.*
Leaf-bases not decurrent, or but slightly decurrent.   2. *S. asperrimum.*

1. Symphytum officinàle L.   Comfrey.   Healing-herb.   Fig. 3547.

Symphytum officinale L. Sp. Pl. 136. 1753.

Roots thick, deep; stem erect, branched, 2°–3° high. Leaves lanceolate, ovate-lanceolate, or the lower ovate, pinnately veined, 3′–10′ long, acute or acuminate at the apex, narrowed into margined petioles, or the uppermost smaller and sessile, decurrent on the stem; petioles of the basal leaves sometimes 12′ long; flowers numerous, in dense racemes or clusters; pedicels 2″–4″ long; calyx-segments ovate or ovate-lanceolate, acute or acuminate, much shorter than the corolla; corolla yellowish or purplish, 6″–10″ long; nutlets brown, shining, slightly wrinkled, 2″ high.

In waste places, Newfoundland to Minnesota, south to Virginia and North Carolina. Naturalized or adventive from Europe. Native also of Asia. June–Aug. Back- or black-wort. Bruisewort. Knitback. Boneset. Consound. Gum-plant.

Symphytum tuberòsum L., with thickened tuberous roots, the nutlets granular-tuberculate, not shining, has been found in sandy meadows in Connecticut.

2. Symphytum aspérimum Donn.   Rough Comfrey.   Fig. 3548.

S. asperrimum Donn; Sims, Bot. Mag. **24**: *pl. 929.* 1806.

Similar to *S. officinale,* but the pubescence rougher, the hairs stiff and reflexed. Leaves ovate-lanceolate to oblong-lanceolate, long-acuminate at the apex, narrowed at the base, all but the uppermost petioled, slightly or not at all decurrent, the lower often 8′ long; flower-clusters rather loose; calyx about half as long as the corolla-tube, its segments hispid; corolla bluish-purple.

Waste grounds, Massachusetts to Maryland. Adventive or naturalized from Europe. June–Aug.

15. BORÀGO [Tourn.] L. Sp. Pl. 137. 1753.

Hirsute or hispid annual or biennial branching herbs, with alternate entire leaves, and showy blue flowers, in terminal loose leafy racemes. Calyx deeply 5-cleft or 5-parted. Corolla rotate, the tube very short, the throat closed by scales, the limb 5-lobed, the lobes imbricated, acute. Stamens 5, inserted on the corolla-tube; filaments dilated below, narrowed above into a slender appendage; anthers linear, erect, and connivent into a cone. Ovary 4-divided; style filiform. Nutlets 4, ovoid, erect, attached by their bases to the flat receptacle, the scar of attachment large, concave. [Middle Latin, *burra,* rough hair, alluding to the foliage.]

Three species, natives of the Mediterranean region, the following typical.

1. Borago officinàlis L. Borage.
Fig. 3549.

Borago officinalis L. Sp. Pl. 137. 1753.

Stem erect, branched, 1°–2½° high, the branches spreading or ascending. Leaves oblong to obovate, acute or obtuse at the apex, 2′–5′ long, narrowed into margined petioles, or the upper smaller, ovate-lanceolate, sessile or partly clasping; flowers 8″–10″ broad, pedicels rather stout, 1½′–2′ long, spreading or recurving; calyx-segments lanceolate, nearly erect in fruit; corolla bright blue, the lobes ovate-lanceolate; the cone of anthers darker, about 3″ long; nutlets 2″ high.

In waste places, escaped from gardens, Nova Scotia to Ontario and Pennsylvania, but probably not persistent within our range. Native of southern Europe. June–Sept.

16. LYCÓPSIS L. Sp .Pl. 138. 1753.

Annual bristly-hispid branched erect or diffuse herbs, with alternate leaves, and small blue or bluish flowers, in dense leafy-bracted terminal spike-like scorpioid racemes. Calyx 5-parted. Corolla slightly irregular, salverform, the tube curved, the limb somewhat unequally 5-lobed, the lobes obtuse, imbricated, the throat closed by hispid scales. Stamens 5, included, inserted on the tube of the corolla; filaments short; anthers obtuse at each end. Ovary 4-divided; style filiform. Nutlets 4, wrinkled, erect, attached by their bases to the flat receptacle, the scar of attachment concave. [Greek, wolf-face.]

About 4 species, natives of the Old World, the following typical.

1. Lycopsis arvénsis L. Small Bugloss.
Fig. 3550.

Lycopsis arvensis L. Sp. Pl. 139. 1753.

Stem erect or ascending, at length divergently or diffusely branched, 1°–2° high, the branches becoming procumbent. Leaves lanceolate, narrowly oblong or the lower oblanceolate, obtuse, 1′–2′ long, undulate or dentate, sessile, or the lower narrowed into petioles, the upper much smaller and acute or acutish; flowers numerous, crowded, 2″–3″ broad, very short-pedicelled; calyx-segments lanceolate, acute, nearly as long as the curved corolla-tube; nutlets shorter than the calyx.

In fields and waste places, Nova Scotia to Ontario, Minnesota, Pennsylvania and Virginia. Naturalized or adventive from Europe. Native also of Asia. June–Sept.

17. ÉCHIUM [Tourn.] L. Sp. Pl. 139. 1753.

Biennial or perennial mostly bristly-hirsute branching herbs, with alternate leaves, and rather large blue violet or rarely white flowers, in leafy-bracted scorpioid spikes. Calyx 5-parted, the segments narrow. Corolla tubular-funnelform, irregular, the limb unequally 5-lobed, the lobes rounded, spreading, the throat not appendaged. Stamens 5, inserted low down on the tube of the corolla, unequal, at least the longer ones exserted; filaments slender, dilated at the base; anthers ovate or oblong. Ovary 4-divided; style filiform, 2-cleft at the summit. Nutlets 4, erect, ovoid, rugose, attached by their bases to the flat receptacle, the scar of attachment not concave. [Greek, a viper.]

About 30 species, natives of the Old World. Type species: *Echium itálicum* L.

1. Echium vulgàre L. Viper's Bugloss. Blue-weed. Fig. 3551.

Echium vulgare L. Sp. Pl. 140. 1753.

Bristly-hairy, biennial; stem erect, at length much branched, 1°–2½° high. Leaves oblong, linear-oblong, or linear-lanceolate, obtuse or acute, entire, 2'–6' long, sessile, or the lower and basal ones narrowed into petioles; flower-buds pink; flowers bright blue, varying to violet purple, 8"–12" long, numerous in short 1-sided spikes, forming a narrow thyrsus; calyx-segments much shorter than the corolla; limb of the corolla oblique, the lobes very unequal.

In fields and waste places, Nova Scotia to North Carolina, Ontario and Nebraska. A troublesome weed in some sections of the North. Naturalized from Europe. Native also in Asia. June–July. Viper's-herb. Viper's-grass. Snake-flower. Blue thistle. Blue stem or cat's-tails. Blue devils. Adder's-wort.

Family 25. **VERBENÀCEAE** J. St. Hil. Expos. Fam. **1**: 245. 1805.

VERVAIN FAMILY.

Herbs, shrubs or some tropical genera trees, with opposite verticillate or rarely alternate leaves, and perfect, more or less irregular, or sometimes regular flowers, in terminal or axillary spikes, racemes, cymes or panicles. Calyx inferior, mostly persistent, usually 4–5-lobed or 4–5-cleft. Corolla gamopetalous, regular, or 2-lipped, the tube usually cylindric and the limb 4–5-cleft. Stamens 4, didynamous, rarely only 2, or as many as the corolla-lobes, inserted on the corolla and alternate with its lobes; anthers 2-celled, the sacs longitudinally dehiscent. Ovary superior, 2–4-celled (rarely 8–10-celled), composed of 2 carpels, each carpel with 2 anatropous or amphitropous ovules, thus in 4-celled ovaries 1 ovule in each cavity; style terminal, simple; stigmas 1 or 2. Fruit dry, separating at maturity into 2–4 nutlets, or a drupe containing the 2–4 nutlets. Endosperm little or none, or rarely fleshy; embryo straight.

About 75 genera and 1300 species, of wide geographic distribution in temperate and warm regions.

Flowers in heads or spikes; ovary 2–4-celled; fruit of 2 or 4 erect nutlets; ours herbs.
　Corolla-limb 5-lobed, regular or nearly so; nutlets 4.　　　　　　　　　　　　1. *Verbena.*
　Corolla-limb 4-lobed, 2-lipped; nutlets 2.　　　　　　　　　　　　　　　　　2. *Lippia.*
Flowers in axillary cymes; shrubs; fruit drupaceous.　　　　　　　　　　　　　3. *Callicarpa.*

1. VERBÈNA [Tourn.] L. Sp. Pl. 18. 1753.

Herbs (some exotic species shrubby), mostly with opposite leaves, and variously colored bracted flowers, in terminal solitary corymbed or panicled spikes. Calyx usually tubular, 5-angled, more or less unequally 5-toothed. Corolla salverform or funnelform, the tube straight or somewhat curved, the limb spreading, 5-lobed, slightly 2-lipped or regular. Stamens 4, didynamous, or very rarely only 2, included; connective of the anthers unappendaged, or sometimes provided with a gland. Ovary 4-celled; ovule 1 in each cavity; style usually short, 2-lobed at the summit, one of the lobes stigmatic. Fruit dry, mostly enclosed by the calyx, at length separating into 4, 1-seeded linear or linear-oblong crustaceous smooth papillose or rugose nutlets. [Latin name of a sacred herb.]

About 100 species, natives of America, or a single one indigenous in the Mediterranean region. Besides the following, some 15 others occur in the southern and western parts of North America. Type species: *Verbena officinalis* L.

Flowers 2"–5" long, in narrow spikes; anthers unappendaged.
　Spikes filiform or slender; bracts shorter than the flowers.
　　Spikes filiform; fruit scattered; corolla usually white.
　　　Leaves incised or pinnatifid; diffuse annual; fruit short.　　　　　1. *V. officinalis.*
　　　Leaves serrate (rarely incised); erect perennial; fruit oblong.　　2. *V. urticifolia.*
　　Spikes slender; fruit densely imbricated; corolla blue.
　　　Plants glabrous or sparingly rough-pubescent; corolla 2"–3" long.
　　　　Leaves lanceolate, acuminate, petioled.　　　　　　　　　　　3. *V. hastata.*
　　　　Leaves linear or spatulate-lanceolate, mostly obtuse and sessil　4. *V. angustifolia.*
　　　Plants densely soft-pubescent; corolla 4"–5" long.　　　　　　　5. *V. stricta.*
　Spikes thick, dense; bracts longer than the flowers.　　　　　　　　6. *V. bracteosa.*
Flowers 7"–12" long, in short dense elongating spikes; connective of the longer stamens appendaged.
　Corolla-limb 6"–12" broad; bracts mostly shorter than the calyx.　　7. *V. canadensis.*
　Corolla-limb 4"–7" broad; bracts equalling or exceeding the calyx.　8. *V. bipinnatifida.*

1. Verbena officinàlis L. European Vervain. Herb-of-the-Cross. Berbine. Fig. 3552.

Verbena officinalis L. Sp. Pl. 20. 1753.

Annual; stem 4-sided, slender, glabrous or nearly so, ascending or spreading, diffusely branched, 1°–3° high. Leaves minutely pubescent, the lower deeply incised or 1–2 pinnatifid, ovate, oblong, or obovate in outline, 1'–3' long, narrowed into margined petioles, the teeth acute; upper leaves linear or lanceolate, acute, entire, sessile; spikes several or numerous, filiform, at length 4'–5' long; fruits less than 1" high, scattered along the spikes, not at all imbricated; bracts ovate, acuminate, shorter than the 5-toothed calyx; corolla purplish or white, the limb 1"–2" broad.

In waste and cultivated ground, Maine to Florida, Tennessee and Texas. Also on the Pacific Coast and in the West Indies. Naturalized from the Old World. Sometimes a troublesome weed. Herb-grace. Holy-herb. Enchanter's-plant. Juno's-tears. Pigeon's-grass. Simpler's-joy. June–Sept.

2. Verbena urticifòlia L. White or Nettle-leaved Vervain. Fig. 3553.

Verbena urticifolia L. Sp. Pl. 20. 1753.
V. urticifolia riparia Britton, Mem. Torr. Club **5**: 276. 1894.
V. riparia Raf.; Small & Heller, Mem. Torr. Club **3**: 12. 1892.

Perennial, usually pubescent; stem slender, strict, erect, 4-sided, paniculately branched above, 3°–5° high, the branches upright. Leaves ovate, oblong, or oblong-lanceolate, all petioled, or the uppermost sessile, serrate-dentate all around, or incised, sometimes 3-cleft near the base, thin, acute or acuminate, mostly rounded at the base, 1½'–5' long; spikes numerous, filiform, erect, or spreading, at length 4'–6' long; fruits oblong, scattered, not at all imbricated, about 1" high; bracts ovate, acuminate, shorter than the calyx; corolla white, blue or pale purple, its limb about 1" broad.

In fields and waste places, New Brunswick to South Dakota, Kansas, Florida and Texas. Hybridizes with *V. bracteosa*, *V. hastata* and *V. stricta*. June–Sept.

Verbena carolinénsis (Walt.) Gmel., with sessile, spatulate to oblong leaves and larger bluish flowers, native of the Southeastern States, is recorded as occurring north to Virginia.

3. Verbena hastàta L. Blue or False Vervain. Wild Hyssop. Fig. 3554.

Verbena hastata L. Sp. Pl. 20. 1753.
Verbena pinnatifida Lam. Tabl. Encycl. **1**: 57. 1791.
Verbena paniculata Lam. Encycl. **8**: 548. 1808.
Verbena hastata pinnatifida Britton, Mem. Torr. Club **5**: 276. 1894.

Perennial, roughish-puberulent; stem erect, strict, 4-sided, usually branched above, 3°–7° high. Leaves oblong-lanceolate or lanceolate, petioled, acute or acuminate at the apex, narrowed at the base, serrate or incised-dentate with acute teeth, sometimes pinnatifid, 3'–6' long, the lower sometimes hastately 3-lobed at the base; spikes numerous, panicled, slender, usually peduncled, 2'–6' long; fruits densely imbricated on the spikes, 1"–1½" high; bracts ovate, acuminate, shorter than the calyx; corolla blue, white, or sometimes pink, its limb about 1½" broad.

In moist fields, meadows and in waste places, Nova Scotia to British Columbia, Florida, Nebraska and Arizona. Hybridizes with *V. stricta* and *V. bracteosa*. American vervain. Purvain. Iron-weed. June–Sept.

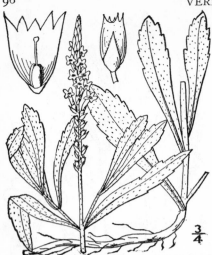

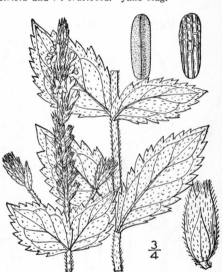

4. Verbena angustifòlia Michx. Narrow-leaved Vervain. Fig. 3555.

V. angustifolia Michx. Fl. Bor. Am. **2**: 14. 1803.

Perennial, roughish-puberulent or pubescent; stem slender, simple or branched, 4-sided above, 1°–2° high. Leaves linear, spatulate or lanceolate, obtuse or subacute at the apex, cuneate at the base and tapering into short petioles, serrate or serrulate, veiny, 1½′–3′ long, 2″–5″ wide; spikes mostly solitary at the ends of the branches, usually peduncled, slender, dense, 2′–5′ long; fruits overlapping or the lower somewhat distant, 1½″ high; bracts lanceolate, acuminate, equalling or shorter than the calyx; corolla purple or blue, about 3″ long, its limb about as broad, the lobes obovate or oblong.

In dry fields, Massachusetts to Florida, west to Minnesota, Kansas and Arkansas. Hybridizes with *V. stricta* and *V. bracteosa*. June–Aug.

5. Verbena strícta Vent. Hoary or Mullen-leaved Vervain. Fig. 3556.

V. stricta Vent. Descr. Pl. Jard. Cels. *pl. 53.* 1800.
Verbena rigens Michx. Fl. Bor. Am. **2**: 14. 1803.

Perennial, densely soft-pubescent all over with whitish hairs; stem stout, obtusely 4-angled, simple, or branched above, strict, very leafy, 1°–2½° high. Leaves ovate, oval, or oblong, very short-petioled, acute or obtuse at the apex, narrowed at the base, prominently veined, incised-serrate or laciniate, 1′–4′ long; spikes solitary, or several, mostly sessile, dense, stout, becoming 6′–12′ long in fruit; fruits much imbricated, 2″–2½″ high; bracts lanceolate-subulate, nearly as long as the calyx; corolla purplish blue, 4″–5″ long, its limb nearly as broad.

In dry soil, Ontario and Ohio to Minnesota, South Dakota and Wyoming, south to Tennessee, Texas and New Mexico. Naturalized as a weed further east. Hybridizes with *V. bracteosa*. June–Sept.

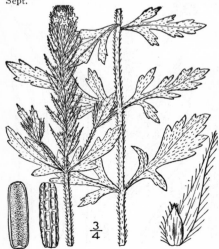

6. Verbena bracteòsa Michx. Large-bracted Vervain. Fig. 3557.

V. bracteosa Michx. Fl. Bor. Am. **2**: 13. 1803.

Perennial, hirsute-pubescent; stem 4-sided, much branched from the base, the branches decumbent or ascending, slender, 6′–15′ long. Leaves ovate, oval, or obovate in outline, pinnately incised or pinnatifid, 1′–3′ long, more or less cuneate at the base and narrowed into short petioles, the lobes mostly dentate; spikes sessile, stout, dense, becoming 4′–6′ long in fruit; bracts conspicuous, linear-lanceolate, rather rigid, longer than the flowers and fruits, the lower ones often incised; corolla purplish blue, about 2″ long.

On prairies and in waste places, Minnesota and Illinois to Virginia, Alabama and Florida, west to British Columbia, Arizona and California. In ballast grounds at Atlantic seaports. Hybridizes with *V. canadensis*. Prostrate vervain. May–Aug.

7. Verbena canadénsis (L.) Britton.
Large-flowered Verbena. Fig. 3558.

Buchnera canadensis L. Mant. 88. 1767.
V. Aubletia Jacq. Hort. V. **2**: 82. *pl. 176.* 1772.
Glandularia carolinensis J. G. Gmel. Syst. **2**: 920.
1796.
Verbena canadensis Britton, Mem. Torr. Club
5: 276. 1894.

Perennial, pubescent or glabrate; stem slender, usually branched, 8′–20′ high, the branches ascending. Leaves membranous, ovate in outline, petioled, 1′–3′ long, truncate or broadly cuneate at the base, irregularly toothed, or pinnately incised, often 3-cleft, the lobes dentate; spikes peduncled, solitary at the ends of the branches, dense, short and capitate when in early flower, becoming 2′–4′ long in fruit; bracts linear-subulate, mostly shorter than the calyx; calyx-teeth filiform-subulate; corolla 10″–12″ long, blue, purple, white or in cultivation variegated, its limb 7″–12″ broad, the lobes oblong or obovate, emarginate or obcordate; fruit 2½″–3″ high.

In dry soil, Illinois to Tennessee, Virginia and Florida, west to Kansas and Texas. This and the next the source of many garden and other hybrids. Cut-leaved races have been referred to *V. Drummondii* (Lindl.) Baxter. May–Aug.

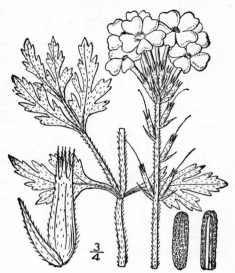

8. Verbena bipinnatífida Nutt. Small-flowered Verbena. Fig. 3559.

Verbena bipinnatifida Nutt. Journ. Acad. Phila.
2: 123. 1821.
Glandularia bipinnatifida Nutt. Trans. Am. Phil.
Soc. (II.) **5**: 184. 1833–37.
Verbena ambrosiaefolia Rydb.; Small, Fl. SE.
U. S. 1011. 1903.

Perennial, producing suckers, hirsute or hispid; stems rather stout, mostly branched, erect, 6′–18′ high. Leaves firm, petioled or the uppermost sessile, broadly ovate in outline, deeply 1–2-pinnatifid into linear or linear-oblong, obtuse or subacute lobes and segments; spikes peduncled or sessile, solitary at the ends of the branches, thick, dense, at first short and capitate, becoming 2′–4′ long in fruit; bracts linear-subulate, about as long as or somewhat exceeding the calyx; calyx-teeth filiform-subulate; corolla 6″–9″ long, purple or lilac, the limb 4″–7″ broad, the lobes emarginate or obcordate; fruit 1½″–2″ long.

On dry plains and prairies, South Dakota to Missouri, Texas and Chihuahua, west to Colorado and Arizona. May–Sept.

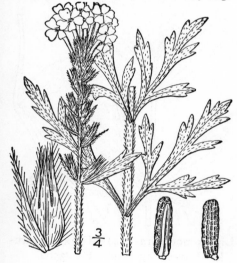

2. LÍPPIA Houst.; L. Sp. Pl. 633. 1753.

Perennial herbs, or shrubs, with opposite, sometimes verticillate, or rarely alternate leaves, and small bracted flowers, in axillary or terminal, mostly peduncled spikes or heads. Calyx small, membranous, ovoid, campanulate or compressed and 2-winged, 2-4-toothed or 2-4-cleft. Corolla-tube straight or incurved, cylindric, the limb oblique, spreading, somewhat 2-lipped, 4-cleft, the lobes broad, often retuse or eroded. Stamens 4, didynamous, included or exserted; anthers ovate, not appendaged, the sacs nearly parallel. Ovary 2-celled; ovules 1 in each cavity; style short; stigma oblique or recurved. Fruit dry, with a membranous exocarp, at length separating into 4 nutlets. [In honor of Auguste Lippi, 1678–1703, French naturalist.]

About 110 species, most abundant in tropical and subtropical America, a few African. Besides the following, which by some authors are separated as a distinct genus (PHYLA Lour.), about 6 others occur in the southern United States. Type species: *Lippia americana* L.

Leaves linear-cuneate to spatulate, 2–8-toothed; peduncles little exceeding leaves.　　1. *L. cuneifolia.*
Leaves sharply serrate; peduncles much longer than leaves.
　　Leaves oblong or lanceolate, mostly acute.　　2. *L. lanceolata.*
　　Leaves spatulate or obovate, mostly obtuse.　　3. *L. nodiflora.*

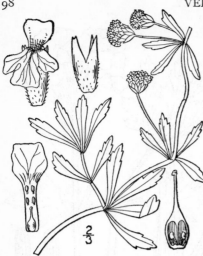

1. Lippia cuneifòlia (Torr.) Steud. Wedge-leaved Fog-fruit. Fig. 3560.

Zapania cuneifolia Torr. Ann. Lyc. N. Y. **2**: 234. 1827.

Lippia cuneifolia Steud.; Torr. in Marcy's Rep. 293. *pl. 17.* 1853.

Pale, minutely puberulent with forked hairs or glabrous, diffusely branched from the woody base; branches terete, slender, rigid, procumbent, somewhat zigzag, with short erect branchlets at the nodes. Leaves linear-cuneate, sessile, obscurely veined, rigid, 1'–1½' long, 2"–3" wide, with 2–8 sharp teeth above the middle or rarely entire, acutish at the apex; peduncles shorter than or somewhat exceeding the leaves; head at first globose, becoming cylindric and 6"–8" long; bracts cuneate, abruptly acuminate from the truncate or retuse summit; calyx flattened, 2-cleft, the lobes 2-toothed or emarginate; corolla-tube longer than the calyx; fruit oblong.

On plains, South Dakota, Nebraska and Colorado to Texas, Mexico and Arizona. May–Aug.

2. Lippia lanceolàta Michx. Fog-fruit. Fig. 3561.

L. lanceolata Michx. Fl. Bor. Am. **2**: 15. 1803.

Green, glabrous, or very sparingly pubescent with forked hairs; stems slender, weak, procumbent or ascending, sometimes rooting at the nodes, simple, or little branched, 1°–2° long. Leaves thin, oblong, ovate, or oblong-lanceolate, pinnately veined, short-petioled, acute or subacute at the apex, sharply serrate to below the middle, narrowed to the somewhat cuneate base, 1'–3' long, 3"–15" wide; peduncles slender, some or all of them longer than the leaves; heads at first globose, becoming cylindric and about ½' long in fruit; bracts acute; calyx flattened, 2-cleft; corolla pale blue, scarcely longer than the calyx; fruit globose.

In moist soil, Ontario to Minnesota, New Jersey, Illinois, Kansas, Florida, Texas and northern Mexico. Also in California. Frog-fruit. June–Aug.

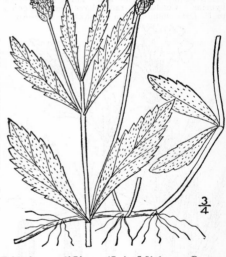

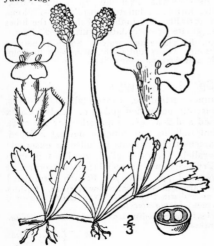

3. Lippia nodiflòra (L.) Michx. Spatulate-leaved Fog-fruit. Fig. 3562.

Verbena nodiflora L. Sp. Pl. 20. 1753.

Lippia nodiflora Michx Fl. Bor. Am. **2**: 15. 1803.

Minutely and rather densely puberulent with short appressed hairs, creeping, or some of the branches ascending, 1°–3° long. Leaves thickish, spatulate, oblanceolate, or obovate, 6"–2½' long, 3"–12" wide, mostly obtuse at the apex, narrowed into a long or short cuneate entire base, sharply serrate above the middle; peduncles slender, 1'–6' long, much longer than the leaves; heads at length cylindric and 5"–12" long, 3"–4" thick; calyx flattened, 2-cleft; corolla purple to white.

In wet or moist soil, South Carolina to southern Missouri, Florida and Texas. Also in California, Central America, the West Indies, and apparently the same species in the warmer regions of the Old World. May–Sept.

3. CALLICÀRPA L. Sp. Pl. 111. 1753.

Shrubs or trees, with opposite leaves, and small blue purple or white flowers in axillary cymes. Calyx short, campanulate, 4-toothed (rarely 5-toothed), or truncate. Corolla-tube short, expanded above, the limb spreading, 4-cleft (rarely 5-cleft), the lobes equal, imbricated in the bud. Stamens 4, equal, exserted; anthers ovate or oval, their sacs parallel. Ovary incompletely 2-celled; ovules 2 in each cavity, laterally attached, amphitropous; style slender; stigma capitate, or 2-lobed. Fruit a berry-like drupe, much longer than the calyx, containing 1–4 nutlets. [Greek, handsome fruit.]

About 45 species, the following typical one of southeastern North America, the others Asiatic, African and tropical American.

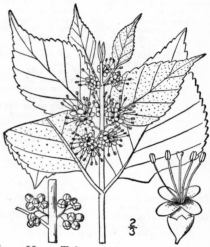

1. Callicarpa americàna L.   French or Bermuda Mulberry. Fig. 3563.

Callicarpa americana L. Sp. Pl. 111. 1753.

A shrub, 2°–5° high, the twigs, petioles and young leaves stellate-scurfy, the mature leaves becoming nearly glabrous and glandular-dotted. Twigs terete; leaves thin, ovate, pinnately veined, slender-petioled, acute or acuminate at the apex, crenate-dentate nearly to the entire base, 3′–6′ long, 1¼′–3′ wide; cymes many-flowered, short-peduncled; pedicels very short; calyx-teeth much shorter than the tube; corolla pale blue, about 1½″ long; fruit violet-blue, globose, 1½″ in diameter, very conspicuous in autumn.

In moist thickets, Virginia to Florida, Missouri, Arkansas and Texas. Bermuda. June–July. Sour-bush.

Callicarpa purpùrea Juss., an Asiatic shrub often planted for ornament, with pink flowers and long leaves, has been observed in a swamp at Wilmington, Delaware.

$\frac{2}{3}$

Family 26. LABIÀTAE B. Juss. Hort. Trian. 1759.
MINT FAMILY.

Aromatic punctate herbs, or shrubs (a few tropical species trees), mostly with 4-sided stems and simple opposite leaves; stipules none. Flowers irregular, perfect, variously clustered, the inflorescence typically cymose, usually bracteolate. Calyx inferior, persistent, regular or 2-lipped, 5-toothed or 5-lobed (rarely 4-toothed), mostly nerved. Corolla with a short or long tube, the limb 4–5-lobed, mostly 2-lipped, regular in a few genera; upper lip 2-lobed, or sometimes entire; lower lip mostly 3-lobed. Stamens borne on the corolla-tube, typically 4 and didynamous, sometimes 2 with or without staminodia, rarely equal; filaments separate, mostly slender, alternate with the corolla-lobes; anthers 2-celled, introrse, or confluently 1-celled, or sometimes of a single sac. Disk usually present, fleshy. Ovary 4-lobed or 4-parted, superior, each lobe or division with 1 mostly anatropous ovule; style arising from the center of the lobed or parted ovary, 2-lobed at the summit. Fruit of 4 1-seeded nutlets. Seed erect (transverse in *Scutellaria*); endosperm scanty, or usually none; embryo mostly straight; radicle short, inferior.

About 160 genera and 3200 species, of wide distribution in temperate and tropical regions. The family is also known as LAMIACEAE. The foliage abounds in volatile oils.

A. **Ovary 4-lobed, the style not basal; nutlets laterally attached.** (AJUGEAE.)
Corolla-limb very irregular, apparently 1-lipped, or the other lip very short; stamens exserted.
    Upper lip of corolla short, truncate.   1. *Ajuga.*
    Upper lip of corolla 2-lobed, or all the lobes united into the lower lip.   2. *Teucrium.*
Corolla-limb nearly equally 5-lobed.
    Corolla-lobes spreading; stamens short-exserted.   3. *Isanthus.*
    Corolla-lobes declined; stamens long-exserted.   4. *Trichostema.*

B. **Ovary 4-parted, the style basal; nutlets basally attached.**
           * Calyx with a protuberance on the upper side. (SCUTELLARIEAE.)   5. *Scutellaria.*
            ** Calyx not gibbous on the upper side.   (STACHYEAE.)
Stamens and style very short, included in the corolla-tube.
    Anther-sacs parellel.   6. *Marrubium.*
    Anther-sacs, at least of the upper stamens, divergent.   7. *Sideritis.*
Stamens longer, not included in the corolla-tube.

† *Corolla strongly 2-lipped; lips unlike, the upper concave.*

    (a.) Anther-bearing stamens 4.

Posterior (upper) pair of stamens longer than the anterior.
  Anther-sacs parallel or nearly so.
    Tall erect herbs; posterior stamens declined, anterior ascending.   8. *Agastache.*
    Trailing herb; stamens all ascending under upper lip of corolla.   9. *Meehania.*
  Anther-sacs divergent.
    Calyx tubular, nearly equally 5-toothed, not 2-lipped; plant erect.   10. *Nepeta.*
    Calyx distinctly 2-lipped, or unequally 5-toothed.
      Trailing herb; calyx unequally 5-toothed.   11. *Glecoma.*
      Erect herbs; calyx 2-lipped.   12. *Moldavica.*
Posterior pair of stamens shorter than the anterior.
  Calyx distinctly 2-lipped, closed in fruit.   13. *Prunella.*
  Calyx 3–10-toothed, not distinctly 2-lipped, open in fruit.
    Calyx membranous, inflated in fruit, faintly nerved.
      Calyx nearly equally 5-toothed, or 5-lobed.   14. *Dracocephalum.*
      Calyx 4-lobed.   15. *Synandra.*
    Calyx not membranous, not inflated in fruit, distinctly 5–10-nerved.
      Style-branches very unequal.   16. *Phlomis.*
      Style-branches equal, or nearly so.
        Anther-sacs transversely 2-valved.   17. *Galeopsis.*
        Anther-sacs not transversely 2-valved, parallel or divergent.
          Nutlets 3-sided, truncate.
            Calyx-teeth not spiny-tipped.   18. *Lamium.*
            Calyx-teeth spiny-tipped.   19. *Leonurus.*
          Nutlets ovoid, rounded above.
            Calyx with a spreading 5-toothed limb.   20. *Ballota.*
            Calyx-limb not spreading.
              Corolla-tube not longer than calyx; anther-sacs divergent.
                      21. *Stachys.*
              Corolla-tube exserted; lower petioles very long; anther-sacs parallel.
                      22. *Betonica.*

    (b.) Anther-bearing stamens 2.

Connective of the anther very long, bearing a perfect sac at one end, and a rudimentary one, or
  none, at the other.   23. *Salvia.*
Connective very short, the anther-sacs confluent.
  Calyx tubular, 15-nerved, equally 5-toothed.   24. *Monarda.*
  Calyx ovoid-tubular, 13-nerved, 2-lipped.   25. *Blephilia.*

†† *Corolla 2-lipped, or regular; upper lip, when present, flat, or only slightly concave.*

    (a.) Flowers in axillary whorls or clusters, or these forming terminal spikes.

Corolla 2-lipped.
  Stamens curved, often converging, or ascending under the upper lip of the corolla.
    Anther-bearing stamens 2.   26. *Hedeoma.*
    Anther-bearing stamens 4.
      Corolla-tube upwardly curved, exserted.   27. *Melissa.*
      Corolla-tube straight.
        Calyx 10-nerved, campanulate, about equally 5-toothed.   28. *Satureia.*
        Calyx mostly 13-nerved, tubular, 2-lipped.   29. *Clinopodium.*
  Stamens straight, often diverging.
    Calyx 15-nerved.   30. *Hyssopus.*
    Calyx 10–13-nerved.
      Anther-bearing stamens 4.
        Anther-sacs divergent.
          Calyx equally 5-toothed; erect herbs.   31. *Origanum.*
          Calyx 2-lipped; creeping herbs.   32. *Thymus.*
        Anther-sacs parallel.   33. *Koellia.*
      Anther-bearing stamens 2.   34. *Cunila.*
Corolla regular, 4–5-lobed.
  Anther-bearing stamens 2; plants not aromatic.   35. *Lycopus.*
  Anther-bearing stamens 4; aromatic fragrant herbs.   36. *Mentha.*

    (b.) Flowers in terminal panicled racemes or spikes; corolla 2-lipped.

Anther-bearing stamens 2; lower lip of corolla long, fimbriate; native.   37. *Collinsonia.*
Anther-bearing stamens 4; lower lip of corolla not fimbriate; introduced.
  Flowers racemose.   38. *Perilla.*
  Flowers densely spiked.   39. *Elsholtzia.*

1. AJUGA L. Sp. Pl. 561. 1753.

    Annual or perennial, often stoloniferous herbs, mostly with dentate leaves, and rather large verticillate-clustered flowers in terminal spikes, or in the upper axils. Calyx ovoid or campanulate, 10–many-nerved, 5-toothed or 5-lobed, the teeth or lobes nearly equal. Corolla-limb 2-lipped, the upper lip short, truncate or emarginate, the lower spreading, with 2 small lateral lobes and a much larger emarginate or 2-cleft middle one. Stamens 4, didynamous, somewhat exserted beyond the upper lip of the corolla, the anterior pair the longer; anther-sacs divergent, only slightly confluent at the base. Ovary not deeply 4-lobed. Nutlets obovoid, rugose-reticulate. [Greek, without a yoke; from the seeming absence of the upper lip of the corolla.]

About 40 species, natives of the Old World. Type species: *Ajuga reptans* L.

Sparingly pubescent, or glabrous, stoloniferous. 1. *A. reptans.*
Pubescent with long hairs, not stoloniferous. 2. *A. genevensis.*

1. Ajuga réptans L. Bugle. Fig. 3564.

Ajuga reptans L. Sp. Pl. 561. 1753.

Perennial, sparingly pubescent or glabrous, producing slender creeping stolons sometimes 1° long; stem erect, rather stout, 6'–15' tall. Basal leaves tufted, obovate, rounded at the apex, crenate or undulate, 1'–3' long, tapering into margined petioles; leaves of the stem oblong or oblanceolate, much smaller, sessile or nearly so, those of the stolons mostly petioled; upper flower-clusters often forming a short spike, the lower commonly distant and axillary; corolla blue or nearly white, about ½' long.

In fields, Quebec and Maine to southern New York, locally naturalized from Europe. Brown bugle. Middle comfrey. Carpenter's herb. Sicklewort. May–June.

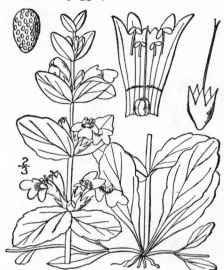

2. Ajuga genevénsis L. Erect Bugle. Fig. 3565.

Ajuga genevensis L. Sp. Pl. 561. 1753.

Perennial, not stoloniferous; stems usually several from the root, ascending, rather stout, long-pubescent. Leaves long-pubescent, the basal ones long-petioled, spatulate or oblanceolate, obtuse, sparingly toothed or entire, 2'–4' long, the upper sessile, obovate to ovate, usually coarsely dentate, much smaller; bracts pubescent, the upper shorter than the flowers; inflorescence mostly dense; corolla ½'–¾' long.

In fields and waste grounds, Maine to New York and Pennsylvania. Adventive from Europe. May–July.

2. TEÙCRIUM [Tourn.] L. Sp. Pl. 562. 1753.

Herbs or shrubs, with dentate entire or laciniate leaves, and rather small pink, white or purplish flowers, in terminal bracted spikes or heads, or verticillate in the upper axils. Calyx tubular-campanulate, 10-nerved, equally or unequally 5-toothed. Corolla-tube short, the limb irregularly 5-lobed, the 2 short upper lobes oblong, declined or erect, the lateral lobe declined, more or less united with the upper ones, the lower lobe broader, also declined. Stamens 4, didynamous, exserted between the 2 upper lobes of the corolla, the anterior pair the longer; anther-sacs divergent, confluent at the base. Ovary 4-lobed; style 2-cleft at the summit. Nutlets obovoid, rugose-reticulated. [Named from the Trojan king, Teucer.]

Over 100 species, of wide distribution in temperate and tropical regions. Besides the following, 2 or 3 others occur in the southern and southwestern United States. Type species: *Teucrium fruticans* L.

*PERENNIAL SPECIES.

1. Leaves toothed.
   † Flowers in terminal dense spike-like panicles.
       Calyx and bracts canescent, without long hairs.
           Leaves, at least the lower, obtuse or rounded at the base, not rugose-veined.

                                                      1. *T. canadense.*

        Leaves narrowed at the base, rugose-veined, mostly narrowly lanceolate.

                                              2. *T. littorale.*
                                              3. *T. occidentale.*

    Calyx and bracts villous and often glandular-pubescent.   4. *T. Scorodonia.*
    †† Flowers in secund terminal spikes.   5. *T. laciniatum.*
2. Leaves laciniate; flowers axillary.
        ** ANNUAL SPECIES, with pinnatifid leaves and axillary flowers.   6. *T. Botrys.*

$\frac{1}{2}$

1. Teucrium canadénse L. American Germander or Wood Sage. Fig. 3566.

Teucrium canadense L. Sp. Pl. 564. 1753.

Teucrium virginicum L. Sp. Pl. 564. 1753.

Perennial, appressed-pubescent or canescent; stem erect, simple or somewhat branched, rather slender, 1°–2° tall. Leaves lanceolate, oblong-lanceolate or ovate-lanceolate, acuminate at the apex, irregularly dentate, mostly rounded at the base, short-petioled, 1½′–5′ long, ½′–2′ wide, glabrous or sparingly pubescent above, densely canescent beneath; spike usually dense, becoming 6′–12′ long in fruit, bracts canescent, the lower sometimes foliaceous, the upper commonly not longer than the canescent calyx; flowers 6″–10″ long, very short-pedicelled; calyx about 3″ long in fruit, its three upper teeth obtuse or subacute.

In moist thickets or along marshes, New England to Ontario, Minnesota, Florida, Kansas and Texas. Ascends to 2600 ft. in Virginia. Ground-pine. June–Sept.

2. Teucrium littoràle Bicknell. Narrow-leaved Germander. Fig. 3567.

Teucrium littorale Bicknell, Bull. Torr. Club 28: 169. 1901.

T. canadense var. *littorale* Fernald, Rhodora 10: 84. 1908.

Pale and canescent, 2° high or less, erect or assurgent, often with ascending branches. Leaves thickish and rugose-veiny, narrowly oblong or sometimes broader, narrowed into the petiole, closely fine-serrate or becoming unequally dentate-serrate, 2½′–4′ long, ½′–1½′ wide; petioles 2½″–5″ long; spikes narrow, often interrupted; bracts about the length of the calyx; calyx small, 2″–2½″ high, becoming somewhat gibbous-urceolate, the teeth short, the upper ones obtuse; corolla pale pink, about 8″ long, loosely pilose without.

On or near the coast, Maine to Florida and Texas, north to Arkansas and Oklahoma. Included in our first edition in *T. canadense* L., and there figured for that species. July–Aug.

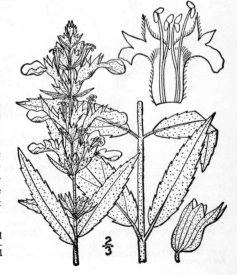

$\frac{2}{3}$

3. Teucrium occidentàle A. Gray. Hairy Germander. Fig. 3568.

Teucrium occidentale A. Gray, Syn. Fl. **2**: 349. 1878.

T. boreale Bicknell, Bull. Torr. Club **28**: 171. 1901.

Perennial, villous or pubescent; stem erect, rather stout, usually much branched, 1°–3° high, the branches ascending. Leaves lanceolate or ovate-lanceolate, thin, acute or acuminate at the apex, sharply dentate, mostly rounded at the base, usually slender-petioled, 1′–3½′ long, ½′–1½′ wide; spikes dense, becoming 3′–8′ long in fruit; bracts lanceolate-subulate or the lower sometimes larger, villous and often glandular; calyx and axis of the spike villous-pubescent and often glandular, the 3 upper calyx-teeth acute or acutish; corolla 4″–6″ long.

In moist soil, Maine and Ontario to eastern Pennsylvania, British Columbia, Ohio, Nebraska, New Mexico and California. July–Sept.

4. Teucrium Scorodònia L. Wood Germander. Fig. 3569.

Teucrium Scorodonia L. Sp. Pl. 564. 1753.

Perennial, villous-pubescent, 2° high or less, the branches erect-ascending. Leaves mostly ovate, 1′–2½′ long, obtuse or acute at the apex, cordate or subtruncate at the base, crenate, the petioles 2½″–7″ long; racemes narrow, rather loosely flowered, often 5′ long, the flowers secund, mostly in pairs, the pedicels shorter than the calyx, equalling or shorter than the ovate, acute or acuminate bracts; calyx veiny, the upper tooth broad; corolla light yellow, 3″–4″ long.

A weed in cultivated fields, Ontario, and reported from Ohio. Adventive from Europe. June–Sept.

5. Teucrium laciniàtum Torr. Cut-leaved Germander. Fig. 3570.

Teucrium laciniatum Torr. Ann. Lyc. N. Y. **2**: 231. 1828.

Melosmon laciniatum Small, Fl. SE. U. S. 1019. 1903.

Diffusely branched from a woody perennial root, 1° high or less, densely leafy, glabrous, or nearly so. Leaves 1½′ long or less, pinnately parted into 3–7 stiff, linear, entire, toothed or lobed segments; flowers solitary in the upper axils, as long as the subtending leaves or shorter, short-peduncled; calyx deeply 5-parted, the lobes narrowly lanceolate, nearly equal; corolla pale blue or lilac, about 9″ long, its lower lobes much longer than the calyx.

Plains, Kansas and Colorado to Texas and Arizona. May–Aug.

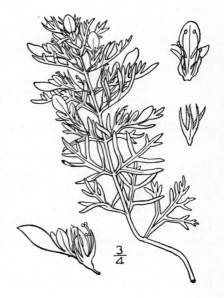

$\frac{3}{5}$

6. **Teucrium bòtrys** L. Cut-leaved Annual Germander. Fig. 3571.

Teucrium botrys L. Sp. Pl. 562. 1753.

Annual, villous-pubescent, branched, 1° high or less. Leaves slender-petioled, deeply pinnatifid into oblong, entire or toothed lobes, the basal ones with petioles longer than the blades; flowers whorled in the upper axils; pedicels shorter than the calyx; calyx campanulate, gibbous, about 8″ long in fruit, veiny, its teeth triangular-ovate, acute, nearly equal; corolla purplish, about 6″ long.

Waste grounds, Massachusetts and Ohio. Naturalized from Europe. July–Sept.

3. **ISÁNTHUS** Michx. Fl. Bor. Am. **2**: 3. *pl. 30.* 1803.

An annual erect finely viscid-pubescent much-branched herb, with narrow entire or few-toothed leaves, and small blue flowers in loose axillary cymes. Calyx broadly campanulate, 10-nerved, nearly equally 5-toothed, the teeth lanceolate. Corolla-tube not longer than the calyx, enlarged into the throat, the limb nearly equally 5-cleft into obovate somewhat spreading lobes. Stamens 4, didynamous, incurved-ascending, not longer than the corolla, the anterior pair slightly the longer; anther-sacs divergent at maturity. Ovary deeply 4-lobed; style minutely 2-cleft at the summit. Nutlets rugose-reticulated. [Greek, equal-flower, the corolla-lobes being nearly equal.]

A monotypic genus of eastern North America.

1. **Isanthus brachiàtus** (L.) B.S.P. False Pennyroyal. Fig. 3572.

Trichostema brachiatum L. Sp. Pl. 598. 1753.
Isanthus coeruleus Michx. Fl. Bor. Am. **2**: 4. *pl. 30.* 1803.
I. brachiatus B.S.P. Prel. Cat. N. Y. 44. 1888.

Stem slender, much branched, 6′–20′ tall, the branches spreading. Leaves oblong or elliptic-lanceolate, acute at each end, entire, or with a few sharp teeth, 3-nerved, short-petioled, 1′–2′ long, 2″–6″ wide; axillary cymes 1–3-flowered; pedicels very slender, some of them as long as the fruiting calyx; calyx-lobes acute or acuminate, longer than or equalling the tube; corolla 2″–3″ long, the fruiting calyx 3″ long.

In sandy soil, especially along streams, Quebec and Ontario to Minnesota, Vermont, Georgia, Kansas and Texas. Flux-weed. Blue gentian. July–Sept.

$\frac{3}{5}$

4. **TRICHOSTÈMA** [Gronov.] L. Sp. Pl. 598. 1753.

Annual or perennial erect branching herbs, some western species shrubby, with lanceolate oblong or linear entire or slightly repand leaves. Flowers small, or middle-sized, pink, blue, purple, or white, paniculate, or in axillary loose or dense cymes. Calyx campanulate, very unequally 5-lobed in our species, the lobes ovate or lanceolate, the 3 upper much longer than the 2 lower. Corolla-tube slender, exserted or included, the limb somewhat oblique and deeply 5-cleft into oblong more or less declined segments. Stamens 4, didynamous, ascending, curved, the anterior pair the longer, the filaments filiform, spirally coiled in the bud, long-exserted; anther-sacs divaricate, more or less confluent at the base. Ovary deeply 4-lobed; style 2-cleft at the summit. Nutlets obovoid, reticulated. [Greek, hair-stamen, referring to the slender filaments.]

About 10 species, natives of North America. Type species: *Trichostema dichotomum* L.

Leaves oblong or lanceolate; plant minutely viscid-pubescent.   1. *T. dichotomum.*
Leaves linear; plant puberulent or glabrous.   2. *T. lineare.*

1. Trichostema dichótomum L. Blue Curls. Bastard Pennyroyal. Fig. 3573.

Trichostema dichotomum L. Sp. Pl. 598. 1753.

Annual, minutely viscid-pubescent; stem slender, rather stiff, much branched, 6′–2° high, the branches spreading or ascending. Leaves oblong or oblong-lanceolate, membranous, obtuse or subacute at the apex, narrowed at the base into short petioles, 1′–3′ long, 3″–10″ wide, the upper gradually smaller; flowers paniculate, 6″–9″ long, borne 1–3 together on 2-bracteolate peduncles; calyx oblique, very unequally 5-lobed, the 3 upper lobes much longer and more united than the 2 lower ones; corolla blue, pink or rarely nearly white, the limb longer than the tube; stamens blue or violet.

In dry fields, Maine to Florida, Vermont, Pennsylvania, Missouri and Texas. The lateral flowers become inverted by torsion of the pedicels. July–Oct.

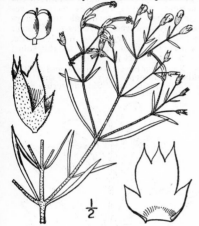

2. Trichostema lineàre Nutt. Narrow-leaved Blue Curls. Fig. 3574.

T. brachiatum Lam. Encycl. **8**: 84. 1808. Not L. 1753.
Trichostema lineare Nutt. Gen. **2**: 39. 1818.

Puberulent or glabrous, not viscid or scarcely so; stem very slender, at length widely branched, 6′–18′ high, the branches ascending. Leaves linear, obtuse or subacute, sessile or very short-petioled, ½′–2′ long, 1″–2″ wide, sometimes with smaller ones or short leafy branches in their axils; flowers very similar to those of the preceding species, sometimes larger.

In sandy fields and dry pine barrens, Connecticut to Georgia and Louisiana, mostly near the coast. July–Aug.

5. SCUTELLÀRIA [Rivin.] L. Sp. Pl. 598. 1753.

Annual or perennial bitter herbs, some species shrubby. Flowers blue to violet, in terminal or axillary bracted mostly secund spike-like racemes, or solitary or 2–3 together in the axils. Calyx campanulate, gibbous, 2-lipped, the lips entire, the upper one with a cres′ or protuberance upon its back and often deciduous in fruit, the lower one persistent. Coroila much exserted, recurved-ascending, dilated above into the throat, glabrous within, the limb 2-lipped; upper lip arched, entire or emarginate; lower lip spreading or deflexed, its lateral lobes small and somewhat connected with the upper, its middle lobe broad, sometimes emarginate, the margins mostly recurved. Stamens 4, didynamous, all anther-bearing, ascending under the upper lip, the upper pair somewhat the shorter, their anthers 2-celled, ciliate; anthers of the lower pair of stamens 1-celled, also ciliate. Style unequally 2-cleft at the apex; ovary deeply 4-parted. Nutlets subglobose or depressed, papillose or tuberculate, borne on a short or elongated gynobase. [Latin, a dish, from the appendage to the fruiting calyx.]

About 100 species of wide geographic distribution. Besides the following, some 15 others occur in the southern and western parts of North America, all known as Skullcap, or Helmet-flower. Type species: *Scutellaria peregrina* L.

* Nutlets wingless, very slightly elevated on the short gynobase.

Flowers 3″–5″ long, in axillary and sometimes terminal secund racemes.	1. *S. lateriflora.*
Flowers 6″–15″ long, in terminal often panicled racemes.	
Plant glabrous or very nearly so; leaves broad.	2. *S. serrata.*
Plants pubescent, puberulent or pilose.	
Leaves all except the floral crenate or dentate, broad.	
Canescent, not glandular; corolla canescent.	3. *S. incana.*
Densely glandular-pubescent; corolla puberulent.	4. *S. cordifolia.*
Pubescent below, glandular above; corolla nearly glabrous.	5. *S. pilosa.*
Leaves all except the lowest entire, narrow.	6. *S. integrifolia.*
Flowers solitary in the axils or sometimes also in terminal bracted racemes.	
Perennial from a thick woody root.	7. *S. resinosa.*
Fibrous-rooted; perennial by rootstocks or stolons.	
Flowers 2″–4″ long.	8. *S. parvula.*
Flowers 8″–13″ long.	

Minutely and densely glandular-pubescent, resiniferous.   9. *S. Brittonii.*
Densely cinerous-pubescent, pale.   10. *S. Bushii.*
Glabrous or merely slightly puberulent.
 Leaves ovate, slender-petioled, cordate, obtuse.   11. *S. saxatilis.*
 Leaves lanceolate or ovate-lanceolate, nearly sessile, acute.   12. *S. galericulata.*
** Nutlets membranous-winged, elevated on the slender gynobase; flowers axillary.
                 13. *S. nervosa.*

1. Scutellaria lateriflòra L. Mad-dog or Blue Skullcap. Fig. 3575.

Scutellaria lateriflora L. Sp. Pl. 598. 1753.

Perennial by slender stolons, glabrous throughout or puberulent above; stem slender, leafy, erect or ascending, commonly branched, 4′–2½° high. Leaves ovate, ovate-oblong or ovate-lanceolate, thin, slender-petioled, acute or acuminate at the apex, coarsely dentate-serrate, obtuse, rounded or subcordate at the base, 1′–3′ long, the upper gradually smaller, the uppermost sometimes entire; racemes narrow, secund, axillary or often also terminal and leafy-bracted, several–many-flowered; flowers 3″–5″ long; calyx short; corolla blue, varying to nearly white, its lips about equal, one-fifth as long as the tube; nutlets borne on a very short gynobase.

In wet places, Newfoundland to Ontario and British Columbia, Florida, Mississippi, New Mexico and Oregon. July–Sept. Blue pimpernel. Side-flowering scullcap. Madweed. Hoodwort.

2. Scutellaria serràta Andr. Showy Skullcap. Fig. 3576.

Scutellaria serrata Andr. Bot. Rep. *pl. 494.* 1809.
S. laevigata Aiken; Eaton, Man. Ed. 6, 333. 1833.

Perennial, glabrous, or puberulent above; stem slender, erect, simple or branched, 1°–2° high. Leaves ovate or elliptic, slender-petioled, acute at the apex, narrowed, or the lowest rounded or subcordate at the base, crenate or dentate, 2′–4′ long, the uppermost reduced to small floral bracts; racemes almost always simple and terminal, loosely flowered, the flowers opposite; fruiting calyx about 3″ long; corolla 1′ long, blue, minutely puberulent, its tube narrow, gradually expanded above into the throat, its rather narrow upper lip a little shorter than the lower; nutlets borne on a short gynobase.

In woods, southern New York and Pennsylvania to South Carolina, Illinois, Kentucky and Tennessee. One of the handsomest of the American species. Ascends to 3000 ft. in Virginia. May–June.

3. Scutellaria incàna Muhl. Downy Skullcap. Fig. 3577.

Scutellaria incana Muhl. Cat. 56. 1813.

Scutellaria canescens Nutt. Gen. 2: 38. 1818.

S. serrata Spreng. Syst. 2: 703. 1825. Not Andr. 1809.

Perennial, finely and densely whitish downy, or the upper surfaces of the leaves glabrous; stem rather strict, erect, usually much branched above, 2°–4° high. Leaves ovate, oval, or oblong, rather firm, slender-petioled, acute at the apex, crenate-dentate, narrowed, rounded or the lower subcordate at the base, 3′–4½′ long; racemes terminal, usually numerous and panicled, several–many-flowered; fruiting calyx 1½′–2′ long; corolla 9″–10″ long, canescent; upper lip of the corolla slightly longer than the lower; gynobase very short.

In moist woods and thickets, Ontario to Michigan, Kansas, Georgia, Florida and Alabama. June–Aug.

4. Scutellaria cordifòlia Muhl.   Heart-leaved Skullcap.   Fig. 3578.

Scutellaria cordifolia Muhl. Cat. 56. 1813.
Scutellaria versicolor Nutt. Gen. **2**: 38. 1818.

Perennial, densely glandular-pubescent; stem erect, usually stout, 1°–3° high, often simple. Leaves prominently veined, slender-petioled, broadly ovate, crenate-dentate all around, 2′–4′ long, all but the uppermost cordate at the base; racemes terminal, narrow, solitary or panicled; bracts ovate, mostly entire, commonly longer than the pedicels; fruiting calyx nearly 3″ long; corolla puberulent, 10″–12″ long, blue with the lower side lighter or white, its tube narrow, its throat moderately dilated, its lateral lobes about as long as the upper lip; gynobase short.

In woods and thickets, especially along streams, Pennsylvania to Florida, west to Minnesota, Kansas, Arkansas and Texas. Includes several races. June–Aug.

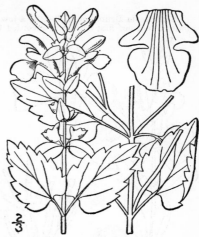

5. Scutellaria pilòsa Michx.   Hairy Skullcap.   Fig. 3579.

Scutellaria pilosa Michx. Fl. Bor. Am. **2**: 11. 1803.
Scutellaria ovalifolia Pers. Syn. **2**: 136. 1807.
S. hirsuta Short, Transyl. Journ. Med. **8**: 582. 1836.
Scutellaria pilosa hirsuta A. Gray, Syn. Fl. **2**: Part 1, 379. 1878.

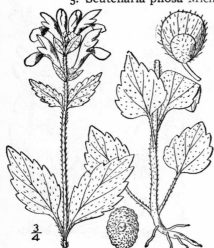

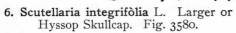

Perennial, stem slender, simple or branched, hairy or downy below, sometimes hirsute, glandular-pubescent above, 1°–3° high. Leaves ovate, oval, or oblong, petioled, obtuse, or the upper subacute at the apex, crenate, 1′–3′ long, narrowed or rounded at the base or the lower subcordate; racemes terminal, solitary or panicled, sometimes also in the upper axils; bracts oblong or spatulate, entire, longer than the pedicels; fruiting calyx about 3″ long; corolla blue, 6″–8″ long, minutely puberulent or glabrous, its lower lip and lateral lobes somewhat shorter than the arched upper one; gynobase short.

In dry sandy woods and thickets, southern New York and Pennsylvania to Michigan, Missouri, Florida and Texas. Races differ in size and in pubescence. Ascends to 4000 ft. in North Carolina. May–July.

6. Scutellaria integrifòlia L.   Larger or Hyssop Skullcap.   Fig. 3580.

Scutellaria integrifolia L. Sp. Pl. 599. 1753.
Scutellaria hyssopifolia L. Sp. Pl. 599. 1753.

Perennial, hoary with a minute down; stem slender, erect, rather strict, simple or branched, 6′–2½° high. Leaves thin, linear to oblong, petioled, or the upper sessile, obtuse at the apex, entire, 1′–2′ long, 2″–6″ wide, or the lower ovate, lanceolate or nearly orbicular, obtuse and sometimes subcordate at the base, often crenate-dentate or incised; racemes solitary or several, terminal; bracts linear-oblong, subacute, longer than the pedicels; fruiting calyx 2″–3″ long; corolla blue, or whitish underneath, 10″–15″ long, its large lips nearly equal; gynobase short.

In fields, woods and thickets, Massachusetts to West Virginia, Tennessee, Arkansas, Florida, Louisiana and Texas. Consists of several races. May–Aug.   Large-flowered scullcap.

Scutellaria Drummóndii Benth., a low annual villous species with flowers solitary in the axils, admitted into our first edition as recorded from Kansas, is not definitely known north of Texas.

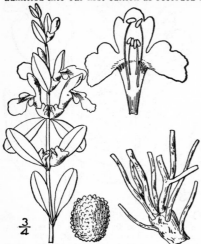

7. Scutellaria resinòsa Torr. Resinous Skullcap. Fig. 3581.

S. resinosa Torr. Ann. Lyc. N. Y. **2**: 232. 1827.

S. Wrightii A. Gray, Proc. Am. Acad. **8**: 370. 1872.

Perennial from a thick woody root, minutely canescent or puberulent and usually resiniferous; stems rather slender, rigid, tufted, leafy, ascending, 6'–10' high. Leaves ovate, oval, or oblong, sessile, or the lower short-petioled, entire, obtuse at the apex, mostly narrowed at the base, 3''–6'' long; flowers solitary in the axils; fruiting calyx nearly 3'' long; corolla violet or nearly white, very pubescent, 6''–8'' long, its tube narrow and lips nearly equal; gynobase short.

On dry plains, Nebraska and Kansas to Texas and Arizona. May–Aug.

8. Scutellaria pàrvula Michx. Small Skullcap. Fig. 3582.

S. parvula Michx. Fl. Bor. Am. **2**: 11. 1803.
Scutellaria ambigua Nutt. Gen. **2**: 37. 1818.
S. parvula var. mollis A. Gray, Syn. Fl. **2**¹: 380. 1878.
S. campestris Britton, Mem. Torr. Club **5**: 283. 1894.

Glabrous, or quite densely pubescent, sometimes slightly glandular, perennial by slender tuberous-thickened rootstocks; stems erect or ascending, very slender, usually branched, 3'–12' tall. Leaves ovate, oval or lanceolate, or the lower nearly orbicular, entire and sessile, or the lower dentate and petioled, 3''–12'' long; flowers solitary in the axils; fruiting calyx about 2'' long; corolla 2''–4'' long, violet, pubescent; gynobase short.

In sandy soil, Quebec to Ontario, South Dakota, Florida, Nebraska and Texas. Races differ in pubescence and in leaf-form. April–July. Little scullcap.

9. Scutellaria Brittònii Porter. Britton's Skullcap. Fig. 3583.

Scutellaria resinosa A. Gray, Syn. Fl. **2**: Part 1, 381. 1878. Not Torr. 1827.

Scutellaria Brittonii Porter, Bull. Torr. Club **21**: 177. 1894.

Perennial by tuberous-thickened rootstocks, viscidly glandular, pubescent or puberulent, branched from the base; stems erect, 4'–8' high, leafy. Leaves oblong or oval, sessile and entire or the lowest short-petioled and slightly crenulate, obtuse at the apex, rather prominently veined on the lower surface, 6''–12'' long, the upper scarcely smaller; flowers solitary in the axils; pedicels mostly shorter than the calyx; corolla pubescent, blue, 10''–15'' long, the tube narrow below, enlarged above into the throat; gynobase short.

Nebraska (according to Coulter); Colorado and Wyoming. June–July.

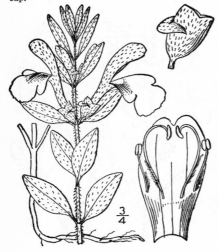

10. Scutellaria Búshii Britton.   Bush's Skullcap. Fig. 3584.

Scutellaria Bushii Britton, Manual, 785. 1901.

Roots thick-fibrous; stems several or numerous, tufted, erect or nearly so, finely cinereous-pubescent, 6′–12′ high. Leaves oblanceolate to oblong-oblanceolate, obtuse, entire, sessile, ¾′–1½′ long, 2″–4″ wide, the veins rather prominent; flowers solitary in the axils, short-peduncled, about 1′ long; corolla short-pubescent.

Barrens, southern Missouri. May–June.

11. Scutellaria saxátilis Riddell.   Rock Skullcap. Fig. 3585.

S. saxatilis Riddell, Suppl. Cat. Pl. Ohio, 14. 1836.

Perennial by filiform runners or stolons, glabrate or sparingly puberulent; stem slender, weak, ascending or reclining, simple or branched, 6′–20′ long. Leaves ovate, slender-petioled, thin, coarsely crenate, obtuse at the apex, cordate at the base, 1′–2′ long, or the lower nearly orbicular, and the upper lanceolate, subacute and entire; flowers solitary in the upper axils, or clustered in a terminal leafy-bracted loose raceme; bracts longer than the pedicels; fruiting calyx about 2″ long; corolla light blue, very nearly glabrous, 8″–10″ long; gynobase short.

On moist banks and in thickets, Delaware to South Carolina, Ohio and Tennessee. Ascends to 3000 ft. in Virginia. May–July.

12. Scutellaria galericulàta L.   Hooded Willow-herb.   Marsh or European Skullcap. Fig. 3586.

Scutellaria galericulata L. Sp. Pl. 599. 1753.

Perennial by filiform stolons, not tuber-bearing, puberulent or pubescent; stem erect, usually branched, 1°–3° high. Leaves oblong-lanceolate to ovate-oblong, thin, short-petioled, or the upper sessile, acute at the apex, dentate with low teeth or the upper entire, subcordate or rounded at the base, 1′–2½′ long, the uppermost usually much smaller and bract-like; flowers solitary in the axils; peduncles shorter than the calyx; corolla blue, puberulent, nearly or quite 1′ long, with a slender tube and slightly enlarged throat; gynobase short.

In swamps and along streams, Newfoundland to Mackenzie, Alaska, New Jersey, the mountains of North Carolina, Ohio, Nebraska, Arizona and Washington. Also in Europe and Asia. June–Sept.

Scutellaria Churchilliàna Fernald, of Maine and New Brunswick, has smaller flowers, solitary in the axils, its leaves much like those of *S. lateriflora*, and is, perhaps, a hybrid.

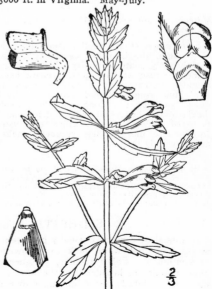

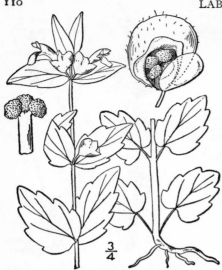

3/4

13. Scutellaria nervòsa Pursh. Veined Skullcap. Fig. 3587.

S. nervosa Pursh, Fl. Am. Sept. 412. 1814.
Scutellaria teucriifolia J. E. Smith in Rees' Cycl.
32: no. 15. 1816.

Perennial by filiform stolons; stem glabrous or sparingly pubescent, erect, slender, simple or sometimes branched, 6'–2° high. Leaves thin, glabrous, or sometimes decidedly pubescent, the lower slender-petioled, nearly orbicular, crenate, often subcordate at the base, the middle ones larger, ovate, 1'–2' long, sessile or nearly so, obtuse or acute, coarsely dentate or crenate, the upper lanceolate or ovate-lanceolate, acute, commonly entire; flowers solitary in the axils; fruiting calyx about 2'' long; corolla blue, 4''–5'' long, puberulent, the lower lip longer than the concave upper one; nutlets membranous-winged, borne on a slender gynobase.

In moist woods and thickets, southern Ontario, New York and New Jersey to Illinois and Missouri, North Carolina and Tennessee. Ascends to 3000 ft. in Virginia. May–Aug.

6. MARRÙBIUM [Tourn.] L. Sp. Pl. 582. 1753.

Perennial branching mostly woolly herbs, with petioled dentate rugose leaves, and small white or purplish flowers in dense axillary clusters, the juice bitter. Calyx tubular, 5–10-nerved, regularly 5–10-toothed, the teeth nearly equal, or the alternate ones shorter, acute or aristate, spreading or recurved in fruit. Corolla-limb 2-lipped, the upper lip erect, entire or emarginate, the lower spreading, 3-cleft, its broader middle lobe commonly emarginate. Stamens 4, didynamous, included, the posterior pair the shorter; anthers 2-celled. Style 2-cleft at the summit, the lobes short. Ovary deeply 4-lobed. Nutlets ovoid, smooth. [Name Middle Latin, perhaps from the Hebrew, referring to its bitter qualities.]

About 40 species, natives of the Old World, the following typical.

1. Marrubium vulgàre L. White or Common Hoarhound. Fig. 3588.

Marrubium vulgare L. Sp. Pl. 583. 1753.

Stem erect, stout, woolly, especially below, 1°–3° high, the branches ascending. Leaves oval, broadly ovate or nearly orbicular, rugose-veined, obtuse at the apex, crenate-dentate, rounded, narrowed or subcordate at the base, 1'–2' long, rough, whitish above, woolly beneath; petioles ½'–1' long, usually exceeding the flowers; clusters all axillary, densely many-flowered; flowers whitish; calyx-teeth usually 10, subulate, more or less recurved, glabrous above, woolly below.

In waste places, Maine and Ontario to Minnesota and British Columbia, North Carolina, Alabama, Texas, Mexico and California. Also in South America. Naturalized from Europe. Native also of Asia. Old names, houndbene, marrube, marvel.

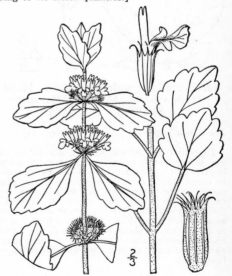

2/3

7. SIDERÌTIS [Tourn.] L. Sp. Pl. 574. 1753.

Annual or perennial, usually pubescent herbs or shrubs, with opposite, entire or toothed leaves, and small white, pink, or yellowish flowers in axillary clusters which are separated or contiguous. Calyx tubular, 5–10-nerved, slightly 2-lipped, the teeth spine-tipped, the upper middle one broader than the others. Corolla-limb exceeding the calyx, 2-lipped; upper lip entire or lobed; lower lip longer than the upper, with a broad middle lobe. Stamens 4, didynamous, included, the anterior pair the longer; anthers 2-celled, or those of the anterior pair more or less imperfect, the sacs divaricate. Style 2-cleft at the summit, lobes obtuse. Ovary 4-lobed. Nutlets obovoid, smooth. [Greek, iron, referring to its medicinal use.]

About 45 species, native of the Old World, most numerous in the Orient. Type species: *Sideritis hirsùta* L.

1. Sideritis romàna L.   Simple-beaked Iron-wort.
Fig. 3589.

Sideritis romana L. Sp. Pl. 575.   1753.

Annual, softly pubescent with spreading hairs, the stem erect, 4′–1° tall, usually branched. Leaves obovate to oblong, ½′–1¼′ long, crenate, sessile or nearly so, ultimately somewhat veiny; clusters few-flowered; flowers white or pinkish; calyx-teeth shorter than the tube, the middle one of the upper lip ovate, all subulate-tipped; corolla slightly exceeding the calyx, the middle lobe of the lower lip reniform.

In fields, southeastern Pennsylvania. Naturalized from the Mediterranean region. Also in Bermuda. June–Aug.

8. AGÁSTACHE Clayt.; Gron. Fl. Virg. 88.   1762.
[VLECKIA Raf. Med. Rep. (II.) **5**: 308.   1808.]
[LOPHANTHUS Benth. Bot. Reg. **15**: under *pl. 1282.*   1829.   Not Adans. 1763.]

Tall erect perennial herbs, with serrate, mainly ovate, petioled leaves, and yellowish purplish or blue flowers, verticillate-clustered in thick dense or interrupted bracted terminal spikes. Calyx narrowly campanulate, somewhat oblique, slightly 2-lipped, 5-toothed, the teeth of the upper lip somewhat larger than those of the lower, or all about equal. Corolla strongly 2-lipped, the tube as long as the calyx; upper lip erect, 2-lobed; lower lip spreading, 3-lobed, its middle lobe broader than the lateral ones and crenulate. Stamens 4, all anther-bearing, didynamous, the upper pair the longer; anthers 2-celled, their sacs nearly parallel. Ovary deeply 4-parted; style 2-cleft at the summit. Nutlets ovoid, smooth. [Greek, many spikes.]

About 8 species, natives of North America. Besides the following, 2 or 3 others occur in the western United States. Type species: *Hyssopus nepetoides* L.

Glabrous or very nearly so, stout; corolla greenish-yellow.	1. *A. nepetoides.*
Pubescent, stout; corolla purplish; leaves green both sides.	2. *A. scrophulariaefolia.*
Glabrous or slightly pubescent, slender; corolla blue; leaves pale beneath.	3. *A. anethiodora.*

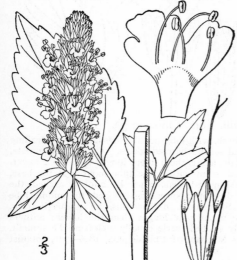

1. Agastache nepetoìdes (L.) Kuntze.
Catnep Giant-Hyssop.   Fig. 3590.

Hyssopus nepetoides L. Sp. Pl. 569.   1753.
Lophanthus nepetoides Benth. Bot. Reg. under *pl. 1282.* 1829.
Vleckia nepetoides Raf. Fl. Tell. **3**: 89.   1836.
A. nepetoides Kuntze, Rev. Gen. Pl. 511.   1891.

Glabrous, or slightly puberulent; stem stout, 2°–5° high, branched, at least above, sharply 4-angled. Leaves ovate or ovate-oblong, acuminate or acute at the apex, rounded, cordate or the upper narrowed at the base, mostly thin, coarsely dentate, 2′–6′ long; spikes 3′–18′ long, usually very dense; petioles of the lowest leaves often 2′ long; bracts ovate, acute or acuminate; calyx-teeth oblong or ovate-oblong, obtuse or sub-acute, sometimes purplish; corolla greenish yellow, about 4″ long, scarcely exceeding the calyx.

In woods and thickets, Quebec and Vermont to South Dakota, south to Georgia, Tennessee, Kansas and Arkansas. July–Sept.

3. Agastache anethiodòra (Nutt.) Britton. Fragrant Giant-Hyssop. Fig. 3592.

Hyssopus anethiodorus Nutt. Fras. Cat. 1813.
Hyssopus anisatus Nutt. Gen. 2: 27. 1818.
Lopanthus anisatus Benth. Bot. Reg. under *pl. 1282.* 1829.
Vleckia anisata Raf. Fl. Tell. 3: 89. 1836.
Vleckia anethiodora Greene, Mem. Torr. Club 5: 282. 1894.
A. anethiodora Britton in Britt. & Brown, Ill. Fl. 3: 85. 1898.

Glabrous, or minutely puberulent; stem rather slender, leafy, usually branched, 2°–4° high. Leaves ovate or triangular-ovate, firm, mostly short-petioled, acute or acuminate at the apex, truncate, obtuse or sometimes subcordate at the base, sharply serrate, green above, pale and minutely canescent beneath, 2′–3′ long, anise-scented; spikes dense or interrupted, seldom 6′ long; bracts broadly ovate, abruptly acuminate; calyx-teeth ovate to lanceolate, acute, purple; corolla blue, 4″–5″ long, somewhat exceeding the calyx.

On prairies and plains, Minnesota to Manitoba, Alberta, Illinois, Nebraska and Colorado. July–Sept. Anise-hyssop.

2. Agastache scrophulariaefòlia (Willd.) Kuntze. Figwort Giant-Hyssop. Fig. 3591.

Hyssopus scrophulariaefolius Willd. Sp. Pl. 3: 48. 1801.
Lophanthus scrophulariaefolius Benth. Bot. Reg. under *pl. 1282.* 1829.
Vleckia scrophulariaefolia Raf. Fl. Tell. 3: 89. 1836.
Agastache scrophulariaefolia Kuntze, Rev. Gen. Pl. 511. 1891.

Similar to the preceding species, but commonly taller, strong-scented, the obtusely 4-angled stem, the petioles and lower surfaces of the leaves more or less pubescent, sometimes villous. Leaves nearly identical with those of *V. nepetoides* in size and outline; spike sometimes interrupted, 3′–18′ long; bracts broadly ovate, abruptly acuminate; calyx-teeth lanceolate or ovate-lanceolate, very acute or sometimes acuminate, whitish or purplish; corolla purplish, 5″–6″ long, considerably exceeding the calyx.

In woods and thickets, New Hampshire to Ontario, Wisconsin, North Carolina, Kentucky and Missouri. July–Oct.

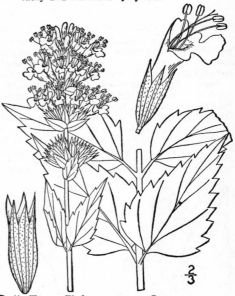

9. MEEHÁNIA Britton, Bull. Torr. Club 21: 32. 1894.

A low pubescent spreading or decumbent herb, with long-petioled cordate leaves, trailing leafy stolons, and large blue flowers in terminal secund bracted spikes. Calyx campanulate, 15-nerved, slightly 2-lipped, its teeth all lanceolate, acute, the 3 upper longer than the 2 lower. Corolla much exserted, puberulent without, pubescent within, the tube narrow at the base, gradually widely ampliate into the throat, the limb 2-lipped; upper lip 2-lobed, arched, the lobes ovate, obtuse; lower lip about equalling the upper, spreading, 3-lobed, the middle lobe emarginate, broader than the lateral ones. Stamens 4, didynamous, all anther-bearing, included, ascending under the upper lip, the upper pair longer than the lower; anthers 2-celled, the sacs nearly parallel. Ovary deeply 4-lobed; style equally 2-cleft at the summit. Nutlets oblong, smooth. [Named for Thomas Meehan, of Philadelphia, 1826–1901, botanist and horticulturist.]

A monotypic genus of eastern North America.

1. Meehania cordàta (Nutt.) Britton. Meehania. Fig. 3593.

Dracocephalum cordatum Nutt. Gen. **2**: 35. 1818.
Cedronella cordata Benth. Lab. 502. 1834.
Meehania cordata Britton, Bull. Torr. Club **21**: 33.
pl. 173. 1894.

Flowering stems ascending, 3′–8′ high; stolons very slender, leafy throughout, sometimes 2° long. Leaves all broadly ovate or ovate-orbicular, thin, obtuse or subacute at the apex, crenate all around, cordate at the base, sparingly pubescent with scattered hairs on both surfaces, or nearly glabrous beneath, green on both sides, 1′–2′ long, the basal sinus broad; spikes 1′–4′ long; bracts ovate or oblong, acute, membranous, the lower sometimes crenulate and surpassing the calyx; bractlets small, lanceolate; calyx about 5″ long, puberulent, its longer teeth about one-half the length of the tube; corolla 1′–1¼′ long, showy.

In rich moist woods and thickets, southwestern Pennsylvania to Illinois, Tennessee and North Carolina. May–July.

$\frac{1}{2}$

10. NÉPETA [Rivin.] L. Sp. Pl. 570. 1753.

Herbs, with dentate or incised leaves, and mostly white or blue rather small flowers in verticillate clusters, usually crowded in terminal spikes, or axillary and cymose. Calyx tubular, somewhat oblique at the mouth, 15-nerved, usually incurved, 5-toothed, scarcely 2-lipped, but the upper teeth usually longer than the lower. Corolla-tube enlarged above, the limb strongly 2-lipped; upper lip erect, emarginate or 2-lobed; lower lip spreading, 3-lobed, the middle lobe larger than the lateral ones. Stamens 4, all anther bearing, didynamous, ascending under the upper lip, the lower pair the shorter; anthers 2-celled, the sacs divaricate. Ovary deeply 4-parted; style 2-cleft at the summit. Nutlets ovoid, compressed, smooth. [Ancient Latin name of catnep.]

About 150 species, natives of Europe and Asia, the following typical.

$\frac{3}{5}$

1. Nepeta Catària L. Catmint. Catnep. Nep. Fig. 3594.

Nepeta Cataria L. Sp. Pl. 570. 1753.

Perennial, densely canescent, pale green; stem rather stout, erect, branched, 2°–3° high, the branches straight, ascending. Leaves ovate to oblong, petioled, acute at the apex, coarsely crenate-dentate, mostly cordate at the base, 1′–3′ long, greener above than beneath; flower-clusters spiked at the ends of the stem and branches, the spikes 1′–5′ long; bracts small, foliaceous; bractlets subulate; calyx puberulent, its teeth subulate, the upper about one-half the length of the tube; corolla nearly white, or pale purple, dark-dotted, puberulent without, 5″–6″ long, its lobe a little longer than the calyx, the broad middle lobe of its lower lip crenulate.

In waste places, New Brunswick and Quebec to South Dakota, Oregon, South Carolina, Kansas and Utah; also in Cuba. Naturalized from Europe. Native also of Asia. July–Nov.

11. GLECÒMA L. Sp. Pl. 578. 1753.

Low diffuse creeping herbs, with long-petioled nearly orbicular or reniform crenate leaves, and rather large blue or violet flowers in small axillary verticillate clusters. Calyx oblong-tubular, 15-nerved, oblique at the throat, not 2-lipped, unequally 5-toothed. Corolla-tube exserted, enlarged above, the limb 2-lipped; upper lip erect, 2-lobed or emarginate; the lower lip spreading, 3-lobed, the middle lobe broad, emarginate, the side lobes small. Stamens 4, didynamous, all anther-bearing, ascending under the upper lip of the corolla, not exserted, the upper pair the longer; anther-sacs divergent. Ovary deeply 4-parted. Nutlets ovoid, smooth. [Greek name for thyme or pennyroyal.]

About 6 species of Europe and Asia, the following typical.

8

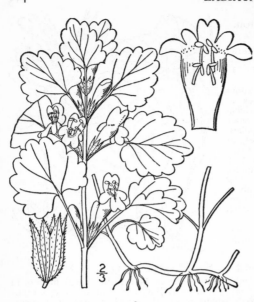

$\frac{2}{3}$

1. **Glecoma hederàcea** L.   Ground
   Ivy.   Gill-over-the-Ground.
   Field Balm.   Fig. 3595.

Glecoma hederacea L. Sp. Pl. 578. 1753.
N. Glechoma Benth. Lab. Gen. & Sp. 485.
   1834.
N. hederacea B.S.P. Prel. Cat. N. Y. 43.
   1888.

. Perennial, pubescent, the creeping stems
leafy, sometimes 18′ long, the branches
ascending.   Lower petioles commonly
longer than the leaves; leaves green
both sides, $\frac{1}{2}′-1\frac{1}{2}′$ in diameter; clusters
few-flowered, the flowers 7″–10″ long,
short-pedicelled;   bractlets   subulate,
shorter than the calyx; calyx puberulent,
its teeth acute or lanceolate-acuminate,
about one-third as long as the tube;
corolla-tube 2–3 times as long as the
calyx; upper pair of stamens much
longer than the lower.

   In waste places, woods and thickets, Newfoundland to Ontario, Minnesota, Oregon,
Georgia, Tennessee, Kansas and Colorado.
Old names, alehoof, cat's-foot, gill, gill-ale,
gill-go-by-the-ground, hayhofe, haymaids,
hove, tunhoof, creeping charlie, robin-run-
away, gill-run-over, crow-vituals, wild
snake-root, hedge-maids. March–May.

12. **MOLDÁVICA** [Tourn.] Adans. Fam. Pl. **2**: 190. 1763.

   Herbs, with dentate entire or incised leaves, and blue or purple flowers in axillary and
terminal bracted clusters, the bracts pectinate in our species.   Calyx tubular, 15-nerved,
straight or incurved, 5-toothed, the upper tooth much larger than the others, or 2-lipped with
the 3 upper teeth more or less united.   Corolla expanded above, its limb 2-lipped; upper lip
erect, emarginate; lower lip spreading, 3-lobed, the middle lobe larger than the lateral ones,
sometimes 2-cleft.   Stamens 4, didynamous, ascending under the upper lip, the upper pair
longer than the lower; anthers 2-celled, the sacs divaricate; style 2-cleft at the summit; ovary
deeply 4-parted.   Nutlets ovoid, smooth.   [From Moldavia.]

   About 35 species, natives of the northern hemisphere.   Only the following are known in North
America.   Type species: *Dracocephalum Moldávica* L.

Corolla 2–3 times as long as the calyx; clusters mostly axillary.   1. *M. parviflora.*
Corolla scarcely exceeding the calyx; clusters mostly terminal, dense.   2. *M. Moldavica.*

1. **Moldavica parviflòra** (Nutt.) Britton.   American Dragon-head.
   Fig. 3596.

Dracocephalum parviflorum Nutt. Gen. **2**: 35.
   1818.

   Annual or biennial, somewhat pubescent,
or glabrous; stem rather stout, usually
branched, 6′–2½° high.   Leaves lanceolate,
ovate, or oblong, slender-petioled, serrate,
or the lower incised, acute or obtuse at the
apex, rounded or narrowed at the base,
thin, 1′–3′ long; clusters dense, many-flowered, crowded in dense terminal spikes,
and sometimes also in the upper axils;
bracts ovate to oblong, pectinate with awn-
pointed teeth, shorter than or equalling the
calyx; pedicels 1″–2″ long; upper tooth of
the calyx ovate-oblong, longer than the
narrower lower and lateral ones, all acuminate; corolla light blue, scarcely longer
than the calyx.

   In dry gravelly or rocky soil, Quebec and
Ontario to Alaska, New York, Iowa, Missouri
and Arizona.   May–Aug.

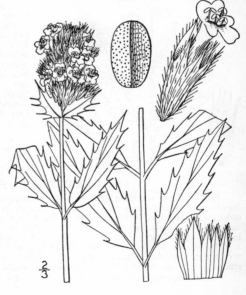

$\frac{2}{3}$

2. Moldavica Moldávica (L.) Britton. Moldavian Dragon-head or Balm. Fig. 3597.

Dracocephalum Moldavica L. Sp. Pl. 595. 1753.

Annual, puberulent; stem erect, usu-ally widely branched, 1°–$2\frac{1}{2}^\circ$ high. Leaves oblong or linear-oblong, dentate or some-what incised, obtuse at the apex, usually narrowed at the base, $1'$–$2'$ long, $2''$–$6''$ wide; clusters loose, few-flowered, com-monly numerous, mostly axillary; bracts narrowly oblong, usually shorter than the calyx, deeply pectinate with aristate teeth; pedicels $2''$–$4''$ long; calyx slightly curved, the 2 lower teeth somewhat shorter than the 3 broader equal upper ones; corolla 2–3 times as long as the calyx.

In a cañon near Spring View, Nebr. Also in northern Mexico. Introduced from central Europe. June–Aug.

13. PRUNÉLLA L. Pl. 600. 1753.

Perennial simple or sometimes branched herbs, with petioled leaves, and rather small clustered purple or white flowers, in terminal and sometimes also axillary, dense bracted spikes or heads. Calyx oblong, reticulate-veined, about 10-nerved, deeply 2-lipped, closed in fruit; upper lip nearly truncate, or with 3 short teeth; lower lip 2-cleft, its teeth lanceolate. Corolla-tube inflated, slightly narrowed at the mouth, its limb strongly 2-lipped; upper lip entire, arched; lower lip spreading, 3-lobed. Stamens 4, didynamous, ascending under the upper lip of the corolla, the lower pair the longer; filaments of the longer stamens 2-toothed at the summit, one of the teeth bearing the anther, the other sterile; anthers 2-celled, the sacs divergent or divaricate. Ovary deeply 4-parted. Nutlets ovoid, smooth. [Origin of name doubtful; often spelled *Brunella,* the pre-Linnaean form.]

About 5 species, of wide geographic distribution. Only the following typical one occurs in North America.

1. Prunella vulgàris L. Self-heal. Heal-all. Dragon-head. Fig. 3598.

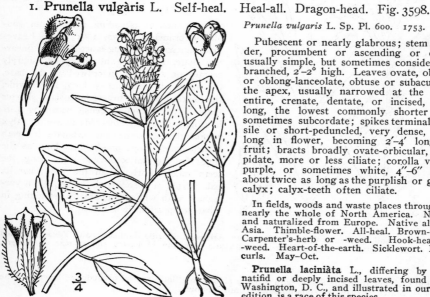

Prunella vulgaris L. Sp. Pl. 600. 1753.

Pubescent or nearly glabrous; stem slen-der, procumbent or ascending or erect, usually simple, but sometimes considerably branched, $2'$–2° high. Leaves ovate, oblong or oblong-lanceolate, obtuse or subacute at the apex, usually narrowed at the base, entire, crenate, dentate, or incised, $1'$–$4'$ long, the lowest commonly shorter and sometimes subcordate; spikes terminal, ses-sile or short-peduncled, very dense, $\frac{1}{2}'$–$1'$ long in flower, becoming $2'$–$4'$ long in fruit; bracts broadly ovate-orbicular, cus-pidate, more or less ciliate; corolla violet, purple, or sometimes white, $4''$–$6''$ long, about twice as long as the purplish or green calyx; calyx-teeth often ciliate.

In fields, woods and waste places throughout nearly the whole of North America. Native and naturalized from Europe. Native also of Asia. Thimble-flower. All-heal. Brown-wort. Carpenter's-herb or -weed. Hook-heal or -weed. Heart-of-the-earth. Sicklewort. Blue-curls. May–Oct.

Prunella laciniàta L., differing by pin-natifid or deeply incised leaves, found near Washington, D. C., and illustrated in our first edition, is a race of this species.

14. DRACOCÉPHALUM [Tourn.] L. Sp. Pl. 594. 1753.

[PHYSOSTEGIA Benth. Lab. Gen. & Sp. 504. 1834.]

Erect perennial glabrous or puberulent herbs, with serrate, dentate or entire leaves, and large or middle-sized, bracted purple violet, pink or white flowers in terminal spikes, or spike-like racemes. Calyx campanulate or oblong, membranous, swollen and remaining open in fruit, faintly reticulate-veined and 10-nerved, equally 5-toothed. Corolla much longer than the calyx, its tube gradually much enlarged upward, its limb strongly 2-lipped; upper lip concave, rounded, nearly or quite entire; lower lip spreading, 3-lobed, the middle lobe commonly emarginate. Stamens 4, didynamous, ascending under the upper lip of the corolla, the lower pair the longer; filaments pubescent; anthers all alike, 2-celled, the sacs nearly parallel, the margins of their valves commonly spinulose or denticulate. Ovary 4-parted. Nutlets ovoid-triquetrous, smooth. [Greek, dragon-head.]

About 7 species, natives of North America, known as False Dragon-head or Lion's-heart. Type species: *Dracocephalum virginianum* L.

Flowers 9″ long, or more; leaves firm.
    Spike dense, many-flowered.   1. *D. virginianum.*
    Spike loose, few-flowered; leaf-serrations mostly blunt.   2. *D. denticulatum.*
Flowers 5″–7″ long; leaves thin.
    Spike loose; 4′–8′ long; leaves few and distant.   3. *D. intermedium.*
    Spike dense, 1′–4′ long; stem leafy.   4. *D. Nuttallii.*

1. Dracocephalum virginiànum L. Dragon-head. Obedient Plant. Lion's Heart. Fig. 3599.

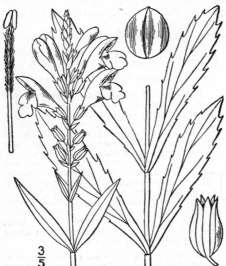

Dracocephalum virginianum L. Sp. Pl. 594. 1753.

Dracocephalum speciosum Sweet, Brit. Fl. Gard. *pl. 93.* 1825.

Physostegia virginiana Benth. Lab. Gen. & Sp. 504. 1834.

Stem erect or ascending, rather stout, simple or branched above, 1°–4° tall. Leaves firm, lanceolate, oblong-lanceolate, or linear-lanceolate, acuminate at the apex, sharply serrate or serrulate, narrowed at the base, the upper all sessile, 2′–5′ long, 2″–7″ wide, the lowest petioled; bracts lanceolate, shorter than the calyx; spikes dense, becoming 4′–8′ long in fruit, many-flowered; flowering calyx campanulate or somewhat turbinate, its teeth ovate, acute, about one-half as long as the tube; fruiting calyx oblong, 4″–5″ long, the teeth much shorter than the tube; corolla pale purple or rose, about 1′ long, often variegated with white, temporarily remaining in whatever position it is placed.

In moist soil, Quebec to Ontario, Minnesota, Arkansas, Florida, Louisiana and Texas; escaped from gardens eastward. Races differ in width and serration of leaves and in size of flowers. July–Sept.

Physostegia Digitàlis Small, with broadly oblong to elliptic repand or undulate leaves, of the Southern States, perhaps extends northward into Missouri.

2. Dracocephalum denticulàtum Ait. Few-flowered Lion's Heart. Fig. 3600.

Prasium purpureum Walt. Fl. Car. 166. 1788?
Drac. denticulatum Ait. Hort. Kew. **2**: 317. 1789.
P. virginiana var. *denticulata* A. Gray, Syn. Fl. **2**[1]: 383. 1878.
P. denticulata Britton, Mem. Torr. Club **5**: 284. 1894.

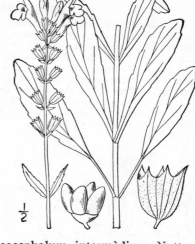

Stem slender, ascending or erect, simple, or little branched, 1°–2° high. Leaves firm or rather thin, oblong, linear-oblong, or oblanceolate, obtuse or acute at the apex, narrowed at the base, crenulate, obtusely dentate, or entire, 1′–3′ long, 2″–6″ wide, the upper sessile, the lower slender-petioled; spike loosely few–several-flowered; bracts lanceolate, little longer than the fruiting pedicels; flowering calyx oval-campanulate, its teeth acute, about one-third as long as the tube; fruiting calyx oblong, 3″–4″ long; corolla rose-pink, nearly or quite 1′ long.

In moist soil, Pennsylvania to Illinois, Florida and Texas. June–Aug.

3. Dracocephalum intermèdium Nutt. Slender Lion's Heart. Fig. 3601.

Dracocephalum intermedium Nutt. Trans. Am. Phil. Soc. (II.) **5**: 187. 1833–37.
Physostegia intermedia A. Gray, Proc. Am. Acad. **8**: 371. 1872.

Stem very slender, usually quite simple, 1°–3° high. Leaves usually few pairs, remote, thin, mostly shorter than the internodes, narrowly lanceolate or linear, acute or acuminate at the apex, repand-denticulate, little narrowed at the base, all sessile, or the lowest petioled, 2′–3′ long, 2″–4″ wide; spikes very slender, remotely many-flowered, 4′–8′ long in fruit; lower bracts often nearly as long as the campanulate calyx; calyx-teeth acute, shorter than the tube; fruiting calyx broadly oval, 2″–2½″ long; corolla much dilated above, 5″–7″ long.

On prairies, western Kentucky to Missouri, Louisiana, Arkansas and Texas. May–July.

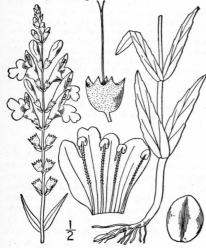

4. Dracocephalum Nuttallii Britton. Purple or Western Lion's Heart. Fig. 3602.

Physostegia parviflora Nutt.; Benth. in DC. Prodr. **12**: 434. As synonym. 1848. A. Gray, Proc. Am. Acad. **8**: 371. Not *Dracocephalum · parviflorum* Nutt.

Stem rather stout, usually simple, 1°–3° high. Leaves lanceolate, oblong-lanceolate, or ovate-lanceolate, acute, acuminate, or the lower obtuse at the apex, sharply serrate or dentate, somewhat narrowed at the base, all sessile or the lowest petioled, thin, 3′–4′ long, 3″–10″ wide; spikes densely several–many-flowered, 1′–4′ long; bracts ovate or ovate-lanceolate, acute, shorter than the calyx; flowering calyx campanulate, its teeth ovate, obtuse or subacute, about one-third as long as the tube; fruiting calyx globose-oblong, 2″–3″ long; corolla purple, 5″–7″ long.

In moist soil, Wisconsin and Minnesota to Nebraska, North Dakota, Saskatchewan, British Columbia and Oregon. June–Aug.

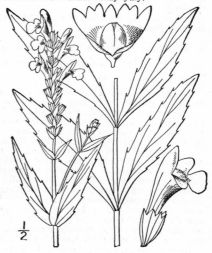

15. SYNÁNDRA Nutt. Gen. 2: 29. 1818.

An annual or biennial, somewhat hirsute, simple or little branched herb, with long-petioled ovate cordate crenate leaves, and large whitish flowers in a terminal leafy-bracted spike. Calyx campanulate-oblong, membranous, deeply 4-cleft, inflated in fruit, faintly and irregularly veined, the lobes narrowly lanceolate, the two upper shorter than the lower. Corolla much longer than the calyx, its tube narrow below, much expanded above, 2-lipped; upper lip concave, entire; lower lip spreading, 3-lobed. Stamens 4, didynamous, ascending under the upper lip of the corolla; filaments villous; anthers glabrous, 2-celled, the sacs divaricate, the contiguous ones of the upper pair of stamens sterile and connate. Ovary deeply 4-lobed; style unequally 2-cleft at the summit. Nutlets ovoid, smooth, sharply angled. [Greek, stamens-together.]

A monotypic genus of southeastern North America.

1. Synandra hispídula (Michx.) Britton.
Synandra. Fig. 3603.

Lamium hispidulum Michx. Fl. Bor. Am. 2: 4. 1803.
Synandra grandiflora Nutt. Gen. 2: 29. 1818.
Torreya grandiflora Raf. Am. Month. Mag. 3: 356. 1818.
S. hispidula Britton, Mem. Torr. Club 5: 285. 1894.

Stem rather slender, erect or ascending, weak, 1°–2½° long, striate. Leaves thin, the lower and basal ones broadly ovate, or nearly orbicular, palmately veined, acute or obtuse at the apex, deeply cordate at the base, the blade 2′–4′ long, and commonly shorter than the petiole; floral leaves sessile, ovate or ovate-lanceolate, acute or acuminate, the flowers solitary in their axils, the uppermost leaves very small; calyx hirsute, its lobes about as long as the tube; corolla 1′–1½′ long, showy, the lower lip with purple lines.

Along streams and in wet woods, Ohio to Illinois, Virginia and Tennessee. Ascends to 3,500 ft. in Virginia. May–June.

16. PHLÒMIS [Tourn.] L. Sp. Pl. 584. 1753.

Tall perennial herbs, or shrubs. Calyx tubular or tubular-campanulate, 5–10-nerved, the limb mostly equally 5-toothed. Corolla-tube usually with a woolly ring within, shorter than or exceeding the calyx, the limb strongly 2-lipped; upper lip erect, concave, arched or sometimes keeled, entire or emarginate; lower lip spreading, 3-cleft. Stamens 4, didynamous, ascending under the upper lip of the corolla, the anterior pair the longer and their filaments with hooked appendages at the base; anther-sacs divergent. Ovary deeply 4-lobed; style subulate, 2-cleft at the summit, one of the lobes smaller than the other. Nutlets ovoid, glabrous, or pubescent above. [Greek, mullen, in allusion to the thick woolly leaves of some species.]

About 50 species, natives of the Old World. Type species: *Phlomis fruticosa* L.

1. Phlomis tuberòsa L. Jerusalem Sage.
Sage-leaf Mullen. Fig. 3604.

Phlomis tuberosa L. Sp. Pl. 586. 1753.

Herbaceous from a thickened root; stem stout, purplish, glabrous or loosely pubescent above, usually much branched, 3°–6° tall, the branches nearly erect. Lower leaves triangular-ovate, long-petioled, acuminate or acute at the apex, coarsely dentate or incised-dentate, rather thick, deeply cordate at the base, strongly veined, 5′–10′ long, 3′–6′ wide; upper leaves lanceolate, short-petioled or sessile, truncate or sometimes narrowed at the base, the uppermost (floral) very small; clusters densely many-flowered; bractlets subulate, ciliate-hirsute or nearly glabrous; calyx 5″–6″ long, its teeth setaceous with a broader base, spreading; corolla 10″–12″ long, pale purple or white, twice as long as the calyx, densely pubescent, and the margins of its upper lip fringed with long hairs.

In waste places, south shore of Lake Ontario. Naturalized from southern Europe. June–Sept.

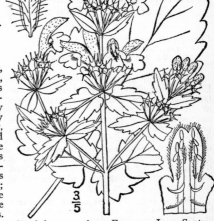

17. GALEÓPSIS L. Sp. Pl. 579. 1753.

Erect annual branching herbs, with broad or narrow leaves, and rather small yellow red purple or mottled verticillate-clustered flowers in the upper axils, or forming terminal dense or interrupted spikes. Calyx campanulate or tubular-campanulate, 5-10-nerved, 5-toothed, the teeth nearly equal, spinulose. Corolla-tube narrow, the throat expanded, the limb strongly 2-lipped; upper lip erect, concave, entire; lower lip spreading, 3-cleft, the middle lobe obcordate or emarginate. Stamens 4, didynamous, ascending under the upper lip of the corolla, the anterior pair the longer; anthers 2-celled, the sacs transversely 2-valved, the inner valve ciliate, the outer smooth, larger. Ovary deeply 4-parted; style 2-cleft at the summit. Nutlets ovoid, slightly flattened, smooth. [Greek, weasel-like.]

About 6 species, natives of Europe and Asia. Type species: *Galeopsis Tetrahit* L.

Plant puberulent; leaves linear to ovate-lanceolate.   1. *G. Ladanum.*
Plant hispid; leaves ovate.   2. *G. Tetrahit.*

1. Galeopsis Ládanum L. Red Hemp-Nettle. Ironwort. Fig. 3605.

Galeopsis Ladanum L. Sp. Pl. 579. 1753.

Puberulent, sometimes glandular above; stem slender, erect, much branched, 6'–18' high, the branches ascending. Leaves linear, ovate-oblong, or ovate-lanceolate, mostly short-petioled, acute at both ends, 'dentate, or nearly entire, 1'–2' long, 2''–8'' wide; flower-clusters mainly axillary, distant; calyx-teeth subulate-lanceolate, shorter than or equalling the tube; corolla 6''–8'' long, red or purple to white and variegated, twice the length of the calyx.

In waste places and on ballast, New Brunswick to Michigan, New Jersey and Indiana. Adventive or naturalized from Europe. Native also of Asia. Dog-nettle. July–Oct.

2. Galeopsis Tetràhit L. Hemp-Nettle. Hemp Dead Nettle. Ironwort. Fig. 3606.

Galeopsis Tetrahit L. Sp. Pl. 579. 1753.

A coarse and rough-hairy herb, the stem rather stout, branched, 1°–3° high, swollen under the joints. Leaves ovate, membranous, slender-petioled, acuminate at the apex, rounded or narrowed at the base, coarsely dentate, 2'–5' long, ½'–2½' wide; flower-clusters axillary, dense, or in a short leafy-bracted spike; calyx-teeth needle-pointed, bristly, as long as or longer than the tube; corolla 8''–12'' long, pink or pale purple variegated with white, about twice the length of the calyx.

In waste places, Newfoundland to British Columbia and Alaska, south to North Carolina, West Virginia and Michigan. Naturalized from Europe. Native also of Asia. Bee-, dog- or blind-nettle. Stinging or flowering nettle. Nettle-, wild- or bastard-hemp. Simon's-weed. June–Sept.

18. LEONÙRUS L. Pl. 584. 1753.

Tall erect herbs, with palmately cleft, parted or dentate leaves, and small white or pink flowers verticillate in dense axillary clusters. Calyx tubular-campanulate, 5-nerved, nearly regular and equally 5-toothed, the teeth rigid, subulate or aristate. Tube of the corolla included or slightly exserted, its limb 2-lipped; upper lip erect, concave or nearly flat, entire; lower lip spreading or deflexed, 3-lobed, the middle lobe broad, obcordate or emarginate. Stamens 4, didynamous, the anterior pair the longer, ascending under the upper lip of the corolla; anthers 2-celled, the sacs mostly parallel. Ovary deeply 4-parted; style 2-cleft at the summit. Nutlets 3-sided, smooth. [Greek, lion's-tail.]

About 10 species, natives of Europe and Asia. Type species: *Leonurus Cardiaca* L.

Lower leaves palmately 2–5-cleft, the upper 3-cleft.   1. *L. Cardiaca.*
Leaves deeply 3-parted, the segments cleft and incised.   2. *L. sibiricus.*
Leaves coarsely dentate or incised-dentate.   3. *L. Marrubiastrum.*

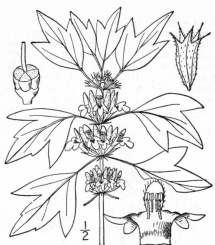

1. **Leonurus Cardìaca** L.   Motherwort.
Fig. 3607.

Leonurus Cardiaca L. Sp. Pl. 584. 1753.

Perennial, puberulent; stem rather stout, strict, commonly branched, 2°–5° tall, the branches straight and ascending.   Leaves membranous, slender-petioled, the lower nearly orbicular, palmately 3–5-cleft, 2′–4′ broad, the lobes acuminate, incised or dentate; upper (floral) leaves narrower, oblong-lanceolate or rhombic, 3-cleft, or the uppermost merely 3-toothed; flower-clusters numerous, exceeded by the petioles; calyx-teeth lanceolate, subulate, somewhat spreading, nearly as long as the tube; corolla pink, purple or white, 3″–5″ long, its tube with an oblique ring of hairs within, its upper lip slightly concave, densely white-woolly without, the lower lip mottled; anther-sacs parallel.

In waste places, Nova Scotia to North Carolina, South Dakota, Montana, Kansas and Utah.   Naturalized from Europe.   Native also of Asia.   Also called cowthwort.   Lion's-ear.   June–Sept.

2. **Leonurus sibíricus** L.   Siberian Mother-
wort or Lion's-tail.   Fig. 3608.

Leonurus sibiricus L. Sp. Pl. 584. 1753.

Biennial, puberulent or glabrate; stem stout, branched, 2°–6° high, the branches slender.   Leaves long-petioled, deeply 3-parted into ovate or lanceolate, more or less cuneate, acute or acuminate deeply cleft and incised segments, the lobes lanceolate or linear, acute; lower leaves sometimes 6′ wide, the uppermost linear or lanceolate, slightly toothed or entire; clusters numerous, dense, usually all axillary; calyx campanulate, 3″ long, glabrous or minutely puberulent, its bristle-shaped teeth slightly spreading, shorter than the tube; corolla purple or red, densely puberulent without, 4″–6″ long, its tube naked within, the upper lip arched; anther-sacs divergent.

In waste and cultivated soil, southern Pennsylvania and Delaware.   Bermuda.   Naturalized from eastern Asia.   Widely distributed in tropical America as a weed.   May–Sept.

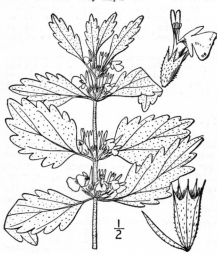

3. **Leonurus Marrubiástrum** L.   Hoar-
hound Motherwort or Lion's-tail.
Fig. 3609.

Leonurus Marrubiastrum L. Sp. Pl. 584. 1753.

Biennial, puberulent or pubescent; stem stout, branched, 2°–5° high.   Leaves petioled, ovate or ovate-oblong, acute or obtuse at the apex, narrowed at the base, coarsely crenate or incised-dentate, 1′–3′ long, ½′–1½′ wide, the upper narrower; flower-clusters dense, numerous, axillary; calyx finely puberulent or glabrate, its bristle-shaped somewhat spreading teeth mostly shorter than the tube; corolla nearly white, glabrate, about 4″ long, its tube scarcely exceeding the calyx, naked within, its lower lip ascending.

In waste places, southern Pennsylvania and Delaware.   Recorded from New Jersey.   Naturalized from Europe.   Native also of Asia.   June–Sept.

19. LÀMIUM [Tourn.] L. Sp. Pl. 579. 1753.

Annual or perennial mostly diffuse herbs, with crenate dentate or incised, usually cordate leaves, and rather small flowers, verticillate in axillary and terminal clusters. Calyx tubular-campanulate, about 5-nerved, 5-toothed, the teeth sharp, equal or the upper ones longer. Tube of the corolla mostly longer than the calyx, dilated above, its limb 2-lipped; upper lip concave, erect, usually entire, narrowed at the base; lower lip spreading, 3-cleft, the middle lobe emarginate, contracted at the base, the lateral ones sometimes each with a tooth-like appendage. Stamens 4, didynamous, ascending under the upper lip of the corolla, the anterior pair the longer; anthers 2-celled, the sacs divaricate, often hirsute on the back. Ovary deeply 4-parted; style 2-cleft at the summit. Nutlets smooth or tuberculate. [Greek, throat, from the ringent corolla.]

About 40 species, natives of the Old World, known as Dead-Nettle or Hedge Dead-Nettle. Type species: *Lamium purpureum* L.

Upper leaves sessile or clasping.	1. *L. amplexicaule.*
Leaves all petioled.	
Flowers red or purple.	
Corolla 6″–9″ long; leaves not blotched.	2. *L. purpureum.*
Corolla 10″–12″ long; leaves commonly blotched.	3. *L. maculatum.*
Flowers white.	4. *L. album.*

1. Lamium amplexicaùle L. Henbit. Greater Henbit. Henbit Dead Nettle. Fig. 3610.

Lamium amplexicaule L. Sp. Pl. 579. 1753.

Biennial or annual, sparingly pubescent; stems branched from the base or also from the lower axils, slender, ascending or decumbent, 6′–18′ long. Leaves orbicular or nearly so, coarsely crenate, ½′–1½′ wide, rounded at the apex, the lower slender-petioled, mostly cordate, the upper sessile and more or less clasping; flowers rather few in axillary and terminal clusters; calyx pubescent, its teeth erect, nearly as long as the tube; corolla purplish or red, 6″–8″ long, its tube very slender, the lateral lobes of its lower lip very small, the middle one spotted; upper lip somewhat pubescent; flowers sometimes cleistogamous.

In waste and cultivated ground, New Brunswick to Ontario, Minnesota, British Columbia, Florida, Arkansas and California and in Bermuda and Jamaica. Naturalized from Europe. Native also of Asia. Feb.–Oct.

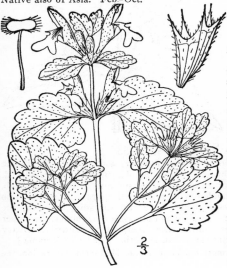

2. Lamium purpùreum L. Red Dead Nettle. Fig. 3611.

Lamium purpureum L. Sp. Pl. 579. 1753.

Annual, slightly pubescent, branched from the base and sometimes also above; stems stout or slender, decumbent, 6′–18′ long. Leaves crenate or crenulate, the lower orbicular or broadly ovate, slender-petioled, rounded at the apex, cordate at the base, the upper ovate, short-petioled, sometimes acute at the apex, ½′–1¼′ long; flowers in axillary and terminal clusters; calyx teeth narrowly lanceolate, acuminate, slightly longer than the tube, spreading, at least in fruit; corolla purple-red, rarely exceeding ½′ long, its tube rather stout, the lateral lobes of its lower lip reduced to 1 or 2 short teeth, its middle lobe spotted; upper lip very pubescent.

In waste and cultivated soil, Newfoundland to Pennsylvania, North Carolina and Missouri. Also in ballast about the northern seaports. Naturalized or adventive from Europe. Native also of Asia. Old names, red or sweet archangel; day-, dog-, french- or deaf-nettle. Rabbit-meat. April–Oct.

Lamium hybridum Vill., occasionally found in waste and cultivated grounds, introduced from Europe, differs by its more deeply and incisely toothed leaves.

3. Lamium maculàtum L.   Spotted Dead Nettle.   Variegated Dead Nettle.   Fig. 3612.

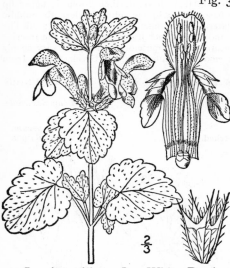

L. maculatum L. Sp. Pl. Ed. 2, 809. 1763.

Perennial, somewhat pubescent; stems mostly slender, commonly branched, decumbent or ascending, 8'–1½° long. Leaves crenate or incised-crenate, all petioled, usually longitudinally blotched along the midrib, broadly ovate or triangular-ovate, acute or obtuse, truncate or cordate at the base, 1'–2' long, or some of the lower ones much smaller and nearly orbicular; clusters few-flowered, mainly axillary; calyx-teeth lanceolate-subulate, as long as or longer than the tube, spreading; corolla 10"–12" long, purple-red, its tube short, contracted near the base, with a transverse ring of hairs within, the lateral lobes of its lower lip very small.

Along roadsides, escaped from gardens, Maine and Vermont to Virginia. Native of Europe and Asia. May–Oct.

4. Lamium álbum L.   White Dead Nettle.   Fig. 3613.

Lamium album L. Sp. Pl. 579. 1753.

Perennial, pubescent; stems decumbent or ascending, rather stout, simple or branched, 1°–1½° long. Leaves ovate, crenate, dentate or incised, all petioled, acute or acuminate at the apex, cordate or truncate at the base, 1'–3' long, or the lower shorter and obtuse; clusters mostly axillary; calyx-teeth very slender, subulate, spreading, usually longer than the tube; corolla white, about 1' long, its tube short, stout, contracted near the base, with an oblique ring of hairs within, the lateral lobes of its lower lip each with a slender tooth.

In waste places, Ontario to Massachusetts and Virginia. Also in ballast about the northern seaports. Naturalized or adventive from Europe. Old names, white archangel; day-, blind-, dumb- or bee-nettle; snake-flower, suck-bottle. April–Oct.

20. BALLÒTA L. Sp. Pl. 582. 1753.

Perennial pubescent or tomentose herbs, some species shrubby, with dentate or crenate leaves, and small bracted flowers in axillary clusters. Calyx tubular-funnelform, 10-nerved, 5–10-toothed, the teeth dilated at the base, or sometimes connate into a spreading limb. Corolla-tube about as long as the calyx, provided with a ring of hairs within, the limb strongly 2-lipped; upper lip erect, concave, emarginate, lower lip spreading, 3-lobed, the middle lobe emarginate or obcordate. Stamens 4, didynamous, ascending under the upper lip of the corolla, the anterior pair the longer; anther-sacs divergent at maturity. Ovary deeply 4-lobed; style 2-cleft at the summit. Nutlets ovoid, smooth. [The Greek name.]

About 30 species, natives of the Old World, most numerous in the Mediterranean region, the following typical.

1. Ballota nìgra L.   Black or Fetid Hoarhound.
Fig. 3614.

Ballota nigra L. Sp. Pl. 582.   1753.

Herbaceous, puberulent or pubescent, ill-scented; stem usually branched, erect, $1\frac{1}{2}°$–$3°$ high, its hairs mostly reflexed. Leaves slender-petioled, ovate, or the lower nearly orbicular, acute or obtuse at the apex, coarsely dentate, thin, narrowed, truncate or subcordate at the base, $1'$–$2'$ long; clusters numerous, several-flowered, dense; bractlets subulate, somewhat shorter than the calyx; calyx about $4''$ long, its teeth lanceolate, sharp, bristle-pointed, spreading in fruit; corolla $6''$–$9''$ long, reddish-purple to whitish, its upper lip pubescent on both sides; nutlets shining.

In waste places, eastern Massachusetts to Pennsylvania. Naturalized from Europe. June–Sept. Black archangel. Hairhound. Henbit. Bastard-hoarhound.

$\frac{1}{2}$

21. STÀCHYS [Tourn.] L. Sp. Pl. 580.   1753.

Annual or perennial glabrous pubescent or hirsute herbs, with small or rather large purple yellow red or white flowers, loosely verticillate-clustered in terminal dense or interrupted spikes, or also in the upper axils. Calyx mostly campanulate, 5–10-nerved, 5-toothed, the teeth nearly equal in our species. Corolla purple in our species, its tube narrow, not exceeding the calyx, the limb strongly 2-lipped; upper lip erect, concave, entire or emarginate; lower lip spreading, 3-cleft, the middle lobe broader than the lateral ones, sometimes 2-lobed. Stamens 4, didynamous, ascending under the upper lip of the corolla, the anterior pair the longer, sometimes deflexed or twisted after anthesis; anthers contiguous in pairs, 2-celled, the sacs mostly divergent. Ovary deeply 4-lobed; style 2-cleft at the summit into subulate lobes. Nutlets ovoid or oblong. [Greek, a spike, from the spicate inflorescence.]

About 160 species, of wide geographic distribution in the north temperate zone, a few in South America and South Africa. Besides the following, some 20 others occur in the southern and southwestern United States. Type species: *Stachys germanica* L.

A. Perennials.

1. Leaves, or some of them narrowed or acute at the base, linear to oblong or lanceolate.
    Stem glabrous or nearly so, sometimes sparingly hirsute on the angles; leaves nearly or quite entire.
        Leaves $1''$–$2\frac{1}{2}''$ long, linear.   1. *S. hyssopifolia*.
        Leaves $2''$–$5''$ wide, oblong to linear-oblong.   2. *S. atlantica*.
    Stem retrorsely hirsute; leaves serrate.
        Stem hirsute only on the angles; leaves slightly pubescent.   3. *S. ambigua*.
        Stem densely hirsute all over; leaves densely pubescent.   4. *S. arenicola*.

2. Leaves rounded, cordate or truncate at the base, oblong, ovate or lanceolate.
    Leaves all subsessile or short-petioled.
        Glabrous or very nearly so, the stem-angles sparsely bristly.   5. *S. latidens*.
        Stem retrorsely hirsute; leaves pubescent.
            Leaves lanceolate to oblong-lanceolate.   6. *S. palustris*.
            Leaves ovate to oblong-ovate.   7. *S. Nuttallii*.
    Leaves, at least the lower, slender-petioled.
        Stem glabrous, or nearly so.   8. *S. tenuifolia*.
        Stem retrorsely hirsute or villous.
            Petioles less than one-fourth as long as the blades.
                Pubescence hirsute.   9. *S. aspera*.
                 Pubescence short, dense.   10. *S. salvioides*.
            Petioles, at least those of lower leaves, about one-half as long as the deeply cordate blades.   11. *S. cordata*.

B. Annuals.

Plant low, hirsute.   12. *S. arvensis*.
Plant tall, densely villous.   13. *S. germanica*.

1. Stachys hyssopifòlia Michx. Hyssop Hedge Nettle. Fig. 3615.

S. palustris Walt. Fl. Car. 162. 1788. Not L. 1753.
S. hyssopifolia Michx. Fl. Bor. Am. 2 : 4. 1803.

Perennial, glabrous or very nearly so throughout, sometimes slightly hirsute at the nodes; stem slender, usually branched, erect or nearly so, $1°-1\frac{1}{2}°$ long. Leaves thin, linear, acute at both ends, or the uppermost rounded at the base, short-petioled or sessile, entire, or sparingly denticulate with low teeth, $1'-2'$ long, $1''-2\frac{1}{2}''$ wide, the uppermost reduced to short floral bracts; clusters few–several-flowered, forming an interrupted spike; calyx glabrous or slightly hirsute, $2''-3''$ long, its teeth lanceolate-subulate, nearly as long as the tube; corolla about $7''$ long, light purple, glabrous.

In fields and thickets, Massachusetts to Florida, Indiana, Michigan and Virginia. July–Sept.

2. Stachys atlántica Britton. Coast Hedge Nettle. Fig. 3616.

Stachys atlantica Britton, Man. 792. 1901.

Perennial, glabrous or with a few hairs at the nodes of the stem. weak, diffuse, $8'-16'$ long. Leaves thin, oblong or linear-oblong, obtuse or obtusish at the apex, narrowed, or the upper sometimes rounded at the base, remotely denticulate or entire, spreading, $2'$ long or less, $2''-5''$ wide; fruiting calyx glabrous, broadly campanulate, about $2\frac{1}{2}''$ long, its teeth triangular-ovate, acuminate, more than half as long as the tube; corolla purplish.

In wet meadows and marshes, Long Island to eastern Pennsylvania. Perhaps a wet-ground race of the preceding species. Aug.–Sept.

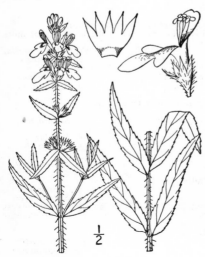

3. Stachys ambígua (A. Gray) Britton. Dense-flowered Hedge Nettle. Fig. 3617.

Stachys hyssopifolia var. *ambigua* A. Gray, Syn. Fl. 2 : Part 1, 387. 1878.
S. ambigua Britton, Mem. Torr. Club 5 : 285. 1894.

Perennial; stem slender, erect, retrorsely hispid, at least below, simple or sparingly branched, $1°-2°$ high. Leaves oblong, oblong-lanceolate, or linear, pubescent or glabrate, acuminate or acute at the apex, narrowed at the base, $2''-10''$ wide, $1'-3'$ long, serrulate; clusters in a terminal rather dense spike, and usually also in the upper axils; calyx more or less hirsute, $2\frac{1}{2}''$ long, its lanceolate-subulate teeth more than one-half as long as the tube; corolla nearly as in *S. hyssopifolia*.

In moist soil, Massachusetts to Pennsylvania, Georgia, Wisconsin and Kentucky. July–Aug.

4. Stachys arenícola Britton. Sand Hedge Nettle. Fig. 3618.

Stachys arenicola Britton, Man. 792. 1901.

Perennial, very densely pubescent, 1½°–3° high. Leaves lanceolate, narrowed at the base, sessile, finely crenate-dentate, acuminate, 2½′–4′ long, the upper much smaller; upper bracts little longer than the flowers; calyx-teeth subulate-acicular, about as long as the tube; corolla about 5″ long.

In sandy soil, southern New York to Illinois and Michigan. July–Sept.

5. Stachys látidens Small. Broad-toothed Hedge Nettle. Fig. 3619.

Stachys latidens Small; Britton, Man. 793. 1901.

Perennial; stem glabrous almost to the inflorescence, erect, 1°–2½° tall, mostly simple, slightly rough on the angles. Leaves thin, various, the lower oval or ovate to oblong, 1½′–4′ long, the upper longer, narrower, oblong-elliptic to lanceolate, acute or acuminate, all crenate-serrate with rather small teeth, rather abruptly narrowed or subcordate at the base, short-petioled; inflorescence closely pubescent, the clusters few; calyx minutely pubescent, often sessile, 2½″–3″ long, the tube campanulate, the teeth triangular, very short; corolla about 5″ long, purplish.

On mountain slopes and summits, Virginia to North Carolina and Tennessee. June–Aug.

6. Stachys palústris L. Hedge Nettle. Marsh or Clown's Woundwort. Fig. 3620.

Stachys palustris L. Sp. Pl. 580. 1753.

Perennial, hirsute or pubescent all over; stem erect, strict, simple or somewhat branched, commonly slender, and retrorse-hispid on the angles, 1°–4° high. Leaves firm, lanceolate, oblong, or oblong-lanceolate, sessile, or very short-petioled, acuminate or acute at the apex, truncate, cordate or subcordate at the base, 2′–5′ long, ½′–1′ wide, crenulate or dentate; flower-clusters forming an elongated interrupted spike, sometimes also in the upper axils; flowers 6–10 in a whorl; calyx pubescent, its subulate teeth more than one-half as long as the tube; corolla purplish to pale red, purple spotted, 6″–8″ long, its upper lip pubescent.

In moist soil, Newfoundland to Oregon, south to southern New York, Illinois, Michigan, and in the Rocky Mountains to New Mexico. Also in Europe and Asia. June–Sept. Old names, clown's-heal or all-heal. Cock-head. Dead nettle. Rough weed. June–Sept.

3/5

8. Stachys tenuifòlia Willd. Smooth Hedge Nettle. Fig. 3622.

Stachys tenuifolia Willd. Sp. Pl. 3 : 100. 1801.
S. glabra Ridd. Suppl. Cat. Ohio Pl. 16. 1836.
S. cincinnatensis Kuntze, Rev. Gen. Pl. 531.
1891.

Perennial; stem quite smooth, or slightly
scabrous on the angles, slender, erect or
ascending, usually branched, 1°–2½° high.
Leaves lanceolate, oblong, or ovate-lanceolate, slender-petioled, thin, acuminate at the
apex, obtuse or subcordate at the base,
sharply dentate or denticulate, dark green,
2′–5′ long, ½′–2′ wide; clusters several or
numerous in terminal spikes, or also in the
upper axils; calyx glabrous, or sparingly
hirsute, 2″ long, its teeth lanceolate, acute,
one-half as long as the tube or more; corolla about 6″–8″ long, pale red and purple.

    In moist fields and thickets, New York to Illinois, Kansas, North Carolina and Louisiana.
Ascends to 4000 ft. in North Carolina. June–Aug.

7. Stachys Nuttállii Shuttlw. Nuttall's Hedge Nettle. Fig. 3621.

Stachys Nuttallii Shuttlw.; DC. Prodr. **12**: 469. 1848.

Perennial, conspicuously hirsute, bright green; stem
stiff, erect, 1½°–3½° tall, simple. Leaves thinnish, oblong, oblong-lanceolate or ovate-lanceolate, acuminate,
serrate-dentate, rounded or truncate at the base, short-
petioled, 2′–4′ long; spike interrupted; bracts surpassing the calyx, the upper ones with 3 tooth-like lobes;
calyx-teeth triangular-lanceolate, acuminate, about ½
as long as the tube; corolla purple, about 5″ long, pubescent; nutlets about 1″ long.

    In woods and on mountain slopes, Maryland and Vir-
ginia to Tennessee. June–Aug.

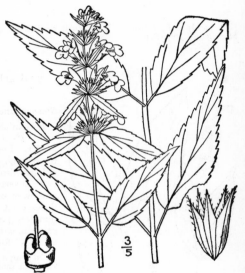

3/5

9. Stachys áspera Michx. Rough Hedge Nettle or Woundwort. Fig. 3623.

Stachys aspera Michx. Fl. Bor. Am. **2**: 5. 1803.

Stachys hispida Pursh, Fl. Am. Sept. 407. 1814.

S. palustris var. *aspera* A. Gray, Man. Ed. 2, 317. 1856.

    Perennial, rough hairy; stem erect or ascending,
simple or branched, 2°–4° high, commonly retrorse-
hispid on the angles. Leaves firm, oblong, oblong-
lanceolate, or ovate-oblong, crenate-dentate, acute or
acuminate at the apex, truncate, rounded or cordate
at the base, 2′–6′ long, ½′–2½′ wide, the lower slender-
petioled, the upper short-petioled; spike terminal,
mostly interrupted; clusters sometimes also in the
upper axils; calyx about 3″ long, hirsute or glabrate,
its teeth triangular-lanceolate, acuminate, about one-
half as long as the tube; corolla red-purple, about ½′
long, its upper lip pubescent.

    In moist soil, Ontario to Massachusetts, Florida, Min-
nesota and Louisiana. Ascends to 5300 ft. in Virginia.
June–Sept. Base hoarhound.

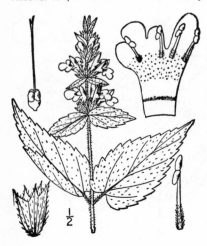

½

10. Stachys salvioìdes Small.　Sage-like Hedge Nettle.　Fig. 3624.

Stachys salvioides Small; Fl. SE. U. S. 1032.　1903.

Perennial, rather finely and often closely puberulent or puberulent-pubescent; stem erect, 1°–3° tall, commonly branched, rough along the angles. Leaves firm, oblong to oblong-ovate or rarely ovate, 2½′–5′ long, or shorter on the lower part of the stem, pubescent on both sides, acute or acutish, crenate, truncate or cordate at the base, slender-petioled; clusters several; calyx sessile or nearly so, becoming 2½″ long, finely pubescent, the tube campanulate-turbinate, the teeth triangular at mutarity; corolla nearly 5″ long, pale purple.

In stony soil, Virginia and West Virginia to Tennessee.　June–Aug.

11. Stachys cordàta Riddell.　Light-green Hedge Nettle.　Fig. 3625.

Stachys cordata Riddell, Suppl. Cat. Ohio Pl. 15.　1836.
Stachys palustris var. *cordata* A. Gray, Man. Ed. 2, 317. 1856.

Perennial, hirsute, pale green; stem slender, weak, mostly simple, ascending or reclining, 2°–3° long. Leaves membranous, flaccid, ovate, oblong or ovate-lanceolate, acuminate, acute or the lowest obtuse at the apex, nearly all of them cordate at the base, dentate or crenate all around, long-petioled, 3′–6′ long, 1′–3′ wide, the lowest petioles nearly as long as the blades; spike interrupted; calyx-teeth subulate-lanceolate, about half the length of the tube; corolla purplish, pubescent or puberulent, about 5″ long.

In woods and thickets, Ohio, Indiana and Illinois to North Carolina and Tennessee.　Ascends to 2100 ft. in Virginia.　July–Aug.

12. Stachys arvénsis L.　Corn or Field Woundwort.　Fig. 3626.

Stachys arvensis L. Sp. Pl. Ed. 2, 814.　1763.

Annual, hirsute; stem very slender, diffusely branched, decumbent or ascending, 3′–2° long. Leaves ovate or ovate-oblong, thin, long-petioled, obtuse at the apex, crenate all around, cordate or the upper rounded at the base, about 1′ long; lower petioles commonly as long as the blades; clusters 4–6-flowered, borne in the upper axils and in short terminal spikes; calyx about 3″ long, its teeth lanceolate, acuminate, nearly as long as the tube; corolla purplish, 3″–5″ long.

In waste places, Maine and Massachusetts to Nebraska, in ballast about the eastern seaports, and in the West Indies, Mexico and South America.　Naturalized from Europe.　July–Oct.

Stachys ánnua L., another European annual species, the stems erect, glabrous or nearly so, the yellow corolla much longer than the calyx, has been found in ballast and waste grounds about the seaports.

1/2

13. Stachys germánica L. Downy Wound-wort. Mouse-ear. Fig. 3627.

Stachys germanica L. Sp. Pl. 581. 1753.

Annual; stem erect, simple, or somewhat branched, 1°–3° high, densely villous. Leaves oval, ovate or lanceolate, crenate-dentate, villous, the lower round-ed or subcordate at the base, long-petioled, mostly obtuse at the apex, the upper short-petioled or ses-sile, narrowed at both ends; clusters of flowers dense, borne in most of the axils; bracts lanceolate, half as long as the calyx; calyx-teeth ovate, acumi-nate, awned; corolla purple, its tube about as long as the calyx.

Roadsides near Guelph, Ontario. Adventive from Europe. July–Sept.

22. BETÓNICA [Tourn.] L. Sp. Pl. 573. 1753.

Annual or perennial herbs, similar to *Stachys.* Lower leaves very long-petioled. Ver-ticils many-flowered, in terminal spikes. Calyx nearly equally 5-toothed, 5–10-nerved. Corolla purple, the tube exceeding the calyx, the limb strongly 2-lipped; upper lip concave; lower 3-cleft, spreading. Stamens and pistil as in *Stachys,* but the anther-sacs parallel in some species. Nutlets ovoid, rounded above. [The classical Latin name of wood betony.]

Ten species, or more, natives of Europe and Asia, the following typical.

1. Betonica officinàlis L. Betony. Wood Betony. Bishop's-wort. Fig. 3628.

Betonica officinalis L. Sp. Pl. 573. 1753.
Stachys Betonica Benth. Lab. Gen. & Sp. 532. 1834.

Perennial, pilose or glabrate, deep green; stem slen-der, erect, usually simple, 1°–3° tall. Leaves oblong or ovate, obtuse at the apex, crenate all around, firm, cor-date or truncate at the base, the basal and lower ones very long-petioled, the blades 3'–6' long, the upper dis-tant, short-petioled or nearly sessile; lower petioles 1½–4 times as long as the blades; spike short, dense; bracts ovate, mucronate, about as long as the calyx; calyx-teeth acicular, half as long as the tube or more; corolla-tube exserted.

In a thicket at Newton, Mass. Fugitive from Europe. Herb christopher. Wild hop. July–Sept.

1/2

23. SÁLVIA [Tourn.] L. Sp. Pl. 23. 1753.

Herbs, or some species shrubs, with clustered usually showy flowers, the clusters mostly spiked, racemed, or panicled. Calyx ovoid, tubular or campanulate, mostly naked in the throat, 2-lipped; upper lip entire or 3-toothed; lower lip 2-cleft or 2-toothed. Corolla strongly 2-lipped; upper lip usually concave, sometimes arched, entire, emarginate or 2-lobed; lower lip spreading or drooping, 3-cleft or 3-lobed. Anther-bearing stamens 2 (the posterior pair wanting or rudimentary); filaments usually short; connective of the anthers transverse, linear or filiform, bearing a perfect anther-sac on its upper end, its lower end dilated, capi-tate or sometimes bearing a small or rudimentary one. Ovary deeply 4-parted; style 2-cleft at the summit. Nutlets smooth, usually developing mucilage and spiral tubes when wetted. [Latin, salvus, safe from its healing virtues.]

About 500 species, of wide distribution in temperate and tropical regions. Besides the follow-ing, some 25 others occur in southern and western North America. Type species: *Salvia officinalis* L.

Leaves mostly basal, only 1–3 pairs on the stem.
   Leaves lyrate-pinnatifid or repand; upper corolla-lip short.   1. *S. lyrata.*
   Leaves crenulate; upper lip arched, longer than the lower.   2. *S. pratensis.*
Stem leafy, bearing several pairs of leaves.
   Leaves narrowly oblong, or lanceolate.
      Corolla 10"–15" long, its tube exserted.   3. *S. Pitcheri.*
      Corolla 4"–6" long, its tube not exserted.   4. *S. lanceifolia.*
   Leaves ovate, or broadly oval.
      Upper corolla-lip short, not exceeding the lower.

Leaves merely crenate or crenulate; fruiting calyx spreading.   5. *S. urticifolia.*
Leaves pinnatifid, sinuate or incised; fruiting calyx deflexed.   6. *S. Verbenaca.*
Upper lip of corolla arched, longer than the lower.   7. *S. Sclarea.*

1. Salvia lyràta L. Lyre-leaved Sage.
Wild Sage. Cancer-weed. Fig. 3629.

Salvia lyrata L. Sp. Pl. 23. 1753.

Perennial or biennial, hirsute or pubescent;
stem slender, simple, or sparingly branched, erect,
1°–3° high, bearing 1 or 2 distant pairs of small
leaves (rarely leafless), and several rather distant
whorls of large violet flowers. Basal leaves tufted,
long-petioled, obovate or broadly oblong, lyrate-
pinnatifid or repand-dentate, thin, 3′–8′ long; stem-
leaves similar, or narrower and entire, sessile, or
short-petioled; clusters distant, about 6-flowered;
calyx campanulate, the teeth of its upper lip subulate, those of the lower longer, aristulate; corolla
about 1′ long, the tube very narrow below, the
upper lip much smaller than the lower; filaments slender; anther-sacs borne on both the
upper and lower ends of the connective, the lower
one often smaller.

In dry, mostly sandy woods and thickets, Connecticut to Florida, west to Illinois, Arkansas and
Texas. Corolla rarely undeveloped. May–July.

2. Salvia praténsis L. Meadow Sage.
Fig. 3630.

Salvia pratensis L. Sp. Pl. 25. 1753.

Perennial, pubescent or puberulent; stem erect,
rather stout, simple or little branched, sparingly
leafy. Basal leaves long-petioled, ovate, oblong or
ovate-lanceolate, irregularly crenulate, obtuse at
the apex, rounded or cordate at the base, thick,
rugose, 2′–7′ long; stem-leaves much smaller, narrower, commonly acute, sessile or nearly so;
clusters spicate, the spike elongated, interrupted;
calyx campanulate, glandular-pubescent, the teeth
of the upper lip minute, those of the lower long,
subulate; corolla purple, minutely glandular, its
upper lip strongly arched, mostly longer than the
lower; lower end of the connective with a small
or imperfect anther-sac.

Atlantic Co., N. J. Fugitive or adventive from
Europe. May–July.

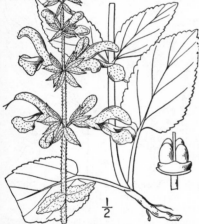

3. Salvia Pítcheri Torr. Pitcher's Sage.
Tall Sage. Fig. 3631.

Salvia Pitcheri Torr.; Benth. Lab. 251. 1833.
Salvia azurea var. *grandiflora* Benth. in DC.
Prodr. 12: 302. 1848.

Perennial, downy; stem stout, branched or
simple, erect, 2°–5° high; branches nearly erect.
Leaves linear or linear-oblong, dentate to entire, sessile, or narrowed at the base into short
petioles, firm, 2′–5′ long, 2″–8″ wide; the uppermost reduced to small bracts; clusters in
long dense terminal spikes, or the lower ones
distant; calyx oblong-campanulate, densely and
finely woolly, about 3″ long, its upper lip entire, obtuse, the lower with 2 ovate acute teeth;
corolla blue or white, finely pubescent without,
1′ long, its lower lip broad, sinuately 3-lobed,
longer than the concave upper one; lower ends
of the connective dilated, often adherent to
each other, destitute of anther-sacs.

On dry plains, Illinois and Minnesota to Missouri, Kansas, Colorado and Texas. July–Sept.

4. Salvia lanceifòlia Poir.   Lance-leaved Sage.   Fig. 3632.

S. lanceolata Brouss. App. Elench. Pl. Hort. Monsp. 15. 1805. Not Lam.
?*S. reflexa* Hornem. Enum. Pl. Hort. Hafn. 34.   1807.
S. lanceifolia Poir. in Lam. Encycl. Suppl. **5** : 49.   1817.

Annual, puberulent or glabrous; stem leafy, usually much branched, erect or diffuse, 6′–18′ high. Leaves oblong, linear-oblong or oblong-lanceolate, petioled, mostly obtuse at the apex and narrowed at the base, crenulate-dentate or entire, 1′–2′ long, 2″–5″ wide, the upper reduced to lanceolate-subulate, rather persistent bracts; flowers mostly opposite, but sometimes 3–4 together in the terminal spike-like racemes; pedicels shorter than the campanulate calyx; calyx 2″–3″ long, its upper lip ovate, entire, the lower 2-cleft, the teeth ovate, mucronate; corolla blue, about 4″ long, its lower lip narrow, twice as long as the upper; lower ends of the connectives dilated; style nearly or quite glabrous.

On plains, Indiana to Nebraska, Montana, Colorado, Texas, Arizona and Mexico. Recorded as introduced in Ohio. May–Sept.

5. Salvia urticifòlia L.   Nettle-leaved or Wild Sage.   Fig. 3633.

Salvia urticifolia L. Sp. Pl. 24.   1753.

Perennial, pubescent, or nearly glabrous; stem glandular above, rather slender, ascending or erect, 1°–2° high. Leaves thin, ovate, 2′–4′ long, irregularly dentate or crenate-dentate, usually acute at the apex, abruptly contracted below into margined petioles; clusters several-flowered, in terminal interrupted spikes; bracts early deciduous; pedicels about as long as the calyx; calyx oblong-campanulate, about 3″ long, the upper lip minutely 3-toothed, the lower 2-cleft, its teeth triangular-lanceolate, acuminate, spreading in fruit; corolla puberulent without, 6″–8″ long, blue and white, the lower lip broad, 3-lobed, twice as long as the upper; lower ends of the connectives dilated; style bearded.

In woods and thickets, Pennsylvania to Kentucky, south to Georgia and Louisiana. April–June.

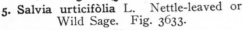

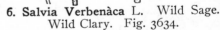

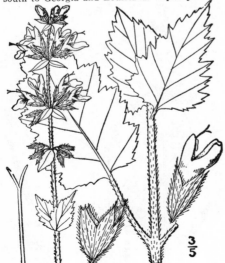

6. Salvia Verbenàca L.   Wild Sage. Wild Clary.   Fig. 3634.

Salvia Verbenaca L. Sp. Pl. 25.   1753.

Perennial; stem glandular-pubescent, erect, simple or sparingly branched, 1°–2° high. Leaves ovate, ovate-lanceolate or oblong, coarsely and irregularly incised-dentate or pinnatifid, petioled, or the uppermost sessile, the lower 3′–8′ long, obtuse at the apex, cordate at the base, nearly glabrous, the upper acute, much smaller; floral bracts broadly ovate, short; clusters several-flowered in elongated terminal interrupted spikes; pedicels shorter than the calyx; calyx deflexed in fruit, 3″–4″ long, its upper lip recurved-spreading, with 3 minute connivent teeth, the lower one with 2 lanceolate acuminate mucronate teeth; corolla blue, about 4″ long, its upper lip nearly straight, scarcely longer than the lower; lower ends of the connectives dilated and adnate to each other.

In waste places, Ohio to South Carolina and Georgia. Naturalized from Europe. Native also of Asia. Eye-seed. Vervain. June–Aug.

Salvia verticillàta L., a European species, with rough lyrate leaves, has been found wild in Pennsylvania.

3/5

7. Salvia Sclàrea L. Clary. Clear-eye. See-bright. Fig. 3635.

Salvia Sclarea L. Sp. Pl. 27. 1753.

Biennial or annual, glandular-pubescent; stem stout, erect, 2°–3½° high. Leaves broadly ovate, rugose, acute or obtuse at the apex, cordate at the base, irregularly denticulate, the lower long-peti-oled, 6'–8' long, the upper small, short-petioled or sessile; clusters several-flowered, numerous in terminal spikes; bracts broad, ovate, acuminate, commonly longer than the calyx, pink or white; calyx campanulate, deflexed-spreading in fruit, 4''–5'' long, its lips about equal, the teeth all subu-late-acicular; corolla blue and white, about 1' long; upper lip arched, laterally compressed, longer than the lower one.

In fields, Pennsylvania to South Carolina, escaped from gardens. Naturalized from Europe. The mucilage of the seeds used to clear specks from the eye. June–Aug.

24. MONÀRDA L. Sp. Pl. 22. 1753.

Perennial or annual erect aromatic herbs, with dentate or serrate leaves, and rather large white red purple yellowish or mottled flowers, in dense capitate clusters, mostly bracteate and bracteolate, terminal and sometimes also axillary, the bracts sometimes brightly colored. Calyx tubular, narrow, 15-nerved, nearly or quite equally 5-toothed, mostly villous in the throat. Corolla glabrous within, usually puberulent or glandular without, the tube slightly dilated above, the limb 2-lipped; upper lip erect or arched, emarginate or entire; lower lip spreading, 3-lobed, the middle lobe larger or longer than the others. Anther-bearing stamens 2, ascending, usually exserted, the posterior pair (staminodia) rudimentary or wanting; anthers linear, versatile, 2-celled, the sacs divaricate, more or less confluent at the base. Ovary deeply 4-parted; style 2-cleft at the apex; nutlets ovoid, smooth. [In honor of Nicolas Monardes, a Spanish physician and botanist of the sixteenth century.]

About 12 species, natives of North America and Mexico. Type species: *Monarda fistulosa* L.
Flower-clusters solitary, terminal (rarely also in the uppermost axils); stamens exserted.
  Leaves manifestly petioled, the petioles commonly slender.
    Corolla scarlet, 1½'–2' long; bracts red.   1. M. didyma.
    Corolla white, pink, or purple, 1'–1½' long.
      Leaves membranous; corolla slightly pubescent, 10''–12'' long.   2. M. clinopodia.
      Leaves thin or firm; corolla pubescent, 1'–1½' long.
        Pubescence spreading; leaves thin.
          Corolla cream-color, pink, or purplish.   3. M. fistulosa.
          Corolla or bracts deep purple or purple-red.   4. M. media.
        Pubescence short, canescent; leaves firm, pale.   5. M. mollis.
  Leaves sessile, or very short-petioled.   6. M. Bradburiana.
Flower-clusters both axillary and terminal; stamens not exserted.
  Calyx-teeth triangular; corolla yellowish, mottled.   7. M. punctata.
  Calyx-teeth subulate-aristate; corolla white or purple, not mottled.
    Bracts lanceolate, gradually acuminate.   8. M. pectinata.
    Bracts oblong to oval, not acuminate.   9. M. dispersa.

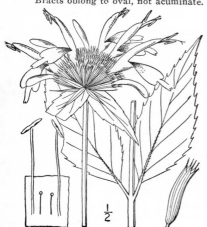

1. Monarda dídyma L. Oswego Tea. American Bee Balm. Fig. 3636.

Monarda didyma L. Sp. Pl. 22. 1753.

Perennial; stem stout, villous-pubescent, or glabrate, 2°–3° high. Leaves thin, ovate or ovate-lanceolate, petioled, dark green, usually pubescent beneath, hairy above, acuminate at apex, rounded or narrowed at the base, sharply serrate, 3'–6' long, 1'–3' wide; lower petioles often 1' long; flower-clusters almost always solitary, terminal; bracts commonly red; calyx glabrous without, glabrous or slightly hirsute in the throat, its teeth subulate, nearly erect, about as long as the diam-eter of the tube; corolla scarlet, 1½'–2' long, puberulent; stamens exserted.

In moist soil, especially along streams, Quebec to Ontario and Michigan, south to Georgia and Ten-nessee. Ascends to 5200 ft. in North Carolina. Red, low or fragrant balm. Horse- or mountain-mint. In-dian's-plume. July–Sept.

2. Monarda clinopòdia L. Basal Balm.
Fig. 3637.

Monarda clinopodia L. Sp. Pl. 22. 1753.
Pycnanthemum Monardella Michx. Fl. Bor. Am. **2**: 8.
  pl. 34. 1803.

Perennial; stem slender, simple, or with few long
ascending branches, glabrous or somewhat villous,
1°–3° high. Leaves lanceolate, ovate or ovate-lanceolate, membranous, bright green, mostly slender-
petioled, more or less villous on the veins beneath
and puberulent above, acuminate at the apex, rounded, narrowed or rarely subcordate at the base, sharply
serrate, 2′–4′ long, ½′–2′ wide; clusters solitary, terminal; bracts pale or white; calyx curved, nearly
glabrous without, somewhat hirsute in the throat,
its teeth subulate, slightly spreading, about as long
as the diameter of the tube; corolla whitish or yel-
lowish-pink, slightly pubescent, 10″–12″ long; stamens
exserted.

In woods and thickets, New York to Illinois, Georgia
and Kentucky. Recorded from Ontario. Ascends to
5000 ft. in North Carolina. June–Aug.

3. Monarda fistulòsa L. Wild Bergamot.
Fig. 3638.

Monarda fistulosa L. Sp. Pl. 22. 1753.

Perennial, villous-pubescent or glabrate; stem
slender, usually branched, 2°–3° high. Leaves
thin but not membranous, green, usually slender-
petioled, lanceolate, ovate or ovate-lanceolate,
acuminate at the apex, serrate, rounded, narrowed
or sometimes cordate at the base, 1½′–4′ long,
½′–2½′ wide; clusters solitary and terminal, or
rarely also in the uppermost axils; bracts whitish
or purplish; calyx puberulent or glabrous, densely
villous in the throat, its subulate teeth rarely
longer than the diameter of the tube; corolla
pubescent, especially on the upper lip, yellowish-
pink, lilac or purplish, 1′–1½′ long; stamens exserted.

On dry hills and in thickets, Maine and Ontario
to Minnesota, Florida, Louisiana and Kansas. Ascends to 2500 ft. in Virginia. Oswego-tea. June–
Sept.

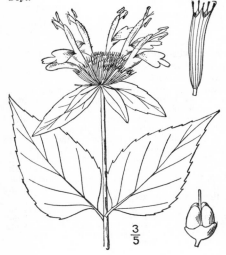

4. Monarda mèdia Willd. Purple Ber-
gamot. Fig. 3639.

Monarda media Willd. Enum. 32. 1809.
Monarda fistulosa var. *rubra* A. Gray, Syn. Fl. **2**:
  Part 1, 374. 1878.
M. fistulosa var. *media* A. Gray, loc. cit. 1878.

Perennial, sparingly hairy or glabrate; stem
stout, commonly branched, 2°–3½° high. Leaves
thin, but not membranous, dark green, ovate
or ovate-lanceolate, or the uppermost lanceolate, slender-petioled, acuminate at the apex,
sharply toothed, usually rounded at the base,
3′–5′ long, 1′–3′ wide; flower-clusters terminal,
solitary, large; bracts deep purple, very conspicuous; calyx curved, glabrous, or very
nearly so, slightly villous in the throat, teeth
slightly spreading; corolla purple or purple-
red, about 1½′ long, its upper lip pubescent;
stamens exserted.

In moist thickets, Maine and Ontario to Pennsylvania and Virginia. June–Aug.

5. Monarda móllis L.   Pale Wild Berga-mot.   Fig. 3640.

Monarda mollis L. Amoen. Acad. **3**: 399. 1787.
Monarda scabra Beck, Am. Journ. Sci. **10**: 260. 1826.

Perennial; stem slender, puberulent at least above, usually branched, 1°–3½° high.   Leaves thick or firm, pale, usually short-petioled, acuminate or acute at the apex, rounded, narrowed or cordate at the base, sharply or sparingly serrate, canescent or puberulent, rarely nearly glabrous, sometimes with a few scattered spreading hairs on the veins or petiole, 1′–3½′ long, ½′–1′ wide; flower-clusters terminal, solitary; bracts green or slightly pink, calyx puberulent, often hairy at the summit, densely villous in the throat, its short pointed teeth nearly erect; corolla yellowish, lilac or pink, about 1¼′ long, pubescent, sometimes glandular; stamens exserted.

On prairies and plains, mostly in dry soil, Maine to Ontario, British Columbia, Alabama, Nebraska, Missouri, Texas and Colorado. June–Aug.

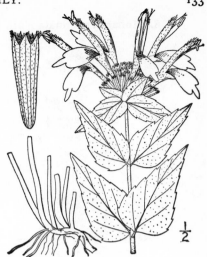

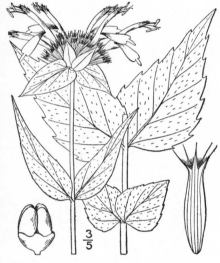

6. Monarda Bradburiàna Beck.   Bradbury Monarda.   Fig. 3641.

Monarda Bradburiana Beck, Am. Journ. Sci. **10**: 260. 1826.

Perennial, sparingly villous or glabrate; stem slender, often simple, 1°–2° high.   Leaves rather thin, bright green, ovate or ovate-lanceolate, sessile, or very nearly so, or partly clasping by the cordate or subcordate base, acuminate at the apex, serrate, 2′–3½′ long; flower-clusters solitary and terminal; bracts green or purplish; calyx glabrous outside, hirsute within and narrowed at the throat, its teeth long, bristle-pointed, divergent, longer than the diameter of the tube; corolla pink or nearly white, about 1′ long, its upper lip pubescent or puberulent, the lower commonly purple-spotted; stamens exserted.

On dry hills or in thickets, Indiana to Alabama, west to Kansas and Arkansas. May–July.

7. Monarda punctàta L.   Horse-mint.   Fig. 3642.

Monarda punctata L. Sp. Pl. 22. 1753.

Perennial, usually rather densely pubescent or downy; stem usually much branched, 2°–3° high. Leaves lanceolate, linear-lanceolate or narrowly oblong, serrate with low teeth, or nearly entire, usually acute at both ends, green, manifestly petioled, 1′–3′ long, 2″–7″ wide, often with smaller ones fascicled in their axils; flower-clusters axillary and terminal, numerous; bracts white or purplish, conspicuous, acute; calyx puberulent, villous in the throat, its teeth short, triangular-lanceolate, acute, not longer than the diameter of the tube; corolla yellowish, purple-spotted, about 1′ long, the stamens equalling or slightly surpassing its pubescent upper lip.

In dry fields, southern New York to Florida, west to Minnesota, Kansas and Texas. Rignum. July–Oct.

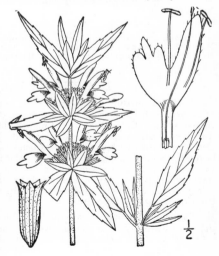

8. Monarda pectinàta Nutt.   Plains Lemon Monarda.   Fig. 3643.

M. *pectinata* Nutt. Proc. Acad. Phila. (II.) **1**: 182.
1848.

Annual, puberulent; stem stout, simple or branched, 1°–2° high. Leaves lanceolate or oblong-lanceolate, mostly sharply serrate or serrulate, acute
at the apex, narrowed at the base, 1′–3′ long,
2″–6″ wide; flower-clusters axillary and terminal,
several or numerous; bracts pale, usually grayish
or brownish, gradually awned at the tip; calyx-
tube nearly glabrous, the throat densely villous,
the teeth bristle-pointed, barbed, lax, nearly half
as long as the tube; corolla pink or nearly white,
not spotted, nearly or quite glabrous, 8″–10″
long; stamens not exserted.

On dry plains, Nebraska and Colorado to Texas
and Arizona. Prairie-bergamot. June–Sept.

Monarda citriodòra Cerv., of Mexico, to which
this was referred in our first edition, differs by long
narrow reflexed bracts and shorter calyx-teeth.

Monarda tenuiaristàta (A. Gray) Small [*M. aristata* Nutt., not Hook.] of the south-central States,
with narrower bracts and longer plumose calyx-teeth,
ranges northward into Kansas.

9. Monarda dispérsa Small.   Purple Lemon Monarda.   Fig. 3644.

M. *dispersa* Small, Fl. SE. U. S. 1038. 1903.

Annual, puberulent; stem stout, usually
branched, 1°–2½° high. Leaves oblong-spatulate, oblanceolate or oblong, or narrowly oblong to linear on the upper part of the stem,
1′–4¼′ long, shallowly serrate; flower clusters
axillary and terminal, often numerous and
usually conspicuous by the broad abruptly
bristle-tipped purple bracts; calyx-tube longer
than in *M. pectinata,* the teeth usually longer
and more slender, fully half as long as the
tube; corolla pale, usually pink, puberulent,
11″–13″ long; stamens mostly not exceeding
the upper lip.

On plains and prairies and in cultivated grounds,
Missouri and Kansas to Georgia, Florida, Texas,
New Mexico and adjacent Mexico. May–Aug.

25. BLEPHÍLIA Raf. Journ. Phys. **89**: 98.   1819.

Perennial hirsute or pubescent erect herbs, with axillary and terminal dense glomerules
of rather small purplish or bluish flowers, or the glomerules in terminal more or less interrupted spikes. Calyx tubular, 13-nerved, not villous in the throat, 2-lipped, the upper lip
3-toothed, the lower 2-toothed, the teeth all aristate or those of the lower lip subulate.
Corolla glabrous within, the tube expanded above, the limb 2-lipped; upper lip erect, entire;
lower lip 3-lobed, the notched middle lobe narrower than the lateral ones. Anther-bearing
(anterior) stamens 2, ascending, exserted or included; posterior stamens reduced to filiform
staminodia, or none; anthers 2-celled, the sacs divaricate, somewhat confluent at the base.
Ovary deeply 4-parted; style 2-cleft at the apex. Nutlets ovoid, smooth. [Greek, eyelash,
from the fringed calyx-teeth.]

Two species, natives of eastern North America. Type species: *Blephilia ciliata* (L.) Raf.

Upper leaves lanceolate or oblong, sessile or short-petioled, slightly serrate.   **1.** *B. ciliata.*
Leaves ovate or ovate-lanceolate, long-petioled, sharply serrate.   **2.** *B. hirsuta.*

1. Blephilia ciliàta (L.) Raf. Downy Blephilia. Fig. 3645.

Monarda ciliata L. Sp. Pl. 23. 1753.
Blephilia ciliata Raf. Journ. Phys. **89** : 98. 1819.

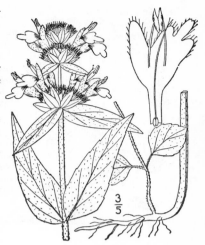

Stem puberulent, or short-villous above, commonly simple, 1°–2° high. Lower leaves and those of sterile shoots ovate or oval, slender-petioled, crenate-denticulate, 1'–2' long, the upper lanceolate or oblong-lanceolate, short-petioled or sessile, mostly acute, longer and narrower, nearly entire; flower-clusters in a terminal spike and in the uppermost axils; outer bracts ovate to lanceolate, acuminate, usually purplish, pinnately veined, ciliate; calyx hirsute, the teeth of the upper lip about one-third longer than those of the lower; corolla purple, villous-pubescent, 5''–6'' long; stamens exserted or included.

In dry woods and thickets, Vermont to Minnesota, south to Georgia, Alabama and Missouri. June–Aug. Ohio horse-mint.

2. Blephilia hirsùta (Pursh) Torr. Hairy Blephilia. Fig. 3646.

Monarda hirsuta Pursh, Fl. Am. Sept. 19. 1814.

Blephilia nepetoides Raf. Journ. Phys. **89** : 98. 1819.

Blephilia hirsuta Torr. Fl. U. S. 27. 1824.

Stem villous-pubescent, or glabrous, usually branched, 1½°–3° high. Leaves membranous, ovate or ovate-lanceolate, acuminate at the apex, rounded, cordate or narrowed at the base, sharply serrate, slender-petioled, 2'–4' long, or the lower shorter and broader; flower-clusters axillary, or in a short terminal spike; outer bracts lanceolate or linear-lanceolate, long-acuminate, hirsute; calyx-tube nearly glabrous, its teeth very villous, those of the upper lip much exceeding the lower; corolla pubescent, pale purple, rather conspicuously darker-spotted, 4''–5'' long.

In woods and thickets, Quebec and Vermont to Minnesota, Kansas, Georgia and Texas. Ascends to 4000 ft. in North Carolina. June–Sept.

26. HEDEÒMA Pers. Syn. **2** : 131. 1807.

Annual or perennial, strongly aromatic and pungent herbs, with small entire or crenulate leaves, and small blue or purple flowers in axillary clusters, these crowded into terminal, leafy-bracted spikes or racemes. Calyx tubular, 13-nerved, villous in the throat, the mouth mostly contracted in fruit, gibbous on the lower side at the base, or nearly terete, 2-lipped, or nearly equally 5-toothed, the upper lip 3-toothed, the lower 2-cleft. Corolla-limb 2-lipped, the upper lip erect, entire, emarginate or 2-lobed, the lower spreading, 3-cleft. Perfect stamens 2, ascending under the upper lip, their anthers 2-celled, the sacs divergent or divaricate. Sterile stamens (staminodia) 2, minute, or none, very rarely anther-bearing. Ovary deeply 4-parted; style 2-cleft at the summit, glabrous. Nutlets ovoid, smooth. [Greek, sweet smell.]

About 15 species, natives of America. Besides the following, some 8 others occur in the southern and southwestern States. Sometimes called Mock Pennyroyal. Type species here taken as *Hedeoma pulegioides* (L.) Pers.

Teeth of the upper lip of the calyx triangular; leaves serrate; annual.   1. *H. pulegioides.*
Teeth of both lips of the calyx subulate; leaves entire.
    Calyx-teeth all nearly equal; annual.   2. *H. hispida.*
    Teeth of the lower lip nearly twice as long as the upper; perennial.   3. *H. longiflora.*

1. Hedeoma pulegioìdes (L.) Pers. American Pennyroyal. Fig. 3647.

Melissa pulegioides L. Sp. Pl. 593. 1753.
Cunila pulegioides L. Sp. Pl. Ed. 2, 30. 1762.
Hedeoma pulegioides Pers. Syn. **2**: 131. 1807.

Annual; stem very slender, erect, much branched, finely soft-pubescent, 6′-18′ high, the branches ascending. Leaves ovate to obovate-oblong, petioled, sparingly serrate, mostly obtuse at the apex and narrowed at the base, glabrous, or sparingly pubescent, thin, ½′-1½′ long, 2″-8″ wide, the upper smaller; clusters few-flowered, axillary, rather loose; pedicels pubescent, shorter than or equalling the calyx; calyx pubescent, gibbous, oblong-ovoid in fruit, its 3 upper teeth triangular, acute, not exceeding the 2 subulate hispid lower ones; corolla bluish-purple, about 3″ long; rudimentary stamens manifest, capitate at the summit, or rarely anther-bearing.

In dry fields, Cape Breton Island to Ontario and Minnesota, Florida, Alabama, Arkansas and Nebraska. Also called tick-weed and squaw-mint. July–Sept.

2. Hedeoma híspida Pursh. Rough Pennyroyal. Fig. 3648.

Hedeoma hispida Pursh, Fl. Am. Sept. 414. 1814.
Hedeoma hirta Nutt. Gen. **1**: 16. 1818.

Annual; stem erect, branched, slender, 3′-8′ high, pubescent, the branches erect-ascending, very leafy and copiously flowered. Leaves linear, entire, firm, sessile, or the lower short-petioled, blunt or subacute at the apex, narrowed at the base, more or less hispid-ciliate but otherwise mostly glabrous, ½′-1′ long, about 1″ wide, the lower much shorter and smaller; clusters axillary, numerous, crowded, several-flowered; pedicels pubescent, shorter than the calyx; bracts subulate, very hispid, about equalling the calyx; calyx oblong, gibbous, hispid, its teeth all subulate, nearly equal in length, upwardly curved in fruit, about one-half as long as the tube, the 2 lower ones somewhat narrower and more hispid than the upper; corolla about 3″ long, bluish-purple; sterile stamens rudimentary or none.

On dry plains, Ontario and New York to Saskatchewan, Illinois, Louisiana, Arkansas and Colorado. May–Aug.

3. Hedeoma longiflòra Rydb. Long-flowered Pennyroyal. Fig. 3649.

Hedeoma longiflora Rydb. Bull. Torr. Club **36**: 685. 1909.

Perennial from a woody base, with an ashy down nearly all over; stems much branched, slender, erect, 6′-18′ high, the branches ascending. Leaves oblong or linear-oblong, entire, short-petioled or sessile, obtuse at the apex, narrowed at the base, spreading, 5″-10″ long, 1″-2½″ wide, the lowest shorter; clusters axillary, loosely few-flowered; pedicels puberulent, about one-half as long as the calyx and equalling or longer than the subulate bracts; calyx oblong, hirsute, slightly gibbous, its teeth all subulate, upwardly curved and connivent in fruit, the 2 lower nearly twice as long as the 3 upper; corolla purple, 4″-6″ long; sterile stamens rudimentary or none.

In dry soil, South Dakota to Nebraska, Kansas and Texas. Included, in our first edition, in *H. Drummondii* Benth. of the Southwest. April–Aug.

27. MELÍSSA [Tourn.] L. Sp. Pl. 592. 1753.

Leafy branching herbs, with broad dentate leaves, and rather small white or yellowish axillary clustered somewhat secund flowers. Calyx oblong-campanulate, deflexed in fruit, 13-nerved, nearly naked in the throat, 2-lipped; upper lip flat, 3-toothed, the lower 2-parted. Corolla exserted, its tube curved-ascending, enlarged above, naked within, the limb 2-lipped; upper lip erect, emarginate; lower lip 3-cleft, spreading. Stamens 4, didynamous, connivent and ascending under the upper lip of the corolla; anthers 2-celled, their sacs divaricate. Ovary deeply 4-parted; style 2-cleft at the summit, the lobes subulate. Nutlets ovoid, smooth. [Greek, bee.]

About 4 species, natives of Europe and western Asia, the following typical.

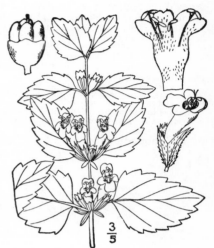

1. Melissa officinàlis L.   Garden or Lemon Balm. Bee-balm. Fig. 3650.

Melissa officinalis L. Sp. Pl. 592. 1753.

Perennial, pubescent; stem rather stout, erect or ascending, 1°–2½° high.   Leaves ovate, petioled, mostly obtuse at both ends, sometimes cordate, pinnately veined, coarsely dentate or crenate-dentate, 1′–2½′ long; flowers several in the axillary clusters; pedicels shorter than the calyx; calyx about 3″ long, the teeth of its lower lip slightly exceeding those of the upper; corolla white, 5″–7″ long.

In waste places, thickets and woods, Maine to Georgia, West Virginia, Missouri and Arkansas, also in Oregon and California. Naturalized from Europe. Plant lemon-scented. Balm-leaf or -mint. Honey-plant. Pimentary. Goose-tongue. Dropsy-plant. Lemon-lobelia. Sweet-mary. June–Aug.

28. SATUREÌA [Tourn.] L. Sp. Pl. 567. 1753.

Herbs or shrubs, with small entire leaves, sometimes with smaller ones fascicled in their axils, and bracted purple flowers in dense terminal or axillary clusters. Calyx campanulate, mostly 10-nerved, 5-toothed, naked or rarely villous in the throat. Corolla-limb 2-lipped, the upper lip erect, flat, entire or emarginate, the lower spreading, 3-cleft. Stamens 4, connivent under the upper lip of the corolla; anthers 2-celled, the sacs parallel or divaricate. Ovary deeply 4-parted; style 2-cleft at the summit. Nutlets oblong or oval. [The classical Latin name of the plant.]

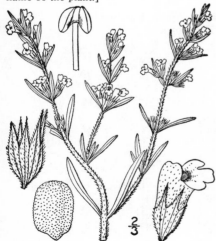

About 18 species, the following typical one introduced as a garden herb from Europe, one of doubtful affinity in Florida, the others of the Mediterranean region.

1. Satureia horténsis L. Savory. Summer Savory. Fig. 3651.

Satureia hortensis L. Sp. Pl. 568. 1753.

Annual, puberulent; stems erect, slender, much branched, 6′–18′ high. Leaves linear or linear-oblong, short-petioled, entire, acute at both ends, ½′–1½′ long, 1″–2″ wide; clusters 3″–5″ in diameter, terminal and in many of the upper axils; bracts linear, small, minute, or wanting; calyx about equalling the corolla-tube, somewhat pubescent, its teeth subulate, about as long as the tube, ciliate; corolla little longer than the calyx; stamens scarcely exserted.

In waste places, New Brunswick and Ontario to Kentucky, west to Nevada. Naturalized or adventive from Europe. July–Sept.

29. CLINOPÒDIUM L. Sp. Pl. 587. 1753.

[CALAMINTHA Moench, Meth. 408. 1794.]

Herbs, or low shrubs, with entire or sparingly dentate leaves, and rather large flowers variously clustered. Calyx tubular or oblong, mostly gibbous at the base, about 13-nerved, 2-lipped, naked or villous in the throat, the upper lip 3-toothed, the lower 2-cleft. Corolla usually expanded at the throat, the tube straight, mostly longer than the calyx, the limb 2-lipped; upper lip erect, entire or emarginate; lower lip spreading, 3-cleft. Stamens 4, all anther-bearing, didynamous, ascending under the upper lip of the corolla, somewhat connivent in pairs, the longer mostly exserted; anthers 2-celled, the sacs divergent or divaricate. Ovary deeply 4-parted; style glabrous, 2-cleft at the summit. Nutlets ovoid, smooth. [Greek, bed-foot, the flowers likened to a bed-castor.]

About 60 species, natives of the north temperate zone. Besides the following, 4 others occur in the southeastern United States and in California. The genus has been included in *Satureia* by authors. Type species: *Clinopodium vulgare* L.

   * **Flower-clusters dense, axillary and terminal, setaceous-bracted.** 1. *C. vulgare.*

  ** **Flower-clusters loose, axillary, or forming terminal thyrses; bracts small.**

Plants pubescent; introduced species.

Clusters peduncled; calyx not gibbous; upper leaves very small; perennial.	2. *C. Nepeta.*
Clusters sessile; calyx very gibbous; plant leafy, annual.	3. *C. Acinos.*

Plants glabrous; native species.

Leaves linear or the lower spatulate, entire; corolla 4″ long.	4. *C. glabrum.*
Leaves oblong or oblong-lanceolate, serrate; corolla 6″–7″ long.	5. *C. glabellum.*

1. Clinopodium vulgàre L.   Field or Wild Basil.   Basil-weed.   Fig. 3652.

Clinopodium vulgare L. Sp. Pl. 587. 1753.
Melissa Clinopodium Benth. Lab. Gen. & Sp. 393. 1834.
Calamintha Clinop. Benth. in DC. Prodr. 12: 233. 1848.

Perennial by short creeping stolons, hirsute; stem slender, erect from an ascending base, usually branched, sometimes simple, 1°–2° high. Leaves ovate or ovate-lanceolate, petioled, obtuse or acutish, entire, undulate or crenate-dentate, rounded, truncate or sometimes narrowed at the base, thin, 1′–2½′ long; flowers in dense axillary and terminal capitate clusters about 1′ in diameter; bracts setaceous, hirsute-ciliate, usually as long as the calyx-tube; calyx pubescent, somewhat gibbous, the setaceous teeth of its lower lip rather longer than the broader ones of the upper; corolla purple, pink, or white, little exceeding the calyx-teeth.

In woods and thickets, Newfoundland to North Carolina, Tennessee, Minnesota and Manitoba, in the Rocky Mountains to New Mexico and Arizona. Ascends to 4000 ft. in Virginia. Also in Europe and Asia. Stone-basil. Bed's-foot. Field- or horse-thyme. Dog-mint. June–Oct.

2. Clinopodium Népeta (L.) Kuntze.   Field Balm.   Field or Lesser Calamint.   Basil-thyme.   Fig. 3653.

Melissa Nepeta L. Sp. Pl. 593. 1753.
Cal. Nepeta Link & Hoffmansg. Fl. Port. 1: 14. 1809.
Clinopodium Nepeta Kuntze, Rev. Gen. Pl. 515. 1891.
Satureia Nepeta Scheele, Flora 26: 577. 1843.

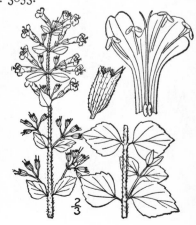

Perennial by a woody root and short rootstocks, villous or pubescent; stem rather stout, at length much branched, the branches nearly straight, ascending. Leaves broadly ovate, petioled, obtuse or acute, crenulate with few low teeth, rounded or narrowed at the base, the lower ½′–1′ long, the upper much smaller and bract-like; flowers few in the numerous loose peduncled axillary cymes, forming an almost naked elongated thyrsus; bracts very small, linear; calyx not gibbous, villous in the throat, about 1½″ long, the teeth of its lower lip twice as long as those of the upper; corolla light purple or almost white, about 4″ long.

In fields and waste places, Maryland to South Carolina, Alabama, Indiana, Kentucky and Arkansas. Bermuda. Naturalized from Europe. Native also of Asia. June–Sept.

Clinopodium Calamíntha (L.) Kuntze, the calamint of the Old World, with larger leaves and flowers, admitted into our first edition, is not known in
the wild state within our area.

3. Clinopodium Ácinos (L.) Kuntze.
Basil-thyme. Basil Balm. Fig. 3654.

Thymus Acinos L. Sp. Pl. 591. 1753.
Melissa Acinos Benth. Lab. Gen. & Sp. 389. 1834.
Cal. Acinos Benth. in DC. Prodr. **12**: 230. 1848.
Clin. Acinos Kuntze, Rev. Gen. Pl. 513. 1891.

Annual, pubescent; stems branched from the
base, very slender, 6′–8′ high. Leaves oblong
or ovate-oblong, petioled, acutish at both ends
or the lower obtuse, crenulate or entire, 4″–8″
long; flowers about 6 in the axils, the clusters
sessile; bracts shorter than the pedicels; calyx
gibbous on the lower side, rough-hairy, longer
than its pedicel, contracted at the throat, its subulate teeth somewhat unequal in length; corolla
purplish, 1½–2 times as long as the calyx.

In waste places, Ontario to Massachusetts and
New Jersey. Adventive or naturalized from Europe.
Mother-of-thyme. Polly mountain. May–Aug.

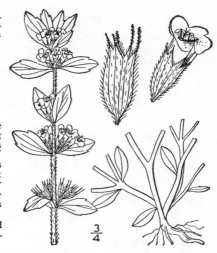

4. Clinopodium glàbrum (Nutt.) Kuntze.
Low Calamint or Bed's-foot. Fig. 3655.

Hedeoma glabra Nutt. Gen. **1**: 16. 1818.
Cal. Nuttallii Benth. in DC. Prodr. **12**: 230. 1848.
Calamintha glabella var. *Nuttallii* A. Gray, Man.
Ed. 2, 307. 1856.
Clin. glabrum Kuntze, Rev. Gen. Pl. 515. 1891.

Perennial, glabrous, stoloniferous; stem very
slender, at length much branched, erect or ascending, 4′–12′ high. Leaves of the flowering
branches linear, entire, very short-petioled or
sessile, mostly obtuse at the apex, obscurely
veined, 4″–9″ long, ½″–1″ wide, the margins
slightly revolute; lower leaves and those of the
stolons shorter and broader, distinctly petioled;
flowers 1–4 in the axils; bracts minute; pedicels
filiform, mostly longer than the calyx; calyx not
gibbous, its throat pubescent in a ring within,
its lower teeth somewhat longer than the upper;
corolla purple, about 4″ long.

On rocks and banks, Ontario to western New
York, Illinois, Minnesota, Missouri, Arkansas and
Texas. May–Aug.

5. Clinopodium glabéllum (Michx.)
Kuntze. Slender Calamint or Bed's-
foot. Fig. 3656.

Cunila glabella Michx. Fl. Bor. Am. **1**: 13. 1803.
Calamintha glabella Benth. in DC. Prodr. **12**: 230.
1848.
Clin. glabellum Kuntze. Rev. Gen. Pl. 515. 1891.

Perennial, stoloniferous, glabrous; stems
weak, spreading or decumbent, at length freely
branched, elongated, slender, 8′–2° long. Leaves
membranous, oblong, short-petioled, obtuse or
the uppermost subacute at the apex, narrowed
to a cuneate base, distinctly serrate with low
teeth, 1′–2′ long, 2″–8″ wide, the lowest and
those of the stolons sometimes proportionately
broader and shorter; axils 2–5-flowered; pedicels filiform, commonly twice as long as the
calyx; bracts minute; calyx not gibbous, its
throat pubescent in a ring within, its teeth
nearly equal; corolla purplish, 6″–7″ long.

On river banks, Indiana to Kentucky, Tennessee and Arkansas. May–July.

30. HYSSÒPUS [Tourn.] L. Sp. Pl. 569. 1753.

A perennial erect herb, the stem somewhat woody at the base, with narrow, entire leaves, and small bracted purple or blue flowers, in dense clusters in the upper axils, and forming elongated terminal more or less interrupted spikes. Calyx tubular, 15-nerved, about equally 5-toothed, not hairy in the throat. Corolla-limb 2-lipped, the upper lip erect, emarginate, the lower spreading, 3-cleft, the middle lobe 2-lobed. Stamens 4, didynamous, the 2 longer ones exserted, divergent; anthers 2-celled, the sacs divaricate. Ovary deeply 4-parted; style 2-cleft at the summit. Nutlets ovoid, somewhat 3-sided, nearly smooth. [Greek, an aromatic herb.]

A monotypic genus of Europe and Asia.

1. Hyssopus officinàlis L.   Hyssop.
Fig. 3657.

Hyssopus officinalis L. Sp. Pl. 569. 1753.

Stems usually several together from the woody base, slender, strict, puberulent, simple or branched, 1°–3° high, the branches upright or ascending. Leaves linear to oblong, sessile or very nearly so, firm, acute at both ends or the lower obtuse at the apex, puberulent or glabrate, faintly veined, 1½′–2′ long, 1″–3″ wide, sometimes with smaller ones or short leafy branches in their axils; spike sometimes 1-sided, dense, ½′–1′ broad; pedicels short, puberulent; outer bracts as long as the calyx; calyx-teeth lanceolate, acute, one-fourth to one-third as long as the tube; corolla 4″–5″ long, its tube exceeding the calyx.

Along roadsides and in waste places, Ontario and Maine to North Carolina, and on the Pacific Coast. Naturalized from Europe. June–Sept.

31. ORÍGANUM [Tourn.] L. Sp. Pl. 588. 1753.

Perennial branching herbs, some species shrubby, with rather small crenate-dentate or entire leaves, and small bracted pink or purple flowers, in dense terminal glomerules. Calyx ovoid or campanulate, villous in the throat, about 13-nerved, 5-toothed or more or less 2-lipped. Corolla-limb 2-lipped, the upper lip erect, emarginate or 2-lobed, the lower longer, spreading, 3-cleft. Stamens 4, didynamous, ascending; anthers 2-celled, the sacs divergent. Style 2-cleft at the summit; ovary deeply 4-parted. Nutlets ovoid or oblong, smooth. [Greek, mountain-joy.]

About 30 species, natives of the Old World, the following typical.

1. Origanum vulgàre L.   Wild Marjoram.
Winter Sweet. Organy. Fig. 3658.

Origanum vulgare L. Sp. Pl. 590. 1753.

Perennial from nearly horizontal rootstocks, villous or hirsute; stem erect, slender, 1°–2½° high. Leaves ovate, petioled, obtuse or subacute at the apex, rounded or subcordate at the base, crenate or entire, 1′–1½′ long, often with smaller ones, or short leafy branches, in their axils; flower-clusters often 2′ broad; bracts purplish, ovate or oval, about equalling the nearly regularly 5-toothed calyx; corolla pink, purple or nearly white, longer than the calyx, the upper lobe broad; all four stamens, or the two longer, exserted.

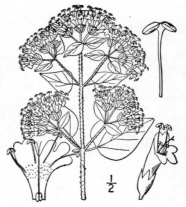

In fields and waste places, Ontario to New Jersey and Pennsylvania. Naturalized from Europe. Native also of Asia. Called also organs, pot-marjoram. July–Sept.

32. THÝMUS [Tourn.] L. Sp. Pl. 590. 1753.

Perennial herbs, or low shrubby creeping plants, with small mostly entire leaves, and small purple flowers clustered in terminal glomerules, or in the axils of the leaves. Calyx ovoid, villous in the throat, 10–13-nerved, 2-lipped, the upper lip erect-spreading, 3-toothed, the lower 2-toothed, its teeth long and slender. Corolla-limb 2-lipped, the upper lip erect, emarginate, the lower spreading, 3-cleft. Stamens 4, more or less didynamous, mostly exserted; anthers 2-celled, the sacs parallel or divergent. Ovary deeply 4-parted; style 2-cleft at the summit. Nutlets ovoid or oblong, smooth. [Greek, incense.]

About 50 species, natives of the Old World, mostly European. Type species: *Thymus vulgaris* L.

1. Thymus Serpýllum L.   Wild or Creeping Thyme.   Mother of Thyme.   Fig. 3659.

Thymus Serpyllum L. Sp. Pl. 590. 1753.

Stems more or less pubescent in lines, very slender, procumbent, tough, much branched, 4′–12′ long, commonly forming dense mats. Leaves oblong or ovate-oblong, petioled, obtuse at the apex, usually narrowed at the base, entire, glabrous, or sometimes ciliate, 2″–5″ long; bracts similar to the leaves, but smaller; flowers numerous in verticillate clusters crowded in dense short terminal spikes, or also in the upper axils; calyx distinctly 2-lipped, the tube usually pubescent and the teeth ciliate; corolla longer than the calyx.

In thickets, woods, and along roadsides, Nova Scotia to southern New York and North Carolina. Naturalized from Europe. Native also of Asia. June–Sept. Old English names, brotherwort, hillwort, penny-mountain, shepherd's-thyme.

33. KOÈLLIA Moench, Meth. 407. 1794.

[BRACHYSTEMON and PYCNANTHEMUM Michx. Fl. Bor. Am. **2**: 5, 7. 1803.]

Perennial erect, mostly branched, glabrous cinereous canescent or pubescent herbs, with small white or purple-dotted flowers, in terminal or sometimes also axillary capitate glomerules or cymose clusters. Calyx ovoid, oblong or tubular, 10–13-nerved, equally or more or less unequally 5-toothed, not villous in the throat, the 2 upper teeth sometimes united below. Corolla 2-lipped, the upper lip emarginate or entire, the lower 3-cleft, its lobes obtuse. Stamens 4, didynamous, nearly equal, or the lower pair a little longer; anther-sacs parallel. Ovary deeply 4-parted; style slender. Nutlets smooth, pubescent, or roughened. [Named for J. L. C. Koelle, a German botanist of the eighteenth century.]

About 17 species, natives of North America. Besides the following, 1 occurs in California and several in the Southern States. Type species: *Koellia capitata* Moench. Mostly very fragrant. Sometimes called Basil, the proper name of Calamint or Ocymum.

* Leaves prevailingly linear, linear-lanceolate or oblong-lanceolate.

Calyx-teeth ovate-triangular, acute, one-fourth as long as the tube.   1. *K. virginiana.*
Calyx-teeth subulate, lance-subulate or bristle-tipped.
    Leaves linear or lanceolate; calyx-teeth subulate or lanceolate.
      Leaves linear or narrowly linear-lanceolate, entire.   2. *K. flexuosa.*
      Leaves lanceolate, entire, or usually serrate.
        Bracts appressed, erect; clusters dense.
         Hirsute or pilose; leaves mainly entire.   3. *K. pilosa.*
         Puberulent, short-pubescent, or glabrate; leaves mostly denticulate.
                   4. *K. verticillata.*
                   5. *K. clinopodioides.*
        Bracts spreading; clusters loose.
    Leaves oblong to linear-oblong, obtuse or subacute; calyx-teeth awn-like. 6. *K. hyssopifolia.*

** Leaves prevailingly ovate, ovate-oblong or ovate-lanceolate.

Calyx-teeth bristle-tipped or subulate.
    Bracts appressed; clusters dense; calyx-teeth bristle-pointed.   7. *K. aristata.*
    Bracts spreading; clusters loose; calyx-teeth subulate.
      Calyx-teeth about one-half as long as the tube.   8. *K. incana.*
      Calyx-teeth as long as the tube.   9. *K. pycnanthemoides.*
Calyx-teeth triangular, triangular-lanceolate or lanceolate.
    Bracts spreading; clusters loose; calyx-teeth short.   10. *K. albescens.*
    Bracts appressed; clusters dense.
      Bracts canescent; leaves firm, acute.   11. *K. mutica.*
      Bracts ciliate or villous; leaves membranous, acuminate.   12. *K. montana.*

1. Koellia virginiàna (L.) MacM.　Virginia Mountain-Mint.　Fig. 3660.

Satureja virginiana L. Sp. Pl. 567.　1753.
P. lanceolatum Pursh, Fl. Am. Sept. 409.　1814.
Koellia virginiana MacM. Met. Minn. 452.　1892.

Stem strict, rather stout, glabrous or pubescent, 1°–3° high. Leaves lanceolate or linear-lanceolate, fragrant, very short-petioled or sessile, firm, entire, acuminate at the apex, glabrous or somewhat puberulent beneath, or the uppermost densely canescent, 1′–2′ long, 2″–5″ wide, often with short leafy branches in their axils; glomerules dense, 4″–6″ in diameter, terminal, corymbed, canescent; bracts appressed, rigid, acute, acuminate or subulate-tipped, not exceeding the clusters; calyx cylindraceous, or expanded above, canescent, its teeth triangular-ovate, equal or nearly so, acute, little longer than wide, about one-fourth as long as the tube; corolla pubescent without, purple-spotted, its tube longer than the calyx.

In dry fields and thickets, Quebec and Ontario to Minnesota, south to Georgia, Alabama and Kansas. Virginia or mountain thyme. Prairie-hyssop. Pennyroyal. Basil. July–Sept.

2. Koellia flexuosa (Walt.) MacM.　Narrow-leaved Mountain-Mint.　Fig. 3661.

Satureja Thymus virginicus L. Mant. 2: 409.　1771.　Not
　S. virginica L. 1753.
Origanum flexuosum Walt. Fl. Car. 165.　1788.
Koellia capitata Moench, Meth. 408.　1794.
Pycnanthemum linifolium Pursh, Fl. Am. Sept. 409.　1814.
P. flexuosum B.S.P. Prel. Cat. U. S. 42.　1888.
Koellia flexuosa MacM. Met. Minn. 452.　1892.

Stem slender, stiff, nearly glabrous throughout, 1½°–2½° high. Leaves linear or linear-lanceolate, entire, glabrous or the uppermost puberulent, sessile, or the lower very short-petioled, 1′–2′ long, ½″–1½″ wide, rather firm, often with short leafy branches in their axils; glomerules dense, terminal, corymbed, 3″–5″ broad, canescent; bracts appressed, acuminate, or subulate-tipped, not longer than the clusters; calyx cylindraceous, canescent, its teeth subulate and rigid, equal or nearly so, 3–4 times longer than broad, about one-third as long as the tube; corolla-tube longer than the calyx.

In fields and thickets, Maine to Florida, New York, Minnesota, Kansas and Texas. Little fragrant. July–Sept.

3. Koellia pilòsa (Nutt.) Britton.　Hairy Mountain-Mint.　Fig. 3662.

Pycnanthemum pilosum Nutt. Gen. 2: 33.　1818.
Pycnanthemum muticum var. *pilosum* A. Gray, Syn. Fl.
　2: Part 1, 355.　1878.
Koellia pilosa Britton, Mem. Torr. Club 5: 279.　1894.

Pubescent, at least above, 1°–2½° high. Leaves lanceolate, very short-petioled or sessile, entire or very sparingly denticulate, 1′–2′ long, 3″–6″ wide, firm, acuminate at the apex, mostly narrowed at the base, commonly with smaller ones, or short leafy shoots in their axils; glomerules dense, numerous, terminal, villous or hirsute-canescent, about 4″ in diameter; bracts lanceolate, acuminate, equalling or exceeding the clusters; calyx cylindraceous, narrow, canescent, its teeth lanceolate-subulate, equal, often ciliate, about 3 times as long as wide and one-fourth the length of the tube; corolla pubescent, its tube little longer than the calyx; stamens exserted.

On prairies and in dry woods, Ontario to Pennsylvania, Georgia, Iowa, Kansas and Arkansas. July–Sept.

Koellia léptodon (A. Gray) Small, of the North Carolina mountains, with bristly-ciliate calyx-teeth, is recorded as extending to Ohio and Missouri.

4. Koellia verticillàta (Michx.) Kuntze. Torrey's Mountain-Mint. Fig. 3663.

Brachystemon verticillatum Michx. Fl. Bor. Am. **2**:
6. *pl. 31.* 1803.
P. Torreyi Benth. Lab. Gen. & Sp. 329. 1834.
Koellia verticillata Kuntze, Rev. Gen. Pl. 520. 1891.

Puberulent, glabrate or pubescent; stem slender,
1°–2½° high. Leaves lanceolate, oblong-lanceolate
or linear-lanceolate (rarely ovate-lanceolate),
short-petioled or sessile, serrulate or entire, acute
or acuminate at the apex, rounded or narrowed
at the base, 1′–3′ long, 3″–10″ wide, the uppermost sometimes canescent; flower-clusters dense,
canescent, 5″–6″ broad, terminal, corymbose and
commonly also in some of the upper axils; bracts
appressed, lanceolate, acuminate, ciliate, equalling
or longer than the clusters; calyx canescent, its
teeth subulate or lance-subulate, ciliate, 2–3 times
as long as wide, one-fourth to one-third as long
as the tube; corolla pubescent, its tube rather exceeding the calyx.

In dry fields and thickets, Vermont to Virginia,
west to Missouri. July–Sept.

5. Koellia clinopodioìdes (T. & G.) Kuntze. Basil Mountain-Mint. Fig. 3664.

Pycnanthemum clinopodioides T. & G.; A. Gray, Am.
Journ. Sci. **42**: 45. 1842.
Koellia clinopodioides Kuntze, Rev. Gen. Pl. 520. 1891.

Pubescent or puberulent; stem slender, 1°–2½° high.
Leaves lanceolate or oblong-lanceolate, rather thin,
short-petioled, sharply serrate, or the upper entire,
1½′–3′ long, 5″–12″ wide, none of them canescent;
flower-clusters loose, terminal and axillary, about 1′
broad; bracts linear-acuminate or subulate-tipped,
not exceeding the clusters, some or all of them
spreading; calyx finely canescent or glabrate, its
teeth subulate, sometimes with a few long hairs,
slightly unequal, about one-third the length of the
tube; corolla-tube longer than the calyx.

In dry soil, Connecticut to Pennsylvania, Virginia and
Tennessee. Ascends to 5000 ft. in Virginia. Aug.–Sept.

6. Koellia hyssopifòlia (Benth.) Britton. Hyssop Mountain-Mint. Fig. 3665.

P. hyssopifolium Benth. Lab. Gen. & Sp. 329. 1834.
Pycnanthemum aristatum var. *hyssopifolium* A. Gray,
Syn. Fl. **2**: Part 1, 354. 1878.

K. hyssopifolia Britton, Mem. Torr. Club **5**: 279. 1894.

Puberulent or glabrate; stem slender, stiff, 1°–3°
high. Leaves oblong, linear-oblong, or lanceolate-
oblong, short-petioled, or the upper sessile, obtuse
or subacute at the apex, narrowed at the base,
entire or denticulate, ½′–1½′ long, 2″–6″ wide,
glabrous or minutely canescent; flower-clusters
dense, minutely canescent, not at all villous, terminal, and usually also in the upper axils, often
1′ broad; bracts linear-oblong, narrowed at each
end, terminated by an awn almost as long as the
body; calyx cylindraceous, glabrous or very nearly
so, prominently nerved, its teeth bristle-pointed,
slightly widened below, nearly as long as the
tube; corolla-tube not longer than calyx.

In dry soil, Virginia to Florida. June–Aug.

7. Koellia aristàta (Michx.) Kuntze. Awned Mountain-Mint. Fig. 3666.

Pyc. aristatum Michx. Fl. Bor. Am. **2**: 8. *pl. 33.* 1803.
Koellia aristata Kuntze, Rev. Gen. Pl. 520. 1891.

Similar to the preceding species; stem slender, stiff, minutely canescent, $1\frac{1}{2}°–2\frac{1}{2}°$ high. Leaves ovate, or some of them ovate-lanceolate, short-petioled, sharply serrate, serrulate, or the upper entire, acute at the apex, rounded at the base, $1'–2'$ long, $4''–12''$ wide, the uppermost usually minutely canescent; inflorescence as in the preceding species; bracts long-awned, appressed, the awn about one-third the length of the body; calyx canescent, its teeth equal, bristle-pointed, widened below, one-third to one-half as long as the tube; corolla-tube about equalling the calyx.

In dry pine barrens, New Jersey to Florida and Louisiana, mostly near the coast. Wild basil. July–Sept.

8. Koellia incàna (L.) Kuntze. Hoary Mountain-Mint. Fig. 3667.

Clinopodium incanum L. Sp. Pl. 588. 1753.
Pycnanthemum incanum Michx. Fl. Bor. Am. **2**: 7. 1803.
Koellia incana Kuntze, Rev. Gen. Pl. 520. 1891.

Stem pubescent, or glabrous below, stout, $1\frac{1}{2}°–3°$ high. Leaves thin, ovate to ovate-lanceolate, petioled, acute at the apex, sharply serrate or serrulate, white-canescent beneath, puberulent or glabrous above, $1\frac{1}{2}'–3'$ long, $\frac{1}{2}'–1\frac{1}{2}'$ wide, or the uppermost smaller and sometimes canescent on both sides; clusters loose, terminal and in the upper axils, $1'–1\frac{1}{2}'$ broad, canescent, the flowers sometimes secund on their branches; bracts linear, or the outer broader, canescent or slightly villous, spreading, mostly shorter than the clusters; calyx canescent, slightly 2-lipped, its teeth subulate, somewhat unequal, the longer one-fourth to one-half as long as the tube, rarely villous; corolla-tube equalling or longer than the calyx.

Dry thickets and hillsides, Maine to Ontario, Florida, Alabama and Missouri. Calamint. Wild basil. Aug.–Oct.

9. Koellia pycnanthemoìdes (Leavenw.) Kuntze. Southern Mountain-Mint. Fig. 3668.

Tullia pycnanthemoides Leavenw. Am. Journ. Sci. **20**: 343. *pl. 5.* 1830.
P. Tullia Benth. Lab. Gen. & Sp. 328. 1834.
K. pycnanthemoides Kuntze, Rev. Gen. Pl. 520. 1891.
P. pycnanthemoides Fernald, Rhodora **10**: 86. 1908.

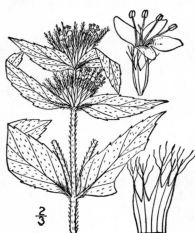

Stem rather stout, pubescent nearly to the base, $2°–3°$ high. Leaves membranous, petioled, mostly ovate-lanceolate, acuminate or acute at the apex, narrowed at the base, sharply serrate, pubescent beneath, puberulent or glabrate and dark green above, $1\frac{1}{2}'–3\frac{1}{2}'$ long, $\frac{1}{2}'–1\frac{1}{2}'$ wide, the lower green, the upper smaller and white-canescent on both sides; clusters loose, villous and canescent; terminal and axillary, $1'–2'$ broad, the flowers often secund; bracts linear-oblong, acuminate or subulate-tipped, villous, spreading; calyx-teeth very unequal, subulate, villous, the longer about equalling the tube; corolla-tube not exceeding the calyx.

In dry woods and on hills, Virginia to Georgia, Kentucky and Tennessee. Calamint. July–Sept.

10. Koellia albéscens (T. & G.) Kuntze.
White-leaved Mountain-Mint. Fig. 3669.

P. albescens T. & G.; A. Gray, Am. Journ. Sci. **42** :
    45. 1842.
Koellia albescens Kuntze, Rev. Gen. Pl. 520. 1891.

Stem slender, soft-pubescent nearly to the base.
1°-2° high.   Leaves ovate to ovate-lanceolate, petioled, acute or subacuminate at the apex, narrowed
or sometimes rounded at the base, sharply serrate
or nearly entire, 1'-2½' long, ½'-1¼' wide, white-
canescent beneath, green above, or the upper canescent on both sides; clusters loose, terminal and
axillary, densely canescent, not at all villous, at
length about 1' broad; bracts linear, or the outer
broader, spreading, sometimes exceeding the clusters; calyx densely canescent, its teeth triangular,
obtuse or acute, slightly unequal, one-fifth to one-
fourth as long as the tube; corolla-tube longer
than the calyx.

In dry woods and thickets, southern Virginia to
Kentucky, Missouri, Arkansas, Florida and Texas.
July–Sept.

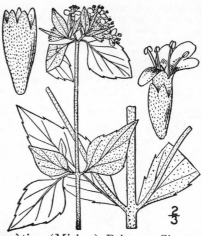

11. Koellia mùtica (Michx.) Britton. Short-
toothed Mountain-Mint. Fig. 3670.

Brachystemon muticum Michx. Fl. Bor. Am. **2** : 6. *pl. 32.*
    1803.
Koellia mutica Britton, Mem. Torr. Club **4** : 145. 1894.

Stem stiff, puberulent, or glabrous below, 1°-2½°
high.   Leaves short-petioled or sessile, strongly veined,
ovate or ovate-lanceolate, firm, acute at the apex,
rounded or sometimes subcordate at the base, sharply
serrate or serrulate, 1½'-3' long, ½'-1½' wide, the lower
mostly glabrous, the uppermost white-canescent on both
sides, much smaller; flower-clusters capitate, dense, terminal, corymbose, often also in the upper axils, pubescent or canescent, 4''-6'' broad; bracts appressed, lanceolate-subulate or the outer broader; calyx pubescent,
its teeth nearly equal, triangular-ovate or triangular-
lanceolate, not much longer than wide, about one-fifth
as long as the tube; corolla-tube exceeding the calyx.

In sandy soil, Maine to Virginia and Florida, west to
Pennsylvania and Missouri. Calamint. July–Sept.

12. Koellia montàna (Michx.) Kuntze. Thin-
leaved Mountain-Mint. Fig. 3671.

Pycnanthemum montanum Michx. Fl. Bor. Am. **2** : 8. 1803.
Monardella montana Benth. Lab. Gen. & Sp. 331. 1834.
Koellia montana Kuntze, Rev. Gen. Pl. 520. 1891.

Stem slender, glabrous or nearly so throughout, 2°-3°
high.   Leaves distinctly petioled, membranous, glabrous,
ovate-lanceolate or the upper lanceolate, long-acuminate
at the apex, usually narrowed at the base, sharply serrate,
2'-5' long, ½'-2' wide, none of them canescent; flower-
clusters rather dense, terminal and in the upper axils, ½'-1'
broad; bracts appressed, lanceolate or linear-lanceolate,
acuminate, not exceeding the fully developed clusters,
bearded more or less with long hairs; calyx tubular,
glabrous or puberulent, or villous above, its teeth triangular-
subulate, equal, one-fifth to one-fourth as long as the
tube; corolla-tube longer than the calyx.

In woods, mountains of southern Virginia to Georgia, Tennessee and Alabama. July–Sept.

34. CUNÌLA L. Syst. Ed. 10, 1359. 1759.

Perennial branching herbs, or low shrubs, with dentate or entire leaves, and rather
small, clustered, purple or white flowers. Calyx tubular-ovoid, 10-13-nerved, villous in the
throat, equally 5-toothed. Corolla 2-lipped, longer than the calyx, the upper lip erect, emarginate, the lower spreading, 3-cleft. Anther-bearing stamens 2, long-exserted, straight, the

posterior pair rudimentary, or wanting; anther-sacs parallel. Ovary deeply 4-parted; style slender, 2-cleft at the summit. Nutlets smooth; scar of attachment basal and small. [Latin name of some plant.]

About 15 species, natives of America. The following typical species is the only one known in the United States.

1/2

1. Cunila origanoìdes (L.) Britton.   Stone Mint. Sweet Horse-Mint.   American Dittany. Wild Basil.   Fig. 3672.

Satureia origanoides L. Sp. Pl. 568.   1753.
Cunila Mariana L. Syst. Ed. 10, 1359.   1759.
Hedyosmos origanoides Kuntze, Rev. Gen. Pl. 520.   1891.
Cunila origanoides Britton, Mem. Torr. Club 5 : 278.   1894.

Plant very aromatic; stem slender, stiff, branched, glabrous, or pubescent at the nodes, erect, 8'–20' high, the branches ascending. Leaves ovate, sessile or very short-petioled, acute at the apex, sharply serrate, rounded, truncate or subcordate at the base, ½'–1½' long, densely punctate; flowers nearly ½' long, numerous in terminal loose cymose clusters; corolla purple-pink, one-half as long as the stamens; posterior pair of stamens usually rudimentary.

In dry woods and thickets, southern New York to Florida, west to Ohio, Missouri, Arkansas and Texas. Aug.–Sept.

35. LÝCOPUS [Tourn.] L. Sp. Pl. 21.   1753.

Herbs, perennial by slender stolons or suckers, with erect or diffuse stems, petioled or sessile leaves, and small white or purple flowers, bracted and verticillate in dense axillary clusters. Calyx campanulate, regular or nearly so, 4–5-toothed, not bearded in the throat, the teeth obtuse or acute. Corolla funnelform-campanulate to cylindric, equalling or longer than the calyx, the limb nearly equally 4-cleft, or one of the lobes broader and emarginate. Perfect stamens 2, anterior, the posterior pair rudimentary, or altogether wanting; anther-sacs parallel. Ovary deeply 4-parted; style slender, 2-cleft at the summit. Nutlets truncate at the summit, narrowed below, trigonous, smooth, their margins thickened. [Greek, wolf-foot.]

About 15 species of the north temperate zone. Besides the following, two or three others occur in western North America. Type species: *Lycopus europaeus* L.

Calyx-teeth 4 or 5, ovate, shorter than the nutlets.
  Base of the stem not tuberous; leaves ovate or ovate-lanceolate.   1. *L. virginicus.*
  Base of the stem tuberous-thickened; leaves oblong to lanceolate.   2. *L. uniflorus.*
Calyx-teeth mostly 5, lanceolate or subulate, longer than the nutlets.
  Bracts minute; corolla twice as long as the calyx.
    Leaves sessile.   3. *L. sessilifolius.*
    Leaves narrowed into a manifest petiole.   4. *L. rubellus.*
  Bracts lanceolate or subulate; corolla not twice as long as the calyx.
    Leaves pinnatifid or deeply incised.   5. *L. americanus.*
    Leaves merely coarsely dentate or serrate (lower rarely incised).
      Leaves oblong or oblong-lanceolate, serrate.   6. *L. asper.*
      Leaves ovate, coarsely dentate.   7. *L. europaeus.*

1. Lycopus virgínicus L.   Bugle-weed.   Bugle-wort.   Fig. 3673.

Lycopus virginicus L. Sp. Pl. 21.   1753.

Perennial by long filiform leafy stolons, glabrous or puberulent; stem slender, erect or ascending, simple or branched, 6'–2° high. Leaves ovate or ovate-lanceolate, acuminate at the apex, sharply dentate, narrowed or cuneate at the base, petioled, or the upper sessile, dark green or purple, 1½'–3' long, ½'–1½' wide; bracts short, oblong; calyx-teeth 4, or sometimes 5, ovate or ovate-lanceolate, obtuse or subacute; corolla about 1'' broad, narrow, nearly twice as long as the calyx, or longer; rudimentary posterior stamens minute; nutlets longer than or about equalling the calyx.

In wet soil, New Hampshire to Florida, Alabama, Missouri and Nebraska. Northern Asia. Sometimes called wood betony. July–Sept.

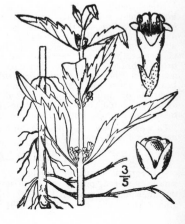

3/5

2. Lycopus uniflòrus Michx.   Northern Bugle-weed.   Fig. 3674.

Lycopus uniflorus Michx. Fl. Bor. Am. 1 : 14.   1803.
Lycopus communis Bicknell, Britton Man. 803.   1901.

Green or sometimes purplish, mostly less puberulent than *L. virginicus;* stems rather slender, simple or sparingly branched, 4'–2⅔° tall, rather acutely angled, tuberous at the base, the stolons rarely tuber-bearing.   Leaves oblong or oblong-lanceolate, acute or acuminate at the apex, serrate, ¾'–3¾' long, sessile or nearly so; calyx-teeth 4 or 5, triangular, ovate or ovate-oblong, rather obtuse; corolla about 1½'' long, less than twice as long as the calyx; rudimentary posterior stamens obsolete or minute; nutlets about as long as the calyx, oblique at the apex.

In low grounds, Newfoundland to British Columbia, North Carolina, Nebraska and Oregon.   Summer and fall.

Lycopus membranàceus Bicknell, with thinner, often coarsely-toothed, longer-petioled and larger leaves, appears to be a race of this species.

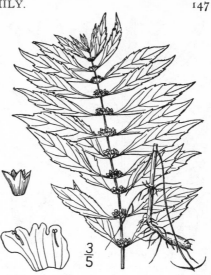

3. Lycopus sessilifòlius A. Gray.   Sessile-leaved Water Hoarhound.   Fig. 3675.

Lycopus europaeus var. *sessilifolius* A. Gray, Man. Ed. 5, 345.   1867.
Lycopus sessilifolius A. Gray, Proc. Am. Acad. 8 : 285.   1870.

Glabrous, or puberulent above, tuberiferous, perennial by stolons and suckers; stem simple, or at length branched, erect, 1°–2° high.   Leaves ovate-lanceolate or oblong-lanceolate, closely sessile, acute or acuminate at the apex, sharply serrate, somewhat narrowed at the base, 1'–2' long; bracts very small, acute; calyx-teeth usually 5, subulate, rigid, nearly as long as the tube; corolla twice as long as the calyx; rudimentary posterior stamens oval; nutlets shorter than the calyx.

In wet soil, eastern Massachusetts to Florida and Mississippi, near the coast.   Aug.–Oct.

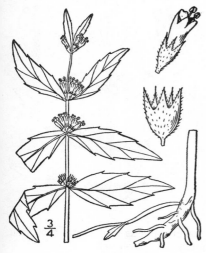

4. Lycopus rubéllus Moench.   Stalked Water Hoarhound.   Fig. 3676.

Lycopus rubellus Moench, Meth. Suppl. 146.   1802.
Lycopus europaeus var. *integrifolius* A. Gray, Man. Ed. 5, 346.   1867.

Glabrous or minutely puberulent, perennial by leafy stolons; stem erect or ascending, simple or at length freely branched, 1°–3° high.   Leaves ovate to oblong-lanceolate, or narrower, acuminate at the apex, usually dentate, narrowed or cuneate at the base, 2'–5' long, ¾'–1½' wide, usually tapering into a conspicuous petiole; bracts minute, acute or acuminate; calyx-teeth triangular-subulate, herbaceous, one-half as long as the tube or more; corolla longer than the calyx; rudimentary posterior stamens oval or oblong; nutlets much shorter than the calyx.

In wet soil, southern Vermont to Florida, Minnesota, Arkansas and Louisiana.   Gipsywort.   July–Oct.

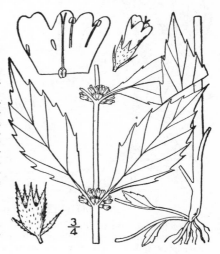

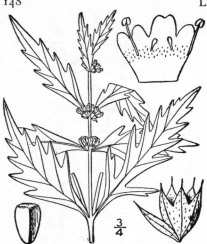

5. Lycopus americànus Muhl. Cut-leaved
Water Hoarhound. Fig. 3677.

L. americanus Muhl.; Bart. Fl. Phil. Prodr. 15. 1815.
Lycopus sinuatus Ell. Bot. S. C. & Ga. 1 : 26. 1817.
Lycopaeus europaeus var. *sinuatus* A. Gray, Man.
Ed. 5, 346. 1867.

Puberulent or glabrous, perennial by suckers;
stem stiff, erect, simple or branched, 1°–2° high.
Leaves lanceolate or ovate-lanceolate in outline,
petioled, acuminate at the apex, incised, pinnatifid or the uppermost merely serrate, 2′–4′ long;
bracts subulate, the outer ones sometimes exceeding the calyx; calyx-teeth triangular-subulate,
cuspidate, rigid; corolla little exceeding the calyx; rudimentary posterior stamens thickened at
their tips; nutlets much shorter than the calyx.

In wet soil, Newfoundland to British Columbia,
south to Florida, Texas, Utah and California. Gipsy-
wort. Bitter bugle. Paul's-betony. June–Oct.

6. Lycopus asper Greene.   Western Water Hoarhound.   Fig. 3678.

Lycopus asper Greene, Pittonia 3 : 339. 1898.
Lycopus lucidus var. *americanus* A. Gray, Proc. Am.
Acad. 8 : 286. 1870.

Pubescent or glabrate, perennial by stolons; stem
usually stout, erect, strict, leafy, simple, or sometimes branched, 1°–3° high. Leaves oblong-lanceolate, acute at the apex, narrowed or rounded at the
base, sessile, or very short-petioled, 2′–6′ long, ¼′–1½′
wide, sharply serrate with acute ascending teeth;
bracts ovate or lanceolate, acuminate-subulate, the
outer ones often as long as the flowers; calyx-teeth
3, subulate-lanceolate, nearly as long as the tube;
corolla little longer than the calyx; rudimentary
stamens slender, thickened at the tips; nutlets much
shorter than the calyx.

In wet soil, Michigan to Kansas, west to Manitoba,
British Columbia, California and Arizona. Regarded in
the first edition of this work as the same as *L. lucidus*
Turcz. of NW. America and NE. Asia. July–Sept.

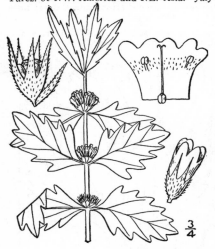

7. Lycopus europaèus L.   Water or Marsh
Hoarhound. Gipsy-wort. Gipsy-herb.
Fig. 3679.

Lycopus europaeus L. Sp. Pl. 21. 1753.

Puberulent or pubescent, perennial by suckers;
stems stout, at length widely branched, 1°–2½°
high. Leaves ovate, ovate-oblong, or oblong-
lanceolate, short-petioled, or the upper sometimes
sessile, coarsely dentate, or the lower incised at
the base, 1′–3′ long, ½′–1′ wide; bracts subulate-
lanceolate, the outer shorter than or equalling
the flowers; calyx-teeth subulate-spinulose; corolla scarcely longer than the calyx; rudimentary
posterior stamens obsolete; nutlets shorter than
the calyx.

In waste places, Massachusetts to Virginia. Naturalized from Europe. Green archangel. Bitter bugle-
weed. July–Sept.

36. **MÉNTHA** [Tourn.] L. Sp. Pl. 576. 1753.

Erect or diffuse odorous herbs, with simple sessile or petioled mostly punctate leaves,
and small whorled purple pink or white flowers, the whorls axillary or in terminal dense or
interrupted spikes.   Calyx campanulate to tubular, 10-nerved, regular, or slightly 2-lipped,

5-toothed. Corolla-tube shorter than the calyx, the limb 4-cleft, somewhat irregular, the posterior lobe usually somewhat broader than the others, entire or emarginate. Stamens 4, equal, erect, included or exserted, sometimes imperfect; filaments glabrous; anthers 2-celled, the sacs parallel. Ovary 4-parted; style 2-cleft at the summit. Nutlets ovoid, smooth. [Name used by Theophrastus; from the nymph Minthe.]

About 30 species, natives of the north temperate zone. The more or less characteristic odors of the species change during the progress of the life of the plant. Type species: *Mentha spicata* L.

* Whorls of flowers in terminal spikes, or some in the upper axils.

Plants glabrous or very nearly so.	
Spikes slim, narrow, mostly interrupted; leaves sessile, or nearly so.	1. *M. spicata.*
Spikes thick, mostly dense, at first short; leaves petioled.	
Leaves lanceolate or oblong, acute.	2. *M. piperita.*
Leaves ovate, obtuse, or the upper acute, subcordate.	3. *M. citrata.*
Plants villous, hirsute or canescent, at least at the nodes.	
Spikes slim or narrow, often interrupted.	
Leaves lanceolate or ovate-lanceolate, acute.	4. *M. longifolia.*
Leaves elliptic or ovate-oblong, obtuse, reticulated beneath.	5. *M. rotundifolia.*
Spikes thick (6″), dense, elongated or short.	
Leaves sessile; spikes 1′–3′ long; plant canescent.	6. *M. alopecuroides.*
Leaves distinctly petioled, or the uppermost sessile; spikes short.	
Leaves simply serrate.	7. *M. aquatica.*
Leaves mostly incised, the margins crisped and wavy.	8. *M. crispa.*

** Whorls of flowers all axillary.

Upper leaves much smaller than the lower.	9. *M. Cardiaca.*
Upper leaves not conspicuously reduced.	
Stem pubescent.	
Leaves rounded or obtuse at the base.	10. *M. arvensis.*
Leaves narrowed, mostly cuneate at the base.	11. *M. canadensis.*
Stem glabrous or nearly so.	12. *M. gentilis.*

1. Mentha spicàta L. Spearmint. Lamb or Common Mint. Our Lady's Mint. Fig. 3680.

Mentha spicata L. Sp. Pl. 576. 1753.
Mentha spicata var. *viridis* L. loc. cit. 1753.
Mentha viridis L. Sp. Pl. Ed. 2, 804. 1763.

Glabrous, perennial by leafy stolons; stem erect, branched, 1°–1½° high. Leaves lanceolate, sessile or short-petioled, sharply serrate, acute or acuminate at the apex, narrowed at the base, the largest about 2½′ long; whorls of flowers in terminal narrow acute usually interrupted spikes, which become 2′–4′ long in fruit, the one terminating the stem surpassing the lateral ones; bracts subulate-lanceolate, ciliate, some of them usually longer than the flowers; calyx campanulate, its teeth hirsute or glabrate, subulate, nearly as long as the tube; corolla glabrous.

In moist fields or waste places, Nova Scotia to Ontario, Minnesota, Washington, Florida, Texas and California. Naturalized from Europe. Also in Bermuda. Native also of Asia. Garden-, brown- or mackerel-mint. Sage-of-bethlehem. July–Sept.

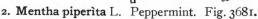

$\frac{3}{5}$

2. Mentha piperìta L. Peppermint. Fig. 3681.

Mentha piperita L. Sp. Pl. 576. 1753.

Perennial by subterranean suckers; stems glabrous, mostly erect, branched, 1°–3° high. Leaves lanceolate, petioled, dark green, acute at the apex, rounded or narrowed at the base, rather firm, sharply serrate, glabrous on both sides, or pubescent on the veins beneath, the larger 1½′–3′ long, 1′–1½′ wide; whorls of flowers in terminal dense or interrupted spikes, which are thick and obtuse, and become 1′–3′ long in fruit, the middle one at length overtopped by the lateral ones; bracts lanceolate, acuminate, not longer than the flowers, or the lower occasionally foliaceous; calyx tubular-campanulate, glabrous below, its teeth subulate, ciliate, one-half as long as the tube or more; corolla glabrous; style occasionally 3-cleft.

In wet soil, Nova Scotia to Ontario and Minnesota, south to Florida, Tennessee and Arkansas. Also in California, Bermuda and Jamaica. Naturalized from Europe. Lamb- or brandy-mint. July–Sept.

$\frac{3}{5}$

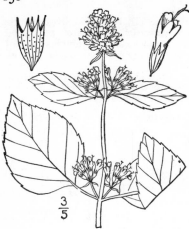

3/5

3. Mentha citràta Ehrh. Bergamot Mint.
Fig. 3682.

Mentha citrata Ehrh. Beitr. **7** : 150. 1792.

Perennial by leafy stolons, glabrous throughout; stem weak, branched, decumbent or ascending, 1°–2° long. Leaves petioled, thin, ovate or ovate-orbicular, obtuse or the upper acute at the apex, rounded or subcordate at the base, sharply serrate with low teeth, the larger about 2′ long, 1′–1½′ wide; whorls of flowers in terminal dense thick obtuse spikes, and commonly also in the uppermost axils; spikes scarcely more than 1′ long in fruit; calyx glabrous, its teeth subulate, one-half as long as the tube, or longer; corolla glabrous.

In wet soil, Connecticut to New York, Ohio, Michigan and Missouri. Naturalized from Europe. Also in Bermuda and Porto Rico. July–Sept.

4. Mentha longifòlia (L.) Huds. Horse Mint. Fig. 3683.

Mentha spicata var. *longifolia* L. Sp. Pl. 576. 1753.
Mentha longifolia Huds. Fl. Angl. 221. 1762.
Mentha sylvestris L. Sp. Pl. Ed. 2, 804. 1763.

Perennial by suckers, canescent or puberulent nearly all over; stems mostly erect, branched, or simple, 1°–2½° high. Leaves lanceolate, ovate-lanceolate, or oblong-lanceolate, sessile, or very short-petioled, acute at the apex, usually rounded at the base, sharply serrate, 1′–3′ long, ½′–1¼′ wide, sometimes glabrous above; whorls of flowers in terminal narrow dense or interrupted acute spikes, which become 2′–5′ long in fruit; bracts lanceolate-subulate, the lower equalling or longer than the flowers; calyx tomentose or canescent, its teeth subulate, one-half as long as the campanulate tube; corolla puberulent.

In waste places, Connecticut to Delaware, New Jersey and Ohio. Naturalized from Europe. Brook- or fish-mint. Water or European horse-mint. July–Oct.

2/3

5. Mentha rotundifòlia (L.) Huds. Round-leaved Mint. Fig. 3684.

Mentha spicata var. *rotundifolia* L. Sp. Pl. 576. 1753.

Mentha rotundifolia Huds. Fl. Angl. 221. 1762.

Perennial by leafy stolons, canescent or tomentose-puberulent, somewhat viscid; stems ascending or erect, simple or branched, usually slender, 1½°–2½° high. Leaves elliptic, or ovate-oblong, short-petioled, or sessile and somewhat clasping by the subcordate or rounded base, obtuse at the apex, crenate-serrate with low teeth, 1′–2′ long, 9″–15″ wide, more or less rugose-reticulated beneath; whorls of flowers in terminal dense or interrupted spikes which elongate to 2′–4′ in fruit; bracts lanceolate, acuminate, commonly shorter than the flowers; calyx-teeth setaceous, usually about one-half as long as the tube; corolla puberulent.

In waste places, Maine to Florida, Ohio, Arkansas, Texas and Mexico. Bermuda. Patagonia- or apple-mint. Horse mint. Wild mint. Naturalized from Europe. July–Sept.

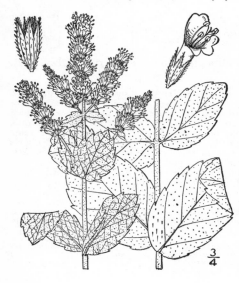

3/4

6. Mentha alopecuroìdes Hull. Woolly Mint. Fig. 3685.

Mentha alopecuroides Hull. Brit. Fl. 221. 1799.

Perennial by suckers, white-woolly; stem stout, leafy, erect or ascending, simple or branched, 1½°–3° high. Leaves broadly oval, sessile, or partly clasping by a subcordate or rarely rounded base, obtuse at the apex, sharply and rather coarsely serrate, pinnately-veined, the lower 2′–3′ long, 1½′–2′ wide; spikes rather thick, dense, stout, obtuse, 2′–3′ long in fruit; bracts lanceolate, shorter than the flowers; calyx-teeth setaceous, one-half as long as the campanulate tube, or more; corolla pubescent.

Along roadsides, Connecticut to New Jersey, Pennsylvania, Wisconsin and Missouri. Naturalized from Europe. July–Oct.

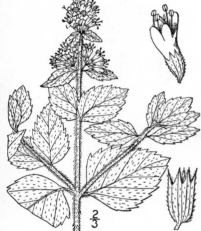

7. Mentha aquática L. Water Mint. Fish Mint. Fig. 3686.

Mentha aquatica L. Sp. Pl. 576. 1753.

Perennial by suckers, hirsute or pubescent, rarely glabrate; stem stout, erect, leafy, usually branched, 1½°–2½° high, its hairs reflexed. Leaves broadly ovate, petioled, acute, subacute or the lower obtuse at the apex, rounded, subcordate or rarely narrowed at the base, sharply serrate, the larger 1½′–3′ long and nearly as wide; whorls of flowers in terminal dense short thick rounded spikes, and usually also in the upper axils; spikes seldom more than 1′ long in fruit; bracts lanceolate, shorter than the flowers; calyx hirsute, its teeth lanceolate-subulate or triangular-lanceolate, one-third to one-half as long as the nearly cylindric tube; corolla sparingly pubescent.

In wet places, Nova Scotia to Pennsylvania and Georgia. Naturalized from Europe. Aug.–Oct.

8. Mentha críspa L. Crisped-leaved, Curled or Cross Mint. Fig. 3687.

Mentha crispa L. Sp. Pl. 576. 1753.
Mentha aquatica var. *crispa* Benth. Lab. Gen. & Sp. 177. 1833.

Sparingly pilose-pubescent at least at the nodes, petioles and veins of the lower surfaces of the leaves; stem rather weak, usually much branched, 1½°–3° long. Leaves distinctly petioled, or the uppermost sessile, ovate in outline, mostly acute at the apex, rounded, truncate or subcordate at the base, their margins crisped, wavy and incised, or the uppermost merely sharply serrate; whorls of flowers in dense thick rounded terminal spikes, which become 1′–1½′ long in fruit; calyx sparingly pubescent or glabrous, its teeth subulate, more than one-half as long as the campanulate tube; corolla glabrous.

In swamps and roadside ditches, Connecticut to New Jersey and Pennsylvania. Balm-mint. Aug.–Oct.

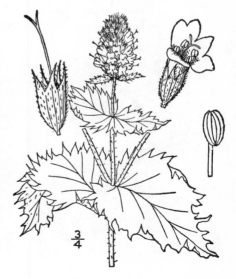

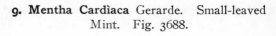

9. Mentha Cardìaca Gerarde. Small-leaved Mint. Fig. 3688.

Mentha Cardiaca Gerarde; Baker, Journ. Bot. **2** : 245. 1865.

Pubescent or glabrate, often much branched, $1\frac{1}{2}°-2\frac{1}{2}°$ high, the upper branches ascending. Leaves lanceolate to oblong-lanceolate or ovate-lanceolate, acuminate or acute at the apex, the lower 2′–3′ long, the upper much smaller, all sharply serrate; flowers whorled in the upper axils; calyx-teeth subulate, about one-half as long as the tube.

Wet grounds, mostly along rivers and streams, Nova Scotia to New Jersey, Pennsylvania and District of Columbia. July–Sept.

10. Mentha arvénsis L. Corn Mint. Field Mint. Fig. 3689.

Mentha arvensis L. Sp. Pl. 577. 1753.

Perennial by suckers, pubescent or glabrate; stems erect or ascending, simple or branched, 6′–2° high, slender. Leaves oblong-lanceolate, oval or ovate, petioled, acute at the apex or the lower obtuse, crenate-serrate with bluntish teeth, rounded at the base, 1′–2½′ long, ½′–1′ wide, the upper not much smaller than the lower; whorls of flowers all axillary, usually about equalling the petioles; calyx pubescent, campanulate, its teeth triangular, about as long as the width of their base, acute or sometimes obtuse, one-third as long as the tube.

In dry waste places, New Brunswick to northern New York, New Jersey, Pennsylvania and Florida. Also in California and Mexico. Naturalized from Europe. Lamb's-tongue. Wild pennyroyal. July–Sept.

11. Mentha canadénsis L. American Wild Mint. Fig. 3690.

Mentha canaaensis L. Sp. Pl. 577. 1753.
Mentha borealis Michx. Fl. Bor. Am. **2** : 2. 1803.
Mentha canadensis var. *glabrata* Benth. in DC. Prodr. **12** : 173. 1848.

Perennial by suckers; stem more or less pubescent with spreading or scarcely reflexed hairs, or glabrate, erect, simple, or branched, usually slender, 6′–2½° high. Leaves oblong or ovate-oblong, or oblong-lanceolate, slender-petioled, acute at the apex, or the lower obtuse, sharply serrate, narrowed to a somewhat cuneate acute or obtuse base, glabrous or very sparingly pubescent, the larger 2′–3′ long, ½′–1′ wide; whorls of flowers all axillary, often shorter than the petioles; calyx oblong-campanulate, densely or sparingly pubescent all over, its teeth one-fourth to one-third as long as the tube.

In moist soil, New Brunswick to Manitoba, British Columbia, Virginia, Nebraska, New Mexico and Nevada. Consists of several races, differing in pubescence, leaf-form and size. Odor like Pennyroyal. July–Oct.

12. Mentha gentìlis L. Creeping or Downy
Whorled Mint. Spearmint. Fig. 3691.

Mentha gentilis L. Sp. Pl. 577. 1753.
Mentha sativa L. Sp. Pl. Ed. 2, 805. 1763.

Perennial by suckers; stem rather stout, ascending or erect, branched, glabrous or puberulent
with short reflexed hairs, 1°–2° high. Leaves
ovate or oval, short-petioled, sparingly pubescent
with scattered hairs on both surfaces, often blotched, pinnately veined, acute at both ends, sharply
serrate, the larger 1½–2′ long, the upper sometimes much smaller than the lower; whorls of
flowers all axillary; pedicels glabrous; calyx campanulate, glabrous below, its teeth subulate, ciliate,
one-half as long as the tube; corolla glabrous.

In waste places and along streams, Nova Scotia to
northern New York, Iowa, North Carolina and Tennessee. Naturalized from Europe. Aug.–Oct.

37. COLLINSÒNIA L. Sp. Pl. 28.
1753.

Tall perennial aromatic herbs, with large membranous petioled leaves, and terminal
loosely panicled racemes of small, yellowish, mostly opposite flowers. Calyx campanulate,
short, 10-nerved, 2-lipped, declined in fruit, usually pubescent in the throat; upper lip 3-toothed;
lower 2-cleft. Corolla much longer than the calyx, obliquely campanulate, 5-lobed, 4 of the
lobes nearly equal, the 5th pendent or declined, fimbriate or lacerate, much larger, appearing
like a lower lip. Anther-bearing stamens 2, not declined, much exserted, coiled before
antithesis; bases of the filaments connected by a woolly ring; anthers 2-celled, or the sacs
at length partially confluent. Ovary deeply 4-parted. Nutlets smooth, globose. [Named for
Peter Collinson, 1693–1768, an English botanist, and correspondent of Linnaeus.]

Three species, natives of eastern North America, the following typical.

1. Collinsonia canadénsis L. Horse- or Ox-balm. Citronella. Rich-weed.
Fig. 3692.

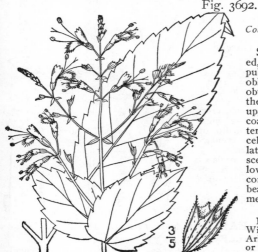

Collinsonia canadensis L. Sp. Pl. 28. 1753.

Stem stout, erect or ascending, branched, 2°–5° high, glabrous, or glandular-
pubescent above. Leaves ovate or ovate-
oblong, acuminate at the apex, narrowed,
obtuse or sometimes cordate at the base,
the lower slender-petioled, 6′–10′ long, the
upper nearly sessile, much smaller, all
coarsely dentate; racemes numerous, in
terminal panicles sometimes 1° long; pedicels ascending, 3″–6″ long in fruit, subu-
late-bracteolate at the base; flowers lemon-
scented; calyx-teeth subulate, those of the
lower lip much longer than the upper;
corolla light yellow, 5″–7″ long; anther-
bearing stamens 2, the upper pair rudimentary; fruiting calyx ribbed, 3″–4″ long.

In moist woods, Quebec and Ontario to
Wisconsin, south to Florida, Alabama and
Arkansas. Root large, thick, woody. Knob-
or horse-weed. Knob-grass or -root. Collinson's-flower. Rich leaf. Stone-root. July–Oct.

38. PERÍLLA Ard.; L. Gen. Pl. Ed. 6, Add. 578. 1764.

Annual herbs, with petioled purple or discolored leaves, and small flowers in loose
bracted racemes. Calyx campanulate, 10-nerved, 5-cleft, nearly regular in flower, enlarging,
declined and becoming 2-lipped in fruit, the upper lip 3-toothed, the lower 2-cleft, the throat
not bearded. Corolla-tube not longer than the calyx, the throat obliquely campanulate, the
limb 5-cleft, the lower lobe slightly the larger. Stamens 4, nearly equal, or the posterior

pair shorter, erect, divergent; anthers 2-celled.   Style deeply 2-cleft; ovary 4-parted.   Nutlets globose, reticulated.   [The native name in India.]

One or 2 species, natives of Asia, the following typical.

1. Perilla frutéscens (L.) Britton.
Perilla.   Beef-steak Plant.   Fig. 3693.

Ocimum frutescens L. Sp. Pl. 597. 1753.
Perilla ocimoides L. Gen. Ed. 6, Add. 578. 1764.
P. frutescens Britton, Mem. Torr. Cl. 5 : 277. 1894.

Purple or purple-green, sparingly pubescent; stem stout, erect, much branched, 1°–3° high, leafy. Leaves long-petioled, broadly ovate, acuminate at the apex, narrowed at the base, coarsely dentate or incised, 3′–6′ long and nearly as wide; racemes terminal and axillary, many-flowered, 3′–6′ long; pedicels spreading, 1½″–3″ long in fruit; calyx minute in flower, much enlarged, gibbous at the base and densely pilose-pubescent in fruit; corolla purple or white, 1½″ long, with a woolly ring within.

In waste places, escaped from gardens, Connecticut to Florida, Illinois, Missouri and Texas. Native of India. July–Oct.

39. ELSHÓLTZIA Willd. in Roem. & Ust. Mag. Bot. 11 : 3.   1790.

Herbs, with thin mostly petioled leaves, and small or minute clustered flowers, in terminal bracted spikes. Calyx campanulate or ovoid, 10-nerved, scarcely oblique, enlarging in fruit, not bearded in the throat, 5-toothed, the teeth nearly equal. Corolla-tube little longer than the calyx, straight, or a little curved, the limb oblique, or slightly 2-lipped, 4-lobed; upper lobe erect, concave, emarginate, the 3 others spreading. Stamens 4, divergent, didynamous, ascending, exserted, the upper pair shorter; anthers 2-celled, or the sacs more or less confluent. Style 2-cleft at the summit. Ovary 4-parted. Nutlets ovoid or oblong, tuberculate, or nearly smooth. [Named in honor of J. S. Elsholtz, a Prussian botanist.]

About 20 species, natives of Asia.   Type species: *Elsholtzia cristata* Willd.

1. Elsholtzia Patrínii (Lepech.) Garcke.
Elsholtzia.   Fig. 3694.

Mentha Patrinii Lepech. Nov. Act. Petrop. 13 : 336. 1802.
E. cristata Willd. in Roem. & Ust. Mag. Bot. 11 : 3. 1790.
Elsholtzia Patrinii Garcke, Garcke, Fl. Deutsch. Ed. 4, 257. 1858.

Annual, glabrous or nearly so; stems, weak, erect or ascending, at length widely branched, 1°–2° high. Leaves long-petioled, ovate or oblong, acute or acuminate at the apex, narrowed at the base, crenate-dentate, 1′–3′ long; spikes terminal, very dense, 1′–3′ high, about ½′ thick; flowers several in the axils of each of the broadly ovate membranous green reticulated mucronate bracts; calyx hirsute, shorter than the bract; corolla 1″ long, pale purple.

Notre Dame du Lac, Temiscouata Co., Quebec. Naturalized from Asia. July–Aug.

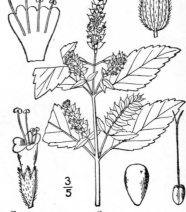

Family 27. SOLANÀCEAE Pers. Syn. 1 : 214.   1805.
POTATO FAMILY.

Herbs, shrubs, vines, or some tropical species trees, with alternate or rarely opposite, exstipulate entire dentate lobed or dissected leaves, and perfect regular or nearly regular cymose flowers. Calyx inferior, gamosepalous, mostly 5-lobed. Corolla gamopetalous, rotate, campanulate, funnelform, salverform or tubular, mostly 5-lobed, the lobes induplicate-valvate or plicate in the bud. Stamens as many as the lobes of the corolla and alternate with them, inserted on the tube, all equal and perfect in the following genera, except in *Petunia*, where 5 are didynamous and the fifth smaller or obsolete; anthers 2-celled, apically or longitudinally dehiscent. Disk present, or none. Ovary entire, superior, 2-celled (rarely 3–5-celled); ovules numerous on the axile placentae, anatropous or amphitropous; style slender, simple; stigma terminal; fruit a berry or capsule. Seeds numerous, the testa sometimes roughened; embryo terete, spiral, curved, or nearly straight; endosperm fleshy; cotyledons semiterete.

About 75 genera and 1750 species, widely distributed, most abundant in tropical regions.
* **Fruit a pulpy berry; corolla plicate, its lobes generally induplicate.**
Anthers unconnected, destitute of terminal pores, dehiscent.
  Fruiting calyx bladdery-inflated.
    Fruiting calyx 5-angled and deeply 5-parted; ovary 3–5-celled.   1. *Physalodes.*
    Fruiting calyx 5-lobed, not parted, 10-ribbed, often 5–10-angled, reticulated, wholly enclosing
      the berry; ovary 2-celled.
      Corolla open-campanulate, yellowish or whitish, often with a dark center; seeds with a
        thin margin, finely pitted.   2. *Physalis.*
      Corolla flat-rotate, violet or purple; seeds thick, rugose-tuberculate.   3. *Quincula.*
  Fruiting calyx somewhat enlarged, but closely fitted to the fruit, thin, obscurely veiny, open at
    the mouth.
    Corolla rotate, whitish; lobes of fruiting calyx much exceeding the berry. 4. *Leucophysalis.*
    Corolla rotate, whitish, sometimes tinged with purple; fruiting calyx not exceeding the
      berry.   5. *Chamaesaracha.*
Anthers connivent or slightly connate; fruiting calyx not enlarged.
  Anthers short or oblong, opening by a terminal pore or short slit in our species. 6. *Solanum.*
  Anthers long, tapering from base to summit, longitudinally dehiscent. 7. *Lycopersicon.*
** **Fruit a nearly dry berry; corolla campanulate, little or not at all plicate, its lobes imbricated.**
        8. *Lycium.*
        *** **Fruit a capsule; corolla funnelform.**
Capsule circumscissile toward the top, which separates as a lid; corolla irregular. 9. *Hyoscyamus.*
Capsule opening by valves.
  Capsule generally prickly.   10. *Datura.*
  Capsule not prickly.
    Flowers paniculate or racemose; stamens nearly uniform in length. 11. *Nicotiana.*
    Flowers solitary; stamens very unequal. 12. *Petunia.*

1. PHYSALÒDES Boehm. in Ludwig, Def. 41. 1760.
[Nicandra Adans. Fam. Pl. **2**: 219. 1763.]

An annual erect branching glabrous herb, with alternate petioled thin sinuate-dentate or lobed leaves, and large light blue peduncled nodding flowers, solitary in the axils. Calyx 5-parted, 5-angled, much inflated in fruit, its segments ovate, connivent, cordate or sagittate at the base, strongly reticulated. Corolla broadly campanulate, plicate in the bud, slightly 5-lobed. Stamens 5, included, inserted on the corolla near its base; filaments filiform, dilated and pilose below; anthers oblong, the sacs longitudinally dehiscent. Ovary 3–5-celled; style slender; stigma 3–5-lobed. Berry globose, nearly dry, enclosed in the calyx. [Greek, Physalis-like.]

A monotypic Peruvian genus.

1. Physalodes physalòdes (L.) Britton.
Apple-of-Peru. Fig. 3695.
Atropa physalodes L. Sp. Pl. 181. 1753.
Physalodes peruvianum Kuntze, Rev. Gen. Pl. 452.
  1891.
P. physalodes Britton, Mem. Torr. Club **5**: 287. 1894.

Stem angled, 2°–5° high. Leaves ovate or oblong, acuminate but blunt-pointed, narrowed at the base, 3′–8′ long, 1′–4′ wide; petioles longer than the peduncles; flowers 1′–1½′ long and broad; corolla-limb almost entire; fruiting calyx 1′–1½′ long and thick, its segments acute at the apex, their basal auricles acute or cuspidate; berry about ½′ in diameter, loosely surrounded by the calyx.

In waste places, escaped from gardens, Nova Scotia to Ontario, Florida, Tennessee and Missouri. Adventive from Peru. Plant with the aspect of a large *Physalis*. Leaves similar to those of *Stramonium*. July–Sept.

2. PHÝSALIS L. Sp. Pl. 182. 1753.*

Annual or perennial herbs, sometimes a little woody below, with entire or sinuately toothed leaves. Peduncles slender, in ours solitary from the axils of the leaves. Calyx campanulate, 5-toothed, in fruit enlarged and bladdery-inflated, membranous, 5-angled, or prominently 10-ribbed and reticulate, wholly inclosing the pulpy berry, its teeth mostly

* Text contributed to our first edition by Dr. P. A. Rydberg, here somewhat revised.

connivent. Corolla yellowish or whitish, often with a darker brownish or purplish center, open-campanulate, or rarely campanulate-rotate, plicate. Stamens inserted near the base of the corolla; anthers oblong, opening by a longitudinal slit. Style slender, somewhat bent; stigma minutely 2-cleft. Seeds numerous, kidney–shaped, flattened, with a thin edge, finely pitted. [Greek, bladder, referring to the inflated calyx.]

Probably 75 species, or more. Two are of European origin, and about half a dozen are natives of India and Australia, the rest are American; about 30 occur in the United States. Type species: *Physalis Alkekéngi* L.

* Annuals with branched fibrous roots.

† Plants more or less pubescent (except in races of *P. barbadensis*.)

Fruiting calyx sharply 5-angled, more or less acuminate at the summit and sunken at the base; calyx-lobes (at flowering time) lanceolate or acuminate, as long as the tube or longer.

Leaves ovate, oblique, acute or acuminate, subentire at the base; upper part repand or subentire; fruiting calyx small and short; stem slender, diffuse, sharply angled.   1. *P. pubescens*.

Leaves cordate, oblique, strongly sinuate to the base; stem stout, obtusely angled; fruiting calyx rounded.   2. *P. pruinosa*.

Leaves cordate, scarcely oblique, more or less abruptly acuminate, acutely repand-dentate; stem tall, acutely angled; fruiting calyx larger, long-acuminate.   3. *P. barbadensis*.

Fruiting calyx obtusely or indistinctly 5–10-angled; calyx-lobes (at flowering time) triangular, generally shorter than the tube.   4. *P. missouriensis*.

†† Plants glabrous, or the upper part sparingly beset with short hairs, or a little puberulent when young; fruiting calyx obtusely 5–10-angled, not sunken at the base.

Corolla yellow, sometimes with the center a little darker but never brown or purple.

Peduncles generally much longer than the fruiting calyx; leaves sinuately toothed or subentire.   5. *P. pendula*.

Peduncles scarcely exceeding the fruiting calyx; leaves sharply dentate.   6. *P. angulata*.

Corolla yellow, with a brown or purple center.   7. *P. ixocarpa*.

** Perennial by thick roots and rootstocks.

† Pubescence not stellate (although in *P. pumila* of branched hairs).

Pubescence on the leaves none, on the upper part of the stem and the calyx sparse and short, if any.

Fruiting calyx ovoid, nearly filled by the berry, scarcely sunken at the base.

Leaves ovate-lanceolate to broadly ovate, usually thin.   8. *P. subglabrata*.

Leaves lanceolate, oblanceolate, or linear.   9. *P. longifolia*.

Fruiting calyx pyramidal, very much inflated and deeply sunken at the base; leaves broadly ovate, usually coarsely dentate.   10. *P. macrophysa*.

Pubescence sparse, consisting of flat, sometimes jointed, and in *P. pumila* branched hairs; in *P. virginiana* sometimes a little viscid.

Fruiting calyx ovoid, scarcely angled and scarcely sunken at the base; leaves thick, obovate or spatulate to rhomboid, subentire.

Leaves obovate or spatulate; hairs all simple.   11. *P. lanceolata*.

Leaves broader, often rhomboid; hairs on the lower surface branched.   12. *P. pumila*.

Fruiting calyx pyramidal, more or less 5-angled and deeply sunken at the base; leaves ovate to lanceolate, generally more or less dentate.   13. *P. virginiana*.

Pubescence dense, short, more or less viscid or glandular, often mixed with long flat jointed hairs.

Leaves large; blade generally over 2' long and more or less cordate.   14. *P. heterophylla*.

Leaves less than 2' long, rounded ovate or rhombic, scarcely at all cordate at the base; calyx, peduncles and younger branches with long white flat and jointed hairs.   15. *P. comata*.

Leaves small, 1'–1⅔' in diameter, nearly orbicular, sometimes a little cordate at the base, not coarsely toothed; stem diffuse or prostrate.   16. *P. rotundata*.

†† Pubescence dense, cinereous, beautifully stellate.   17. *P. viscosa*.

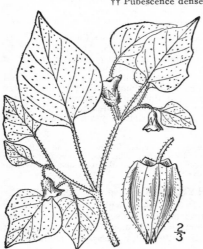

2/3

1. Physalis pubéscens L. Low Hairy Ground-Cherry. Fig. 3696.

Physalis pubescens L. Sp. Pl. 183. 1753.

Annual; stem generally diffuse or spreading, much branched, angled, often a little swollen at the nodes, villous-pubescent or sometimes nearly glabrous; leaves thin, 1'–2½' long, ovate, acute or acuminate, at the base oblique, slightly cordate and generally entire, upward repand-denticulate or entire, pubescent, sometimes becoming nearly glabrous except along the veins; peduncles short, 1"–2" long or in fruit about 5"; calyx-lobes narrow but not with a subulate tip; corolla 3"–5" in diameter, yellow with dark centre; anthers usually purplish; fruiting calyx membranous, 10"–15" long, pyramidal, ovoid-acuminate, more or less retuse at the base.

In sandy soil, Pennsylvania to Florida, Arkansas and California; also in Mexico, the West Indies, Central and South America and India. Called also dwarf cape-gooseberry and strawberry-tomato. July–Sept.

2. Physalis pruinòsa L.　Tall Hairy Ground-Cherry.　Fig. 3697.

Physalis pruinosa L. Sp. Pl. 184. 1753.
P. pubescens Dunal, in DC. Prodr. 13: part 1, 446. 1852.

Annual; stem stout, generally erect, and more hairy than the preceding and the two following species; stem obtusely angled, finely villous or somewhat viscid; leaves firm, 1½′–4′ long, finely pubescent, ovate, cordate, generally very oblique at the base, and deeply sinuately toothed with broad and often obtuse teeth; peduncles 1″–2″ long, in fruit about 5″; calyx villous or viscid; lobes as long as the tube, narrow but not subulate-tipped; corolla 2″–4″ in diameter; anthers yellow, or tinged with purple; fruiting calyx a little firmer and more pubescent than in the preceding, reticulate, 10″–15″ long, ovoid, sunken at the base; berry yellow or green.

In cultivated soil, Massachusetts to Ontario, Florida, Iowa, Missouri and Colorado. July–Sept.

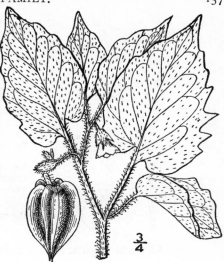

3. Physalis barbadénsis Jacq.　Barbadoes Ground-Cherry.　Fig. 3698.

Physalis barbadensis Jacq. Misc. 2: 359. 1781.
Physalis obscura Michx. Fl. Bor. Am. 1: 149. 1803.
Physalis barbadensis obscura Rydberg, Mem. Torr. Club 4: 327. 1896.

Annual; stem tall and erect or widely spreading, acutely 3–4-angled, pubescent, viscid, or nearly glabrous; leaves 1½′–2½′ long, ovate or heart-shaped, acute, or abruptly acuminate, sharply repand-dentate, pubescent with short hairs; peduncles short, 1½″–2″ long, but in fruit sometimes 10″, calyx generally densely viscid-hirsute, lobes lanceolate, acuminate, but not subulate-tipped; corolla 2½″–5″ in diameter; anthers generally purplish; fruiting calyx longer than in the two preceding species, 1′–1¼′ long, acuminate and reticulate, retuse at the base.

Sandy soil, Pennsylvania to Illinois, Missouri, Florida, Mexico, the West Indies and South America. July–Sept.

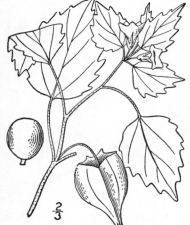

4. Physalis missouriénsis Mack. & Bush.　Missouri Ground-Cherry.　Fig. 3699.

Physalis missouriensis Mack. & Bush, Fl. Jackson Co. 167. 1902.

Annual; stem spreading, often zigzag, branched, striate, or slightly angled, villous with short hairs. Leaves ½′–3½′ long, ovate, oblique and cuneate, obtuse, or cordate at the base, acute but not acuminate, repand or sinuately dentate, hairy, at least on the veins; peduncles ½″–2½″ long, erect, in fruit 2½″–5″, reflexed, shorter than the fruiting calyx; calyx villous, lobes shorter than the tube, triangular; corolla 1½″–4″ in diameter, yellow; fruiting calyx 7″–10″ long, round-ovoid, nearly filled by the berry, scarcely sunken or commonly rounded at the base.

Missouri and Kansas to Arkansas and Oklahoma. July–Sept. Referred in our first edition to the tropical *P. Lagáscae* R. & S.

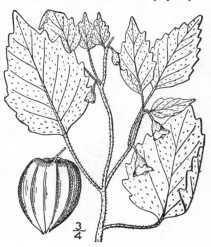

5. Physalis péndula Rydberg.   Lance-leaved Ground-Cherry.   Fig. 3700.

Physalis pendula Rydberg; Small, Fl. SE. U. S. 983.   1903.

Annual, stem erect, generally 1½° high, branched, angled, glabrous; leaves lanceolate or ovate-lanceolate, thin, usually coarsely toothed; calyx cylindric-campanulate, its lobes broadly triangular, shorter than the tube; peduncles filiform, about 1′ long, erect with nodding flower, in fruit 1¾′–2′ long and reflexed; corolla 3″–4″ in diameter, campanulate, yellow, without a dark spot; anthers yellow, more or less tinged with purple, fruiting calyx about 10″ long, rounded ovoid, indistinctly 10-angled and purple veined, nearly filled by the berry.

Illinois to Kansas and Texas.   July–Sept.   Referred in the first edition to *P. lanceifolia* Nees.

6. Physalis angulàta L.   Cut-leaved Ground-Cherry.   Fig. 3701.

Physalis angulata L. Sp. Pl. 183.   1753.

Annual, erect, 1½°–3° high, glabrous; stem angular, usually much branched; leaves ovate, with more or less cuneate base, somewhat sinuately toothed with long-acuminate teeth; blades 2′–2½′ long, on slender petioles 1′–2′ long, thin, the veins not prominent; peduncles slender, 10″–15″ long, erect, in fruit often reflexed but seldom exceeding the fruiting calyx in length; calyx smooth, lobes triangular to lanceolate, generally shorter than the tube; corolla 2½″–5″ in diameter; anthers more or less purplish tinged; fruiting calyx about 1¼′ long, ovoid, 5–10-angled, sometimes purple-veined, nearly filled by the yellow berry.

In rich soil, Pennsylvania to Illinois, Minnesota, Missouri, Texas, Central America, Brazil and the West Indies.   Also in India.   July–Sept.

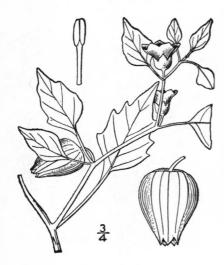

7. Physalis ixocàrpa Brot.   Tomatillo. Mexican Ground-Cherry.   Strawberry Tomato.   Fig. 3702.

Physalis ixocarpa Brot.; Horneman, Hort. Hafn. Suppl. 26.   1819.
P. aequata Jacq. f.; Nees, Linnaea **6**: 470.   1831.

Annual, stem at first erect, later widely spreading, much branched, angled, glabrous, or the younger parts sparingly hairy; leaves from cordate to ovate, with a cuneate base which is somewhat oblique, sinuately dentate or entire, 1′–2½′ long; peduncles short, 1″–2½″ long; calyx sparingly hairy, its lobes short, broadly triangular, shorter than the tube; corolla bright yellow with purple throat, 5″–7″ (sometimes nearly 10″) in diameter; fruiting calyx round-ovoid, obscurely 10-angled, often purple-veined, filled by the purple berry, which sometimes bursts it.

Native of Mexico.   It is often cultivated for its fruit and frequently escapes from cultivation, New York to Texas and California.

8. Physalis subglabràta Mackenzie and Bush.   Smooth Ground-Cherry. Fig. 3703.

?Physalis philadelphica Lam. Encycl. **2**: 101.   1786.
P. subglabrata Mackenzie & Bush, Trans. Acad. St. Louis **12**: 86.   1902.

Perennial from a deep rootstock, tall, erect, 2½°–5° high; stem angled, dichotomously branched, glabrous, or sometimes slightly pubescent with sparse and short hairs on the upper parts; blades ovate to ovate-lanceolate, often very oblique at the base and more or less acuminate, entire or repand-denticulate, 2½′–4′ long, on petioles 1½′–2½′ long, often in pairs; peduncles slender, 5″–10″ long, generally longer than the flower; calyx glabrous, or minutely ciliolate, lobes ovate-lanceolate or trangular, sometmes broadly ovate and unequal, generally equalling the tube; corolla yellow or greenish yellow with purplish throat, ¾′–1′ in diameter; anthers tinged with purple; fruiting calyx at first somewhat 10-angled and sunken at the base, at last often filled with or burst by the large red or purple berry.

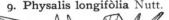

Ontario and Rhode Island to Georgia, Minnesota, Kentucky and Colorado.   July–Sept.

9. Physalis longifòlia Nutt.   Long-leaved Ground-Cherry.   Fig. 3704.

Physalis longifolia Nutt. Trans. Am. Phil. Soc. (II.) **5**: 193.   1833–37.

Physalis lanceolata var. *laevigata* A. Gray, Proc. Am. Acad. **10**: 68.   1874.

Perennial from a thick rootstock; stem in the common form stout and tall, 1½°–3° high, slightly angled, branched above, the branches strict, glabrous. Leaves lanceolate, oblanceolate, or linear, tapering into a short stout petiole 5″–10″ long, subentire or repand; peduncles 5″–10″ long, in fruit often recurved; calyx generally glabrous, its lobes triangular-lanceolate, about the length of the tube; corolla 6″–10″ in diameter, yellow with a dark, commonly brownish center; anthers yellow, tinged with purple; fruiting calyx ovoid, about 1¼′ long, not sunken at the base; berry yellow, the lower portion and the stipe glutinous.

In rich soil, Iowa to South Dakota, Montana, Arkansas, Utah and Mexico.   July–Sept.

10. Physalis macróphysa Rydb.   Large-bladder Ground-Cherry.   Fig. 3705.

P. macrophysa Rydberg, Bull. Torr. Club **22**: 308.   1895.

Perennial; rootstock rather thick and fleshy; stem erect, 1½°–3° high, comparatively slender, angled, perfectly glabrous, or the upper parts sparingly pubescent with very short hairs. Leaves large, thin, 1½′–3½′ long, 1′–2′ wide, the lower obtuse, the upper acute or acuminate; petioles slender, 10″–20″ long; peduncles 5″–8″ long, erect, in fruit reflexed; calyx smooth, its lobes ovate-triangular or broadly lanceolate, generally a little shorter than the tube; corolla yellow with a dark center, about 10″ in diameter; anthers generally yellow, sometimes tinged with purple; fruiting calyx large, 1¼′–1½′ long, 1′–1¼′ in diameter, pyramidal to ovoid-conic, indistinctly 10-angled, deeply sunken at the base; berry small, in the center of the calyx.

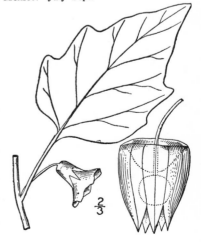

In rich soil, Kansas and Arkansas to Texas.   May–July.   Rare.

11. Physalis lanceolàta Michx. Prairie Ground-Cherry. Fig. 3706.

Physalis lanceolata Michx. Fl. Bor. Am. 1 : 149. 1803.

Physalis pennsylvanica var. *lanceolata* A. Gray, Man. Ed. 5, 382. 1867.

Perennial; rootstock apparently slender and creeping; stem about 1½° high, first erect, later spreading or diffuse, only slightly angled, sparingly hirsute with flat hairs. Leaves lanceolate, oblanceolate or spatulate, tapering into the petiole, acute or obtuse, nearly always entire, rarely wavy, but never sinuately toothed, thickish, sparingly hairy with short hairs; peduncles 5″–10″ long, in fruit reflexed; calyx strigose or villous, rarely glabrous, its lobes triangular-lanceolate; corolla dullish yellow with a brownish center, about 8″ in diameter; fruiting calyx round-ovoid, not sunken at the base, indistinctly 10-angled; berry yellow or greenish yellow.

On dry prairies, South Carolina to Illinois, South Dakota, Wyoming, Kansas and New Mexico. July–Sept.

12. Physalis pùmila Nutt. Low Ground-Cherry. Fig. 3707.

Physalis pumila Nutt. Trans. Am. Phil. Soc. (II.) **5** : 193. 1834.

Physalis lanceolata var. *hirta* A. Gray, Proc. Am. Acad. 10 : 68. 1874.

Perennial from a slender rootstock, 1½°–3° high; stem hirsute, obscurely angled. Leaves thick, broadly ovate to oblong, acute at both ends and somewhat rhomboid, the lower often obtuse and obovate, generally much larger than in the preceding; blades 2′–4′ long, entire or seldom sinuate, on petioles 10″–15″ long, strigose with many-branched hairs especially on the lower surface; peduncles 5″–10″ long, in fruit reflexed and 1½′–2′ long; calyx densely hirsute, not stellate-pubescent, its lobes triangular, generally a little shorter than the tube; corolla yellow with brown center, 8″–10″ in diameter; fruiting calyx usually more elongated than in the preceding, 1½′–2′ long, oblong-ovoid, a little sunken at the base, indistinctly 10-angled.

Plains and prairies, Illinois to Colorado and Texas. July–Sept.

13. Physalis virginiàna Mill. Virginia Ground-Cherry. Fig. 3708.

Physalis virginiana Mill. Gard. Dict. Ed. 8, no. 4. 1768.

Physalis pennsylvanica A. Gray, Man. Ed. 5, 382. 1867. Not L. 1753.

Physalis virginiana intermedia Rydberg, Mem. Torr. Club 4 : 345. 1896.

Perennial; rootstock thick and somewhat fleshy; stem 1½°–3° high, erect, dichotomously branched, somewhat angular, more or less strigose-hairy with flat hairs, sometimes a little glandular, or sometimes nearly glabrous. Leaves ovate-lanceolate, tapering to both ends, 1½′–2½′ long, generally more or less sinuately dentate, often yellowish green; peduncles 5″–10″ long, generally erect, in fruit curved but scarcely reflexed; calyx strigose, hirsute, or at least puberulent, its lobes triangular or broadly lanceolate, nearly equalling the tube; corolla sulphur-yellow with purplish spots, ⅜′–1′ in diameter; anthers yellow; fruiting calyx pyramidal-ovoid, 5-angled, sunken at the base; berry reddish.

Rich soil, especially in open places, Ontario to Manitoba, Connecticut, Florida, Louisiana and Texas. Consists of numerous races, differing in pubescence. July–Sept. Wild cherry.

14. Physalis heterophýlla Nees.   Clammy Ground-Cherry.   Fig. 3709.

Physalis viscosa Pursh, Fl. Am. Sept. 157. 1814. Not
L. 1753.
Physalis heterophylla Nees, Linnaea 6: 463. 1831.
Physalis virginiana A. Gray, Syn. Fl. 2: Part 1, 235.
1878. Not Mill. 1768.

Perennial from a slender creeping rootstock, 1½°–3°
tall, at first erect, later generally decumbent and
spreading, viscid and glandular, and villous with
long spreading jointed flat hairs; leaves large, blade
generally over 2′ long, usually broadly cordate, often
acute and very rarely with an elongated tip, thick,
more or less sinuately toothed, or sometimes subentire; calyx long-villous, lobes triangular, generally
shorter than the tube; corolla 8″–10″ in diameter,
greenish yellow with a brownish or purplish center;
anthers mostly yellow; berry yellow.

In rich soil, especially where the surface has been
disturbed, New Brunswick to Saskatchewan, Florida,
Colorado and Texas. The most common of our species,
and includes several races.

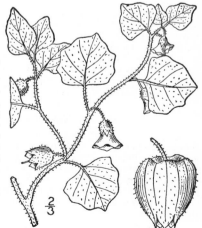

Physalis peruviàna L., a native of South America, is cultivated for its fruit and often escapes.
It resembles *P. heterophylla,* but differs in the leaves, which have a long tip, and in the pubes-
cence, which is shorter, denser, and not at all viscid.   Cape gooseberry.   Strawberry tomato.
Peruvian ground-cherry.   Husk tomato.

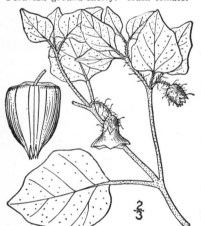

15. Physalis comàta Rydberg.   Hillside
Ground-Cherry.   Fig. 3710.

P. comata Rydberg, Bull. Torr. Club 22: 306. 1895.

Perennial, erect, about 1½° high; pubescence fine
and short, that on the calyx, peduncles and upper
branches mixed with long white flat jointed hairs.
Like *P. heterophylla* Nees (*P. virginiana* Gray, not
Mill.), but leaves smaller, blade not over 2′ long,
round-ovate, scarcely at all cordate at the base, about
2′ long, thin, somewhat repand-dentate, or nearly
entire; petioles as long as the leaves; peduncles as
long as the fruiting calyx, or longer; corolla greenish yellow, with brown center, 6″–10″ in diameter;
fruiting calyx of thin texture, round-ovoid, somewhat 10-angled, scarcely sunken at the base.

Hillsides of Nebraska, Kansas, Colorado and Texas.
Rare.

16. Physalis rotundàta Rydberg.   Round-
leaved Ground-Cherry.   Fig. 3711.

Physalis hederaefolia Holzinger, Cont. U. S. Nat. Herb.
1: 212. 189 . Not Gray.
P. rotundata Rydberg, Mem. Torr. Club 4: 352. 1896.

Diffuse and spreading, zigzag, generally dichotomously much branched, from a perennial rootstock,
densely and finely viscid-pubescent, usually more
glandular than the preceding.   Leaves nearly orbicular with more or less cordate base, 1′–1⅔′ in diameter, with small teeth; petioles short, more or less
winged; peduncles short, in fruit scarcely more
than half the length of the calyx; corolla 8″ in
diameter, greenish yellow with a brownish center;
fruiting calyx ovoid, slightly angled, scarcely sunken
at the base.

Dry plains, South Dakota to Texas and New Mexico.
July–Sept.

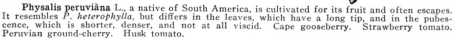

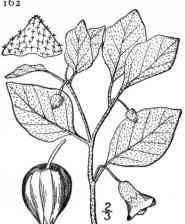

17. Physalis viscòsa L. Stellate Ground-Cherry. Yellow-henbane. Fig. 3712.

Physalis viscosa L. Sp. Pl. 183. 1753.
Physalis pennsylvanica L. Sp. Pl. Ed. 2, 1670. 1763.

Perennial from a slender creeping rootstock; stems slender, creeping, with a dense ashy stellate pubescence, or in age rarely glabrate. Leaves elliptic, oval or ovate, obtuse, thinish, entire or undulate, in the typical South American race often cordate at the base, but rarely so in our plant; peduncles ½′–1′ long; calyx stellate-pubescent, its lobes triangular, generally shorter than the tube; corolla greenish yellow with a darker center, 8″–10″ in diameter; fruiting calyx 10″–15″ long, round-ovoid, scarcely sunken at the base; berry orange or yellow.

On sea beaches, or in sand near the coast, Virginia and North Carolina to Florida. Eastern South America.

Physalis Alkekéngi L., Strawberry tomato or Winter cherry, is a native of Europe and Asia, often cultivated for its fruit and sometimes escapes from cultivation. The flowers are whitish, the limb distinctly 5-lobed; leaves broadly deltoid, acute at both ends, repand or angulately toothed.

3. QUÍNCULA Raf. Atl. Journ. 145. 1832.

A low and diffuse somewhat scurfy herb, with a stout perennial root. Leaves from sinuate to pinnatifid, somewhat fleshy. Peduncles most commonly in pairs from the axils of the leaves, sometimes solitary, or in fascicles of 3–5. Calyx campanulate, 5-toothed, in fruit inflated, sharply 5-angular and reticulate, enclosing the fruit, the lobes connivent. Corolla flat-rotate, pentagonal in outline, veiny, violet or purplish. Anthers opening by a longitudinal slit. Seeds comparatively few, kidney-shaped, somewhat flattened, with thick margins, rugose-tuberculate. [Name unexplained.]

A monotypic genus of central North America.

1. Quincula lobàta (Torr.) Raf. Purple-flowered Ground-Cherry. Fig. 3713.

Physalis lobata Torr. Ann. Lyc. N. Y. **2**: 226. 1827.
Quincula lobata Raf. Atl. Journ. 145. 1832.

Perennial, low, spreading or prostrate, more or less scurfy-puberulent; stem obtusely angled and striate, much branched. Leaves oblanceolate or spatulate to oblong, sinuately toothed, or pinnatifid with rounded lobes, or rarely subentire, cuneate at the base, thickish and veiny, tapering into margined petioles; peduncles 1′–2′ long, in fruit reflexed; calyx-lobes triangular, acute, shorter than the tube; corolla purplish, 10″–15″ in diameter; anthers yellow, tinged with purple; fruiting calyx about as wide as long, sharply 5-angled, sunken at the base.

On high plains, Kansas to California, Texas, New Mexico and Mexico. May–Sept.

4. LEUCOPHÝSALIS Rydberg, Mem. Torr. Club **4**: 365. 1896.

A tall erect viscid and villous annual, with entire leaves, the blade decurrent on the petiole. Peduncles generally in fascicles of 2–4 in the axils. Calyx campanulate, 5-lobed, at first a little inflated, but soon filled by and closely fitted to the berry, thin, neither angled nor ribbed, faintly veiny, open at the mouth, the lobes exceeding the fruit. Corolla rotate, white, sometimes tinged with purple and generally creamy or yellow in the center, the limb plicate. Stamens inserted near the base of the corolla; filaments long and slender; anthers oblong, opening by a longitudinal slit. Style and stigmas as in *Physalis*. Seeds kidney-shaped, flattened, punctate. [Greek, white *Physalis*.]

A monotypic genus of northern North America.

1. Leucophysalis grandiflòra (Hook.) Rydberg. Large White Ground-Cherry. Fig. 3714.

P. grandiflora Hook. Fl. Bor. Am. **2**: 90. 1834.
Leucophysalis grandiflora Rydberg, Mem. Torr.
    Club **4**: 366. 1896.

Erect, tall, $1\frac{1}{2}°-3°$ high; stem somewhat angled, striate, more or less villous. Leaves large, $4'-8'$ long, ovate to lanceolate-ovate, generally acute and entire, somewhat decurrent on the petiole, more or less villous and viscid, especially on the veins of the lower surface; peduncles several from each axil, $\frac{2}{3}'-\frac{4}{5}'$ long, villous; calyx villous, its lobes lanceolate, equalling the tube; corolla large, $1\frac{1}{4}'-1\frac{1}{2}'$ in diameter, rotate, white with a more or less yellowish center; filaments slender; anthers short, yellow, often tinged with purple; fruiting calyx ovoid, early filled by the berry.

Sandy soil, Quebec to Saskatchewan, Michigan and Minnesota. May–July.

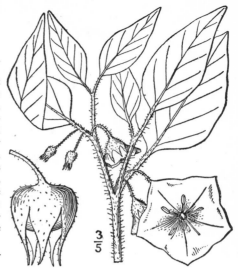

5. CHAMAESÁRACHA A. Gray, Bot. Cal. **1**: 540. 1876.

Perennials, with entire to pinnatifid leaves, the blades decurrent on the petioles. Peduncles solitary, or in fascicles of 2–4 in the axils. Calyx campanulate, 5-lobed, in fruit somewhat enlarged, but not bladdery-inflated, close-fitting to the berry, thin, not angled nor ribbed, and faintly if at all veiny, open at the mouth, not exceeding the berry. Corolla rotate, white or cream-colored, often tinged with purple, the limb plicate. Stamens inserted near the base of the corolla; filaments long and slender; anthers oblong, opening by longitudinal slits; style and stigma as in *Physalis*. Seeds kidney-shaped, flattened, rugose-favose or punctate. [Ground-*Saracha*, the latter a genus named in honor of Isidore Saracha, a Spanish Benedictine botanist.]

An American genus, consisting of half a dozen species, natives of Mexico and the southwestern United States. Type species: *Chamaesaracha Coronopus* (Dunal) A. Gray.

Pubescence dense, puberulent and hirsute.   1. *C. conioides.*
Pubescence sparse, puberulent or stellate, hirsute (if at all) only on the calyx.   2. *C. Coronopus.*

1. Chamaesaracha conioìdes (Moricand) Britton. Hairy Chamaesaracha. Fig. 3715.

Solanum conioides Moric.; Dunal in DC. Prodr. **13**: Part
    1, 64. 1852.
Withania (?) *sordida* Dunal, *loc. cit.,* 456. 1852.
Chamaesaracha sordida A. Gray, Bot. Cal. **1**: 540. 1876.
Chamaesaracha conioides Britton, Mem. Torr. Club **5**:
    287. 1895.

Much branched from a perennial base, at first upright, at length spreading, cinereous-puberulent with short branched somewhat glutinous or viscid hairs, generally also viscidly hirsute or villous with long and branched hairs, especially on the calyx; leaves oblanceolate to obovate-rhombic, usually acutish and tapering into a short petiole, generally deeply lobed, but varying from subentire to pinnatifid; calyx-lobes triangular, generally acutish; corolla about $\frac{1}{2}'$ in diameter, white, cream-colored or sometimes violet-purplish; berry $2\frac{1}{2}''-4''$ in diameter.

In dry clayey soil, southern Kansas to California and Mexico. May–Sept.

$\frac{3}{4}$

2. Chamaesaracha Corónopus (Dunal)
A. Gray. Smoothish Chamaesaracha.
Fig. 3716.

Solanum Coronopus Dunal in DC. Prodr. 13: Part 1, 64. 1852.

C. Coronopus A. Gray, Bot. Cal. 1: 540. 1876.

Branched and diffuse from a perennial base; stem obtusely angled; pubescence on the stem and leaves more or less roughish pruinose or stellate, often scarcely any; on the calyx stellate or sometimes hirsute. Leaves linear or lanceolate, tapering at the base, more or less sinuately lobed, occasionally subentire, sometimes pinnatifid; calyx-lobes triangular, acute; corolla white or ochroleucous, the appendages of the throat often protuberant; berry $2\frac{1}{2}''$–$4''$ in diameter, nearly white.

In clayey soil, Kansas to Utah, California and Mexico. May–Sept.

6. SOLÀNUM [Tourn.] L. Sp. Pl. 184. 1753.

Herbs or shrubs, often stellate-pubescent, sometimes climbing. Flowers cymose, umbelliform, paniculate, or racemose, white, blue, purple, or yellow. Calyx campanulate or rotate, mostly 5-toothed or 5-cleft. Corolla rotate, the limb plaited, 5-angled or 5-lobed, the tube very short. Stamens inserted on the throat of the corolla; filaments short; anthers linear or oblong, acute or acuminate, connate or connivent into a cone, the cells dehiscent by a terminal pore, or sometimes by a short introrse terminal slit, or sometimes also longitudinally. Ovary usually 2-celled; stigma small. Berry mostly globose, the calyx either persistent at its base or enclosing it. [Name, according to Wettstein, from *solamen,* quieting.]

About 1000 species, of wide geographic distribution, most abundant in tropical America. Besides the following, some 20 others occur in the southern and western United States. Type species: *Solanum nigrum* L.

* Glabrous or pubescent herbs, not prickly.

Plants green; pubescence simple, or some of it stellate; flowers white.	
Leaves repand or entire; ripe berries black.	1. *S. nigrum.*
Leaves deeply pinnatifid; ripe berries green.	2. *S. triflorum.*
Plant silvery stellate-canescent; flowers violet.	4. *S. elaeagnifolium.*

** Stellate-pubescent and prickly herbs.

Berry not enclosed by the calyx; perennials.	
Hirsute; leaves ovate or oblong, sinuate or pinnatifid.	3. *S. carolinense.*
Densely silvery-canescent; leaves linear or oblong, repand or entire.	4. *S. elaeagnifolium.*
Pubescent; leaves ovate, 5–7-lobed.	5. *S. Torreyi.*
Berry partly or wholly invested by the spiny calyx; annuals.	
Lowest anther larger than the other four.	
Plant densely stellate-pubescent; corolla yellow.	6. *S. rostratum.*
Plant glandular-pubescent, with few stellate hairs; corolla violet.	7. *S. citrullifolium.*
Anthers all equal.	8. *S. sisymbrifolium.*

*** Climbing vine, not prickly; leaves hastate or 3-lobed. 9. *S. Dulcamara.*

1. Solanum nìgrum L. Black, Deadly or
Garden Nightshade. Morel. Fig. 3717.

Solanum nigrum L. Sp. Pl. 186. 1753.

Annual, glabrous, or somewhat pubescent with simple hairs, green; stem erect, branched, $1°$–$2\frac{1}{2}°$ high. Leaves ovate, petioled, more or less inequilateral, $1'$–$3'$ long, entire, undulate, or dentate, thin, acute, acuminate or acutish at the apex, narrowed or rounded at the base; peduncles lateral, umbellately 3–10-flowered, $\frac{1}{2}'$–$1\frac{1}{2}'$ long; pedicels $3''$–$7''$ long; flowers white, $4''$–$5''$ broad; calyx-lobes oblong, obtuse, spreading, much shorter than the corolla, persistent at the base of the berry; filaments somewhat pubescent; anthers obtuse; berries black when ripe, smooth and glabrous, globose, $4''$–$5''$ in diameter, on nodding peduncles.

In waste places, commonly in cultivated soil, Nova Scotia to the Northwest Territory, south to Florida and Texas. Widely distributed in nearly all countries as a weed, and includes numerous races, differing principally in leaf-form and pubescence. Petty-morel. Duscle. Hound's-berry. July–Oct.

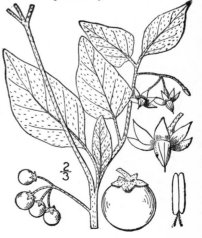

$\frac{2}{3}$

Solanum villòsum (L.) Mill., with coarsely dentate leaves, the pubescence villous and somewhat viscid, has been found in ballast about the seaports.

2. Solanum triflòrum Nutt.　Cut-leaved Nightshade.　Fig. 3718.

Solanum triflorum Nutt. Gen. **1**: 128. 1818.

Annual, sparingly pubescent with simple hairs, or glabrous; stem branched, 1°–3° high. Leaves pinnatifid or some of them pinnately lobed, acute at the apex, petioled, 2′–4′ long, the lobes triangular-lanceolate, acute or obtuse, entire or dentate, the sinuses rounded; peduncles lateral, 1–3-flowered, ½′–1′ long; pedicels 3″–7″ long, reflexed in fruit; calyx-segments lanceolate, shorter than the corolla, persistent at the base of the berry; corolla white, 4″–5″ broad; anthers obtuse; berries green when mature, globose, smooth and glabrous, about 5″ in diameter.

On prairies and in waste places, Ontario to Manitoba, Kansas, New Mexico and Arizona. Introduced in Missouri. May–Oct.

3. Solanum carolinénse L.　Horse-Nettle.　Sand-Brier.　Fig. 3719.

Solanum carolinensis L. Sp. Pl. 184. 1753.

Perennial, green, finely stellate-pubescent with 4–8-rayed hairs; stem erect, branched, 1°–4° high, the branches, petioles, midveins and sometimes the lateral veins of the leaves armed with straight subulate yellow prickles. Leaves oblong or ovate, repand, lobed, or pinnatifid, 2′–6′ long, the lobes obtuse or acutish; petioles 3″–10″ long; flowers cymose-racemose, appearing terminal, but really lateral, as is manifest in fruit; pedicels 3″–7″ long, recurved in fruit; calyx-lobes lanceolate, acuminate, about one-half the length of the corolla, persistent at the base of the berry; corolla-lobes ovate-lanceolate, acute; anthers elongated; berries orange-yellow, smooth and glabrous, 8″–10″ in diameter.

In dry fields and in waste places, southern Ontario to Vermont, Massachusetts and Florida, west to Illinois, Nebraska and Texas. Adventive in its northeastern range. Apple-of-sodom. Radical-weed. Bull-nettle. Tread-softly. May–Sept.

4. Solanum elaeagnifòlium Cav.　Silver-leaved Nightshade.　Fig. 3720.

S. elaeagnifolium Cav. Icon. **3**: 22. *pl. 243.* 1794.

Perennial, densely and finely stellate-pubescent, silvery-canescent all over; stem branched, 1°–3° high, armed with very slender sharp prickles, or these wanting. Leaves lanceolate, oblong, or linear, petioled, 1′–4′ long, 3″–12″ wide, mostly obtuse at the apex, narrowed or rounded at the base, repand-dentate or entire; flowers cymose, 8″–12″ broad, violet or blue; peduncles short and stout, appearing terminal, but soon evidently lateral; calyx-lobes lanceolate or linear-lnceolate, acute; anthers linear; ovary white-tomentose; berries globose, yellow or darker, smooth and glabrous, 4″–6″ in diameter.

On dry plains and prairies, Missouri and Kansas to Texas and Arizona. Trompillos. May–Sept.

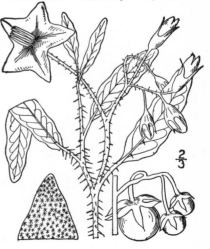

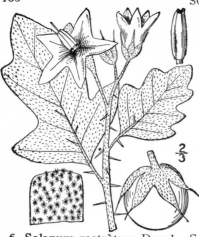

5. Solanum Tórreyi A. Gray. Torrey's Nightshade. Fig. 3721.

S. *Torreyi* A. Gray, Proc. Am. 'Acad. 6: 44. 1862.

Perennial, hoary with a stellate pubescence of 8–12-rayed hairs, more or less armed with small subulate prickles. Leaves ovate in outline, 3′–6′ long, sinuately 5–7-lobed, the lobes entire or undulate, obtuse; cymes appearing terminal, soon evidently lateral, branched, loosely several-flowered; flowers showy, large; calyx-lobes ovate, abruptly long-acuminate, persistent at the base of the berry; corolla violet, 1′–1½′ broad, its lobes ovate, acute; berry globose, smooth and glabrous, 1′ or more in diameter, yellow when ripe.

On dry plains and prairies, Kansas to Texas. Reported from Missouri.

6. Solanum rostràtum Dunal. Sand Bur. Beaked or Prickly Nightshade. Buffalo-bur. Fig. 3722.

Solanum rostratum Dunal, Sol. 234. *pl. 24.* 1813.
S. *heterandrum* Pursh, Fl. Am. Sept. 156. *pl. 7.* 1814.

Annual, densely stellate-pubescent with 5–8-rayed hairs, usually copiously armed with yellow subulate prickles; stem erect, branched, 1°–2½° high. Leaves ovate or oval in outline, irregularly pinnately 5–7-lobed or 1–2-pinnatifid, 2′–5′ long, petioled, the lobes mostly oblong, obtuse; flowers racemose, yellow, about 1′ broad; racemes lateral; pedicels stout, 3″–6″ long, erect both in flower and fruit; calyx densely prickly, surrounding and wholly enclosing the berry, the prickles becoming as long as the fruit, or longer; calyx-lobes lanceolate, acuminate; corolla about 1′ broad, slightly irregular, its lobes ovate, acute; stamens and style declined, the lowest stamen longer with an incurved beak; fruit, including its prickles, 1′ in diameter or more.

On prairies, South Dakota to Texas and Mexico. Occasional in waste places, Ontario to New Hampshire, Tennessee and Florida, adventive from the west. Texas-nettle. Prickly potato. May–Sept. The original food of the Colorado beetle.

7. Solanum citrullifòlium Braun. Melon-leaved Nightshade. Fig. 3723.

Solanum citrullifolium Braun, Ind. Sem. Frib. 1849.

Annual, glandular-pubescent, or a few 4–5-rayed hairs on the leaves, copiously armed with slender yellow subulate prickles, diffusely branched, 1°–3° high. Leaves irregularly bipinnatifid, resembling in outline those of the watermelon, 2′–6′ long; racemes lateral, several-flowered; flowers 1′–1½′ broad, violet; stamens and style declined; lowest anther violet, larger than the four other yellow ones; corolla somewhat irregular, its lobes ovate, acuminate; fruit similar to that of the preceding species.

In dry soil, Iowa and Kansas to Texas, Mexico and New Mexico. Referred in our first edition, to S. *heterodoxum* Dunal. July–Sept.

8. Solanum sisymbriifòlium Lam. Viscid Nightshade. Fig. 3724.

Solanum sisymbriifolium Lam. Ill. **2**: 25. 1793.

Annual, branched, 2°–4° high, villous-pubescent with long viscid hairs and armed all over with bright yellow prickles. Leaves thin, deeply pinnatifid into oblong toothed or sinuate lobes; flowers 1¼′–1½′ broad, light blue or white; stamens and style nearly erect; anthers all equal, yellow; corolla slightly irregular, its lobes deltoid or ovate-deltoid, acute or obtusish; fruit included in the accrescent calyx which has a prickly tube and thinnish lobes.

In waste places and on ballast, especially about sea-ports, from Massachusetts to the Gulf States. Introduced from tropical America. June–Sept.

9. Solanum Dulcamàra L. Climbing or Bitter Nightshade. Blue Bindweed. Fellenwort. Bittersweet. Dogwood. Fig. 3725.

Solanum Dulcamara L. Sp. Pl. 185. 1753.

Perennial, pubescent with simple hairs or glabrate, stem climbing or straggling, somewhat woody below, branched, 2°–8° long. Leaves petioled, ovate or hastate in outline, 2′–4′ long, 1′–2½′ wide, acute or acuminate at the apex, usually slightly cordate at the base, some of them entire, some with a lobe on one side near the base, some deeply 3-lobed or 3-divided, with the terminal segment much the largest; cymes compound, lateral; pedicels slender, articulated at the base, spreading or drooping; flowers blue, purple or white, 5″–7″ broad; calyx-lobes short, oblong, obtuse, persistent at the base of the berry; corolla deeply 5-cleft, its lobes triangular-lanceolate, acuminate; berry oval or globose, red.

In waste places or in moist thickets, sometimes appearing as if indigenous, Nova Scotia to Minnesota, Washington, New Jersey, Pennsylvania and Kansas. May–Sept. Woody nightshade. Poison-flower. Poison- or snake-berry. Scarlet berry. Naturalized from Europe. Native also of Asia.

Solanum trìquetrum Cav., a Texan and Mexican nearly glabrous herb, scarcely climbing, with somewhat ridged stems, 3-lobed deltoid-cordate or hastate leaves, lateral few-flowered cymes and globose red berries, is reported from Kansas.

Solanum Melongèna L., the egg-plant, with blue or purplish flowers, and an ovoid or obovoid berry up to 6′ long, is occasionally found in waste grounds. It is native of Asia. species: *Lycium afrum* L.

7. LYCOPÉRSICON Mill. Gard. Dict. Abr. Ed. 4. 1754.

Annual, or rarely perennial, coarse branching or feebly climbing herbs, with 1–2-pinnately divided leaves, and lateral irregular raceme-like cymes of small yellowish flowers opposite the leaves. Calyx 5-parted, or rarely 6-parted, the segments linear or lanceolate. Corolla rotate, the tube very short, the limb 5-cleft or rarely 6-cleft, plicate. Stamens 5 (rarely 6), inserted on the throat of the corolla; filaments very short; anthers elongated, connate or connivent, introrsely longitudinally dehiscent. Ovary 2–3-celled; style simple; stigma small, capitate. Berry in the wild plants globose or pyriform, much modified in cultivation, the calyx persistent at its base. [Greek, wolf-peach.]

About 4 species, natives of North America, the following typical.

1. Lycopersicon Lycopérsicon (L.)
Karst. Tomato. Love Apple.
Cherry Tomato. Fig. 3726.

Solanum Lycopersicum L. Sp. Pl. 185. 1753.
L. esculentum Mill. Gard. Dict. Ed. 8. 1768.
Lycopersicum Lycopersicum Karst. Deutsch. Fl.
966. 1880–83.

Viscid-pubescent, much branched, 1°–3°
high, the branches spreading. Leaves petioled, pinnately divided, 6′–18′ long, the segments stalked, the larger 7–9, ovate or ovate-
lanceolate, mostly acute, dentate, lobed or
again divided, 2′–4′ long, with several or
numerous smaller, sometimes very small ones
interspersed; clusters several-flowered; peduncles 1′–3′ long; flowers 5″–8″ broad; calyx-
segments about equalling the corolla; berry
the well-known tomato or love-apple.

Escaped from cultivation and occasionally
spontaneous from New York and Pennsylvania
southward. Jews' ear. June–Sept.

8. LÝCIUM L. Sp. Pl. 191. 1753.
Shrubs, or woody vines, often spiny, with small alternate entire leaves, commonly with
smaller ones fascicled in their axils, and white greenish or purple, axillary or terminal, solitary or clustered flowers. Calyx campanulate, 3–5-lobed or -toothed, not enlarged in fruit,
persistent at the base of the berry. Corolla funnelform, salverform, or campanulate, the tube
short or slender, the limb 5-lobed (rarely 4-lobed), the lobes obtuse. Stamens 5 (rarely 4),
exserted, or included; filaments filiform, sometimes dilated at the base; anther-sacs longitudinally dehiscent. Ovary 2-celled; style filiform; stigma capitate or 2-lobed. Berry globose, ovoid, or oblong. [Named from the country Lycia.]

About 75 species, widely distributed in temperate and warm regions. Besides the following,
introduced from Europe, some 17 native species occur in the western parts of North America. Type
species: *Lycium afreum* L.

1. Lycium halimifòlium Mill. Matrimony Vine. Box-thorn. Fig. 3727.

Lycium halimifolium Mill. Gard. Dict. Ed. 8, no. 6. 1768.
Lycium Barbarum var. *vulgare* Ait. f. Hort. Kew. Ed. 2,
2 : 3. 1811.
Lycium vulgare Dunal in DC. Prodr. 13 : Part 1, 509. 1852.

Glabrous, spiny or unarmed; stems slender, climbing or trailing, branched, 6°–25° long, the branches
somewhat angled, the spines, when present, slender,
about ½′ long. Leaves lanceolate, oblong, or spatulate, acute or obtuse at the apex, narrowed into short
petioles, firm, ½′–1½′ long, 2″–4″ wide; flowers 2–5
together in the axils, or solitary; peduncles filiform,
spreading, 6″–12″ long; calyx-lobes ovate, acute, or
obtuse, 1½″ long; corolla funnelform, purplish changing to greenish, 4″–6″ broad, its lobes ovate-oblong;
stamens slightly exserted; berry oval, orange-red.

In thickets and waste places, escaped from gardens,
Ontario to Virginia, Minnesota and Kansas. Introduced
from Europe. Bastard jessamine. Jasmine. Jackson-vine.
May–Aug.

9. HYOSCÝAMUS [Tourn.] L. Sp. Pl. 179. 1753.
Erect coarse viscid-pubescent narcotic annual biennial or perennial herbs, with alternate
mostly lobed or pinnatifid leaves, and large nearly regular flowers, the lower solitary in the
axils, the upper in a more or less 1-sided spike or raceme. Calyx urn-shaped or narrowly
campanulate, 5-cleft, striate, enlarged and enclosing the capsule in fruit. Corolla funnelform, the limb somewhat oblique, 5-cleft, the lobes more or less unequal, spreading. Stamens
declined, mostly exserted; filaments filiform; anthers oblong or ovate, their sacs longitudi-

nally dehiscent. Ovary 2-celled; style slender; stigma capitate. Capsule 2-celled, circumscissile above the middle. [Greek, hog-bean.]

About 15 species, natives of the Mediterranean region, the following typical.

1. Hyoscyamus nìger L. Black Hen bane. Hog's-bean. Fig. 3728.

Hyoscyamus niger L. Sp. Pl. 179. 1753.

Annual or biennial, villous and viscid, of an ill odor; stem stout, 1°–2½° high. Leaves ovate, lanceolate, or oblong in outline, 3'–7' long, acute or acuminate at the apex, sessile, or the upper clasping the stem, irregularly lobed, cleft or pinnatifid; flowers very short-pedicelled, 1'–2' broad; calyx-lobes triangular-ovate, acute; corolla greenish-yellow, strongly reticulated with purple veins, its lobes ovate, obtusish; capsule globose-oblong, about 5" high.

In waste places, Nova Scotia to Ontario, New York and Michigan. Naturalized from Europe. Fetid nightshade. Insane-root. Belene. Chenile. Poison-tobacco. June–Sept.

10. DATÙRA L. Sp. Pl. 179. 1753.

Annual or perennial erect tall branching narcotic herbs, some tropical species shrubs or trees, with alternate petioled entire sinuate-dentate or lobed leaves, and large solitary erect short-peduncled white purple or violet flowers. Calyx elongated-tubular or prismatic, its apex 5-cleft or spathe-like, in the following species circumscissile near the base which is persistent and subtends the globose ovoid prickly capsule. Corolla funnelform, the limb plaited, 5-lobed, the lobes broad, acuminate. Stamens included or little exserted; filaments filiform, very long, inserted at or below the middle of the corolla-tube. Ovary 2-celled, or falsely 4-celled; style filiform; stigma slightly 2-lobed. Capsule 4-valved from the top, or bursting irregularly. [The Hindoo name, dhatura.]

About 12 species, of wide geographic distribution. Type species: *Datura Stramonium* L. The following are introduced weeds.

Glabrous or very sparingly pubescent; leaves lobed, calyx prismatic.   1. *D. Stramonium.*
Finely glandular-pubescent; leaves entire or undulate; calyx tubular.   2. *D. Metel.*

1. Datura Stramònium L. Stramonium. Jamestown or Jimson-weed. Thorn-Apple. Fig. 3729.

Datura Stramonium L. Sp. Pl. 179. 1753.

Datura Tatula L. Sp. Pl. Ed. 2, 256. 1762.

Annual, glabrous or the young parts sparingly pubescent; stem green to purple, stout, 1°–5° high. Leaves thin, ovate in outline, acute or acuminate at the apex, mostly narrowed at the base, 3'–8' long, irregularly sinuate-lobed, the lobes acute; petioles 1'–4' long; flowers white or violet, about 4' high, the limb 1½'–2' broad; calyx prismatic, less than one-half the length of the corolla; capsule ovoid, densely prickly, about 2' high, the lower prickles shorter than the upper or all about equal.

In fields and waste places, Nova Scotia to Florida, west to Minnesota and Texas. Naturalized from tropical regions. June–Sept. Peru-, mad- or devil's-apple. Devil's-trumpet. Jamestown-lily. Fire-weed. Dewtry. Races differ in color of flowers and in length of the prickles on the pods.

2. Datura Mètel L. Entire-leaved Thorn-Apple. Fig. 3730.

Datura Metel L. Sp. Pl. 179. 1753.

Annual, densely and finely glandular-pubescent; stem stout, much branched, 4°–8° high. Leaves broadly ovate, acute at the apex, inequilateral, rounded or subcordate at the base, 4′–10′ long, entire or merely undulate; petioles 1′–3′ long; flowers white, 6′–7′ high, the limb 3′–4′ broad; calyx tubular, about one-half as long as the corolla; capsule globose or ovoid-globose, obtuse, prickly and pubescent, 1′–1½′ in diameter.

In waste places, escaped from gardens, Rhode Island to Florida. Native of tropical America. July–Sept.

11. NICOTIÀNA [Tourn.] L. Sp. Pl. 180. 1753.

Annual or perennial viscid-pubescent acrid narcotic herbs or shrubs, with large alternate entire or slightly undulate leaves, and rather large white yellow greenish or purplish flowers, in terminal, often bracted, racemes or panicles. Calyx tubular-campanulate or ovoid, 5-cleft. Corolla funnelform, salverform, or nearly tubular, the tube usually longer than the limb, 5-lobed, the lobes spreading. Stamens 5, inserted on the tube of the corolla; filaments filiform; anthers ovate or oblong, their sacs longitudinally dehiscent. Ovary 2-celled (rarely 4-celled); style slender; stigma capitate. Capsule 2-valved, or sometimes 4-valved at the summit. Seeds very numerous, small. [Named for John Nicot, French ambassodar to Portugal, who sent some species to Catherine de Medici, about 1560.]

About 50 species, mostly natives of America. Besides the following, some 10 others occur in the southern and western United States. Type species: *Nicotiana Tabacum* L.

Corolla 1′ long, the tube cylindric; calyx-lobes triangular.   1. *N. rustica.*
Corolla 4′–6′ long. the tube very slender; calyx-lobes linear or narrowly
  lanceolate.   2. *N. longiflora.*

1. Nicotiana rústica L. Wild Tobacco. Fig. 3731.

Nicotiana rustica L. Sp. Pl. 180. 1753.

Annual; stem rather slender, 2°–4° high. Leaves broadly ovate, thin, entire, slender-petioled, 2′–8′ long, 1′–6′ wide; petioles ½′–5½′ long; flowers greenish-yellow, about 1′ long, panicled; pedicels 3″–6″ long, rather stout; calyx-lobes broadly triangular, acute, shorter than the tube; corolla-tube cylindric, somewhat enlarged above, the lobes short, obtuse, slightly spreading; capsule globose, glabrous, about 5″ in diameter, 2-valved, longer than the calyx.

In fields and waste places, escaped from gardens, Ontario to Minnesota, south to southern New York and Florida. Cultivated by the Indians, its origin unknown. Leaves greenish when dry. Indian, Syrian or real-tobacco. June–Sept.

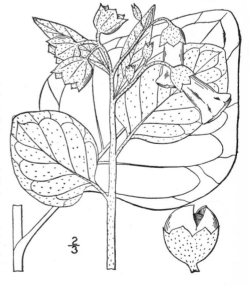

2. Nicotiana longiflòra Cav.　Long-flowered
Tobacco.　Fig. 3732.

Nicotiana longiflora Cav. Descr. Pl. 106.　1802.

　Annual, minutely rough-puberulent and viscid; stem
erect, slender, branched, 1½°–3° high.　Basal leaves
ovate-lanceolate or broadly oblanceolate, obtuse, 6′–10′
long, 1′–3′ wide, tapering into slender winged petioles;
stem-leaves linear or lanceolate, sessile, 2′–4′ long;
flowers in terminal racemes, short-pedicelled, 4′–6′ long;
calyx oblong, pubescent, its narrow lobes nearly as long
as the tube; corolla white or purplish, viscid, the tube
slender, 1″–1½″ in diameter, expanding above, the lobes
ovate-lanceolate, acute; capsule oblong.

　Near Harrisburg and Easton, Pa.　Escaped from gardens.　Native of South America.　Aug.–Oct.

12. PETÙNIA Juss. Ann. Mus. Paris 2 : 215. *pl. 47.* 1803.

　Viscid-pubescent annual or perennial branching herbs,
with entire leaves, and axillary or terminal solitary white
violet or purple flowers (in cultivation sometimes variegated).　Calyx deeply 5-cleft or
5-parted, the segments narrow.　Corolla funnelform or salverform, the limb plicate, spreading, slightly irregular.　Stamens 5, inserted on the throat of the corolla, 4 of them didynamous, perfect, the fifth smaller or obsolete; filaments slender; anthers ovoid, 2-lobed.
Disk fleshy.　Ovary 2-celled; ovules numerous in each cavity; style filiform; stigma 2-lamel-
late.　Capsule 2-celled, 2-valved, the valves entire.　Seeds small, the testa rugose.　[*Petun,*
an Indian name of tobacco.]

　About 12 species, natives of South America.　Type species : *Petunia parviflora* Juss.

Corolla white, its tube cylindric.	1. *P. axillaris.*
Corolla violet-purple, its tube campanulate.	2. *P. violacea.*

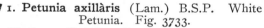

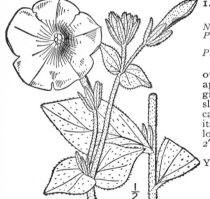

1. Petunia axillàris (Lam.) B.S.P.　White Petunia.　Fig. 3733.

Nicotiana axillaris Lam. Encycl. 4 : 480.　1797.
Petunia nyctaginiflora Juss. Ann. Mus. Paris, 2 : 215.
pl. 47. f. 2.　1803.
Petunia axillaris B.S.P. Prel. Cat. N. Y. 38.　1888.

　Very viscid; stem stout, about 1° high.　Leaves
ovate to obovate, obtuse or blunt-pointed at the
apex, sessile, or the lower narrowed into short margined petioles, 3′–4′ long, 1′–2′ wide; peduncles
slender, 2′–4′ long, often longer than the leaves;
calyx-segments linear-oblong, obtuse; corolla white,
its tube cylindric, slightly enlarged above, 1′–1½′
long, 3–4 times as long as the calyx, its limb about
2′ broad, the lobes rounded.

　In waste places, escaped from gardens, southern New
York and Pennsylvania.　Native of Brazil.　July–Sept.

2. Petunia violàcea Lindl.　Violet Petunia Fig. 3734.

Petunia violacea Lindl. Bot. Reg. *pl. 1626.*　1833.

　Similar to the preceding species, but usually
rather lower, and the stem slender.　Leaves ovate
or obovate, all but the uppermost petioled, mostly
obtuse, 1′–2½′ long; peduncles slender, 1′–2′ long;
calyx-segments linear, subacute, or obtuse; corolla violet-purple, its tube campanulate, 9″–15″
long, the limb less abruptly spreading, 1′–1½′
broad, the lobes subacute.

　In waste places, escaped from gardens, southern
New York and Pennsylvania.　Native of southern
Brazil and Paraguay.　June–Sept.

　Petunia parviflòra Juss., a prostrate pubescent
annual, with small linear to spatulate leaves, and a
funnelform corolla 4″–5″ long, has been found on
ballast about the eastern seaports, from South
America.

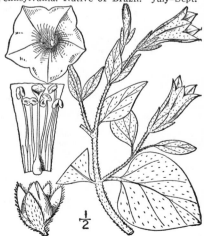

Family 28. **SCROPHULARIÀCEAE** Lindl. Nat. Syst. Ed. 2, 288. 1826.

Figwort Family.

Herbs, shrubs or trees, with opposite or alternate exstipulate leaves, and perfect mostly complete and irregular flowers (corolla wanting in one species of *Synthyris*). Calyx inferior, persistent, 4–5-toothed, -cleft, or -divided, or sometimes split on the lower side, or on both sides, the lobes or segments valvate, imbricate or distinct in the bud. Corolla gamopetalous, the limb 2-lipped, or nearly regular. Stamens 2, 4 or 5, didynamous, or nearly equal, inserted on the corolla and alternate with its lobes; anthers 2-celled, the sacs equal, or unequal, or sometimes confluent into one. Disk present or obsolete. Pistil 1, entire or 2-lobed; ovary superior, 2-celled, or rarely 1-celled; ovules mostly numerous, rarely few, anatropous or amphitropous, borne on axile placentae; style slender, simple; stigma entire, 2-lobed or 2-lamellate. Fruit mostly capsular and septicidally or loculicidally dehiscent. Seeds mostly numerous, the testa reticulated, pitted, striate, ribbed, or nearly smooth; endosperm fleshy; embryo small, straight or slightly curved; cotyledons little broader than the radicle.

About 165 genera and 2700 species, widely distributed, most abundant in temperate regions.

1. **Anther-bearing stamens 5; corolla rotate; leaves alternate.** (VERBASCEAE.) 1. *Verbascum.*
2. **Anther-bearing stamens 2 or 4; leaves opposite, verticillate or alternate.**
   * **Corolla spurred, saccate or gibbous on the lower side at the base.** (ANTIRRHINEAE.)
   Corolla spurred at the base.
     Leaves palmately 3–5-veined.   2. *Cymbalaria.*
     Leaves pinnately veined.
       Flowers solitary in the axils.
         Throat of the corolla closed by the palate; leaves broad.   3. *Kickxia.*
         Throat of the corolla not closed by the palate; leaves narrow.   4. *Chaenorrhinum.*
       Flowers in terminal racemes.   5. *Linaria.*
   Corolla saccate or gibous at the base.   6. *Antirrhinum.*
   ** **Corolla neither spurred, saccate nor gibbous on the lower side.**
   † *Anther-bearing stamens 4, the fifth sterile or rudimentary.* (CHELONEAE.)
   Sterile stamen a scale adnate to the upper side of the corolla.   7. *Scrophularia.*
   Sterile stamen elongated, longer or shorter than the others.
     Corolla tubular, 2-lipped, the lobes of the lower lip flat.
       Sterile stamens shorter than the others; seeds winged.   8. *Chelone.*
       Sterile stamen about equalling the others; seeds wingless.   9. *Pentstemon.*
     Corolla 2-cleft, declined; middle lobe of the lower lip conduplicate.   10. *Collinsia.*
   †† *Stamens 4, all anther-bearing, or 2 sterile, or 2 only.*
   a. Stamens 4, all anther-bearing; large Asiatic tree. (PAULOWNIEAE.)
                                                                              11. *Paulownia.*
   b. Stamens 4, all anther-bearing; herbs; corolla 2-lipped; stamens not enclosed
            in upper lip of corolla. (MIMULEAE.)
   Calyx prismatic, 5-angled, 5-toothed.   12. *Mimulus.*
   Calyx 5-parted, not prismatic.
     Calyx-segments equal; leaves pinnatifid in our species.   13. *Conobea.*
     Calyx-segments unequal, the upper one the largest.
       Corolla nearly regular, about equally 5-lobed.   14. *Bramia.*
       Corolla manifestly 2-lipped.
         Leaves palmately nerved; flower with an hypogynous disk.   15. *Hydrotrida.*
         Leaves pinnately veined; no hypogynous disk.   16. *Mecardonia.*
   c. Stamens 4, 2 anther-bearing and 2 sterile, or 2 only; corolla obviously 2-lipped. (GRATIOLEAE.)
   Calyx 5-parted; upper lip of corolla present.
     Sterile filaments short or none.
       Anther-sacs transverse, separated.   17. *Gratiola.*
       Anther-sacs parallel, contiguous.   18. *Sophrononthe.*
     Sterile filaments slender, 2-lobed.   19. *Ilysanthes.*
   Calyx 4-toothed; upper lip of corolla obsolete; low mud plant.   20. *Micranthemum.*
   d. Stamens 4, all anther-bearing; corolla nearly regular; flowers on scapes. (LIMOSELLEAE.)
                                                                              21. *Limosella.*
   e. Stamens 2 only (rarely 4 in *Synthyris*); corolla rotate, salverform, tubular, or
            none. (VERONICEAE.)
   Leaves alternate; flowers spicate; corolla 2–3-lobed or none.   22. *Synthyris.*
   Leaves, at least the lower, opposite or verticillate; corolla 4-lobed.
     Corolla rotate or salverform; capsule obcordate or emarginate, compressed. 23. *Veronica.*
     Corolla tubular-funnelform; capsule ovoid, not compressed.   24. *Leptandra.*
   f. Stamens 4, all anther-bearing; corolla campanulate, salverform or funnelform, scarcely 2-lipped.
   Leaves alternate; flowers in 1-sided racemes. (DIGITALEAE.)   25. *Digitalis.*
   Leaves, at least the lower, opposite. (BUCHNEREAE.)
     Corolla salverform; flowers in a long spike.   26. *Buchnera.*
     Corolla campanulate or funnelform.
       Stamens nearly equal; calyx-lobes as long as the tube.   27. *Afzelia.*

Stamens strongly didynamous, unequal; calyx-teeth shorter than the tube.
 Anthers awned at the base; corolla yellow.   28. *Dasystoma.*
 Anthers awnless; corolla purple, pink or rarely white.
  Anthers all alike; flowers pedicelled; leaves not auricled.   29. *Agalinis.*
  Anthers of the shorter stamens smaller; leaves auricled at base.   30. *Otophylla.*
g. Stamens 4, all anther-bearing, ascending under the upper lip of the corolla. (EUPHRASIEAE.)
Ovules several or numerous; capsule many-seeded.
 Anther-sacs dissimilar, the inner one pendulous by its apex; leaves mostly alternate.
  Upper lip of the corolla much longer than the lower.   31. *Castilleja.*
  Upper lip of the corolla scarcely longer than the lower.   32. *Orthocarpus.*
 Anther-sacs similar and parallel; leaves mostly opposite.
  Calyx 2-bracteolate at the base, oblique, 5-toothed.   33. *Schwalbea.*
  Calyx not bracteolate, 4–5-toothed, or cleft or split.
   Upper lip of the corolla 2-lobed, its margins recurved; calyx 4-cleft.   34. *Euphrasia.*
   Margins of the upper lip of the corolla not recurved.
    Calyx scarcely or not at all inflated in fruit; galea entire.
     Calyx 4-toothed or 4-cleft; capsule straight.
      Seeds spreading, numerous.   35. *Bartsia.*
      Seeds pendulous, few.   36. *Odontites.*
     Calyx split on the lower side or on both sides; capsule oblique.
      Galea short-beaked or beakless.   37. *Pedicularis.*
      Galea filiform-beaked.   38. *Elephantella.*
    Calyx ovoid, much inflated and veiny in fruit.   39. *Rhinanthus.*
Ovules only 1 or 2 in each cell of the ovary; capsule 1–4-seeded; leaves opposite.   40. *Melampyrum.*

1. **VERBASCUM** [Tourn.] L. Sp. Pl. 177. 1753.

Biennial or rarely perennial, mostly tall and erect herbs, with alternate dentate pinnatifid or entire leaves, and rather large yellow purple red or white flowers, in terminal spikes, racemes or panicles. Calyx deeply 5-cleft or 5-parted. Corolla flat-rotate or slightly concave, 5-lobed, the lobes a little unequal, the upper exterior, at least in the bud. Stamens 5, inserted on the base of the corolla, unequal, all anther-bearing; filaments of the 3 upper stamens, or of all 5, pilose; anther-sacs confluent into one. Ovules numerous; style dilated and flattened at the summit. Capsule globose to oblong, septicidally 2-valved, many-seeded, the valves usually 2-cleft at the apex. Seeds rugose, not winged. [The Latin name of the great mullen; used by Pliny.]

About 125 species, natives of the Old World. Besides the following, another is naturalized in the western United States. Type species: *Verbascum Thapsus* L.

Plants densely woolly; flowers in dense terminal spikes, or spike-like racemes.
 Leaves strongly decurrent on the stem.   1. *V. Thapsus.*
 Leaves not decurrent, or but slightly so.   2. *V. phlomoides.*
Leaves white-tomentose beneath; flowers in large terminal panicles.   3. *V. Lychnitis.*
Plant glabrous or sparingly glandular; flowers racemose.   4. *V. Blattaria.*

1. **Verbascum Thápsus** L.   Great Mullen.   Velvet or Mullen Dock.   Fig. 3735.

Verbascum Thapsus L. Sp. Pl. 177. 1753.

Erect, stout, simple or with some erect branches, densely woolly all over with branched hairs; stem 2°–7° high, wing-angled by the bases of the decurrent leaves. Leaves oblong, thick, acute, narrowed at the base, dentate or denticulate, 4′–12′ long, the basal ones borne on margined petioles; flowers yellow, 8″–12″ broad, sessile, numerous in dense elongated spikes rarely branched above; stamens unequal, the three upper shorter with white hairy filaments and short anthers, the two lower glabrous or nearly so with larger anthers; capsule about 3″ high, slightly longer than the woolly calyx.

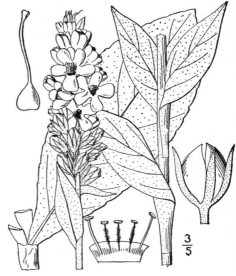

In fields and waste places, Nova Scotia to South Dakota, California, Florida and Kansas. Often a troublesome weed. Naturalized from Europe. Native also of Asia. Among some 40 English names are hedge-, hig- or high-taper. Candlewick. Cow's or bullock's-lungwort. Aaron's- or Adam's-rod or -flannel. Feltwort. Hare's-beard. Jacob's-, Jupiter's- or Peter's-staff. Ice-leaf. Torches. Flannel-leaf. Old man's-flannel or blanket-leaf. Woolen, *i. e.,* mullen. Shepherd's-club. Velvet-plant. June–Sept.

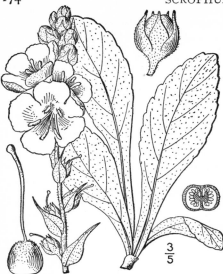

$\frac{3}{5}$

2. Verbascum phlomoìdes L. Clasping-leaved Mullen. Fig. 3736.

Verbascum phlomoides L. Sp. Pl. 1194. 1753.

Stem rather stout, usually simple, 1°–4° high. Leaves oblong to ovate-lanceolate, crenate, crenulate, or entire, woolly-tomentose on both sides, sessile or somewhat clasping, or slightly decurrent on the stem, or the lower often petioled with truncate or subcordate bases; flowers yellow, or cream-color, 1' broad or more, usually in a solitary elongated tomentose spike-like raceme; pedicels clustered, shorter than the calyx; stamens as in *V. Thapsus;* capsule 4''–5'' long, exceeding the tomentose calyx.

Eastern Massachusetts to Kentucky. Adventive or fugitive from Europe or eastern Asia. June–Aug.

3. Verbascum Lychnìtis L. White Mullen. Fig. 3737.

Verbascum Lychnitis L. Sp. Pl. 177. 1753.

Stem angled, rather stout, paniculately branched above, 2°–4½° high, densely covered, as well as the lower surfaces of the leaves, with a white canescent nearly stellate pubescence. Leaves oblong, ovate or oblong-lanceolate, crenate-dentate, 2'–7' long, the upper acute, sessile, but not decurrent on the stem, the lower obtuse or acute at the apex and narrowed into margined petioles; flowers in a large terminal panicle, racemose on its branches, white or cream-color, 5''–6'' broad, nearly sessile; pilose hairs of the 3 shorter filaments white; capsule about 2'' high, equalling or exceeding the calyx.

In fields and waste places, Ontario to New Jersey and Pennsylvania. Reported from Kansas. Naturalized from Europe. Native also of Asia. June–Sept. Its down once used for lighting, hence *Lichnitis,* lamp.

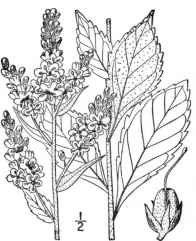

$\frac{1}{2}$

4. Verbascum Blattària L. Moth Mullen. Fig. 3738.

Verbascum Blatteria L. Sp. Pl. 178. 1753.

Stem erect, strict, slender, terete, glabrous or sparingly glandular-pubescent, usually quite simple, 2°–6° high. Leaves oblong, ovate or lanceolate, dentate, laciniate, or pinnatifid, acute or acuminate, the upper ½'–2½' long, truncate or cordate-clasping at the base, the lower and basal ones sessile or somewhat petioled, sometimes 1° long, seldom present at flowering time; raceme 1°–2° long, loose; pedicels spreading, ½'–1' long, bracted at the base; corolla yellow or white, about 1' broad, with brown marks on the back; filaments all pilose with violet hairs; capsule depressed-globose, 3'' in diameter, longer than the calyx.

In fields and waste places, Quebec to Florida, west to Oregon and California. Naturalized from Europe. Native also of Asia. June–Nov. Said to repel the cockroach (*Blatta*), whence the name *Blattaria;* frequented by moths, hence moth-mullen.

Verbascum virgàtum Stokes, a similar Old World species, found on Cape Breton Island and naturalized in the Pacific States, has fruiting pedicels shorter than the capsules.

$\frac{1}{2}$

2. CYMBALÀRIA Medic. Phil. Bot. 2: 70. 1791.

Perennial creeping or spreading herbs, with long-petioled, mostly lobed, palmately veined leaves, and solitary axillary white to violet flowers. Calyx 5-parted. Corolla irregular, 2-lipped, short-spurred; upper lip 2-lobed, lower lip 3-lobed; throat nearly or quite closed by the palate. Stamens 4, didynamous, ascending, included, the filaments filiform. Style very slender. Capsule dehiscent by 2 terminal 3-toothed pores. Seeds numerous, small. [From the Greek for cymbal.]

About 9 species, natives of the Old World, the following typical.

1. Cymbalaria Cymbalària (L.) Wettst. Kenilworth or Coliseum Ivy. Fig. 3739.

Antirrhinum Cymbalaria L. Sp. Pl. 612. 1753.
Linaria Cymbalaria Mill. Gard. Dict. Ed. 8, no. 17. 1768.
Cymbalaria Cymbalaria Wettst. in Engl. & Prantl, Nat. Pfl. Fam. 4: Abt. 3b, 58. 1891.

Perennial, glabrous; stem trailing, branched, often rooting at the nodes, 3'-12' long. Leaves slender-petioled, reniform-orbicular, palmately 3–5-veined, 3–5-lobed, ¼'–1' in diameter, the lobes broad and obtuse; petioles usually as long as the blade; flowers axillary, solitary, blue or lilac, 4"–5" long; peduncles slender, recurved, shorter than the petioles; calyx-segments lanceolate, acute; palate yellowish; capsule globose, several-seeded; seeds rugose, wingless.

Waste places and roadsides, adventive from Europe, Ontario to New Jersey and Pennsylvania, and in seaport ballast. Other English names are ivy-leaved toadflax, ivy-weed. Climbing or roving sailor. Aaron's-beard. Wandering jew. Mother-of-thousands. Oxford-weed. Pennywort. June–Aug.

3. KÍCKXIA Dumort. Fl. Belg. 35. 1827.

[ELATINOIDES Wettst. in Engl. & Prantl, Nat. Pfl. Fam. 4: Abt. 3b, 58. 1891.]

Mostly annual spreading or creeping herbs, with pinnately veined, short-petioled entire toothed or lobed leaves, and solitary axillary white yellow or variegated flowers. Calyx 5-parted. Corolla irregular, spurred, 2-lipped, the throat closed by the palate. Stamens 4, didynamous, included; filaments filiform. Capsule subglobose, or ovoid, opening by 1 or 2 terminal slits, pores or valves. Seeds numerous, ovoid, mostly rough or tubercled. [In honor of Jean Kickx, 1775–1831, professor in Brussels.]

About 25 species, natives of the Old World. Type species: *Antirrhinum Elatìne* L.

Leaves ovate-orbicular, cordate or rounded at the base. 1. *K. spuria.*
Leaves hastate. 2. *K. Elatine.*

1. Kickxia spùria (L.) Dumort. Round-leaved Toad-Flax. Fig. 3740.

Antirrhinum spurium L. Sp. Pl. 613. 1753.
Linaria spuria Mill. Gard. Dict. Ed. 8, no. 15. 1768.
Kickxia spuria Dumont. Fl. Belg. 35. 1827.
Elatinoides spuria Wettst. in Engl. & Prantl, Nat. Pfl. Fam. 4: Abt. 3b, 58. 1891.

Annual, pubescent all over; stems prostrate, branched or simple, 3'–2° long. Leaves short-petioled, ovate-orbicular, entire, or sometimes dentate, mucronulate at the apex, cordate or rounded at the base, ¼'–1' in diameter; petioles 1"–2" long; flowers solitary in the axils, small; peduncles filiform, very pubescent, often much longer than the leaves; calyx-segments ovate, acute at the apex, cordate or rounded at the base, one-half as long as the corolla; corolla yellowish with a purple upper lip, the spur curved, about as long as the tube; capsule subglobose, shorter than the calyx; seeds rugose, not winged.

In waste places and ballast, New York to North Carolina and Missouri. Adventive from Europe. This and the next called also cancerwort and female-fluellin. June–Sept.

2/3

2. **Kickxia Elatìne** (L.) Dumort.　Sharp-pointed
Fluellin or Toad-Flax.　Fig. 3741.

Antirrhinum Elatine L. Sp. Pl. 612. 1753.
Linaria Elatina Mill. Gard. Dict. Ed. 8, no. 16. 1768.
Kickxia Elatine Dumort. Fl. Belg. 35. 1827.
Elatinoides Elatine Wettst. in Engl. & Prantl, Nat. Pfl.
　Fam. 4: Abt. 3b, 58. 1891.

Annual, pubescent; stems prostrate, usually branched,
slender, 6′–2° long.　Leaves short-petioled, ovate, ½′–1′
long, acute or acutish at the apex, triangular, hastate,
truncate, or subcordate at the base, the basal auricles
divergent, acute; petioles 1″–3″ long; flowers solitary
in the axils, about 3″ long; peduncles filiform, glabrous, or somewhat hairy, usually longer than the
leaves; calyx-segments narrowly lanceolate, acute; corolla yellowish, purplish beneath, its spur slender,
straight, declined; capsule subglobose, shorter than the
calyx; seeds wingless.

In sandy waste places, Canada (?); Massachusetts to
Georgia and Missouri.　Naturalized from Europe.　Native also of Asia.　Called also canker-root.　June–Sept.

4. **CHAENORRHÌNUM** [DC.] Lange; Willk. & Lange, Prodr. Fl. Hisp. **2**:
577.　1870.

Herbs with alternate, usually entire leaves, and violet, blue or white axillary flowers.
Calyx 5-parted, the segments narrow.　Corolla similar to that of *Linaria,* but with an open
throat and a straight upper lip.　Stamens 4, didynamous; filaments slender.　Style filiform.
Capsule inequilateral, one carpel larger than the other.　Seeds ovoid or cuneate, ribbed.
[Greek, open nose, referring to the open corolla-throat.]

About 20 species, chiefly in the Mediterranean region and Asia.　Type species: *Antirrhinum
minus* L.

1. **Chaenorrhinum mìnus** (L.) Lange.　Small Snap-dragon.　Fig. 3742.

Antirrhinum minus L. Sp. Pl. Ed. 2, 852. 1763.
Linaria minor Desf. Fl. Atlant. **2**: 46. 1800.
Chaenorrhinum minus Lange; Willk. & Lange, Prodr. Fl. Hisp.
　2: 579. 1870.

Annual, glandular-pubescent all over; stem 5′–13′ tall,
often branched.　Leaves linear-spatulate to linear, mostly
obtuse, 5″–15″ long, narrowed at the base; flowers shorter
than the pedicels, blue or bluish, 2½″–4″ long; calyx-segments linear to linear-spatulate, somewhat shorter than the
corolla; spur short and stout, much shorter than the body
of the corolla; capsule globose-ovoid.

Waste grounds and ballast, New Brunswick to New York,
Pennsylvania and Michigan.　Adventive from Europe.

5. **LINÀRIA** [Tourn.] Mill. Gard. Dict. Abr.
Ed. 4.　1754.

Herbs, some exotic species shrubby, with alternate entire
dentate or lobed leaves, or the lower and those of sterile
shoots opposite or verticillate, and yellow white blue purple
or variegated flowers, in terminal bracted racemes or spikes.
Calyx 5-parted, the segments imbricated.　Corolla irregular, spurred at the base, or the spur rarely obsolete, 2-lipped,
the upper lip erect, 2-lobed, covering the lower in the bud,
the lower spreading, 3-lobed, its base produced into a palate
often nearly closing the throat.　Stamens 4, didynamous,
ascending, included; filaments and style filiform.　Capsule ovoid or globose, opening by 1
or more mostly 3-toothed pores or slits below the summit.　Seeds numerous, wingless or
winged, angled or rugose.　[Latin, *linum,* flax, which some species resemble.]

About 150 species, of wide geographic distribution, most abundant in the Old World.　Besides
the following, another species occurs in Florida.　The corolla, especially the terminal one of the
raceme, occasionally has 5 spurs and is regularly 5-lobed, and is then said to be in the Peloria state.
Type species: *Antirrhinum Linaria* L.

Flowers yellow, 8″–15″ long; leaves linear; flowers 12″–15″ long.
　Leaves alternate.　　　　　　　　　　　　　　　　　　　　　　1. *L. Linaria.*
　Lower leaves whorled.　　　　　　　　　　　　　　　　　　　　2. *L. supina.*
Flowers blue to white, 3″–6″ long.
　Spur of corolla filiform, curved; native species.　　　　　　　　3. *L. canadensis.*
　Spur of the corolla short, conic; European adventive species.　　4. *L. repens.*

1. Linaria Linària (L.) Karst. Ranstead. Butter-and-Eggs. Fig. 3743.

Antirrhinum Linaria L. Sp. Pl. 616. 1753.
Linaria vulgaris Hill, Brit. Herb. 108. 1756.
Linaria Linaria Karst. Deutsch. Fl. 947. 1880–83.

Perennial by short rootstocks, pale green and slightly glaucous; stems slender, erect, very leafy, glabrous, or sparingly glandular-pubescent above, simple or with few erect branches, 1°–3° high. Leaves linear, sessile, entire, acute at both ends, mostly alternate, $\frac{3}{4}$′–1$\frac{1}{2}$′ long, 1″–1$\frac{1}{2}$″ wide; flowers densely racemose, light yellow, 1′–1$\frac{1}{4}$′ long, the spur of the erect corolla somewhat darker, the palate orange-colored; pedicels 2″–4″ long, nearly erect; calyx-segments oblong, acutish, about 1$\frac{1}{2}$″ long; spur subulate, nearly as long as the body of the corolla; middle lobe of the lower lip shorter than the other two; capsule ovoid, the seeds rugose, winged.

In fields and waste places, Newfoundland to Oregon, Virginia and New Mexico. Naturalized from Europe. Native also of Asia. June–Oct. Brideweed. Flaxweed. Eggs and bacon. Yellow toad-flax. Impudent lawyer. Jacob's-ladder. Rancid. Wild flax or tobacco. Devil's flax. Snap-dragon. Devil's-flower. Dead men's bones. Bread and butter. Continental weed. Gallwort. Rabbit-flower. Widely distributed in temperate regions as a weed.

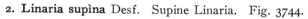

Linaria genistaefòlia (L.) Mill., found many years ago at the northern part of New York Island, and admitted into the first edition of this work, has not recently been collected in America.

2. Linaria supìna Desf. Supine Linaria. Fig. 3744.

Linaria supina Desf. Fl. Atlant. **2**: 44. 1800.

Perennial by short rootstocks, bright green; stems few or numerous, decumbent, 4′–9′ long, glabrous or slightly glandular-pubescent. Leaves linear-spatulate to narrowly linear, the lower ones whorled, the upper alternate, mostly 5″–15″ long; flowers few together in short racemes, nearly similar to those of *L. Linaria*, but smaller; capsule globose or ovoid-globose.

Waste places and ballast, northern Atlantic seaboard. Naturalized from Europe.

3. Linaria canadénsis (L.) Dumort. Blue or Wild Toad-Flax. Fig. 3745.

Antirrhinum canadense L. Sp. Pl. 618. 1753.

Linaria canadensis Dumont, Bot. Cult. **2**: 96. 1802.

Biennial or annual, glabrous, green, sometimes fleshy; flowering stems erect or ascending, very slender, simple, or branched, 4′–2$\frac{1}{2}$° high, the sterile shoots spreading or procumbent, very leafy. Leaves linear or linear-oblong, 4″–15″ long, $\frac{1}{2}$″–1″ wide, entire, sessile, those of the sterile shoots, or some of them, usually opposite; flowers 3″–4″ long, in slender long racemes; pedicels 2″–3″ long, erect and appressed in fruit, minutely bracted at the base; calyx-segments lanceolate, acute or acuminate, about as long as the capsule; spur of the corolla filiform, curved, as long as the tube or longer; palate a white convex 2-ridged projection; capsule opening by 2 apical valves, each valve becoming 3-toothed; seeds angled, wingless.

In dry soil, Nova Scotia to Florida, west to Minnesota, Oregon, Texas and California. Also in Central and South America. A dwarf form with no corolla is frequent. May–Sept.

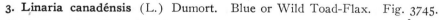

4. Linaria rèpens (L.) Mill. Pale-blue Toad-Flax. Fig. 3746.

Antirrhinum repens L. Sp. Pl. 614. 1753.

Linaria repens Mill. Gard. Dict. Ed. 8, no. 6. 1768.

L. striata DC. Fl. France, 3: 586. 1805.

Glabrous, perennial by a horizontal or creeping rootstock; stem erect, or the base decumbent, 8′–30′ high, usually branched, the branches slender. Leaves linear, entire, short-petioled or sessile, ½′–2′ long, 1″–2″ wide, narrowed to both ends, the lower crowded, sometimes whorled, the upper more scattered; flowers in slender terminal elongating racemes; pedicels 2″–5″ long; bracts narrowly linear, acute; corolla nearly white, but striped with blue or purple, about 6″ long; spur short, conic; capsule subglobose; seeds wrinkled, wingless.

Newfoundland, and in ballast about the Atlantic seaports. Adventive from Europe. Summer.

$\frac{3}{5}$

6. ANTIRRHÌNUM [Tourn.] L. Sp. Pl. 612. 1753.

Annual or perennial herbs, with alternate leaves, or the lower and those of sterile shoots opposite, and mostly large red purple yellow or white flowers, in terminal racemes, or solitary in the upper axils. Calyx 5-parted, the segments imbricated. Corolla irregular, gibbous, or saccate, but not spurred at the base, 2-lipped, the upper lip erect, 2-lobed, the lower spreading, 3-lobed, its base produced into a palate nearly or quite closing the throat. Stamens 4, didynamous, included; filaments filiform, or dilated at the summit. Style filiform. Capsule ovoid or globose, opening by chinks or pores below the summit. Seeds numerous, oblong, truncate, rugose or smooth, not winged. [Greek, nose-like.]

About 40 species, natives of Europe, Asia and western North America. Besides the following introduced species, some 18 others inhabit the western United States. Type species: *Antirrhinum majus* L.

Flowers 1′–1½′ long; calyx-segments ovate, much shorter than the corolla.   1. *A. majus.*
Flowers 5″–7″ long; calyx-segments linear, as long as the corolla.   2. *A. Orontium.*

1. Antirrhinum màjus L. Great Snap-dragon. Lion's-mouth. Fig. 3747.

Antirrhinum majus L. Sp. Pl. 617. 1753.

Perennial, glabrous below, usually more or less glandular-pubescent above; stem branched or simple, 1°–3° high. Leaves lanceolate, linear or oblong-lanceolate, entire, short-petioled, acute at both ends, rather firm, glabrous, 1′–3′ long, 1″–5″ wide; flowers racemose, purplish-red (of a variety of colors in cultivated forms), 1′–1½′ long; pedicels rather stout, 3″–6″ long, erect in fruit; calyx-segments oval to ovate, obtuse, 2″–3″ long; capsule obliquely ovoid, 4″–5″ high, opening by 2 pores just below the summit or at length apically 2-valved, much longer than the calyx.

In waste places, sparingly escaped from gardens in the Atlantic States. Adventive from Europe. Other English names are rabbit's mouth, bonny rabbits, calf-snout, dragon's-, tiger's-, dog's- or toad's-mouth. Bulldogs. Lion's-snap. June–Sept.

$\frac{3}{5}$

2. Antirrhinum Oróntium L. Lesser Snapdragon. Fig. 3748.

Antirrhinum Orontium L. Sp. Pl. 617. 1753.

Annual, glabrous or pubescent; stem erect, simple, or branched, slender, about 1° high. Leaves narrowly linear, or the lower linear-spatulate, almost sessile, narrowed at both ends, 1′–2′ long, 1″–2″ wide; flowers solitary in the upper axils, purple, mostly distant, 5″–7″ long; peduncles shorter than the flowers; calyx-segments linear, somewhat unequal, as long as the corolla, elongated in fruit so as much to exceed the pubescent capsule.

Fields and waste places, Ontario, New England, New York, Vancouver Island and Jamaica. Adventive from Europe. Native also of Asia. Corn-snapdragon. June–Aug.

7. SCROPHULÀRIA [Tourn.] L. Sp. Pl. 619. 1753.

Perennial strong-smelling herbs, some exotic species shrubby, with mostly opposite large leaves, and small purple greenish or yellow proterogynous flowers, in terminal panicled cymes or thyrses. Calyx 5-parted or 5-cleft, the segments or lobes mostly obtuse. Corolla irregular, the tube globose to oblong, not gibbous nor spurred at the base, the limb 5-lobed, the 2 upper lobes longer, erect, the lateral ones ascending, the lower spreading or reflexed. Stamens 5, 4 of them anther-bearing and didynamous, declined, mostly included, their anther-sacs confluent into one, the fifth sterile, reduced to a scale on the roof of the corolla tube. Style filiform; stigma capitate or truncate. Capsule ovoid, septicidally dehiscent. Seeds rugose, not winged. [Named for its repute as a remedy for scrofula.]

About 120 species, natives of the northern hemisphere, most abundant in southern Europe. Besides the following, 2 or 3 others occur in the western United States. Type species: *Scrophularia nodòsa* L.

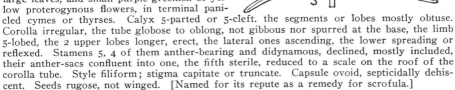

Corolla dull outside; sterile stamen deep purple. 1. *S. marylandica.*
Corolla shining outside; sterile stamen greenish yellow.
  Upper lip of the corolla as long as the tube; panicle-branches sparingly glandular; leaf-blades
    not hastate-incised at the base. 2. *S. leporella.*
  Upper lip of the corolla much shorter than the tube; panicle-branches densely glandular; leaf-
    blades, especially the lower ones, incised-hastate at the base. 3. *S. occidentalis.*

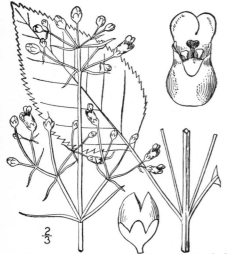

1. Scrophularia marylándica L. Maryland Figwort, Heal-all or Pilewort. Fig. 3749.

Scrophularia marylandica L. Sp. Pl. 619. 1753.
Scrophularia nodosa var. *marylandica* A. Gray, Syn. Fl. **2**: Part 1, 258. 1878.

Glabrous below, somewhat glandular-pubescent above; stem slender, 4-angled with grooved sides, usually widely branched, erect, 3°–10° high. Leaves membranous, slender-petioled, usually puberulent beneath, ovate or ovate-lanceolate, acuminate at the apex, sharply serrate, narrowed, truncate or subcordate at the base, 3′–12′ long; flowers greenish-purple, 3″–4″ long, very numerous in the nearly leafless thyrses; bractlets mostly opposite, pedicels slender, ascending, 4″–12″ long; calyx-lobes broadly ovate, obtuse, about the length of the tube; corolla green, dull without, brownish purple and shining within, little contracted at the throat, the two lateral lobes slightly spreading, the upper lip erect, its lobes short, rounded; capsule subglobose, with a slender tip; sterile stamen deep purple.

In woods and thickets, Maine to South Dakota, North Carolina, Georgia and Tennessee. Scrofula-plant. Carpenter's-square. Ascends to 4000 ft. in North Carolina. July–Sept.

Scrophularia neglécta Rydb. differs from *S. marylandica* in the pubescent leaves, the larger corollas and larger capsules; it occurs in the western part of our range.

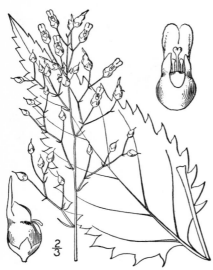

2. **Scrophularia leporélla** Bicknell.   Hare Figwort.   Fig. 3750.

Scrophularia leporella Bicknell, Bull. Torr. Club **23**: 317. 1896.

Stem puberulent below, viscid-glandular above, sharply 4-angled with flat sides, 3°–8° tall, simple, or somewhat branched.   Leaves short-petioled, ovate to lanceolate, acuminate at the apex, mostly narrowed at the base, but sometimes subcordate, glabrous on both sides when mature, usually incised-dentate, 2′–10′ long; flowers 4″–5″ long, in elongated narrow thyrses; bractlets mostly alternate; calyx-lobes ovate, obtuse, or acute; corolla contracted at the throat, green to purple and shining without, dull within, the two lateral lobes erect; lobes of the upper lip often narrowly oblong; sterile stamen greenish yellow; capsule ovoid-conic.

In woods and along roadsides, Vermont to Minnesota, Virginia and Kansas. Ascends to 3500 ft. in Virginia.   May–July.

3. **Scrophularia occidentàlis** (Rydb.) Bicknell.   Western Figwort.   Fig. 3751.

Scrophularia nodosa occidentalis Rydb. Contr. Nat. Herb. 3: 517. 1896.
Scrophularia occidentalis Bicknell, Bull. Torr. Club **23**: 315. 1896.

Similar to *S. leporella* in habit, but with densely glandular panicle-branches.   Stem more or less glandular; leaf-blades ovate to lanceolate or oblong-lanceolate, incised or incised-serrate, at least those of the lower leaves hastately incised at the base; corolla 4″–5″ long, the upper lip much shorter than the tube; capsules ovoid, 3½″–4″ long.

In low grounds and thickets, South Dakota to Oklahoma, Washington and California.   June–Aug.

8. **CHELÒNE** [Tourn.] L. Sp. Pl. 611. 1753.

Perennial, mostly glabrous branched or simple herbs, with opposite serrate petioled leaves, and large white red or purple flowers, in terminal and axillary dense spikes.   Calyx 5-parted, bracted at the base, the segments ovate or lanceolate.   Corolla irregular, the tube elongated, enlarged above, the limb 2-lipped; upper lip concave, emarginate or entire, exterior in the bud; lower lip spreading, woolly within, 3-lobed, its lateral lobes sometimes longer than the middle one.   Stamens 5, included, 4 of them antheriferous, didynamous, the fifth sterile, smaller; filaments slender, woolly; anthers woolly, cordate.   Style filiform; stigma small, capitate. Capsule ovoid, septicidally dehiscent.   Seeds numerous, compressed, winged.   [Greek, tortoise, the head of which the corolla resembles.]

Three species, natives of eastern North America.   Type species: *Chelone glabra* L.

Corolla white to purplish ; bracts not ciliolate.   1. *C. glabra.*
Corolla red or rose-purple ; bracts ciliolate.
    Leaves oblong or lanceolate.   2. *C. obliqua.*
    Leaves ovate, acuminate ; mountain plant.   3. *C. Lyoni.*

1. Chelone glàbra L. Snake-head.
Turtle-head. Fig. 3752.

Chelone glabra L. Sp. Pl. 611. 1753.

Stem slender, erect, obtusely 4-sided, simple or sometimes branched, strict, 1°–3° high, the branches erect. Leaves linear-lanceolate to ovate-lanceolate, sharply serrate with low appressed teeth, acuminate at the apex, narrowed at the base, short-petioled, 3′–6′ long, ½′–1¼′ wide, the principal veins about 10 on each side of the midvein; flowers white or faintly pink, about 1′ long; bracts glabrous, not ciliolate; calyx-segments ovate-oblong, obtuse; capsule ovoid, obtuse, about ½′ high, twice as long as the calyx.

In swamps and along streams, Newfoundland to Florida, Manitoba, Alabama and Kansas. Ascends to 3000 ft. in the Adirondacks. Shell-flower. Cod-head. Bitter-herb. Balmony. Salt-rheum weed. Turtle-bloom. Fish-mouth. Lower leaves sometimes broadly oval. July–Sept.

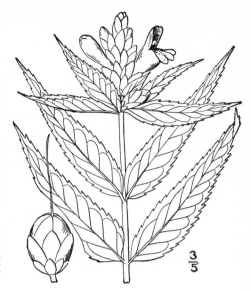

2. Chelone oblìqua L. Red Turtle-head.
Fig. 3753.

Chelone obliqua L. Syst. Ed. 11, no. 4. 1767.

Stem slender, ascending, 1°–2° high, usually branched, the branches spreading or ascending. Leaves oblong, or broadly lanceolate, acuminate at the apex, mostly narrowed at the base, petioled, sharply serrate with somewhat spreading teeth, or laciniate, 2′–6′ long, ½′–2½′ wide; petioles 2″–6″ long; principal veins about 10 on each side; flowers red or rose-purple, about 1′ long; bracts and calyx-segments ciliolate and usually puberulent; capsule similar to that of the preceding species.

In wet thickets and along streams, Virginia to Illinois, south to Florida. July–Sept.

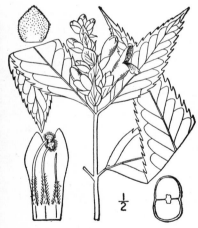

3. Chelone Lỳoni Pursh. Lyon's Turtle-head.
Fig. 3754.

Chelone Lyoni Pursh, Fl. Am. Sept. 737. 1814.

Stem slender, erect or nearly so, simple or branched, 1°–3° high. Leaves ovate, acuminate at the apex, rounded, truncate or subcordate at the base; 3′–7′ long, 1′–4′ wide, usually slender-petioled, sharply serrate with divergent teeth, the principal veins 8–10 on each side; flowers red or rose-purple, about 1′ long; bracts and calyx-segments ciliolate and puberulent.

In swamps and wet thickets, mountains of Virginia, North Carolina and Tennessee to Georgia. July–Sept.

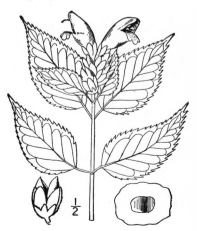

9. PENTSTÈMON Mitchell; Soland. in Ait. Hort. Kew. 3: 511.   1789.

Perennial herbs, mostly branched from the base only, with opposite or rarely verticillate leaves, or the upper occasionally alternate, and large, usually showy, blue purple red or white flowers, in terminal thyrses, panicles, or racemes. Calyx 5-parted, the segments imbricated. Corolla irregular, the tube elongated, more or less enlarged above, the limb 2-lipped; upper lip 2-lobed; lower lop 3-lobed.   Stamens 5, included, 4 of them antheriferous and didynamous, the fifth sterile, as long as or shorter than the others; anther-sacs divergent or connivent.   Style filiform; stigma capitate.   Capsule ovoid, oblong, or globose, septicidally dehiscent.   Seeds numerous, angled or even, wingless.   [Greek, five stamens.]

About 100 species, natives of North America and Mexico.   Type species: *Pentstemon pubescens* Soland.

* More or less pubescent or glandular, at least the calyx and pedicels.

Corolla ½′–1¼′ long; leaves entire, serrate, or denticulate.
  Stem pubescent or puberulent nearly or quite to the base.
    Thyrsus open, panicle-like.
      Corolla large, over 10″ long; stem hirsute or canescent, often glandular.
        Corolla densely bearded in the throat; outer calyx-segments ovate or ovate-lanceolate.
                        1. *P. hirsutus.*
        Corolla scarcely bearded in the throat; outer calyx-segments lanceolate or linear-
          lanceolate.   2. *P. canescens.*
      Corolla small, less than 10″ long; stem puberulent.   3. *P. pallidus.*
    Thyrsus narrow, raceme-like or spike-like.
      Corolla-tube abruptly enlarged; sterile filament densely woolly.   4. *P. erianthera.*
      Corolla-tube gradually enlarged; sterile filament slightly bearded.   5. *P. albidus.*
  Only the inflorescence, or pedicels, or calyx pubescent.
    Thyrsus open, panicle-like.
      Stem leaves oblong, ovate, or lanceolate.
        Calyx-lobes lanceolate to ovate, much less than 5″ long.
          Corolla purplish; tube not gibbous above the point of enlargement.
                            6. *P. Pentstemon.*
          Corolla white or pinkish; tube gibbous above the point of enlargement.
                            7. *P. Digitalis.*
        Calyx-lobes linear-subulate from a narrowly lanceolate base, becoming fully 5″ long.
                            8. *P. calycosus.*
      Stem leaves linear-lanceolate.   9. *P. gracilis.*
    Thyrsus narrow, interrupted; calyx viscid.   10. *P. tubiflorus.*
Corolla 2′ long, the tube much enlarged above; leaves dentate.   11. *P. Cobaea.*

** Completely glabrous throughout, mostly glaucous.

Leaves lanceolate, oblong, ovate, obovate, or orbicular.
  Stem leaves rounded, clasping; flowers 2′ long.   12. *P. grandiflorus.*
  Stem leaves acute or acuminate; flowers 9″–15″ long.
    Corolla 9″–10″ long; stem leaves lanceolate.   13. *P. acuminatus.*
    Corolla 1′–1½′ long; stem leaves mostly oblong.   14. *P. glaber.*
Leaves linear or linear-lanceolate; flowers densely thyrsoid.
  Bracts lanceolate, small; flowers 6″–8″ long.   15. *P. angustifolius.*
  Bracts ovate, acuminate, large; flowers 1′ long or more.   16. *P. Haydeni.*

½

1. Pentstemon hirsùtus (L.) Willd.
Hairy Beard-tongue.   Fig. 3755.

Chelone hirsuta L. Sp. Pl. 611.   1753.
P. pubescens Soland. in Ait. Hort. Kew. 3: 360.   1789.
Pentstemon hirsutus Willd. Sp. Pl. 3: 227.   1801.

Stem slender, erect, downy nearly or quite to the base, 1°–3° high. Leaves puberulent or glabrous, denticulate or the uppermost entire, the basal oblong or ovate, obtusish at the apex, 2′–4½′ long, ½′–2′ wide, narrowed into petioles, the upper sessile, lanceolate, mostly acuminate, sessile or slightly clasping; inflorescence thyrsoid, rather loose, glandular-pubescent; pedicels mostly short; corolla purplish or violet, the tube gradually dilated above, 2-grooved on the lower side, about 1′ long, the throat nearly closed by the villous palate at the base of the lower lip; sterile filament densely bearded for about one-half its length.

In dry woods and thickets, Maine to Ontario, Florida, Minnesota, Alabama and Missouri. Erroneously recorded from Texas. May–July.

2. Pentstemon canéscens Britton. Gray Beard-tongue. Fig. 3756.

Pentstemon laevigatus var. *canescens* Britton, Mem. Torr. Club **2**: 30. 1890.
P. canescens Britton, Mem. Torr. Club **5**: 291. 1894.

Densely and finely canescent or puberulent, or the leaves sometimes nearly glabrous; stem rather stout, 1°–3° high. Leaves denticulate, the lower and basal ones oval, obtuse, narrowed into long margined petioles, the next 1 or 2 pairs contracted below the middle and somewhat fiddle-shaped, 3'–6' long, the upper ovate or ovate-lanceolate, acuminate, clasping; thyrsus elongated, open, glandular-pubescent, leafy-bracted below; pedicels very short; calyx-segments lanceolate, acuminate; corolla purple or nearly white, about 1' long, slightly or not at all bearded in the throat; sterile filament slightly bearded for about one-third its length; capsule ovoid, glabrous, longer than the calyx.

In dry woods, Virginia, West Virginia and North Carolina to western Kentucky. Recorded from Missouri. May–June.

3. Pentstemon pállidus Small. Pale Beard-tongue. Fig. 3757.

Pentstemon pallidus Small, Fl. SE. U. S. 1060. 1903.

Puberulent, or sometimes canescent-puberulent; stem 3°–9° tall. Basal and lower stem-leaves with oblong elliptic or nearly spatulate blades; upper stem-leaves few, the blades oblong, lanceolate or linear-lanceolate, rather remotely toothed, partly clasping; panicles narrow; calyx-segments ovate, becoming triangular, acute, closely puberulent or pubescent during anthesis; corolla white or purplish, 9''–10'' long, the tube rather gradually dilated, the throat sparingly bearded; sterile filament bearded with very short hairs.

In sandy soil or swamps, Connecticut and New York to Missouri, Florida and Oklahoma. June–July.

4. Pentstemon eriánthera Pursh. Crested Beard-tongue. Fig. 3758.

Pentstemon erianthera Pursh, Fl. Am. Sept. 737. 1814.
Pentstemon cristatus Nutt. Gen. **2**: 52. 1818.

Puberulent below, glandular-villous and viscid above; stem rather stout, leafy, 6'–18' high. Leaves firm, entire or repand, the lower and basal ones oblong or spatulate, obtuse or acutish, 2'–4' long, narrowed into margined petioles, the upper sessile or somewhat clasping, acute or acuminate; thyrsus dense, narrow, leafy-bracted; flowers almost sessile; calyx-segments linear-lanceolate, acuminate, villous when young; corolla about 1' long, rather abruptly dilated above, red or purple, its lower lip villous within; sterile filament densely long-bearded.

On plains and bluffs, South Dakota to Manitoba, Montana, Nebraska and Nevada. May–July.

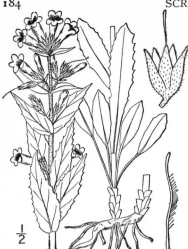

$\frac{1}{2}$

6. Pentstemon Pentstèmon (L.) Britton. Smooth Beard-tongue. Fig. 3760.

Chelone Pentstemon L. Sp. Pl. Ed. *2*, 850. 1763.
Pentstemon laevigatus Soland. in Ait. Hort. Kew. **2**: 300. 1789.
P. Pentstemon Britt. Mem. Torr. Club **5**: 291. 1894.

Glabrous, except the somewhat glandular-pubescent inflorescence; stem slender, 2°–3° high. Basal and lower leaves oblong or oval, obtuse, 3′–6′ long, narrowed into margined petioles, denticulate; upper leaves sessile or slightly clasping, acute, oblong, or lanceolate, denticulate; thyrsus open, usually many-flowered; flowers nearly sessile; calyx-segments lanceolate, acute, short; corolla purple or purplish, 8″–10″ long, the tube gradually enlarged above, the throat wide open, scarcely or not at all bearded, the lobes spreading; sterile filament thinly bearded for about one-half its length, or more densely bearded above.

Woods and thickets, Pennsylvania to Florida, Kentucky and Louisiana. Occasionally escaped from cultivation further north and east. May–July.

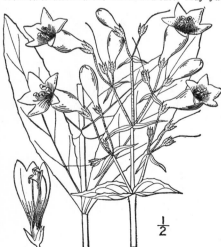

$\frac{1}{2}$

5. Pentstemon álbidus Nutt. White-flowered Beard-tongue. Fig. 3759.

Pentstemon albidus Nutt. Gen. **2**: 53. 1818.

Stems puberulent below, densely glandular-pubescent above, rather stout, 6′–10′ high. Basal and lower leaves spatulate or oblong, obtusish, mostly entire, the upper lanceolate or oblong, sessile, denticulate, acute or acuminate, 1½′–2½′ long, 3″–6″ wide; thyrsus narrow, raceme-like, leafy-bracted, interrupted; calyx-segments lanceolate, acuminate, viscid, one-half as long as the corolla-tube; corolla white or nearly so, 8″–10″ long, funnelform, the tube gradually dilated upward, the limb nearly equally 5-lobed, the lobes spreading; sterile filament slightly bearded with short hairs.

On plains, Minnesota and South Dakota to Assiniboia, Colorado, Nebraska and Texas. June–Aug.

$\frac{1}{2}$

7. Pentstemon Digitàlis (Sweet) Nutt. Foxglove Beard-tongue. Fig. 3761.

Chelone Digitalis Sweet, Brit. Fl. Gard. **2**: *pl. 120.* 1825–27.
Pentstemon Digitalis Nutt. Trans. Am. Phil. Soc. (II.) **5**: 181. 1833–37.
Pentstemon laevigatus var. *Digitalis* A. Gray, Syn. Fl. **2**: Part 1, 268. 1878.

Glabrous, except the glandular-pubescent inflorescence; stem rather stout, 2°–5° high. Lower and basal leaves oblong or oval, obtuse or acutish, entire or repand, 2′–7′ long, narrowed into margined petioles; upper leaves ovate, lanceolate or ovate-lanceolate, sessile and more or less cordate-clasping at the base, acuminate, sharply denticulate; thyrsus open, many-flowered; pedicels 1″–3″ long; calyx-segments lanceolate, corolla white, 1′–1¼′ long, the tube abruptly dilated, the limb moderately 2-lipped, the throat open; sterile filament bearded above.

In fields and thickets, Maine to New York, Illinois, Kansas, Virginia and Arkansas. Doubtless escaped from cultivation in its northeastern range. May–July.

8. Pentstemon calycòsus Small. Long-sepaled Beard-tongue. Fig. 3762.

Pentstemon calycosus Small, Bull. Torr. Club **25**: 470. 1898.

Glabrous except a few scattered hairs in the inflorescence and lines of puberulence on the internodes, deep green. Leaves various, the basal spatulate, 2′–6′ long, entire or undulate, with winged petioles; lower stem-leaves similar to the basal, the upper lanceolate or ovate-lanceolate, acute or acuminate, finely repand or distinctly serrate, rounded at the base and clasping; calyx-segments with a lanceolate base and a slender tip, often becoming 5″ long; corolla purple, 12″–15″ long; sterile filament sparingly bearded.

On bluffs, Kentucky to Missouri, Alabama and Arkansas. May–June.

9. Pentstemon grácilis Nutt. Slender Beard-tongue. Fig. 3763.

Pentstemon gracilis Nutt. Gen. **2**: 52. 1818.

Glabrous or very nearly so up to the glandular-pubescent inflorescence; stem slender, strict, 6′–18′ high. Basal and lower leaves linear-oblong or spatulate, mostly obtuse, denticulate, or entire, 1′–3′ long, narrowed into margined petioles; upper leaves sessile, linear-lanceolate or the uppermost lanceolate, acuminate, denticulate; thyrsus open, several–many-flowered; pedicels 2″–4″ long; calyx-segments lanceolate, acute or acuminate; corolla purple, 9″–12″ long, its tube gradually enlarged above, its throat wide open; sterile filament bearded for about one-half its length; capsule one-third longer than the calyx.

On moist prairies, Manitoba to Minnesota, Missouri, Athabasca, Oklahoma and Colorado. May–July.

10. Pentstemon tubiflòrus Nutt. Funnelform Beard-tongue. Fig. 3764.

Pentstemon tubiflorus Nutt. Trans. Am. Phil. Soc. (II.) **5**: 181. 1833–37.

Glabrous, except the viscid-pubescent calyx and pedicels; stem slender, strict, 2°–3½° high, leafless above. Leaves oblong, ovate, or lanceolate, 1′–4′ long, obtusish or acute, the basal narrowed into broad margined petioles, the upper sessile or clasping, entire or merely undulate; thyrsus narrow, interrupted, the clusters several-flowered; pedicels 1″–3″ long, calyx-segments ovate, acuminate, short, striate-nerved; corolla nearly funnelform, white or purplish, less than 1′ long, nearly as broad when expanded, its tube gradually enlarged, the limb nearly equally five-lobed, the lobes spreading; sterile filament short bearded above; capsule ovoid, acute, about twice as long as the calyx.

In moist soil, Missouri, Kansas and Arkansas. May–July.

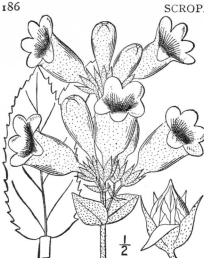

11. Pentstemon Cobaèa Nutt. Cobaea Beard-tongue. Fig. 3765.

Pentstemon Cobaea Nutt. Trans. Am. Phil. Soc. (II.) 5: 182. 1833–37.

Stem stout, densely and finely pubescent below, glandular-pubescent above, 1°–2° high. Leaves oblong to ovate, firm, 3′–5′ long, dentate, the lower mostly glabrous and narrowed into margined petioles, the upper sessile or cordate-clasping, usually pubescent; thyrsus short, several–many-flowered, open; flowers about 2′ long; calyx-segments lanceolate, acuminate, 5″–7″ long; corolla purple, puberulent without, glabrous within, its tube narrow up to the top of the calyx, then abruptly dilated and campanulate, the limb scarcely 2-lipped, the lobes short, rounded, spreading; sterile filament sparingly bearded; capsule ovoid, acute, pubescent, reticulate-veined, as long as the calyx.

On dry prairies, Missouri and Kansas to Texas. Recorded from Ohio. May–July.

12. Pentstemon grandiflòrus Nutt. Large-flowered Beard-tongue. Fig. 3766.

P. grandiflorus Nutt. in Fras. Cat. 1813.

Glabrous and somewhat glaucous; stem stout, 2°–4° high. Leaves all entire and obtuse, the basal ones obovate, narrowed into broad petioles, those of the lower part of the stem sessile, oblong or oval, 1′–2½′ long, the upper nearly orbicular, cordate-clasping, shorter; thyrsus open, leafy-bracted, the bracts orbicular, cordate; pedicels 2″–6″ long; flowers nearly 2′ long; calyx-segments lanceolate, acute, 3″–4″ long; corolla lavender-blue, the tube rather abruptly dilated above the calyx, the limb somewhat 2-lipped; sterile filament incurved, villous and capitate at the summit; capsule acute, 8″–10″ high, three times as long as the calyx.

On prairies, Illinois to Minnesota, North Dakota, Kansas and Colorado. Locally introduced eastward. June–Aug.

13. Pentstemon acuminàtus Dougl. Sharp-leaved Beard-tongue. Fig. 3767.

Pentstemon acuminatus Dougl.; Lindl. Bot. Reg. *pl. 1285.* 1829.

Glabrous and glaucous; stem rather stout, strict, 6′–2° high, leafy. Leaves firm, entire, the lower and basal ones oblong or spatulate, obtuse or acute, narrowed into petioles, the upper sessile or clasping, lanceolate or ovate-lanceolate, 2′–3′ long; thyrsus narrow, sometimes 1-sided, usually leafy-bracted below; pedicels becoming 4″–8″ long in fruit; calyx-segments lanceolate, acute, 2″–3″ long; corolla blue, 9″–10″ long, its tube rather gradually dilated, the limb 2-lipped; sterile filament bearded along the dilated summit; capsule acute, twice as long as the calyx.

In dry soil, Minnesota to Nebraska, Texas, Manitoba, Alberta, Oregon and New Mexico. St. Joseph's-wand. May–Aug.

14. Pentstemon glàber Pursh. Large Smooth Beard-tongue. Fig. 3768.

Pentstemon glaber Pursh, Fl. Am. Sept. 738. 1814.

Glabrous, somewhat glaucous; stem ascending or erect, rather stout, leafy, 1°–2° high. Leaves entire, firm, the basal and lower ones narrowed into petioles, the middle ones oblong or oblong-lanceolate, acuminate, sessile, the upper lanceolate, acuminate, sessile, scarcely clasping; thyrsus narrow, elongated, densely many-flowered; pedicels 3″–7″ long in fruit; calyx-segments ovate-lanceolate, scarious-margined, abruptly acuminate, 3″–4″ long, their margins commonly eroded; corolla blue or purple, 1′–1½′ long, rather abruptly expanded above the calyx, the limb somewhat 2-lipped, the lobes rounded; sterile filament bearded at the slightly enlarged summit; capsule narrowly ovoid, acute, about twice as long as the calyx.

In moist soil, South Dakota to Nebraska, Wyoming and Arizona. Far western plants previously referred to this species prove to be distinct. May–Aug.

$\frac{1}{2}$

15. Pentstemon angustifòlius Pursh. Pale-blue Beard-tongue. Fig. 3769.

P. angustifolius Pursh, Fl. Am. Sept. 738. 1814.
Pentstemon coeruleus Nutt. Gen. 2 : 52. 1818.

Glabrous and glaucous; stem slender, erect, leafy, 6′–15′ high. Leaves all linear or linear-lanceolate, entire, the lower narrowed into petioles, obtusish at the apex, the upper sessile, acute, 1½′–2½′ long, 1½″–2″ wide; thyrsus narrow, spike-like, mostly dense; bracts lanceolate, acuminate; pedicels very short; calyx-segments linear-lanceolate, acuminate, 2″–3″ long; corolla blue, or nearly white, 8″–10″ long, the tube gradually enlarged, the limb somewhat 2-lipped; sterile filament bearded at the summit.

In dry soil, western Nebraska to North Dakota and Montana. May–July.

$\frac{3}{5}$

16. Pentstemon Hàydeni S. Wats. Hayden's Beard-tongue. Fig. 3770.

P. Haydeni S. Wats. Bot. Gaz. 16 : 311. 1891.

Glabrous, not at all glaucous or slightly so; stem decumbent, simple or branched, leafy, 1°–2° high. Leaves linear or elongated-lanceolate, entire, sessile and slightly clasping, acute, acuminate, or the lowest obtusish at the apex, 2′–5′ long, 1″–5″ wide; thyrsus narrow, dense; bracts ovate or ovate-lanceolate, large, cordate-clasping, acute, or acuminate; fruiting pedicels 2″–3″ long; calyx-segments lanceolate, striate-nerved, acuminate, 3″–5″ long; corolla blue, 1′ long or more, the tube broadly dilated above the calyx, the limb nearly equally 5-lobed; capsule acute, twice as long as the calyx.

In moist soil, Nebraska, Kansas and Wyoming.

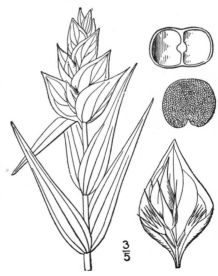

$\frac{3}{5}$

10. COLLÍNSIA Nutt. Journ. Acad. Phil. 1: 190. *pl. 9.* 1817.

Winter-annual or biennial herbs, with opposite or verticillate leaves, and blue pink white or variegated flowers, verticillate, or solitary in the axils. Calyx campanulate, 5-cleft. Corolla irregular, the tube short, the limb 2-lipped; upper lip 2-cleft, the lobes erect or recurved; lower lip larger, 3-lobed, the lateral lobes spreading or drooping, flat, the middle one conduplicate, keel-like, enclosing the 4 declined stamens and the filiform style. Stamens didynamous. Corolla with a gland on the upper side of the tube near the base. Filaments filiform; anther-sacs confluent at the apex. Stigma small, capitate or 2-lobed. Capsule ovoid or globose, septicidally 2-valved, the valves 2-cleft. Seeds few, large, peltate, concave on the inner side. [Named for Zaccheus Collins, botanist, of Philadelphia, 1764–1831.]

About 20 species, natives of North America. Type species: *Collinsia verna* Nutt.

Corolla 5"–8" long, the throat shorter than the limb.
 Leaves, at least the lower, ovate or oblong; corolla-lobes notched.   1. *C. verna.*
 Leaves lanceolate; corolla-lobes obcordate.   2. *C. violacea.*
Corolla 2"–3" long, the throat longer than the limb.   3. *C. tenella.*

1. Collinsia vérna Nutt. Blue-eyed Mary. Innocence. Broad-leaved Collinsia. Fig. 3771.

C. verna Nutt. Journ. Acad. Phil. 1: 190. *pl. 9.* 1817.

Glabrous or puberulent; stem slender, weak, 6'–2° long, simple or branched. Leaves thin, opposite, the lower broadly ovate or orbicular, obtuse at the apex, rounded, narrowed or subcordate at the base, crenate or entire, slender-petioled; middle leaves sessile or cordate-clasping, ovate or oblong, obtuse, dentate, 1'–2' long, floral leaves ovate to spatulate, mostly acute, dentate or entire; upper whorls 4–6-flowered; peduncles ¾'–1' long; corolla 6"–8" long, its throat equalling or shorter than the calyx, its lower lip blue, the upper purple or nearly white, the lobes emarginate or truncate; capsule globose, 2½"–3" in diameter, shorter than the linear calyx-lobes.

In moist woods and thickets, Ontario and western New York to Wisconsin, south to Pennsylvania, Kentucky and Kansas. April–June.

The California **Collinsia bicolor** Benth., which differs from this by short-peduncled flowers, is recorded as found introduced in Illinois.

2. Collinsia violàcea Nutt. Violet or Narrow-leaved Collinsia. Fig. 3772.

Collinsia violacea Nutt. Trans. Am. Phil. Soc. (II.) 5: 179. 1833–37.

Similar to the preceding species, stem slender, erect, usually branched, 6'–15' high. Leaves lanceolate or oblong-lanceolate, rather thick, entire or denticulate, obtuse or obtusish, the lower opposite, petioled, the middle similar, sessile, 1'–2' long, 3"–5" wide, the floral linear or linear-lanceolate, opposite or verticillate; upper whorls 2–5-flowered: corolla 5"–6" long, violet, its lobes obcordate or emarginate; capsule globose, about 2" in diameter, shorter than the lanceolate acute calyx-lobes.

Rich soil, Missouri, Kansas and Arkansas to Texas. April–May.

3. Collinsia tenélla (Pursh) Piper.
Small-flowered Collinsia. Fig. 3773.

Antirrhinum tenellum Pursh, Fl. Am. Sept.
  421. 1814.
Collinsia parviflora Dougl.; Lindl. Bot. Reg.
  pl. 1082. 1827.
C. tenella Piper, Contr. Nat. Herb. 11: 496.
  1906.

Puberulent, at length diffusely branched;
stems very slender, 3′-15′ long. Leaves
oblong or lanceolate, mostly obtuse at the
apex and narrowed at the base, ½′-1′ long,
entire, or sparingly toothed, the lower opposite, sometimes broader, petioled, the
floral sessile, opposite or verticillate; upper
whorls 2-6-flowered; pedicels commonly
longer than the flowers; corolla 3″-4″ long,
about twice as long as the calyx, blue or
whitish, the throat longer than the limb;
capsule globose, 1″-1½″ in diameter, little
shorter than the lanceolate calyx-lobes.

In moist places, Ontario to British Columbia, Michigan, Colorado, Arizona and Utah.
April–June.

$\frac{3}{4}$

11. PAULÒWNIA Sieb. & Zucc. Fl. Jap. 1: 25. *pl. 10.* 1835.

A large tree, with the aspect of *Catalpa,* with broad opposite entire or 3-lobed, petioled
pubescent leaves, and large violet flowers in terminal panicles. Calyx deeply 5-cleft, the
lobes short. Corolla irregular, the tube elongated, enlarged above, the 5 lobes spreading,
somewhat unequal. Stamens 4, didynamous, included; anther-sacs divaricate. Style slender,
slightly thickened toward the summit, stigmatic on the inner side. Capsule coriaceous, ovoid,
acute, loculicidally dehiscent. Seeds numerous, striate, winged. Flowers expanding before
the leaves appear. [Named for Anna Paulowna, daughter
of the Czar Paul I.]

A monotypic Japanese genus.

1. Paulownia tomentòsa (Thunb.) Baill. Paulownia. Fig. 3774.

Bignonia tomentosa Thunb. Fl. Jap. 252. 1784.
Paulownia imperialis Sieb. & Zucc. Fl. Jap. 1: 27. 1835.
Paulownia tomentosa Baill. Hist. Pl. 9: 434. 1888.

A tree with thin flaky bark, reaching a maximum
height of about 70° and a trunk diameter of 4°, the
branches stout, spreading. Leaves broadly ovate, 6′-15′
long, 4′-8′ wide, long-petioled, canescent on both sides
when young, glabrate above when old, the petioles
terete; flowers about 2½′ long, numerous in large erect
terminal panicles; pedicels stout, densely tomentose;
calyx 5-lobed, the lobes thick, tomentose; corolla slightly
irregular, puberulent without; capsule 2′ high, 1′ in
diameter.

Escaped from cultivation, southern New York and New
Jersey to Georgia. May–July.

$\frac{2}{5}$

12. MÍMULUS L. Sp. Pl. 634. 1753.

Erect or decumbent herbs, with opposite mostly dentate leaves. Flowers axillary, solitary, peduncled, pink, violet, or yellow, usually showy. Calyx prismatic, 5-angled, 5-toothed,
the upper tooth usually the largest. Corolla irregular, its tube cylindric with a pair of ridges
on the lower side within, its limb 2-lipped; upper lip erect or reflexed, 2-lobed; lower lip
spreading, 3-lobed, the lobes rounded. Stamens 4, didynamous, inserted on the corolla-tube;
anther-sacs divergent, or sometimes confluent at the summit. Style filiform; stigma 2-lamel-
late. Capsule oblong or linear, loculicidally dehiscent, many-seeded, enclosed by the calyx.
[Diminutive of *mimus,* a mimic actor.]

About 40 species, natives of America. Besides the following, some 30 others occur in the
western United States and British Columbia. Type species: *Mimulus ringens* L.
Corolla violet, or rarely white; eastern species.
  Leaves sessile, clasping; peduncles longer than the calyx.   1. *M. ringens.*
  Leaves petioled; peduncles shorter than the calyx.   2. *M. alatus.*
Corolla yellow; western; two species adventive in the East.
  Plants glabrous or glabrate.

Erect; branches spreading; leaves ovate; flowers 1′ long.   3. *M. Langsdorffii.*
Diffuse; leaves nearly orbicular; flowers about 6″ long.   4. *M. Geyeri.*
Plant villous and viscid, diffuse, musk-scented.   5. *M. moschatus.*

1. Mimulus ríngens L.   Square-stemmed Monkey-flower.   Fig. 3775.

Mimulus ringens L. Sp. Pl. 634.   1753.

Glabrous, perennial by rootstocks; stem erect, 4-sided or somewhat 4-winged, usually much branched, 1°–3° high.   Leaves oblong, lanceolate, or oblong-lanceolate, pinnately veined, acuminate or acute at apex, serrate, auriculate-clasping at the base, or the lower merely sessile, 2′–4′ long, ½′–1′ wide; peduncles slender, 1′–2′ long in fruit, 2–4 times as long as the calyx; calyx-teeth lanceolate-subulate; corolla violet, rarely white, about 1′ long, the throat narrow, exceeding the calyx; base of lower lip puberulent within; fruiting calyx oblong, 6″–8″ long; seeds oblong, minute, reticulated.

In swamps and along streams, Nova Scotia to Virginia, Tennessee, Manitoba, Nebraska and Texas. Ascends to 3000 ft. in Virginia.   June–Sept.

2. Mimulus alàtus Soland.   Sharp-winged Monkey-flower.   Fig. 3776.

Mimulus alatus Soland. in Ait. Hort. Kew. **2:** 361. 1789.

Similar to the preceding species, glabrous; stem sharply 4-angled, the angles more or less winged. Leaves ovate, ovate-lanceolate, or oblong, acute or acuminate at the apex, dentate-serrate, narrowed at the base, petioled, 2′–5′ long, 9″–18″ wide; petioles ¼′–1′ long, narrowly margined; peduncles stout, shorter than the calyx; corolla violet, 1′ long; calyx-teeth short, broad, abruptly mucronulate; seeds smooth.

In swamps, Ontario to Connecticut, Illinois, Kansas, Georgia and Texas.   June–Sept.

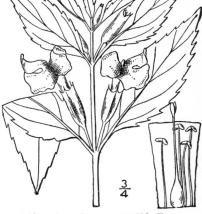

3. Mimulus Langsdŏrffii Donn.   Langsdorff's Yellow Monkey-flower.   Fig. 3777.

Mimulus Langsdorffii Donn; Sims. Bot. Mag. *pl. 1501. 1812.*
Mimulus guttatus DC. Cat. Hort. Monsp. 127. 1818.

Perennial by stolons, glabrous or puberulent; stem rather stout, simple or branched; branches spreading.   Leaves ovate to obovate, dentate or denticulate, obtuse at the apex, rounded, narrowed or cordate at the base, 1′–3′ long, the lower short-petioled, the upper smaller, sessile or clasping; peduncles shorter than or equalling the flowers; calyx oblique; corolla yellow, 1′–2′ long, the lower lip bearded at the base, often blotched with red or purple; fruiting calyx about ½′ long; seeds longitudinally striate.

In wet meadows, Norfolk, Conn., and southern New York.   Introduced from California. Summer.

4. Mimulus Géyeri Torr. Geyer's Yellow Monkey-flower. Fig. 3778.

M. Geyeri Torr. in Nicollet, Rep. Up. Miss. 157. 1843.
Mimulus Jamesii T. & G.; Benth. in DC. Prodr. 10: 371. 1846.
M. glabratus var. *Jamesii* A. Gray, Syn. Fl. N. A. Ed. 2, 2¹: 447. 1886.

Perennial by stolons, glabrous or nearly so; stems slender, creeping, diffusely branched, rooting at the nodes, 6′–18′ long. Leaves broadly ovate, orbicular or broader, very obtuse, denticulate or entire, membranous, truncate, subcordate, or rarely narrowed at the base, palmately veined, short-petioled or the upper sessile, ¼′–1′ in diameter; peduncles slender, longer than the oblique calyx in fruit, about as long as the leaves; calyx scarcely toothed, 3″–4″ long at maturity; corolla yellow, 4″–6″ long, the lower lip bearded at the base within, the throat broad; seeds nearly smooth.

In brooks and swamps, Ontario to Illinois, Nebraska, Wyoming and Colorado. June–Sept.

5. Mimulus moschàtus Dougl. Musk-flower. Musk-plant. Fig. 3779.

M. moschatus Dougl.; Lindl. Bot. Reg. *pl. 1118.* 1827.

Perennial, villous-pubescent, viscid, musk-scented; stems creeping and ascending, branched, slender, 6′–12′ long. Leaves ovate or oblong, short-petioled, acute or obtuse at the apex, denticulate, rounded or subcordate at the base, 1′–2′ long, 5″–12″ wide; peduncles slender, longer than the calyx; flowers 1′ long or less; calyx-teeth lanceolate, acuminate, nearly equal; corolla yellow, 2–3 times as long as the calyx.

In wet places, Newfoundland to Pennsylvania, Ontario and Michigan; apparently introduced from western North America, where it is widely distributed. June–Sept.

Mazus japónicus (Thunb.) Kuntze, a low herb with mostly basal, slender-petioled obovate repand leaves and racemose violet flowers, the campanulate calyx not angled, has been found at Washington, D. C., and abundantly about New Orleans, La. It is native of eastern Asia.

13. CONÒBEA Aubl. Pl. Guian. 2: 639. *pl. 258.* 1775.

Herbs, with opposite pinnately parted pinnatifid incised or serrate leaves, and small blue or white peduncled flowers, solitary or two together in the axils. Calyx 5-parted, the segments narrow, equal. Corolla irregular, the tube cylindric, the limb 2-lipped; upper lip emarginate or 2-lobed; lower lip 3-lobed. Stamens 4, didynamous, ascending, included; filaments filiform; anther-sacs parallel, not confluent. Style incurved at the summit; stigma 2-lamellate. Capsule globose, oblong or linear, septicidally dehiscent, the valves entire or 2-cleft. Seeds numerous, oblong, striate. [Guiana name.]

About 8 species, natives of America. Besides the following, another occurs in the southwestern United States. Type species: *Conobea aquatica* Aubl.

1. Conobea multífida (Michx.) Benth. Conobea. Fig. 3780.

Capraria multifida Michx. Fl. Bor. Am. 2: 22. *pl. 35.* 1803.
Conobea multifida Benth. in DC. Prodr. 10: 391. 1846.

Annual, finely viscid-pubescent; stem at length diffusely branched, 4′–8′ high, very leafy. Leaves petioled, ½′–1′ long, pinnately parted into 3–7 linear or linear-oblong obtuse entire or incised segments; flowers greenish-white, 2″–2½″ long, mostly solitary in the opposite axils, about as long as their filiform peduncles; calyx-segments linear-subulate, slightly shorter than the corolla; capsule narrowly ovoid, glabrous, about equalling the calyx.

Along streams and rivers, Ontario to Ohio, Iowa, Kansas, Kentucky, Alabama and Texas. Introduced along the Delaware below Philadelphia. June–Sept.

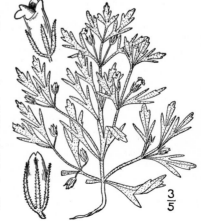

14. BRÀMIA Lam. Encycl. 1: 459. 1783.

[MONNIERA P. Br. Civ. & Nat. Hist. Jam. 269. *pl. 28. f. 3.* Hyponym. 1755. Not L. 1759.]

[HERPESTIS Gaertn. Fruct. & Sem. 3: 186. *pl. 214.* 1805.]

Diffuse or prostrate herbs, with opposite mostly entire obtuse leaves, and small peduncled flowers, mostly solitary in the axils. Calyx subtended by 2 bracts, 5-parted, the upper segment the broadest. Corolla blue or white, nearly regular, the tube cylindric, the limb nearly equally 5-lobed. Stamens 4, didynamous, included. Style slender; stigma capitate, or 2-lobed. Capsule globose or ovoid, septicidally dehiscent. Seeds numerous. [From Brami, a Malabar name.]

About 20 species, natives of warm and tropical regions. Type species: *Bramia indica* Lam.

Leaves spatulate or cuneate; capsules acuminate at the apex.   1. *B. Monniera.*
Leaves obovate or orbicular-obovate; capsules blunt at the apex.   2. *B. rotundifolia.*

1. Bramia Monnièra (L.) Drake.   Monnier's Hedge-Hyssop.   Fig. 3781.

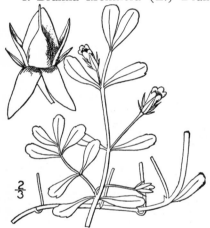

Gratiola Monniera L. Cent. Pl. 2. 1756.
Herpestis Monniera H.B.K. Nov. Gen. 2: 366. 1817.
M. Monniera Britton, Mem. Torr. Club 5: 292. 1894.
Bacopa Monniera Wettst. in Engler & Prantl, Nat. Pfl. 4^{sb}: 77. 1891.
Bramia Monniera Drake, Fl. Polyn. Franc. 142. 1892.

Perennial, glabrous, fleshy; stem creeping, rooting at the nodes, branched, 6'–18' long. Leaves spatulate or cuneate-obcordate, sessile, rounded at the apex, entire, or sparingly denticulate, 3''–10'' long, 1''–2½'' wide; peduncles mainly in alternate axils, slender, 2-bracteolate at the summit, in fruit longer than the leaves; flowers pale blue, 5'' long; upper calyx-segment ovate, acute; corolla obscurely 2-lipped; stamens nearly equal; stigma slightly 2-lobed; capsule ovoid, acute, shorter than the calyx.

On shores, near the coast, Maryland to Florida, Texas and Mexico. Widely distributed in tropical regions of the Old World and the New. June–Oct. Herb-grace. Water-hyssop.

2. Bramia rotundifòlia (Michx.) Britton.   Round-leaved Hedge-Hyssop. Fig. 3782.

M. rotundifolia Michx. Fl. Bor. Am. 2: 22. 1803.
Herpestis rotundifolia Pursh, Fl. Am. Sept. 418. 1814.
Bacopa rotundifolia Wettst. in Engler & Prantl, Nut. Pfl. 4^{sb}: 76. 1891.

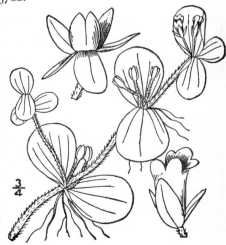

Perennial by stolons, succulent; stems creeping and spreading, branched or simple, villous-pubescent, 1°–2° long. Leaves obovate or orbicular, palmately veined, entire, or slightly undulate, narrowed to a sessile or clasping base, ½'–1' broad, glabrous, not punctate; peduncles stout, shorter than the leaves, solitary or 2 together in the axils, longer than the flowers; flowers blue, 3''–4'' long; upper calyx-segment oval; corolla 2-lipped, longer than the calyx; stamens approximate in pairs; stigma 2-lobed; disk obsolete; capsule oblong, obtuse, 1½'' high, at length 4-valved, shorter than the calyx.

On muddy shores, Illinois to South Dakota, Tennessee and Texas. June–Sept.

15. HYDROTRÌDA Willd.

Perennial aromatic succulent creeping herbs, usually pubescent. Stems terete. Leaves opposite; blades broadest below the middle, palmately nerved, entire or shallowly toothed, punctate, partly clasping. Flowers solitary on short axillary peduncles. Calyx subtended by 2 small bractlets. Sepals nearly distinct, the outer ones cordate, the upper one broadest. Corolla blue or white, manifestly 2-lipped, the upper lip merely notched. Hypogynous disk present. Stamens 4, included; filaments adnate to near the throat of the corolla-tube. Capsule ovoid to conic, septicidally dehiscent, the valves cleft. Seeds numerous. [Name unexplained.]

About 8 species, of the southeastern United States, the West Indies and South America. Type species: *Herpestis obovata* Poepp. (See Linnaea **5**: 107.)

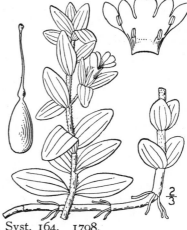

1. Hydrotrida caroliniàna (Walt.) Small. Blue Hedge-Hyssop. Fig. 3783.

Obolaria caroliniana Walt. Fl. Car. 166. 1788.
Monniera amplexicaulis Michx. Fl. Bor. Am. **2**: 22. 1803.
Herpestis amplexiculis Pursh, Fl. Am. Sept. 418. 1814.
M. caroliniana Kuntze, Rev. Gen. Pl. 463. 1891.
Septilia caroliniana Small, Fl. SE. U. S. 1064. 1903.

Perennial by stolons, more or less pubescent, fleshy; stems creeping and ascending, rooting at the lower nodes, 6'-2° long, simple, or sparingly branched, leafy. Leaves ovate to oval, sessile and clasping by a subcordate base, obtuse at the apex, parallel-veined, punctate, entire, the margins ciliolate or naked; peduncles shorter than the leaves; usually shorter than the calyx; upper calyx-segment ovate, cordate; flowers blue, 4"-5" long, ephemeral; disk 10-12-toothed; stamens approximate in pairs.

In wet pine barrens, New Jersey to Florida and Louisiana. June–Oct.

16. MECARDÒNIA R. & P. Syst. 164. 1798.

Perennial relatively rigid herbs, with 4-angled erect or diffuse, but rarely creeping stems. Leaves opposite; blades toothed, narrowed at the base, pinnately nerved. Flowers solitary on slender axillary peduncles subtended by two small bractlets. Calyx not subtended by bractlets; sepals 5, unequal, the upper one broadest; corolla white, purple or yellow, 2-lipped, the upper lip notched. Stamens adnate to near the corolla-throat. Hypogynous disk wanting. Stigma 2-lobed. Capsule septicidal, the valves cleft. Seeds numerous. [Named in honor of Anton Meca y Cardona, a founder of the botanical garden at Barcelona.]

About 10 species, natives of warm and tropical America. Besides the following, some 4 others inhabit the southeastern United States. Type species: *Mecardonia ovata* R. & P.

1. Mecardonia acuminàta (Walt.) Small. Purple Hedge-Hyssop. Fig. 3784.

Gratiola acuminata Walt. Fl. Car. 61. 1788.
Matourea nigrescens Benth. Comp. Bot. Mag. **1**: 173. 1835.
Herpestis nigrescens Benth. Comp. Bot. Mag. **2**: 56. 1836.
Monniera acuminata Kuntze, Rev. Gen. Pl. 463. 1891.
Mecardonia acuminata Small, Fl. SE. U. S. 1065. 1903.

Perennial, glabrous; stem erect, branched above, 1°-2° high, very leafy. Leaves oblong or oblong-lanceolate, serrate, at least above the middle, short-petioled or sessile, obtuse or acute at the apex, narrowed or somewhat cuneate at the base, 1'-2' long, faintly veined; peduncles in alternate and opposite axils, ascending, in fruit longer than the leaves, not bracteolate at the summit; flowers 5"-6" long, purple or purplish; upper calyx-segment lanceolate, acute; corolla 2-lipped, the lower lip longer than the upper; stamens approximate in pairs; capsule oblong, 4-valved, 3"-4" high, about equalling the calyx.

In wet soil, Maryland to Florida, Missouri and Texas, mostly near the coast. Plant blackening in drying. June–Sept.

17. GRATÌOLA L. Sp. Pl. 17. 1753.

Erect or diffuse glabrous or glandular-pubescent herbs, with opposite sessile, entire or dentate leaves, and yellow or whitish peduncled flowers solitary in the axils. Peduncles 2-bracteolate at the summit in the following species. Calyx 5-parted, the segments narrow, slightly unequal. Corolla irregular, its tube cylindric, its limb more or less 2-lipped; upper lip entire, emarginate, or 2-cleft; lower lip 3-lobed. Perfect stamens 2, the anterior pair wanting, or represented by rudiments; filaments filiform; anther-sacs distinct, transverse and separated by a broad connective. Style filiform; stigma dilated, slightly 2-lobed. Capsule loculicidally and septicidally dehiscent, ovoid or globose, 4-valved. Seeds numerous, longitudinally and transversely striate. [Latin, grace or favor, from its reputed healing properties.]

About 30 species, of wide geographic distribution in temperate and warm regions. Besides the following, some 9 others occur in southern and western North America. Type species: *Gratiola officinalis* L.

Sterile filaments minute or none.
    Glandular-puberulent; flowers 4″–5″ long; capsule ovoid.   1. *G. virginiana.*
    Glabrous; flowers 7″ long; capsule globose.   2. *G. sphaerocarpa.*
Sterile filaments 2, slender, capitate at the summit.
    Leaves lanceolate, entire or remotely denticulate.   3. *G. aurea.*
    Leaves ovate or oblong, sharply serrate.   4. *G. viscosa.*

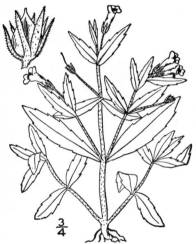

1. Gratiola virginiàna L. Clammy Hedge-Hyssop. Fig. 3785.

Gratiola virginiana L. Sp. Pl. 17. 1753.

Annual; stem erect, at length widely branched, glandular-puberulent, at least above, 3′–12′ high. Leaves oblong or oblong-lanceolate, sessile, narrowed to both ends, denticulate, 1′–2′ long, 2″–5″ wide, glabrous or nearly so; peduncles slender, glandular, shorter than or equalling the leaves; flowers 4″–5″ long; bractlets as long as the calyx, or longer; calyx about one-half as long as the yellowish corolla-tube; limb of the corolla short, white; sterile filaments minute or none; anther-sacs transverse, separated by a broad connective; capsule broadly ovoid, 2″ high, as long as the calyx.

In wet places, Quebec to British Columbia, south to Florida, Texas and California. Ascends to 3000 ft. in Virginia. Water jessamine. May–Oct.

2. Gratiola sphaerocàrpa Ell. Round-fruited Hedge-Hyssop. Fig. 3786.

Gratiola sphaerocarpa Ell. Bot. S. C. & Ga. 1: 14. 1816.

Annual (or perennial?), glabrous; stem ascending or erect, rather stout, simple or branched, 6′–12′ high. Leaves oblong or obovate-oblong, sessile, dentate or denticulate, 3–5-nerved, acute or obtusish at the apex, narrowed at the base, 1′–2′ long, 3″–8″ wide; peduncles stout, little cr not at all longer than the calyx; bractlets about equalling the calyx; calyx-lobes linear; calyx nearly one-half as long as the corolla; flowers about 7″ long; corolla-tube yellow, the limb paler; anther-sacs broad, transverse; sterile filaments wanting; capsule globose, 3″ in diameter.

In wet places, southern New Jersey to Florida, Illinois, Missouri, Texas and Mexico. June–Sept.

3. Gratiola aùrea Muhl. Goldenpert. Golden Hedge-Hyssop. Fig. 3787.

Gratiola aurea Muhl. Cat. 2. 1813.

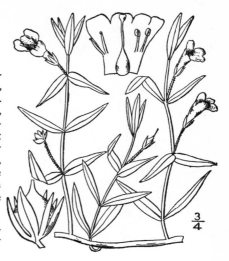

Perennial, glandular-puberulent above, or glabrate; stems decumbent, creeping or ascending, simple or branched, 4'–12' long, somewhat 4-sided. Leaves lanceolate to linear-oblong, ½'–1' long, 1''–3'' wide, sparingly denticulate, scarcely narrowed to the sessile and somewhat clasping base; peduncles filiform, in fruit equalling or longer than the calyx; corolla bright yellow, 6''–7'' long, 3 times as long as the calyx; sterile filaments 2, capitate at the summit; anther-sacs of the fertile stamens broad, transverse; capsule globose-ovoid, shorter than or equalling the calyx.

In sandy wet places, Quebec and Ontario to New Jersey and Virginia. Recorded from Florida. June–Sept.

4. Gratiola viscòsa Schwein. Viscid Hedge-Hyssop. Fig. 3788.

Gratiola viscosa Schwein.; LeConte, Ann. Lyc. N. Y. 1: 106. 1823.

Annual; stem weak, finely viscid-pubescent, slender, commonly simple, 6'–18' long. Leaves ovate, ovate-oblong or ovate-lanceolate, sharply serrate, acute at the apex, sessile, cordate-clasping at the base, ½'–1' long; peduncles slender, shorter than or exceeding the leaves; bractlets and calyx-segments foliaceous, entire or dentate, one-third to one-half as long as the yellow or purplish corolla; flowers 5''–6'' long; sterile filaments 2, capitate at the summit; anther-sacs of the fertile stamens transverse, separated by the broad connective; capsule subglobose, shorter than the calyx.

In brooks and swamps, Delaware to Kentucky, Georgia and Tennessee. May–Sept.

18. SOPHRONÁNTHE Benth.; Lindl. Introd. Nat. Syst. Ed. 2, 445   1836.

Annual or perennial rigid caulescent herbs, with rough-pubescent foliage. Leaves opposite; blades entire or toothed, sessile. Flowers solitary in the axils, short-peduncled or nearly sessile. Calyx sessile in 2 bractlets; sepals 5, nearly distinct. Corolla white or purplish, 2-lipped. Stamens 2, included. Staminodia filiform, capitate at the apex. Anther-sacs parallel, contiguous. Capsule somewhat elongated, acuminate. [Greek, referring to the included anthers.]

Two known species of eastern North America. Type species: *Sophronanthe hispida* Benth.

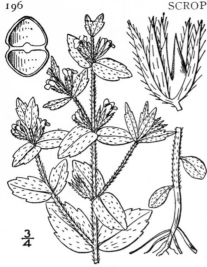

¾

1. Sophronanthe pilòsa (Michx.) Small. Hairy Hedge-Hyssop. Fig. 3789.

Gratiola pilosa Michx. Fl. Bor. Am. 1: 7. 1803.

Sophronanthe pilosa Small, Fl. SE. U. S. 1067. 1903.

Perennial (?), hirsute; stems slender, erect, strict, simple, or branched, 6′–2° high. Leaves ovate or ovate-lanceolate, dentate or denticulate, sessile, acute or obtusish at the apex, rounded, truncate or subcordate at the base, ½′–1′ long; flowers nearly or quite sessile, commonly numerous, about 4″ long; corolla purplish or white, slightly longer than the calyx and bractlets; sterile filaments 2, capitate at the summit; anther-sacs of the fertile stamens parallel, contiguous; capsule oblong-conic, acuminate, about the length of the calyx, or shorter.

In dry soil, southern New Jersey to Florida, Arkansas and Texas. May–Aug.

19. ILYSÁNTHES Raf. Ann. Nat. 13. 1820.

Annual or biennial glabrous slender branching herbs, with opposite, mostly dentate and sessile leaves, and small purplish peduncled flowers solitary in the axils. Peduncles not bracteolate. Calyx 5-parted, the segments linear. Corolla irregular, the tube somewhat expanded above, the limb 2-lipped; upper lip 2-cleft, erect; lower lip larger, 3-lobed, spreading. Fertile stamens 2, included, their anther-sacs divergent; sterile stamens 2, 2-lobed, one of the lobes capitate, glandular, the other glabrous, shorter. Style slender; stigma slightly 2-lobed. Capsule oblong or ovoid, septicidally dehiscent. Seeds numerous, wrinkled. [Greek, mud-flower.]

About 10 species, of wide geographic distribution. Besides the following, 2 or 3 others occur in the southeastern United States. Type species: *Ilysanthes riparia* Raf.

Peduncles longer than the leaves; calyx-segments shorter than the capsule.   1. *I. dubia.*
Peduncles shorter than the leaves; calyx-segments mostly as long as the capsule, or longer.
                                                         2. *I. attenuata.*

1. Ilysanthes dùbia (L.) Barnhart. Long-stalked False Pimpernel. Fig. 3790.

Gratiola dubia L. Sp. Pl. 17. 1753.
Capraria gratioloides L. Sp. Pl. Ed. 2, 876. 1763.
Ilysanthes riparia Raf. Ann. Nat. 13. 1820.
Ilysanthes gratioloides Benth. in DC. Prodr. 10: 419. 1846.
Ilysanthes dubia Barnhart, Bull. Torr. Club 26: 376. 1899.

Stem slender, mostly erect, at length diffusely branched, 3′–8′ long. Leaves ovate, ovate-oblong, or the lower obovate, sessile or very nearly so, or slightly clasping at the base, remotely denticulate or entire, obtuse or acutish at the apex, thickish, 3–7-nerved, ¼′–1′ long, the upper ones commonly much smaller; peduncles slender, considerably exceeding the leaves; flowers 3″–5″ long; calyx-segments linear, about one-half the length of the corolla, shorter than the capsule; capsule narrowly ovoid-oblong, bluntish, 2″–3″ high; seeds 1″ long, reddish, the ends usually truncate.

In wet places, Massachusetts to Florida, west to Ontario, Minnesota and Texas. Also on the Pacific Coast, and in the West Indies and South America. July–Sept.

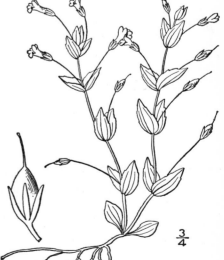

¾

2. Ilysanthes attenuàta (Muhl.) Small.
Short-stalked False Pimpernel.
Fig. 3791.

Lindernia attenuata Muhl. Cat. 59. 1813.
Ilysanthes gratioloïdes curtipedicellata Bush, Bull.
     Torr. Club 21: 494. 1894.
I. attenuata Small, Bull. Torr. Club 23: 297. 1896.

Stem erect or ascending, 3'–16' long, the branches spreading. Leaves oblong to ovate, or sometimes obovate, ½'–1½' long, thinnish, obtuse, serrate with a few low teeth, 3–5-nerved, narrowed into short petioles, or sessile; peduncles shorter than the leaves; calyx-segments linear-subulate, as long as the capsule, or longer; corolla 2''–6'' long; capsule narrowly ovoid, about 2'' long, pointed; seeds slightly curved, 1½''–2'' long, yellowish brown, the ends usually rounded.

In wet places, New Brunswick and Ontario to Wisconsin, south to Florida and Arkansas. Ascends to 2000 ft. in Virginia. Perhaps not specifically distinct from the preceding. May–Oct.

20. MICRÁNTHEMUM Michx. Fl. Bor. Am. 1: 10. *pl. 12.* 1803.

Creeping or ascending, branched small leafy annual glabrous herbs, with opposite obovate oval or orbicular sessile entire leaves, and minute white or purplish short-peduncled flowers, solitary in some of the axils. Calyx 4–5-lobed or 4–5-parted. Corolla very irregular, the tube short, the upper lip shorter than the lower, or wanting, the lower 3-lobed, spreading or ascending, the middle lobe the largest. Stamens 2, anterior; filaments short, somewhat dilated or appendaged at the base; anthers small, their sacs distinct, parallel, or slightly divergent. Style short; stigma 2-lobed. Capsules globose, 2-celled by a membranous partition or becoming 1-celled. Seeds several or numerous, minute. [Greek, small flower.]

About 16 species, natives of America. Besides the following, another occurs in the southern United States. Type species: *Micranthemum orbiculatum* Michx.

1. Micranthemum micranthemoïdes (Nutt.) Wettst. Nuttall's Micranthemum.
Fig. 3792.

Hemianthus micranthemoides Nutt. Journ. Acad. Phil.
     1: 119. *pl. 6.* 1817.
Micranthemum Nuttallii A. Gray, Man. Ed. 5, 331. 1867.
Micranthemum micranthemoides Wettst. in Engl. &
     Prantl, Nat. Pfl. Fam. 4: Abt. 3b. 77. 1891.

Somewhat fleshy; stem filiform, creeping, the branches ascending, ½'–2½' high. Leaves obovate to oval, obtuse, 1''–2½'' long; flowers about ½'' long, borne on peduncles of about the same length; calyx campanulate in flower, obovoid in fruit, 4-lobed, usually split along one side; peduncles recurved in fruit; upper lip of the corolla nearly obsolete; middle lobe of the lower lip longer than the lateral ones; appendages at the bases of the stamens nearly as long as the filaments; capsule obovoid-globose, ½'' in diameter, as long as the calyx.

In tidal mud, New Jersey to Florida. Aug.–Oct. This species is the type of the genus *Hemianthus* Nutt., which has been regarded as distinct from *Micranthemum* by other authors, a view which may be maintained.

21. LIMOSÉLLA L. Sp. Pl. 631. 1753.

Low glabrous succulent floating or creeping, tufted annual herbs (or perennial by stolons?), with filiform stems rooting at their nodes, basal slender-petioled entire leaves, and filiform 1-flowered scape-like peduncles, the flowers small, white, pink, or purple. Calyx campanulate, 5-lobed. Corolla nearly regular, open-campanulate, the tube short, the limb 5-cleft. Stamens 4, inserted on the corolla-tube, scarcely exserted; filaments short; anther-sacs confluent. Style short; stigma capitate. Ovary 2-celled at the base, 1-celled above. Capsule globose or oblong, becoming 1-celled, many-seeded. [Greek, mud seated.]

About 6 species, of wide geographic distribution. Type species: *Limosella aquatica* L.

1. Limosella aquática L. Mudweed. Mudwort. Fig. 3793.

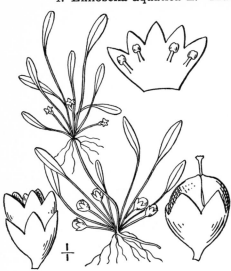

Limosella aquatica L. Sp. Pl. 631. 1753.
Limosella tenuifolia Hoffm. Deutsch. Fl. 29.
  1804.
Limosella australis R. Br. Prodr. Fl. Nov. Holl.
  1 : 443. 1810.

Leaves 1'–5' long, the blade oblong, linear-oblong, narrowly linear or spatulate, obtuse, one-fourth or one-third as long as the filiform petiole. Peduncles shorter than the leaves, arising with the petioles from the base of the plant or from nodes of the creeping or floating stem; corolla pink or white, about 1" broad, scarcely longer than the calyx; calyx-lobes ovate, acute or acutish, about the length of the tube; stamens inserted high up on the corolla-tube; filaments somewhat longer than the anthers; capsule globose or oblong-globose, obtuse, 1½" high, longer than the calyx.

On muddy shores and in brooks, Labrador and Hudson Bay to the Northwest Territory, south to New Jersey, and in the Rocky Mountains to Colorado, and in the Sierra Nevada to California. Also in Europe, Australia and South America. June–Aug.

22. SÝNTHYRIS Benth. in DC. Prodr. 10: 454. 1846.

Perennial herbs, with thick rootstocks, simple erect stems, large petioled basal leaves, those of the stem much smaller, alternate, sessile, or clasping, bract-like. Flowers small, pink or purple, in terminal dense elongated spikes or racemes. Calyx 4–5-parted, the segments oblong or linear. Corolla oblong or campanulate, 2–4-lobed, or parted, or wanting, the lateral lobes, when present, exterior in the bud. Stamens 2 (occasionally 4), posterior, inserted on the corolla, or on the outer side of the hypogynous disk, exserted; filaments slender; anther-sacs parallel or divergent, not confluent. Ovary 2-celled or rarely 3-celled; style filiform; stigma capitate. Capsule compressed, obtuse, or emarginate, many-seeded, loculicidally dehiscent. Seeds flat, oval, or orbicular. [Greek, closed doors, referring to the capsule-valves.] Our species were referred, in the first edition of this work, to the Old World genus *Wulfenia*.

About 10 species, natives of North America and Europe. Type species: *Synthyris reniformis* Benth.

Corolla present, usually 2-lobed.   1. *S. Bullii.*
Corolla none.   2. *S. rubra.*

1. Synthyris Bullii (Eaton) Heller. Bull's Synthyris. Fig. 3794.

Gymnandra Bullii Eaton; Eaton & Wright, 259. 1840.
Synthyris Houghtoniana Benth. in DC. Prodr. 10: 454.
  1846.
Wulfenia Houghtoniana Greene, Erythea 2 : 83. 1894.
Gymnandra Bullii Barnhart, Bull. Torr. Club 26 : 378. 1899.
Synthyris Bullii Heller, Muhlenbergia 1 : 4. 1900.

Pubescent; stem stout, 1°–2½° high. Basal leaves ovate or orbicular, rounded at the apex, truncate, cordate or reniform at the base, crenulate all around, 2'–5' long, 5–7-nerved, petioled, the petiole usually shorter than the blade; stem-leaves small, ½'–1' long, sessile or slightly clasping, crenulate, obtuse, or acute, passing gradually into the bracts of the dense spike; flowers greenish yellow, 2"–3" long; corolla present, variously 2–4-lobed (commonly 2-lobed), little, if any, longer than the calyx, its lobes obtuse, the stamens inserted on its base; spike much elongated in fruit; capsule emarginate, slightly exceeding the calyx.

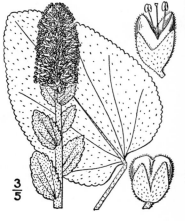

3/5

On dry prairies, Ohio to Minnesota, Michigan and Iowa. May–July.

2. Synthyris rùbra (Hook.) Benth. Western Synthyris. Fig. 3795.

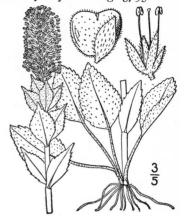

Gymnandra rubra Hook. Fl. Bor. Amer. **2**: 103. *pl. 172.* 1838.

Synthyris rubra Benth. in DC. Prodr. **10**: 455. 1846.

Wulfenia rubra Greene, Erythea **2**: 83. 1894.

Besseya rubra Rydberg, Bull. Torr. Club **30**: 280. 1903.

Similar to the preceding species but lower, pubescent or tomentose, seldom over 1° high. Basal leaves ovate or oblong, obtuse or acute at the apex, narrowed, truncate or cordate at the base, 1½′–3′ long, crenulate, petioled, indistinctly nerved; stem-leaves ovate or lanceolate, acute, sessile, crenulate, or entire, ¾′–1′ long; spike very dense, 1′–2′ long in flower, 2′–5′ long in fruit, its bracts purplish; corolla none; stamens inserted on the outer side of the hypogynous disk; capsule little compressed, emarginate, slightly longer than the calyx.

In dry soil, South Dakota to Nebraska, British Columbia and Utah. May–June.

23. VERONICA [Tourn.] L. Sp. Pl. 9. 1753.

Annual or perennial herbs (some exotic species shrubs or trees), with opposite and alternate, rarely verticillate leaves, and mostly small blue purple pink or white flowers, terminal or axillary, racemose, spicate, or solitary. Calyx mostly 4-parted, sometimes 5-parted, the segments oblong or ovate. Corolla rotate, its tube very short, deeply and more or less unequally 4-lobed (rarely 5-lobed), the lower lobe commonly the narrowest. Stamens 2, divergent, inserted on either side and at the base of the upper corolla-lobe; anthers obtuse, their sacs confluent at the summit; filaments slender. Ovary 2-celled; style slender; stigma capitate; ovules few or numerous in each cavity. Capsule more or less compressed, sometimes very flat, emarginate, obcordate, or 2-lobed, loculicidally dehiscent. Seeds smooth or rough, flat, plano-convex, or excavated on the inner side. [Named for St. Veronica.]

About 200 species, of wide geographic distribution. Besides the following, 3 others occur in northwest America. Type species: *Veronica officinalis* L.

* Flowers racemose in the axils of the leaves, bracteolate.

Glabrous, or minutely glandular above (No. 3 rarely hairy); brook or swamp plants.
  Leaves ovate, oval, oblong, or oblong-lanceolate; capsule compressed.
    Stem leaves sessile, partly clasping, serrulate or entire.　　　　　1. *V. Anagallis-aquatica.*
    All the leaves petioled, serrate.　　　　　　　　　　　　　　2. *V. americana.*
  Leaves linear or linear-lanceolate; capsule very flat.　　　　　　3. *V. scutellata.*
Pubescent, dry soil plants; leaves crenate or dentate.
  Leaves oval or obovate, petioled; pedicels shorter than the calyx.　4. *V. officinalis.*
  Leaves ovate, nearly or quite sessile; pedicels longer than the calyx.　5. *V. Chamaedrys.*

** Flowers in terminal spikes or racemes, or solitary in the axils.

Flowers in terminal spikes or racemes.
  Leaves all sessile; capsule elliptic, emarginate.　　　　　　　　6. *V. Wormskioldii.*
  Lower leaves petioled; capsule orbicular, obcordate.　　　　　　7. *V. serpyllifolia.*
Flowers solitary in most of the axils; peduncles shorter than the leaves.
  Erect; glabrous or glandular; capsule emarginate.　　　　　　　8. *V. peregrina.*
  Diffuse; pubescent; capsule obcordate.　　　　　　　　　　　9. *V. arvensis.*
Flowers solitary in the axils; peduncles as long as the leaves, or longer.
  Leaves ovate or oblong, crenate or dentate.
    Corolla not longer than the calyx; capsule narrowly emarginate.　10. *V. agrestis.*
    Corolla longer than the calyx; capsule broadly emarginate.　　　11. *V. Tournefortii.*
  Leaves orbicular, or broader, 3–5-lobed or -crenate.　　　　　　12. *V. hederaefolia.*

$\frac{2}{3}$

1. Veronica Anagállis-aquática L.   Water Speedwell or Pimpernel.   Fig. 3796.

Veronica Anagallis-aquatica L. Sp. Pl. 12. 1753.

Perennial by stolons or leafy shoots developed in autumn; stem rather stout, glabrous, or glandular-puberulent above, erect or decumbent, often rooting at the lower nodes, usually branched, 1°–3° high. Leaves of sterile autumn shoots orbicular to obovate, obtuse, serrulate, narrowed into margined petioles, those of the flowering stems ovate, oblong, or lanceolate, sessile and more or less clasping or the lowest short-petioled, serrulate or entire, 1½′–4′ long, ¼′–2′ wide; racemes peduncled, borne in most of the axils, 2′–5′ long; bractlets shorter than or exceeding the pedicels; flowers blue, or purplish striped, 2″ broad; capsule compressed, not very flat, nearly orbicular, 2-lobed, emarginate, 1½″ high; seeds flat.

In brooks and swamps, Nova Scotia to British Columbia, south to North Carolina, Nebraska and New Mexico. Also in Europe and Asia. The plant of the Atlantic Coast appears as if introduced.   Ascends to 4000 ft. in Virginia. May–Sept.

2. Veronica americàna Schwein.   American Brooklime.   Fig. 3797.

V. americana Schwein.; Benth. in DC. Prodr. **10**: 468. 1846.

Similar to the preceding species, perennial by stolons or leafy shoots, glabrous throughout; stem decumbent, usually branched, rooting at the lower nodes, 6′–3° long. Leaves oblong, ovate or oblong-lanceolate, all distinctly petioled, sharply serrate, truncate, rounded, or subcordate at the base, obtuse or acutish at the apex, 1′–3′ long, ¼′–1′ wide; racemes peduncled, borne in most of the axils, loose, elongated, sometimes 6′ long; bractlets shorter than the pedicels; flower blue or nearly white, usually striped with purple, 2″ broad; capsule nearly orbicular, compressed, but not very flat, emarginate, 1½″ high; seeds flat.

In brooks and swamps, Anticosti to Alaska, south to Pennsylvania, Nebraska, New Mexico and California. Ascends to 2600 ft. in the Catskills. Wallink. Blue-bells. April–Sept.

Veronica Beccabúnga L., a European brooklime, similar to *V. americana* in habit, but with crenate or low-serrate broad tipped leaves, is naturalized about Quebec and has been found on ballast about seaports in New York and New Jersey.

$\frac{3}{5}$

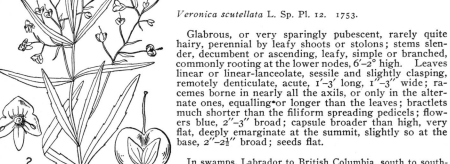

$\frac{2}{3}$

3. Veronica scutellàta L.   Marsh or Skullcap Speedwell.   Fig. 3798.

Veronica scutellata L. Sp. Pl. 12. 1753.

Glabrous, or very sparingly pubescent, rarely quite hairy, perennial by leafy shoots or stolons; stems slender, decumbent or ascending, leafy, simple or branched, commonly rooting at the lower nodes, 6′–2° high. Leaves linear or linear-lanceolate, sessile and slightly clasping, remotely denticulate, acute, 1′–3′ long, 1″–3″ wide; racemes borne in nearly all the axils, or only in the alternate ones, equalling or longer than the leaves; bractlets much shorter than the filiform spreading pedicels; flowers blue, 2″–3″ broad; capsule broader than high, very flat, deeply emarginate at the summit, slightly so at the base, 2″–2½″ broad; seeds flat.

In swamps, Labrador to British Columbia, south to southern New York, Minnesota and California. Also in Europe and Asia. May–Sept.

4. Veronica officinàlis L. Common Speedwell. Fluellin. Gipsy-weed. Fig. 3799.

Veronica officinalis L. Sp. Pl. 11. 1753.

Perennial by stolons, pubescent all over; stem ascending, 3′–10′ high. Leaves oblong, oval, or obovate, petioled, ½′–2′ long, obtuse at the apex, serrate, narrowed into the petioles; racemes spike-like, narrow, dense, elongated, often borne only in alternate axils, much longer than the leaves; subulate bractlets and the calyx longer than the pedicels; flowers pale blue, 2″–3″ broad; capsule obovate-cuneate, compressed, broadly emarginate, 2″ high, 1½″ broad; seeds numerous, flat.

In dry fields and woods, Nova Scotia to Ontario, South Dakota, North Carolina and Tennessee. Ascends to 5600 ft. in Virginia. Also in Europe and Asia. Appears, in most places, as if introduced. Paul's-betony. Ground-hele. Upland speedwell. May–Aug.

Veronica longifòlia L., an erect European species with lanceolate petioled acuminate sharply serrate leaves, has been found in waste grounds and fields from Nova Scotia to New York.

5. Veronica Chamaèdrys L. Germander Speedwell. Fig. 3800.

Veronica Chamaedrys L. Sp. Pl. 13. 1753.

Perennial; stem ascending, simple or branched, slender, pubescent in two lines, 4′–12′ high. Leaves ovate, sessile, or very nearly so, pubescent, truncate, rounded or cordate at the base, incised-dentate, obtuse at the apex, ½′–1¾′ long; racemes borne in opposite or alternate axils, peduncled, more or less pubescent, loose, 2′–6′ long; pedicels filiform, longer than the calyx and usually longer than the bractlets; flowers light blue, 3″–4″ broad; capsule obcordate, narrowed at the base; seeds numerous, flattish.

In fields and waste places, Nova Scotia and Quebec to southern New York and Pennsylvania. Naturalized from Europe. Other English names are blue eye, eyebright, angel's-eye, god's-eye, bird's-eye, cat's-eye, base vervain. Forget-me-not. May–July.

Veronica Teùcrium L., also European, similar to *V. Chamaedrys*, but with oblong to lanceolate crenate leaves, has been found in New England and Ohio.

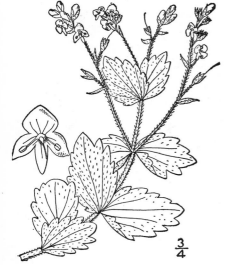

6. Veronica Wormskioldii R. & S. Wormskiold's Speedwell. Fig. 3801.

V. Wormskioldii R. & S. Syst. 1: 101. 1817.

Perennial, pubescent or nearly glabrous; stems ascending or erect, slender, usually simple, 2′–12′ high. Leaves oblong, ovate, or elliptic, sessile, mostly rounded at both ends, crenulate or entire, ½′–1′ long; flowers in a short narrow raceme at the end of the stem, light blue, 2″–3″ broad; pedicels shorter than the calyx in flower, much shorter than the bractlets, 2″–3″ long in fruit; capsule ellipsoid or slightly obovoid, moderately compressed, emarginate, 2″–3″ high; seeds numerous, flattish.

Labrador; mountains of Quebec and New England to Alaska, south in the Rocky Mountains to Colorado and in the Sierra Nevada to Nevada. Summer. Referred in our first edition to *V. alpina* L.

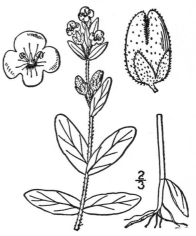

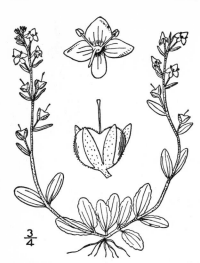

3/4

7. Veronica serpyllifòlia L.   Thyme-leaved Speedwell.   Fig. 3802.

Veronica serpyllifolia L. Sp. Pl. 12. 1753.

Perennial, puberulent or glabrous; stems slender, decumbent, branched, the branches ascending or erect, 2'–10' high. Leaves all opposite and petioled, or the uppermost sessile, oblong, oval, or ovate, ¼'–½' long, crenulate or entire; flowers in short narrow racemes at the end of the stem and branches; pedicels equalling or longer than the calyx, usually shorter than the bractlets; corolla blue with darker stripes, sometimes white, 2''–4'' broad; capsule broader than high, broadly obcordate or emarginate at the summit, about 1'' long, about equalling the calyx; seeds flat, numerous.

In fields and thickets, Labrador to Alaska, south to Georgia, New Mexico and California. Also in Europe, Asia and South America. Ascends to 2600 ft. in the Catskills. Paul's-betony. April–Aug.

Veronica humifùsa Dickson, differing in larger flowers and more pubescent inflorescence, is apparently a northern race of this species, also occurring in Europe.

8. Veronica peregrìna L.   Purslane Speedwell. Neckweed.   Fig. 3803.

Veronica peregrina L. Sp. Pl. 14. 1753.

Annual, glabrous, or glandular-puberulent; stem erect or ascending, simple or branched, 3'–12' high. Leaves oblong, oval, linear or slightly spatulate, 3''–10'' long, obtuse or acutish, the lowest opposite, short-petioled, or sessile, broader than the upper and usually denticulate, the upper alternate, sessile, mostly entire, each with a short-pedicelled flower in its axil; flowers nearly white, about 1'' broad; pedicels much shorter than the calyx; capsule nearly orbicular, obcordate, usually a little shorter than the calyx, 1''–1½'' high, many-seeded, the seeds flat.

In moist places, and common as a weed in cultivated soil, Nova Scotia to British Columbia, south to Florida, Mexico and California. Also in Central and South America, distributed as a weed in the Old World. May–Oct.

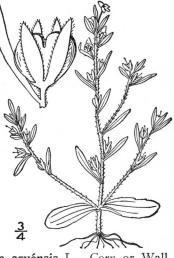

3/4

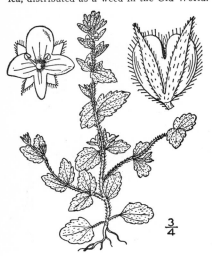

3/4

9. Veronica arvénsis L.   Corn or Wall Speedwell.   Fig. 3804.

Veronica arvensis L. Sp. Pl. 13. 1753.

Annual, pubescent; stem slender, at first simple and erect, at length much branched and diffuse, 3'–10' long. Lower leaves ovate or oval, opposite, obtuse at both ends, crenate or crenulate, 2''–6'' long, the lowest petioled; upper leaves sessile, alternate, ovate or lanceolate, acute or acutish, commonly entire, each with a short pedicelled minute flower in its axil; pedicels shorter than the calyx; corolla blue, or nearly white, 1'' broad or less; capsule broadly obovate, obcordate, 1'' high.

In fields, woods and waste places and in cultivated soil, Nova Scotia to Ontario and Minnesota, south to Florida, Kansas and Texas. Also in Bermuda. Naturalized from Europe. Native also of Asia. March–Sept.

10. Veronica agréstis L.   Procumbent, Field or Garden Speedwell.   Fig. 3805.

Veronica agrestis L. Sp. Pl. 13.   1753.

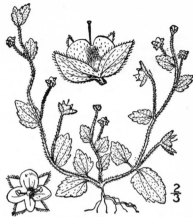

Annual, pubescent; stems creeping or procumbent, very slender, branched, 3′–8′ long, the branches ascending or spreading.   Leaves broadly ovate or oval, obtuse at the apex, rounded, truncate or subcordate at the base, crenate, all short-petioled, or the uppermost sessile, the lowest opposite, the upper alternate and each with a slender-peduncled small blue flower in its axil; peduncles equalling or longer than the leaves; corolla not exceeding the calyx; capsule broader than high, compressed, but not very flat, narrowly emarginate at the summit, 1″ high, 2″ broad; seeds few, hollowed out on the inner side.

In fields and waste places, Nova Scotia to New Jersey and Louisiana.   Naturalized from Europe.   Native also of Asia.   Other English names are germander-chickweed and winter-weed.   May–Sept.

11. Veronica Tournefortii Gmelin.   Tournefort's Speedwell.   Byzantine Speedwell.   Fig. 3806.

Veronica Tournefortii Gmelin, Fl. Bad. **1** : 39.   1805.
Veronica agrestis var. *byzantina* Sibth. & Smith, Fl. Graec. **1** : *pl. 8.*   1806.
V. Buxbaumii Tenore, Fl. Nap. **1** : 7. *pl. 1.*   1811.
V. byzantina B.S.P. Prel. Cat. N. Y. 40.   1888.

Annual, pubescent; stems diffusely branched, spreading or ascending, 6′–15′ long.   Leaves ovate or oval, short-petioled, obtuse or acutish, sometimes narrowed at the base, crenate-dentate or somewhat incised, 4″–12″ long, the lowest opposite, the upper all alternate and each with a slender-peduncled rather large blue flower in its axil; peduncles filiform, as long as the leaves or longer; corolla exceeding the calyx, 3″–4″ broad; capsule twice as broad as high, 3″ broad, with a wide and shallow emargination at the summit; seeds few or several, hollowed out on the inner side.

In waste places, Nova Scotia to southern New York and Ohio; Colorado to California.   Adventive or naturalized from Europe.   Native also of Asia.   Bird's-eye.   Cat's-eye.   May–Sept.

12. Veronica hederaefòlia L.   Ivy-leaved Speedwell.   Fig. 3807.

Veronica hederaefolia L. Sp. Pl. 13.   1753.

Annual, pubescent; stems slender, diffusely branched, 3′–18′ long.   Leaves orbicular or broader, truncate or subcordate at the base, 3–5-lobed or 3–5-crenate, petioled, ¼′–1′ in diameter, the lower opposite, the upper all alternate and with slender peduncled small blue flowers in their axils; peduncles filiform, often longer than the leaves; corolla 2″ broad, scarcely longer than the calyx; capsule little compressed, 2-lobed, broader than high, shorter than the densely ciliate sepals, 2–4-seeded; seeds excavated on the inner side.

In thickets, fields and waste places, southern New York, Pennsylvania and New Jersey to South Carolina.   Naturalized from Europe.   Native also of Asia.   April–Oct.   Ivy-chickweed.   Mother-of-wheat.   Small henbit.   Winter-weed.   Morgeline.

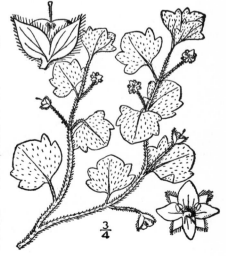

24. LEPTÁNDRA Nutt. Gen. 1 : 7. 1818.

Tall stout erect perennial herbs, with verticillate or opposite leaves, and small minutely bracted white or blue flowers, in dense peduncled spike-like racemes, terminal, or in the upper axils. Calyx 4-parted, short. Corolla tubular, or salverform, nearly regular, 4-lobed, the tube cylindric, longer than the lobes. Stamens 2, exserted, inserted low down on the corolla-tube; filaments filiform; anthers obtuse, short. Style about as long as the stamens, stigma minute. Capsule narrowly ovoid, scarcely compressed, not emarginate nor obcordate, 4-valved at the apex. Seeds numerous, oval, minutely reticulated. [Greek, slender stamens, referring to the filaments.]

Two species, the following typical one native of eastern North America, the other of north-eastern Asia.

1. Leptandra virgínica (L.) Nutt. Beaumont's-, Bowman's- or Culver's-root. Fig. 3808.

Veronica virginica L. Sp. Pl. 9. 1753.
Leptandra virginica Nutt. Gen. 1 : 7. 1818.

Stem glabrous, or very nearly so, simple, strict, 2°–7° high. Leaves verticillate in 3's–9's or some of the upper-most opposite, lanceolate or oblong-lanceolate, short-petioled, long-acuminate at the apex, sharply serrulate, narrowed at the base, pinnately veined, glabrous both sides, or pubescent beneath, 3′–6′ long, ½′–1′ wide; spike-like racemes several or rarely solitary, 3′–9′ long, very dense, the terminal one first developing; pedicels and bractlets about as long as the calyx; calyx-segments ovate-lanceolate, acute; corolla tubular, white or bluish, 2″ long; capsule ovoid-oblong, 1″–1½″ long, 2–3 times as long as the calyx.

In meadows, moist woods and thickets, Ontario to Mani-toba, Massachusetts, Alabama and Texas. Recorded from Nova Scotia. Ascends to 2700 ft. in Virginia. Black-root. Culver's-physic. Brinton's-root. Oxadaddy. Quitch. Tall speedwell. June–Sept.

25. DIGITÀLIS [Tourn.] L. Sp. Pl. 621. 1753.

Tall biennial or perennial herbs, the stems simple, or branched at the base, with alternate dentate or entire leaves, and large showy purple yellowish or white flowers, in long terminal commonly 1-sided racemes. Calyx 5-parted, the segments imbricated. Corolla declined, somewhat irregular, the tube contracted above the ovary, then rather abruptly expanded, longer than the 4–5-lobed slightly 2-lipped limb; upper lip broadly emarginate or 2-cleft; lower lip 3-lobed, the middle lobe largest, the lateral ones exterior in the bud. Stamens 4, didynamous, ascending, mostly included; anthers approximate in pairs. Style slender; stigma 2-lobed. Capsule ovoid, septicidally dehiscent. Seeds numerous, rugose. [Latin, *digitale,* the finger of a glove, which the flowers resemble.]

About 20 species, native of Europe and Asia, the following typical.

1. Digitalis purpùrea L. Purple Fox-glove. Thimbles. Fairy Cap. Fig. 3809.

Digitalis purpurea L. Sp. Pl. 621. 1753.

Usually biennial, pubescent; stem stout, erect, 2°–5° high. Basal and lower leaves ovate or ovate-lanceolate, 6′–10′ long, slender-petioled, acute at the apex, narrowed at the base, dentate; upper leaves similar, smaller, sessile; racemes 1° long or more, dense, 1-sided; flowers purple to white, 1½′–2′ long, drooping; upper calyx-segment narrower than the four other foliaceous ones; corolla spotted within.

Cape Breton Island and New York, naturalized from Europe, sparingly escaped from cultivation; also from Washington to California. June–Aug. Among some 60 English names are folk's-glove [by corruption fox-glove], *i. e.,* fairy's-glove, fairy-thimbles, -fingers, fairy-bells, pop-dock, or -glove. Rabbit's-flower. Cottagers. Lion's-mouth. Scotch mercury. Throatwort. Lady-fingers, -glove, -thimble. Dog's-finger. Witches' thimbles.

26. BÚCHNERA L. Sp. Pl. 630. 1753.

Erect, perennial or biennial, simple or branched, strict hispid or scabrous herbs, blackening in drying, the lower leaves opposite, the upper sometimes alternate. Flowers rather large, white, blue, or purple, in dense terminal bracted spikes, the lower commonly distant. Calyx tubular, or oblong, 5–10-nerved, 5-toothed. Corolla salverform, its tube cylindric, somewhat curved, its limb deeply and nearly equally 5-cleft, spreading, the lateral lobes exterior in the bud. Stamens 4, didynamous, included; anther-sacs confluent into 1. Style slender, thickened or club-shaped above; stigma small, entire or emarginate. Capsule oblong or ovoid, loculicidally dehiscent. Seeds numerous, reticulated. [Named for J. G. Buchner.]

About 30 species, natives of warm and temperate regions. Besides the following, another occurs in the southern United States. Type species: *Buchnera americana* L.

1. Buchnera americàna L.   Blue-hearts.
Fig. 3810.

Buchnera americana L. Sp. Pl. 630. 1753.

Hispid and rough; stem slender, stiff, 1°–2½° high. Leaves usually all opposite, prominently veined, the lowest obovate or oblong, obtuse, narrowed into very short petioles, the middle ones oblong or oblong-lanceolate, dentate, obtuse, or acute at the apex, narrowed at the base, sessile, the upper lanceolate or linear-lanceolate, entire or nearly so; spike peduncled, 6'–10' long in fruit, the flowers mostly opposite, nearly 1' long; bractlets shorter than the calyx; calyx strigose; corolla purple, its lobes obovate, obtuse, 3''–4'' broad; capsule slightly oblique, 4'' high, a little longer than the calyx.

In sandy or gravelly soil, New Jersey to western New York, southern Ontario and Minnesota, south to Virginia, Louisiana, Kansas and Arkansas. June–Sept.

27. AFZÈLIA J. G. Gmel. Syst. Nat. 2: 927. 1796.
[Seymeria Pursh, Fl. Am. Sept. 736. 1814.]

Erect stout branched annual or perennial herbs, mostly with opposite leaves, at least the lower 1–2-pinnately parted or dissected, and yellow flowers solitary in the axils, or in terminal bracted spikes or racemes. Calyx campanulate, 5-cleft or 5-parted. Corolla slightly irregular, campanulate or rotate, the tube short, broad, the limb 5-lobed, the lower lobe exterior in the bud. Stamens 4, slightly unequal, scarcely or not at all exserted; filaments short, villous, at least near the base; anthers 2-celled, the sacs parallel, distinct. Style short or slender. Capsule globose or ovoid, acute and more or less compressed at the summit. Seeds numerous, reticulated. [Named for Adam Afzelius, 1750–1812, professor at Upsala.]

About 10 species, natives of North America, Mexico and Madagascar. Besides the following, 4 others inhabit the southern United States. Type species: *Anonymos cassioides* Walt.

1. Afzelia macrophýlla (Nutt.) Kuntze.
Mullen Foxglove. Fig. 3811.

Seymeria macrophylla Nutt. Gen. 2: 49. 1818.
Gerardia macrophylla Benth. Comp. Bot. Mag. 1: 205. 1835.
A. macrophylla Kuntze, Rev. Gen. Pl. 457. 1891.

Annual (?), puberulent or glabrate; stem sparingly branched, or simple, 4°–6° high. Lower leaves long-petioled, pinnately parted, 6'–15' long, their segments lanceolate, coarsely dentate, irregularly incised, or pinnatifid; upper leaves short-petioled or sessile, oblong or lanceolate, 1'–3' long, entire, obtuse or acutish at the apex, narrowed at the base, each with a sessile flower in its axil; flowers 5''–7'' long; calyx-lobes lanceolate or ovate, acute, about as long as the tube; corolla 2–3 times as long as the calyx, woolly in the throat; style short, club-shaped; capsule globose-ovoid, 3''–4'' high, twice as long as the calyx.

In moist thickets and along streams, Ohio to Iowa, Nebraska, Kentucky and Texas. Aug.–Oct.

28. DASÝSTOMA Raf. Journ. Phys. 89: 99. 1819.

Large erect simple or branched, glandular-puberulent, pubescent or glabrous, annual or perennial herbs, partly parasitic on the roots of other plants, with opposite whorled or some alternate leaves, and large showy yellow flowers, in terminal mostly leafy-bracted racemes or panicles. Calyx campanulate or turbinate, 5-lobed, the lobes longer than or equalling the tube, sometimes foliaceous. Corolla slightly irregular, funnelform, or campanulate-funnelform, the tube villous or pubescent within, the limb spreading, 5-lobed. Stamens 4, didynamous, included, villous or pubescent; filaments slender; anthers all alike, their sacs distinct, parallel, awned at the base. Style filiform. Capsule oblong, acute, loculicidally dehiscent, longer than the calyx. [Greek, thick or hairy mouth, referring to the corolla.]

Six species, natives of eastern North America. Type species: *Dasistoma aurea* Raf.

Plant glandular-pubescent; corolla pubescent without.   1. *D. pedicularia.*
Puberulent, cinereous or glabrous; corolla glabrous without.
  Cinereous-puberulent.
    Leaves entire, dentate, or some of the lower pinnatifid, firm.
      Leaves entire, undulate or the lower pinnatifid, with entire lobes.   2. *D. flava.*
      Upper leaves mostly serrate, lower pinnatifid with toothed lobes.   3. *D. serrata.*
    Leaves, at least all but the uppermost, pinnatifid.   4. *D. grandiflora.*
  Glabrous or very nearly so throughtout.
    Leaves all pinnatifid, thin.   5. *D. laevigata.*
    Leaves entire, or the lowest dentate or incised.   6. *D. virginica.*

1. Dasystoma pedicularia (L.) Benth. Fern-leaved or Lousewort False Foxglove. Fever-weed. Fig. 3812.

Gerardia pedicularia L. Sp. Pl. 611. 1753.
Dasystoma pedicu'aria Benth. in DC. Prodr. 10: 521. 1846.

Annual or biennial, more or less glandular-pubescent, viscid, and with some longer hairs; stem rather slender, much branched, leafy, 1°–4° high. Leaves sessile, or the lower petioled, 1–2-pinnatifid, ovate or ovate-lanceolate in outline, usually broadest at the base, 1'–3' long, the segments incised or crenate-dentate; pedicels slender, ascending, mostly longer than the calyx, 1'–2' long in fruit; calyx-lobes oblong, foliaceous, usually incised or pinnatifid, 3"–4" long, corolla 1'–1½' long, pubescent without, limb about 1' broad; capsule pubescent, 5"–6" long, beak flat.

In dry woods and thickets, Maine and Ontario to Minnesota, Florida and Missouri. Races differ in pubescence and in leaf-division. Lousewort. Bushy gerardia. Aug.–Sept.

Dasystoma pectinàta (Nutt.) Benth., of the Southern States, ranging north to Kentucky and Missouri, appears to be a very glandular race.

2. Dasystoma flàva (L.) Wood. Downy False Foxglove. Fig. 3813.

Gerardia flava L. Sp. Pl. 610. 1753.
D. aurea Raf. Journ. Phys. 89: 99. 1819.
D. pubescens Benth. in DC. Prodr. 10: 520. 1846.
Dasystoma flava Wood, Bot. & Flor. 230. 1873.

Perennial, downy, grayish; stem strict, erect, simple, or with a few nearly erect branches, 2°–4° high. Leaves oblong, lanceolate or ovate-lanceolate, usually opposite, rarely whorled in 3's, firm, entire, or the lower sinuate-dentate or sometimes pinnatifid, 3'–6' long, short-petioled, the lobes obtuse; the upper much smaller and sessile, passing into the bracts of the raceme; pedicels stout, usually shorter than the calyx even in fruit; calyx-lobes lonceolate, entire, about as long as the tube; corolla 1½'–2' long, glabrous outside, its tube much expanded above; capsule 8"–10" long, pubescent, twice as long as the calyx.

In dry woods and thickets, Maine to Ontario and Wisconsin, south to southern New York, Georgia and Mississippi. Yellow foxglove. July–Aug.

3. Dasystoma serràta (Benth.) Small. Serrate False Foxglove. Fig. 3814.

D. *Drummondii serrata* Benth. in DC. Prodr. **10**: 521. 1846.

G. *grandiflora integriuscula* A. Gray, Syn. Fl. N. Am. **2¹**: 291. 1871.

Gerardia grandiflora serrata Robinson, in A. Gray, Man. Ed. 7, 730. 1908.

Dasystoma serrata Small, Bull. Torr. Club **28**: 451. 1901.

Perennial, grayish puberulent or finely pubescent; stems 1°–3½° tall, often widely branched. Leaf-blades of the lower part of the stem pinnatifid and their segments entire, merely serrate and much smaller above; calyx-tube 2½″–3″ broad during anthesis; calyx-lobes linear-lanceolate to lanceolate, entire; corolla 1′–1⅜′ long; capsules ovoid or globose-ovoid, 5″–7½″ long, short-beaked.

In dry soil, Missouri to Louisiana and Texas. July–Sept.

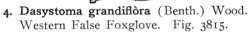

$\frac{3}{5}$

4. Dasystoma grandiflòra (Benth.) Wood. Western False Foxglove. Fig. 3815.

Gerardia grandiflora Benth. Comp. Bot. Mag. **1**: 206. 1835.

D. *Drummondii* Benth. in DC. Prodr. **10**: 520. 1846.

D. *grandiflora* Wood, Bot. & Flor. 231. 1873.

Perennial, cinereous-puberulent and roughish; stem much branched, very leafy to the top, 2°–3° high, the branches ascending or spreading. Leaves short-petioled, thin, ovate or ovate-lanceolate in outline, all pinnatifid or deeply incised, 2′–4′ long, the lobes acute or obtuse, serrate, or nearly entire; upper leaves smaller, sessile; pedicels, even in fruit, shorter than the calyx; calyx-lobes oblong or ovate, dentate or entire, about as long as the tube; corolla 1½′–2′ long, glabrous without, its tube much expanded above.

In dry woods and thickets, Minnesota and Wisconsin to Tennessee, Kansas and Texas. July–Aug.

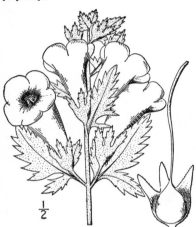

$\frac{1}{2}$

5. Dasystoma laevigàta Raf. Entire-leaved False Foxglove. Fig. 3816.

Gerardia laevigata Raf. Ann. Nat. 13. 1820.

Dasystoma quercifolia var. *integrifolia* Benth. in DC. Prodr. **10**: 520. 1846.

Dasystoma laevigata Raf.; Chapm. Fl. S. States, Ed. 2, 636. 1883.

Perennial, glabrous or very nearly so, not glaucous; stem strict, simple, or sparingly branched, 1°–3° high, the branches ascending. Leaves usually all petioled, lanceolate or ovate-lanceolate, 1½′–4′ long, entire, or the lowest dentate or incised; pedicels shorter than the calyx, or in fruit longer; calyx-lobes ovate-lanceolate, equalling or shorter than the tube; corolla glabrous without, 1′–1½′ long, the limb fully as broad, the tube much expanded above; capsule glabrous, twice as long as the calyx.

In dry thickets, Pennsylvania to Michigan, Georgia and Missouri. July–Aug.

$\frac{1}{2}$

6. Dasystoma virgínica (L.) Britton.   Smooth False Foxglove.   Fig. 3817.

Rhinanthus virginicus L. Sp. Pl. 603.   1753.
Gerardia quercifolia Pursh, Fl. Am. Sept. 423. pl. 19. 1814.
D. quercifolia Benth. in DC. Prodr. 10: 520. 1846.
D. virginica Britton, Mem. Torr. Club 5: 295. 1894

Perennial, glabrous and glaucous; stem strict, rather stout, usually branched, 3°-6° high, the branches ascending. Leaves usually all petioled, ovate or ovate-lanceolate in outline, the lower 1-2-pinnatifid, 4'-6' long, the upper pinnatifid or deeply incised, the lobes lanceolate or oblong, acute, entire, or dentate; fruiting pedicels longer than the calyx; calyx-lobes ovate or ovate-lanceolate, acute, entire, about equalling the tube; corolla 1½'-2' long, glabrous outside, its tube not widely expanded above; capsule glabrous, twice as long as the calyx.

In dry or moist woods, Maine to Minnesota, south to Florida and Illinois. Golden-oak. July–Sept.

Dasystoma calycòsa Mackenzie & Bush has been distinguished from D. virginica by its more finely divided leaves, its elongate spreading branches and its long and narrow calyx-lobes; it occurs in Missouri and Arkansas.

29. AGALÌNIS Raf. N. Fl. 2: 61. 1836.

Erect branching annual or perennial herbs, some South American species shrubby, mainly with opposite and sessile leaves. Flowers showy, usually large, purple, violet, yellow, red, or rarely white, racemose, or paniculate, or solitary and axillary. Calyx campanulate, 5-toothed, or 5-lobed. Corolla somewhat irregular, campanulate, or funnelform, the tube broad, short, or elongated, the limb 5-lobed, slightly 2-lipped, the lower lobes exterior in the bud. Stamens 4, didynamous, included; filaments more or less pubescent; anthers 2-celled, their sacs obtuse or mucronate at the base, style filiform. Capsule globose or ovoid, loculicidally dehiscent, many-seeded. Seeds numerous, mostly angled. [Greek, remarkable flax.]

About 45 species, natives of America. Besides the following, some 10 others occur in the southern United States. Most of the species blacken in drying. Type species: Agalinis palustris Raf. (Gerardia purpurea L.) The generic name Gerardia (Plumier) L., used for these plants in our first edition,. is typified by Gerardia tuberosa L. (Stenandrim tuberosum (L.) Britton, of the Acanthaceae).

Pedicels in flower shorter than the calyx, or but 1-2 times as long.
  Corolla 10"-13" long.
    Calyx-teeth minute; root perennial.   1. A. linifolia.
    Calyx-teeth triangular, lanceolate or oblong, acute; annuals.
      Leaves very scabrous, filiform; capsule oblong.   2. A. aspera.
      Leaves slightly scabrous, linear; capsule globose.
        Calyx-teeth as long as the tube or longer.   3. A. heterophylla.
        Calyx-teeth shorter than the tube.
          Stem smooth or nearly so; branches spreading; leaves rarely clustered in axils.
            4. A. purpurea.
          Stem scabrous; branches virgate; leaves much clustered in axils.
            5. A. fasciculata.
  Corolla 5"-8" long.
    Calyx-teeth triangular-subulate, acute.   6. A. paupercula.
    Calyx-teeth broad, short, obtuse.   7. A. maritima.
Pedicels in flower 2-6 times as long as the calyx.
  Leaves flat, linear, spreading or ascending; capsule globose.
    Leaves ¼"-1" wide, ½'-1½' long; pedicels spreading.   8. A. tenuifolia.
    Leaves 1"-2" wide, 1½'-3' long; pedicels ascending.   9. A. Besseyana.
  Leaves subulate, or filiform, the margins often revolute.
    Corolla-lobes obcordate or deeply emarginate.   10. A. parvifolia.
    Corolla-lobes rounded or slightly emarginate.
      Corolla 5"-7" long; pedicels erect-ascending.   11. A. Skinneriana.
      Corolla 8"-10" long; pedicels spreading-ascending.
        Calyx-teeth broadly triangular.   12. A. setacea.
        Calyx-teeth linear-subulate.   13. A. Gattingeri.

1. Agalinis linifòlia (Nutt.) Britton   Flax-leaved Agalinis.   Fig. 3818.

Gerardia linifolia Nutt. Gen. **2** : 47. 1818

Perennial, glabrous and smooth; stem branched, 2°–3° high. Leaves narrowly linear, rather thick, 1′–2′ long, 1″–1½″ wide, erect the upper much smaller and subulate; pedicels erect, in flower equalling or a little longer than the calyx, longer in fruit; calyx campanulate, truncate, its teeth minute; corolla purple, about 1′ long, narrower than that of the two following species, villous within, the lobes ciliate; filaments and anthers densely villous; anther-sacs mucronate at the base; capsule globose, 2″–3″ in diameter, but little longer than the calyx.

In moist pine barrens, Delaware to Florida. Aug.–Sept.

2. Agalinis áspera (Dougl.) Britton.   Rough Purple Agalinis.   Fig. 3819.

G aspera Dougl.; Benth. in DC. Prodr. **10** : 517. 1846.

Annual, 1°–2° high, hispidulous-scabrous with rough stiff short whitish hairs, branched, the branches nearly erect. Leaves narrowly linear, 1′–1½′ long, less than 1″ wide, erect or ascending; pedicels equalling or becoming a little longer than the turbinate calyx; calyx-teeth triangular-ovate or triangular-lanceolate, acute, one-fourth to one-third as long as the tube; corolla deep purple, about 1′ long, nearly or quite glabrous within, very pubescent without, the lobes ciliate; filaments villous; anthers all alike, obtuse at the base; capsule oblong, 3″–4″ high, considerably longer than the calyx.

On dry plains and prairies, Indiana to North Dakota, Missouri, Kansas and Arkansas. Recorded from Colorado. Aug.–Oct.

3. Agalinis heterophýlla (Nutt.) Small.   Prairie Agalinis.   Fig. 3820.

Gerardia heterophylla Nutt. Trans. Am. Phil. Soc. **5** : 180. 1837.

Gerardia crustata Greene, Leaflets **2** : 108. 1910.

Annual, more or less scabrous; stems 6′–2° tall, branching above, the branches ascending. Leaves linear, or narrowly linear-lanceolate below, ½′–2′ long, rarely 2″ broad, acute, erect or erect-ascending, decidedly scabrous along the margins; pedicels shorter than the calyx, enlarged upward; calyx campanulate, 3″–4″ long, its teeth lanceolate or subulate-lanceolate, about as long as the tube; corolla rose-purple, about ¾′ long; capsules globose-oblong or globose-ovoid, 2½″–3″ long.

On prairies and in wet woods, Missouri and Arkansas to Texas. Aug.–Oct.

4. Agalinis purpùrea (L.) Britton. Large Purple Agalinis. Fig. 3821.

Gerardia purpurea L. Sp. Pl. 610. 1753.
G. racemulosa Pennel, Torreya 11 : 15. 1911.

Annual, glabrous, smooth, or roughish; stem slender, branched, 1°–2½° high, the branches spreading. Leaves narrowly linear, usually widely spreading, 1'–1½' long, about 1" wide, rarely with smaller ones fascicled in their axils; flowers racemose on the branches, purple (rarely white), about 1' long and broad; pedicels shorter than or but little longer than the campanulate calyx, even in fruit; calyx-teeth triangular-lanceolate or ovate-oblong, acute, one-third to one-half the length of the tube; corolla much expanded above, villous or nearly glabrous within, pubescent without, the lobes ciliolate; anthers all alike, the sacs mucronulate at the base; filaments villous; capsule globose, 2"–3" in diameter, longer than the calyx.

In moist fields and meadows, Maine to Florida, Indiana, Wisconsin, Missouri and Texas. Aug.–Oct.

5. Agalinis fasciculàta (Ell.) Raf. Fascicled Agalinis. Fig. 3822.

Gerardia fasciculata Ell. Bot. S. C. & Ga. 2 : 115.

Gerardia fasciculata Ell. Bot. S. C. & Ga. 2 : 115. 1824.

Annual, finely pubescent and scabrous; stems 1½°–4° tall, often puberulent or sparingly hispidulous, fastigiately branched. Leaves numerous, the larger ones with conspicuous clusters of smaller ones in their axils, linear, mostly less than 1" wide, acute, very scabrous; flowers racemose, often numerous, purple, about 1' long; pedicel shorter than the calyx; calyx-tube campanulate, 1¼"–1¾" high, the teeth usually minute, often less than ¼" long; corolla abruptly expanded above the short tube, minutely pubescent without, copiously pubescent within, the lobes ciliolate; capsule 2"–2½" in diameter, becoming longer than the calyx.

In marshes or sometimes in dry soil, Virginia to Florida and Texas. Aug.–Oct.

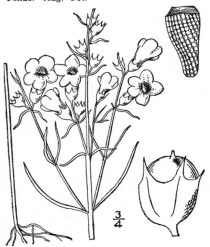

6. Agalinis paupércula (A Gray) Britton. Small-flowered Agalinis. Fig. 3823.

Gerardia purpurea var. *paupercula* A. Gray, Syn. 2 : Part 1, 293. 1878.
Gerardia intermedia Porter; A. Gray, loc. cit. As synonym. 1878.
G. paupercula Britton, Mem. Torr. Club 5 : 295. 1894.

Annual, glabrous and smooth or very nearly so; stems strict, branched above, 6'–18' high, the branches nearly erect. Leaves narrowly linear, ½'–1' long, ½"–1" wide, spreading or ascending; pedicels equalling the calyx, or longer in fruit; calyx campanulate, its teeth about one-half the length of the tube, triangular-lanceolate, acute, or acuminate; corolla 6"–10" long, puberulent, rose purple, its limb about as broad, somewhat villous in the throat within, the lobes ciliate; stamens very villous; anther-sacs mucronulate at the base; capsule globose-oblong, 3" high, longer than the calyx.

In bogs and low meadows, Nova Scotia to New Jersey, Tennessee, Manitoba and Wisconsin; apparently also in Georgia and South Carolina. July–Sept.

7. Agalinis marítima Raf.   Sea-side or Salt-marsh Agalinis.   Fig. 3824.

Gerardia maritima Raf. Med. Rep. (II.) **5** : 361.   1808.

A. maritima Raf. New Fl. N. A. **2** : 62.   1836.

Annual, smooth and glabrous, fleshy; stem erect, usually branched, 4′–16′ high, the branches ascending. Leaves linear, thick, ½′–1′ long, 1″ wide or less, obtuse, spreading; the upermost very small and subulate, scarcely longer than the pedicels; pedicels in flower shorter than or equalling the calyx, about twice as long in fruit; calyx-teeth broad, short, obtuse; corolla rose purple, 5″–9″ long, glabrous, the limb about as broad; anther-sacs mucronulate at the base; capsule globose-oblong, 2″–3″ high.

In salt marshes, Maine to Florida.   July–Aug.   The southern plant has larger flowers than the northern. Flowers before the upland species.

$\frac{2}{3}$

8. Agalinis tenuifòlia (Vahl.) Raf.   Slender Agalinis.   Fig. 3825.

Gerardia tenuifolia Vahl, Symb. Bot. **3** : 79.   1794.

A. tenuifolia Raf. New Fl. N. A. **2** : 64.   1836.

Annual, glabrous; stem very slender, paniculately branched, 6′–24′ high, the branches spreading or ascending. Leaves very narrowly linear, flat, acute, ½′–1¼′ long, ¼″–1″ wide, spreading; pedicels mostly equalling or longer than the flowers; calyx campanulate, its teeth very short, pointed; corolla light purple, spotted, rarely white, 6″–9″ long, vertically compressed when fully expanded, minutely puberulent, or glabrous; anther-sacs mucronate at the base; capsule globose or slightly obovoid, 2″–2½″ in diameter, longer than the calyx.

In dry woods and thickets, Quebec to Georgia, west to western Ontario, Kansas and Texas.   Aug.–Oct.

$\frac{3}{4}$

9. Agalinis Besseyàna Britton.   Bessey's Agalinis.   Fig. 3826.

Gerardia tenuifolia var. *macrophylla* Benth. Comp. Bot. Mag. **1** : 209.   1835.   Not *G. macrophylla* Benth.

Gerardia Besseyana Britton, Mem. Torr. Club **5** : 295. 1894.

Annual; stem glabrous, rather stout, strict, branched, 1°–2° high, the branches ascending or nearly erect. Leaves linear, 1′–2′ long, 1″–2″ wide, scabrous, acute, ascending; pedicels ascending, longer than the flowers; calyx campanulate, its teeth triangular-subulate, one-third to one-half the length of the tube; corolla purple, 5″–6″ long; capsule globose, 2″–3″ in diameter, exceeding the calyx.

On dry hills and prairies, Ontario to South Dakota, Colorado, Louisiana and Kansas.   Recorded east to Connecticut.   July–Sept.

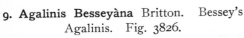

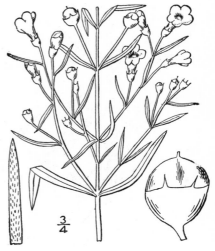

$\frac{3}{4}$

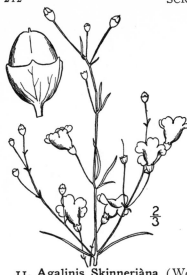

10. Agalinis parvifòlia (Chapm.) Small.
Ten-lobed Agalinis. Fig. 3827.

Gerardia parvifolia Chapm. Fl. S. U. S. 300. 1860.

Gerardia decemloba Greene, Pittonia 4: 51. 1899.

Annual, pale green, usually roughish; stems 8′–2½° tall, simple or branched above, striate-angled, the branches mostly ascending. Leaves few, remote, linear-spatulate, to narrowly linear, more or less revolute, 2½″–10″ long or rarely slightly longer, scabrous-pubescent above; pedicel exceeding the calyx; calyx campanulate or turbinate-campanulate, 1″–1½″ high, the lobes minute, triangular; corolla light rose-colored, about ½′ long, or less, the lobes obcordate or deeply emarginate; capsules globose-ovoid, or oval, 1½″–2″ long, apiculate.

In dry sandy soil or on banks, Massachusetts to Florida and Louisiana. Aug.–Oct.

11. Agalinis Skinneriàna (Wood) Britton. Skinner's Agalinis. Fig. 3828.

Gerardia Skinneriana Wood, Classbook 408. 1847.
?*G. tenuifolia asperula* A. Gray, Bot. Gaz. 4: 153. 1879.

Annual, roughish; stem strict, striate, branched, or sometimes simple, 6′–18′ high, very slender, the branches erect or ascending. Leaves setaceous, ascending or commonly erect and appressed, ½′–1′ long, ½″ wide or less, the uppermost minute; pedicels longer than the calyx, scarcely longer than the flowers, 2–4 times the length of the capsule; calyx-teeth minute; corolla light purple, 5″–6″ long and about as broad, glabrous without, its lobes ciliolate; capsule oblong, 2″–3″ high, considerably longer than the calyx.

In dry sandy woods and thickets, Indiana to Minnesota and Kansas. Recorded from Ontario. Aug.–Oct.

Gerardia víridis Small, differing by more widely spreading pedicels and narrow calyx-teeth often one-half as long as the tube, inhabits the western Gulf States and is recorded as extending northward into Missouri.

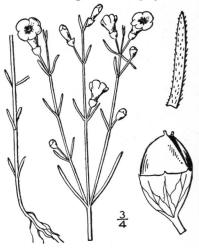

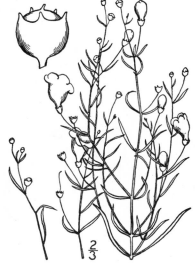

12. Agalinis setàcea (Walt.) Raf. Thread-leaved Agalinis. Fig. 3829.

Gerardia setacea Walt. Fl. Car. 170. 1788.

A. setacea Raf. New Fl. N. A. 2: 64. 1836.

G. Holmiana Greene, Pittonia 4: 52. 1899.

Annual, smooth or slightly scabrous; stems 1°–2° tall, loosely branching, the branches mostly ascending. Leaves rather numerous, mostly opposite, linear-filiform or setaceous-filiform, ¾′–1¼′ long, scabrous or nearly smooth; pedicels very slender, ½′–1½′ long; calyx campanulate, 1½″–2″ long, the lobes broadly triangular, about ⅛ as long as the tube; corolla rose purple, 5″–8″ long or rarely larger; capsules subglobose or globose-oval, 2″–2½″ in diameter.

In dry soil or pine barrens, New Jersey to Florida and Texas. Aug.–Oct.

13. Agalinis Gattíngeri Small.   Gattinger's Agalinis.   Fig. 3830.

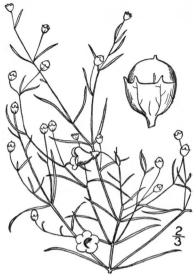

Gerardia Gattingeri Small, Fl. SE. U. S. 1078. 1903.

Annual, smooth or slightly scabrous; stem 8′–2° tall, wiry, with smooth very slender long branches. Leaves numerous, linear-filiform or almost filiform, ½′–1¼′ long, acute, smooth or slightly roughened; pedicels spreading or ascending, filiform, ½′–1′ long, less than twice the length of the leaves; calyx campanulate, its teeth subulate or linear-subulate, much shorter than the tube; corolla rose purple, 8″–10″ long; capsules subglobose, 1¾″–2″ in diameter.

In dry soil or woods, Wisconsin and Iowa to Tennessee and Texas. Aug.–Oct.

30. OTOPHÝLLA Benth. in DC. Prodr. 10: 512. 1846.

Annual caulescent herbs, with hirsute-pubescent foliage. Leaves opposite; blades entire or pinnately divided, all or some of them auricled at the base, sessile. Flowers in terminal spikes. Calyx of 5 partially united sepals. Corolla purple or rarely white; tube broadly dilated at the throat; lobes spreading. Stamens 4, didynamous, included; filaments glabrous or nearly so; anthers awnless, those of the shorter stamens much smaller than those of the longer. Style slender. Stigma entire. Capsule oval or globose-oval, often minutely pointed. Seeds angled. [Greek, meaning lobed leaf.]

Two species in eastern North America. Type species: *Gerardia auriculata* Michx.

Leaf-blades, or some of them, with 2 auricles at the base; corolla 8″–10″ long; capsules over   5″ long.   1. *O. auriculata.*
Leaf-blades parted into 3–7 linear segments; corolla 1′–1¼′ long; capsules about ¼′ long.   2. *O. densiflora.*

1. Otophylla auriculàta (Michx.) Small.   Auricled Gerardia.   Fig. 3831.

G. auriculata Michx. Fl. Bor. Am. **2**: 20. 1803.

Otophylla auriculata Small, Fl. SE. U. S. 1075. 1903.

Annual, scabrous; stem slender, simple, or branched above, hirsute, 1°–2° high. Leaves lanceolate or ovate-lanceolate, acuminate at the apex, sessile, mostly rounded and 2-lobed at the base, or quite entire, 1′–2′ long, the basal lobes oblong or lanceolate, obtusish, short; flowers solitary in the upper axils, sessile, purple, 8″–10″ long; calyx 5-cleft, its lobes lanceolate, acute, slightly unequal, as long as or longer than the tube; corolla densely puberulent outside, glabrous within; filaments glabrous; anther-sacs obtuse at the base; capsule oval-oblong, about ½′ high, a little shorter than the calyx.

In moist open soil, Pennsylvania to Illinois, Minnesota, North Carolina and Kansas. Adventive at Woodbridge, N. J. July–Sept.

2. **Otophylla densiflòra** (Benth.) Small.
Cut-leaved Gerardia.   Fig. 3832.

G. densiflora Benth. Comp. Bot. Mag. 1 : 206. 1835.
Otophylla densiflora Small, Fl. SE. U. S. 1075. 1903.

Annual, scabrous and short-hispid; stems stiff,
erect, branched, or simple, 1°–2½° high, very leafy.
Leaves sessile, ovate in outline, ascending, about
1' long, pinnately parted nearly to the midvein
into 3–7 narrowly linear acute rigid segments less
than 1″ wide; flowers 1'–1¼' long, rose purple,
sessile in the upper axils; calyx 5-cleft, its lobes
linear, acuminate, ciliate, about as long as the
tube; corolla glabrous both outside and within;
filaments glabrous or villous; capsule about ½'
high, shorter than the calyx.

On dry prairies, Kansas to Texas. Aug.–Oct.

31. **CASTILLÈJA** Mutis; L. f. Suppl. 47. 1781.

Herbs, parasitic on the roots of other plants, with alternate leaves, and red yellow purple
or white flowers, in dense leafy-bracted spikes, the bracts often brightly colored and larger
than the flowers. Calyx tubular, laterally compressed, cleft at the summit on the upper side,
or also on the lower, the lobes entire or 2-toothed. Corolla very irregular, its tube not longer
than the calyx, its limb 2-lipped; upper lip (galea) arched, elongated, concave or keeled.
laterally compressed, entire, enclosing the 4 didynamous stamens; lower lip short, 3-lobed.
Anther-sacs oblong or linear, unequal, the outer one attached to the filament by its middle,
the inner one pendulous from its apex. Style filiform; stigma entire or 2-lobed. Capsule
ovoid or oblong, loculicidally dehiscent, many-seeded. Seeds reticulated. [Named for Cas-
tillejo, a Spanish botanist.]

About 50 species, mostly natives of the New World. In addition to the following, about 30
others occur in the western parts of North America. Type species: *Castilleja fissifòlia* L. f.

* Plants villous-pubescent.

Bracts broad, dilated, lobed, or entire.
  Stem leaves deeply and irregularly cleft into narrow segments.   1. *C. coccinea.*
  Leaves linear-lanceolate, entire, or rarely with a few lobes.   2. *C. indivisa.*
Bracts linear or linear-lanceolate, entire.   3. *C. minor.*

** Plants glabrous, woolly at the summit, or cinereous-puberulent.

Glabrous, or tomentòse at the summit; leaves lanceolate, mostly entire.   4. *C. acuminata.*
Cinereous-puberulent, pale; stem leaves cleft.   5. *C. sessiliflora.*

1. **Castilleja coccínea** (L.) Spreng.   Scarlet
Painted-cup.   Indian Paint-brush or
Pink.   Prairie-fire.   Fig. 3833.

Bartsia coccinea L. Sp. Pl. 602. 1753.
Castilleja coccinea Spreng. Syst. 2 : 775. 1825.

Annual or biennial, villous-pubescent; stem
rather slender, simple, or with few erect branches,
1°–2° high. Leaves sessile, parallel-veined, the
basal oblong, obovate, or linear, tufted, mostly
entire, 1'–3' long, those of the stem deeply 3–5-
cleft into linear obtusish segments, the bracts
broader and shorter, 3–5-lobed or cleft, bright
red or scarlet, conspicuous; flowers sessile, 10″–
12″ long, usually not exceeding the bracts; calyx
cleft both above and below into 2 dilated entire
or retuse oblong and obtuse lobes, sometimes
scarlet; corolla greenish-yellow, its tube shorter
than the calyx, its upper lip much longer than
the lower; capsule oblong, acute, 5″–6″ long.

In meadows and moist thickets, Maine and Ontario
to Manitoba, North Carolina, Tennessee, Kansas and
Texas. Ascends to 4000 ft. in Virginia. Bracts and
calyx rarely yellow. Red indians. Election-posies.
Wickawee. Bloody-warrior. Nose-bleed. May–July.

2. Castilleja indivìsa Engelm. Entire-leaved Painted-cup. Fig. 3834.

Castilleja indivisa Engelm.; Engelm. & Gray, Bost. Journ. Nat. Hist. **5**: 255. 1845.

Winter-annual; stem villous-pubescent, usually simple, 8′–18′ high. Leaves sessile, parallel-veined, linear to linear-lanceolate, 1′–4′ long, 1½″–3″ wide, entire, or rarely with 2–4 lateral lobes; no tuft of basal leaves; bracts dilated, obovate to spatulate, bright red; flowers sessile, about 1′ long or less, not longer than the bracts; calyx cleft as in the preceding species, and corolla similar.

In sandy soil, Kansas to Texas. Spring.

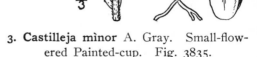

3. Castilleja mìnor A. Gray. Small-flowered Painted-cup. Fig. 3835.

Castilleja affinis var. *minor* A. Gray, Bot. Mex. Bound. Surv. 119. 1859.
Castilleja minor A. Gray, in Brew. & Wats. Bot. Cal. **1**: 573. 1876.

Annual, villous-pubescent; stem slender, strict, simple, or with 1 or 2 erect branches, 1°–2½° high. Leaves all linear-lanceolate and entire, parallel-veined, sessile, acuminate, 2′–3′ long, the bracts similar, smaller, red or red-tipped, very narrow, equalling or longer than the short-pedicelled flowers; calyx green, cleft on both sides to about the middle, the lobes lanceolate, acute, entire, or 2-toothed; corolla yellow, 6″–10″ long, its upper lip much longer than the small lower one; capsule oblong, acute, 6″–8″ long.

In moist soil, Nebraska to New Mexico, west to Nevada and Arizona. Indian-pink. May–July.

4. Castilleja acuminàta (Pursh) Spreng. Lance-leaved Painted-cup. Fig. 3836.

Bartsia acuminata Pursh, Fl. Am. Sept. 429. 1814.
Castilleja acuminata Spreng. Syst. **2**: 775. 1825.
C. septentrionalis Lindl. Bot. Reg. *pl. 925.* 1825.
Castilleja pallida var. *septentrionalis* A. Gray, in Brew. & Wats. Bot. Cal. **1**: 575. 1876.

Perennial, glabrous or loosely tomentose above; stems slender, commonly clustered, 6′–2° high, usually simple. Leaves sessile, 3–5-nerved, mostly quite entire, the lower linear, the upper lanceolate, acuminate or acute at the apex, somewhat narrowed at the base, 2′–4′ long; bracts oblong, oval, or obovate, obtuse, dentate, or entire, yellowish, greenish-white or purple, as long as the sessile flowers; calyx cleft on both sides to about the middle, the lobes lanceolate, usually again 2-cleft; corolla 6″–8″ long, its upper lip 2–4 times as long as the lower; capsule oblong, 6″–8″ high.

In moist soil, Newfoundland and Labrador to Hudson Bay, the mountains of New England, Ontario and Minnesota. Plants previously referred to this species from farther west are now regarded as distinct from it. June–Aug. Pale painted-cup.

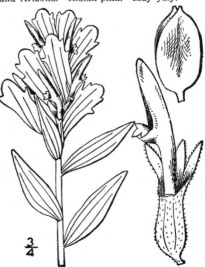

3/4

5. Castilleja sessiliflòra Pursh. Downy Painted-cup. Fig. 3837.

Castilleja sessiliflora Pursh, Am. Sept. 738. 1814.

Perennial, cinereous-puberulent all over; stems stout, simple, or branched from near the base, 6′–15′ high, densely leafy. Leaves sessile, 1′–2′ long, the lowest commonly linear, obtuse and entire, the others laciniate into narrow, entire or cleft segments; bracts green, similar to the upper leaves, shorter than the sessile flowers; calyx deeper cleft on the lower side than on the upper, its lobes linear-lanceolate, acute; corolla yellowish, 1½′ long, the upper lip about twice as long as the lower, the lobes of the latter linear; capsule oblong-lanceolate, acute, 6″–8″ long.

On dry prairies, Illinois to Manitoba, Saskatchewan, Nebraska, Wyoming and Texas. May–July.

32. ORTHOCÀRPUS Nutt. Gen. 2: 56. 1818.

Annual or rarely perennial herbs, mostly with alternate leaves, and yellow white or purplish flowers, in bracted usually dense spikes, the bracts sometimes brightly colored. Calyx tubular or tubular-campanulate, 4-cleft, or sometimes split down both sides. Corolla very irregular, the tube slender, the limb 2-lipped; upper lip little if any longer than the 3-lobed 1–3-saccate lower one. Stamens 4, didynamous, ascending under the upper lip; anther-sacs dissimilar, the outer one affixed by its middle, the inner pendulous from its upper end, commonly smaller. Style filiform; stigma entire. Capsule oblong, loculicidally dehiscent, many-seeded. Seeds reticulated. [Greek, erect-fruit.]

About 30 species, natives of America, mostly of the western United States, 1 or 2 Andean. Type species: *Orthocarpus luteus* Nutt.

1. Orthocarpus lùteus Nutt. Yellow Orthocarpus. Fig. 3838.

Orthocarpus luteus Nutt. Gen. 2: 57. 1818.

Annual, rough-pubescent or puberulent; stem strict, erect, branched above, or simple, 6′–18′ high, densely leafy. Leaves erect or ascending, linear or lanceolate, entire, or sometimes 3-cleft, 1′–1½′ long, 1″–2″ wide, sessile, long-acuminate; bracts of the dense spike lanceolate, broader and shorter than the leaves, entire or 3-cleft, acute, green, mostly longer than the flowers; flowers bright yellow, 4″–5″ long; calyx-teeth acute, shorter than the tube; corolla about twice as long as the calyx, puberulent without, its upper lip ovate, obtuse, about as long as the saccate 3-toothed lower one; capsule about as long as the calyx-tube.

2/3

On dry plains and prairies, Manitoba to Minnesota and Nebraska, west to British Columbia and California. July–Sept.

33. SCHWÀLBEA [Gronov.] L. Sp. Pl. 606. 1753.

A perennial erect finely pubescent and minutely glandular, simple or sparingly branched, leafy herb, with sessile entire 3-nerved leaves, and rather large yellowish-purple flowers in a terminal bracted spike. Calyx tubular, somewhat oblique, 10–12-ribbed, 2-bracteolate at the base, 5-toothed, the upper tooth much the smallest, the 2 lower ones partly connate; corolla very irregular, the tube cylindric, the limb 2-lipped; upper lip arched, concave, entire; lower lip somewhat shorter, 3-lobed, 2-plaited. Stamens 4, didynamous, ascending within the upper lip of the corolla; anther-sacs equal. Style filiform. Capsule oblong, many-seeded. Seeds linear, with a loose reticulated testa. [Named for C. G. Schwalbe, of Holland, who wrote (1719) on Farther India.]

A monotypic genus of eastern North America.

1. Schwalbea americàna L.　Chaff-seed.
Fig. 3839.

Schwalbea americana L. Sp. Pl. 606. 1753.

Stem slender, strict, 1°–2° high. Leaves oblong or ovate-oblong, entire, acute at both ends, 1′–1½′ long, 2″–8″ wide, the upper gradually smaller and passing into the bracts of the rather loose spike; flowers very nearly sessile, 1′–1½′ long, longer than the bracts; bractlets at the base of the calyx linear, shorter than its tube; corolla-tube slightly exceeding the lower lobes of the calyx, these connate to near their apices; capsule enclosed by the calyx.

In wet sandy soil, eastern Massachusetts to Florida and Louisiana, near the coast. May–July.

34. EUPHRÀSIA [Tourn.] L. Sp. Pl. 604. 1753.

Annual or perennial low mostly branched herbs, parasitic on other plants, with opposite dentate or incised leaves, and small blue purplish yellow or white often variegated flowers in terminal leafy-bracted spikes. Calyx not bracteolate at the base, campanulate or tubular, 4-cleft (rarely 5-cleft with one of the lobes much smaller than the others). Corolla very irregular, 2-lipped, the upper lip erect, scarcely concave, 2-lobed, its margins recurved; lower lip larger, 3-lobed, spreading, its lobes either emarginate or obtuse. Stamens 4, didynamous, ascending under the upper lip of the corolla; anther-sacs equal and parallel, mucronate at the base. Capsule oblong, loculicidally dehiscent, many-seeded. Seeds oblong, longitudinally ribbed. [Greek, delight.]

About 110 species, natives of temperate and cold regions of both the northern and southern hemispheres. Besides the following, another occurs in northwestern North America. Type species: *Euphrasia officinalis* L.

Flowers 2½″–4″ long.
　　Leaves markedly pubescent on both sides.　　　　　　　　　　　1. *E. arctica.*
　　Leaves glabrate or sparingly pubescent.　　　　　　　　　　　2. *E. americana.*
Flowers 1½″–2″ long.
　　Stem 3′–20′ long; flowers racemose-spicate.　　　　　　　　　3. *E. Randii.*
　　Stem 1′–2′ high; flowers subcapitate.　　　　　　　　　　　　4. *E. Oakesii.*

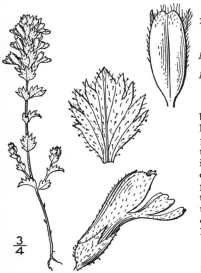

1. Euphrasia àrctica Lange.　Glandular Eyebright.　Fig. 3840.

Euphrasia arctica Lange, Bot. Tidskr. **4**: 47. 1870.

Euphrasia latifolia Pursh, Fl. Am. Sept. 430. 1814. Not Willd.

Annual; stem erect, simple, or with a few erect branches from near the base, pubescent with crisped hairs, 2′–7′ high. Leaves ovate to obovate, obtuse, pubescent on both sides, 2–5-toothed on each margin, the teeth sharp or blunt; spike 1′–4′ long, the bracts imbricated, at least above, broadly oval to orbicular, cuneate at the base, sharply toothed, glandular-pubescent beneath or also on the margins; calyx-teeth acute; corolla 3″–4″ long, lilac, or variegated, the lobes of its lower lip nearly parallel; capsule oblong-elliptic, ciliate on the margins, and more or less pubescent or pilose, about as long as the calyx.

Bluffs and slopes, Greenland to Hudson Bay, New Brunswick, Maine and Minnesota. Has been erroneously referred to *E. hirtella* Jordan. Summer.

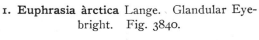

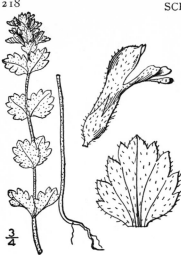

2. Euphrasia americàna Wettst. Hairy Eyebright. Eyebright. Fig. 3841.

Euphrasia americana Wettst. Mon. Euph. 127. 1896.
E. canadensis Townsend, Journ. Bot. **36**: 1. 1898.

Annual, stem pubescent with crisped hairs, often at length much branched, 4′–10′ high. Leaves glabrous, or sparingly pubescent, ovate to oval, obtuse or acutish at the apex, 3–5-toothed on each side, 3″–6″ long, 1½″–4½″ wide, the teeth acute; bracts cuneate or obtuse at the base, dentate, pubescent, not glandular, slightly shorter than the flowers; calyx-teeth lanceolate, acuminate; corolla 3″–4″ long, purplish or nearly white, the somewhat spreading lobes of its lower lip emarginate; capsule narrow, cuneate, 2″–3″ high, about equalling the calyx.

In fields and on hills, Maine and New Hampshire to New Brunswick and Newfoundland. Summer.

Euphrasia officinàlis L. is not known from North America.

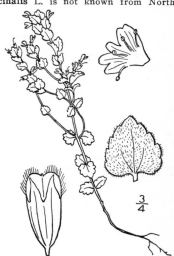

3. Euphrasia Rándii Robinson. Rand's Eyebright. Fig. 3842.

Euphrasia Randii Robinson, Rhodora **3**: 273. 1901.

E. Randii var. (?) *Farlowii* Robinson, loc. cit. 274. 1901.

Annual; stem simple or branched, erect or decumbent, 3′–20′ long, purplish, covered with whitish hairs. Leaves ovate to suborbicular or fan-shaped, finely pubescent on both sides, 4–6-toothed on each margin, the teeth obtuse or acutish; flowers spicate-racemose; calyx pubescent, its teeth triangular-lanceolate; corolla about 2″ long, violet to nearly white, with a yellow eye; capsule oblong-elliptic, ciliate, notched.

Thickets and grassy places, Newfoundland to Quebec and Maine. July–Aug.

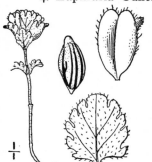

4. Euphrasia Oakesii Wettst. Oakes' Eyebright. Fig. 3843.

Euphrasia Oakesii Wettst. Mon. Euph. 142. 1896.

Stem erect or ascending, very slender or filiform, simple, 1′–2′ high, somewhat pubescent. Stem leaves 2 or 3 pairs. Bracts and leaves orbicular or broadly oval, narrowed or nearly truncate at the base, 2″–3″ long, obtuse, pubescent, not glandular, with 2–5 blunt teeth on each side; spike short, capitate, only 3″–5″ long, the bracts densely imbricated; calyx-teeth triangular-lanceolate, short; corolla 1½″–2″ long, purplish-white with a yellow eye; capsule oblong-elliptic, ciliate, longer than the calyx.

White Mountains of New Hampshire and on Mt. Katahdin, Maine. July–Aug.

Euphrasia Williamsii Robinson, found on Mt. Washington, N. H., differs in being nearly glabrous, with a brown-purple corolla.

35. BÁRTSIA L. Sp. Pl. 602. 1753.

Annual or perennial herbs, partly parasitic on the roots of other plants, with opposite leaves, and purple pink red or yellow flowers, in terminal leafy-bracted spikes. Calyx campanulate or tubular, 4-toothed or 4-cleft. Corolla very irregular, the tube straight or recurved, the limb 2-lipped; upper lip erect, concave, entire, the margins not recurved; lower lip spreading, 3-lobed. Stamens 4, didynamous, ascending under the upper lip of the corolla; anther-sacs similar, parallel. Capsule globose, oblong, or ovoid, loculicidally dehiscent, several–many-seeded. Seeds horizontal, striate, or ribbed. [Named for John Bartsch, a Prussian botanist, died 1738.]

About 6 species of the northern hemisphere. Only the following is known to occur in North America. Type species: *Bartsia viscòsa* L.

1. **Bartsia alpìna** L. Alpine Bartsia.
Fig. 3844.

Bartsia alpina L. Sp. Pl. 602. 1753.

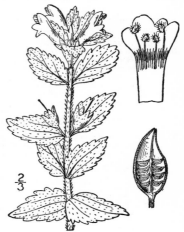

Perennial by short rootstocks, pubescent; stem erect, leafy, simple, or rarely with 1 or 2 short branches, 4′–10′ high. Leaves sessile, ovate, or ovate-oblong, crenate-dentate, obtuse or acutish, rounded and sometimes slightly clasping at the base, ½′–1′ long; bracts similar, smaller, mostly shorter than the flowers; spike 1′–2′ long; flowers 8″–10″ long, sometimes borne also in the upper axils; calyx 4-cleft nearly to the middle; corolla purple, its tube much longer than the calyx; anthers pubescent, at least on the back; capsule ovoid-oblong, equalling or longer than the calyx.

Labrador to Greenland and the Arctic Sea. Also in Europe. Summer.

36. **ODONTÌTES** Gmel. Fl. Sib. 3: 213. 1768.

Annual erect herbs, half parasitic on the roots of other plants, with small opposite leaves, and yellow or red flowers in terminal bracted spikes or racemes. Calyx 4-toothed. Corolla with a narrow tube and a strongly 2-lipped limb, the upper lip concave, entire, or 2-lobed, the lower 3-lobed, spreading. Stamens 4, didynamous, ascending; anther-sacs similar. Capsule mostly subglobose, loculicidally dehiscent. Seeds few, pendulous. [Greek, referring to its supposed value as a cure for toothache.]

About 20 species, mostly natives of the Mediterranean region, the following typical.

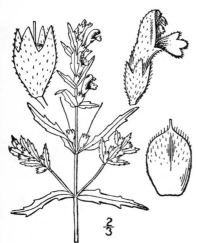

1. **Odontites Odontìtes** (L.) Wettst. Red Bartsia. Red Eyebright. Fig. 3845.

Euphrasia Odontites L. Sp. Pl. 604. 1753.
Bartsia Odontites Huds. Fl. Angl. Ed. 2268. 1778.
Odontites Odontites Wettst. in Engl. & Prantl, Nat. Pfl. Fam. 4: Abt. 3b, 102. 1891.

Annual, appressed-pubescent, roughish; stem slender, at length much branched, 6′–15′ high, the branches erect or ascending. Leaves sessile, lanceolate, or oblong-lanceolate, acute or acuminate at the apex, serrate with low distant teeth, slightly narrowed at the base, ½′–1½′ long, 1″–3″ wide; spikes slender, becoming 2′–5′ long in fruit, somewhat 1-sided; bracts similar to the leaves, but smaller; flowers numerous, 4″–5″ long; calyx 4-cleft; corolla red or pink, its tube somewhat longer than the calyx; anthers slightly pubescent; capsule oblong, shorter than the calyx.

In fields and waste places, coast of Maine to Nova Scotia. Naturalized from Europe. Native also of Asia. June–Sept.

37. **PEDICULÀRIS** [Tourn.] L. Sp. Pl. 607. 1753.

Herbs, with alternate opposite or rarely verticillate, pinnately lobed cleft or pinnatifid leaves, and yellow red purple or white flowers, in terminal spikes or spike-like racemes. Calyx tubular, cleft on the lower side or sometimes also on the upper, or 2–5-toothed. Corolla strongly 2-lipped, the tube cylindric, the upper lip (galea) laterally compressed, concave or conduplicate, sometimes short-beaked; lower lip erect or ascending, 3-lobed, the lobes spreading or reflexed, the middle one the smallest. Stamens 4, didynamous, ascending within the upper lip of the corolla; anthers approximate in pairs, their sacs transverse, equal, parallel, obtuse or rarely mucronate at the base. Capsule compressed, oblique or curved, beaked, many-seeded, loculicidally dehiscent. Seeds reticulate, pitted, striate or ribbed. [Latin, pertaining to lice, long supposed to breed lice in sheep that feed on these plants.]

About 125 species, mostly natives of the northern hemisphere. In addition to the following, some 30 others inhabit the western parts of North America. Mostly known as Lousewort; a few red-flowered species are called Red Rattle. Type species: *Pedicularis sylvática* L.

Beak of the galea conic, decurved, 1″ long.   1. *P. lapponica.*
Beak of the galea very short, or none.
  Annuals or biennials; stems leafy, freely branching; northern.
    Puberulent; upper leaves crenulate, lower pinnatifid.   2. *P. euphrasioides.*
    Glabrous or very nearly so; leaves all pinnatifid.   3. *P. palustris.*
  Perennials; stems leafy, simple (rarely branched in No. 6).
    Corolla yellow, or the galea red; plants 6′–3° high; eastern species.
      Leaves pinnately lobed; capsule ovate, scarcely longer than the calyx.
                                                                       4. *P. lanceolata.*
    Leaves pinnately parted; capsule lanceolate, 3 times as long as the calyx.
                                                                         5. *P. canadensis.*
      Lower leaves pinnately divided; capsule ovate.   6. *P. Furbishiae.*
    Galea crimson or purple; plant 1′–4′ high; arctic.   7. *P. flammea.*
  Perennial; stem scapose, or 1-leaved; flowers capitate; arctic.   8. *P. capitata.*

1. Pedicularis lappónica L. Lapland Pedicularis. Fig. 3846.

Pedicularis lapponica L. Sp. Pl. 609. 1753.

Perennial, puberulent; stems simple, or sparingly branched, leafy, 4′–8′ high. Leaves sessile, or very short-petioled, alternate or the lowest opposite, lanceolate or linear-lanceolate, obtuse or acutish at the apex, $\frac{1}{2}′$–$1\frac{1}{2}′$ long, pinnately incised into numerous approximate oblong serrulate lobes; spike short, the flowers almost capitate, light yellow, 6″–7″ long; calyx cleft on the lower side, 2-toothed on the upper; galea erect, arched, tipped by an abruptly spreading or recurved conic beak about 1″ long.

In open places, Labrador and Greenland to the Arctic Sea. Also in Arctic Europe and Asia. Summer.

Pedicularis pedicellàta Bunge, an Alaskan species is recorded by Bunge from Labrador. It is distinguished from the above by its scapose stem, deeply pinnatifid leaves and pedicellate lower flowers. We have not seen specimens from the eastern side of the continent.

2. Pedicularis euphrasioìdes Steph. Eyebright Pedicularis. Fig. 3847.

P. euphrasoides Steph.; Willd. Sp. Pl. **3**: 204. 1801.

Biennial or annual, puberulent; stem branched, 6′–15′ high, the branches ascending. Lower leaves petioled, lanceolate or oblong-lanceolate in outline, 2′–4′ long, 4″–8″ wide, pinnatifid into oblong, obtuse, crenate-dentate segments; upper leaves sessile, linear or linear-oblong, smaller, merely crenulate; flowers in a short terminal spike and solitary in the upper axils; calyx cleft on the lower side, 2–3-toothed on the upper, shorter than the corolla-tube; corolla yellow, or the galea purplish, about 6″ long; galea as long as the tube, tipped with a very short truncate beak, minutely 2-toothed on the lower side at the apex; capsule apparently shorter than the calyx.

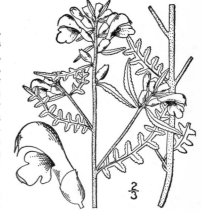

Labrador to Greenland, the Arctic Sea, Alaska and British Columbia. Also in northern Asia. Summer.

3. Pedicularis palústris L.   Purple Pedicularis.   Marsh Lousewort.   Fig. 3848.

Pedicularis palustris L. 607. 1753.
?*P. parviflora* J. E. Smith in Rees' Cyclop. **26**: No. 4.
   1814.
Pedicularis Wlassoviana Stev. Mem. Soc. Nat. Mosc.
   6: 29. *pl. 9, f. 1.* 1823.
Pedicularis palustris var. *Wlassoviana* Bunge;
   Ledeb. Fl. Ross. **3**: 283. 1847–49.

Biennial or annual, glabrous throughout, or the
petiole-bases ciliate; stem erect, much branched,
1°–2° high. Leaves 1′–2′ long, alternate, or some
of them opposite, oblong-lanceolate in outline,
all pinnatifid into oblong crenate or incised segments; flowers in terminal rather loose spikes
and solitary in the upper axils, 7″–9″ long; calyx
2-cleft, the lobes with an incised crested border;
corolla purple, sometimes white, its tube twice as
long as the calyx, the lip much shorter, the galea
arched at the top, blunt or apiculate, not beaked,
bearing a pair of minute teeth below its summit;
capsule obliquely ovate, twice as long as the calyx
when mature.

In wet situations, Labrador to Alaska, south to
Quebec, the Northwest Territory and Oregon. Europe. Summer. Red rattle, Cow's-wort.

4. Pedicularis lanceolàta Michx.   Swamp Lousewort.   Fig. 3849.

Pedicularis lanceolata Michx. Fl. Bor. Am. **2**: 18. 1803.
Pedicularis auriculata J. E. Smith in Rees' Cyclop. **26**:
No. 4. 1814.

Perennial, glabrous or very nearly so throughout;
stem stout, simple, or branched above, 1°–3° high, the
branches erect. Leaves alternate and opposite, lanceolate, or linear-lanceolate, 2′–5′ long, pinnately lobed,
the lower petioled, the upper sessile, the lobes oblong,
obtuse, short, crenate-dentate, the margins cartilaginous; spikes short; calyx 2-lobed, the lobes with foliaceous margins; corolla yellow, 8″–10″ long, the galea
arched, terminated by a very short truncate beak, the
lower lip erect-ascending; capsule ovate, little exceeding the calyx, about 5″ high.

In swamps, Ontario to Connecticut, North Carolina,
Manitoba, Ohio, Michigan, South Dakota and Nebraska.
Aug.–Oct.

5. Pedicularis canadénsis L.   Wood or Head Betony.   Lousewort.   Fig. 3850.

Pedicularis canadensis L. Mant. 86. 1767.

Perennial, hirsute, pubescent, or glabrate below; stems commonly tufted, ascending or
erect, 6′–18′ high, simple. Leaves alternate or
some of them opposite, oblong-lanceolate, 3′–5′
long, all but the uppermost slender-petioled,
pinnately parted into oblong, obtuse, incised or
dentate lobes; flowers spicate, the spike short
in flower, 5′–8′ long in fruit, the lower bracts
usually foliaceous, the others small; calyx cleft
on the lower side; 2–3-crenate on the upper,
oblique; corolla yellow, or reddish (rarely
white), 7″–10″ long, the tube much longer than
the calyx, the galea arched, incurved, not beaked,
minutely 2-toothed below the apex; capsule
lanceolate, oblique, 7″–8″ long, 2″ wide, about
3 times as long as the calyx.

In dry woods and thickets, Nova Scotia to Manitoba, Florida, Mississippi, Kansas, Colorado and
North Mexico. Ascends to 3000 ft. in Virginia.
High heal-all. Beefsteak-plant. Lousewort-fox-
glove. Snaffles. April–June.

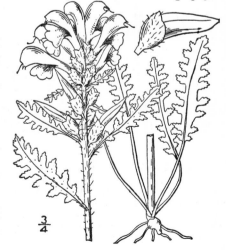

$\frac{3}{4}$

6. Pedicularis Furbíshiae S. Wats. Miss Furbish's Pedicularis. Fig. 3851.

P. Furbishiae S. Wats. Proc. Am. Acad. **17**: 375. 1882.

Perennial, pubescent, at least above; stem strict, simple, 2°–3° high. Leaves lanceolate, alternate, or some of them opposite, the lower long-petioled, 4′–6′ long, pinnately divided into ovate or oblong, pinnatifid or incised segments, the upper sessile, pinnately parted or lobed; calyx 5-lobed, the lobes entire or dentate; narrow, somewhat unequal; corolla yellow, 7″–9″ long, the galea arched, truncate, not beaked, 2-cuspidate at the apex; capsule ovate.

In swamps and along streams, Maine and New Brunswick. July–Sept.

7. Pedicularis flámmea L. Red-tipped Pedicularis. Fig. 3852.

Pedicularis flammea L. Sp. Pl. 609. 1753.

Perennial, glabrous, or somewhat woolly; stem simple, 2′–4′ high, with several linear-oblong, pinnately parted leaves. Basal and lower leaves slender-petioled, 1′–1½′ long, the uppermost sessile, the lobes ovate or oblong, incised-serrate; flowers about 6″ long, pedicelled in a short spike-like raceme, longer than the narrow bracts; calyx 5-toothed, the teeth lanceolate, acute, unequal; corolla-tube and the lower lip greenish yellow, the galea slightly arched, very blunt, much longer than the lower lip, its summit crimson or purple; capsule lanceolate, 6″–8″ long, 2–3 times as long as the calyx; fruiting pedicels 3″–5″ long.

Labrador and Greenland, west to Alaska. Also in arctic and alpine Europe. Summer.

$\frac{3}{4}$

8. Pedicularis capitàta Adams. Capitate Pedicularis. Fig. 3853.

Pedicularis capitata Adams, Mem. Soc. Nat. Mosc. **5**: 100. 1817.

Perennial, pubescent or glabrous; stem scapose, leafless, or 1-leaved, 1′–5′ high. Leaves slender-petioled, often shorter than the scape, pinnately divided, the segments ovate or oblong, incised; flowers several in a capitate cluster at the end of the scape, 1′–1½′ long; calyx 5-cleft, the lobes foliaceous, incised or crenate; corolla described as white; galea scarcely broadened above, slightly curved, very obtuse, twice as long as the lower lip; capsule oblong, a little longer than the calyx, beaked on the outer side near the summit.

Arctic America; Hudson Bay to Alaska. Summer.

$\frac{1}{2}$

38. ELEPHANTÉLLA Rydb. Mem. N. Y. Bot. Gard. 1: 362. 1900.

Herbs similar to *Pedicularis* in habit. Leaves pinnately parted or pinnately divided. Inflorescence erect. Calyx 2-lipped, often campanulate. Corolla very strongly 2-lipped, the tube short, the upper lip (galea) produced into an elongated slender beak which is soon turned upward, the lower lip very broad. [Greek, little elephant, referring to the resemblance of the galea of the corolla to an elephant's head.]

Two or three species of northern regions. Type species: *Elephantella groenlándica* (Retz.) Rydb.

1. Elephantella groenlándica (Retz.) Rydb. Long-beaked Pedicularis. Fig. 3854.

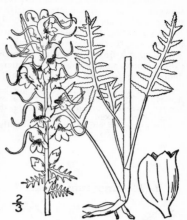

P. groenlandica Retz. Fl. Scand. Ed. 2, 145.　1795.
Elephantella groenlandica Rydb. Mem. N. Y. Bot. Gard. 1 : 363.　1900.

Perennial, glabrous; stem simple, erect, 1°–1½° high. Leaves alternate, lanceolate in outline, acute or acuminate, pinnately parted or the lower pinnately divided into lanceolate, acute, crenulate or incised segments, the upper sessile, the lower slender-petioled, 2′–6′ long; spike 1′–6′ long, very dense; calyx 5-toothed, nearly as long as the corolla-tube, the teeth short, acutish; corolla red or purple, the galea produced into a filiform beak 6″–8″ long, which is decurved against the lower lip and upwardly recurved beyond it; body of the corolla 2½″–3″ long; capsule obliquely ovate, about 3″ long.

In wet soil, Labrador, Greenland and Hudson Bay to Athabasca, British Columbia, south in the Rocky Mountains to New Mexico, and in the Sierra Nevada to California. Summer.

39. RHINÁNTHUS L. Sp. Pl. 603.　1753.

Annual erect mostly branched herbs, with opposite leaves, and yellow blue violet or variegated flowers, in terminal 1-sided leafy-bracted spikes, or solitary in the upper axils. Calyx compressed, 4-toothed, much inflated, membranous and conspicuously veiny in fruit. Corolla very irregular, 2-lipped, the upper lip (galea) compressed, arched, minutely 2-toothed below the entire apex, the lower lip 3-lobed, shorter, the lobes spreading. Stamens 4, didynamous, ascending under the galea; anthers pilose, the sacs obtuse at the base, transverse, distinct. Capsule orbicular, flat, loculicidally dehiscent, several-seeded. Seeds nearly orbicular, winged. [Greek, nose-flower, from the beaked corolla.]

About 3 species, natives of the northern hemisphere. Type species: *Rhinanthus Crista-galli* L.

1. Rhinanthus Crísta-gálli L. Rattle. Rattle-box. Yellow or Penny Rattle. Fig. 3855.

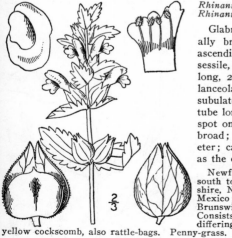

Rhinanthus Crista-galli L. Sp. Pl. 603.　1753.
Rhinanthus minor Ehrh. Beitr. 6: 44.　1791.

Glabrous, or pubescent above; stem slender, usually branched, 6′–18′ high, the branches erect or ascending. Leaves lanceolate or oblong-lanceolate, sessile, coarsely serrate-dentate, acute or obtuse, 1′–2′ long, 2″–4″ wide; bracts broader, ovate, or ovate-lanceolate, incised-dentate, the teeth acuminate or subulate-tipped; flowers yellow, 6″–8″ long; corolla-tube longer than the calyx, commonly with a purple spot on one or both lips, the teeth of the upper lip broad; fruiting calyx ovate-orbicular, 4″–6″ in diameter; capsule orbicular, or broader, nearly as broad as the calyx, very flat, not oblique.

Newfoundland and Labrador to Alaska and Oregon, south to Quebec, the White Mountains of New Hampshire, New York and in the Rocky Mountains to New Mexico; on the Atlantic Coast from Connecticut to New Brunswick. Common in northern Europe and Asia. Consists of several races, sometimes regarded as species, differing in habit and in color of the corolla-lips. Called yellow cockscomb, also rattle-bags. Penny-grass. Money-grass. June–Aug.

Rhinthus màjor Ehrh., found many years ago in fields at Plymouth, Massachusetts, differs in having the teeth of the upper corolla-lip elongated, its flowers somewhat larger. It is a native of Europe.

40. MELAMPŶRUM [Tourn.] L. Sp. Pl. 605. 1753.

Annual branching herbs, with opposite leaves, and small white yellow violet or variegated flowers, solitary in the upper axils, or in terminal bracted spikes. Calyx 4-toothed, the 2 upper teeth somewhat the longer. Corolla irregular, 2-lipped, the tube narrow, gradually enlarged above, the upper lip compressed, obtuse or emarginate with a groove behind the margins, or these recurved or with a tooth on each side; lower lip spreading or ascending, 3-toothed, 2-grooved beneath. Stamens 4, didynamous, ascending under the upper lip; anther-sacs distinct, parallel obtuse or mucronulate at the base. Capsule flat, oblique, loculicidally dehiscent, 2-4-seeded. Seeds smooth, strophiolate. [Greek, black wheat.]

About 10 species, all of the northern hemisphere. Only the following are known in North America. Type species: *Melampyrum arvense* L.

Leaves lanceolate or linear-lanceolate, the floral 2–4-toothed at the base; capsule much longer than wide, long-beaked.   1. *M. lineare.*
Leaves ovate, all entire; capsule slightly longer than wide, short-beaked.   2. *M. latifolium.*

1. Melampyrum lineàre Lam. Narrow-leaved Cow-Wheat. Fig. 3856.

Melampyrum lineare Lam. Encycl. 4: 22. 1797.
M. americanum Michx. Fl. Bor. Am. 2: 16. 1803.

Puberulent; stem slender, obscurely 4-sided above, at length widely branched, 6'–1½° high. Leaves lanceolate or linear-lanceolate to ovate, short-petioled, acuminate or acute at the apex, narrowed, obtuse, or the upper truncate at the base, 1'–2½' long, 1½''–6'' wide, the lower entire, the upper floral ones ovate or lanceolate, with 2–6 bristle-pointed teeth near the base or entire; flowers short-peduncled, 4''–6'' long; calyx about one-third the length of the corolla, its subulate teeth longer than its tube; corolla white or whitish, puberulent, the lower lip yellow; capsule 4''–5'' long, about 2'' wide, twice as long as the calyx, long-beaked.

In dry woods and thickets, Nova Scotia to British Columbia, south to Georgia, Tennessee, Iowa, Montana and Idaho. Consists of several races, differing in leaf-form. Ascends 3500 ft. in Virginia. May– Aug.

2. Melampyrum latifòlium Muhl. Broad-leaved Cow-Wheat. Fig. 3857.

Melampyrum latifolium Muhl. Cat. 57. 1813.

Similar to the preceding, widely branched, 1°–1½° high; but the leaves all entire, short-petioled, the lowest small, spatulate, obtuse, the middle ones lanceolate or ovate, acuminate, narrowed at the base, 2'–3' long, the floral ovate or ovate-lanceolate, shorter, acute, mostly rounded at the base; corolla purple, veiny, larger; capsule little longer than wide, short-beaked.

In dry woods, Delaware (according to Muhlenberg); mountains of Virginia to Georgia and Tennessee. June– Aug.

Family 29. **LENTIBULARIÀCEAE** Lindl. Veg. Kingd. 686. 1847.*

Bladderwort Family.

Aquatic plants, or if terrestrial usually on moist ground, the leaves in a basal rosette, or borne along floating stems, or more or less root-like, or wanting. Inflorescence scapose. Flowers solitary or racemose, perfect, irregular. Calyx inferior, 2–5-lobed or parted, persistent. Corolla hypogynous, gamopetalous, 2-lipped, the upper lip entire or 2-lobed, the lower lip entire or 3-lobed, usually with a nectariferous spur. Stamens 2, borne upon the base of the corolla; anthersacs confluent into 1. Ovary superior, usually subglobose, 1-celled; placenta subglobose, central, free; ovules 2 or more, usually very numerous; style short or obsolete; stigma 2-lipped, the anterior lip larger, lamelliform. Fruit a capsule, bursting irregularly, or dehiscent by valves. Seeds variously appendaged or sculptured; embryo in the axis, often imperfectly developed; endosperm none.

About 16 genera and 300 species, of world-wide distribution.

Scapes without bracts or scales, 1-flowered; leaves in a basal rosette.   1. *Pinguicula.*
Scapes with one or more bracts, with or without scales; leaves usually alternate, often dissected
    or root-like and bladder-bearing.
    Bracts at the base of the pedicels without bractlets; calyx not enclosing the fruit.
        Branches verticillate, and verticillately or oppositely decompound; lateral lobes of lower lip
            of corolla saccate.   2. *Vesiculina.*
        Branches alternate or none; lateral lobes of lower lip of corolla not saccate.
            Bracts, and scales if present, flat, basally attached; aquatic.   3. *Utricularia.*
            Bract solitary, tubular, surrounding the scape; scales none.   4. *Lecticula.*
            Bracts and scales peltate; terrestrial.   5. *Setiscapella.*
    Bracts at the base of the pedicels accompanied by a pair of bractlets; calyx enclosing the fruit;
        terrestrial.   6. *Stomoisia.*

1. **PINGUÍCULA** [Tourn.] L. Sp. Pl. 17. 1753.

Acaulescent herbs, with fibrous roots, naked 1-flowered circinate scapes, and leaves in a basal rosette, the upper surface commonly glandular and covered with a viscid secretion. Calyx 5-lobed, more or less 2-lipped, the upper lip 3-lobed, the lower 2-lobed. Corolla 5-lobed, more or less 2-lipped, the upper lip 2-lobed, the lower 3-lobed; base of the corolla saccate and contracted into a nectariferous spur. Capsule 2-valved. Seeds oblong, reticulate. [Latin, *pinguis*, fat, from the apparent greasiness of the leaves of several species.]

About 35 species, of wide distribution in the northern hemisphere, and southward along the Andes to Patagonia. Besides the following, 4 others are found in the southeastern United States. Type species: *Pinguicula vulgaris* L.

Scapes villous, 1′–2′ high; corolla pale violet, less than 6″ long.   1. *P. villosa.*
Scapes glabrous or nearly so, 2′–6′ high; corolla violet-purple, more than 7″ long.   2. *P. vulgaris.*

1. **Pinguicula villòsa** L. Hairy Butterwort.
Fig. 3858.

Pinguicula villosa L. Sp. Pl. 17. 1753.

Pinguicula acutifolia Michx. Fl. Bor. Am. 1: 11. 1803.

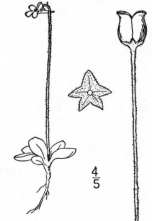

Scapes slender, densely villous, 1′–2′ high, becoming 1½′–5′ high in fruit. Leaves 3–6, the blades oval, obtuse or emarginate, 3″–6″ long, the margins usually inrolled; calyx minute, ¼″–¾″ long, the lobes acute, the 2 lower ones smaller and partly united; corolla pale violet with a yellowish striped throat, 4″–6″ long including the slender obtuse spur, 2-lipped, the upper lip erect, 2-lobed, the lower spreading, 3-lobed; capsule subglobose, about 1″ in diameter.

$\frac{4}{5}$

In bogs, circumpolar; southward in America to Labrador, Hudson Bay, Yukon, and Alaska. June–July.

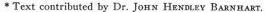

* Text contributed by Dr. John Hendley Barnhart.

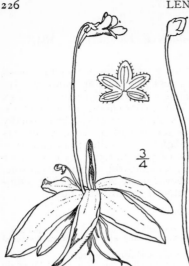

2. Pinguicula vulgàris L.　Common Butterwort.　Bog Violet.　Fig. 3859.

Pinguicula vulgaris L. Sp. Pl. 17.　1753.

Scapes glabrous or nearly so, 1'–6' high, little if at all elongating in fruit. Leaves 3–7, the blades ovate to elliptic, obtuse, ½'–1½' long, the margins usually inrolled; calyx 1½"–2½" long, the lobes obtuse, the 2 lower ones more or less united; corolla violet-purple, 7"–10" long including the subulate acute spur, 2-lipped, the lips equally spreading, the upper 2-lobed, the lower 3-lobed; capsule ovoid, 3"–4" long, 2½"–3" in diameter.

On wet rocks or gravelly places, circumpolar; southward in America to Newfoundland, New Brunswick, Vermont, northern New York, Michigan, Minnesota, Montana, and British Columbia.　June–July.　Beanweed. Yorkshire sanicle. Sheep-root or -rot. Rot-grass. Sheepweed.　Steep or earning-grass, from its use in curdling milk.

Pinguicula alpìna L., reported from Labrador on account of a single specimen said to have been collected there many decades ago, probably does not occur in North America.　It has flowers about the size of those of *P. vulgaris,* but nearly white and with a very much shorter obtuse spur.

2.　VESICULÌNA Raf. Fl. Tellur. 4: 109.　1838.

Aquatic herbs, with horizontal submerged stems, the branches verticillate, and verticillately or oppositely decompound.　Leaves, at least in the adult plant, none.　Bladders terminal on the ultimate branches, the mouth naked or with a single median hairy proboscis.　Inflorescence racemose, 1–4-flowered; scales on the lower portion of the scape none; pedicels from the axils of bracts, without bractlets, erect in fruit.　Calyx 2-lobed, the lobes herbaceous, concave.　Corolla strongly 2-lipped, the upper lip not lobed, the lower lip 3-lobed, the lateral lobes saccate and together constituting a prominent 2-lobed palate, the middle lobe flat, comparatively inconspicuous.　Anthers not lobed.　Capsule many-seeded.　Seeds tuberculate.　[Latin, *vesicula,* a little bladder.]

About 6 species, confined to the New World.　Type species: *Utricularia saccàta* LeConte.

1.　Vesiculina purpùrea (Walt.) Raf.　Purple Bladderwort.　Fig. 3860.

Utricularia purpurea Walt. Fl. Car. 64.　1788.
U. saccata LeConte; Ell. Bot. S. C. & Ga. 1 : 21.　1816.
Vesiculina saccata Raf. Fl. Tellur. 4 : 109.　1838.
Vesiculina purpurea Raf. loc. cit.　1838.

Stems 1°–3° long, the internodes 1'–2' long, the primary branches in whorls of 5–7, verticillately decompound.　Leaves none; bladders 1"–1½" long, without bristles, the exterior surface of the valve with a central tuft of glandular hairs; scape 2'–6' high, 2–4-flowered; bracts membranaceous, basally attached, or more commonly with a free portion below the line of insertion; pedicels 3"–4" long, half longer in fruit; calyx-lobes subequal, 1½" long; corolla red-purple, the upper lip subrhomboid, about 4" long and 6" broad, concave, the lower lip 4"–6" long and broad, with a yellow spot at the base; spur conic, appressed to and shorter than the lower lip; capsule 1½" in diameter; seeds numerous, minute, tuberculate-spiny, especially toward each end.

In ponds, Maine to Florida and Louisiana, near the coast, also Michigan and Indiana to Minnesota. Hooded or horned milfoil.　July–Sept.

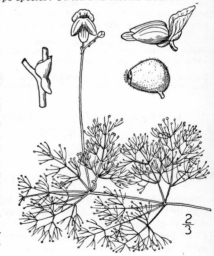

3.　UTRICULÀRIA L. Sp. Pl. 18.　1753.

Aquatic herbs, with horizontal submerged leafy stems.　Leaves alternate, dissected, sometimes root-like, 2–8-parted from the very base, and thus often appearing opposite or verticillate, the segments dichotomously or pinnately dissected, some or all of them bladder-bearing. Bladders with a pair of bristles and more or less other armature about the mouth.　Inflorescence racemose, the raceme sometimes reduced to a single flower; scales on the lower portion of the scape, when present, basally attached, sometimes replaced by a whorl of inflated floats; pedicels from the axils of basally inserted but sometimes auriculate bracts, without bractlets. Calyx 2-lobed, the lobes concave, herbaceous, usually spreading under the mature capsule. Corolla strongly 2-lipped, the palate at the base of the lower lip, prominent, usually 2-lobed. Anthers not lobed. Capsule few–many-seeded. Seeds more or less peltate, flat-topped, the margin variously winged or wingless. [Latin, *utriculus,* a little bag.]

About 75 species, of world-wide distribution. Besides the following, 2 or 3 others occur in the southeastern United States. Type species: *Utricularia vulgaris* L.

Scape without floats.
  Stems creeping on the bottom in shallow water; some or all of the leaves root-like.
    Spur and palate conspicuous; pedicels ascending in fruit.
      Segments of leaves all capillary; lips of corolla nearly equal in length.
        Spur stout, conic, shorter than the lower lip.   1. *U. gibba.*
        Spur slender, equalling or exceeding the lower lip.
          Spur tapering from base to apex; leaves all alike, bladder-bearing; scapes 2'–5' high.   2. *U. pumila.*
          Spur conic at base, linear above; leaves not all bladder-bearing; scapes 4'–16' high.   3. *U. fibrosa.*
      Segments of some leaves linear, flat, bristly-serrulate; upper lip of corolla about half length of lower lip.   4. *U. intermedia.*
    Spur a mere sac; palate obsolete; pedicels recurved in fruit.   5. *U. minor.*
  Stems free-floating, except for a single point of attachment.
    Scapes 2–5-flowered, without scales; cleistogamous flowers also present.   6. *U. geminiscapa.*
    Scapes 6–20-flowered, with 1–5 scales; cleistogamous flowers none.   7. *U. macrorhiza.*
Scape with a whorl of more or less united, conspicuous, inflated floats.
  Scape from floats to lowest pedicel 2' long or less; corolla 6″–9″ long.   8. *U. radiata.*
  Scape from floats to lowest pedicel 3' long or more; corolla 10″–12″ long.   9. *U. inflata.*

1. Utricularia gíbba L. Humped or Swollen-spurred Bladderwort. Fig. 3861.

Utricularia gibba L. Sp. Pl. 18. 1753.
U. fornicata LeConte, Ann. Lyc. N. Y. 1: 76. 1824.

Stems creeping on the bottom in shallow water, radiating from the base of the scape; leaves alternate, mostly once or twice dichotomous, the segments few, capillary, bladder-bearing. Scapes solitary, ¾'–4' high, 1–3-flowered, with 1 scale or none; pedicels 1″–5″ long, erect-ascending; calyx-lobes 1″–1¼″ long; corolla yellow, the upper lip subtriangular, 2″–3″ long, 3″–4″ broad, the lower about 3″ long and broad, the palate prominent, 2-lobed; spur conic, very obtuse, shorter than the lower lip; capsule globose, 1½″–2″ in diameter.

In shallow water, Maine to Michigan, south to Florida and Texas. June–Sept.

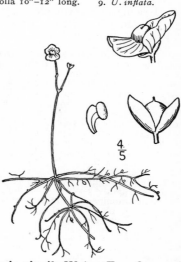

$\frac{4}{5}$

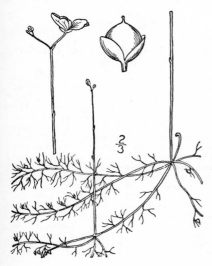

$\frac{2}{3}$

2. Utricularia pùmila Walt. Two-flowered Bladderwort. Fig. 3862.

Utricularia pumila Walt. Fl. Car. 64. 1788.
Utricularia biflora Lam. Ill. 1: 50. 1791.
U. longirostris LeConte; Ell. Bot. S. C. & Ga. 1: 21. 1816.
U. macrorhyncha Barnh. Bull. Torr. Club 25: 515. 1898.

Stems creeping on the bottom in shallow water, radiating from the base of the scape; leaves alternate, dichotomously divided, the segments capillary, bladder-bearing. Scapes solitary or two together, 2'–5' high, 1–4-flowered, with 1–3 scales or none; pedicels 2½″–8″ long, erect-ascending; corolla yellow, the upper lip more or less 3-lobed, 4″–5″ long, 6″–7″ broad, the lower 4″–5″ long and broad, the palate prominent, 2-lobed; spur slenderly conic or subulate, obtuse or emarginate, equalling or usually exceeding the lower lip; capsule globose, 1½″ in diameter.

In shallow water, Massachusetts to Florida and Louisiana, near the coast. July–Aug.

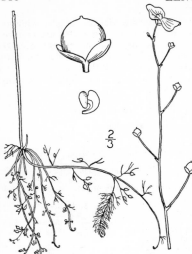

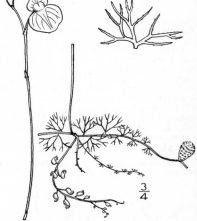

3. Utricularia fibròsa Walt. Fibrous Bladderwort. Fig. 3863.

Utricularia fibrosa Walt. Fl. Car. 64. 1788.
U. striata LeConte; Torr. Cat. Pl. N. Y. 89. 1819.

Stems creeping on the bottom in shallow water, radiating from the base of the scape. Leaves alternate, trichotomous at base, the rays di- or trichotomously divided, the segments capillary, fewer and more or less root-like when bladder-bearing. Scapes solitary or 2 together, 4′–16′ high, 2–6-flowered, with 1 or 2 scales or none; pedicels 2½″–6″ long, one-half longer in fruit, ascending; calyx-lobes 1″–1½″ long, a little longer in fruit; corolla yellow, the upper lip broadly triangular, faintly 3-lobed, 3½″–5″ long, 6″–7″ broad, radiately striate, the lower slightly 3-lobed, 3½″–5″ long and broad, the palate prominent, 2-lobed; spur conic at base, slender above, obtuse or emarginate, appressed to and equalling or exceeding the lower lip; capsules 2″–2½″ in diameter.

In shallow water, Long Island to Florida and Mississippi, near the coast. June–Aug.

4. Utricularia intermèdia Hayne. Flat-leaved Bladderwort. Fig. 3864.

U. intermedia Hayne, in Schrad. Journ. Bot. 1800[1]: 18. 1801.

Stems creeping on the bottom in shallow water, radiating from the base of the scape. Leaves alternate, those on some branches or portions of branches 2½″–7″ long, trichotomous at base, the rays di- or trichotomously divided, with linear, flat, bristly-serrulate segments, and without bladders, those on other branches or portions of branches shorter and root-like, with fewer capillary segments and a few large bladders up to 2½″ long; scapes solitary, 2′–8′ high, 1–4-flowered, with 1 or several auriculate scales; pedicels 4″–10″ long, erect-ascending; calyx-lobes 1″–1½″ long, a little larger in fruit; corolla yellow, the upper lip broadly triangular, 2½″–3″ long, 3½″–4″ broad, the lower slightly 3-lobed, 5″–6″ long and broad, the palate prominent; spur conic at base, cylindric above, acute, about ¾ as long as the lower lip; capsules 1½″ in diameter.

In shallow water, Newfoundland to British Columbia, south to New Jersey, Indiana and California. Also in Europe. May–Aug. Commonly propagated by the velvety winter-buds.

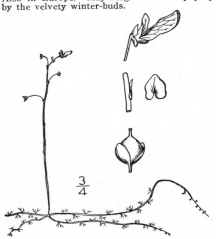

5. Utricularia mìnor L. Lesser Bladderwort. Fig. 3865.

Utricularia minor L. Sp. Pl. 18. 1753.

Stems creeping on the bottom in shallow water, 4′–12′ long, sparingly branched. Leaves alternate, with few divisions, usually only ½″–1½″ long, and bearing 1–5 bladders each; larger bladderless leaves, 1″–3″ long, with flat segments, sometimes occurring on special branches or segments of the main stem; scapes solitary, filiform, 2′–6′ high, 3–6-flowered, with 2–5 minute auriculate scales; pedicels capillary, 1″–4″ long, recurved in fruit; calyx-lobes about ½″ long; corolla pale yellow, the upper lip minute, 1″–2″ long and half as wide, the lower 2″–4″ long, the palate nearly obsolete; spur very short and saccate; capsules about ½″ in diameter.

In shallow water, circumpolar, southward in America to Connecticut, New York, Pennsylvania, Ohio, Indiana, Colorado and California. June–July. Often propagated by winter-buds.

6. Utricularia geminiscàpa Benj. Hidden-fruited Bladderwort. Fig. 3866.

Utricularia geminiscapa Benj. Linnaea **20**: 305. 1847.
U. clandestina Nutt.; A. Gray, Man. 287. 1848.

Stems about 8'–10' long, floating horizontally beneath the surface of the water, sparingly branched. Leaves alternate, about 10'' long, 4–7-dichotomous, and bladderless, or more or less reduced and bladder-bearing; scape slender, 2'–5' high, 2–5-flowered, without scales; cleistogamous flowers solitary on short peduncles, 1 at the base of the scape and others scattered along the stems, often in pairs; pedicels of conspicuous flowers 2''–3'' long; corolla yellow, 3'' long or more, the lower lip longer and broader than the upper, 3-lobed, with a prominent palate; spur a little shorter than the lower lip, obtuse.

In shallow water, New Brunswick to Virginia. Hooded or horned milfoil. July–Aug.

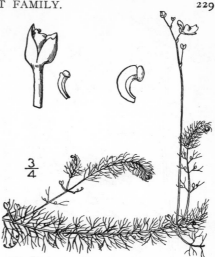

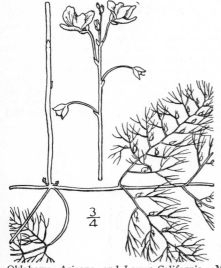

7. Utricularia macrorhìza LeConte. Greater Bladderwort. Hooded Water Milfoil. Pop-weed. Fig. 3867.

Utricularia macrorhiza LeConte, Ann. Lyc. N. Y. **1**: 73. 1824.
U. vulgaris var. *americana* A. Gray, Man. Ed. 5, 318. 1867.

Stems 1°–3° long, floating horizontally beneath the surface of the water, sparingly branched. Leaves alternate, ¾'–2' long, dichotomous at the base, each ray pseudo-pinnately divided, bladder-bearing, the bladders sometimes 2'' long; scape stout, 3'–24' high, 6–20-flowered, with 1–5 auriculate scales; pedicels 3''–8'' long, becoming 5''–10'' long and recurved at maturity of fruit; corolla yellow, 7''–10'' long, the lower lip a little longer and much broader than the upper, with a spreading, undulate, slightly 3-lobed border, and a prominent palate; spur shorter than the lower lip, subulate, upwardly curved, acute or obtuse.

In stagnant water or sluggish streams, Newfoundland to Yukon, south to Maryland, Missouri, Oklahoma, Arizona, and Lower California. May–Aug. Often propagated by winter-buds. Very variable, but appears to differ constantly from the related and equally variable European species, *Utricularia vulgaris* L., by the longer stems, the shape and direction of the spur, and the minuteness of the appendages (rudimentary stolons) at the base of the scape.

8. Utricularia radiàta Small. Small Swollen Bladderwort. Fig. 3868.

Utricularia inflata var. *minor* Chapm. Fl. S. U. S. 282. 1860. Not *U. minor* L. 1753.
Utricularia radiata Small, Fl. SE. U. S. 1090. 1903.

Stems long, floating horizontally beneath the surface of the water. Leaves alternate, 6–10-dichotomous, usually less than 1½' long, bladder-bearing; scape 1–4-flowered, with a whorl of 4–7 inflated floats, above the floats 1½'–4' high; floats up to 1½' long, lobed and finely dissected toward the apex; pedicels 4''–10'' long; corolla 6''–9'' long, the upper lip suborbicular, undulate, the lower lip about the same length but broader, 3-lobed, with a prominent 2-lobed palate; spur conic, obtuse, appressed to and shorter than the lower lip.

In ponds, Maine to Florida and Texas, near the coast. March–Sept.

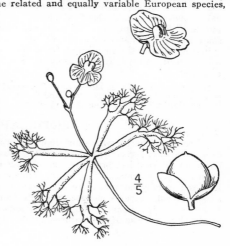

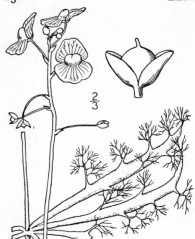

9. Utricularia inflàta Walt. Large Swollen Bladderwort. Fig. 3869.

Utricularia inflata Walt. Fl. Car. 64. 1788.

U. ceratophylla Michx. Fl. Bor. Am. 1: 12. 1803.

Stems long, floating horizontally beneath the surface of the water. Leaves alternate, 10–12-dichotomous, usually more than 2′ long, bladder-bearing; scape 4–12-flowered, with a whorl of 4–9 inflated floats, above the floats 5′–12′ high; floats 2′–3½′ long, lobed and dissected from near the middle; pedicels 8″–16″ long; corolla 10″–12″ long, the upper lip subtriangular, undulate, often emarginate, the lower lip a little longer and much broader, undulate, faintly 3-lobed, with a prominent 2-lobed palate; spur conic, toothed, appressed to and about half the length of the lower lip.

In ponds, Delaware to Florida, near the coast. May.

4. LECTÍCULA Barnhart.

Herbs, with horizontal usually submerged leafy stems. Leaves alternate, 3-parted from the very base, the middle lobe erect and linear, the lateral lobes capillary and root-like, bladder-bearing; bladders slightly beaked, but without bristles. Inflorescence strictly 1-flowered, the pedicel continuous with the scape, its point of origin marked by the solitary bract, appearing like a scale above the middle of the scape; true scales none; bract basally inserted, amplexicaul and tubular, the free margin truncate, more or less deeply 2-notched, without bractlets. Calyx 2-parted, the lobes concave, herbaceous, appressed to the mature capsule. Corolla very strongly 2-lipped, the palate a mere convexity at the base of the lower lip. Anthers not lobed. Capsule many-seeded. [Latin, a couch, from the transverse position of the corolla.]

Two species, the following, and another, in tropical South America. Type species: *Utricularia resupinata* B. D. Greene.

1. Lecticula resupinàta (B. D. Greene) Barnhart. Reclined Bladderwort. Fig. 3870.

Utricularia resupinata B. D. Greene; Bigel. Fl. Bost. Ed. 3, 10. 1840.
Utricularia Greenei Oakes, Hovey's Mag. Hort. 7: 180. 1841.

Scape and pedicel slender, 1′–4′ high, becoming much elongated, 4′–6′ high in fruit, the bract ½″–1″ long. Flower half-reversed so as to rest transversely upon the summit of the pedicel; calyx-lobes subequal, about 1″ long; corolla purple, 4″–6″ long, the upper lip narrowly oblong-spatulate, the lower spreading, entire; spur conic-cylindric, obtuse, the tip distant from the lower lip and bent upward; capsule globose, 1½″–2″ in diameter.

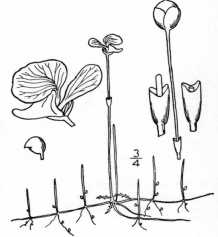

Margins of ponds and lakes, New Brunswick to western Ontario and Pennsylvania, and South Carolina to Florida. Rare and local. July–Aug.

5. SETISCAPÉLLA Barnhart.

Terrestrial herbs, with short root-like branches from the base of the scape. Leaves delicate, some basal, erect, with linear blades, usually evanescent and rarely seen, others root-like, borne on the root-like branches, and bladder-bearing; bladders minute, 2-horned at the apex. Inflorescence racemose, the raceme sometimes reduced to a single flower, when several-flowered usually becoming zig-zag above at maturity; scales on the lower portion of the scape several, scarious, peltate; pedicels from the axils of scarious peltate bracts, without bractlets. Calyx 2-parted, the lobes scarious, strongly longitudinally ribbed, spreading under or clasping the base of the mature capsule. Corolla strongly 2-lipped, the lower lip commonly strongly and divergently 3-lobed; palate usually very prominent and 2-lobed. Anthers not lobed. Capsule many-seeded. Seeds prismatic, reticulate. [Latin, *seta*, a bristle, and *scapus*, scape, from the stiff, bristle-like scapes.]

About 12 species, chiefly American; 1 species in tropical Africa and 2 in Asia. Type species: *Utricularia subulata* L.

Corolla yellow, the lower lip conspicuous, 2″–4″ long; spur conic.   1. *S. subulata.*
Corolla white or purplish, both lips minute, less than ¼″ long; spur saccate.   2. *S. cleistogama.*

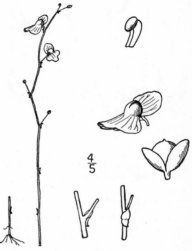

1. Setiscapella subulàta (L.) Barnhart. Zig-zag or Tiny Bladderwort. Fig. 3871.

Utricularia subulata L. Sp. Pl. 18. 1753.
Utricularia setacea Michx. Fl. Bor. Am. 1: 12. 1803.

Scape filiform, stiff, 1½′–9′ high, 1–12-flowered, the pedicels capillary, 1½″–3″ long. Calyx-lobes minute, about ½″ long, becoming 1″ long in fruit; corolla pale yellow, 3″–6″ long, the upper lip ovate, obtuse, the lower 2″–4″ long, strongly and divergently 3-lobed, with a prominent faintly 2-lobed palate; spur flattened-conic, obtuse, but acute in outline when viewed from the side, appressed to and about equalling the lower lip; capsule globose, 1″ in diameter; seeds irregularly ovoid.

In wet sandy soil, Nantucket to Florida, west to Arkansas and Texas. Also in the West Indies. July–Aug.

$\frac{4}{5}$

$\frac{3}{4}$

2. Setiscapella cleistógama (A. Gray) Barnhart. Pin or Closed Bladderwort. Fig. 3872.

U. subulata var. *cleistogama* A. Gray, Syn. Fl. 2¹: 317. 1878.
U. cleistogama Britton, Trans. N. Y. Acad. Sci. 9: 12. 1889.

Scape filiform, stiff, ½′–2½′ high, 1–3-flowered, the pedicels 1″–3″ long. Calyx-lobes minute, the upper faintly 7-nerved, the lower strongly 5-nerved; corolla dirty-white or purplish, ½″ in diameter or less, cleistogamous (?), consisting chiefly of the large saccate spur, the lips minute, obtuse, the lower faintly 3-lobed; capsule globose, ½″ in diameter; seeds very minute and numerous, irregularly ovoid, $\frac{1}{10}$″ long.

In wet soil, eastern Massachusetts, Long Island and New Jersey. Rare and local. July–Aug.

6. STOMOISIA Raf. Fl. Tellur. 4: 108. 1838.

Terrestrial scapose herbs, with tufts of root-like bladder-bearing branches arising from near the base of the scape, and few extremely delicate leafy branches radiating horizontally from the same region, so inconspicuous and so readily detached from the scape that they are rarely seen. Leaves delicate, some linear, erect, and grass-like, others colorless and root-like, bladder-bearing. Bladders minute and rudimentary, beaked but without bristles. Inflorescence racemose or subspicate, sometimes reduced to a single flower; scales on the lower portion of the scape several, basally inserted; pedicels from the axils of basally inserted bracts, with an inner pair of opposite lateral bractlets. Calyx 2-parted, the lobes thin, veiny, appressed to and exceeding the mature capsule. Corolla 2-lipped, the lips nearly distinct, the upper narrow, erect, with a distinct claw, the lower with a prominent galeate palate, the edge of the aperture of the spur usually ciliate. Anthers vertically 2-lobed. Capsule many-seeded. Seeds very minute, subglobose or prismatic, reticulate or areolate. [Greek, hairy mouth, from the ciliate aperture of the spur in the type species.]

About 50 species, of wide geographic distribution. Type species: *Utricularia cornuta* Michx.
Corolla much exceeding the calyx.
    Lower lip of corolla 6″–8″ long; spur 3½″–6″ long.   1. *S. cornuta.*
    Lower lip of corolla 4″–5″ long; spur 2½″–4″ long.   2. *S. juncea.*
Corolla shorter than or about equalling the calyx.   3. *S. virgatula.*

1. Stomoisia cornùta (Michx.) Raf. Horned Bladderwort. Fig. 3873.

Utricularia cornuta Michx. Fl. Bor. Am. 1: 12. 1803.
Stomoisia cornuta Raf. Fl. Tellur. 4: 108. 1838.

Scape strict, brownish, $1\frac{1}{2}'$–12' high, 1–5-flowered, the flowers usually approximate near the summit, the scales several or numerous, rarely exceeding 1″ in length, acute. Bracts $\frac{1}{2}$″–1″ long, acute; bractlets the same length, but narrower; pedicels rarely exceeding the bracts; calyx yellowish, the upper lobe acuminate, 2″–3″ long, the lower acute, often much shorter; corolla yellow, the lower lip 6″–8″ long, with a spreading margin around the galeate palate; spur subulate, pendent, $3\frac{1}{2}$″–6″ long; capsule subglobose, $1\frac{1}{2}$″–2″ in diameter, closely invested by the calyx-lobes and slightly beaked by the upper one.

On wet borders of ponds, or in bogs, Newfoundland to Minnesota, south to Florida and Texas. Also in the Bahamas and Cuba. July–Aug.

2. Stomoisia júncea (Vahl.) Barnhart. Rush Bladderwort. Fig. 3874.

Utricularia juncea Vahl, Enum. 1: 202. 1804.
Utricularia personata LeConte; Ell. Bot. S. C. & Ga. 1: 23. 1816.

Scape strict, brownish, 4′–16′ high, 1–12-flowered, the flowers usually scattered along the upper part, the scales several or numerous, $\frac{1}{2}$″–$\frac{3}{4}$″ long, acute. Bracts $\frac{3}{4}$″–1″ long, acute; bractlets the same length, but narrower; pedicels rarely exceeding the bracts; calyx yellowish, the upper lobe acuminate, 2″–$2\frac{1}{2}$″ long, the lower acute, much shorter; corolla yellow, the lower lip 4″–5″ long, with little or no spreading margin; spur subulate, pendent, $2\frac{1}{2}$″–4″ long; capsule subglobose, 1″–$1\frac{1}{2}$″ in diameter, closely invested by the calyx-lobes and slightly beaked by the upper one.

On wet borders of ponds, and in bogs, New York to Florida to Mississippi, chiefly near the coast. Also in the West Indies and South America. July–Aug.

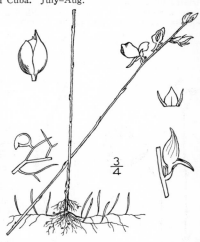

2. Stomoisia virgátula Barnhart. Fairy-wand Bladderwort. Fig. 3875.

Utricularia simplex C. Wright, in Sauvalle, Anal. Acad. Habana 6: 315. 1870. Not R. Br. 1812.
U. virgatula Barnh. Bull. Torr. Club 34: 580. 1908.

Scape wiry, strict, brownish, 1′–8′ high, 1–6-flowered, the scales several, minute, acute. Bracts less than $\frac{1}{2}$″ long, acute; bracts the same length, but narrower; pedicels shorter than the bracts; calyx purplish, the upper lobe acuminate, 2″ long, the lower acute, much shorter; corolla yellow, the upper lip barely if at all exceeding the upper calyx-lobe, the lower of about the same length, with a minute tuft of hairs in the throat; spur conic, pendent, acute, 1″–$1\frac{1}{2}$″ long; capsule globose, 1″ in diameter or less, closely invested by the calyx-lobes and beaked by the upper one.

On wet borders of ponds, New York, New Jersey, Florida and Mississippi. Also in Cuba. Rare and local. Sept.

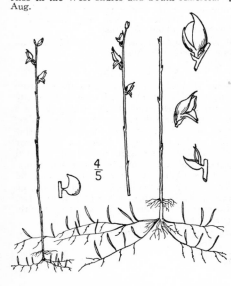

Family 30. **OROBANCHÀCEAE** Lindl. Nat. Syst. Ed. 2, 287. 1830.

BROOM-RAPE FAMILY.

Erect, simple or branched, brown yellowish purplish or nearly white root-parasites, the leaves reduced to alternate appressed scales, the flowers perfect, irregular (rarely cleistogamous), sessile in terminal bracted spikes, or solitary and peduncled in the axils of the scales. Calyx inferior, gamosepalous, 4–5-toothed, 4–5-cleft, or split nearly or quite to the base on one or both sides. Corolla gamopetalous, more or less oblique, the tube cylindric, or expanded above, the limb 2-lipped, 5-lobed. Stamens 4, didynamous, inserted on the tube of the corolla and alternate with its lobes, a fifth rudimentary one occasionally present; filaments slender; anthers 2-celled, the sacs parallel, equal. Ovary superior, 1-celled, the four placentae parietal; ovules numerous, antropous; style slender; stigma discoid, 2-lobed, or sometimes 4-lobed. Capsule 1-celled, 2-valved. Seeds numerous, reticulated, wrinkled or striate; embryo minute; cotyledons scarcely differentiated.

About 11 genera and over 200 species, of wide geographic distribution, mostly in the northern hemisphere.

Flowers all complete and perfect.
  Calyx 2–5-lobed or toothed.
    Calyx about equally 5-cleft; no bractlets on peduncles nor calyx.   1. *Thalesia.*
    Calyx unequally toothed, or split on both sides; flowers bracteolate.   2. *Orobanche.*
  Calyx spathe-like, split on the lower side, 3–4-toothed on the upper.   3. *Conopholis.*
Lower flowers cleistogamous, fertile; upper complete, mostly sterile.   4. *Leptamnium.*

1. **THALÈSIA** Raf. Am. Month. Mag. **2**: 267. 1818.

[ANOPLANTHUS Endl. Icon. Gen. Pl. 12. *pl. 72.* 1838.]

[APHYLLON A. Gray, Man. 290. 1848.]

Glandular or viscid-pubescent simple-stemmed herbs, parasitic on the roots of various plants, with scattered scales, and long-peduncled yellowish white or violet, complete and perfect flowers without bractlets. Calyx campanulate or hemispheric, nearly equally 5-cleft, the lobes acute or acuminate. Corolla oblique, the tube elongated, curved, the limb slightly 2-lipped, the upper lip erect-spreading, 2-lobed, the lower spreading, 3-lobed, the lobes all nearly equal. Stamens included; anther-sacs mucronate at the base. Ovary ovoid; placentae equidistant, or contiguous in pairs; style slender, deciduous; stigma peltate, or transversely 2-lamellate. [Dedicated to Thales.]

About 5 species, natives of North America. Type species: *Orobanche uniflora* L.

Stem very short; peduncles 1–4, erect; calyx-lobes lanceolate, acuminate.   1. *T. uniflora.*
Stem erect, 2′–5′ high; peduncles several; calyx-lobes broad, acute.   2. *T. fasciculata.*

1. **Thalesia uniflòra** (L.) Britton. Pale or Naked Broom-rape. One-flowered Broom-rape. Squaw-drops. Cancer-root. Fig. 3876.

Orobanche uniflora L. Sp. Pl. 633. 1753.
Anoplanthus uniflorus Endl. Icon. Gen. Pl. 12. *pl. 72.* 1838.
Aphyllon uniflorum T. & G.; A. Gray, Man. 290. 1848.
T. uniflora Britton, Mem. Torr. Club **5**: 298. 1894.

Stem usually less than 1′ long, nearly subterranean, bearing several ovate-oblong scales and 1–4 slender erect scape-like glandular-puberulent naked 1-flowered peduncles 3′–8′ high. Calyx campanulate, pubescent, glandular, 4″–5″ high, less than one-half the length of the corolla, its lobes as long as the tube or longer, lanceolate, acuminate; corolla white or violet, puberulent without, 8″–12″ long, the curved tube about 3 times as long as the limb, the short lobes oval or obovate, obtuse; placentae nearly equidistant; capsule ovoid, longer than the calyx.

In woods and thickets, parasitic on the roots of various herbs, Newfoundland to Ontario, South Carolina and Texas. Far western plants, formerly referred to this species, prove to be distinct. Pipes. April–June.

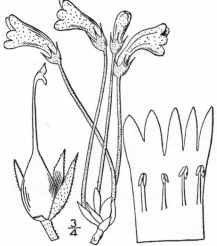

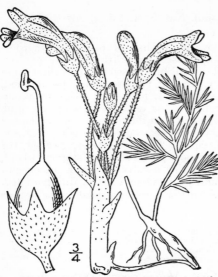

2. Thalesia fasciculàta (Nutt.) Britton.
Clustered or Yellow Cancer-root.
Fig. 3877.

Orobanche fasciculata Nutt. Gen. **2**: 59. 1818.
Anoplanthus fasciculatus Walp. Rep. **3**: 480. 1844–45.
Aphyllon fasciculatum A. Gray, Syn. Fl. **2**: Part 1, 312. 1878.
Thalesia fasciculata Britton, Mem. Torr. Club **5**: 298. 1894.

Stem erect, 2′–4′ high, densely glandular-pubescent, bearing several scales and 3–15 naked 1-flowered peduncles 1′–4′ long. Calyx glandular, broadly campanulate, 3″–5″ high, about one-third the length of the corolla, its lobes triangular-lanceolate or triangular-ovate, acute, equalling or shorter than the tube; corolla nearly 1′ long, purplish to yellow, puberulent without, the curved tube 3 times as long as the limb, the lobes oblong, obtuse, the limb more manifestly 2-lipped than in the preceding species; capsule ovoid to globose.

In sandy soil, parasitic on the roots of various plants, mostly composites, northern Indiana to Minnesota, Yukon and British Columbia, Nebraska, Arizona and California. April–Aug.

Thalesia lutea (Parry) Rydb. [*Thalesia fasciculata lutea* (Parry) Britton] is a race with light yellow flowers, growing on grasses in western Nebraska and Wyoming.

2. OROBÁNCHE [Tourn.] L. Sp. Pl. 632. 1753.

Glandular-pubescent, erect simple or branched, reddish yellowish violet or nearly white herbs, parasitic on the roots of various plants, with scattered scales, and spicate or racemose, complete and perfect, bracted and sometimes bracteolate flowers. Calyx split both above and below, nearly or quite to the base, the divisions 2-cleft or rarely entire, or more or less unequally 2–5-toothed. Corolla oblique, strongly 2-lipped; upper lip erect, emarginate or 2-lobed; lower lip spreading, 3-lobed. Stamens included; anther-sacs mostly mucronate at the base. Placentae equidistant, or approximate in pairs. Style slender, commonly persistent until after the dehiscence of the capsule; stigma peltate to funnelform, entire, or laterally 2-lamellate. [Greek, Choke-vetch.]

About 90 species, natives of the Old World and western America. Besides the following some 6 others occur in the western parts of North America. Type species: *Orobanche major* L.

Calyx 4-toothed; stems mostly branched. 1. *O. ramosa.*
Calyx split on both sides; stem simple. 2. *O. minor.*
Calyx 5-cleft; stem simple. 3. *O. ludoviciana.*

1. Orobanche ramòsa L. Hemp or Branched Broom-rape. Fig. 3878.

Orobanche ramosa L. Sp. Pl. 633. 1753.

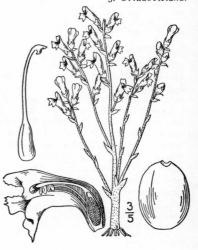

Plant yellowish; stem rather slender, branched, or rarely simple, 3′–15′ high, the scales few and distant, 2″–5″ long. Spike loosely many-flowered, denser above than below, the lowest flowers short-pedicelled; bracts usually 3, the longest about equalling the calyx; calyx 4-toothed, the teeth triangular-ovate, acute, or acuminate, about as long as the tube; corolla 5″–9″ long, the tube yellow, slightly constricted above the ovary, the limb bluish.

Parasitic on the roots of tomato, hemp and tobacco, New Jersey, Illinois, Kentucky. Adventive or naturalized from Europe. Summer. Strangle-tare.

Orobanche purpùrea Jacq., another European species, with violet flowers and a 5-toothed calyx, is recorded as found on *Achillea* in lawns at Wingham, Ontario.

2. Orobanche mìnor J. E. Smith. Lesser or Clover Broom-rape. Herb-bane. Fig. 3879.

O. minor J. E. Smith, Engl. Bot. *pl. 422.* 1797.

Plant yellowish-brown; stem rather stout, simple, 4′–20′ high; lower scales numerous, ovate-oblong, the upper lanceolate, acute, scattered, 3″–10″ long. Spike dense, or the lower flowers separated, 3′–8′ long; bracts 1 or 2, lanceolate, equalling or longer than the flowers; flowers 5″–9″ long; calyx split both above and below, each of the lateral segments 2-cleft, the teeth lanceolate-subulate; corolla-tube yellowish, scarcely constricted above the ovary, the limb bluish.

Parasitic on the roots of clover, New Jersey to Virginia. Naturalized from Europe. Called also devil's-root and hell-root. Strangle-tare. May–July.

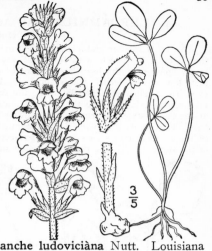

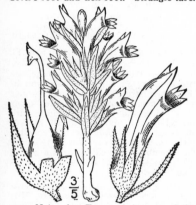

3. Orobanche ludoviciàna Nutt. Louisiana Broom-rape. Fig. 3880.

Orobanche ludoviciana Nutt. Gen. **2**: 58. 1818.
Aphyllon ludovicianum A. Gray, Bot. Cal. **1**: 585. 1876.
Myzorrhiza ludoviciana Rydb.; Small, Fl. SE. U. S. 1093. 1903.

Stems stout, simple, solitary or clustered, viscid-puberulent, 4′–12′ high, scaly. Flowers 6″–8″ long, very numerous in dense terminal spikes, 1–2-bracteolate under the calyx; calyx 5-cleft, the lobes somewhat unequal, linear-lanceolate, acuminate, longer than the corolla-tube, or shorter; corolla 2-lipped, purplish, its tube narrow, about twice as long as the limb, the teeth of its lips acute; anthers woolly; capsule ovoid-oblong, shorter than the calyx.

In sandy soil, Illinois to South Dakota, Saskatchewan, Nebraska, Texas, Arizona and California. Strangle tare. June–Aug.

3. CONÓPHOLIS Wallr. Orobanch. 78. 1825.

An erect stout simple glabrous, densely scaly, light brown herb, parasitic on the roots of trees, with yellowish flowers 2-bracteolate under the calyx, in a thick dense bracted spike, the bracts similar to the scales of the stem. Calyx oblique, deeply split on the lower side, 3–4-toothed on the upper. Corolla strongly 2-lipped, the tube slightly curved, the upper lip concave, nearly erect, emarginate, the lower spreading, 3-lobed. Stamens exserted; anthersacs bristly pubescent. Placentae about equidistant; stigma capitate, obscurely 2-lamellate. Capsule ovoid-globose. [Greek, signifying a scaly cone.]

Three known species, the following typical one of eastern North America, the other southwestern and Mexican.

1. Conopholis americàna (L. f.) Wallr. Squaw-root. Fig. 3881.

Orobanche americana L. f. Suppl. 88. 1767.
Conopholis americana Wallr. Orobanch. 78. 1825.

Plants 3′–10′ high from a thickened base, light brown, usually clustered, covered all over with stiff imbricated scales. Upper scales lanceolate or ovate, acute, 6″–10″ long, the lowest much shorter; flowers about ½′ long, exceedingly numerous in the dense spike which is 6″–10″ thick; corolla pale yellow, somewhat exceeding the calyx; anthers sagittate; capsule ovoid-globose, 4″–5″ high.

In rich woods at bases of trees, Maine to Ontario, Michigan, Florida, Alabama and Tennessee. Cancer-root. Earth-club. Clap-wort. April–Aug.

4. LEPTÁMNIUM Raf. Am. Month. Mag. 2: 267. Feb. 1818.

[EPIFAGUS Nutt. Gen. 2: 60. 1818.]

An erect slender glabrous, purplish or yellowish, rather stiff branching herb, parasitic on the roots of the beech, with few small scattered scales, and sessile dimorphous flowers, distantly spicate on the branches, the lower cleistogamous, abundantly fertile, the upper complete but mostly sterile. Calyx short, nearly equally 5-toothed. Corolla of the upper flowers cylindric, slightly flattened laterally, the tube much longer than the 4-lobed limb, the upper lobe concave, larger than the 3 lower ones; stamens not exserted; anther-sacs mucronulate at the base; style filiform, 2-lobed; ovary with an adnate gland on the upper side near the base. Corolla of the lower flowers minute, not unfolding, borne like a hood on the summit of the ovoid ovary; style very short; placentae contiguous in pairs; capsule at length 2-valved at the summit. [Greek, referring to the small calyx.]

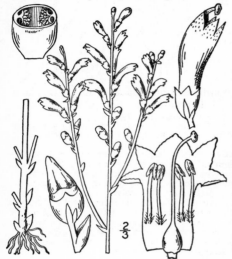

A monotypic genus of eastern North America.

1. Leptamnium virginiànum (L.) Raf.
Beech-drops. Fig. 3882.

Orobanche virginiana L. Sp. Pl. 633. 1753.
Epifagus americana Nutt. Gen. 2: 60. 1818.
Epiphegus virginiana Bart. Comp. Fl. Phil. 2: 50. 1818.
Leptamnium virginianum Raf.; A. Gray, Syn. Fl. 2: Part 1, 314. 1878. As synonym.

Plant 6′–2° high from a thick scaly base, the roots brittle, fibrous. Stem paniculately branched, the branches straight, ascending; scales very few and small; corolla of the upper flowers 4″–5″ long and 1″ thick, striped with purple and white, the limb not at all spreading; stamens about as long as the corolla; style slightly exserted; lower flowers 1½″ long, the corolla resembling the calyptra of a moss; capsule 3″ high, somewhat oblique and compressed, many-seeded.

In beech woods, Nova Scotia and New Brunswick to Florida, west to Ontario, Michigan, Missouri and Louisiana. Cancer-root or -drops. Virginia brown-rape. Aug.–Oct.

Family 31. BIGNONIÁCEAE Pers. Syn. 2: 168. 1807.
TRUMPET-CREEPER FAMILY.

Trees, shrubs or woody vines, a few exotic species herbs, with opposite (rarely alternate) compound or simple leaves, and mostly large and showy, clustered terminal or axillary, more or less irregular flowers. Calyx inferior, gamosepalous. Corolla gamopetalous, funnelform, campanulate, or tubular, 5-lobed, somewhat 2-lipped, at least in the bud. Anther-bearing stamens 2 or 4, inserted on the tube of the corolla and alternate with its lobes; anthers 2-celled, the sacs longitudinally dehiscent. Disk annular or cup-like. Ovary sessile or stalked, mostly 2-celled; placentae parietal, or on the partition-wall of the ovary; ovules very numerous, horizontal, anatropous; style slender; stigma terminal, 2-lobed. Capsule 2-valved, loculicidally, septifragally or septicidally dehiscent. Seeds flat, transverse, winged in our genera; endosperm none; cotyledons broad and flat, emarginate or 2-lobed; radicle short, straight.

About 60 genera and over 500 species of wide geographic distribution in tropical regions, a few in the north and south temperate zones.

Leaves compound; anther-bearing stamens 4; our species vines.
  Calyx-limb undulate; capsule flattened parallel with its partition-wall.   1. *Anisostichus.*
  Calyx 5-toothed; capsule compressed at right angles to its partition-wall.   2. *Bignonia.*
Leaves simple; anther-bearing stamens mostly 2; trees.   3. *Catalpa.*

1. ANISÓSTICHUS Bureau, Mon. Bignon. 43. 1864.

Woody vines, with opposite 2-foliolate leaves, the terminal leaflet reduced to a tendril, and large flowers in axillary cymes. Calyx campanulate, the limb merely undulate, truncate or slightly 5-toothed. Tube of the corolla much expanded above the calyx, the limb somewhat 2-lipped, 5-lobed, the lobes rounded. Anther-bearing stamens 4, didynamous, included, inserted near the base of the corolla; anther-sacs glabrous, divergent. Capsule linear, flat-tened parallel with the thin partition, septifragally dehiscent, the margins of the valves more or less thickened. Seeds in 2 unequal rows on both margins of the partition, winged, much broader than high, the wing entire, or erose at the end. [Greek, unequal-ranked.]

An apparently monotypic genus. Type species: *Bignonia capreolata* L.

1. Anisostichus capreolàta (L.) Bureau.
Tendrilled Trumpet-flower.   Cross-vine.   Fig. 3883.

? *Bignonia crucigera* L. Sp. Pl. 624. 1753.
Bignonia capreolata L. loc. cit. 1753.
Doxantha capreolata Miers. Proc. Roy. Hort. Soc. **3**: 190. 1863.
Anisostichus capreolata Bureau, Mon. Bigon. Atlas 8, pt. 6. 1864.

A glabrous woody vine, often climbing to the height of 40°–60°, the stems sometimes 4' in diameter, exhibiting a conspicuous cross in the transverse section. Leaves petioled, commonly with small, simple, stipule-like ones in their axils, 2-foliolate, terminated by a branched tendril; leaflets stalked, oblong or ovate, entire, acute or acuminate at the apex, cordate at the base, pinnately veined, 3'–7' long; cymes numerous, short-peduncled, 2–5-flowered; pedicels 1'–2' long; calyx membranous; corolla 2' long, orange and puberulent without, yellow within; capsule 5'–7' long, 8"–10" broad, very flat, each valve longitudinally 1-nerved; seeds broadly winged laterally, narrowly winged above and below, 1½' broad.

In moist woods, Virginia to Florida, Louisiana, Ohio and southern Illinois. Quarter vine. April–June.

2. BIGNONIA L. Sp. Pl. 633. 1753.

Climbing woody vines with aerial rootlets, with opposite pinnately compound leaves, and large showy red or orange flowers, in terminal corymbs. Calyx tubular-campanulate, somewhat unequally 5-toothed. Corolla-tube elongated, enlarged above the calyx, narrowly campanulate, the limb slightly 2-lipped, 5-lobed, the lobes spreading. Anther-bearing stamens 4, didynamous, ascending under the upper lip; anther-sacs divergent, glabrous or slightly pubescent. Capsule elongated, slightly compressed at right angles to the partition, loculicidally and septicidally dehiscent. Seeds in several rows on each side of the margins of the partition, flat, winged, the wing translucent. [Named after the Abbe Bignon, 1662–1743, librarian to to Louis XV.]

Two known species, the following typical one, the other Japanese. The name *Tecoma*, used for this vine in our first edition, belongs properly to a genus of pinnate-leaved yellow-flowered shrubs of tropical America, typified by *Tecoma stans* (L.) H.B.K.

1. Bignonia radìcans L. Trumpet-flower.
Trumpet-creeper. Foxglove. Fig. 3884.

Bignonia radicans L. Sp. Pl. 624. 1753.
Tecoma radicans DC. Prodr. **9**: 223. 1845.
Campsis radicans Seem. Journ. Bot. **5**: 362. 1867.

A woody vine, climbing to the height of 20°–40° or prostrate if meeting no support. Leaves petioled, odd-pinnate, not tendril-bearing, 8'–15' long; leaflets 7–11, ovate to lanceolate, short-stalked, sharply serrate, reticulate-veined, glabrous, or pubescent on the veins beneath, acute or acuminate at the apex, narrowed at the base, 1½'–3' long; flowers corymbose, short-pedicelled; corymbs 2–9-flowered; calyx coriaceous, about 1' long; corolla scarlet, 2¼' long, the tube veined within, 3 times as long as the limb, anther-sacs glabrous; stigma spatulate; capsule stalked, 4'–6' long, 10" in diameter, narrowed at both ends, little flattened, ridged above and below by the margins of the valves; seeds in several rows on each surface of the broad partition, broadly winged laterally, the wing eroded.

In moist woods and thickets, southern New Jersey and Pennsylvania to Florida and Texas, north to Illinois and Iowa. Escaped from cultivation further north. Trumpet-vine. Trumpet-ash. Cow-itch. Cross-vine. Aug.–Sept.

3. CATÁLPA Scop. Introd. 170. 1771.

Trees, or some exotic species shrubs, with opposite or rarely verticillate simple petioled leaves, and large showy white or mottled flowers in terminal panicles or corymbs. Calyx closed in the bud, splitting irregularly or into 2 lips in opening. Corolla-tube campanulate or obconic, oblique, expanded above, 2-lipped, 5-lobed, the lobes all spreading, their margins crisped. Anther-bearing stamens 2, ascending under the upper lip of the corolla; anther-sacs glabrous, linear or oblong, divergent; sterile stamens (staminodia) 3, short (or occasionally 4 perfect didynamous stamens and 1 staminodium). Disk obsolete. Ovary sessile, 2-celled; ovules in 2–several rows on the sides of the partition. Capsule elongated-linear, terete, loculicidally dehiscent. Seeds flat, the large lateral wings dissected into capillary processes. [The American Indian name of the first species below.]

About 5 species, the following in eastern North America, 2 in eastern Asia. Type species: *Bignonia Catalpa* L. West Indian trees referred to this genus prove to be distinct.

Corolla thickly spotted within, 1'–1½' long, the lobes crimped.   1. *C. Catalpa.*
Corolla little spotted, but purple-lined, 2' long, the lobes nearly flat.   2. *C. speciosa.*

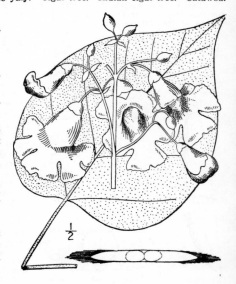

1. Catalpa Catálpa (L.) Karst. Catalpa. Indian or Smoking Bean. Candle-tree. Bean-tree. Fig. 3885.

Bignonia Catalpa L. Sp. Pl. 622. 1753.
Catalpa bignonioides Walt. Fl. Car. 64. 1788.
Catalpa Catalpa Karst. Deutsch. Fl. 927. 1880–83.

A tree, with thin flaky bark, reaching a maximum height of about 60° and a trunk diameter of 4°, the branches spreading. Leaves strong-scented, broadly ovate, entire, or 3-lobed, acute or acuminate at the apex, densely pubescent beneath, becoming glabrous above, obtuse at the base, 6'–12' long, the lobes, when present, acuminate; petioles stout, nearly as long as the blade; flowers white, numerous, mottled with yellow and purple within, 1'–1½' long, in large terminal erect panicles; capsules 6'–18' long, 4"–7" thick, thin-walled, drooping, the partition narrow.

In woods in the Gulf States. Escaped from cultivation northward as far as Pennsylvania and southern New York. Wood brown, soft, weak, durable in contact with the soil. Weight per cubic foot 28 lbs. June–July. Cigar-tree. Indian cigar-tree. Catawba.

2. Catalpa speciòsa Warder. Catawba Tree. Larger Indian Bean. Western Catalpa. Fig. 3886.

Catalpa cordifolia Duham. Nouveau 2: *pl. 5.* 1802. Not Moench, 1794.

Catalpa speciosa Warder; Engelm. Coult. Bot. Gaz. 5: 1. 1880.

A tree, with thick rough bark, reaching a maximum height of 120° and a diameter trunk of 4½°, similar to the preceding species. Leaves not unpleasantly scented, broadly ovate, commonly entire, long-acuminate at the apex; panicles few-flowered; corolla faintly mottled within; capsule thick-walled, 8'–20' long, nearly 10" in diameter.

In woods, southern Indiana to Tennessee, west to Missouri and Arkansas. Wood brown, soft, weak, durable. Weight per cubic foot 26 lbs. May–June. Cigar-tree. Hardy catalpa. Shawnee-wood.

Family 32. **MARTYNIÀCEAE** Link, Handb. 1 : 504. 1829.
UNICORN-PLANT FAMILY.

Herbs, with opposite leaves, or the upper sometimes alternate, and perfect irregular flowers, racemose in our species.   Calyx inferior, 4–5-cleft or 4–5-parted or sometimes split to the base on the lower side.   Corolla gamopetalous, irregular, the tube oblique, often decurved, the limb slightly 2-lipped, 5-lobed, the lobes nearly equal, the 2 upper ones exterior in the bud.   Anther-bearing stamens 4, didynamous, or the posterior pair sterile; anthers 2-celled, the sacs longitudinally dehiscent.   Ovary 1-celled, with 2 parietal placentae expanded into broad surfaces, or 2–4-celled by the intrusion of the placentae or by false partitions; ovules numerous or few in each cavity of the ovary, anatropous; style slender; stigma 2-lobed or 2-lamellate.   Fruit various in the different genera.   Seeds oblong or orbicular, compressed, wingless or narrowly winged; endosperm none; embryo large; cotyledons fleshy, flat; radicle short and straight.

Three genera and about 12 species, mainly tropical.

1. **MARTÝNIA** L. Sp. Pl. 618.   1753.

Coarse diffusely branched glandular-pubescent and viscid strong-scented herbs, with opposite or alternate long-petioled leaves, and large violet purple whitish or mottled flowers in short terminal racemes.   Calyx 1–2-bracteolate at the base, campanulate, inflated, unequally 5-cleft, deciduous.   Corolla funnelform-campanulate, oblique, decurved, the 5 lobes nearly equal, spreading.   Stamens 4 in our species; anthers gland-tipped, their sacs divergent.   Ovary 1-celled, the 2 parietal placentae intruded and expanded in the center of the cavity into broad surfaces bearing the ovules in 1 or 2 rows.   Fruit an incurved beaked loculicidally 2-valved capsule, the exocarp somewhat fleshy, the endocarp fibrous, woody, crested below or also above, 4-celled by the extension of the placentae.   Seeds numerous, tuberculate.   [Named for John Martyn, 1693–1768, professor of botany at Cambridge, England.]

About 8 species, natives of America. Besides the following, 2 others occur in the southwestern United States. Type species: *Martynia annua* L.

1. Martynia louisiàna Mill. Unicorn-Plant. Elephant's Trunk.   Martinoe.   Double-claw.   Fig. 3887.

M. louisiana Mill. Gard. Dict. Ed. 8, no. 3.   1768.
Martynia proboscidea Glox. Obs. 14.   1785.

Annual, densely glandular-pubescent all over; stem stout, much branched, the branches prostrate or ascending, 1°–3° long. Leaves broadly ovate to orbicular, rounded at the apex, cordate at the base, repand, undulate or entire, 3′–12′ in diameter, the petiole stout, mostly longer than the blade; bractlets at the base of the calyx oblong or linear, deciduous; calyx somewhat cleft on the lower side; racemes several-flowered; pedicels slender; corolla whitish or yellowish, mottled with purple or yellow within, 1½′–2′ long, the limb nearly as broad as long, the lobes obtuse; stamens all anther-bearing; fruit strongly curved, 4′–6′ long when mature, the beak longer than the body, splitting into 2 elastically diverging segments, the endocarp crested on the under side only.

In waste places, escaped from gardens, Maine to western New York, New Jersey and Georgia. Native from Indiana to Iowa, Utah, Texas and New Mexico. July–Sept.

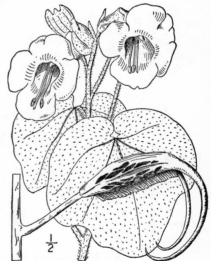

Family 33. **ACANTHÀCEAE** J. St. Hil. Expos. Fam. 1 : 236. 1805.
ACANTHUS FAMILY.

Herbs, or some tropical genera shrubs or small trees, with opposite simple exstipulate leaves, and irregular or nearly regular perfect flowers.   Calyx inferior, persistent, 4–5-parted or 4–5-cleft, the sepals or segments imbricated, equal or unequal.   Corolla gamopetalous, nearly regularly 5-lobed with the lobes convolute in the bud, or conspicuously 2-lipped.   Anther-bearing stamens 4, didynamous, or 2 only; anthers mostly 2-celled, the sacs longitudinally dehiscent. Disk annular or cup-like. Ovary 2-celled; ovules 2–10 in each cavity, anatropous or amphitropous; style filiform, simple; stigmas 1 or 2. Capsule dry, 2-celled, loculicidally elastically 2-valved. Seeds globose or orbicular, not winged, borne on curved projections (retinacula) from the placentae, the testa close, mostly roughened, often developing spiral threads and mucilage when wetted. Endosperm in the following genera none; cotyledons flat, commonly cordate.

About 175 genera and 2000 species, natives of temperate and tropical regions of the Old World and the New.

Corolla convolute in the bud, nearly regular; stamens 4.
  Ovules 2 in each cavity; capsule 2–4-seeded.   1. *Dyschoriste.*
  Ovules 3–10 in each cavity; capsule 6–20-seeded.   2. *Ruellia.*
Corolla imbricated in the bud, strongly 2-lipped; stamens 2.
  Lower lip of the corolla 3-cleft; flowers bracted, not involucrate.   3. *Dianthera.*
  Lower lip of the corolla entire or 3-toothed; flowers involucrate.   4. *Diapedium.*

1. DYSCHORÍSTE Nees, in Wall. Pl. As. Rar. **3**: 81. 1832.

[CALÓPHANES Don; Sweet, Brit. Fl. Gard. (II). *pl. 181.* 1833.]

  Erect or procumbent perennial herbs or shrubs, with entire leaves (smaller ones sometibes fascicled in their axils), and blue or purple, rather large, bracted flowers, axillary, solitary or clustered. Calyx deeply 5-cleft, the lobes setaceous. Corolla funnelform, the tube slightly curved, enlarged above, or cylindraceous, the limb spreading, 5-lobed, somewhat 2-lipped, the lobes rounded, sinistrorsely convolute in the bud. Stamens 4, didynamous, included, all anther-bearing in our species; anther-sacs mucronate at the base. Ovules 2 in each cell of the ovary; summit of the style recurved; stigma simple, or of 2 unequal lobes. Capsule oblong, linear, narrowed at the base, 2–4-seeded, in some species coherent. Seeds flat, orbicular, attached by their edges to the retinacula. [Greek, referring to the coherent capsule-valves.]

  About 30 species, of wide distribution in warm and tropical regions. Besides the following, 4 others occur in the southern and southwestern United States. Type species: *Dyschoriste depressa* Nees.

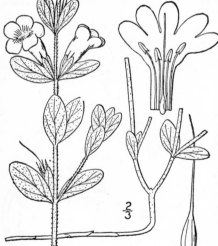

1. Dyschoriste oblongifòlia (Michx.) Kuntze. Dyschoriste. Fig. 3888.

Ruellia biflora L. Sp. Pl. 633. 1753?
R. oblongifolia Michx. Fl. Bor. Am. **2**: 23. 1803.
Calophanes oblongifolia Don; Sweet, Brit. Fl. Gard. (II.) *pl. 181.* 1833.
Dipteracanthus biflorus Nees, Linnaea **16**: 294. 1842.
D. oblongifolia Kuntze, Rev. Gen. Pl. 486. 1891.

  Rootstocks horizontal, slender; stems slender, erect, simple, or branched below, pubescent or puberulent, obtusely 4-angled, 6′–15′ high, rather stiff. Leaves ascending or erect, oblong or oval, rounded at the apex, somewhat narrowed at the base, very short-petioled, or sessile, pubescent or glabrate, 8″–15″ long; flowers commonly solitary in the axils; calyx-segments filiform, hirsute, exceeding the oblong, obtuse bractlets; corolla blue or mottled with purple, 8″–12″ long, slightly 2-lipped, its tube enlarged above; capsule oblong, about one-half the length of the calyx; anther-sacs mucronate-aristate at the base.

  In sandy pine barrens, Virginia to Florida, mainly near the coast. June–Sept.

2. RUÉLLIA [Plumier] L. Sp. Pl. 634. 1753.

  Perennial herbs, or shrubs, mostly pubescent, with entire or rarely dentate leaves, and large violet blue white or yellow flowers, solitary or clustered in the axils, or cymose in terminal panicles, or sometimes cleistogamous and inconspicuous. Calyx 5-cleft, or 5-parted, the segments narrow. Corolla funnelform or salverform, the tube usually narrow, slightly enlarged above, the limb spreading, 5-lobed, the lobes obtuse, mostly nearly equal, sinistrorsely convolute in the bud. Stamens 4, included or exserted; anther-sacs not mucronate at the base. Ovules 3–10 in each cavity of the ovary; apex of the style recurved; stigma simple, or of 2 unequal lobes. Capsule oblong or club-shaped, compressed or terete, 6–20-seeded. Seeds compressed, ovate or orbicular, attached by their edges to the retinacula. [Named for I. Ruel or de la Ruelle, 1474–1537, an early French herbalist.]

About 200 species, mainly of tropical America, a few in Africa, Asia and Australia. Besides the following, some 6 others occur in the southern and southwestern United States. Type species: *Ruellia tuberosa* L.

Flowers sessile or nearly so.
  Calyx-segments linear-lanceolate, scarcely exceeding the capsule.   1. *R. strepens.*
  Calyx-segments filiform-linear, exceeding the capsule.
    Leaves sessile or very short-petioled.   2. *R. ciliosa.*
    Leaves slender-petioled.   3. *R. parviflora.*
Flowers peduncled; peduncle with two large bracts at the summit.   4. *R. pedunculata.*

1. Ruellia strèpens L. Smooth or Short-tube Ruellia. Fig. 3889.

Ruellia strepens L. Sp. Pl. 634. 1753.

Dipteracanthus micranthus Engelm. & Gray, Bost. Journ. Nat. Hist. **5**: 257. 1845.

Ruellia strepens var. *cleistantha* A. Gray, Syn. Fl. **2**: Part 1, 327. 1878.

Glabrate or somewhat pubescent; stem erect, slender, simple, or branched, 4-sided, 1°–4° high, the branches ascending. Leaves oblong, oval, or ovate, petioled, acute or subacute at the apex, narrowed at the base, 3′–6′ long; petioles 2″–10″ long; flowers solitary or several together in the axils, some often cleistogamous; calyx-segments linear-lanceolate, shorter than the corolla-tube, slightly pubescent or ciliate, 8″–12″ long, ½″–1″ wide; corolla blue, 1½–2′ long, the limb nearly as broad, the tube about as long as the throat and limb; capsule club-shaped, longer than or equalling the calyx.

In dry woods, Pennsylvania to Wisconsin, Kansas, Florida and Texas. May–July.

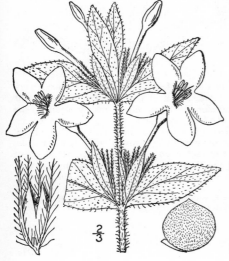

2. Ruellia ciliòsa Pursh. Hairy or Long-tube Ruellia. Fig. 3890.

Ruellia ciliosa Pursh, Fl. Am. Sept. 420. 1814.

Hirsute or pubescent, at least above; stem ascending, rather stout, 1°–2½° high. Leaves hairy, ciliate, oblong, oval, or ovate, sessile or nearly so, obtuse or subacute at the apex, narrowed at the base, 1½′–3′ long; flowers clustered or solitary in the axils, sometimes cleistogamous; calyx-segments filiform, hirsute, 10″–12″ long, about ¼″ wide; corolla blue, 1½′–2′ long, the tube equalling or longer than the obconic throat and nearly regular limb and about twice as long as the calyx; capsule shorter than the calyx.

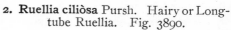

In dry soil, southern New Jersey and Pennsylvania to Florida, west to Michigan, Kansas and Louisiana. June–Sept.

$\frac{2}{3}$

3. Ruellia parviflòra (Nees) Britton.
Slender Hairy Ruellia.   Fig. 3891.

Dipteracanthus ciliosus parviflorus Nees, Linnaea
   16: 294. 1842.
R. cilosa ambigua A. Gray, Syn. Fl. **2**[1]: 326. 1878.
Ruellia ciliosa parviflora Britton, in Britton &
   Brown, Ill. Fl. Ed. 1, **3**: 203. 1898.
R. parviflora Britton, Man. 854. 1901.

Sparingly finely pubescent, but green; stem
erect, 6'–18' tall, simple or sometimes sparingly branched. Leaves minutely pubescent,
ciliolate, oblong to oblong-lanceolate, or sometimes oval, 1¼'–4' long, obtuse or acutish, undulate, rather acuminately narrowed at the
base, manifestly petioled; flowers solitary or
clustered in the axils, the pedicels very short;
calyx-segments linear-filiform, bristly, 8''–12''
long; corolla blue, 1½'–2' long, the tube somewhat longer than the throat and somewhat exceeding the calyx, the limb ¾'–1¼' broad; capsule oblong above the stipe-like base, barely
8'' long.

In sandy soil, Maryland to Indiana, Florida and
Texas. May–Aug.

4. Ruellia pedunculàta Torr.   Stalked
Ruellia.   Fig. 3892.

Ruellia pedunculata Torr.; A. Gray, Syn. Fl.
   2: Part 1, 326. 1878.

Finely pubescent; stem erect, 1°–2½° tall,
the branches spreading. Leaves ovate to
oblong-lanceolate, acute or acuminate at
the apex, narrowed at the base, short-
petioled, the larger 2'–3' long; peduncles
slender, spreading, 1'–3' long, with 2 leaf-
like bracts at the summit which subtend a
solitary flower, or 2 or 3 slender-pedicelled
ones with pedicels similarly bracted; calyx-
segments awn-like, equalling the narrow
corolla-tube, or shorter; corolla funnelform, 1½'–2' long; capsule about 10'' long,
puberulent, longer than the calyx.

In dry soil, Illinois and Missouri to Arkansas and Louisiana. June–Sept.

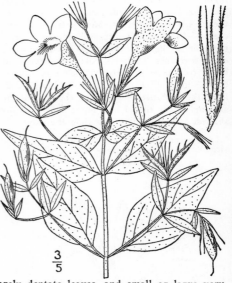

$\frac{3}{5}$

3.   DIANTHÈRA L. Sp. Pl. 27.
1753.

Herbs mostly perennial, with entire or rarely dentate leaves, and small or large very
irregular flowers, variously clustered or solitary in the axils. Calyx deeply 4–5-parted, the
segments narrow. Corolla-tube slender, short or elongated, curved or nearly straight, the
limb conspicuously 2-lipped; upper lip interior in the bud, erect or ascending, concave, entire,
or 2-dentate; lower lip spreading, 3-cleft. Stamens 2, inserted on the throat of the corolla,
not exceeding the upper lip; anther-sacs ovate or oblong, slightly divergent, not mucronate,
separated by a rathed broad connective. Ovules 2 in each cavity of the ovary; style slender;
stigma entire, or 2-lobed. Capsule contracted at the base into a long stipe, about 4-seeded.
Seeds flat, orbicular or ovate, the placentae not separating from the walls of the capsule.
[Greek, double anthers.]

About 100 species, natives of tropical America, a few in tropical Asia and Africa. Besides the
following, 4 others occur in the southern and southwestern United States. Type species: *Dianthera
americana* L.

Flowers capitate, the heads dense, at length oblong.   1. *D. americana.*
Flowers in loose spikes.
   Flowers in pairs in the spikes.   2. *D. ovata.*
   Flowers scattered singly along the spikes.   3. *D. lanceolata.*

1. Dianthera americàna L. Dense-flowered Water Willow. Fig. 3893.

Dianthera americana L. Sp. Pl. 27. 1753.

Perennial, glabrous; stem erect, grooved and angled, slender, usually simple, 1°–3° high, or sometimes 6° long when growing in water. Leaves lanceolate or linear-lanceolate, gradually acuminate, 3′–6′ long, 3″–8″ wide, entire, narrowed at the base into short petioles, or sessile; flowers violet, or nearly white, capitate-spicate at the ends of slender axillary peduncles which are shorter than or equal to the leaves; bractlets linear-subulate, shorter than the flowers; corolla 5″–6″ long, its tube shorter than the lips, the base of the lower lip rough and palate-like; capsule 6″ long, exceeding the calyx, its stipe about the length of the slightly compressed body.

In water and wet places, Quebec to Ontario, Michigan, Georgia and Texas. May–Aug.

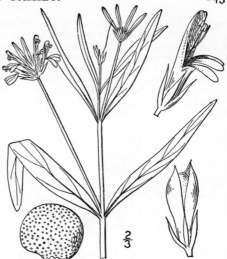

2. Dianthera ovàta Walt. Loose-flowered Water Willow. Fig. 3894.

Dianthera ovata Walt. Fl. Car. 63. 1788.

Perennial, glabrous; stem ascending or erect from a horizontal base, slender, 6′–20′ high, simple, or sparingly branched. Leaves short-petioled, or sessile, ovate, oblong or oval, 1′–3′ long, 8″–18″ wide; flowers opposite in loose slender-peduncled axillary spikes, which become 1′–3′ long; peduncles shorter than or but little exceeding the leaves; calyx-segments narrowly linear, much longer than the bracts and bractlets; corolla pale purple, 4″–5″ long; capsule about 6″ long.

In wet soil, especially along streams, southern Virginia to Florida. Records from Arkansas and Texas apply to the following species. June–Aug.

3. Dianthera lanceolàta (Chapm.) Small. Narrow-leaved Water Willow. Fig. 3895.

D. ovata lanceolata Chapm. Fl. S. States 304. 1860.
D. lanceolata Small; Britton, Man. 855. 1901.

Perennial, puberulent; stem erect or spreading, more or less branched, 4′–12′ long. Leaves linear to linear-elliptic or elliptic-lanceolate, 1¼′–4′ long, more or less acuminate, undulate, sessile or nearly so; flowers in interrupted slender spikes 1¼′–4′ long; calyx-segments narrowly linear, 2½″–3½″ long; corolla whitish or pale-purple, about 5″ long, the lips nearly as long as the tube, which is saccate near the base, the upper lip truncate or retuse, the lower one 3-lobed, the middle lobe truncate or retuse, the lateral ones obtuse; capsule 7″–8″ long, the body as long as the stipe-like base or shorter.

In swamps and low grounds, Missouri to Tennessee, Florida and Texas. June–Sept.

4. DIAPEDIUM Konig; Konig & Sims, Ann. Bot. 2: 189. 1806.

[DICLIPTERA Juss. Ann. Mus. Paris, 9: 267. 1807.]

Erect or diffuse branched pubescent or glabrous herbs, with entire petioled leaves, and blue red or violet flowers, subtended by involucres of 2–4 distinct or connate bracts, the inflorescence mostly cymose or spicate, the involucres subtending 1 flower or several. Calyx 4–5-cleft, the lobes linear or subulate. Corolla-tube slender, slightly enlarged above, the limb conspicuously 2-lipped; upper lip erect, concave, interior in the bud, entire or 2–3-toothed; lower lip spreading, entire or 3-toothed. Stamens 2; anther-sacs parallel, sometimes unequal, separated by a narrow connective. Style filiform; ovules 2 in each cavity of the ovary. Capsule flattened, ovate or suborbicular, sessile or stipitate, 2–4-seeded. Placentae separating elastically from the walls of the capsule. Seeds compressed, nearly orbicular.

About 60 species, natives of warm and tropical regions. Besides the following, 4 others occur in the southern and southwestern United States. Type species: *Justicia chinensis* L.

3/5

1. Diapedium brachiàtum (Pursh) Kuntze. Diapedium. Fig. 3896.

Justicia brachiata Pursh, Fl. Am. Sept. 13. 1814.
Dicliptera brachiata Spreng. Syst. 1: 86. 1825.
D. brachiatum Kuntze, Rev. Gen. Pl. 485. 1891.

Annual (?), glabrate, or pubescent; stem slender, 6-grooved, erect, much branched, 1°–2° high. Leaves ovate, membranous, long-petioled, acuminate or acute at the apex, narrowed or rounded at the base, 3′–6′ long, 1′–3′ wide; inflorescence paniculate, the involucres 1–4-flowered, each of 2 oblong or obovate, obtuse or mucronate, opposite bracts; corolla 8″–12″ long, pink or purple, the lips about as long as the slender tube; upper lip 2–3-toothed, the lower entire; capsule oblong, 2″–3″ high, a little longer than the involucre, the valves slightly divergent in dehiscence, the placentae remaining attached to their summits.

In moist thickets, North Carolina to Florida, Missouri, Kansas and Texas. July–Oct.

Family 34. PHRỲMACEAE Schauer in DC. Prodr. 11: 520. 1847.

LOPSEED FAMILY.

An erect perennial herb with divaricate branches, opposite membranous simple leaves, and small irregular purplish flowers, distant in slender elongated spikes. Calyx cylindric, 2-lipped; upper lip 2-cleft, the teeth setaceous; lower lip much shorter, 3-toothed, the teeth subulate. Corolla-tube cylindric, the limb 2-lipped; upper lip erect, concave, emarginate; lower lip larger, spreading, convex, 3-lobed, the lobes obtuse. Stamens 4, didynamous, included. Ovary oblique, 1-celled; ovule 1, orthotropous, ascending; style slender; stigma 2-lobed. Calyx reflexed in fruit, enclosing the dry achene, becoming prominently ribbed, closed and its teeth hooked at the ends. Cotyledons convolute; radicle superior.

Consists of the following monotypic genus of eastern North America, eastern and central Asia. [Name unexplained.]

1. PHRỲMA L. Sp. Pl. 601. 1753.

1. Phryma Leptostàchya L. Lopseed. Fig. 3897.

Phyrma Leptostachya L. Sp. Pl. 601. 1753.

Puberulent; stem 1½°–3° high, somewhat 4-sided, sometimes constricted above the nodes, branched above, the branches slender, elongated, divergent. Leaves ovate, acute or acuminate at the apex, obtuse of narrowed at the base, very thin, coarsely dentate, 2′–6′ long, the lower petioled, the upper often nearly sessile; spikes very narrow, 3′–6′ long; flowers about 3″ long, mostly opposite, distant, borne on very short minutely 2-bracteolate pedicels, at first erect, soon spreading, the calyx, after flowering, abruptly reflexed against the axis of the spike.

In woods and thickets, New Brunswick to Manitoba, Florida and Kansas. Bermuda. Eastern Asia. June–Aug.

2/3

Family 35. **PLANTAGINÀCEAE** Lindl. Nat. Syst. Ed. 2, 267. 1836.

PLANTAIN FAMILY.

Annual or perennial, mostly acaulescent or short-stemmed, rarely stoloniferous herbs, with basal, or, in the caulescent species, opposite or alternate leaves, and small perfect polygamous or monoecious flowers, bracteolate in dense terminal long-scaped spikes or heads, or rarely solitary. Calyx 4-parted, inferior, persistent, the segments imbricated. Corolla hypogynous, scarious or membranous, mostly marcescent, 4-lobed. Stamens 4 or 2 (only 1 in an Andean genus), inserted on the tube or throat of the corolla; filaments filiform, exserted or included; anthers versatile, 2-celled, the sacs longitudinally dehiscent. Ovary sessile, superior, 1–2-celled, or falsely 3–4-celled. Style filiform, simple, mostly longitudinally stigmatic. Ovules 1–several in each cavity of the ovary, peltate, amphitropous. Fruit a pyxis, circumscissile at or below the middle, or an indehiscent nutlet. Seeds 1–several in each cavity of the fruit; endosperm fleshy; cotyledons narrow.

Three genera and over 225 species, of wide geographic distribution.

Flowers in terminal spikes or heads; fruit a pyxis.   1. *Plantago.*
Flowers monoecious, the staminate solitary, peduncled, the pistillate sessile among the linear leaves;
    fruit indehiscent.   2. *Littorella.*

1. **PLANTÀGO** [Tourn.] L. Sp. Pl. 112. 1753.

Acaulescent or leafy-stemmed herbs, the scapes arising from the axils of the basal or alternate leaves, bearing terminal spikes or heads of small greenish or purplish flowers (flowers solitary in a few exotic species). Calyx-segments equal, or two of them larger. Corolla salverform, the tube cylindric, or constricted at the throat, the limb spreading in anthesis. Fruit a pyxis, mostly 2-celled. Seeds various. [The Latin name.]

Over 200 species, of wide distribution. Besides the following, some 6 others occur in western North America. Known as Plantain, Ribwort or Roadweed. Type species: *Plantago major* L.

    * **Plants acaulescent; flowers spicate or capitate at the ends of scapes.**
Corolla-lobes spreading or reflexed in fruit, not closed over the top of the pyxis.
    Leaves ovate, lanceolate or oblong.
        Seeds several or many in each pyxis.
            Pyxis ovoid, circumscissile at about the middle.   1. *P. major.*
            Pyxis oblong, circumscissile much below the middle.   2. *P. Rugelii.*
        Seeds 2–4 in each pyxis.
            Leaves all narrowed at the base, parallel-ribbed.
                Seeds excavated on the inner side.   3. *P. lanceolata.*
                Seeds flat or but slightly concave on the inner side.
                    Spike very dense; leaves pubescent.   4. *P. media.*
                    Lower flowers scattered; leaves glabrous or very nearly so.   5. *P. eriopoda.*
            Leaves, or some of them, cordate; veins starting from the midrib.   6. *P. cordata.*
    Leaves linear or filiform.
        Leaves fleshy; plant maritime.   7. *P. maritima.*
        Leaves not fleshy; plants not maritime.
            Spike densely tomentose; bracts usually not longer than flowers.   8. *P. Purshii.*
            Plant green and glabrate; bracts much longer than the flowers.   9. *P. aristata.*
Corolla-lobes erect and closed over the top of the pyxis.
    Leaves spatulate to obovate; stamens 4.   10. *P. virginica.*
    Leaves linear-filiform; stamens 2.
        Capsule about 4-seeded, slightly exceeding the calyx.   11. *P. pusilla.*
        Capsule 7–30-seeded, twice as long as the calyx.   12. *P. heterophylla.*
    ** **Stem erect, leafy; flowers capitate at ends of axillary peduncles.**   13. *P. arenaria.*

1. **Plantago màjor** L. Common or Greater Plantain. Dooryard Plantain. Fig. 3898.

Plantago major L. Sp. Pl. 112. 1753.

Perennial, glabrous or somewhat pubescent; rootstock short, thick, erect. Leaves long-petioled, rather firm in texture, mostly ovate, obtuse or acutish, entire, or coarsely dentate, 1'–10' long, 3–11-ribbed; scapes 2'–3° high; spike linear-cylindric, usually very dense, commonly blunt, 2'–10' long, 3''–4'' thick; flowers perfect, proterogynous; sepals broadly ovate to obovate, scarious on the margins, one-half to two-thirds as long as the ovoid obtuse or subacute, 5–16-seeded pyxis, which is circumscissile at about the middle; stamens 4.

In waste places, nearly throughout North America. In part naturalized from Europe, but indigenous in the North and on salt meadows. Also in the West Indies. Small leaves are occasionally borne near the bases of the spikes; spikes rarely branched. May–Sept. Way-side or round-leaf plantain. Broadleaf. Hen-plant. Lamb's-foot. Way-bread. Healing-blade.

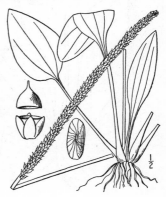

Plantago halóphila Bicknell, of saline situations along the Atlantic coast, is densely pubescent, but otherwise like this species. A similar race occurs about lakes in northern New York.

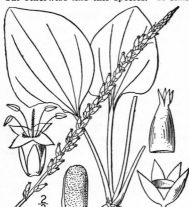

Plantago asiática L. is a boreal race with thinner nearly erect leaves.

2. Plantago Rugèlii Dcne. Rugel's or Pale Plantain. Fig. 3899.

Plantago Rugelii in DC. Prodr. **13**: Part 1, 700. 1852.

Similar to the preceding species, the spikes less dense, at least toward the base, usually long-attenuate at the summit, sometimes 5″ thick. Sepals oblong, prominently keeled on the back, the margins green or scarious; pyxis oblong-cylindric, 2″-3″ long, twice as long as the sepals, circumscissile much below the middle and entirely within the calyx, 4-10-seeded; stamens 4.

In fields, woods and waste places, New Brunswick to Ontario, South Dakota, Florida, Kansas and Texas. Petioles commonly purple at the base. Usually brighter green and with thinner leaves than *P. major*. This species, or the preceding one, was known to the Indians as "White-man's-foot." Silk-plant. June–Sept.

3. Plantago lanceolàta L. Ribwort. Ribgrass. English, Buck, or Buckthorn Plantain. Snake, Lance-leaved, or Ripple Plantain. Fig. 3900.

Plantago lanceolata L. Sp. Pl. 113. 1753.

Perennial or biennial, more or less pubescent; rootstock mostly short, with tufts of brown hairs at the bases of the leaves. Leaves narrowly oblong-lanceolate, mostly erect, shorter than the scapes, entire, acute or acuminate at the apex, gradually narrowed into petioles, 3–5-ribbed, 2′–12′ long, 3″–12″ wide; scapes slender, channelled, sometimes 2½° tall; spikes very dense, at first short and ovoid, becoming cylindric, blunt and ½′–4′ long in fruit, 4″–6″ thick; flowers perfect, proterogynous; sepals ovate, with a narrow green midrib and broad scarious margins, the two lower ones commonly united; corolla glabrous, its tube very short; filaments white; pyxis oblong, very obtuse, 2-seeded, slightly longer than the calyx; seeds deeply excavated on the face.

In fields and waste places, New Brunswick to Northwest Territory, British Columbia, Florida and Kansas. Bermuda. Naturalized from Europe; native also of Asia. Sepals rarely metamorphosed into small leaves. April–Nov. Black-jacks. Jack-straws. Dog's-ribs. Cocks. Kemps. Leechwort. Ram's-tongue. Rattail. Windles. Long-plantain. Ripplegrass. Kempseed. Headsman. Hen-plant. Clock. Chimney-sweeps. Cats'-cradles.

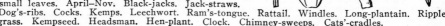

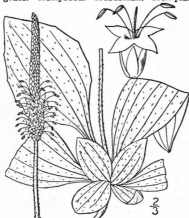

4. Plantago mèdia L. Hoary Plantain. Lamb's-tongue. Healing herb. Fig. 3901.

Plantago media L. Sp. Pl. 113. 1753.

Perennial, intermediate in aspect between *P. major* and *P. lanceolata,* the short rootstock clothed with brown hairs among the bases of the leaves. Leaves spreading, ovate, broadly oblong or elliptic, obtuse or acute at the apex, entire, or repand-dentate, densely and finely canescent, 5-7-ribbed, narrowed at the base into margined, usually short petioles; scapes slender, much longer than the leaves, 1°-2° tall; spikes very dense, cylindric and 1′-3′ long in fruit, about 3″ thick; flowers perfect, white; sepals all distinct, oblong, with a narrow green midrib and broad scarious margins; corolla glabrous; stamens pink or purple; pyxis oblong, obtuse, about as long as the calyx, 2-4-seeded, the seeds merely concave, stamens 4.

In waste places, Maine, Rhode Island, Ontario and New York. Adventive from Europe. Native also of Asia. May–Sept. Fire-leaves. Fire-weed. Lamb's-lettuce.

5. Plantago eriópoda Torr. Saline Plantain. Fig. 3902.

Plantago glabra Nutt. Gen. **1**: 100. 1818?
Plantago eriopoda Torr. Ann. Lyc. N. Y. **2**: 237. 1827.

Perennial, succulent; rootstock long, usually densely covered with long brown hairs among the bases of the leaves. Leaves oblong, oblong-lanceolate, or ob-lanceolate, entire, or repand-dentate, acute at the apex, narrowed into petioles, 5-9-ribbed, 3'-12' long, ½'-1½' wide, glabrous or very nearly so; scapes stout, more or less pubescent, longer than the leaves, 6'-18' high; spikes 1'-5' long, dense above, the lower flowers scattered; flowers perfect; sepals distinct, oblong-obovate with a narrow green midrib and broad scari-ous margins; corolla glabrous, its lobes spreading or reflexed; pyxis ovoid-oblong, very obtuse, one-third longer than the calyx, 2-4-seeded, circumscissile be-low the middle, seeds nearly flat.

In maritime or saline soil, Nova Scotia and Quebec to Minnesota, Athabasca, Colorado and California. June–Sept.

Plantago sparsiflòra Michx., of the Southeastern States, admitted into our first edition as found in southern Illinois, is not known to range north of North Carolina.

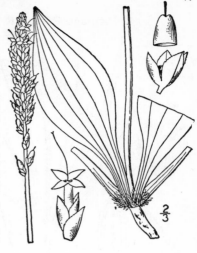

6. Plantago cordàta Lam. Heart-leaved Plantain. Water Plantain. Fig. 3903.

Plantago cordata Lam. Tabl. Encycl. **1**: 338. 1791.

Perennial, glabrous, purple-green; rootstock short, very stout. Leaves broadly ovate or nearly orbicu-lar, pinnately veined, obtuse or acute at the apex, entire or dentate, rounded, abruptly narrowed, or cordate at the base, often 10' long; petioles stout, margined above; scapes stout, longer than the leaves; spikes loosely flowered, sometimes 1° long, with interrupted clusters; flowers perfect; sepals ovate to obovate, obtuse, green; corolla-lobes spread-ing; pyxis ovoid-globose, obtuse, circumscissile at or slightly below the middle, 1-4-seeded; seeds not excavated on the face.

In swamps and along streams, Ontario to New York, Alabama, Minnesota, Missouri and Louisiana. March–July.

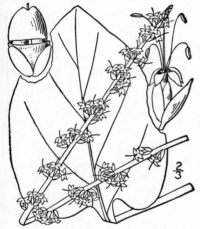

7. Plantago marítima L. Sea or Seaside Plantain. Fig. 3904.

Plantago maritima L. Sp. Pl. 114. 1753.
Plantago decipiens Barneoud, Mon. Plantag. 16. 1845.
P. borealis Lange, Fl. Dan. *pl. 2707.*

Annual, biennial, or perennial, fleshy; rootstock stout or slender, sometimes with tufts of whitish hairs among the bases of the leaves. Leaves linear, glabrous, very obscurely nerved, sessile, or narrowed into short margined petioles, 2'-10' long, entire, or with a few small teeth, 1"-2½" wide; scapes slender, more or less pubescent, longer than or equalling the leaves; spikes dense, linear-cylindric, blunt, 1'-5' long; flowers perfect; sepals ovate-lanceolate to nearly orbicular, green, somewhat keeled; corolla pubescent without, its lobes spreading; pyxis ovoid-oblong, obtuse, 2-4-seeded, circumscissile at about the middle, nearly twice as long as the calyx; seeds nearly flat on the face.

In salt marshes and on sea-shores, Greenland and Labrador to New Jersey, and on the Pacific Coast from Alaska to California. Also on the coasts of Europe and Asia. Consists of several slightly differing races. Buckshorn. Gibbals. Sea-kemps. June–Sept.

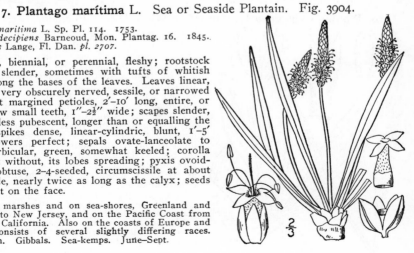

8. Plantago Púrshii R. & S. Pursh's Plantain. Fig. 3905.

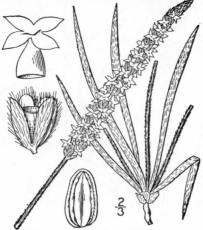

Plantago Purshii R. & S. Syst. 3: 120. 1818.
Plantago gnaphalioides Nutt. Gen. 1: 100. 1818.
Plantago patagonica var. gnaphalioides A. Gray, Man. Ed. 2, 269. 1856.

Annual, woolly or silky all over, pale green; scapes slender, 2′-15′ tall, longer than the leaves. Leaves ascending, linear, acute or acuminate at the apex, narrowed into margined petioles, 1-3-nerved, 1½″-4″ wide, entire, or very rarely with a few small teeth; spikes very dense, cylindric, obtuse, 1′-5′ long, about 3″ in diameter, exceedingly woolly; bracts rigid, equalling or slightly exceeding the flowers; flowers perfect but heterogonous, many of them cleistogamous; sepals oblong, obtuse, scarious-margined; corolla-lobes broadly ovate, spreading; stamens 4; pyxis oblong, obtuse, 1¼″ long, little exceeding the calyx, 2-seeded, circumscissile at about the middle; seeds convex on the back, deeply concave on the face.

On dry plains and prairies, Indiana to western Ontario, British Columbia, Texas and northern Mexico. Locally adventive eastward. May-Aug. Salt-and-pepper-plant.

9. Plantago aristàta Michx. Large-bracted Plantain. Fig. 3906.

Plantago aristata Michx. Fl. Bor. Am. 1: 95. 1803.
Plantago patagonica var. aristata A. Gray, Man. Ed. 2, 269. 1856.

Annual, dark green, villous, or glabrate; scapes stout, erect, 6′-18′ tall, exceeding the leaves. Leaves linear, acuminate at the apex, entire, narrowed into slender petioles, sometimes prominently 3-ribbed 1½″-4″ wide; spikes very dense, cylindric, 1′-6′ long, pubescent but not woolly; bracts puberulent, linear, elongated, ascending, the lower often 10 times as long as the flowers; flowers very similar to those of the preceding species; pyxis 2-seeded; the seeds concave on the face.

On dry plains and prairies, Illinois to South Dakota, Louisiana and Texas, west to British Columbia and New Mexico. Also widely adventive as a weed in the Eastern States from Maine to Georgia, its eastern natural limits now difficult to determine. May-Oct.

Plantago spinulòsa Dcne., differing in its pale green color and mostly shorter involucral bracts, enters the western part of our area in Nebraska.

10. Plantago virgínica L. Dwarf or White Dwarf Plantain. Fig. 3907.

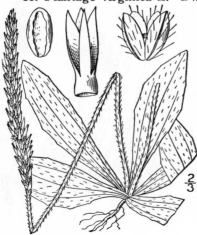

Plantago virginica L. Sp. Pl. 113. 1753.

Annual or biennial, pubescent or villous; scapes erect, slender, 1′-18′ high, much longer than the leaves. Leaves spatulate or obovate, obtuse or acutish, thin, entire, or repand-denticulate, narrowed into margined petioles, or almost sessile, varying greatly in size, 3-5-nerved, ascending or spreading; spikes very dense, or the lower flowers scattered, linear-cylindric, obtuse, 3″-4″ thick, usually 1′-4′ long, but in dwarf forms reduced to 2-6 flowers; flowers imperfectly dioecious; corolla-lobes of the fertile plants erect and connivent on the top of the pyxis, those of the sterile widely spreading; stamens 4; pyxis oblong, about as long as the calyx, appearing beaked by the connivent corolla-lobes, 2-4-seeded, the seeds yellow to brown.

In dry soil, Rhode Island to Florida, Illinois, Michigan, Missouri, Arizona and northern Mexico. Bermuda. March-July.

Plantago rhodospèrma Dcne., of the Southwest, with larger flowers and larger reddish seeds, is recorded as adventive in Missouri.

11. **Plantago pusílla** Nutt.   Slender Plantain.
Fig. 3908.

Plantago pusilla Nutt. Gen. 1 : 100. 1818.

Annual, puberulent; scapes filiform, 2'–7' high, longer than the linear-filiform, mostly entire, blunt-pointed obscurely 1-nerved leaves. Leaves about ½'' wide; spikes slender, linear, rather loosely flowered, ½'–3' long, 1½''–2'' thick; flowers imperfectly dioecious or polygamous; sepals oblong, obtuse, about as long as the bract, scarious-margined; corolla-lobes of the more fertile plants becoming erect over the pyxis; stamens 2; pyxis ovoid-oblong, obtuse, one-fourth to one-third longer than the calyx, about 4-seeded, cir-cumscissile at about the middle; seeds nearly flat on both sides.

In dry sandy soil, Massachusetts to Georgia, Illinois, Kansas and Texas. April–Aug.

Plantago elongàta Pursh, to which this was referred in our first edition, differs by larger seeds and saccate bracts, and enters our western limits in Nebraska.

12. **Plantago heterophýlla** Nutt.   Many-seeded Plantain. Fig. 3909.

Plantago heterophylla Nutt. Trans. Am. Phil. Soc. (II.) 5 : 177. 1833–37.

Annual, similar to the preceding species, but gla-brous or slightly puberulent; scapes ascending or spreading, equalling or exceeding the leaves, 2'–10' long. Leaves narrowly linear or filiform, the larger about 2'' wide, entire or often with several distant small teeth or linear lobes; spikes loose, linear, ½'–5' long; sepals oblong, obtuse, scarious-margined, mostly shorter than the bract, corolla-lobes in the more fer-tile plants becoming erect over the pyxis; stamens 2; pyxis oblong, subacute, about twice as long as the calyx, 7–30-seeded, circumscissile rather below the middle; seeds somewhat angled, scarcely concave on the face.

In moist soil, New Jersey to Florida, Illinois, Arkan-sas, Texas and apparently introduced in California. April–July.

13. **Plantago arenària** W. & K   Sand Plantain.
Fig. 3910.

Plantago arenaria W. & K. Pl. Rar. Hung. 1 : 51. *pl. 51.* 1802.

Annual, pubescent, somewhat viscid; stem simple, or commonly becoming much branched, leafy, 3'–15' high. Leaves opposite, or whorled, narrowly linear, entire, sessile, 1'–3' long, about 1'' wide; peduncles axillary, often umbellate at the ends of the stem and branches, slender, as long as the leaves or longer; heads of flowers conic, oval, or subglobose, 5''–10'' long, about 5'' thick; lower bracts acute or acuminate; calyx-lobes unequal; corolla-lobes ovate to lanceolate, acute; capsule 2-seeded.

Fields, Dayton, Ohio, and Lancaster County, Pennsylva-nia. Adventive from central Europe. Summer.

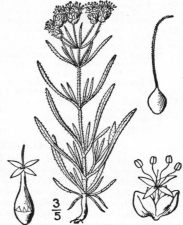

2. LITTORÉLLA L. Mant. 2 : 295. 1771.

A low perennial succulent herb, with linear entire basal leaves and monoecious flowers, the staminate solitary or two together at the summits of slender scapes, the pistillate sessile among the leaves. Sepals 4 Corolla of the staminate flowers with a somewhat urceolate tube, and a spreading 4-lobed limb. Corolla of the pistillate flowers urn-shaped, 3–4-toothed. Staminate flowers with 4 long-exserted stamens, their filaments filiform, the anthers ovate. Pistillate flowers with a single ovary and a long-exserted filiform style. Fruit an indehiscent 1-seeded nutlet. [Latin, shore.]

Two known species, the following typical, the other in southern South America.

3/5

1. Littorella uniflòra (L.) Ascherson. Plantain Shore-weed. Shore-grass. Fig. 3911.

Plantago uniflora L. Sp. Pl. 115. 1753.
Littorella lacustris L. Mant. 2 : 295. 1771.
Littorella uniflora Ascherson, Fl. Brand. 544. 1864.

Tufted, usually growing in mats; leaves bright green, 1′–3′ long, ½″–1″ wide, spreading or ascending, mostly longer than the scapes of the staminate flowers, which bear a small bract at about the middle; sepals lanceolate, mostly obtuse, with a dark green midrib and lighter margins, sometimes only 3 in the fertile flowers; stamens conspicuous, 4″–6″ long; corolla-lobes ovate, subacute; pistillate flowers very small; nutlet about 1″ long.

Borders of lakes and ponds, Maine and Vermont to Minnesota, Newfoundland, Nova Scotia and Ontario. July–Aug.

Family 36. RUBIÀCEAE B. Juss. Hort. Trian. 1759.

MADDER FAMILY.

Herbs, shrubs, or trees, with simple, opposite or sometimes verticillate, mostly stipulate leaves, and perfect, often dimorphous or trimorphous, regular and nearly symmetrical flowers. Calyx-tube adnate to the ovary, its limb various Corolla gamopetalous, funnelform, club-shaped, campanulate, or rotate, 4–5-lobed, often pubescent within. Stamens as many as the lobes of the corolla and alternate with them, inserted on its tube or throat; anthers mostly linear-oblong. Ovary 1–10-celled; style short or elongated, simple or lobed; ovules 1–∞ in each cavity. Fruit a capsule, berry, or drupe. Seeds various; seed-coat membranous or crustaceous; endosperm fleshy or horny (wanting in some exotic genera); cotyledons ovate, cordate, or foliaceous.

About 340 genera and about 6000 species, of very wide geographic distribution, most abundant in tropical regions. Known as Madderworts.

* **Leaves opposite, stipulate (sometimes verticillate in No. 3).**
Ovules numerous in each cavity of the ovary ; herbs.
　Top of the capsule free from the calyx ; seeds few, peltate. 　　　　　　　1. *Houstonia.*
　Capsule wholly adnate to the calyx ; seeds minute, angular. 　　　　　　2. *Oldenlandia.*
Ovules 1 in each cavity of the ovary.
　Shrubs or small trees ; flowers in dense globular heads. 　　　　　　　　3. *Cephalanthus.*
　Low evergreen herbs ; flowers 2 together, their ovaries united. 　　　　　4. *Mitchella.*
　Herbs ; flowers axillary, nearly sessile, distinct.
　　Capsule separating into 2 carpels, one dehiscent, the other indehiscent. 　5. *Spermacoce.*
　　Capsule separating into 2 (or 3) indehiscent carpels. 　　　　　　　　6 *Diodia.*
** **Leaves appearing verticillate; herbs (some of the leaves rarely opposite in No. 7); ovule 1 in each cavity of the ovary.**
Corolla rotate ; calyx teeth minute or none. 　　　　　　　　　　　　　7. *Galium.*
Corolla funnelform.
　Flowers in involucrate heads. 　　　　　　　　　　　　　　　　　8. *Sherardia.*
　Flowers in panicles. 　　　　　　　　　　　　　　　　　　　　9. *Asperula.*

1. HOUSTÒNIA L. Sp. Pl. 105. 1753.

Erect or diffuse, usually tufted herbs, with opposite entire often ciliate leaves, and small blue purple or white, mostly dimorphous flowers. Calyx-tube globose or ovoid, 4-lobed, the lobes distant. Corolla funnelform or salverform, 4-lobed, the lobes valvate, the throat glabrous or pubescent. Stamens 4, inserted on the tube or throat of the corolla; anthers linear or oblong. Ovary 2-celled. Style slender; ovules numerous in each cavity; stigmas 2, linear. Capsule partly inferior, globose-didymous, or emarginate at the apex, loculicidally dehiscent above, its summit free from the calyx. Seeds few or several in each cavity, peltate, more or less concave, not angled.　Seed-coat reticulate or roughened; endosperm horny; embryo club-shaped.　[Named in honor of Dr. William Houston, botanist and collector in South America, died 1733.]

About 25 species, natives of North America and Mexico.　Type species: *Houstonia coerulea* L.

* Plants 1'-7' high; peduncles 1-flowered.
† Peduncles filiform, 1'-2½' long.

Erect; leaves obovate or spatulate, narrowed into petioles.	1. *H. coerulea.*
Diffuse or spreading; leaves nearly orbicular.	2. *H. serpyllifolia.*

†† Peduncles 3"-18" long, stouter.

Calyx-lobes narrow, about equalling the capsule.	3. *H. patens.*
Calyx-lobes broad, much exceeding the capsule.	4. *H. minima.*

** Plants 4'-18' high; flowers cymose.

Calyx-lobes lanceolate-subulate, 2 to 3 times as long as the capsule.	5. *H. lanceolata.*

Calyx-lobes linear-subulate, scarcely longer than the capsule.

Leaves broad, ovate, or ovate-lanceolate.	6. *H. purpurea.*
Leaves oblong or spatulate, ciliate.	7. *H. ciliolata.*
Leaves linear-lanceolate or oblanceolate, not ciliate.	8. *H. longifolia.*

Leaves filiform or narrowly linear.

Flowers loosely cymose on filiform pedicels; leaves not fascicled.	9. *H. tenuifolia.*
Flowers densely cymose on very short pedicels; leaves usually fascicled.	10. *H. angustifolia.*

1. Houstonia coerùlea L.　Bluets.
Innocence.　Eyebright.　Fig. 3912.

Houstonia coerulea L. Sp. Pl. 105.　1753.
Hedyotis coerulea Hook. Fl. Bor. Am. 1: 286. 1833.
Oldenlandia coerulea A. Gray, Man. Ed. 2, 174.　1856.

Erect, 3'-7' high, glabrous, or nearly so, perennial by slender rootstocks and forming dense tufts.　Lower and basal leaves spatulate or oblanceolate, about 6" long, sometimes hirsute or ciliate, narrowed into a petiole, the upper oblong, sessile; flowers solitary on filiform terminal and axillary peduncles; corolla salverform, violet, blue, or white with a yellow center, 4"-6" broad, its tube slender and about the length of the lobes or longer; capsule didymous, compressed, about 2" broad and broader than long, the upper half free from the calyx and shorter than its lobes.

In open grassy places, or on wet rocks, Miquelon and Nova Scotia to Quebec, New York, Michigan, Georgia, Alabama, Tennessee and Missouri.　April-July, or producing a few flowers through the summer.　Includes several races.　Called also quaker-ladies, quaker-

bonnets, Venus'-pride.　Bright eyes.　Angel-eyes.　Blue-eyed-grass or -babies.　Wild forget-me-not.　Nuns.　Star-of-Bethlehem.　Little washerwoman.

2. Houstonia serpyllifòlia Michx.
Thyme-leaved Bluets.
Fig. 3913.

Houstonia serpyllifolia Michx. Fl. Bor. Am. 1: 85.　1803.
Hedyotis serpyllifolia T. & G. Fl. N. A. 2: 39. 1841.

Perennial; stems prostrate or diffuse, slender, glabrous, 4'-10' long.　Leaves orbicular or broadly oval, abruptly petioled, 3"-4" long, sometimes hispidulous; or those of the flowering stems narrower, distant; flowers on terminal and axillary filiform peduncles; corolla usually deep blue, 4"-6" broad, its tube rather shorter than the lobes; capsule similar to that of the preceding species but usually slightly larger, nearly as long as the calyx.

Somerset County, Pennsylvania, to the high mountains of Virginia and West Virginia, South Carolina, Georgia and east Tennessee. May.

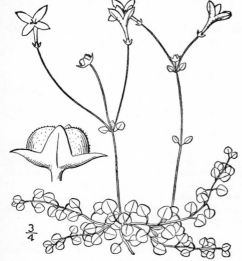

3. Houstonia pàtens Ell.   Small Bluets.   Fig. 3914.

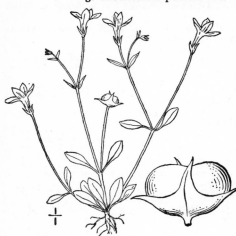

Houstonia Linnaei var. *minor* Michx. Fl. Bor. Am. **1** : 35.   1803.

H. *patens* Ell. Bot. S. C. & Ga. **1** : 191.   1821.

Houstonia *minor* Britton, Mem. Torr. Club **5** : 302.   1894.

Annual, glabrous or nearly so, branched from the base, 1′–6′ high. Lower and basal leaves oval or ovate, 5″–6″ long, narrowed into petioles often of their own length, the upper narrower and sessile; peduncles axillary and terminal, erect-divergent, 3″–18″ long, 1-flowered; corolla violet-blue or purple, 3″–4″ broad, its tube about equalling the lobes or somewhat longer; capsule compressed, didymous, 2½″–3″ broad, its upper part free from the calyx and about equalling or exceeding the subulate lobes.

In dry soil, Virginia to Florida, Illinois, Arkansas and Texas. March–April. Starviolet (Texas).

4. Houstonia mínima Beck.   Least Bluets.
Fig. 3915.

Houstonia *minima* Beck, Am. Journ. Sci. **10** : 262.   1826.

Hedyotis *minima* T. & G. Fl. N. A. **2** : 38.   1841.

Oldenlandia *minima* A. Gray, Man. Ed. 2, 173.   1856.

Annual, spreading or diffuse, 1′–2½′ high, roughish. Lower and basal leaves oval or ovate, with petioles shorter than or equalling the blade, the upper oblong, sessile; peduncles axillary and terminal, rather stout, 3″–12″ long; flowers 4″–5″ broad; corolla violet or purple, the tube about as long as the lobes; capsule didymous, compressed, about 3″ broad, its upper part free from the calyx and considerably exceeded by the lanceolate foliaceous lobes.

In dry soil, Illinois to Kansas, Arkansas, **Tennessee and Texas.** March–April.

5. Houstonia lanceolàta (Poir.) Britton.   Calycose Houstonia.   Fig. 3916.

Hedyotis *lanceolata* Poir. in Lam. Encycl. Suppl. **3** : 14.   1813.

Houstonia *purpurea calycosa* A. Gray, Syn. Fl. **1²** : 26.   1878.

Houstonia *lanceolata* Britton, Man. 861.   1901.

Houstonia *calycosa* Mohr, Contr. Nat. Herb. **6** : 739.   1901.

Perennial, glabrous or pubescent, rather stout, 6′–16′ high. Leaves lanceolate to linear-lanceolate, firm in texture, not ciliate, sessile, or the lowest spatulate and narrowed into petioles; corolla-tube little exceeding the calyx; calyx-lobes lanceolate or lanceolate-subulate, 5″ long or less, much exceeding the capsule; capsule ovoid-globose, about 2″ thick.

In dry soil, Illinois to North Carolina, Alabama, Tennessee and Oklahoma. Recorded from Maine. May–June.

6. Houstonia purpùrea L. Large Houstonia.
Venus'-pride. Fig. 3917.

Houstonia purpurea L. Sp. Pl. 105. 1753.
Hedyotis purpurea T. & G. Fl. N. A. **2**: 40. 1841.
Oldenlandia purpurea A. Gray, Man. Ed. 2, 173. 1856.
Houstonia purpurea pubescens Britton, Mem. Torr. Club **4**:
125. 1894.

Perennial, stout, erect, tufted, branched or simple,
glabrous or pubescent, 4′–18′ high. Leaves ovate or
ovate-lanceolate, sessile, or the lower ones short-
petioled, 3–5-nerved and pinnately veined, obtuse or
acute, ½′–2′ long, 3″–15″ wide, the margins often ciliate;
flowers in terminal cymose clusters; pedicels 1″–4″
long; corolla purple or lilac, funnelform, 3″–4″ long,
the tube at least twice as long as the lobes; capsule
compressed-globose, 1½″ broad, somewhat didymous, its
upper half free and considerably shorter than the subulate-linear calyx-lobes.

In open places, Maryland to Iowa, Kentucky, Missouri,
Georgia, Alabama and Mississippi, especially in the mountains. May–Sept.

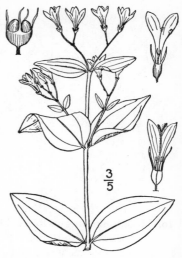

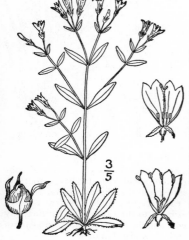

7. Houstonia ciliolàta Torr. Fringed Hous-
tonia. Fig. 3918.

Houstonia ciliolata Torr. Fl. N. U. S. **1**: 173. 1824.
Houstonia purpurea var. *ciliolata* A. Gray, Man. Ed. 5,
212. 1867.

Perennial, tufted, erect or ascending, 4′–7′ high.
Lower and basal leaves petioled, thick, 1-nerved, obovate or oblanceolate, obtuse, 6″–10″ long, their margins
conspicuously ciliate; stem leaves oblong or oblanceolate, sessile or nearly so; flowers in corymbed cymes;
pedicels filiform, 1″–4″ long; corolla funnelform, lilac
or pale purple, about 3″ long, the lobes about one-
third the length of the tube; capsule little compressed,
obscurely didymous, 1½″ wide, overtopped by the linear-
lanceolate calyx-lobes.

On rocks and shores, Ontario to Minnesota, south to
Pennsylvania, West Virginia, Kentucky and Arkansas.
May–Aug.

8. Houstonia longifòlia Gaertn. Long-
leaved Houstonia. Fig. 3919.

Houstonia longifolia Gaertn. Fruct. **1**: 226. *pl. 49. f. 8.*
1788.
Houstonia purpurea var. *longifolia* A. Gray, Man. Ed.
5, 212. 1868.

Perennial, usually tufted, erect, glabrous or puberulent, 5′–10′ high. Basal leaves spatulate or oblanceolate, obtuse, not ciliate, very short-petioled;
stem leaves linear or linear-oblong, acute or obtuse,
1-nerved, 6″–12″ long, 1″–2½″ wide; flowers in
corymbed cymes; calyx-lobes subulate; corolla pale
purple or nearly white, 2½″–3″ long, its lobes about
one-third the length of the tube; capsule little compressed, globose, ovoid, about 1″ in diameter, its upper
half free and much exceeded by the subulate calyx-
lobes.

In dry open places, Maine and Ontario to Manitoba,
Saskatchewan, Georgia, Mississippi and Missouri. May–
Sept.

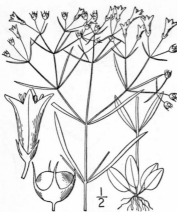

$\frac{1}{2}$

9. Houstonia tenuifòlia Nutt. Slender-leaved Houstonia. Fig. 3920.

Houstonia tenuifolia Nutt. Gen. 1: 95. 1818.
Houstonia purpurea var. *tenuifolia* A. Gray, Syn. Flor. 1: Part 2, 26. 1878.

Perennial, somewhat tufted, very slender and widely branching, erect, glabrous, 6'–1° high, sometimes finely pubescent below. Basal and lowest stem leaves ovate or oval, obtuse, petioled, 4''–6'' long; upper leaves narrowly linear or filiform, blunt-pointed, 6''–15'' long, $\frac{1}{2}$''–1$\frac{1}{2}$'' wide; flowers in loose corymbose cymes; pedicels filiform, 2''–6'' long; corolla purple, narrow, 2''–3'' long, its lobes short; capsule compressed-globose, didymous, about 1'' in diameter, its upper half free and only slightly exceeded by the subulate calyx-lobes.

In dry soil, Pennsylvania to Ohio, North Carolina and Tennessee. May–July.

10. Houstonia angustifòlia Michx. Narrow-leaved Houstonia. Fig. 3921.

Houstonia angustifolia Michx. Fl. Bor. Am. 1: 85. 1803.
Oldenlandia angustifolia A. Gray, Pl. Wright. 2: 68. 1853.

Perennial by a deep root, erect, stiff, glabrous, usually branched, 1°–2° high. Leaves linear, 6''–18'' long, 1''–3'' wide, or the lowest narrowly spatulate, usually with numerous smaller ones fascicled in the axils, or on short axillary branches; flowers in terminal dense cymose clusters; pedicels short; corolla white or purplish, between funnelform and salverform, about 2'' long, its lobes shorter than the tube; capsule compressed-obovoid, 1$\frac{1}{2}$'' wide, its summit free and scarcely exceeded by the calyx-lobes.

In dry open places, Illinois to Kansas, Texas, Tennessee and Florida. May–July. Star-violet. Venus'-pride.

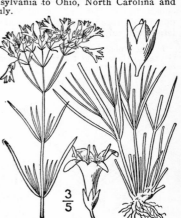

$\frac{3}{5}$

2. OLDENLÁNDIA [Plumier] L. Sp. Pl. 119. 1753.

Erect or diffuse slender herbs, with opposite leaves, and small axillary or terminal solitary or clustered, white or pink flowers. Calyx-tube obovoid or subglobose, the limb 4-toothed. Corolla rotate or salverform, 4-lobed. Stamens 4, inserted on the throat of the corolla; anthers oblong. Ovary 2-celled; ovules numerous in each cavity; style slender, 2-lobed. Capsule small, ovoid, top-shaped, or hemispheric, wholly adnate to the calyx-tube, loculicidally dehiscent at the summit, several or many-seeded. Seeds angular, not peltate; endosperm fleshy; embryo club-shaped. [Named for H. B. Oldenland, a Danish botanist.]

About 175 species, mostly of tropical distribution, most abundant in Asia. Besides the following, 2 others occur in the Southern States and 1 in New Mexico. Type species: *Oldenlandia corymbòsa* L.

$\frac{3}{5}$

1. Oldenlandia uniflòra L. Clustered Bluets. Fig. 3922.

Oldenlandia uniflora L. Sp. Pl. 119. 1753.
Oldenlandia glomerata Michx. Fl. Bor. Am. 1: 83. 1803.

Annual, weak, usually tufted, more or less hirsute-pubescent, diffuse or ascending; stems 1'–15' long. Leaves short-petioled or sessile, mostly thin, entire, 3–5-nerved, ovate, oblong, or oval, acute at the apex, narrowed at the base, $\frac{1}{2}$'–1' long; flowers sessile or nearly so, white, about 1'' broad, terminal and axillary, clustered or solitary; calyx hirsute, hemispheric in fruit, the ovate or oval lobes erect and nearly equalling the tube.

In low grounds, southern New York to Florida and Texas, north to Arkansas and reported from Missouri. Also in Cuba and Jamaica. June–Sept.

3. CEPHALÁNTHUS L. Sp. Pl. 95. 1753.

Shrubs, or small trees, with opposite or verticillate short-petioled entire leaves, and terminal or axillary, densely capitate, bracteolate small white or yellow flowers. Calyx-tube obpyramidal, its limb with 4 obtuse lobes. Corolla tubular-funnelform, with 4 short erect or spreading lobes. Stamens 4, inserted on the throat of the corolla; filaments very short; anthers oblong, 2-cuspidate at the base. Ovary 2-celled; ovules solitary in each cavity, pendulous; style filiform, exserted; stigma capitate. Fruit dry, obpyramidal, 1–2-seeded. Endosperm cartilaginous; cotyledons linear-oblong. [Greek, head-flower.]

About 6 species, natives of America and Asia. The following, here taken as typical, is the only one known to occur in North America, unless the southwestern and Mexican plant proves to be distinct.

1. Cephalanthus occidentàlis L. Button-bush. Button-tree. Honey-balls. Globe-flower. Fig. 3923.

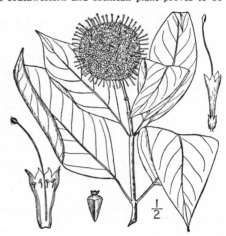

Cephalanthus occidentalis L. Sp. Pl. 95. 1753.

A shrub 3°–12° high, or sometimes a tree, up to 20° high, with opposite or verticillate leaves and branches, glabrous, or somewhat pubescent. Leaves petioled, ovate or oval, entire, acuminate or acute at the apex, rounded or narrowed at the base, 3′–6′ long, 1′–2½′ wide; peduncles 1′–3′ long; heads globose, about 1′ in diameter, the receptacle pubescent; flowers sessile, white, 4″–6″ long; style very slender, about twice the length of the corolla; calyx-tube prolonged beyond the ovary.

In swamps, and low grounds, New Brunswick to western Ontario and Wisconsin, Florida, Texas, Arizona and California. Pond dogwood. Button-wood shrub. Box. Pin-ball. Little snowball. Button- or crane willow. Swamp-wood. River- or crouper-bush. June–Sept.

4. MÍTCHELLA L. Sp. Pl. 111. 1753.

Creeping herbs, with opposite petioled, entire or undulate, evergreen leaves, and white axillary or terminal peduncled geminate dimorphous flowers, their ovaries united. Calyx-tube ovoid, the limb 3–6-lobed (usually 4-lobed). Corolla funnelform, usually 4-lobed, the lobes recurved, bearded on the inner side. Stamens as many as the lobes of the corolla and inserted on its throat; filaments short and style exserted, or filaments exserted and style short. Ovary 4-celled; stigmas 4, short, filiform; ovules 1 in each cavity, erect, anatropous. Fruit composed of 2 united drupes usually containing 8 roundish nutlets. Seed erect; cotyledons short, obtuse; embryo minute. [Named after Dr. John Mitchell, botanist and correspondent of Linnaeus in Virginia.]

Two species, the following typical one North American, the other Japanese.

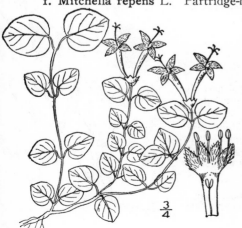

1. Mitchella rèpens L. Partridge-berry. Twin-berry. Fig. 3924.

Mitchella repens L. Sp. Pl. 111. 1753.

Stems slender, trailing, rooting at the nodes, 6′–12′ long, branching, glabrous, or very slightly pubescent. Leaves ovate-or-bicular, petioled, obtuse at the apex, rounded or somewhat cordate at the base, 3″–10″ long, pinnately veined, dark green, shining; peduncles shorter than the leaves, bearing 2 sessile white flowers at the summit; corolla 5″–6″ long; drupes red (rarely white), broader than high, 2″–4″ in diameter, persistent through the winter, edible.

In woods, Nova Scotia to Florida, west to western Ontario, Minnesota, Arkansas and Texas. April–June, sometimes flowering a second time in the autumn. Hive- or squaw-vine. Checker-berry. Deer-berry. Fox- or box-berry. Partridge-vine. Winter-clover. Chicken-, cow-, pigeon-, snake- or tea-berry. Two-eyed- or one-berry. Squaw-plum. Leaves often whitish-veined; flower-buds pink. Ascends to 5000 ft. in Virginia.

5. **SPERMACÒCE** L. Sp. Pl. 102. 1753.

Herbs, with 4-sided stems, opposite pinnately veined stipulate leaves, and small white flowers, in dense axillary and terminal clusters. Calyx-tube obovoid or obconic, its limb 4-toothed. Corolla funnelform, 4-lobed. Stamens 4, inserted on the tube of the corolla; anthers oblong or linear. Ovary 2-celled; ovules 1 in each cavity; style slender; stigma capitate, or slightly 2-lobed. Capsule coriaceous, didymous, of 2 carpels, one dehiscent, the other indehiscent. Seeds oblong, convex on the back; endosperm horny; embryo central; cotyledons foliaceous. [Greek, seed-point, from the sharp calyx-teeth surmounting the carpels.]

Two or three species, natives of America. Type species: *Spermacoce tenuior* L.

1. **Spermacoce glàbra** Michx. Smooth Button-weed. Fig. 3925.

Spermacoce glabra Michx. Fl. Bor. Am. 1: 82. 1803.

Glabrous, decumbent or ascending, rather stout; stems 10′–20′ long. Leaves lanceolate or elliptic-lanceolate, petioled, or the uppermost sessile, 1′–3′ long, 4″–12″ wide, acute at each end, the margins rough; corolla pubescent in the throat, about 1½″ long, scarcely exceeding the ovate-lanceolate acute calyx teeth; stamens and style included; capsule obovoid, about 2″ long, glabrous; seeds black, punctate.

On river-banks and in wet soil, southern Ohio and Kentucky to Florida, Kansas and Texas. June–Sept.

½

6. **DIÒDIA** L. Sp. Pl. 104. 1753.

Decumbent or ascending branching herbs, with opposite, mostly sessile, entire conspicuously stipulate leaves, and small axillary white lilac or purple flowers. Calyx-tube obconic or obovoid, the limb 2–4-lobed (sometimes 1–6-lobed), often with minute teeth between the lobes. Corolla funnelform or salverform, mostly 4-lobed. Stamens usually 4, inserted on the throat of the corolla; filaments slender; anthers versatile, oblong-linear, exserted. Ovary 2-celled (rarely 3–4-celled); ovules 1 in each cavity; style filiform, simple or 2-cleft; stigmas 2. Fruit crustaceous or somewhat fleshy, oblong, obovoid, or subglobose, 2-celled, finally separating into 2 indehiscent carpels. Seeds oblong, convex on the back; endosperm horny; cotylendons foliaceous; embryo straight. [Greek, thoroughfare, where the species are frequently found.]

About 35 species, mostly American. Besides the following, another occurs in the Southern States. Type species: *Diodia virginiana* L.

Leaves linear-lanceolate; style entire; stigmas capitate.   1. *D. teres.*
Leaves lanceolate or oval; style 2-cleft; stigmas filiform.   2. *D. virginiana.*

1. **Diodia tères** Walt. Rough Button-weed. Fig. 3926.

Diodia teres Walt. Fl. Car. 87. 1788.

Spermacoce diodina Michx. Fl. Bor. Am. 1: 82. 1803.

Rigid, usually rough, much branched from near the base, the branches prostrate or ascending, 4-sided above, 4′–30′ long. Leaves linear or linear-lanceolate, very rough, ½′–1½′ long, 1½″–3″ wide, acute, the margins revolute when dry; flowers lilac or purple, 2″–3″ long, usually solitary in the axils; style entire; stigmas capitate; fruit obovoid or top-shaped, hispid, about 2″ high, the usually 4 persistent calyx-lobes ovate to lanceolate.

In dry or sandy soil, Connecticut to Florida, west to Illinois, Kansas, Texas, New Mexico and Sonora. July–Sept. Poverty-, or poor-land-weed. Poor Joe.

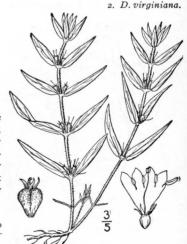

⅗

2. Diodia virginiàna L. Larger Button-weed. Fig. 3927.

Diodia virginiana L. Sp. Pl. 104. 1753.

Hispid-pubescent or glabrate, much branched from near the base, the branches procumbent or ascending, 1°–2° long. Leaves lanceolate to narrowly oval, narrowed at the base, acute, or the lowest obtuse at the apex, 1'–3' long; flowers 1 or 2 in each axil, about 6" long, the corolla-tube very slender; fruit somewhat fleshy, but becoming dry, hirsute or glabrous, oval, 3"–4" high, furrowed, crowned with the 2 or 3 persistent lanceolate calyx-lobes.

In moist soil, southern New Jersey to Florida, west to Arkansas and Texas. June–Aug.

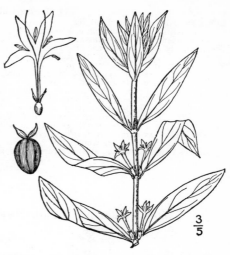

$\frac{3}{5}$

7. GÀLIUM L. Sp. Pl. 105. 1753.

Annual or perennial herbs, with 4-angled slender stems and branches, apparently verticillate leaves, and small white, green, yellow or purple flowers, mostly in axillary or terminal cymes or panicles, the pedicels usually jointed with the calyx. Flowers perfect, or in some species dioecious. Calyx-tube ovoid or globose, the limb minutely toothed, or none. Corolla rotate, 4-lobed (rarely 3-lobed). Stamens 4, rarely 3; filaments short; anthers exserted. Ovary 2-celled; ovules 1 in each cavity. Styles 2, short; stigmas capitate. Fruit didymous, dry or fleshy, smooth, tuberculate, or hispid, separating into 2 indehiscent carpels, or sometimes only 1 of the carpels maturing. Seed convex on the back, concave on the face, or spherical and hollow; endosperm horny; embryo curved; cotyledons foliaceous. [Greek, milk, from the use of *G. verum* for curdling.]

About 250 species, of wide geographic distribution. Besides the following, some 35 others occur in the southern and western parts of North America. The leaves are really opposite, the intervening members of the ventricles being stipules. Type species: *Galium Mollùgo* L.

I. **Fruit dry.**
    A. **Flowers yellow.**   1. *G. verum.*
  B. **Flowers white, green or purple.**
      1. *Annuals.*
Flowers in axillary cymules, or panicled.
  Fruit granular or tubercled, not bristly.
    Fruit slightly granular, or smooth, 1 mm. broad; pedicels not recurved; stem very slender.
                                              2. *G. parisiense.*
    Fruit granular-tubercled, 3 mm. broad; fruiting pedicels recurved; stem stout.
                                              3. *G. tricorne.*
  Fruit densely bristly-hispid.
    Cymes few-flowered; leaves 2–8 cm. long; fruit fully 4 mm. broad.   4. *G. Aparine.*
    Cymes mostly several-flowered; leaves 1–2.5 cm. long; fruit smaller.   5. *G. Vaillantii.*
Flowers solitary in the axils, subtended by 2 foliaceous bracts; fruit bristly. 6. *G. virgatum.*

        2. *Perennials.*
        * Fruit bristly hispid.
        a. Leaves in 4's, 1-nerved.   7. *G. pilosum.*
        b. Leaves in 4's, 3-nerved.
Leaves lanceolate, oval, or ovate; flowers in open cymes.
  Upper leaves lanceolate to ovate-lanceolate, acuminate.   8. *G. lanceolatum.*
  Upper leaves ovate, oblong, oval, ovate-lanceolate or obovate, obtuse.
    Corolla usually hirsute; plant mostly pubescent; leaves oblong to ovate-lanceolate.
                                        9. *G. circaezans.*
    Corolla glabrous; plant little pubescent; some leaves obovate. 10. *G. kamtschaticum.*
Leaves linear to lanceolate; flowers in terminal panicles. 11. *G. boreale.*

        c. Leaves in 6's.   12. *G. triflorum.*
        ** Fruit smooth or warty.
        † Flowers brown-purple.
Leaves lanceolate, 3-nerved; fruit smooth.   13. *G. latifolium.*
Leaves narrowly lanceolate, 1-nerved; fruit warty.   14. *G. arkansanum*

†† Flowers white or greenish.

Stems smooth; introduced species.
  Leaves linear to oblanceolate, cuspidate.   15. *G. Mollugo.*
  Leaves lanceolate, acuminate.   16. *G. sylvaticum.*
Stems mostly more or less retrorsely scabrous; native species.
  Leaves obtuse; stems slightly scabrous.
    Plants of wet soil, not shining.
      Flowers solitary, or few in small simple cymes.
        Corolla-lobes mostly 4, acute.
          Fruit 1½″ in diameter; leaves ascending or spreading.   17. *G. tinctorium.*
          Fruit ½″ in diameter; leaves mostly reflexed.   18. *G. labradoricum.*
        Corolla-lobes mostly 3, obtuse.
          Pedicels rough, curved; flowers mostly solitary.   19. *G. trifidum.*
          Pedicels smooth, straight; flowers 2 or 3 together.   20. *G. Claytoni.*
      Flowers numerous in forked cymes.   21. *G. palustre.*
  Shining plant of dry woodlands.   22. *G. concinnum.*
  Leaves cuspidate-acute; stems retrorsely hispid.   23. *G. asprellum.*

        II. **Fruit fleshy.**   24. *G. bermudense.*

1. **Galium vèrum** L. Yellow Bedstraw. Lady's Bedstraw. Fig. 3928.

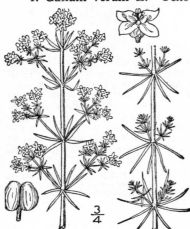

Galium verum L. Sp. Pl. 107. 1753.

Perennial from a somewhat woody base, erect or ascending, simple or branched, 6′–2½° high. Stems smooth or minutely roughened; leaves in 6's or 8's, narrowly linear, 4″–12″ long, about ½″ wide, rough on the margins, at length deflexed; flowers yellow, the cymes in dense narrow panicles; lower branches of the panicles longer than the internodes at anthesis; fruit usually glabrous, less than 1″ broad.

In waste places and fields, Maine and Ontario to Massachusetts, southern New York, New Jersey and Pennsylvania. Adventive or naturalized from Europe. Native also of Asia. May–Sept. Cheese-rennet. Curdwort. Bedflower. Fleawort. Maids'-hair. Yellow cleavers. Our Lady's-bedstraw.

Galium Wírtgeni F. Schultz, differs in having the lower branches of the panicle very short at anthesis, and is recorded as established in a meadow at Norfolk, Connecticut.

2. **Galium parisiénse** L. Wall Bedstraw. Fig. 3929.

Galium parisiense L. Sp. Pl. 108. 1753.

Galium anglicum Huds. Fl. Angl. Ed. 2, 69. 1778.

Annual, erect or ascending, very slender, much branched; stem rough on the angles, 6′–12′ high. Leaves in verticils of about 6 (4–7), linear or linear-lanceolate, cuspidate, minutely scabrous on the margins and midrib, 2″–5″ long; cymes several-flowered, axillary and terminal on filiform peduncles; flowers minute, greenish-white; fruit glabrous, finely granular, less than ½″ wide.

Along roadsides, Virginia and Tennessee. Adventive or naturalized from Europe. June–Aug.

3. Galium tricórne Stokes. Rough-fruited Corn Bedstraw. Fig. 3930.

Galium tricorne Stokes; With. Bot. Arr. Brit. Pl. Ed. **2, 1**: 153. 1787.

Rather stout, decumbent or ascending, 6'–12' high, simple, or little branched. Stem rough with reflexed prickles; leaves in 6's or 8's, linear or narrowly oblanceolate, 1' long or less, 1½''–2'' wide, mucronate, rough on the margins and midrib; peduncles axillary, shorter than the leaves; pedicels thickened and curved downward in fruit; cymes axillary, usually 3- (1–3-) flowered; fruit tuberculate or granular, not hispid, 4''–5'' broad.

In waste places, Ontario, and in ballast about the eastern seaports. May–Aug.

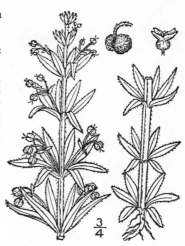

4. Galium Aparìne L. Cleavers. Goose-grass. Cleaver-wort. Fig. 3931.

Galium Aparine L. Sp. Pl. 108. 1753.

Annual, weak, scrambling over bushes, 2°–5° long, the stems retrorsely hispid on the angles. Leaves in 6's or 8's, oblanceolate to linear, cuspidate at the apex, 1'–3' long, 2''–5'' wide, the margins and midrib very rough; flowers in 1–3-flowered cymes in the upper axils; peduncles 5''–12'' long; fruiting pedicels straight; fruit 2''–3'' broad, densely covered with short hooked bristles.

In various situations, New Brunswick to Ontario, South Dakota, Florida and Texas. Bermuda. Apparently in part naturalized from Europe. Widely distributed in temperate regions as a weed. May–Sept. Among some 70 other English names are catchweed, beggar-lice, burhead, claver-grass, cling-rascal, scratch-grass, wild hedge-burs, hairif or airif, stick-a-back, or stickle-back, gosling-grass, gosling-weed, turkey-grass, pigtail, grip or grip-grass, loveman, sweethearts, scratch-weed, poor robin.

5. Galium Vaillántii DC. Vaillant's Goose-grass or Cleavers. Fig. 3932.

Galium Vaillantii DC. Fl. France **4**: 263. 1805.

Galium Aparine var. *Vaillantii* Koch, Fl. Germ. 330. 1837.

Similar to the preceding species but smaller, the stem equally rough-angled. Leaves smaller, 1' in length or less, linear-oblong or slightly oblanceolate, cuspidate-pointed, rough on the margins and midrib; cymes 2–9-flowered; fruit 1''–1½'' broad, usually less hispid.

In low grounds, Ontario to British Columbia, Missouri, Arizona and to California. Europe. The European *G. spurium* L., to which this plant was referred in the first edition, appears to have uniformly smooth fruit.

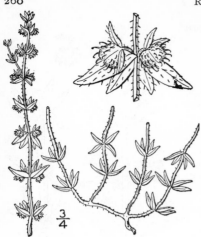

6. Galium virgàtum Nutt. Southwestern Bedstraw. Fig. 3933.

Galium virgatum Nutt.; T. & G. Fl. N. A. **2** : 20. 1841.

Annual, 4′–12′ high, usually hispid, sometimes nearly glabrous; stem very slender, 4-angled, branched from the base, or simple. Leaves in 4's, oblong or linear-oblong, 2½″–5″ long, 1″ wide, or less, obtuse or acutish; peduncles axillary, 1-flowered, less than 1″ long, recurved in fruit; flower white, subtended by 2 large oblong to lanceolate bracts which closely resemble the leaves; fruit about 1″ in diameter, covered with slender barbed bristles.

Dry prairies, barrens and glades, Tennessee and Missouri to Arkansas, Louisiana and Texas. April–June.

7. Galium pilòsum Ait. Hairy Bedstraw. Fig. 3934.

Galium pilosum Ait. Hort. Kew. **1** : 145. 1789.
Galium puncticulosum Michx. Fl. Bor. Am. **1** : 80. 1803.
Galium pilosum puncticulosum T. & G. Fl. N. A. **2** : 24. 1841.

Perennial, more or less hirsute-pubescent or glabrate; stems ascending, branched, 1°–2½° long. Leaves in 4's, oval or oval-ovate, punctate, 1-nerved, obtuse, or obscurely 3-nerved at the base, mucronulate, 4″–12″ long, 2″–5″ wide, the lower usually smaller; peduncles axillary and terminal; cymes numerous but few-flowered; pedicels 1″–6″ long, flowers yellowish-purple; fruit dry, densely hispid, nearly 2″ in diameter.

In dry or sandy soil, New Hampshire to Ontario, Michigan, Kansas, Florida and Texas. June–Aug.

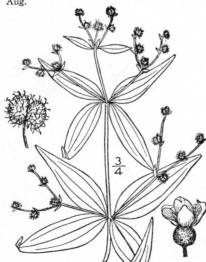

8. Galium lanceolàtum Torr. Torrey's Wild Liquorice. Fig. 3935.

Galium circaezans var. *lanceolatum* Torr. Cat. Pl. N. Y. 23. 1819.

Galium lanceolatum Torr. Fl. U. S. 168. 1824.

Galium Torreyi Bigel. Fl. Bost. Ed. 2, 56. 1824.

Perennial, glabrous or nearly so, the stems minutely roughened, simple or often branched. 1°–2° high. Leaves in 4's, lanceolate or ovate-lanceolate, acutish or acuminate, 3-nerved, more or less ciliate on the margins and nerves, 1′–2½′ long, 5″–11″ wide, the lower smaller and obtuse or obtusish; cymes rather few-flowered, loose, widely branched; flowers sessile or very nearly so; corolla glabrous, yellowish green to purple, its lobes acuminate; fruit dry, hispid with long hairs, 2″–2½″ broad.

In dry woods, Quebec and Ontario to Minnesota, south to North Carolina and Kentucky. Ascends to 4000 ft. in Virginia. June–Aug.

9. Galium circaèzans Michx. Wild Liquorice. Cross-Cleavers. Fig. 3936.

G. circaezans Michx. Fl. Bor. Am. 1: 80. 1803.

Galium circaezans glabellum Britton, Mem. Torr. Club 5: 303. 1894.

Perennial, more or less pubescent, or glabrate, branched, 1°–2° high. Leaves in 4's, oval, oval-lanceolate or ovate, obtuse or obtusish at the apex, 3-nerved, 6″–18″ long, 4″–8″ wide, usually somewhat pubescent on both surfaces, ciliolate, the lower smaller; cymes divaricately branched; flowers sessile or nearly so, greenish; corolla hirsute without, or glabrous, its lobes acute; fruit hispid, similar to that of the preceding species, at length deflexed.

In dry woods, Quebec and Ontario to Minnesota, Florida, Kansas and Texas. May–July.

10. Galium kamtscháticum Steller. Northern Wild Liquorice. Fig. 3937.

Galium kamtschaticum Steller; R. & S. Mant. 3: 186. 1827.
Galium Littellii Oakes, Hovey's Mag. 7: 179. 1841.
Galium circaezans var. montanum T. & G. Fl. N. A. 2: 24. 1841.

Similar to the preceding species, but weak, smaller, stems 4′–15′ long. Leaves in 4's, broadly oval, orbicular, or obovate, thin, 3-nerved, obtuse, mucronulate, 6″–18″ long, 4″–12″ wide, glabrate, or, pubescent with short scattered hairs on the upper surface and on the nerves beneath, sometimes ciliate; flowers few, all on pedicels 2″–6″ long; corolla glabrous, yellowish-green, its lobes acutish; fruit hispid, 2″ broad.

In mountainous regions, Cape Breton Island, Quebec, northern New England and northern New York. Also in northeastern Asia. Summer.

11. Galium boreàle L. Northern Bedstraw. Fig. 3938.

Galium boreale L. Sp. Pl. 108. 1753.

Galium septentrionale R. & S. Syst. 3: 253. 1818.

Erect, perennial, smooth and glabrous, strict, simple, or branched, leafy, 1°–2½° high. Leaves in 4's, lanceolate or linear, 3-nerved, obtuse or acute, 1′–2½′ long, 1″–3″ wide, the margins sometimes ciliate; panicles terminal, dense, many-flowered, the flowers white, panicled, in small compact cymes; fruit hispid, at least when young, sometimes becoming glabrous when mature, about 1″ broad.

In rocky soil or along streams, Quebec to Alaska, south to New Jersey, Pennsylvania, Michigan, Missouri, New Mexico and California. Also in Europe and northern Asia. May–Aug.

$\frac{3}{4}$

13. **Galium latifòlium** Michx.   Purple Bedstraw.   Fig. 3940.

Galium latifolium Michx. Fl. Bor. Am. 1: 79. 1803.

Perennial, erect, smooth and glabrous (rarely hispid), branched, 1°–2° high. Leaves in 4's, lanceolate, 3-nerved, acute or acuminate at the apex, narrowed or rounded at the base, 1′–2′ long, 4″–8″ wide, the midrib sometimes pubescent beneath, the margins minutely roughened, the lower smaller and sometimes opposite; peduncles axillary and terminal, slender, but usually shorter than the leaves; flowers purple; cymes loosely many-flowered, the pedicels slender, 2″–6″ long, smooth or very nearly so; fruit smooth, slightly fleshy, 2″ broad, usually only one of the carpels developing.

In dry woods in mountainous regions, Pennsylvania to Tennessee and Georgia. May–Aug.

12. **Galium triflòrum** Michx.   Sweet-scented or Fragrant Bedstraw. Fig. 3939.

Galium triflorum Michx. Fl. Bor. Am. 1: 80. 1803.

Perennial, diffuse, procumbent, or ascending, glabrous or nearly so, shining, fragrant in drying, the stems and margins of the leaves sometimes roughened.   Leaves in 6's, narrowly oval or slightly oblanceolate, 1-nerved, cuspidate at the apex, narrowed at the base, 1′–3½′ long, 2″–6″ wide; peduncles slender, terminal and axillary, often exceeding the leaves, 3-flowered or branched into 3 pedicels which are 1–3-flowered; flowers greenish; fruit 1½″–2″ broad, hispid with hooked hairs; seed almost spherical, the groove obsolete.

In woods, Greenland to Alaska, south to Florida, Louisiana, Colorado and California. Also in northern Europe, Japan and the Himalayas. June–Aug. Three-flowered bedstraw.

14. **Galium arkansànum** A. Gray.   Arkansas Bedstraw.   Fig. 3941.

Galium arkansanum A. Gray, Proc. Am. Acad. 19: 80. 1883.

Similar to the preceding species but usually lower, the leaves linear-lanceolate, 6″–12″ long, 1″–3″ wide, the lateral nerves obscure or none, the midrib sometimes pubescent beneath, and the margins ciliate; fruiting pedicels roughish, 3″–12″ long; flowers brown-purple, the numerous cymes loosely several–many-flowered; fruit glabrous, warty, somewhat fleshy, each carpel 1″–1½″ in diameter.

Southern Missouri, Arkansas and Oklahoma. June–July.

$\frac{3}{4}$

15. Galium Mollùgo L. Wild Madder. White or Great Hedge Bedstraw. Fig. 3942.

Galium Mollugo L. Sp. Pl. 107. 1753.

Perennial, glabrous or nearly so throughout. Stems smooth, erect, or diffusely branched, 1°–3° long; leaves in 6's or 8's, oblanceolate or linear, cuspidate at the apex, 6″–15″ long, 1″–2″ wide, sometimes roughish on the margins; flowers small, white, very numerous in terminal panicled cymes; pedicels filiform, divaricate; fruit smooth and glabrous, nearly 1″ broad.

In fields and waste places, Newfoundland to Vermont, Ohio, Pennsylvania, New Jersey and Virginia. Naturalized from Europe. Called also whip-tongue. Infants'- or babies'-breath. May–Sept.

Galium eréctum Huds., also European, differs slightly by having somewhat larger flowers on ascending pedicels, and is adventive in fields from Quebec to Connecticut and New York.

16. Galium sylváticum L. Wood Bedstraw. Fig. 3943.

G. sylvaticum L. Sp. Pl. Ed. 2, 155. 1762.

Perennial, erect, 2°–3° tall; stems several or many, shining, obtusely 4-angled, glabrous, or slightly pubescent, not scabrous. Leaves lanceolate or oblong-lanceolate, pale beneath, whorled in 8's or 6's, or those of the branches in 4's, or opposite, the larger sometimes 2′ long; panicles large; pedicels filiform, erect-spreading in fruit; flowers white; corolla-lobes apiculate; fruit smooth.

Fields and thickets, Maine and Vermont, escaped from cultivation. Native of Europe. June–July.

17. Galium tinctòrium L. Stiff Marsh Bed-straw. Wild Madder. Fig. 3944.

Galium tinctorium L. Sp. Pl. 106. 1753.
Galium trifidum var. *latifolium* Torr. Fl. N. & Mid. States, 78. 1826.
Galium tinctorium filifolium Wiegand, Bull. Torr. Club 24: 397. 1897.

Perennial; stem erect, 6′–15′ high, rather stiff, branched almost to the base, the branches commonly solitary, strict (not irregularly diffuse), several times forked; stem 4-angled, nearly glabrous; leaves commonly in 4's, linear to lanceolate, ½′–1′ long, mostly broadest below the middle, obtuse, cuneate at the base, dark green and dull, not papillose, 1-nerved, the margins and midrib roughish; flowers terminal in clusters of 2 or 3; pedicels slender, not much divaricate in fruit; corolla white, large, 1″–1¾″ broad, 4-parted, its lobes oblong, acute; disk large; fruit smooth; seed spherical, hollow, annular in cross-section.

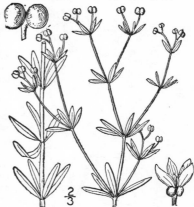

Damp shady places, wet meadows and swamps, Quebec to North Carolina, Florida, Tennessee, Michigan, Nebraska and Arizona. May–July.

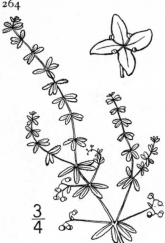

18. Galium labradóricum Wiegand. Labrador Marsh Bedstraw. Fig. 3945.

Galium tinctorium labradoricum Wiegand, Bull. Torr. Club 24: 398. 1897.

Galium labradoricum Wiegand, Rhodora 6: 21. 1904.

Perennial, with very slender rootstocks; stems weak, smooth, slender, more or less branched, 2′-12′ high. Leaves ¼′-¾′ long, linear-oblanceolate, narrowed at the base, becoming reflexed, scabrous on the margins and midvein beneath; flowers solitary, about 1″ broad, or in simple cymes; corolla white, mostly 4-parted; fruit smooth, seed annular in cross-section.

In mossy bogs, Newfoundland to Wisconsin, Connecticut, western Massachusetts and New York. June–Aug.

19. Galium trífidum L. Small Bedstraw. Small Cleavers. Fig. 3946.

Galium trifidum L. Sp. Pl. 105. 1753.

G. trifidum var. *pusillum* A. Gray, Man. Ed. 5, 209. 1867.

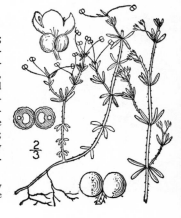

Perennial by slender rootstocks, very slender and weak; stem ascending, 16′ long or less, much branched and inter-tangled; stem sharply 4-angled, rough; branches commonly in 2′s; leaves in 4′s, linear-spatulate, 2½″-7″ long, obtuse, cuneate at the base, 1-nerved, dark green and dull on both surfaces, scarcely papillose, the margins and mid-rib retrorse-scabrous; flowers small, on lateral or terminal pedicels which are capillary and much longer than the leaves, commonly two at each node or three terminal; corolla very small, white, ¼″ long, trifid, its lobes broadly oval, very obtuse; fruit glabrous; seed spherical and hollow, annular in cross-section

Sphagnous bogs and cold swamps, Newfoundland to New York, British Columbia, Ohio, Nebraska and Colorado. Europe and Asia. Summer.

20. Galium Clàytoni Michx. Clayton's Bedstraw. Fig. 3947.

Galium Claytoni Michx. Fl. Bor. Am. 1: 78. 1803.

Galium tinctorium Bigelow, Fl. Bost. Ed. 2, 54. 1824.

Perennial; stem erect or ascending, more diffuse when old, 6′-2° high; stem slender or sometimes quite stout, sharply 4-angled, more or less rough, the diffuse branches in 2′s; leaves of medium size, 4″-8″ long, commonly in 5′s or 6′s, linear-spatulate or spatulate-oblong, obtuse, cuneately narrowed into a short petiole, rather firm in texture, scabrous on the margin and midrib, dark green and dull above, not papillose, dis-colored in drying; flowers in clusters of 2′s or 3′s, terminal, provided with 1 or 2 minute bracts; pedicels straight, in fruit strongly divaricate, glabrous and rather stout; corolla minute, white, 3-parted, the lobes broadly oval, obtuse; fruit glabrous; seed spherical and hollow, annular in cross-section.

Swamps, Quebec to New York, North Carolina, Michigan, Missouri, Nebraska and Texas. May–July.

21. Galium palústre L.   Marsh Bedstraw.   Fig. 3948.

Galium palustre L. Sp. Pl. 105. 1753.
Galium trifidum var. *bifolium* Macoun, Cat. Can.
   Plants 202. 1884?

Perennial, stem erect and rather slender, about
16' high; internodes very long (middle one 2½'-3'
long); short branches mostly in 2's.   Stem sharply
4-angled, glabrous or a little rough; leaves in
typical specimens rather small, in 2's to 6's, linear-
elliptic to spatulate, cuneate at the base, obtuse,
3"-8" long, 1"-2" wide, the rather firm margins
and the midrib slightly scabrous, not papillose;
flowers numerous in terminal and lateral cymes;
bracteoles in the inflorescence minute; pedicels in
flower ascending, 1½"-2½" long, in fruit strongly
divaricate; corolla large, white, 1"-1¾" broad,
4-parted, the lobes oblong, acute; disk almost
obsolete; fruit glabrous; endosperm of the seed
grooved on the inner face, in cross-section lunate.

In damp shady or open places along roadsides and
ditches, or in the margins of swamps.   Newfoundland, Prince Edward Island and Quebec, to Connecticut and New York.   Also in Europe.   Summer.

22. Galium concínnum Torr. & Gray. Shining Bedstraw.   Fig. 3949.

Galium parviflorum Raf. Med. Rep. (II.) 5: 360.
   1808?

Galium concinnum T. & G. Fl. N. A. 2: 23.   1841.

Perennial, glabrous, shining, usually much
branched, the angles of the stem and edges of
the leaves minutely scabrous.   Leaves usually
all in 6's, linear or sometimes broader above
the middle, narrowed at the base, blunt-pointed
or minutely cuspidate, 4"-6" long, 1"-1½" wide,
green in drying; peduncles filiform; pedicels
short; flowers minute, white, numerous in open
cymes; fruit small, glabrous; endosperm deeply
grooved.

In dry woodlands, western New Jersey to Virginia, west to Minnesota, Kansas and Arkansas.
June–Aug.

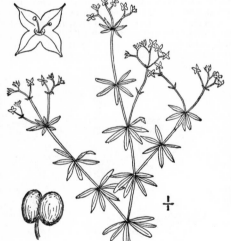

23. Galium aspréllum Michx.   Rough Bedstraw.   Fig. 3950.

Galium asprellum Michx. Fl. Bor. Am. 1: 78.   1803.
Perennial, weak, much branched and reclining
on bushes, or sometimes erect; stem retrorsely
hispid, 2°-6° long.   Leaves in 6's or 5's, or those
of the branches rarely in 4's, narrowly oval or
slightly oblanceolate, cuspidate at the apex, narrowed at the base, sometimes so much so as to
appear petioled, 4"-8" long, 1"-2" wide, their
margins and midribs rough; cymes terminal and
axillary, several–many-flowered; flowers white;
fruit smooth and glabrous, about 1" broad; endosperm with a shallow groove.

In moist soil, Newfoundland to western Ontario,
south to North Carolina, Illinois, Wisconsin and
Nebraska.   Called also pointed cleavers.   Ascends
to 3500 ft. in the Adirondacks.   June–Aug.

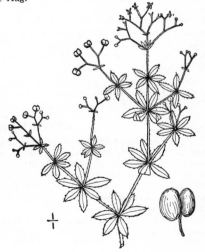

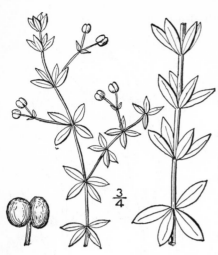

3/4

24. Galium bermudénse L. Coast Bed-straw. Fig. 3951.

Galium bermudense L. Sp. Pl. 105. 1753.
Galium hispidulum Michx. Fl. Bor. Am. 1: 79. 1803.
Relbunium bermudense Britten, Journ. Bot. 47: 42. 1909.

Perennial, much branched, hirsute, hispid or nearly glabrous, 1°–2° high. Leaves in 4's, 1-nerved, oval, mucronate, rather thick, 3″–10″ long, 1½″–4″ wide, the margins more or less revolute in drying; flowers few, terminating the branchlets, white; pedicels 3″–4″ long, rather stout, becoming deflexed in fruit, sometimes 1-bracteolate; fruit fleshy, purplish, minutely pubescent, becoming glabrate, about 2″ broad.

In dry or sandy soil, southern New Jersey to Florida and Georgia. Bermuda; Bahamas. May–Aug.

8. SHERARDIA [Dill.] L. Sp. Pl. 102. 1753.

Slender annual procumbent or diffuse herbs, with verticillate spiny-pointed leaves, and small nearly sessile pink or blue flowers, in terminal and axillary involucrate heads. Calyx-tube obovoid, its limb 4–6-lobed, the lobes lanceolate, persistent. Corolla funnelform, 4–5-lobed, the tube as long as the lobes or longer. Stamens 4 or 5, inserted on the tube of the corolla; filaments slender; anthers small, oblong, exserted. Ovary 2-celled; style 2-cleft at the summit; ovules 1 in each cavity. Fruit didymous, the carpels indehiscent. Seed erect. [Named for Dr. William Sherard, 1659–1728, patron of Dillenius.]

A monotypic genus of the Old World.

1. Sherardia arvénsis L. Blue Field Madder. Herb Sherard. Spurwort. Fig. 3952.

Sherardia arvensis L. Sp. Pl. 102. 1753.

Tufted, roughish, stems numerous, prostrate, ascending, or decumbent, 3′–10′ long. Leaves in 4's, 5's or 6's, the upper linear or lanceolate, acute and sharp-pointed, rough-ciliate on the margins, 3″–8″ long, 1″–2″ wide, the lower often obovate, mucronate; flowers in slender-peduncled involucrate heads, the involucre deeply 6–8-lobed, the lobes lanceolate, sharp-pointed; corolla-lobes spreading; fruit crowned with the 4–6 lanceolate calyx-teeth.

3/4

In waste places, Ontario to eastern Massachusetts and New Jersey. Also in Bermuda. Adventive from Europe. June–July.

9. ASPÉRULA L. Sp. Pl. 103. 1753.

Erect or ascending branching perennial herbs, with 4-angled stems, verticillate leaves, and small white pink or blue flowers in terminal or axillary, mostly cymose clusters. Calyx-tube somewhat didymous, the limb obsolete. Corolla funnelform, 4-lobed. Stamens 4, inserted on the tube or throat of the corolla; anthers linear or oblong. Ovary 2-celled; ovules 1 in each cavity; style 2-cleft. Fruit globose-didymous, the carpels indehiscent. Seed adherent to the pericarp; endosperm fleshy; embryo curved. [Latin diminutive of *asper*, rough, referring to the leaves.]

About 80 species, natives of the Old World. Type species: *Asperula odoràta* L.

Leaves oblong-lanceolate to obovate; fruit hispid.   1. *A. odorata.*
Leaves linear, 1″ wide or less; fruit smooth.   2. *A. galioides.*

1. Asperula odoràta L.   Sweet Woodruff.
Fig. 3953.

Asperula odorata L. Sp. Pl. 103. 1753.

Stems erect, slender, smooth. Leaves usually in 8's (6's–9's), thin, oblong-lanceolate, acute or obtuse, mucronate, 1-nerved, roughish on the margins, 6″–18″ long, the lower smaller, often obovate or oblanceolate; peduncles terminal and axillary, slender; cymes several-flowered; flowers white or pinkish, 1½″ long; pedicels 1″–2″ long; fruit very hispid, about 1″ broad.

In waste places, New Brunswick, N. J. Fugitive from Europe. Other English names are hay-plant, mugwet or mugget, rockweed, sweet hairhoof, woodrip, woodrowel, star-grass, and sweet-grass. May–July.

Asperula arvénsis L., another European species, with terminal capitate flowers, and linear obtuse leaves, has been found in waste places on Staten Island.

2. Asperula galioìdes Bieb.   Bedstraw Asperula.
Fig. 3954.

Asperula galioides Bieb. Fl. Taur. Cauc. **1** : 101.   1808.

Glaucous, glabrous, stems erect or ascending, 2½° high or less. Leaves linear, rigid, involute-margined, ½′–1½′ long, about 1″ wide, whorled in 5's–10's (often in 8's), subulate-tipped or mucronate; cymes panicled; flowers white; fruit smooth.

In fields, Connecticut to Michigan. Adventive from Europe. May–July.

Family 37.   **CAPRIFOLIÀCEAE** Vent. Tabl. **2** : 593.   1799.
Honeysuckle Family

Shrubs, trees, vines, or perennial herbs, with opposite simple or pinnate leaves, and perfect, regular or irregular, mostly cymose flowers. Stipules none, or sometimes present. Calyx-tube adnate to the ovary, its limb 3–5-toothed or 3–5-lobed. Corolla gamopetalous, rotate, campanulate, funnelform, urn-shaped, or tubular, the tube often gibbous at the base, the limb 5-lobed, sometimes 2-lipped. Stamens 5 (very rarely 4), inserted on the tube of the corolla and alternate with its lobes; anthers oblong or linear, versatile. Ovary inferior, 1–6-celled; style slender; stigma capitate, or 2–5-lobed, the lobes stigmatic at the summit; ovules anatropous, 1 or several in each cavity. Fruit a 1–6-celled berry, drupe, or capsule. Seeds oblong, globose, or angular; seed-coat membranous or crustaceous, smooth or cancellate; embryo usually small, placed near the hilum; radicle terete; cotyledons ovate.

About 10 genera and 300 species, mostly natives of the northern hemisphere, a few in South America and Australia.

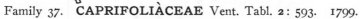

Corolla rotate or urn-shaped; flowers in compound cymes; styles deeply 2–5-lobed; shrubs or trees.
    Leaves pinnate; drupe 3–5-seeded.   1. *Sambucus.*
    Leaves simple; drupe 1-seeded.   2. *Viburnum.*
Corolla tubular or campanulate, often 2-lipped; style slender.
    Erect perennial herbs; leaves connate.   3. *Triosteum.*
    Creeping, somewhat woody herb; flowers long-peduncled, geminate.   4. *Linnaea.*

Shrubs or vines.
    Fruit a few-seeded berry.
        Corolla short, campanulate, regular, or nearly so.   5. *Symphoricarpos.*
        Corolla more or less irregular, tubular or campanulate.   6. *Lonicera.*
    Fruit a 2-celled capsule; corolla funnelform.   7. *Diervilla.*

1. SAMBUCUS [Tourn.] L. Sp. Pl. 269. 1753.

Shrubs or trees (or some exotic species perennial herbs), with opposite pinnate leaves, serrate or laciniate leaflets, and small white or pinkish flowers in compound depressed or thyrsoid cymes. Calyx-tube ovoid or turbinate, 3–5-toothed or 3–5-lobed. Corolla rotate or slightly campanulate, regular, 3–5-lobed. Stamens 5, inserted at the base of the corolla; filaments slender; anthers oblong. Ovary 3–5-celled; style short, 3-parted; ovules 1 in each cavity, pendulous. Drupe berry-like, containing 3–5, 1-seeded nutlets. Endosperm fleshy; embryo nearly as long as the seed. [Latin name of the elder.]

About 25 species, of wide geographic distribution. In addition to the following, about 10 others occur in western North America. Type species: *Sambucus nigra* L.
Cyme convex; fruit purplish black.   1. *S. canadensis.*
Cyme thyrsoid-paniculate, longer than broad; fruit red.   2. *S. racemosa.*

$\frac{3}{5}$

1. Sambucus canadénsis L. American Elder. Sweet or Common Elder. Fig. 3955.

Sambucus canadensis L. Sp. Pl. 269. 1753.

A shrub, 4°–10° high, glabrous or very nearly so, the stems but little woody, the younger ones with large white pith. Leaflets 5–11, usually 7, ovate or oval, acuminate or acute at the apex, short-stalked, glabrous above, sometimes pubescent beneath, 2′–5′ long, sharply serrate, sometimes stipellate; cymes convex, broader than high; flowers white, about 1½″ broad; drupe deep purple or black, nearly 3″ in diameter; nutlets roughened.

In moist soil, Nova Scotia to Florida, west to Manitoba, Kansas and Texas. Also in the West Indies. Ascends to 4000 ft. in North Carolina. Called also elder-blow, elder-berry. The flowers and fruit have strong medicinal properties. Leaves heavy-scented when crushed, those of young shoots often stipulate. June–July.

2. Sambucus racemòsa L. Red-berried Elder. Fig. 3956.

$\frac{3}{5}$

S. racemosa L. Sp. Pl. 270. 1753.
S. pubens Michx. Fl. Bor. Am. 1: 181. 1803.

A shrub, 2°–12° high, the twigs and leaves commonly pubescent; stems woody, the younger with reddish-brown pith. Leaflets 5–7, ovate-lanceolate or oval, acuminate at the apex, often narrowed and usually inequilateral at the base, 2′–5′ long, not stipellate, sharply serrate; cymes thyrsoid, longer than broad; flowers whitish, turning brown in drying; drupe scarlet or red, 2″–3″ in diameter; nutlets very minutely roughened.

In rocky places, Newfoundland to Alaska, British Columbia, Georgia, Michigan, Colorado and California. Fruit rarely white. April–May. Called also mountain elder. Poison-elder (Me.). Boor- or bore-tree. Boutry. Ascends to 5000 ft. in Virginia. *Sambucus pùbens dissécta* Britton, Mem. Torr. Club 5: 304. 1894, is a race with leaflets laciniate. Lake Superior and Pennsylvania.

Sambucus laciniàta Mill., a cut-leaved race of the related European *S. nigra* L., has been found at Cape May, N. J., perhaps escaped from cultivation.

2. VIBÚRNUM [Tourn.] L. Sp. Pl. 267. 1753.

Shrubs or trees, with entire dentate or lobed, sometimes stipulate leaves, and white or rarely pink flowers in compound cymes, the outer flowers sometimes radiant and neutral. Calyx-tube ovoid or turbinate, its limb short, 5-toothed. Corolla rotate or short-campanulate in our species, regular, 5-lobed. Stamens 5, inserted on the tube of the corolla; anthers oblong, exserted. Ovary 1–3-celled; style short, 3-lobed or 3-parted; ovules solitary in each cavity, pendulous. Drupe ovoid or globose, sometimes flattened, 1-seeded. Seed compressed; endosperm fleshy; embryo minute. [The ancient Latin name.]

About 100 species, of wide geographic distribution. Besides the following, about 5 others occur in the southern and western parts of North America. Type species: *Viburnum Tinus* L.

* Outer flowers of the cyme large, radiant; drupe red.

Leaves doubly serrate, pinnately veined.	1. *V. alnifolium.*
Leaves 3-lobed, palmately veined.	2. *V. Opulus.*

** None of the flowers radiant; drupe blue or black (red in no. 3).

1. Leaves palmately veined, or 3-ribbed.

Cymes ½′–1′ broad, the rays short; drupe red.	3. *V. pauciflorum.*
Cymes 1½′–2½′ broad, the rays slender; drupe nearly black.	4. *V. acerifolium.*

2. Leaves pinnately veined.

a. Leaves coarsely dentate, the veins mostly prominent beneath.

Leaves very short-petioled, pubescent.	5. *V. pubescens.*
Petioles 3″–20″ long.	
Leaves glabrous, or with tufts of hairs in the axils beneath.	6. *V. dentatum.*
Leaves pubescent beneath, the pubescence more or less stellate.	
Drupe globose-ovoid; eastern and southern.	
Veins of the leaves not very prominent.	7. *V. scabrellum.*
Veins very prominent on the under sides of the leaves.	8. *V. venosum.*
Drupe oblong, twice as long as thick; western.	9. *V. molle.*

b. Leaves entire, crenulate, or serrulate, the veins not prominent.

Native shrubs; drupes blue or black.

Cymes manifestly peduncled.	
Peduncles shorter than the cyme; leaves crenulate.	10. *V. cassinoides.*
Peduncle equalling or longer than the cyme; leaves mostly entire.	11. *V. nudum.*
Cymes sessile, or nearly so.	
Leaves prominently acuminate.	12. *V. Lentago.*
Leaves obtuse, or merely acute.	
Leaves and scarcely winged petioles glabrous, or nearly so.	13. *V. prunifolium.*
Veins of lower leaf-surfaces and winged petioles tomentose.	14. *V. rufidulum.*
European shrub, or small tree, escaped from cultivation; drupes red.	15. *V. Lantana.*

1. Viburnum alnifòlium Marsh. Hobble-bush. American Wayfaring Tree. Moose-bush or -berry. Fig. 3957.

V. alnifolium Marsh. Arb. Am. 102. 1785.
Viburnum lantanoides Michx. Fl. Bor. Am. 1: 179. 1803.

A shrub, with smooth purplish bark, sometimes reaching a height of 10°, widely and irregularly branching, the branches often procumbent and rooting, the youngest twigs scurfy. Leaves orbicular, or very broadly ovate, strongly pinnately veined, short-acuminate or acute at the apex, usually cordate at the base, finely stellate-pubescent, or at length glabrous above, scurfy with stellate pubescence on the veins beneath, finely serrate all around, 3′–8′ broad; petioles ½′–1½′ long; cymes sessile, 3′–5′ broad, the exterior flowers usually radiant and neutral, about 1′ broad; drupes red, becoming purple, ovoid-oblong, 5″–6″ long; stone 3-grooved on one side, 1-grooved on the other.

In low woods, New Brunswick to North Carolina, Ontario, Tennessee and Michigan. Leaves of shoots from cut stumps thin, ovate, corsely toothed. May–June. Tangle-legs or -foot. Dog-wood. Dogberry. Trip-toe. Witch-hopple or -hobble. Winter-buds naked.

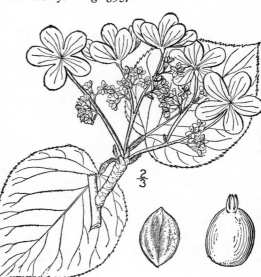

⅔

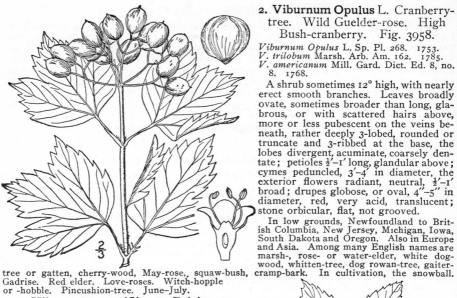

$\frac{2}{3}$

tree or gatten, cherry-wood, May-rose, squaw-bush, Gadrise. Red elder. Love-roses. Witch-hopple or -hobble. Pincushion-tree. June–July.

3. Viburnum pauciflòrum Pylaie. Few-flowered Cranberry-tree. Fig. 3959.

V. pauciflorum Pylaie; T.&G. Fl. N.A. **2**: 17. 1841.
Viburnum Opulus var. *eradiatum* Oakes, Hovey's Mag. **7**: 183. 1841.

A straggling shrub, 2°–6° high, with twigs and petioles glabrous or nearly so. Leaves broadly oval, obovate, or broader than long, 5-ribbed, truncate or somewhat cordate at the base, mostly with 3 rather shallow lobes above the middle, coarsely and unequally dentate, glabrous above, more or less pubescent on the veins beneath, 1½′–3′ broad; cymes peduncled, short-rayed, ½′–1′ broad; flowers all perfect and small; drupes globose to ovoid, light red, acid, 4″–5″ long; stone flat, orbicular, scarcely grooved.

In cold mountain woods, Newfoundland to Alaska, south to Maine, New Hampshire, Vermont, Pennsylvania, in the Rocky Mountains to Colorado, and to Washington. June–July.

2. Viburnum Opulus L. Cranberry-tree. Wild Guelder-rose. High Bush-cranberry. Fig. 3958.

Viburnum Opulus L. Sp. Pl. 268. 1753.
V. trilobum Marsh. Arb. Am. 162. 1785.
V. americanum Mill. Gard. Dict. Ed. 8, no. 8. 1768.

A shrub sometimes 12° high, with nearly erect smooth branches. Leaves broadly ovate, sometimes broader than long, glabrous, or with scattered hairs above, more or less pubescent on the veins beneath, rather deeply 3-lobed, rounded or truncate and 3-ribbed at the base, the lobes divergent, acuminate, coarsely dentate; petioles ½′–1′ long, glandular above; cymes peduncled, 3′–4′ in diameter, the exterior flowers radiant, neutral, ½′–1′ broad; drupes globose, or oval, 4″–5″ in diameter, red, very acid, translucent; stone orbicular, flat, not grooved.

In low grounds, Newfoundland to British Columbia, New Jersey, Michigan, Iowa, South Dakota and Oregon. Also in Europe and Asia. Among many English names are marsh-, rose- or water-elder, white dogwood, whitten-tree, dog rowan-tree, gaiter-cramp-bark. In cultivation, the snowball.

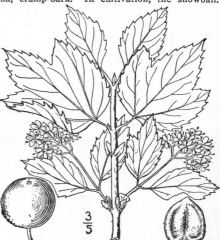

$\frac{3}{5}$

4. Viburnum acerifòlium L. Maple-leaved Arrow-wood. Fig. 3960.

V. acerifolium L. Sp. Pl. 268. 1753.

A shrub 3°–6° high, with smooth gray slender branches, and somewhat pubescent twigs and petioles. Leaves ovate, orbicular, or broader than long, cordate or truncate at the base, pubescent on both sides, or becoming glabrate, 2′–5′ broad, mostly rather deeply 3-lobed, coarsely dentate, the lobes acute or acuminate; petioles ½′–1′ long; cymes long-peduncled, 1½′–3′ broad; flowers all perfect, 2″–3″ broad; drupe nearly black, 3″–4″ long, the stone lenticular, faintly 2-ridged on one side and 2-grooved on the other.

In dry or rocky woods, New Brunswick to Georgia, Alabama, Ontario, Michigan and Minnesota. Upper leaves sometimes merely toothed, not lobed. May–June. Squash-berry. Maple-leaf guelder-rose. Dockmakie.

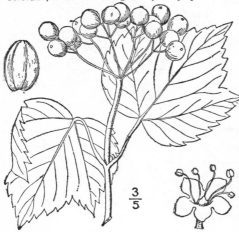

$\frac{3}{5}$

5. Viburnum pubéscens (Ait.) Pursh. Downy-leaved Arrow-wood
Fig. 3961.

Viburnum dentatum var. *pubescens* Ait. Hort. Kew. 1: 372. 1789.
V. pubescens Pursh, Fl. Am. Sept. 202. 1814.

A shrub, 2°–5° high, with numerous straight and slender gray branches. Leaves sessile, or on petioles less than 3″ long, ovate or oval, rounded or slightly cordate at the base, acute or acuminate at the apex, coarsely dentate, 1½′–3′ long, densely velvety-pubescent beneath, glabrous, or with scattered hairs above, or rarely glabrate on both surfaces; cymes peduncled, 1½′–2½′ broad, the flowers all perfect; drupes oval, nearly black, about 4″ long; stone slightly 2-grooved on both faces.

Rocky woods and banks, Quebec and Ontario to Manitoba, south, especially along the Alleghanies to Georgia and to Illinois, Iowa, Michigan and Wyoming. The leaves of shoots are sometimes entire or nearly so. June–July.

6. Viburnum dentàtum L. Arrow-wood. Fig. 3962.

Viburnum dentatum L. Sp. Pl. 268. 1753.

A shrub with slender glabrous gray branches, sometimes reaching a height of 15°. Twigs and petioles glabrous; petioles 3″–12″ long; leaves ovate, broadly oval or orbicular, rounded or slightly cordate at the base, acute or short-acuminate at the apex, prominently pinnately veined, coarsely dentate all around, 1½′–3′ broad, glabrous on both sides, or sometimes pubescent with simple hairs in the axils of the veins beneath; cymes long-peduncled, 2′–3′ broad; flowers all perfect; drupe globose-ovoid, about 3″ in diameter, blue, becoming nearly black; stone rather deeply grooved on one side, rounded on the other.

In moist soil, New Brunswick to Ontario, south along the mountains to Georgia and to western New York, Michigan and Minnesota. Called also mealy-tree. Withe-rod or -wood. May–June.

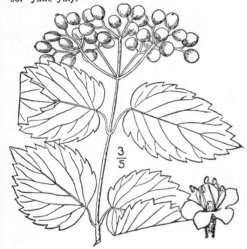

7. Viburnum scabréllum (T. & G.) Chapm. Roughish Arrow-wood. Fig. 3963.

V. dentatum semitomentosum Michx. Fl. Bor. Am. 1: 179. 1803.
V. dentatum var. (?) *scabrellum* T. & G. Fl. N. A. 2: 16. 1841.
V. scabrellum Chapm. Fl. S. States 172. 1860.
V. semitomentosum Rehder, Rhodora 6: 59. 1904.

Similar to the preceding species but the twigs, petioles, rays of the cyme and lower surfaces of the leaves more or less densely stellate-pubescent; petioles short and stouter; leaves usually larger, crenate or dentate, commonly somewhat pubescent above; drupe globose-ovoid, blue, 4″ in diameter, its stone similar to that of *V. dentatum*.

Woodlands and river banks, southern Pennsylvania to Kentucky, Florida and Texas. Referred, in our first edition, following previous authors, to *V. molle* Michx., a species long misunderstood.

$\frac{3}{5}$

sometimes bear reniform, very thin subcordate leaves with minute distant teeth.   June–July.

9. **Viburnum mólle** Michx.   Soft-leaved Arrow-wood.   Fig. 3965.

V. molle Michx. Fl. Bor. Am. **1**: 180. 1803.
Viburnum Demetrionis Deane & Robinson, Bot. Gaz. **22**: 167. *pl. 8.* 1896.

A shrub about 12° high, the older twigs at length grayish black, the bark exfoliating. Bud-scales acutish, ciliolate; leaves broadly ovate or nearly orbicular, short-acuminate at the apex, cordate or truncate at the base, 3′–5′ long, coarsely dentate, glabrous and bright green above, soft-pubescent and paler beneath, some of the pubescence stellate; petioles 8″–20″ long; stipules linear-filiform, 2″–5″ long; cymes terminal, peduncled, 4–7-rayed, glandular-puberulent; calyx-teeth ciliate; drupe oblong, obtuse at both ends, about 5″ long and 2½″ broad, much flattened, with 2 grooves when dry.

Bluffs and rocky woods, Kentucky, Missouri and Iowa. June.

8. **Viburnum venòsum** Britton.   Veiny Arrow-wood.   Fig. 3964.

Viburnum venosum Britton, Man. 871. 1901.
V. venosum Canbyi Rehder, Rhodora **6**: 60. 1904.

A shrub, 6° high or less, the bark grayish-brown, the young twigs stellate-pubescent or stellate-tomentose.   Leaves ovate to orbicular, 2′–5′ long, firm, coarsely and sharply dentate, glabrous or with sparse pubescence on the upper side when young, stellate-pubescent, at least on the veins, beneath, the petioles ¼–¾′ long; cymes long-stalked, often 3′–4′ broad, stellate-pubescent or glabrate; drupe globose or short-oval, 3″–4″ in diameter, nearly black.

Thickets, eastern Massachusetts to New Jersey, Pennsylvania and Virginia.   Shoots

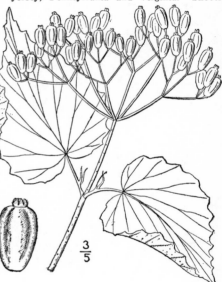

$\frac{3}{5}$

10. **Viburnum cassinoìdes** L.   Withe-rod.   Appalachian Tea.   Fig. 3966.

Viburnum cassinoides L. Sp. Pl. Ed. 2, 384. 1762.
Viburnum nudum var. *cassinoides* T. & G. Fl. N. A. **2**: 14. 1841.

A shrub, 2°–12° high, with ascending gray branches, the twigs somewhat scurfy, or glabrous.   Leaves ovate or oval, thick, pinnately veined, narrowed or sometimes rounded at the base, acute or blunt-acuminate at the apex, 1′–3′ long, usually crenulate, rarely entire, glabrous or very nearly so on both sides; peduncle shorter than or equalling the cyme; drupe pink, becoming dark blue, globose to ovoid, 3″–5″ long; stone round or oval, flattened.

In swamps and wet soil, Newfoundland to Manitoba, Minnesota, New Jersey, Georgia and Alabama.   June–July.   False Paraguay-tea.

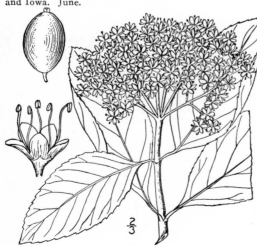

$\frac{2}{3}$

11. Viburnum nùdum L. Larger or Naked Withe-rod. Fig. 3967.

Viburnum nudum L. Sp. Pl. 268. 1753.
Viburnum nudum var. *Claytoni* T. & G. Fl. N. A. 2 : 14. 1841.

Similar to the preceding species, but usually a larger shrub, sometimes 15° high. Leaves oval, oval-lanceolate, or obovate, entire or obscurely crenulate, mostly larger (sometimes 9′ long), narrowed at the base, acute or obtuse at the apex, more prominently veined, sometimes scurfy on the upper surface; peduncle equalling or exceeding the cyme.

In swamps, Connecticut to Florida, west to Kentucky and Louisiana. Blooms a little later than the preceding species. Bilberry. Nannyberry. Possum- or Shawnee-haw.

13. Viburnum prunifòlium L. Black Haw. Stag-bush. Sloe. Fig. 3969.

Viburnum prunifolium L. Sp. Pl. 268. 1753.

A shrub or small tree somewhat similar to the preceding species; but the winter buds smaller, less acute, often reddish-pubescent. Leaves shorter-petioled, ovate or broadly oval, obtuse or acutish but not acuminate at the apex, narrowed or rounded at the base, 1′–3′ long, finely serrulate, glabrous or nearly so; petioles rarely margined; cyme sessile, several-rayed, 2′–4′ broad; flowers expanding with the leaves or a little before them; drupe oval, bluish-black and glaucous, 4″–5″ long; stone very flat on one side, slightly convex on the other, oval.

In dry soil, Connecticut to Georgia, west to Michigan, Kansas and Texas. Wood hard, reddish-brown; weight per cubic foot 52 lbs. April–June. Sheep- or nanny-berry. Fruit ripe in September, sweet and edible. A race, *Viburnum prunifòlium globòsum* Nash, has the drupe globose, about 3″ in diameter, smaller; New Jersey and eastern Pennsylvania.

12. Viburnum Lentàgo L. Nannyberry. Sheep- or Sweet-berry. Sweet Viburnum. Fig. 3968.

Viburnum Lentago L. Sp. Pl. 268. 1753.

A shrub, or often a small tree, sometimes 30° high and with a trunk diameter of 10′. Winter buds acuminate, glabrous; leaves slender-petioled, ovate, mostly rounded at the base, acuminate at the apex, 2′–4′ long, glabrous on both sides, or rarely a little pubescent beneath, sharply serrulate; petioles often broadened and wavy-margined, 9″–12″ long; cyme sessile, several-rayed, 2′–5′ broad; drupes oval to subglobose, bluish-black with a bloom, sweet and edible, 5″–6″ long; stone very flat, circular or oval.

In rich soil, Quebec to Hudson Bay, Manitoba, New Jersey, along the Alleghanies to Georgia, and to Indiana, Kansas and Colorado. Wood orange-brown, hard; weight 45 lbs. to the cubic foot. May–June. Fruit ripe in October. Nanny-bush. Black thorn or haw. Nanny-plum. Tea-plant (Wis.). Wild raisin.

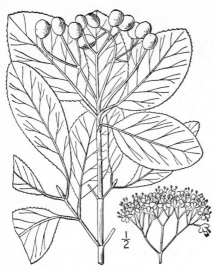

$\frac{1}{2}$

15. Viburnum Lantàna L. Wayfaring Tree. Fig. 3971.

Viburnum Lantana L. Sp. Pl. 268. 1753.

A shrub, or small tree, sometimes 12 ft. high, widely branched, the winter-buds naked. Young twigs, buds and petioles densely stellate-tomentose. Leaves ovate to ovate-elliptic, serrulate, dark green and loosely stellate-pubescent or glabrous above, paler and more or less stellate-tomentose beneath, 2'–4' long, rounded or acutish at the apex, subcordate at the base, the petioles stout and short; cymes short-stalked, stellate-tomentose, densely many-flowered, the flowers all alike, 3"–4" broad; drupe red, oval, 4"–5" long, its stone grooved.

14. Viburnum rufídulum Raf. Southern Black Haw. Fig. 3970.

Viburnum rufidulum Raf. Alsog. Am. 56. 1838.
Viburnum prunifolium var. *ferrugineum* T. & G. Fl. N. A. **2**: 15. 1841. Not *V. ferrugineum* Raf. 1838.
Viburnum rufotomentosum Small, Bull. Torr. Club **23**: 410. 1896.

A small tree, becoming 20° high. Leaves elliptic to obovate, mostly obtuse at the apex, finely and sharply serrate or serrulate, narrowed or obtuse at the base, the veins brown-tomentose beneath; petioles 3"–8" long, winged, brown-tomentose; cymes large, sessile, or very short-peduncled, the principal rays 3–5, mostly 4; flowers 3"–3½" broad; drupe oval, 5"–7" long, blue with a bloom; seed nearly orbicular.

In woods and thickets, New Jersey to Missouri, Kansas, Florida and Texas. Ascends to 3500 ft. in Virginia. April–May. Fruit ripe Aug.–Sept.

Viburnum obovàtum Walt., admitted into our first edition as recorded from Virginia, is not definitely known to range north of South Carolina.

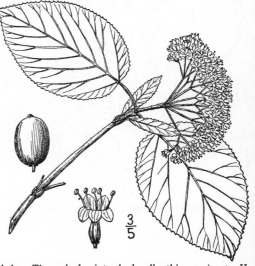

$\frac{3}{5}$

Roadsides, escaped from cultivation in New England. Native of Europe and Asia. The naked winter-buds ally this species to *V. alnifolium* Marsh. May–July.

3. TRIÓSTEUM L. Sp. Pl. 176. 1753.

Perennial herbs, with simple terete stems and opposite connate-perfoliate or sessile leaves narrowed below the middle. Flowers axillary, perfect, solitary or clustered, sessile, yellowish, green, or purplish, 2-bracted. Calyx-tube ovoid, its limb 5-lobed, the lobes elongated, persistent and sometimes foliaceous in our species. Corolla-tube narrow or campanulate, gibbous at the base, the limb oblique. Stamens 5, inserted on the corolla-tube; filaments very short; anthers linear, included. Ovary 3–5-celled; ovules 1 in each cavity; style filiform; stigma 3–5-lobed. Drupe coriaceous, orange or red, enclosing 2–3 (rarely 4–5) 1-seeded nutlets. Endosperm fleshy; embryo minute. [Greek, three-bone, from the 3 bony nutlets.]

Six known species, the following of eastern North America, two Japanese, one Himalayan. Type species: *Triosteum perfoliatum* L.

Leaves ovate or oval; flowers purplish or dull red.
    Leaves, or some of them, connate-perfoliate; fruit orange-yellow.   1. *T. perfoliatum.*
      Leaves narrowed to a sessile base; fruit orange-red.   2. *T. aurantiacum.*
Leaves lanceolate or oval-lanceolate; flowers yellowish.   3. *T. angustifolium.*

1. Triosteum perfoliàtum L. Fever-wort. Horse-Gentian. Fig. 3972.

Triosteum perfoliatum L. Sp. Pl. 176. 1753.

Stem erect, stout, finely glandular-pubescent, 2°–4° high. Leaves ovate to broadly oval, 4′–9′ long, 2′–4′ wide, acute or acuminate at the apex, abruptly narrowed at the base, connate-perfoliate, soft-pubescent beneath, somewhat hairy above, the margins entire or sinuate; bracts linear; corolla dull purplish-brown, greenish below, 6″–10″ long, viscid-pubescent, about the length of the calyx-lobes, the limb nearly regular; filaments bearded; drupe 4″–6″ long, obovoid-globose, orange-yellow, densely and finely pubescent; nutlets usually 3.

In rich soil, Massachusetts to Alabama, Kentucky and Kansas. Fever-root. Wild or wood ipecac. Tinker's-weed. Wild coffee. Horse-ginseng. White gentian. Genson (N. C.). Ascends 3000 ft. in Virginia. May–July.

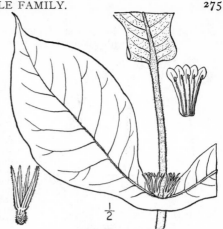

2. Triosteum aurantìacum Bicknell. Scarlet-fruited Horse-Gentian. Fig. 3973.

T. aurantiacum Bicknell, Torreya 1: 26. 1901.

Stems stout, erect, 2°–4° tall, glandular-puberulent to hirsute. Leaves thin, ovate, ovate-oblong or oblong-lanceolate, 5′–10′ long, acuminate at the apex, more or less dilated at the base, but not connate-perfoliate, minutely soft-pubescent beneath, thinly appressed-pubescent above or nearly glabrous; corolla dull-red, 7″–10″ long, often shorter than the calyx-lobes; filaments bearded throughout or nearly so; drupe oblong-ovoid or obovoid, 6″–7″ long, orange-red, densely short-pubescent; nutlets usually 3.

In rich woods and thickets, New Brunswick to Quebec, Minnesota, Missouri and North Carolina. May–June.

Triosteum angustifòlium L. Yellow or Narrow-leaved Horse-Gentian. Fig. 3974.

Triosteum angustifolium L. Sp. Pl. 176. 1753.

Stem slender and hirsute-pubescent, 1°–3° high. Leaves lanceolate or oval-lanceolate, acute or acuminate at the apex, 3′–5′ long, ½′–1½′ wide, rough-pubescent, tapering to the sessile base, or the lower smaller, obtuse and spatulate; corolla yellowish, 6″–7″ long; flowers commonly solitary in the axils.

In rich soil, Connecticut and Long Island to New Jersey, Pennsylvania, Virginia, Alabama, Illinois and Louisiana. May–Aug.

4. LINNAÈA [Gronov.] L. Sp. Pl. 631. 1753.

Creeping, somewhat woody herbs, with opposite evergreen petioled obovate or orbicular leaves, and perfect pink or purplish flowers borne in pairs at the summit of elongated terminal peduncles. Calyx-tube ovoid, the limb 5-lobed. Corolla campanulate or funnelform, 5-lobed, the lobes imbricate. Stamens 4, inserted near the base of the corolla-tube, didynamous, included. Ovary 3-celled, 2 of the cavities with several abortive ovules, the other with 1 perfect pendulous ovule. Fruit nearly globose, 3-celled, 2 of the cells empty, the other with a single oblong seed. Endosperm fleshy; embryo cylindric. [Named by Grono-vius for Linnaeus, with whom the plant was a favorite.]

Three or four species, of the north temperate zone. Type species: *Linnaea boreàlis* L., of Europe.

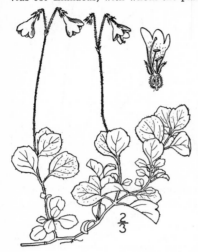

1. Linnaea americàna Forbes. Twin-flower. Ground-vine. Deer-vine. Fig. 3975.

Linnaea americana Forbes, Hort. Woburn. 135. 1825.
L. borealis var. *americana* Rehder, Rhodora 6: 56. 1904.

Branches slender, slightly pubescent, trailing, 6'–2° long. Petioles 1"–2" long; leaves obscurely crenate, thick, 3"–8" wide, sometimes wider than long; peduncles slender, erect, 2-bracted at the summit, 2-flowered (or rarely proliferously 4-flowered); pedicels filiform, 3"–10" long, 2-bracteolate at the summit; flowers nodding, 4"–6" long, fragrant; corolla funnelform; calyx-segments about 1" long; ovary subtended by a pair of ovate glandular scales which are connivent over the fruit or adnate to it.

In cold woods, mountains of Maryland, New Jersey, Long Island, north to Newfoundland, west through Brit-ish America to Alaska and Vancouver, south to Mich-igan, in the Rocky Mountains to Colorado and in the Sierra Nevada to California. Has been considered identical with the similar *L. borealis* L. of Europe and Asia, which has a nearly bell-shaped corolla and longer calyx. Twin sisters. Two-eyed berries. June–Aug.

5. SYMPHORICÀRPOS [Dill.] Ludwig, Def. 35. 1760.

Shrubs, with opposite deciduous short-petioled simple leaves, and small white or pink, perfect flowers, in axillary or terminal clusters. Calyx-tube nearly globular, the limb 4–5-toothed. Corolla campanulate or salverform, regular, or sometimes gibbous at the base, 4–5-lobed, glabrous or pilose in the throat; stamens 4 or 5, inserted on the corolla. Ovary 4-celled, 2 of the cavities containing several abortive ovules, the other two each with a single suspended ovule; style filiform; stigma capitate, or 2-lobed. Fruit an ovoid or globose 4-celled 2-seeded berry. Seeds oblong; endosperm fleshy; embryo minute. [Greek, fruit borne together, from the clustered berries.]

About 10 species, natives of North America and the mountains of Mexico. Known as St. Peter's-wort. Type species: *Lonicera Symphoricarpos* L.

Fruit white; style glabrous.
    Stamens and style included; clusters usually few-flowered.   1. *S. racemosus.*
    Stamens and style somewhat exserted; clusters many-flowered.   2. *S. occidentalis.*
Fruit red; style bearded.   3. *S. Symphoricarpos.*

1. Symphoricarpos racemòsus Michx. Snowberry. Wax-berry. Fig. 3976.

Symphoricarpos racemosus Michx. Fl. Bor. Am. 1: 107. 1803.
S. racemosus var. *pauciflorus* Robbins; A. Gray, Man. Ed. 5, 203. 1867.
S. pauciflorus Britton, Mem. Torr. Club 5: 305. 1894.

An erect or diffuse shrub, 1°–4° high, glabrous, or usually so, the branches slender. Petioles about 2" long; leaves oval, obtuse at each end, sometimes pubescent or whitened beneath, ½'–2' long, entire, undulate, or those of young shoots sometimes dentate; axillary clusters few-flowered, the terminal one mostly interruptedly spicate; corolla campanulate, about 3" long, slightly gib-bous at the base, bearded within; style glabrous; stamens and style included; berry snow-white, globose, loosely cellu-lar, 2½"–5" in diameter.

In rocky places and on river shores, Nova Scotia and Quebec to British Colum-bia, south to Pennsylvania, Kentucky, Min-nesota, South Dakota, Montana and in California. Commonly planted and some-times escaped from cultivation. Races differ in size, habit and pubescence. Snowdrop-berry. Egg-plant. June–Sept.

3. Symphoricarpos occidentàlis
Hook.　Wolfberry.　Fig. 3977.

Symphoricarpos occidentalis Hook. Fl. Bor.
Am. **1**: 285. 1833.

Similar to *S. racemosus* but stouter, with
larger leaves, 1′–3′ long, more or less pubescent beneath, entire, or often undulate-
crenate; petioles 2″–3″ long; axillary
clusters spicate, many-flowered, 6″–12″
long; corolla funnelform-campanulate, 3″
long, lobed to beyond the middle; stamens
and glabrous style somewhat exserted;
berry nearly globular, white, 4″–5″ in
diameter.

Rocky situations, Illinois, Michigan and
Minnesota to British Columbia, Kansas and
Colorado.　Buck-bush.　June–July.

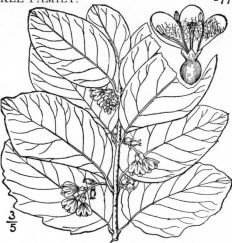

$\frac{3}{5}$

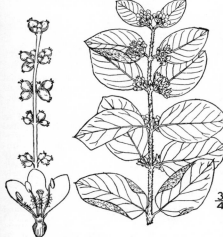

$\frac{3}{4}$

4. Symphoricarpos Symphoricàrpos
(L.) MacM.　Coral-berry.　Indian
Currant.　Fig. 3978.

Lonicera Symphoricarpos L. Sp. Pl. 175. 1753.
S. orbiculatus Moench, Meth. 503. 1794.
Symphoricarpos vulgaris Michx. Fl. Bor. Am. **1**:
106. 1803.
Symphoricarpos Symphoricarpos MacM. Bull.
Torr. Club **19**: 15. 1892.

A shrub, 2°–5° high, the branches erect or
ascending, purplish, usually pubescent. Petioles
1″–2″ long; leaves oval or ovate, entire or
undulate, mostly obtuse at each end, glabrous
or nearly so above, usually soft-pubescent
beneath, 1′–1½′ long; clusters dense, many-
flowered, at length spicate, shorter than the
leaves; corolla campanulate, sparingly pubescent within, pinkish, about 2″ long; style
bearded; stamens included; berry purplish
red, ovoid-globose, 1½″–2″ long.

Along rivers and in rocky places, banks of the Delaware in New Jersey and Pennsylvania, west
to western New York and South Dakota, south to Georgia, Kansas and Texas.　Also sparingly
escaped from cultivation farther east.　Fruit persistent after the leaves have fallen.　Buck-bush.
Turkey- or snap-berry.　July.

6. LONÍCERA L. Sp. Pl. 173.　1753.
Erect or climbing shrubs, with opposite mostly entire leaves; flowers spicate, capitate
or geminate, usually somewhat irregular. Calyx-tube ovoid or nearly globular, the limb
slightly 5-toothed. Corolla tubular, funnelform, or campanulate, often gibbous at the base,
the limb 5-lobed, more or less oblique, or 2-lipped. Stamens 5, inserted on the tube of the
corolla; anthers linear or oblong. Ovary 2–3-celled; ovules numerous in each cavity, pendulous; style slender; stigma capitate. Berry fleshy, 2–3-celled or rarely 1-celled, few-seeded.
Seeds ovoid or oblong with fleshy endosperm and a terete embryo.　[Named for Adam
Lonitzer, 1528–1586, a German botanist.]

About 160 species, natives of the north temperate zone, a few in tropical regions.　Besides the
following, some 10 others occur in the western parts of North America.　Type species: *Lonicera
Caprifòlium* L.

* Climbing or trailing vines; flowers in heads or interrupted spikes;
upper leaves connate-perfoliate.

Corolla 2-lipped, the upper lip 4-lobed, the lower entire.
　　Corolla glabrous within.　　　　　　　　　　　　　　　　　　1. *L. Caprifolium.*
　　Corolla pubescent within.
　　　　Leaves pubescent, at least beneath; corolla yellow.
　　　　　　Leaves pubescent on both sides, at least when young, ciliate; corolla slightly gibbous
　　　　　　　　at base.　　　　　　　　　　　　　　　　　　　2. *L. hirsuta.*
　　　　　　Leaves glabrous above, pubescent beneath; corolla-tube strongly gibbous at the base.
　　　　　　　　　　　　　　　　　　　　　　　　　　　　　3. *L. glaucescens.*

　　　　Leaves glabrous on both sides, very glaucous beneath.
　　　　　　Corolla greenish-yellow, the tube somewhat gibbous.
　　　　　　　　Corolla-tube 3″–5″ long; filaments hirsute at the base.　　4. *L. dioica.*

Corolla-tube 5″–7″ long; filaments nearly glabrous.   5. *L. Sullivantii.*
Corolla bright yellow or orange, its slender tube not gibbous.   6. *L. flava.*
Corolla tubular, the short limb nearly equally 5-lobed.   7. *L. sempervirens.*

** Climbing vine; flowers in pairs on short axillary peduncles.
     8. *L. japonica.*

*** Shrubs; flowers in pairs on axillary bracted peduncles.
Bracts of the peduncle subulate, linear, minute, or none.
   Leaves rarely cordate, more or less pubescent, or ciliate.
     Leaves pale, or glaucous, thick, strongly reticulate-veined.
       Peduncles shorter than the flowers; fruit blue; leaves ciliate.   9. *L. coerulea.*
       Peduncles equalling the flowers; fruit red; leaves not ciliate.   10. *L. oblongifolia.*
     Leaves bright green, thin, ciliate, not strongly reticulate; fruit red.   11. *L. canadensis.*
     Leaves pale, densely pubescent beneath, even when old.   12. *L. Xylosteum.*
   Leaves cordate, glabrous.   13. *L. tatarica.*
Bracts of the peduncle broad, foliaceous.   14. *L. involucrata.*

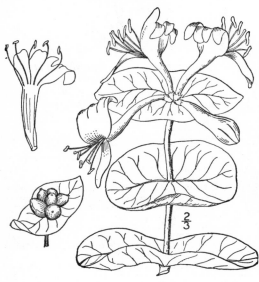

1. Lonicera Caprifòlium L. Italian or Perfoliate Honeysuckle. Fig. 3979.

L. Caprifolium L. Sp. Pl. 173. 1753.
L. grata Ait. Hort. Kew. 1: 231. 1789.
Caprifolium gratum Pursh, Fl. Am. Sept.
   161. 1814.

Climbing high, glabrous and somewhat glaucous. Upper one to three pairs of leaves connate-perfoliate, glaucous beneath, the others sessile or short-petioled, oval or obovate, all rounded at the base, entire; flowers in terminal capitate sessile clusters; corolla glabrous within, 1′–1½′ long, purple without, the limb white within, strongly 2-lipped; upper lip 4-lobed, the lower one narrow, reflexed; tube slightly curved, not gibbous; stamens and style much exserted; berries red.

Thickets, New York, New Jersey and Pennsylvania to Michigan, Missouri, and in the Southern States. Escaped from cultivation and naturalized. Native of Europe. Called also american or fragrant woodbine. May–June.

2. Lonicera hirsùta Eaton. Hairy Honeysuckle. Fig. 3980.

L. hirsuta Eaton, Man. Ed. 2, 307. 1818.

Twining, the branches hirsute and glandular-pubescent. Upper one or two pairs of leaves connate-perfoliate, the others oval or ovate, short-petioled or sessile, softly pubescent beneath, dark green and appressed-pubescent above, ciliate, obtuse or obtusish at the apex, rounded or narrowed at the base, 2′–3½′ long; flowers verticillate in short terminal interrupted spikes; corolla pubescent within, 1′–1½′ long, viscid-pubescent without, orange-yellow, turning reddish, the tube slender, somewhat gibbous at the base, the limb strongly 2-lipped, about as long as the tube or shorter; filaments hirsute below.

In woodlands, Vermont and Ontario to Manitoba, Pennsylvania, Ohio and Michigan. Rough woodbine. June–July.

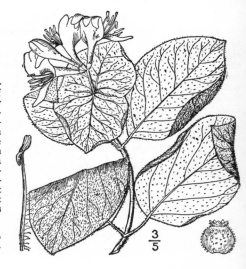

3. Lonicera glaucéscens Rydb. Douglas' Honeysuckle. Fig. 3981.

Lonicera Douglasii Hook. Fl. Bor. Am. 1: 282. 1833. Not *Caprifolium Douglasii* Lind. 1830.
Lonicera glaucescens Rydb. Bull. Torr. Club **24**: 90. 1897.

Similar to the preceding species, the branches glabrous. Leaves glabrous above, pubescent, at least on the veins, beneath, $1\frac{1}{2}'$–$2'$ long, chartaceous-margined, not ciliate, usually only the upper pair connate-perfoliate; flowers verticillate in a short terminal interrupted spike; corolla yellow, changing to reddish, pubescent or puberulent without, pubescent within, $1'$ long, or less, the tube rather strongly gibbous at the base, the 2-lipped limb shorter than the tube; stamens nearly glabrous, or somewhat pubescent; style hirsute; both exserted; ovary sometimes hirsute.

Ontario to Alberta, Pennsylvania, North Carolina, Ohio and Nebraska. May–June.

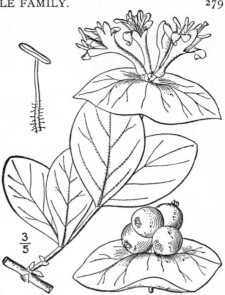

In rocky and usually dry situations, Quebec to Manitoba, south, especially along the mountains to North Carolina, and to Ohio and Missouri. Ascends to 3500 ft. in North Carolina. All the leaves of young shoots are sometimes connate-perfoliate. Small yellow or crimson honeysuckle. Small woodbine. May–June.

5. Lonicera Sullivántii A. Gray. Sullivant's Honeysuckle. Fig. 3983.

Lonicera Sullivantii A. Gray, Proc. Amer. Acad. **19**: 76. 1883.

Similar to the preceding species, very glaucous. Leaves oval or obovate, glaucous and commonly puberulent beneath, obtuse; flowers larger than those of the preceding species, the tube $5''$–$7''$ long, slightly exceeding the limb, pale yellow; stamens usually nearly glabrous; fruit yellow, $3''$ in diameter.

In woodlands, Tennessee, Ohio and western Ontario to Iowa, Wisconsin and Minnesota. Recorded from Manitoba. May–June.

4. Lonicera diòica L. Smoothed-leaved or Glaucous Honeysuckle. Fig. 3982.

Lonicera dioica L. Syst. Ed. 12, 165. 1767.
L. glauca Hill, Hort. Kew. 446. *pl. 18.* 1769.
L. parviflora Lam. Encycl. 1: 728. 1783.

Glabrous throughout, twining or shrubby, $3°$–$10°$ long. Leaves very glaucous beneath, $1\frac{1}{2}'$–$3'$ long, the upper connate-perfoliate, oval, obtuse, the lower sessile or short-petioled, narrower; flowers several in a terminal cluster, yellowish green and tinged with purple, glabrous without, pubescent within, the tube $3''$–$4''$ long, gibbous at the base, scarcely longer than the 2-lipped limb; stamens hirsute below, exserted with the style; berries red, $3''$–$4''$ in diameter.

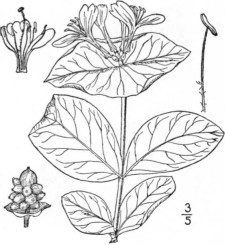

6. Lonicera flàva Sims. Yellow Honeysuckle. Fig. 3984.

Lonicera flava Sims, Bot. Mag. *pl. 1318.* 1810.

Twining to a height of several feet, or trailing, glabrous. Leaves broadly oval, or elliptic, entire, obtuse at the apex, narrowed or rounded at the base, short-petioled, or the upper sessile, green above, glaucous beneath, the pairs subtending flowers connate-perfoliate; flowers bright orange-yellow, fragrant, in a terminal interrupted spike; corolla 1′–1½′ long, the slender tube pubescent above within, not gibbous at the base, the limb strongly 2-lipped, about half as long as the tube; filaments and style glabrous, exserted; fruit about 3″ in diameter.

North Carolina to Kentucky, Missouri, Georgia and Alabama. April–May.

7. Lonicera sempérvirens L. Trumpet or Coral Honeysuckle. Fig. 3985.

Lonicera sempervirens L. Sp. Pl. 173. 1753.
Lonicera sempervirens hirsutula Rehder, Rep. Mo. Bot Gard. 14: 169 1903.

Glabrous or somewhat pubescent, high climbing, evergreen in the South. Leaves oval, obtuse, 2′–3′ long, or the lower ones smaller, narrower and acutish, the upper pairs connate-perfoliate, all conspicuously glaucous and sometimes slightly pubescent beneath, dark green above; flowers verticillate in terminal interrupted spikes; corolla scarlet or yellow, 1′–1½′ long, glabrous or somewhat pubescent, the tube narrow, slightly expanded above, the limb short and nearly regular; stamens and style scarcely exserted; berries scarlet, about 3″ in diameter.

In low grounds, or on hillsides, Maine to Florida, New Hampshire, New York, Nebraska and Texas. Woodbine (N. C.). Scarlet trumpet-honeysuckle. April–Sept.

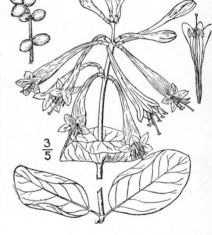

8. Lonicera japónica Thunb. Japanese or Chinese Honeysuckle. Fig. 3986.

Lonicera japonica Thunb. Fl. Jap. 89. 1784

Pubescent, climbing high or trailing. Leaves all short-petioled, ovate, entire, 1′–3′ long, acute at the apex, rounded at the base, dark green and glabrous above, pale and usually sparingly pubescent beneath; flowers in pairs from the upper axils, peduncled, leafy-bracted at the base, white or pink, fading to yellow, pubescent without, the tube nearly 1′ long, longer than the strongly 2-lipped limb; stamens and style exserted; berries black, 3″–4″ in diameter.

Freely escaped from cultivation, Connecticut, New York and Pennsylvania to North Carolina, Florida and West Virginia. Naturalized from eastern Asia. June–Aug.

9. Lonicera coerùlea L.　Blue or Mountain Fly-Honeysuckle.　Fig. 3987.

Lonicera coerulea L Sp. Pl. 174. 1753.
Lonicera villosa Muhl. Cat. 23. 1813.

Erect, shrubby, 1°–3° high, the twigs sometimes slightly pubescent. Leaves oval or obovate, 1′–1½′ long, very obtuse at the apex, rounded or narrowed at the base, thick, conspicuously reticulate-veined, pale and more or less pubescent beneath, glabrous above, at least when mature, ciliate on the margins; flowers in pairs in the axils, short-peduncled, subulate-bracted, yellow, 6″–8″ long; corolla pubescent, or glabrate, the tube gibbous at the base, the limb nearly regular, its lobes oblong, equalling or slightly exceeding the tube; ovaries of the two flowers becoming united and forming an oblong or nearly globose, bluish-black 2-eyed berry, about 2½″ in diameter.

In low grounds, Newfoundland to Alaska, south to Rhode Island, Pennsylvania, Wisconsin, Wyoming and California. Europe and Asia. June.

Lonicera Mórrowi A. Gray, a Japanese species, with red berries and nearly white flowers, has escaped from cultivation in Massachusetts.

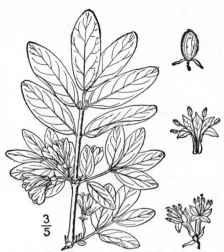

$\frac{3}{5}$

10. Lonicera oblongifòlia (Goldie) Hook.　Swamp Fly-Honeysuckle.　Fig. 3988.

Xylosteum oblongifolium Goldie, Edinb. Phil. Journ. 6: 323. 1822.
Lonicera oblongifolia Hook. Fl. Bor. Am. 1: 284. *pl. 100.* 1833.
L. altissima Jennings, Ann. Carn. Mus. 4: 74. *pl. 20.* 1906.

Leaves 1′–2′ long, reticulate-veined, glabrous or nearly so on both sides when mature, downy-pubescent when young, not ciliate; flowers in pairs on long slender peduncles; corolla yellow, or purplish within, 8″–9″ long, gibbous at the base, deeply 2-lipped; bracts minute or none; ovaries remaining distinct, or becoming united, the berries red or crimson.

In swamps, New Brunswick and Quebec to Manitoba, New York, Pennsylvania, Michigan and Minnesota. May–June.

11. Lonicera canadènsis Marsh.　American Fly-Honeysuckle.　Fig. 3989.

Vaccinium album L. Sp. Pl. 350. 1753. Not *L. alba* L.
Lonicera canadensis Marsh. Arb. 81. 1785.
Lonicera ciliata Muhl. Cat. 23. 1813.

Shrubby, 3°–5° high, the twigs glabrous. Petioles 2″–3″ long, very slender; leaves thin, 1′–4′ long, bright green on both sides, ovate or sometimes oval, acute or acutish at the apex, rounded or cordate at the base, villous-pubescent beneath when young, glabrous or nearly so when mature, but the margins strongly ciliate; flowers in pairs from the axils, greenish-yellow, about 8′ long; peduncles long-filiform; bracts very small, subulate; corolla-limb nearly regular, its lobes short; berries separate, ovoid, light red, about 3″ thick.

In moist woods, Nova Scotia to Saskatchewan, Connecticut, Pennsylvania, Indiana, Michigan and Minnesota. Medaddy-bush. May.

$\frac{2}{3}$

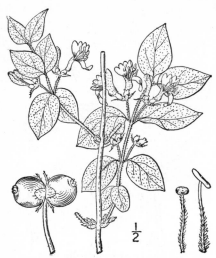

12. Lonicera Xylósteum L. Fly-Honey-suckle. Fig. 3990.

Lonicera Xylosteum L. Sp. Pl. 174. 1753.

A shrub, 3°–7° high, the foliage densely appressed-pubescent when young. Leaves ovate, oval, or obovate, entire, short-petioled, rather pale green, obtuse, or the upper acute at the apex, obtuse, subcordate or narrowed at the base, glabrous above when mature, persistently pubescent beneath, 1′–3′ long; petioles 2″–4″ long; peduncles axillary, 2-flowered, 4″–8″ long, about as long as the flowers, or longer; flowers yellowish white; bracts linear-subulate; berries scarlet.

Escaped from cultivation, Rhode Island, New York and New Jersey. Native of Europe and Asia. May–June.

13. Lonicera tatárica L. Tartarian Bush-Honeysuckle. Fig. 3991.

Lonicera tatarica L. Sp. Pl. 173. 1753.

A glabrous shrub, 5°–10° high. Leaves ovate, rather thin, not conspicuously reticulate-veined, 1′–3′ long, acute or obtusish at the apex, cordate at the base, not ciliate; flowers in pairs on slender axillary peduncles; corolla pink to white, 7″–8″ long, the tube gibbous at the base, the limb irregularly and deeply 5-lobed, somewhat 2-lipped; peduncles 1′ long; bracts linear, sometimes as long as the corolla-tube; stamens and style scarcely exserted; berries separate, red.

Escaped from cultivation, Ontario, Maine and Vermont to southern New York, New Jersey and Kentucky. May. Native of Asia. Garden fly-honeysuckle.

14. Lonicera involucràta (Richards.) Banks. Involucred Fly-Honeysuckle. Fig. 3992.

Xylosteum involucratum Richards. App. Frank. Journ. Ed. 2, 6. 1823.
Lonicera involucrata Banks; Richards. loc. cit. 1823.
Distegia involucrata Rydb. Bull. Torr. Club 33: 152. 1906.

A glabrate or pubescent shrub, 3°–10° high. Leaves short-petioled, ovate, oval, or obovate, 2′–6′ long, acute or acuminate at the apex, narrowed or rounded at the base, more or less pubescent, at least when young; peduncles axillary, 1′–2′ long, 2–3-flowered; bracts foliaceous, ovate or oval, often cordate; bractlets also large, at length surrounding the fruit; flowers yellow; corolla pubescent, funnelform, the limb nearly equally 5-lobed; lobes short, little spreading; stamens and style slightly exserted; berries separate, globose, or oval, nearly black, about 4″ in diameter.

In woodlands, New Brunswick and Quebec to western Ontario and Michigan, west to British Columbia and Alaska, south to Arizona, Utah and California. June–July.

7. DIERVÍLLA [Tourn.] Mill. Gard. Dict. Abr. Ed. 8. 1754.

Shrubs, with opposite leaves, and yellow axillary and terminal cymose or solitary flowers. Calyx-tube slender, elongated, narrowed below, the limb with 5 linear persistent lobes. Corolla narrowly funnelform, the tube slightly gibbous at the base, the limb nearly regular, 5-lobed. Stamens 5, inserted on the corolla; anthers linear. Ovary 2-celled; ovules numerous in each cavity; style filiform; stigma capitate. Fruit a linear-oblong capsule, narrowed or beaked at the summit, septicidally 2-valved, many-seeded. Seed coat minutely reticulated; endosperm fleshy; embryo minute. [Named for Dr. Dierville, who brought the plant to Tournefort.]

Three species, the following typical one, the others in the mountains of the Southern States. The Japanese and Chinese *Weigelas*, often referred to this genus, are here regarded as distinct.

1. Diervilla Diervílla (L.) MacM. Bush-Honeysuckle. Fig. 3993.

Lonicera Diervilla L. Sp. Pl. 175. 1753.
Diervilla Lonicera Mill. Gard. Dict. Ed. 8. 1768.
Diervilla trifida Moench, Meth. 492. 1794.
D. Diervilla MacM. Bull. Torr. Club **19**: 15. 1892.

A shrub, 2°–4° high, glabrous or nearly so throughout, with terete branches. Leaves short-petioled, ovate or oval, acuminate at the apex, usually rounded at the base, 2'–5' long, irregularly crenulate and often slightly ciliate on the margins; peduncles terminal, or in the upper axils, slender, 1–5-flowered; flowers about 9" long; corolla more or less pubescent both without and within, regular or slightly irregular, 3 of its lobes somewhat united; capsule glabrous, linear-oblong, slender, beaked, crowned with the persistent calyx-lobes.

In dry or rocky woodlands, Newfoundland to Manitoba, North Carolina, Michigan and Wisconsin. Gravel-weed. Life-of-man. May–June.

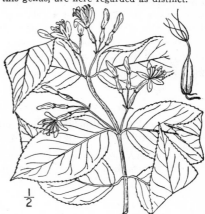

$\frac{1}{2}$

Family 38. ADOXÀCEAE Fritsch; Engl. & Prantl, Nat. Pfl. Fam. **4**⁴: 170. 1891.

MOSCHATEL FAMILY.

A glabrous perennial slender herb, with scaly or tuberiferous rootstocks, basal and opposite ternately compound leaves, and small green flowers in terminal capitate clusters. Calyx-tube hemispheric, adnate to the ovary, its limb 2–3-toothed. Corolla rotate, regular, 4–6-lobed. Stamens twice as many as the lobes of the corolla, inserted in pairs on its tube; filaments short; anthers peltate, 1-celled. Ovary 3–5-celled; style 3–5-parted; ovules 1 in each cavity, pendulous. Fruit a small drupe with 3–5-nutlets. Endosperm cartilaginous.

The family contains only the following monotypic genus of the north temperate zone.

1. ADÓXA L. Sp. Pl. 367. 1753.

Characters of the family. [Greek, without glory, *i. e.*, insignificant.]

1. Adoxa Moschatéllina L. Musk-root. Hollow-root. Moschatel. Fig. 3994.

Adoxa Moschatellina L. Sp. Pl. 367. 1753.

Stems simple, weak, erect, 3'–6' high, bearing a pair of opposite ternate leaves usually above the middle. Basal leaves 1–4, long-petioled, ternately compound, the segments broadly ovate or orbicular, obtuse, thin, 3-cleft or 3-parted, the lobes obtuse and mucronulate; head 3"–4" in diameter, composed of 3–6-flowers; corolla of the terminal flower 4–5-lobed, those of the others usually 5–6-lobed; drupe green, bearing the persistent calyx-lobes above the middle.

In shaded rocky places, Arctic America, south to Iowa, Wisconsin, South Dakota and Colorado. Also in northern Europe and Asia. Other English names are bulbous fumitory, glory-less, musk-crowfoot or wood-crowfoot. Odor musky. May.

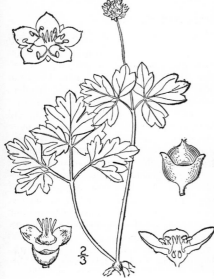

$\frac{2}{3}$

Family 39. **VALERIANÀCEAE** Batsch, Tabl. Aff. *227*. 1802.

VALERIAN FAMILY.

Herbs with opposite leaves, no stipules, and usually small perfect or polygamo-dioecious flowers, in corymbed panicled or capitate cymes. Calyx-tube adnate to the ovary, its limb inconspicuous or none in flower, often becoming prominent in fruit. Corolla gamopetalous, epigynous, somewhat irregular, its tube narrowed, and sometimes gibbous or spurred at the base, its limb spreading, mostly 5-lobed. Stamens 1–4, inserted on the corolla and alternate with its lobes, usually exserted. Ovary inferior, 1–3-celled, one of the cavities containing a single anatropous ovule, the others empty. Fruit indehiscent, dry, containing a single suspended seed. Endosperm little or none; embryo straight; cotyledons oblong.

About 9 genera and 300 species, of wide distribution, most abundant in the northern hemisphere.

Fruit 1-celled; persistent calyx-lobes becoming awn-like; tall herbs.

1. *Valeriana.*

Fruit 3-celled; calyx-lobes minute or none; low herbs.

2. *Valerianella.*

1. **VALERIÁNA** [Tourn.] L. Sp. Pl. 31. 1753.

Perennial, strong-smelling, mostly tall herbs, the leaves mainly basal and the cymose flowers paniculate in our species. Calyx-limb of 5–15 bristle-like plumose teeth, short and inrolled in flower, but elongated, rolled outward and conspicuous in fruit. Corolla funnel-form or tubular, usually more or less gibbous at the base, the limb nearly equally 5-lobed. Stamens commonly 3. Style entire, or minutely 2–3-lobed at the summit. Fruit compressed, 1-celled, 1-nerved on the back, 3-nerved on the front. [Name Middle Latin, from *valere,* to be strong.]

About 175 species, mostly in the temperate and colder parts of the north temperate zone and the Andes of South America. Besides the following, 5 others occur in southern and western North America. Type species: *Valeriana pyrenàica* L.

Corolla-tube very slender, 6″–10″ long; basal leaves cordate. 1. *V. pauciflora.*
Corolla-tube 1″–3″ long; basal leaves not cordate.
  Leaves thick, parallel-veined, entire, or the segments not dentate; roots fusiform.
      2. *V. edulis.*
  Leaves thin, reticulate-veined, the segments dentate; roots fibrous.
    Lower leaves spatulate, often entire; plants glabrous.
      Segments of middle stem-leaves 9–13, usually sinuate-dentate; corolla 1.7″–2.5″ wide.
        3. *V. uliginosa.*
      Segments of middle stem-leaves 5–7, entire; corolla 1.5″ wide or less. 4. *V. septentrionalis.*
    All the leaves pinnately divided; plants pubescent, especially at the nodes.
      5. *V. officinalis.*

1. **Valeriana pauciflòra** Michx. Large-flowered Valerian. Fig. 3995.

V. pauciflora Michx. Fl. Bor. Am. 1 : 18. 1803.

Rootstocks slender, usually horizontal. Stem glabrous, erect or ascending, 1°–3° high, often sending out runners from the base; leaves thin, the basal ones slender-petioled, simple, or sometimes with a pair of small leaflets on the petiole, broadly ovate, cordate, acute at the apex, the margins crenate or dentate; stem leaves pin-nately 3–7-divided, the terminal segment larger than the others; cymes terminal, clustered; flowers few or numerous; co-rolla pink, its tube very slender, 6″–10″ long; bracts linear; fruit oblong or oblong-lanceolate, about 3″ long, glabrous or puberulent; bristles of the calyx at length elongated and plumose.

In moist soil, Pennsylvania to West Virginia, Illinois, Kentucky, Tennessee and Mis-souri. American wild valerian. May–June.

2. Valeriana édulis Nutt. Edible Valerian. Tobacco-root. Fig. 3996.

V. edulis Nutt. in T. & G. Fl. N. A. **2** : 48. 1841.

Erect, 1°–4° high, from a deep fusiform carrot-shaped root. Stem glabrous, or nearly so, the young leaves commonly more or less pubescent and the older ones finely ciliate, sometimes glabrous; basal leaves spatulate or oblanceolate, thick, 3'–12' long, 2"–10" wide, obtuse at the apex, narrowed into a margined petiole, parallel-veined, entire or with a few obtuse entire lobes; stem leaves few, sessile, pinnately-parted into linear or lanceolate segments; flowers yellowish-white, small (2"), polygamo-dioecious, paniculate, the inflorescence at length widely branching; bracts lanceolate, short; fruit narrowly ovate, glabrous or nearly so, 2" long, at length exceeded by the plumose calyx-teeth.

In wet open places, Ontario to British Columbia, south to Ohio, Iowa, Wisconsin, and in the Rocky Mountains to Arizona and New Mexico. Called also oregon tobacco; the root cooked for food. May–Aug.

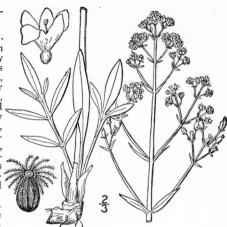

3. Valeriana uliginòsa (T. & G.) Rydb. Marsh or Swamp Valerian. Fig 3997.

Valeriana dioica Pursh, Fl. Am. Sept. 727. 1814. Not L. 1753.
V. sylvatica uliginosa T. & G. Fl. N. A. **2** : 47. 1841.
V. uliginosa Rydb.; Britton, Man. 878. 1901.

Erect, glabrous or very nearly so throughout, 8'–2½° high. Rootstocks creeping or ascending; basal leaves thin, petioled, oblong or spatulate, obtuse, entire, or with a few obtuse lobes, reticulate-veined, 2'–10' long, 3"–18' wide; stem leaves 2–4 pairs, petioled, pinnately parted into 3–15 ovate to lanceolate, dentate or entire, acute or obtuse segments; inflorescence cymose-paniculate, at length loosely branched; flowers pink or nearly white, 3"–4" long, about 2" wide; bracts linear-lanceolate; fruit ovate, glabrous, 1½" long.

In wet soil, Quebec to New York, Ontario and Michigan. American wild valerian. Referred, in our first edition, to the following northern and western species. May–Aug.

4. Valeriana septentrionàlis Rydb. Northern Valerian. Fig. 3998.

Valeriana sylvatica Banks; Richards. App. Frank. Journ. Ed. 2, 2. 1823. Not F. W. Schmidt.

Valeriana septentrionalis Rydb. Mem. N. Y. Bot. Gard. **1** : 376. 1900.

Erect, 8'–16' high, glabrous, or the inflorescence minutely pubescent. Basal leaves spatulate or oval, 4' long or less, entire; stem leaves usually 3 pairs, the segments 5–7, oval to linear-lanceolate, entire, or undulate-margined; inflorescence cymose-paniculate, dense; flowers white, about 1½" wide; fruit 1½"–2" long, glabrous.

In wet soil, Newfoundland to British Columbia, south in the Rocky Mountains to New Mexico. Summer.

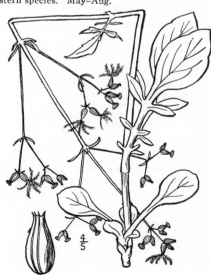

3/4

5. Valeriana officinàlis L. Common, Garden or Great Wild Valerian. Vandal-root. Fig. 3999.

Valeriana officinalis L. Sp. Pl. 31. 1753.

Erect, 2°–5° high, more or less pubescent, especially at the nodes. Leaves all pinnately parted into 7–25 thin reticulate-veined lanceolate acute or acuminate segments, sharply dentate, or those of the upper leaves entire, usually with some scattered hairs beneath; flowers pink or nearly white, about 2″ long; inflorescence of several rather compact corymbed cymes; bracts linear-lanceolate, rather large; fruit glabrous, ovate, about 1½″ long.

Escaped from gardens to roadsides in New York, Ohio and New Jersey. Native of Europe and Asia. Old names, cats'-valerian, setwell, cut-heal, all-heal. Garden-, summer- or hardy-heliotrope. St. George's-herb. June–Aug.

2. VALERIANELLA [Tourn.] Mill. Gard. Dict. Abr. Ed. 4. 1754.

Annual dichotomously branched herbs, the basal leaves tufted, entire, those of the stem sessile, often dentate, the flowers in terminal, compact or capitate, in our species corymbed or panicled cymes. Corolla small, white, blue, or pink, nearly regular. Calyx-limb short or obsolete in flower, in fruit various, not divided into filiform plumose segments, often none. Corolla-tube narrowed at the base, the limb spreading, 5-lobed. Stamens 3; style minutely 3-lobed at the summit. Fruit 3-celled, 2 of the cells empty, and in our species about as large as the fertile one. [Name a diminutive of Valerian.]

About 50 species, natives of the northern hemisphere, most abundant in the Mediterranean region. Besides the following, 8 others occur in the western parts of North America. Type species: *Valeriana Locusta* L.

Corolla funnelform, the short tube not longer than the limb or about equalling it.
   Fruit flattened, twice as broad as thick; corolla blue.   1. *V. Locusta.*
   Fruit triangular-pyramidal; corolla white.   2. *V. chenopodifolia.*
   Fruit oblong-tetragonal or ovoid-tetragonal, grooved.
      Groove of the fruit broad and shallow.   3. *V. radiata.*
      Groove of the fruit narrow.   4. *V. stenocarpa.*
   Fruit globose or saucer-shaped.   5. *V. Woodsiana.*
Corolla salverform, purplish, the slender tube much longer than the limb.   6. *V. longiflora.*

1. Valerianella Locústa (L.) Bettke. European Corn Salad. Fig. 4000.

Valeriana Locusta and var. *olitoria* L. Sp. Pl. 33. 1753.
Valerianella olitoria Poll. Hist. Pl. Palat. 1: 30. 1776.
Valerianella Locusta Bettke. Anim. Val. 10. 1826.

Glabrous, or pubescent at the nodes, 6′–12′ high, usually branched from the base and repeatedly forked. Basal leaves spatulate or oblanceolate, rounded and obtuse at the apex, 1¼′–2′ long, 2″–5″ wide, entire; upper stem leaves oblong-lanceolate, usually dentate; peduncles short; cymes 3″–6″ broad, almost capitate; bracts linear or linear-oblong; corolla blue, about 1″ long; fruit flattened, rounded on the edges, 1″ long, glabrous, twice as broad as thick, depressed-orbicular in outline, the two empty cavities smaller than the fertile one, which has a corky mass at its back.

In fields and waste places, Maine to Ontario, Idaho, Arkansas, New Jersey, Pennsylvania, Virginia and Louisiana. Naturalized from Europe. The plant is cultivated and the leaves used for salad under the name of fètticus. White pot-herb, lamb's-lettuce, milk-grass. April–July.

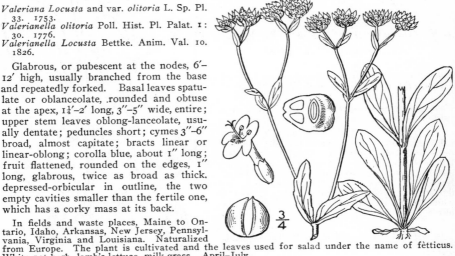

3/4

2. Valerianella chenopodifòlia (Pursh) DC. Goose-foot Corn Salad. Fig. 4001.

Fedia chenopodifolia Pursh, Fl. Am. Sept. 727. 1814.

V. chenopodifolia DC. Prodr. 4: 629. 1830.

Fedia Fagopyrum T. & G. Fl. N. A. 2: 51. 1841.

Glabrous, 1°–2° high. Leaves entire, or the basal and lower ones repand, spatulate, obtuse; upper stem leaves oblong or lanceolate, 1′–3′ long; cymes dense, 6″–8″ broad, at length slender-peduncled; bracts lanceolate or oblong-lanceolate; corolla white, about 1″ long; fruit triangular-pyramidal, 2″ long, 1″ thick, glabrous or minutely pubescent, the two empty cavities narrower than the fertile one but about as deep.

In moist soil, western New York to Virginia, Minnesota and Kentucky. May–July.

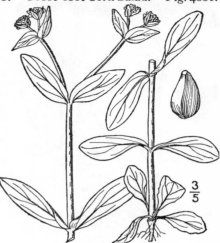

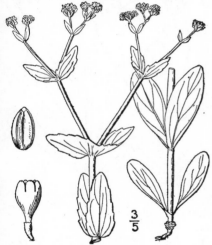

3. Valerianella radiàta (L.) Dufr. Beaked Corn Salad. Fig. 4002.

Valeriana Locusta var. *radiata* L. Sp. Pl. 34. 1753.
Fedia radiata Michx. Fl. Bor. Am. 1: 118. 1803.
Valerianella radiata Dufr. Hist. Val. 57. 1811.

Glabrous, or minutely pubescent below, 6′–18′ high. Basal and lower leaves spatulate, obtuse, entire, the upper lanceolate, usually dentate; cymes 4″–6″ broad, dense; bracts small, lanceolate or oblong-lanceolate; corolla white, 1″ long; fruit narrowly ovate-tetragonal, finely pubescent or sometimes glabrous, 1″ long, ½″ thick, the empty cavities as thick as or thicker than the beaked fertile one and separated from each other by a broad shallow groove.

In moist soil, Massachusetts to Florida, west to Minnesota, Michigan, Missouri and Texas. Called also lamb's-lettuce. May–July.

4. Valerianella stenocàrpa (Engelm.) Krok. Narrow-celled Corn Salad. Fig. 4003.

Fedia stenocarpa Engelm. Bost. Journ. Nat. Hist. 6: 216. 1857.

Valerianella stenocarpa Krok, Kongl. Svensk. Akad. Handl. 5: 64. 1866.

Similar to the preceding species and perhaps better regarded as a race of it. Fruit oblong-tetragonal, slightly smaller, glabrous or sometimes pubescent; sterile cavities not as thick as the oblong seed-bearing one, and separated from each other by a narrow groove.

Prairies and woodlands, Kansas and Missouri to Texas. March–June.

5. Valerianella Woodsiàna (T. & G.) Walp.
Woods' Corn Salad.   Fig. 4004.

Fedia Woodsiana T. & G. Fl. N. A. **2** : 52. 1841.
Valerianella Woodsiana Walp. Rep. **2** : 527. 1843.
F. umbilicata Sulliv. Am. Journ. Sci. **42** : 50. 1842.
Fedia patallaria Sulliv.; A. Gray, Man. 183. 1848.

Usually larger than any of the preceding species, sometimes 3° high, glabrous or very nearly so throughout. Basal and lower leaves spatulate, obtuse, entire; upper leaves lanceolate or linear-oblong, usually dentate; cymes 3″–6″ broad, few-flowered; bracts comparatively large, lanceolate; corolla white, about 1″ long; fruit glabrous, nearly globular to saucer-shaped, about 1″ in diameter, the empty cavities inflated, introrse with a depression or concavity between them, as broad as the fertile one.

In moist soil, New York, Pennsylvania and Ohio to Tennessee and Texas. May–July.

6. Valerianella longiflòra (T. & G.)
Walp. Long-flowered Corn Salad.
Fig. 4005.

Fedia longiflora T. & G. Fl. N. A. **2** : 51. 1841.
Valerianella longiflora Walp. Rep. **2** : 527. 1843.

Glabrous; stem usually several times forked, 6′–12′ high. Leaves very obtuse, the basal ones spatulate, 1′–2½′ long, 4″–8″ wide, those of the stem oblong or spatulate-oblong, smaller, somewhat clasping; cymes dense, corymbed, commonly numerous, several–many-flowered; corolla salverform, pink or purplish, about 6″ long, the almost filiform tube 3–4 times as long as the somewhat irregular 5-parted limb, the lobes linear-oblong; bracts with small gland-tipped teeth; fruit broadly ovate or nearly orbicular in outline, the empty cavities divergent, larger than the oblong seed-bearing one.

In moist rocky situations, Missouri and Arkansas. April–May.

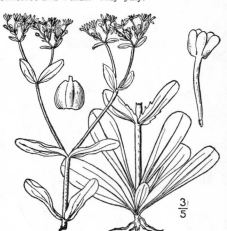

Family 40. DIPSACÀCEAE Lindl. Veg. Kingd. 699. 1847.
TEASEL FAMILY.

Perennial biennial or annual herbs, with opposite or rarely verticillate leaves, and perfect gamopetalous flowers in dense involucrate heads. Stipules none. Flowers borne on an elongated or globose receptacle, bracted and involucellate. Calyx-tube adnate to the ovary, its limb cup-shaped, disk-shaped, or divided into spreading bristles. Corolla epigynous, the tube usually enlarged at the throat, the limb 2–5-lobed. Stamens 2–4, inserted on the tube of the corolla and alternate with its lobes; filaments distinct; anthers versatile, longitudinally dehiscent. Ovary inferior, 1-celled, style filiform; stigma undivided, terminal, or oblique and lateral; ovule 1, anatropous. Fruit an achene, its apex crowned with the persistent calyx-lobes. Seed-coat membranous; endosperm fleshy; embryo straight.

About 7 genera and 140 species, natives of the Old World.

Scales of the elongated receptacle prickly pointed.   1. *Dipsacus.*
Scales of the receptacle not prickly, herbaceous, capillary, or none.
  No receptacular scales.   2. *Scabiosa.*
  Scales of the receptacle about as long as the flowers.   3. *Succisa.*

1. DÍPSACUS [Tourn.] L. Sp. Pl. 97. 1753.

Rough-hairy or prickly tall erect biennial or perennial herbs, with opposite dentate entire or pinnatifid, usually large leaves, and blue or lilac flowers in dense terminal peduncled oblong heads in our species. Bracts of the involucre and scales of the receptacle rigid or spiny pointed. Involucels 4–8-ribbed with a somewhat spreading border. Limb of the calyx cup-shaped, 4-toothed or 4-lobed. Corolla oblique or 2-lipped, 4-lobed. Stamens 4. Stigma oblique or lateral. Achene free from or adnate to the involucel. [Greek, to thirst, the leaves of some species holding water.]

About 15 species, natives of the Old World. Type species: *Dipsacus fullonum* L.

Scales of the receptacle straight-pointed.   1. *D. sylvestris.*
Scales of the receptacle hooked at the apex.   2. *D. fullonum.*

1. Dipsacus sylvéstris Huds. Wild, Common or Card Teasel. Fig. 4006.

Dipsacus fullonum L. Sp. Pl. 97. In part. 1753.
Dipsasus sylvestris Huds. Fl. Angl. 49. 1762.

Biennial, stout, with numerous short prickles on the stem, branches, peduncles, midribs of the leaves and involucre, otherwise glabrous or nearly so, 3°–6° high. Leaves sessile, or the upper ones connate-perfoliate, lanceolate or oblong, the upper acuminate and generally entire, the lower obtuse or obtusish, crenate or sometimes pinnatifid at the base, often 1° long; heads at first ovoid, becoming cylindric, at length 3′–4′ long; flowers lilac, 4″–6″ long; leaves of the involucre linear, curved upward, as long as the head or longer; scales of the receptacle ovate, tipped with a long straight subulate barbed awn, usually exceeding the flowers.

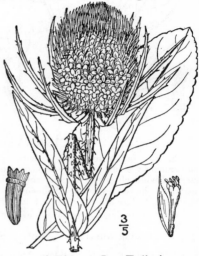

In waste places, Maine and Ontario to North Carolina, west to Michigan. Naturalized from Europe and native also of Asia. July–Sept. Other English names are Venus'-bath or -cup; wood- or church-brooms; shepherds'-staff; card- or water-thistle; gipsy-combs; hutton-weed. Indian's-thistle. Prickly-back. Adam's-flannel.

2. Dipsacus fullònum L. Fuller's or Draper's Teasel. Fuller's Thistle. Fig. 4007.

Dipsacus fullonum L. Sp. Pl. 97. 1753.
Dipsacus fullonum var. *sativus* L. Sp. Pl. Ed. 2, 1677. 1763.

Similar to the preceding species. Leaves of the involucre, or some of them, shorter than the heads, spreading or at length reflexed; scales of the receptacle with hooked tips, about equalling the flowers, which are usually paler than in *D. sylvestris.*

About wool mills, Eastern and Middle States, rare. Fugitive from Europe, and perhaps nowhere permanently established within our range. Other English names are clothiers' brush, Venus'-bath or -cup. Generally regarded as probably a cultivated variety of the preceding species, as it is not found wild, except as an evident escape.

Dipsacus laciniàtus L., with pinnatifid or bipinnatifid ciliate leaves, those of the involucre spreading, has been found at Albany, New York. Fugitive or adventive from Europe.

2. SCABIÒSA [Tourn.] L. Sp. Pl. 98. 1753.

Herbs, with opposite leaves, no prickles, and blue pink or white flowers in peduncled involucrate heads. Bracts of the involucre herbaceous, separate, or slightly united at the base. Scales of the pubescent receptacle none. Involucels compressed, the margins often minutely 4-toothed. Calyx-limb 5–10-awned. Limb of the corolla 4–5-cleft, oblique or 2-lipped. Stamens 4 (rarely 2). Stigma oblique or lateral. Achene more or less adnate to the involucel, crowned with the persistent calyx. [Latin, scale, from its repute as a remedy for scaly eruptions.]

About 75 species, natives of the Old World. Type species: *Scabiosa arvensis* L.

1. Scabiosa arvénsis L. Field Scabious.
Fig. 4008.

Scabiosa arvensis L. Sp. Pl. 99. 1753.
Knautia arvensis T, Coult. Dips. 29. 1823.

Perennial, pubescent, simple or little branched,
1°–3° high. Basal and lower leaves petioled, lan-
seolate, acute or acuminate, entire, lobed, or pinnatifid, 3′–8′ long; upper leaves sessile, often
deeply pinnatifid; heads long-peduncled, depressed-
globose, 1′–1½′ broad; flowers lilac purple, about
6″ long; receptacle depressed-hemispheric, not
scaly, covered with hairs between the flowers;
achene angled, crowned with the 8 or 10 linear-
subulate calyx-teeth.

In cultivated fields and waste places, Quebec to
Massachusetts, Vermont, New York and Pennsylvania. Adventive from Europe. Other English
names are blue buttons, blue caps, gypsy- or egyptian-
rose, pincushion. June–Sept.

3. SUCCÌSA (Vaill.) Moench. Meth. 488. 1794.

Herbs, similar to *Scabiosa*, with opposite leaves, the flowers in long-peduncled heads,
subtended by a several–many-leaved involucre. Scales of the receptacle herbaceous or capillary. Involucels grooved, the margins 4-lobed or 4-toothed. Calyx-limb 5-toothed or 5-awned.
Corolla oblique, 4–5-lobed. Stamens 4. Achene crowned with the persistent calyx. [From
the Latin, to bite off, the rootstock in some species being short and blunt.]

About 4 species, mostly natives of southern Europe. Type species *Scabiosa Succisa* L.

1. Succisa austràlis (Wulf.) Reichenb. Southern Scabious. Fig. 4009.

Scabiosa australis Wulf. in Roem. Arch. 3: Part 3,
316. 1803.
Succisa australis Reichenb. Fl. Germ. Excurs. 196.
1830.

Perennial, puberulent or pubescent, at least
above; stem slender, branched, 1½°–3° high. Basal
leaves oblanceolate to oblong, mostly obtuse, 4′–12′
long, the petiole often as long as the blade or
longer; stem leaves distant, lanceolate or oblong-
lanceolate, entire, or toothed, short-petioled, or
the upper sessile, acute or acuminate; heads of
pale blue-purple flowers long-peduncled, rather
less than 1′ in diameter, oblong-ovoid in fruit;
receptacle scaly, the scales about as long as the
glabrous, 8-ribbed involucels or longer; calyx 5-
toothed; achene crowned with 5 calyx-teeth.

Fields and meadows. Naturalized from Europe in
Pennsylvania, central New York and Massachusetts.
Pincushion-flower. Summer.

Succisa Succìsa (L.) Britton (*Succisa pratensis*
Moench; *Scabiosa Succisa* L.) with villous 4-angled
involucels, the calyx-limb 5-awned, has been found in fields at Louisburg, Cape Breton Island.

Family 41. CUCURBITÀCEAE B. Juss. Hort. Trian. 1759.
GOURD FAMILY.

Climbing or trailing, herbaceous vines, usually with tendrils. Leaves alternate,
petioled, generally palmately lobed or dissected. Flowers solitary or racemose,
monoecious or dioecious. Calyx-tube adnate to the ovary, its limb campanulate
or tubular, usually 5-lobed, the lobes imbricated. Petals usually 5, inserted on
the limb of the calyx, separate, or united into a gamopetalous corolla. Stamens
mostly 3 (sometimes 1), 2 of them with 2-celled anthers, the other with a 1-celled
anther; filaments short, often somewhat monadelphous. Ovary 1–3-celled; style

terminal, simple or lobed; ovules few or numerous, anatropous. Fruit a pepo, indehiscent, or rarely dehiscent at the summit, or bursting irregularly; or sometimes dry and membranous   Seeds usually flat; endosperm none.

About 90 genera and 700 species, mainly natives of tropical regions, a few in the temperate zones.

Flowers large, yellow; prostrate vine.   1. *Pepo.*
Flowers small, white or greenish; climbing vines.
 Fruit glabrous; seeds numerous, horizontal.   2. *Melothria.*
 Fruit prickly; seeds 1 or few, erect or pendulous.
  Fruit dehiscent at the apex or bursting irregularly; several-seeded.
   Leaves 3–7-lobed; anthers 3.   3. *Micrampelis.*
   Leaves digitately compound; anther 1.   4. *Cyclanthera.*
  Fruit indehiscent, 1-seeded.   5. *Sicyos.*

1. PÈPO [Tourn.] Mill. Gard. Dict. Abr. Ed. 4. 1754.

Rough prostrate vines, rooting at the nodes, with branched tendrils, usually lobed leaves mostly cordate at the base, and large yellow axillary monoecious flowers. Calyx-tube campanulate, usually 5-lobed. Corolla campanulate, 5-lobed to about the middle, the lobes recurving. Staminate flowers with three stamens, the anthers linear, more or less united and no pistil. Pistillate flowers with 1 pistil; ovary oblong with 3–5 many-ovuled placentae; style short, thick; stigmas 3–5, each 2-lobed, papillose; staminodia 3. Fruit large, fleshy, with a thick rind, many-seeded, indehiscent. [From the Greek name of some large fruit.]

About 10 species, natives of America, Asia and Africa. Besides the following, some 6 others occur in the southern and southwestern United States. Type species: *Cucurbita Pepo* L.

1. Pepo foetidíssima (H.B.K.) Britton. Missouri Gourd. Calabazilla. Wild Pumpkin. Fig. 4010.

Cucurbita foetidissima H.B.K. Nov. Gen. **2**: 123. 1817.
Cucumis perennis James in Long's Exp. **2**: 20. 1823.
Cucurbita perennis A. Gray, Bost. Journ. Nat. Hist. **6**: 193. 1850.

Stem stout, rough, hirsute, trailing to a length of 15°–25°. Root large, carrot-shaped. Petioles stout, 3'–8' long, very rough; leaves ovate-triangular, thick and somewhat fleshy, cordate or truncate at the base, acute at the apex, 4'–12' long, usually slightly 3–5-lobed, denticulate, rough above, canescent beneath; peduncles 1'–2' long; flowers mostly solitary; corolla 2½'–4' long; pepo globose or globose-ovoid, 2'–3' in diameter, smooth, its pulp fibrous and bitter.

Dry soil, Missouri and Nebraska to Texas and Mexico, west to southern California. May–Sept.

Pumpkins, Squashes, Cucumbers and Melons, cultivated in many races belong to this genus. Gourds belong to *Cucurbita Lagenaria* L., the type of the genus *Cucurbita*.

Citrullus Citrúllus (L.) Karst., the Watermelon, is found escaped from cultivation along river-shores in Virginia and West Virginia, and southward.

2. MELÒTHRIA L. Sp. Pl. 35. 1753.

Slender, mostly climbing vines, with simple or rarely bifid tendrils, lobed or entire thin leaves, and small white or yellow monoecious flowers, the staminate clustered, the pistillate often solitary. Calyx campanulate, 5-toothed. Corolla campanulate, deeply 5-parted. Stamens 3 in the staminate flowers, the anthers distinct or slightly united, the pistil wanting or rudimentary. Fertile flowers with 1 pistil; ovary ovoid, constricted below the corolla; placentae 3; ovules numerous; style short; stigmas 3, linear. Fruit small, berry-like, pulpy, many-seeded. [From the Greek for some vine, probably *Bryonia cretica.*]

About 70 species, natives of warm and tropical regions, most abundant in the Old World. Besides the following typical one, 3 or 4 others occur in the southern United States.

1. Melothria péndula L. Creeping Cucumber.
Fig. 4011.

Melothria pendula L. Sp. Pl. 35. 1753.

Root perennial. Stem slender, climbing to a height of 3°–5°, branched, glabrous, grooved; petioles ½′–2½′ long; leaves nearly orbicular in outline, finely pubescent or scabrous on both sides, cordate at the base, 5-lobed or 5-angled, denticulate or dentate; tendrils puberulent; staminate flowers 4–7, racemose, borne on a peduncle ½′–1′ long; fertile flowers solitary, slender-peduncled; corolla greenish white, about 2″ broad; fruit smooth, ovoid, green, 4″–6″ long.

In thickets, Pennsylvania (Schweinitz, according to Cogniaux); Virginia to Florida, west to Indiana, Kentucky, Missouri and northern Mexico. June–Sept.

3. MICRÁMPELIS Raf. Med. Rep. (II.) 5: 350. 1808.
[ECHINOCYSTIS T. & G. Fl. N. A. 1: 542. 1840.]

Mostly annual climbing herbs, with branched tendrils, lobed or angled leaves, and small white monoecious flowers. Calyx-tube campanulate, 5–6-lobed. Corolla very deeply 5–6-parted. Stamens 3 in the staminate flowers, the anthers more or less coherent. Pistillate flowers with a 2-celled ovary; ovules 2 in each cavity; style very short; stigma hemispheric or lobed. Fruit fleshy, or dry at maturity, densely spiny, 1–2-celled, usually with 2 seeds in each cavity, dehiscent at the summit. Testa of the seed roughened. [Greek, small-vine.]

About 25 species, natives of America. Besides the following typical one, about 10 others occur in the western United States.

1. Micrampelis lobàta (Michx.) Greene. Wild Balsam Apple. Mock Apple.
Mock Orange. Fig. 4012.

Momordica echinata Muhl. Trans. Am. Phil. Soc. 3: 180. Name only. 1793.
Sicyos lobata Michx. Fl. Bor. Am. 2: 217. 1803.
Echinocystis lobata T. & G. Fl. N. A. 1: 542. 1840.
Micrampelis lobata Greene, Pittonia 2: 128. 1890.

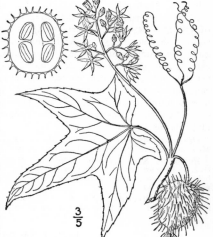

Stem nearly glabrous, angular and grooved, branching, climbing to a height of 15°–25°, sometimes villous-pubescent at the nodes. Petioles 1′–3′ long; leaves thin, roughish on both sides, deeply cordate at the base, 3–7-lobed to about the middle, the lobes triangular-lanceolate, acute or acuminate, the margins remotely serrulate; staminate flowers very numerous in narrow compound racemes; pistillate flowers solitary, or rarely 2 together; fruit ovoid, green, about 2′ long, armed with slender spines; seeds flat.

Along rivers, and in waste places, New Brunswick to Ontario, Manitoba, Montana, Virginia, Pennsylvania, Kentucky, Kansas and Texas. Eastward, mostly occurring as an introduced plant. Wild cucumber. Creeper. Creeping Jenny. July–Sept.

4. CYCLANTHÈRA Schrad. Ind. Sem. Hort. Goett. 1831.

Climbing, annual or perennial, mostly glabrous vines, with forked or simple tendrils, usually digitately compound leaves and small white or greenish monoecious flowers. Calyx cup-shaped, 5-toothed. Corolla rotate, deeply 5-parted. Staminate flowers racemose or panicled, the stamens united into a central column; anther 1, annular in our species. Pistillate flowers solitary; ovary obliquely ovoid, beaked, 1–3-celled, with 2 ovules in each cavity; style short; stigma large, hemispheric. Fruit spiny, obliquely ovoid, beaked, at length irregularly dehiscent, few-seeded. [Greek, circle-anther.]

About 40 species, natives of America. Type species: *Cyclanthera pedàta* Schrad.

1. Cyclanthera dissécta (T. & G.) Arn. Cut-leaved Cyclanthera. Fig. 4013.

Discanthera dissecta T. & G. Fl. N. A. 1: 697. 1840.
Cyclanthera dissecta Arn. in Hook. Journ. Bot. 3: 280. 1841.

Annual; stem grooved and angular, glabrous, branching, climbing to a height of 3°–4°, or straggling. Petioles 1′–2′ long; leaves digitately 3–7-foliolate, the leaflets oval or oblong, usually acute at each end, ½′–2′ long, rough on both sides, dentate, or somewhat lobed; staminate flowers racemose, borne on a peduncle ¼′–2′ long; pistillate flowers solitary, very short-peduncled; fruit narrowed at the base, slightly oblique, about 1′ long, armed with slender spines.

Thickets, Kansas to Texas, Louisiana and northern Mexico. July–Sept.

5. SÍCYOS L. Sp. Pl. 1013. 1753.

Annual climbing vines, with branched tendrils, angled or lobed leaves, and small white or green monoecious flowers. Calyx-tube campanulate or cup-shaped, 5-toothed. Corolla campanulate or rotate, 5-parted nearly to the base. Staminate flowers corymbose or racemose, with 3 stamens, the filaments united into a short column, the anthers coherent; pistil wanting. Pistillate flowers several together in capitate long-peduncled clusters, with no stamens; ovary oblong or fusiform, 1-celled; ovule 1, pendulous; style short, slender; stigmas usually 3. Fruit spiny, indehiscent, 1-seeded. [Greek, a cucumber or gourd.]

About 35 species, natives of America and Australasia. Besides the following typical species, 2 others occur in the southwestern states.

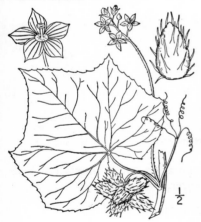

1. Sicyos angulàtus L. One-seeded Bur-Cucumber. Star Cucumber. Fig. 4014.

Sicyos angulatus L. Sp. Pl. 1013. 1753.

Stem angled, more or less viscid-pubescent, climbing to a height of 15°–25°, or trailing. Petioles stout, 1′–4′ long, pubescent; leaves nearly orbicular, rough on both sides, rather thin, deeply cordate at the base, 5-angled or 5-lobed, the lobes acute or acuminate, the margins denticulate; staminate flowers loosely corymbose or racemose, borne on elongated peduncles; fertile flowers capitate, their peduncles shorter; fruits sessile, 3–10 together, yellowish, about ½′ long, pubescent, armed with slender rough spines.

Along river banks and in moist places, Quebec and Ontario to Florida, west to South Dakota, Kansas and Texas. Naturalized in eastern Europe. Called also nimble kate, wild cucumber. Leaves sometimes 10′ across. June–Sept.

Family 42. CAMPANULÀCEAE Juss. Gen. 163. 1789.

BELLFLOWER FAMILY.

Herbs (some tropical species shrubs or even trees), with alternate exstipulate leaves, usually milky juice, and racemose spicate paniculate or solitary perfect flowers. Calyx-tube adnate to the ovary, its limb mostly 5-lobed or 5-parted, the lobes equal or slightly unequal, valvate or imbricate in the bud, commonly persistent. Corolla gamopetalous, regular, inserted at the line where the calyx becomes free from the ovary, its tube entire, or deeply cleft on one side, its limb 5-lobed. Stamens 5, alternate with the corolla-lobes, inserted with the corolla; filaments separate; anthers 2-celled, introrse, separate or connate. Ovary 2–5-celled (rarely 6–10-celled, with the placentae projecting from the axis, or 1-celled with two parietal.placentae; style simple; stigma mostly 2–5-lobed; ovules anatropous. Fruit a capsule. Seeds very numerous and small; embryo minute, straight; endosperm fleshy.

About 40 genera and over 1000 species, of wide geographic distribution.

Capsule opening by lateral pores or valves.
  Corolla campanulate, rarely rotate; flowers all complete.   1. *Campanula.*
  Corolla rotate; earlier flowers cleistogamous.   2. *Specularia.*
Capsule opening by terminal pores or valves.   3. *Jasione.*

1. CAMPÁNULA [Tourn.] L. Sp. Pl. 163. 1753.

Perennial or annual herbs, with alternate or basal leaves. Flowers large or small, solitary, racemose, paniculate, or glomerate, regular, complete, blue, violet, or white. Calyx-tube hemispheric, turbinate, obovoid, or prismatic, adnate to the ovary, the limb deeply 5-lobed or 5-parted (rarely 3–4-parted). Corolla campanulate or rotate, 5-lobed or 5-parted. Stamens 5, free from the corolla; filaments usually dilated at the base; anthers separate. Ovary inferior, 3–5-celled; stigma 3–5-lobed. Capsule wholly or partly inferior, crowned by the persistent calyx-lobes, opening on the sides, either near the top, middle or bottom by 3–5 small valves or perforations, or tending to be indehiscent in some species. [Diminutive of the Latin *campana,* a bell.]

About 250 species, natives of the northern hemisphere. Besides the following, some 8 others occur in the southern and western parts of North America; all known as Bell-flower. Type species: *Campanula latifolia* L.

*Corolla campanulate; flowers solitary, racemose, glomerate, or panicled.

Flower solitary at the end of the stem; arctic and alpine plants.
  Corolla 4″–6″ long; capsule-openings near the summit.   1. *C. uniflora.*
  Corolla 6″–12″ long; capsule-openings near the base.   2. *C. rotundifolia.*
Flowers racemose, glomerate, or paniculate.
  Corolla 7″–15″ long.
    Stem leaves linear, the basal orbicular, mostly cordate.   2. *C. rotundifolia.*
    Leaves all ovate to lanceolate; plants pubescent or scabrous.
      Flowers pedicelled, or clustered.
        Calyx and corolla glabrous, or calyx finely pubescent.   3. *C. rapunculoides.*
        Calyx and corolla bristly-hairy.   4. *C. Trachelium.*
      Flowers sessile in terminal and axillary clusters.   5. *C. glomerata.*
  Corolla 2″–5″ long.
    Plants rough; style not exserted.
      Corolla white, or tinged with blue, 2½″–4″ long; leaves mostly linear-lanceolate,
        crenulate.   6. *C. aparinoides.*
      Corolla blue, 5″–6″ long; leaves linear, denticulate with minute callous teeth.
                                                                              7. *C. uliginosa.*
    Plants smooth, glabrous, slightly viscid; style long-exserted.   8. *C. divaricata.*
      ** Corolla rotate; flowers spicate.   9. *C. americana.*

1. Campanula uniflòra L. Arctic Harebell or Bell-flower. Fig. 4015.

Campanula uniflora L. Sp. Pl. 163. 1753.

Perennial, glabrous or nearly so; stem simple, 1-flowered, 1′–6′ high. Leaves linear or linear-oblong, acute, sessile, thickish, entire or sparingly dentate, 9″–18″ long, or the lower and basal ones spatulate, obtuse and narrowed into petioles; flower erect; calyx-tube turbinate, glabrous or pubescent, shorter than or equalling the lobes; corolla campanulate, 4″–6″ long, blue; capsule cylindric or club-shaped, about 6″ long, erect, opening by valves near the summit.

Labrador and Arctic America to Alaska, south in the Rocky Mountains to Colorado. Also in northern Europe and Asia. Summer.

$\frac{3}{4}$

2. Campanula rotundifòlia L.   Harebell.
Blue Bells of Scotland.   Fig. 4016.

Campanula rotundifolia L. Sp. Pl. 163. 1753.
Campanula rotundifolia velutina DC. Fl. France
   6: 432. 1815.
C. linifolia var. *Langsdorfiana* A. DC. Prodr. 7:
   471. 1839.
Campanula rotundifolia Langsdorfiana Britton,
   Mem. Torr. Club 5: 309. 1894.

Perennial by slender rootstocks, glabrous or
nearly so or sometimes pubescent or canescent; stems erect or diffuse, often several
from the same root, simple or branched, 6′–3°
high. Basal leaves nearly orbicular or broadly
ovate, usually cordate, slender-petioled, ¼–1′
wide, dentate or entire, often wanting at flowering time; stem leaves linear or linear-oblong,
acute, mostly entire, sessile, or the lower narrowed into short petioles and somewhat spatulate; flowers several or numerous, racemose or
sometimes solitary, drooping or spreading,
slender-pedicelled; calyx-lobes subulate to filiform, spreading, longer than the short-turbinate tube; corolla blue, campanulate, 7″–12″
long; capsule obconic or ovoid, pendulous,
ribbed, opening by short clefts near the base.

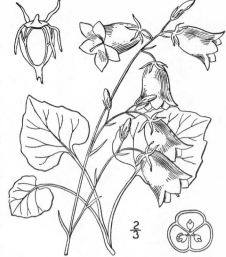

2/3

On moist rocks and in meadows, Labrador to Alaska, south to New Jersey, Pennsylvania, Illinois, Nebraska, in the Rocky Mountains to Arizona and in the Sierra Nevada to California. Also
in Europe and Asia. Consists of many races, differing in pubescence, number and size of flowers;
arctic and alpine plants are usually 1–few-flowered. Other English names are thimbles, lady's-
thimble, heath- or witches'-bells, round-leaved bellflower. June–Sept.

Campanula pátula L., which is retrorse-scabrous on the stems and leaf-margins and nerves,
the basal leaves obovate to spatulate, has been found in fields in Connecticut, introduced from
Europe.

3/5

3. Campanula rapunculoìdes L.   Creep-
ing or European Bellflower.   Fig. 4017.

Campanula rapunculoides L. Sp. Pl. 165. 1753.

Perennial by slender rootstocks; stem glabrous or pubescent, simple or rarely branched,
leafy, erect, rather stout, 1°–3° high. Leaves
pubescent or puberulent, crenate-denticulate,
ovate or ovate-lanceolate, acute or acuminate,
the lower and basal ones mostly cordate, 3′–6′
long, 1′–2′ wide, slender-petioled, the upper
short-petioled or sessile, smaller; flowers short-
pedicelled, drooping, 1′–1½′ long in an elongated bracted 1-sided raceme; corolla campanulate, blue to violet, rather deeply 5-lobed,
much longer than the linear spreading calyx-
lobes; capsule globose, nodding, about 4″ in
diameter, opening by pores near the base.

In fields and along roadsides, New Brunswick
to Ontario, southern New York, Pennsylvania and
Ohio. Naturalized from Europe. July–Sept.

4. Campanula Trachèlium L.   Nettle-leaved Bell-
flower.   Fig. 4018.

C. Trachelium L. Sp. Pl. 166. 1753.

Perennial; stem rather stout, little branched, usually
bearing scattered hairs, 3° high or less; basal leaves
sparingly bristly-pubescent, ovate to reniform, cordate,
slender-petioled; stem leaves ovate-lanceolate to lanceolate, coarsely irregularly serrate, 2½′–5′ long, acute or
acuminate at the apex, narrowed at the base, short-
petioled or the upper sessile; flowers nodding in terminal leafy-bracted racemes; calyx bristly-hairy or glabrate; corolla campanulate, 1′–1½′ long; capsule opening
by basal pores.

Roadsides and thickets, Quebec to southern New York and
Ohio. Naturalized from Europe. Canterbury bells. July–Sept.

1/2

5. Campanula glomeràta L. Clustered Bellflower. Dane's Blood. Fig. 4019.

Campanula glomerata L. Sp. Pl. 166. 1753.

Perennial by short rootstocks; stem stout, simple, erect, pubescent, leafy, 1°–2° high. Leaves pubescent on both sides, crenulate, the lower and basal ones oblong or ovate, mostly obtuse, sometimes cordate, slender-petioled, 2′–4′ long, the upper lanceolate or ovate-lanceolate, acute, sessile or clasping, smaller; flowers about 1′ long, sessile, erect and spreading in terminal and axillary glomerules; corolla campanulate, blue, rather deeply 5-lobed; calyx-lobes lanceolate, acuminate; capsule ovoid or oblong, erect, about 3″ high, opening near the base.

In fields and along roadsides, eastern Massachusetts and Quebec. Naturalized from Europe. Sometimes called canterbury bells, a name more properly belonging to *C. medium* and *C. Trachelium.* June–Aug.

6. Campanula aparinoìdes Pursh. Marsh or Bedstraw Bellflower. Fig. 4020.

Campanula aparinoides Pursh, Fl. Am. Sept. 159. 1814.

Perennial; stems very slender or filiform, weak, reclining or diffuse, rough with short retrorse bristles, leafy, paniculately branched, 6′–2° long. Leaves lanceolate, or linear-lanceolate, sessile, sparingly crenulate with low teeth, or entire, rough on the margins and midrib, acute at both ends, ½′–1½′ long, 1″–3″ wide; flowers leafy-paniculate, 2½″–4″ long; pedicels filiform, divergent; buds drooping; corolla open-campanulate, deeply 5-cleft, white or bluish-tinged, 2½″–4″ long, its tube equalling or longer than the triangular-lanceolate acute calyx-lobes; style included; capsule subglobose, opening near the base.

In grassy swamps, Maine to Georgia, Kentucky and Colorado. Called also slender bellflower. June–Aug.

7. Campanula uliginòsa Rydb. Blue Marsh Bellflower. Fig. 4021.

Campanula uliginosa Rydb.; Britton, Man. 885. 1901.

Perennial; similar to the preceding but branches less spreading; stem 1°–2° long, retrorsely hispidulous on the angles. Leaves linear, 1′–2½′ long, ½″–2½″ wide, retrorsely hispidulous on the margins and midribs, acute, minutely denticulate with callous teeth or entire; corolla blue with darker veins, cleft to below the middle into lanceolate lobes; capsule subglobose, about 2½″ long and nearly as broad, opening near the base.

In wet meadows, New Brunswick to Saskatchewan, New York, Indiana and Nebraska. June–Aug.

8. Campanula divaricàta Michx. Panicled Bellflower. Fig. 4022.

Campanula divaricata Michx. Fl. Bor. Am. **1**: 109. 1803.
Campanula flexuosa Michx. loc. cit. 1803?

Perennial, glabrous but sometimes viscid; stem erect, paniculately branched, slender, 1°–3° high. Leaves lanceolate, ovate or oblong-lanceolate, the uppermost sometimes linear, sharply serrate, acuminate at the apex, narrowed to the base, the upper sessile, the lower petioled, 2′–3′ long, 3″–12″ wide, or the lowest commonly shorter and broader; flowers very numerous in compound panicles, drooping, slender-pedicelled; corolla light blue, campanulate, about 3″ long; calyx-lobes lanceolate, acute, scarcely spreading, often dentate; style long-exserted; capsule turbinate, about 2½″ long, opening near the middle.

On rocky banks, mountains of Virginia and West Virginia to Kentucky, Georgia and Tennessee. Ascends to 2500 ft. in North Carolina. June–Sept.

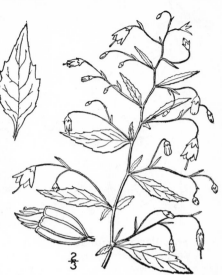

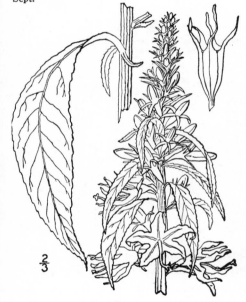

9. Campanula americàna L. Tall Bellflower. Fig. 4023.

Campanula americana L. Sp. Pl. 164. 1753.

Annual or biennial, more or less pubescent; stem erect or nearly so, rather slender, simple or rarely with a few long branches, 2°–6° high. Leaves thin, ovate, oblong, or lanceolate, serrate, acuminate at the apex, narrowed at the base, petioled. or the upper sessile, 3′–6′ long, the lowest sometimes cordate; flowers in a loose or dense terminal sometimes leafy spike, which is often 1°–2° long; lower bracts foliaceous, the upper subulate; corolla rotate, blue, or nearly white, about 1′ broad, deeply 5-cleft; calyx-lobes linear-subulate, spreading, style declined and curved upward, long-exserted; capsule narrowly turbinate, ribbed, erect, 4″–5″ long, opening near the summit.

In moist thickets and woods, New Brunswick to Ontario and South Dakota, south to Florida, Kentucky, Kansas and Arkansas. Rare near the coast in the Middle States and New England. Ascends to 3000 ft. in West Virginia. July–Sept.

2. SPECULÀRIA Heist.; Fabr. Enum. Pl. Hort. Helmst. 225. 1763.

[LEGOUZIA Durand, Fl. Bourg. **2**: 26. 1782.]

Annual herbs, with alternate toothed or entire leaves, the stem and branches long, slender. Flowers axillary, sessile or nearly so, 2-bracted, or the upper panicled in some exotic species, the earlier (lower) ones small, cleistogamous, the later with a blue or purple nearly rotate corolla. Calyx-tube narrow, the lobes in the earlier flowers 3 or 4, in the later 4 or 5. Corolla 5-lobed or 5-parted, the lobes imbricated in the bud. Filaments flat; anthers separate, linear. Ovary 3-celled (rarely 2- or 4-celled); ovules numerous; stigma usually 3-lobed. Capsule prismatic, cylindric, or narrowly obconic, opening by lateral valves. Seeds ovoid, oblong, or lenticular. [From *Speculum Veneris*, the Latin name of the type species.]

About 10 species, natives of the northern hemisphere, one extending into South America. Type species: *Campanula Spéculum* L.; *S. Spéculum* (L.) DC., of Europe, which is adventive in Lancaster County, Pennsylvania.

Capsule narrowly oblong.
    Leaves sessile; capsule-valves near the top. 1. *S. biflora.*
    Leaves cordate-clasping; capsule-valves at about the middle. 2. *S. perfoliata.*
Capsule linear-cylindric; leaves sessile; western. 3. *S. leptocarpa.*

1. Specularia biflòra (R. & P.) F. & M. Small Venus' Looking-glass. Fig. 4024.

Campanula biflora R. & P. Fl. Per. **2**: 55. *pl. 200. f. 6.* 1799.

S. biflora F. & M. Ind. Sem. Hort. Petrop. **1**: 17. 1835.

Legouzia biflora Britton, Mem. Torr. Club **5**: 309. 1894.

Glabrous, or nearly so; stem simple or branched, very slender, roughish on the angles, 6′–2° high. Leaves ovate, oblong, or the upper lanceolate, sessile, acute or obtuse at the apex, crenate with a few teeth, or entire, 4″–10″ long, or the upper smaller; earlier flowers with 3 or 4 ovate to lanceolate calyx-lobes, those of the later flowers 4 or 5, lanceolate-subulate, longer; capsule oblong-cylindric, 3″–5″ long, opening by valves close under the calyx-teeth.

In dry soil, Virginia to Kentucky, Missouri, Kansas, Florida and Texas. Also in Oregon, California and South America. April–July.

2. Specularia perfoliàta (L.) A. DC. Venus' Looking-glass. Fig. 4025.

Campanula perfoliata L. Sp. Pl. 169. 1753.
S. perfoliata A. DC. Mon. Campan. 351. 1830.
L. perfoliata Britton, Mem. Torr. Club **5**: 309. 1894.

More or less pubescent; stem densely leafy, simple or branched from near the base, slender, rather weak, sometimes prostrate, retrorse-hispid on the angles, or nearly smooth, 6′–24′ long. Leaves orbicular or broadly ovate, strongly cordate-clasping or the lower merely sessile, crenate-dentate or sometimes entire, ¼′–1′ wide; flowers solitary or 2–3 together in the axils, sessile, the later (upper) ones with 5 (rarely 4) triangular-lanceolate acuminate rigid calyx-lobes, and a rotate blue or violet corolla 5″–10″ broad, the earlier ones with 3–4 shorter calyx-lobes longer than the rudimentary corolla; capsule oblong, or narrowly turbinate, 2″–3″ long, finally opening at about the middle; seeds lenticular.

In dry woods, Maine and Ontario to British Columbia, south to Florida, Louisiana, Mexico, Arizona and Oregon. Called also clasping bellflower. May–Sept. Also in the mountains of Jamaica and Santo Domingo.

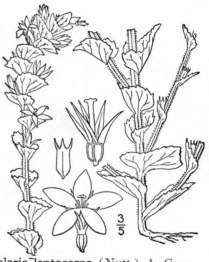

3. Specularia leptocarpa (Nutt.) A. Gray. Western Venus' Looking-glass. Fig. 4026.

Campylocera leptocarpa Nutt. Trans. Am. Phil. Soc. (II.) **8**: 257. 1843.

S. leptocarpa A. Gray, Proc. Am. Acad. **11**: 82. 1876.

L. leptocarpa Britton, Mem. Torr. Club **5**: 309. 1894.

Hirsute, or nearly glabrous; stem slender, simple, or branched from the base, 6′–15′ high. Leaves linear-lanceolate to oblong, sessile, not clasping, acute at both ends, or the lowest obtuse at the apex, entire or sparingly denticulate, ½′–1′ long, 1″–2″ wide; flowers sessile and usually solitary in the axils, the later ones with 4–5 subulate calyx-lobes and a rotate corolla 5″–9″ broad, the earlier ones with 3 shorter calyx-lobes and rudimentary corolla; capsule linear-cylindric, 4″–8″ long, less than 1″ thick; the upper at length opening near the summit; seeds oblong.

In dry soil, western Missouri and Kansas to Montana, Colorado and Texas. May–Aug.

3. JASIONE L. Sp. Pl. 928. 1753.

Herbs, the flowers in terminal heads subtended by a many-leaved involucre. Calyx-tube campanulate, adnate to the ovary, the limb 5-divided. Corolla at first tubular, splitting later to the base into 5 linear or linear-oblanceolate segments. Filaments subulate. Anthers united at the base, free above. Ovary inferior, 2-celled; stigma club-shaped. Capsule 2-celled, dehiscent by two terminal pores. [Greek name of some medicinal plant.]

Five species of central Europe and the Mediterranean Region, the following typical.

1. Jasione montàna L. Sheep's-bit. Fig. 4027.

Jasione montana L. Sp. Pl. 928. 1753.

Annual or biennial, with several stems from a simple root; stems 6'–12' high, branched above, leafy below, decumbent, erect or ascending, the branches spreading. Leaves linear, rough-hairy, sessile; flowers in long-peduncled hemispherical heads; corolla blue, seldom white or pink.

In waste places, Massachusetts to southern New York. Adventive from Europe. June–Sept.

Family 43. LOBELIÀCEAE Dumort. Comm. Bot. 57. 1822.

LOBELIA FAMILY.

Herbs, or in tropical regions rarely shrubs or trees, often with milky sap which contains a narcotic-acid poison, with alternate, exstipulate, simple, entire, toothed or pinnately parted leaves and solitary, spicate, racemose or paniculate flowers. Calyx-tube adnate to the ovary, its limb 5-lobed or 5-parted, the lobes equal or unequal. Corolla gamopetalous, irregular, often bilabiate, its tube open on one side nearly or quite to the base, its limb 5-lobed; stamens 5, inserted with the corolla; filaments sometimes cohering into a tube; anthers united. Ovary 2–5-celled; style single; stigma fringed; ovules numerous, sessile, horizontal, anatropous. Fruit a 1–5-celled capsule, or a berry. Seeds numerous, with a smooth or furrowed testa. Endosperm fleshy.

About 20 genera and 600 species, of wide geographical distribution.

1. LOBÈLIA L. Sp. Pl. 929. 1753.

Herbs (some tropical species shrubs), with alternate or basal leaves and racemose spicate or paniculate, often leafy bracted, red, yellow, blue or white flowers. Calyx-tube turbinate, hemispheric or ovoid, adnate to the ovary. Corolla-tube straight, oblique or incurved, divided to the base on one side, 2-lipped in our species, the lobe on each side of the cleft erect or recurved, turned away from the other three which are somewhat united, the sinuses inclining to extend to the base of the corolla at maturity so as to divide it into 5 petals. Stamens free from the corolla-tube, monadelphous, at least above, two or all the 5 anthers with a tuft of hairs at the tips, three of them usually larger than the other two, all united into a tube or ring around the style. Ovary 2-celled, the 2 placentae many-ovuled; stigma 2-lobed or 2-cleft. Capsule loculicidally 2-valved. [Named after Matthias de L'Obel, 1538–1616, a Flemish botanist.]

About 250 species, of wide geographic distribution. Besides the following, some 16 others occur in the southern and western United States. Type species: *Lobelia Dortmanna* L.

* Aquatic; stem simple, nearly naked; flowers light blue.

Leaves terete, hollow, obtuse, tufted at the base.	1. *L. Dortmanna.*
Leaves flat, linear-oblong or spatulate, entire or glandular-denticulate.	2. *L. paludosa.*

** Terrestrial plants of wet or dry soil; stems leafy.

1. Corolla-tube 5″–12″ long.

Flowers bright scarlet (rarely white); corolla-tube 10″–12″ long.	3. *L. cardinalis.*
Flowers blue, white, or blue and white; corolla-tube 5″–7″ long.	
Leaves ovate, lanceolate, or the lower ones obovate.	
Leaves glabrous or sparingly pubescent.	
Calyx-lobes hirsute; sinuses with large deflexed auricles.	4. *L. syphilitica.*
Calyx-lobes glabrous or glandular, usually without auricles.	
Leaves ovate to ovate-lanceolate, spreading.	5. *L. amoena.*
Leaves linear to narrowly lanceolate, erect.	6. *L. elongata.*
Leaves densely puberulent; calyx hirsute; auricles small.	7. *L. puberula.*
Leaves elongated-linear, strongly glandular-denticulate.	8. *L. glanduosa,*

2. Corolla-tube only 2″–4″ long.

Stems mostly simple; flowers in terminal spike-like racemes.
    Sinuses of the calyx without auricles.   9. *L. spicata.*
    Sinuses of the calyx with reflexed subulate auricles.   10. *L. leptostachys.*
Stems mostly paniculately branched; flowers in loose racemes.
    Stem stout, pubescent; leaves ovate or oblong, dentate.   11. *L. inflata.*
    Stems slender, glabrous; stem-leaves narrow, the basal wider.
        Pedicels mostly longer than flowers, 2-bracteolate near the middle.   12. *L. Kalmii.*
        Pedicels not longer than flowers, not bracteolate, or only so at the base.
            Corolla 2½″–3½″ long; calyx-tube hemispheric in fruit.   13. *L. Nuttallii.*
            Corolla 4½″–5½″ long; calyx-tube turbinate.   14. *L. Canbyi.*

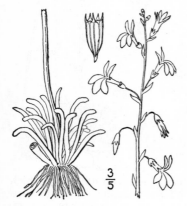

1. **Lobelia Dortmánna** L. Water Lobelia. Water Gladiole. Fig. 4028.

Lobelia Dortmanna L. Sp. Pl. 929. 1753.

Perennial, aquatic, glabrous throughout, somewhat fleshy; roots numerous, white, fibrous; stem slender, simple, erect, hollow, minutely scaly, 6′–18′ high. Leaves all submersed and tufted at the base of the stem, terete, hollow, obtuse, longitudinally divided by a partition, 1′–2′ long, about 2″ thick; flowers in a loose terminal raceme, blue, 6″–8″ long; pedicels filiform, shorter than or equalling the flowers; calyx-lobes subulate or lanceolate, shorter than the tube, the sinuses usually not at all appendaged; corolla-tube 3″–4″ long, its lower lip glabrous or nearly so.

Borders of ponds, usually in sandy soil, sometimes wholly emersed when the water is low, New Jersey and Pennsylvania to Newfoundland, Wisconsin, Washington and British Columbia. Also in Europe. July–Sept.

2. **Lobelia paludòsa** Nutt. Swamp Lobelia. Fig. 4029.

Lobelia paludosa Nutt. Gen. **2**: 75. 1818.

Perennial, aquatic, glabrous throughout; roots few and thick; stem nearly naked, slender, simple, or branched above, 1°–4° high. Leaves mostly tufted at the base, flat, narrowly oblong or spatulate, emersed, obtuse or acutish, entire or repand-denticulate and glandular, those of the stem few, small and sessile, the basal ones 2′–9′ long, 2″–4″ wide, narrowed into petioles; flowers pale blue, racemose, 5″–6″ long; calyx-lobes narrowly lanceolate, about as long as the tube, the sinuses commonly not at all appendaged; corolla-tube 3″–4″ long, its lower lip pubescent at the base.

In swamps and ponds, Delaware to Florida and Louisiana mostly near the coast. May–July.

3. **Lobelia cardinàlis** L. Cardinal-flower. Red Lobelia. Red Betty. Fig. 4030.

Lobelia cardinalis L. Sp. Pl. 930. 1753.

Perennial by offsets; stem slightly pubescent, or glabrous, leafy, simple or rarely branched, 2°–4½° high. Leaves oblong, oval, ovate-lanceolate, or lanceolate, thin, glabrous or sparingly pubescent, 2′–6′ long, ¼′–1½′ wide, acuminate or acute at both ends, crenulate or denticulate, the upper sessile, the lower petioled; flowers racemose, commonly numerous, bright scarlet or red (rarely white), 1′–1½′ long; bracts usually glandular; calyx glabrous or pubescent, its lobes linear, elongated, acute; corolla-tube nearly or quite 1′ long; larger anthers glabrous.

In moist soil, New Brunswick to Florida, Ontario, Kansas, Colorado and Texas. Slink-weed. Hog's-physic. July–Sept.

4. Lobelia syphilítica L.   Great Lobelia.
Blue Cardinal-flower.   Fig. 4031.

Lobelia syphilitica L. Sp. Pl. 931. 1753.
Lobelia syphilitica ludoviciana A. DC. Prodr. **7**: 377. 1839.

Perennial by short offsets; stem sparingly pubescent, rather stout, very leafy, usually simple, 1°–3° high. Leaves glabrous or sparingly puberulent, 2′–6′ long, ½′–2′ wide, oval, oblong, or lanceolate, acute or acuminate at the apex, narrowed at the base, repand-denticulate, irregularly crenate-dentate or nearly entire, sessile, or the lower obovate, obtuse and narrowed into petioles; flowers bright blue, or occasionally white, 10″–12″ long, densely racemose, leafy-bracted; calyx-lobes lanceolate, acuminate, pubescent or ciliate, the sinuses appendaged by large deflexed auricles; corolla-tube 5″–6″ long, about 2″ thick, the lobes of its larger lip oblong-oval, obtuse or acutish, glabrous; larger anthers glabrous.

In moist soil, Maine and Ontario to South Dakota, Colorado, Georgia, Louisiana and Kansas. Hybridizes with the preceding species. High-belia. July–Oct.

5. Lobelia amoèna Michx.   Southern Lobelia.
Fig. 4032.

Lobelia amoena Michx. Fl. Bor. Am. **2**: 152. 1803.
Lobelia amoena glandulifera A. Gray, Syn. Fl. **2**: 4. 1878.

Nearly glabrous throughout, perennial; stem simple, slender, leafy, 1°–4° high. Leaves thin, ovate to oblong-lanceolate, narrowly oblong, or oval, repand-dentate or denticulate, the lower petioled and mostly obtuse, 2′–6′ long, the upper sessile, acute or acutish, smaller; flowers blue, racemose, nearly 1′ long; bracts narrow and small, or the lower foliaceous, glandular; pedicels 1″–2½″ long; calyx-lobes linear-subulate, acuminate, glabrous, glandular, elongated, the sinuses usually not auricled; corolla-tube 5″–7″ long, 1½″–2″ thick; larger anthers glabrous, or puberulent at the tip; lobes of the larger lip of the corolla broadly ovate to oval, obtuse, glabrous.

In swamps, Delaware to Florida and Alabama. July–Sept.

6. Lobelia elongàta Small.   Long-leaved
Lobelia.   Fig. 4033.

L. elongata Small, Fl. SE. U. S. 1144. 1903.

Perennial, at least by offsets, glabrous or nearly so; stem simple, 1°–3° tall, commonly wand-like. Leaves rather few, erect or ascending, linear or nearly so, ¾′–4′ long, mostly acute, serrate or dentate-serrate with gland-tipped teeth, sessile or narrowed into petiole-like bases; flowers deep-blue in rather closely flowered, but not densely flowered, one-sided racemes 4′–12′ long; bracts linear to lanceolate, serrate with gland-tipped teeth, the lower ones sometimes surpassing the corollas; calyx-lobes elongate, linear-subulate or linear-setaceous, entire, as long as the corolla or shorter, without auricles at the sinuses; corolla-tube 5″–8″ long, lobes of the lower lip oval or ovate, glabrous.

In low grounds or swamps, Virginia to Florida and Louisiana. Aug.–Oct.

7. Lobelia pubérula Michx. Downy Lobelia.
Fig. 4034.

Lobelia puberula Michx. Fl. Bor. Am. **2** : 152. 1803.

Perennial, densely and finely puberulent all over, slightly viscid; stem simple, or rarely with a few branches, stout or slender, leafy, 1°–3° high. Leaves oval, oblong, ovate, or obovate, 1′–2′ long, rather thick, the lower petioled, all obtuse or the uppermost acute, denticulate or crenate-dentate, the teeth often glandular; flowers blue, 8″–10″ long, in long spike-like racemes; lower bracts, or sometimes all of them, foliaceous, glandular; pedicels very short; calyx hirsute or pubescent, its lobes narrowly lanceolate, elongated, with small short rounded auricles at the sinuses; corolla-tube about 5″ long, 1″–1½″ thick, the lobes of its larger lip broadly ovate, glabrous; larger anthers minutely bearded.

In moist sandy soil, southern New Jersey to Florida, Illinois, Iowa, Kansas and Texas. Ascends to 3500 ft. in North Carolina. Aug.–Oct.

8. Lobelia glandulòsa Walt. Glandular Lobelia.
Fig. 4035.

Lobelia glandulosa Walt. Fl. Car. 218. 1788.

Perennial; stem slender, leafy below, nearly naked above, simple, glabrous, or sparingly pubescent, 1°–4° high. Leaves elongated-linear to narrowly lanceolate, thick, glabrous, strongly glandular-dentate, the lower petioled, obtuse, often 7′ long and 4″ wide, the upper sessile, acutish, shorter; flowers racemose-spicate, secund, often few and distant, blue, nearly 1′ long; bracts narrow, glandular; peduncles very short, sometimes with a pair of glands near the base; calyx-tube often densely hirsute, its lobes subulate, the sinuses not appendaged; corolla-tube 5″–6″ long, 1″–1½″ thick, about twice as long as the calyx-lobes; larger lip of the corolla pubescent within at the base, its lobes ovate, acutish; anthers all bearded at the tip.

In swamps near the coast, southern Virginia to Florida. July–Sept.

9. Lobelia spicàta Lam. Pale Spiked Lobelia. Fig. 4036.

Lobelia spicata Lam. Encycl. **3** : 587. 1789.
L. spicata hirtella A. Gray, Syn. Fl. **2** : 6. 1878.
L. spicata parviflora A. Gray, Syn. Fl. **2** : 6. 1878.

Perennial or biennial, puberulent, smooth or roughish; stem strict, simple, leafy, 1°–4° high. Leaves thickish, pale green, repand-dentate, crenulate, or entire, the basal ones commonly tufted, broadly oblong, oval, or obovate, very obtuse, narrowed into short petioles, 1′–3½′ long; 1′–2′ wide; stem leaves sessile, oblong, lanceolate, or spatulate, obtuse, the uppermost gradually smaller and acutish; flowers pale blue, 3″–5″ long, densely or distantly racemose-spicate, the inflorescence sometimes 2° long; bracts linear, entire; pedicels very short, ascending; calyx-tube turbinate, usually glabrous, shorter than its subulate spreading, sometimes hirsute or ciliate lobes, the sinuses usually not at all append-aged; corolla-tube about 2½″ long and 1″ thick.

In dry, mostly sandy soil, or in meadows, Prince Edward Island to Saskatchewan, North Carolina, Alabama, Louisiana and Arkansas. Races differ in pubescence and in size of flowers. Ascends to 2500 ft. in Virginia. June–Aug.

10 Lobelia leptóstachys A. DC. Spiked Lobelia. Fig. 4037.

Lobelia leptostachys A. DC. Prodr. 7: 376. 1839.

Similar to the preceding species; stem usually stouter, puberulent or glabrous, 2°–4° high. Basal leaves oval or obovate, obtuse; stem leaves spatulate, oblong, or lanceolate, obtuse, sometimes slightly scabrous, denticulate or entire, or the uppermost narrower and acute; spike-like raceme elongated, usually dense; bracts linear, glabrous; pedicels very short; calyx-lobes linear-subulate, nearly as long as the corolla-tube, the sinuses with subulate deflexed auricles; flowers blue, 4″–5″ long.

In dry soil, Virginia to Ohio and Illinois, Georgia and Kansas. June–Aug.

11. Lobelia inflàta L. Indian or Wild Tobacco. Eyebright. Fig. 4038.

Lobelia inflata L. Sp. Pl. 931. 1753.

Annual, pubescent or hirsute, very acrid; stem leafy, commonly paniculately branched, 1°–3° high. Leaves thin, repand-dentate or denticulate, the lower oval or obovate, obtuse, 1′–2½′ long, narrowed into short petioles, the upper sessile, oval, oblong, ovate, or ovate-lanceolate, obtuse, or the uppermost acute; flowers light blue, 2″–3″ long, usually distant in somewhat spike-like racemes; lower bracts foliaceous, the upper subulate; pedicels 2″–5″ long in fruit; calyx glabrous or nearly so, its subulate lobes about as long as the corolla; capsule inflated, 3″–4″ long, finely transversely veined between the ribs.

In fields and thickets, usually in dry soil, Labrador to Saskatchewan, Georgia, Kansas and Arkansas. Gag-root. Puke-weed. Asthma-weed. Lowbelia. Emetic-weed. Bladder-pod lobelia. July–Nov.

12. Lobelia Kàlmii L. Brook or Kalm's Lobelia. Fig. 4039.

Lobelia Kalmii L. Sp. Pl. 930. 1753.

Perennial by short offsets, glabrous throughout, or sparingly pubescent below; stem leafy, erect, paniculately branched, rarely simple, slender, 6′–20′ high. Lower and basal leaves spatulate, obtuse, narrowed into short petioles, sparingly repand-denticulate, or entire, 6″–12″ long, 1½′–2½′ wide; upper leaves sessile, usually longer and narrower, linear, linear-oblong, or narrowly spatulate, the uppermost acute; flowers light blue, 4″–5″ long, in loose racemes; lower bracts linear-lanceolate, the upper subulate; pedicels nearly filiform, 4″–12″ long, usually 2-glandular or 2-bracteolate near the middle; calyx-lobes lanceolate-subulate, longer than the turbinate tube, the sinuses not appendaged; capsule wholly inferior, not inflated, campanulate or subglobose, about 2″ long.

On wet banks, and in wet meadows, Nova Scotia to New Jersey, west to Ontario, Manitoba, Ohio, Michigan and Iowa. July–Sept.

$\frac{2}{3}$

13. Lobelia Nuttàllii R. & S.   Nuttall's Lobelia.   Fig. 4040.

Lobelia gracilis Nutt. .Gen. **2**: 77. 1818. Not Andr.
Lobelia Nuttallii R. & S. Syst. **5**: 39. 1819.

Annual, or perhaps biennial, glabrous throughout, or puberulent below; stem weak, usually reclining, very slender, loosely branched, at least when old, 1°–3° long. Basal leaves spatulate to oval, obtuse, mostly petioled, 6″–12″ long, slightly repand, or entire; stem leaves distant, linear, linear-oblong, or slightly spatulate, longer and narrower, entire or sparingly glandular-denticulate; flowers 2½″–4″ long, pale blue, loosely racemose; bracts linear or the upper subulate; pedicels filiform, 2″–4″ long, naked, or minutely 2-bracteolate near the base; calyx-lobes subulate, longer than the depressed-hemispheric strongly ribbed tube, the sinuses unappendaged; capsule depressed-globose, half-inferior, about 1″ long.

In sandy soil, Long Island to Pennsylvania, Florida and Georgia. June–Sept.

14. Lobelia Cánbyi A. Gray.   Canby's Lobelia.   Fig. 4041.

Lobelia Canbyi A. Gray, Man. Ed. 5, 284. 1867.

Annual, slightly puberulent, usually glabrous; stem erect, slender, paniculately branched, or simple, 2°–3° high. Stem leaves linear or linear-oblong, ½′–1½′ long, ½″–1½″ wide, the lower obtuse, sometimes slightly repand-denticulate, the upper acute, narrower, entire; flowers racemose, blue, 4″–5½″ long; lower bracts linear, the upper subulate; pedicels erect or ascending, naked, filiform, 1″–3″ long; calyx-lobes linear-subulate, glandular-denticulate, equalling or somewhat longer than the narrowly turbinate tube, mostly shorter than the tube of the corolla; capsule oblong-turbinate, 2″ long, shorter than the calyx-tube.

Swamps, New Jersey to South Carolina. July–Sept.

$\frac{2}{3}$

Family 44. CICHORÌACEAE Reichenb. Fl. Excurs. 248. 1831.

CHICORY FAMILY.

Herbs (two Pacific Island genera trees), almost always with milky, acrid or bitter juice, alternate or basal leaves, and yellow, rarely pink, blue purple or white flowers in involucrate heads (anthodia). Bracts of the involucre in 1 to several series. Receptacle of the head flat or flattish, naked, scaly (paleaceous), smooth, pitted, or honeycombed. Flowers all alike (heads homogamous), perfect. Calyx-tube completely adnate to the ovary, its limb (pappus) of scales, or simple or plumose bristles, or both, or wanting. Corolla gamopetalous, with a short or long tube, and a strap-shaped (ligulate) usually 5-toothed limb (ray). Anthers connate into a tube around the style, the sacs sagittate or auricled at the base, not tailed, usually appendaged at the summit, the simple pollen-grains usually 12-sided. Ovary 1-celled; ovule 1, anatropous; style very slender, 2-cleft, or 2-lobed, the lobes minutely papillose. Fruit an achene. Seed erect; endosperm none; radicle narrower than the cotyledons.

About 70 genera and 1500 species, of wide geographic distribution. The family is also known as LIGULIFLORAE, and is often regarded as a tribe of the COMPOSITAE.

* Pappus of scales, or of scales and bristles, or none.

Flowers blue or white; pappus of blunt scales.   1. *Cichorium.*
Flowers yellow.
     Bracts of the involucre membranous, or herbaceous.
         Pappus none; achenes 20–30-nerved.   2. *Lapsana.*
         Pappus none; achenes 8–10-ribbed.   3. *Serinea.*

Pappus of rounded scales, with or without an inner series of bristles.
  Bracts of the involucre 9–18, reflexed in fruit, not keeled; pappus-bristles usually more
    numerous than the scales.
    Annuals; pappus-scales 5, obovate or rounded.   4. *Krigia.*
    Perennials; pappus-scales 10–15, linear or oblong, minute.   5. *Cynthia.*
  Bracts of the involucre 5–8, erect in fruit, keeled; pappus-bristles and scales 5.
                                                      6. *Cymbia.*
Bracts of the involucre thickened and keeled after flowering; pappus none.   7. *Arnoseris.*

** Pappus, at least some of it, of plumose bristles.

Receptacle chaffy.   8. *Hypochaeris.*
Receptacle naked.
  Plume-branches of the pappus not interwebbed.
    Flowers yellow; plants scapose, the leaves basal.   9. *Apargia.*
    Flowers yellow; plants leafy-stemmed.   10. *Picris.*
    Flowers pink.   11. *Ptiloria.*
  Plume-branches of the pappus interwebbed.   12. *Tragopogon.*

*** Pappus of simple bristles or of some soft white scales.
† *Receptacle with a few deciduous bristles; flowers yellow*   13. *Malacothrix.*
†† *Receptacle naked.*

    1. Achenes spinulose, or with short processes near the summit.
Heads few- (6–15-) flowered, yellow; stem branching.   14. *Chondrilla.*
Heads many-flowered, yellow, solitary on scapes.   15. *Leontodon.*

    2. Achenes smooth, or papillose, not spinulose toward the summit.
                 (a) Achenes flattened.
Achenes truncate, not beaked; flowers yellow.   16. *Sonchus.*
Achenes narrowed at the summit, or beaked; flowers blue or yellow.   17. *Lactuca.*

             (b) Achenes cylindric, or prismatic.
Achenes terete, not narrowed either at the base or summit; flowers pink or purple.   18. *Lygodesmia.*
Achenes narrowed at the base, narrowed or beaked at the summit; flowers yellow.
  Achenes slender-beaked.
    Pappus-bristles not surrounded by a villous ring at base.   19. *Agoseris.*
    Pappus-bristles surrounded by a villous white ring at base.   20. *Sitilias.*
  Achenes merely narrowed above, not beaked.   21. *Crepis.*
Achenes narrowed at the base, otherwise columnar, truncate (except in 2 species of *Hieracium*).
  Flowers yellow, orange, or red.   22. *Hieracium.*
  Flowers white, cream-color, or purple.   23. *Nabalus.*

1. CICHÒRIUM [Tourn.] L. Sp. Pl. 813. 1753.

Erect branching herbs, with alternate and basal leaves, those of the stem and branches
usually very small and bract-like, and large heads of blue purple pink or white flowers,
peduncled, or in sessile clusters along the branches. Involucre of 2 series of herbaceous
bracts, the outer somewhat spreading, the inner erect and subtending, or partly enclosing,
the outer achenes. Receptacle flat, naked, or slightly fimbrillate. Rays truncate and
5-toothed at the apex. Anthers sagittate at the base. Style-branches slender, obtusish.
Achenes 5-angled or 5-ribbed, truncate, not beaked. Pappus of 2 or 3 series of short blunt
scales. [From the Arabic name.]

    About 8 species, natives of the Old World, the following typical.

1. Cichorium Íntybus L. Chicory. Wild Succory.
    Blue Sailors. Bunk. Fig. 4042.

Cichorium Intybus L. Sp. Pl. 813. 1753.
Cichorium Intybus divaricatum DC. Prodr. 7: 84. 1838.

Perennial from a long deep tap-root; stems slightly
hispid, stiff, much branched, 1°–3° high. Basal leaves
spreading, runcinate-pinnatifid, dentate or lobed, spatulate,
3′–6′ long, narrowed into long petioles; upper leaves much
smaller, lanceolate or oblong, lobed or entire, clasping and
auricled at the base; heads numerous, 1′–1½′ broad, 1–4
together in sessile clusters on the nearly naked or bracted
branches, or sometimes peduncled; inner bracts of the
involucre about 8; flowers blue, or sometimes white.

    Roadsides, fields and waste places, Nova Scotia to Minne-
sota, Washington, North Carolina, Kansas, Colorado and California. Bermuda. The ground-up root is used as a substitute
or adulterant for coffee. Heads usually closed by noon. The
flowers sometimes bear supplementary rays. Blue daisy or
dandelion. Coffee-weed. Bachelor's-buttons. Consists of several races. July–Oct.

½

2. LÁPSANA L. Sp. Pl. 811. 1753.

Annual erect branching herbs, with alternate dentate or pinnatifid leaves, and small panicled slender-peduncled heads of yellow flowers. Involucre nearly cylindric, its principal bracts in 1 series, nearly equal, with a few exterior small ones at the base. Receptacle flat, naked. Rays truncate and 5-toothed at the apex. Anthers sagittate at the base. Style-branches slender. Achenes obovate-oblong, 20–30-nerved, somewhat flattened, narrowed below, rounded at the summit. Pappus none. [Greek, *lampsana*, the name of a crucifer.]

About 9 species, natives of the Old World, the following typical.

1. Lapsana commùnis L. Nipplewort. Succory Dock-cress. Fig. 4043.

Lapsana communis L. Sp. Pl. 811. 1753.

Stem paniculately branched, glabrous above, more or less hispid-pubescent below, 1°–3½° high. Lower leaves ovate, repand-dentate, obtuse, thin, pubescent, or glabrate, petioled, 2′–4′ long, often with 2–6 lobes on the petiole; the uppermost oblong or lanceolate, sessile, acute, much smaller, mostly entire; heads very numerous, 3″–6″ broad; involucre oblong-cylindric, 2″–3″ high, and of about 8 linear glaucous principal bracts and several very small outer ones.

Along roadsides and in waste places, Quebec and Ontario to New Jersey, Pennsylvania and Michigan. Also on the Pacific Coast and in Jamaica. Naturalized from Europe. Called also bolgan-leaves, ballogan. June–Sept.

3. SERINIA Raf. Fl. Ludov. 149. 1817.

[APOGON Ell. Bot. S. C. & Ga. 2: 267. 1824.]

Low glaucescent branching annual herbs, with alternate clasping entire or lobed leaves, or those of the stem sometimes appearing as if opposite, and few small long-peduncled heads of yellow flowers. Involucre broadly campanulate, its bracts about 8, equal, membranous, becoming concave after flowering. Receptacle flat, naked. Rays truncate and 5-toothed at the apex. Anthers sagittate at the base. Style-branches slender. Achenes obovoid, 8–10-ribbed, contracted at the base, rounded at the summit. Pappus none, or a mere vestige. [Greek, small chicory.]

Three known species, natives of the southern United States, the following typical.

1. Serinia oppositifòlia (Raf.) Kuntze. Serinia. Fig. 4044.

Krigia oppositifolia Raf. Fl. Ludov. 57. 1817.
Apogon humilis Ell. Bot. S. C. & Ga. 2: 267. 1824.
Serinia oppositifolia Kuntze, Rev. Gen. Pl. 364. 1891.

Glabrous throughout, or slightly glandular-pubescent along the ends of the peduncles, branched from the base, 4′–10′ high. Basal and lower leaves petioled, oblong-lanceolate or spatulate in outline, acute or obtuse, entire, lobed or pinnatifid, 3′–5′ long, 2″–6″ wide; upper leaves mainly sessile, clasping, alternate, or appearing as if opposite, usually entire, smaller; peduncles very slender, sometimes 4′ long; heads 1½″–2″ broad; bracts of the involucre acute or acuminate, about the length of the rays.

Kansas to Texas, east to North Carolina and Florida. Recorded from Missouri. March–May.

4. KRÍGIA Schreb. Gen. Pl. 532. 1791.

An annual herb, with scapose stems, basal, entire, sinuate-dentate or pinnatifid leaves, and a small or middle-sized head of yellow flowers, solitary at the end of the scape. Involucre campanulate, its herbaceous bracts 9–18, reflexed in fruit, in 2 series, with no exterior shorter ones. Receptacle flat, naked. Rays truncate and 5-toothed at the apex. Anthers sagittate at the base. Style-branches slender, obtusish. Achenes turbinate, 15–20-ribbed, truncate. Pappus in 2 series, the outer of 5 thin broad rounded scales, the inner of 10 or more slender naked bristles. [In honor of David Krig, who collected plants in Maryland early in the eighteenth century.]

A monotypic genus of North America. This and the two following genera were included in *Adopogon* Neck., in our first edition, but that genus is not typified, and the name probably belongs to an Old World plant.

1. **Krigia virgínica** (L.) Willd. Carolinia Dwarf Dandelion. Krigia.
Fig. 4045.

Hyoseris virginica L. Sp. Pl. 809. 1753.

Hyoseris caroliniana Walt. Fl. Car. 194. 1788.

Krigia virginica Willd. Sp. Pl. 3: 1618. 1804.

Krigia caroliniana Nutt. Gen. 2: 126. 1818.

Adopogon carolinianum Britton, Mem. Torr. Club 5: 346. 1894.

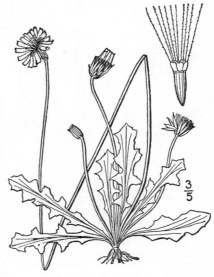

Annual, acaulescent; scapes usually several from the same root, very slender, glabrous or hispidulous, monocephalous, 1'–15' high, simple, or sometimes branched at or near the base. Leaves commonly all basal, tufted, spatulate, lanceolate or linear, pinnatifid, sinuate, lobed, dentate or rarely entire, 1'–6' long, narrowed at the base into usually margined petioles; head 3''–7'' broad; involucre of 9–18 linear-lanceolate bracts, reflexed after the fall of the narrowly turbinate somewhat 5-angled achenes; pappus of 5 round short scales and 10 or more long capillary bristles.

In dry, sandy soil, Maine to Ontario and Minnesota, Florida and Texas; also in Washington. April–Aug.

5. **CÝNTHIA** D. Don, Edinb. Phil. Journ. 12: 305. 1829.

Perennial herbs, with tufted basal leaves, the large many-flowered heads at the ends of simple or branched scapes. Flowers orange or yellow. Bracts of the involucre 9–18, reflexed in fruit, lanceolate to linear-lanceolate, not keeled. Pappus of 10–15 small scales and as many long bristles or more. [Mythological name.]

Four species, natives of North America. Type species: *Cynthia virginica* (L.) D. Don.

Caulescent, branched above. 1. *C. virginica.*
Acaulescent, monocephalous. 2. *C. Dandelion.*

1. **Cynthia virgínica** (L.) D. Don. Cynthia. Virginia Goatsbeard.
Fig. 4046.

Tragopogon virginicum L. Sp. Pl. 789. 1753.
Krigia amplexicaulis Nutt. Gen. 2: 127. 1818.
Cynthia virginica D. Don, Edinb. Phil. Journ. 12: 309. 1829.
Adopogon virginicum Kuntze, Rev. Gen. Pl. 304. 1891.
Cynthia falcata Standley, Contr. Nat. Herb. 13: 356. 1911.

Perennial, glabrous and glaucous; stem 1°–2½° high, 1-leaved and branched above, bearing 1–6 long-peduncled heads at its summit and usually 1 oblong, entire or toothed clasping leaf below the middle. Basal leaves tufted, runcinate, sinuate, denticulate, or entire, 2'–7' long, narrowed into margined petioles, heads about 1½' broad; involucre of 9–15 lanceolate nerveless bracts, 3''–4'' long, reflexed in fruit; achenes nearly oblong; pappus of 10–15 small oblong scales and an equal or greater number of capillary bristles; flowers orange to reddish orange.

In moist woods and meadows, Massachusetts to southern Ontario and Manitoba, Georgia, Kentucky, Missouri and Colorado. Ascends to 4000 ft. in Virginia. False dandelion. May–Oct.

$\frac{3}{5}$

2. Cynthia Dandèlion (L.) DC. Dwarf Dandelion or Goatsbeard. Fig. 4047.

Tragopogon Dandelion L. Sp. Pl. Ed. 2, 1111. 1763.

Krigia Dandelion Nutt. Gen. **2**: 127. 1818.

Cynthia Dandelion DC. Prodr. **7**: 89. 1838.

Adopogon Dandelion Kuntze, Rev. Gen. Pl. 304. 1891.

Perennial, acaulescent, glabrous and some-what glaucous; scape 6′–18′ high, slender leaf-less, with a single head. Stolons filiform, bear-ing globose tubers; leaves all basal, tufted, linear-lanceolate to spatulate, entire, denticulate, sinuate, or pinnatifid, narrowed at the base, 3′–6′ long, 2″–5″ wide; head about 1′ broad; involucre nearly ½′ high; pappus similar to that of the preceding species.

In moist soil, Maryland to Florida, Illinois, Mis-souri, Kansas and Texas. April–June.

6. CÝMBIA (T. & G.) Standley, Contr. Nat. Herb. **13**: 354. 1911.

An annual acaulescent herb, the leaves forming rosettes, the scapes monocephalous. Involucral bracts 5–8, ovate to ovate-lanceolate, remaining erect in fruit. Receptacle naked. Rays yellow, 5-toothed. Achenes turbinate, ribbed. Pappus an outer series of 5 obovate scales, and an inner series of 5 bristles, much longer than the scales. [Greek, referring to the cup-shaped fruiting involucre.]

A monotypic genus of the south-central United States.

1. Cymbia occidentàlis (Nutt.) Standley. Western Dwarf Dandelion. Fig. 4048.

Krigia occidentalis Nutt. Journ. Acad. Phila. **7**: 104. 1834.

Adopogon occidentale Kuntze, Rev. Gen. Pl. 304. 1891.

C. occidentalis Standley, Contr. Nat. Herb. **13**: 354. 1911.

Scapes tufted, 2′–8′ high, usually glandular, at least above, sometimes glabrous, bearing a single head 5″–10″ broad. Leaves basal, lanceolate to obovate, entire to pinnatifid, mostly shorter than the scapes; involucre 2″–3″ high, firm and keeled at maturity, remaining erect; achenes transversely wrinkled; pappus of 5 obovate scales and 5 alter-nating bristles, or these wanting.

Prairies, southern Missouri and Kansas to Texas. April–May.

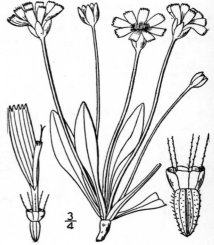

$\frac{3}{4}$

7. ARNÓSERIS Gaertn. Fr. & Sem. **2**: 355. *pl. 157.* 1791.

A low annual scapose herb, glabrous, or nearly so, with tufted basal dentate or nearly entire leaves. Scapes several or numerous, simple or branched, upwardly thickened below the solitary heads of yellow flowers. Involucre campanulate, its bracts in 1 series, narrow, equal, thickened and keeled on the back after flowering, rarely with a few outer minute ones. Receptacle flat, pitted, not chaffy. Anthers sagittate. Style-branches obtuse. Achenes oval, 8–10-ribbed, narrowed below, truncate, or with a denticulate margin. Pappus none. [Greek, lamb-succory.]

A monotypic genus of western Europe.

1. Arnoseris mínima (L.) Dumort. Lamb Succory. Fig. 4049.

Hyoseris minima L. Sp. Pl. 879. 1753.
Arnoseris pusilla Gaertn. Fr. & Sem. **2**: 355. 1791.
Arnoseris minima Dumort. Fl. Belg. 63. 1827.

Scapes slender, 3'–12' high, leafless, simple, or with 1–4 branches mostly above the middle, gradually thickened and hollow upward for a space of an inch or more below the heads. Leaves oblanceolate, obovate, or oblong, 1'–3' long, 3"–10" wide, usually coarsely and sharply toothed, narrowed into margined petioles; heads 8" broad, or less; bracts of the involucre linear-lanceolate, acuminate, strongly keeled after flowering, 2"–4" long, curving over the achenes.

Fields and waste grounds, Maine to Ontario and Michigan. Adventive from Europe. Also called dwarf swine's- or hog's-succory; dwarf nipplewort. Summer.

8. HYPOCHAÈRIS [Vaill.] L. Sp. Pl. 810. 1753.

Mostly perennial herbs, with scapose, often branched stems and mostly basal tufted leaves, pinnatifid to entire, those of the scapes few, scale-like, very small, or none. Heads mostly large, long-peduncled. Flowers yellow. Involucre oblong-cylindric to campanulate, its bracts herbaceous, imbricated in several series, appressed, the outer smaller. Receptacle flat, chaffy. Anthers sagittate. Style-branches slender, obtusish. Achenes oblong to linear, 10-ribbed, somewhat narrowed below, contracted above into a long or short beak, or the outer truncate. Pappus of 1 row of plumose bristles, sometimes with some shorter simple ones. [Greek, for pigs, which are fond of its roots.]

About 50 species, natives of Europe, Asia and South America. Type species: *Hypochaeris glàbra* L.

1. Hypochaeris radicàta L. Long-rooted Cat's-ear. Gosmore. Fig. 4050.

Hypochaeris radicata L. Sp. Pl. 811. 1753.

Perennial; stems several together, glabrous, slender, 1°–2° high, bracted, or rarely simple, bearing a few scales. Leaves spreading on the ground, oblanceolate to obovate in outline, pinnatifid-lobed to dentate, 2'–6' long, hirsute on both sides; involucre oblong-cylindric, about 1' high, its bracts glabrous, or sparingly pubescent; heads 1' broad or more; chaff of the receptacle of narrow soft pointed scales; achenes rough, all with very slender beaks longer than the body; flowers longer than the involucre.

In waste places, Ontario to New York, southern New Jersey and Pennsylvania. Also in California, Oregon, Washington and Colorado. Jamaica. Adventive or naturalized from Europe. Native also of Asia. May–Oct.

Hypochaeris glàbra L., the smooth cat's-ear, a smaller species, with nearly or quite glabrous leaves, flowers scarcely longer than the involucre, and the outer achenes truncate, has been found in Maine and Ontario, and in Georgia, and is naturalized on the Pacific Coast.

9. APÁRGIA Scop. Fl. Carn. Ed. 2, **2**: 113. 1772.

Perennial scapose herbs, with tufted basal, mostly pinnatifid leaves, branched and scaly, or simple and naked scapes, and large heads of yellow flowers, solitary at the end of the scape or of its branches. Involucre ovoid or oblong, its principal bracts in 1 or 2 series, nearly equal, with several series of short outer ones. Receptacle flat, fimbrillate, villous, or somewhat honeycombed. Rays truncate and 5-toothed at the apex. Anthers sagittate at the base. Style-branches slender. Achenes oblong or linear, finely striate, contracted or beaked at the summit. Pappus of 1 or 2 series of plumose persistent brownish bristles, somewhat broadened at the base, or the outer scale-like and simple. [Greek, from the growth of these plants on unused land.]

About 45 species, natives of the Old World. Type species: *Apargia incana* Scop.

Scape scaly-bracted, mostly branched; pappus of plumose bristles only.   1. *A. autumnalis.*
Scape bractless, monocephalous; pappus of two kinds.
    Pappus of marginal and inner flowers dissimilar.   2. *A. nudicaulis.*
    Pappus of all flowers alike.   3. *A. hispida.*

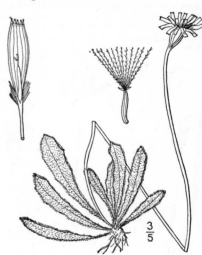

1. Apargia autumnàle (L.) Hoffm. Fall Dandelion. Autumnal Hawkbit. Lion's-tooth. Fig. 4051.

Leontodon autumnale L. Sp. Pl. 798. 1753.

Apargia autumnalis Hoffm. Deutsch. Fl. Ed. 2, 2: 113. 1800.

Plant glabrous or nearly so, or the involucres and ends of peduncles black-pubescent; scape slender, usually branched and scaly, 6′–2° high. Leaves narrowly oblong to linear-lanceolate, pinnatifid into narrow lobes, or some of them coarsely dentate, 3′–8′ long, 3″–12″ wide, acuminate at the apex, narrowed into rather short petioles; heads several, rarely solitary, about 12″–15″ broad; involucre oblong; achenes tapering into a short beak, or the outer ones nearly terete; pappus-bristles all plumose.

In fields and along roadsides, Newfoundland to Ontario, New Jersey, Pennsylvania and Michigan. Naturalized from Europe. Native also of Asia. June–Nov. Dog-dandelion. Arnica-bud.

2. Apargia nudicaulis (L.) Britton. Rough or Hairy Hawkbit. Fig. 4052.

Crepis nudicaulis L. Sp. Pl. 805. 1753.
Leontodon hirtum L. Sp. Pl. Ed. 2, 1123. 1763.
Leontodon nudicaule Banks; Lowe, Trans. Camb. Phil. Soc. 4: 28. 1831.

Plant more or less hirsute; scape simple, slender, 4′–12′ high, minutely scaly, or naked. Leaves linear-oblong to narrowly spatulate, acute or obtuse, not acuminate, nearly entire, coarsely sinuate-dentate or sometimes pinnatifid, 2′–5′ long, 3″–8″ wide, narrowed into petioles; head solitary at the end of the scape, 5″–10″ broad; involucre canescent or pubescent; outer achenes with a pappus of simple narrow scales, the inner ones with a pappus of plumose bristles.

In ballast and waste places, Connecticut to New Jersey and Pennsylvania, and on Vancouver Island. Adventive from Europe. June–Oct.

3. Apargia hispida (L.) Willd. Common Hawkbit. Fig. 4053.

Leontodon hispidus L. Sp. Pl. 799. 1753.

Leontodon hastile L. Sp. Pl. Ed. 2, 1123. 1763.

A. hispida Willd. Sp. Pl. 3: 1552. 1804.

Glabrous, or bristly-hispid. Scape stout or slender, 4′–25′ high, bractless, monocephalous; head nodding before flowering; leaves 2′–6′ long, oblong-lanceolate in outline, coarsely and sharply dentate or subpinnatifid; head 1½′ broad or less; involucre 6″–7″ high; pappus an outer series of short and an inner series of long plumose bristles, alike in all the flowers.

Fields and waste grounds, Rhode Island to Pennsylvania, Ontario and Ohio. July–Sept.

10. **PÌCRIS** L. Sp. Pl. 792. 1753.

Erect hispid, mostly branching, leafy herbs, with alternate leaves (in our species), and rather large, usually corymbose or paniculate heads of yellow flowers. Involucre campanulate or cup-shaped, its principal bracts in 1 series, nearly equal, with 2-3 series of small or large exterior spreading ones. Receptacle flat, short-fimbrillate. Rays truncate and 5-toothed at the apex. Anthers sagittate at the base. Style-branches slender. Achenes linear or oblong, somewhat incurved, terete or angled, 5-10-ribbed and transversely wrinkled, narrowed at the base and summit, or beaked in some species. Pappus of 1 or 2 series of slender plumose bristles. [Greek, bitter.]

About 35 species, natives of the Old World, one perhaps indigenous in Alaska. Type species: *Picris asplenioides* L.

Outer involucral bracts linear; achenes not beaked.   1. *P. hieracioides.*
Outer involucral bracts ovate, foliaceous; achenes short-beaked.   2. *P. echioides.*

1. Picris hieracioìdes L. Hawkweed Picris. Fig. 4054.

Picris hieracioides L. Sp. Pl. 792. 1753.

Biennial, more or less hispid, much branched, 1°–3° high. Leaves lanceolate or oblong-lanceolate, dentate, or nearly entire, the basal narrowed into petioles, acute, often 6′ long, those of the stem mostly sessile and smaller; heads numerous, ½–1′ broad; involucre 4″–6″ high, its outer bracts linear, subulate, spreading, the inner linear-lanceolate, acuminate; pappus copious, nearly white.

In waste places, Illinois, Pennsylvania, New Jersey, and in ballast about the seaports. Adventive from Europe. Native also of Asia. Very bitter. June–Sept. Old name lang-de-beef.

2. Picris echioìdes L. Bristly Ox-tongue. Fig. 4055.

Picris echiọides L. Sp. Pl. 792. 1753.

Helmintha echinoides Gaertn. Fruct. & Sem. **2**: 368. 1802.

Annual or biennial, branched, hispid; stem about 2½° high. Basal and lower leaves spatulate or oblong, obtuse, repand-dentate, 2′–6′ long, narrowed into petioles; upper leaves sessile and clasping, oblong or lanceolate, smaller, the uppermost mainly acute and entire; heads numerous, rather crowded, short-peduncled, about ½′ broad; outer bracts of the involucre 4 or 5, foliaceous, ovate, acute, hispid-ciliate, the inner ones lanceolate, membranous; achenes beaked.

In waste places, Nova Scotia and Ontario to Pennsylvania, and in ballast about the seaports. Also in California. Adventive from Europe. July–Sept. Called also bugloss and bugloss-picris.

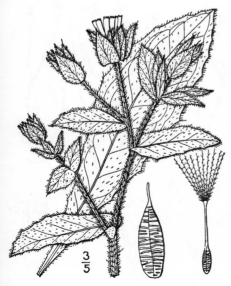

11. PTILÒRIA Raf. Atl. Journ. 145. 1832.

[STEPHANOMERIA Nutt. Trans. Amer. Phil. Soc. (II.) 7: 427. 1841.]

Annual or perennial, mostly glabrous, often glaucous herbs, with erect, simple or branched, usually rigid stems, alternate or basal, entire dentate or runcinate-pinnatifid leaves, those of the stem and branches often reduced to subulate scales, and small erect heads of pink flowers, paniculate, or solitary at the ends of the branches, opening in the morning. Involucre cylindric or oblong, its principal bracts few, equal, scarious-margined, slightly united at the base, with numerous short exterior ones and sometimes a few of intermediate length. Receptacle flat, naked. Anthers sagittate at the base. Style-branches slender. Achenes oblong or linear, terete or columnar, 5-ribbed, truncate or beaked at the summit, the ribs smooth or rugose. Pappus of 1 series of rather rigid plumose bristles. [Greek, referring to the feathery pappus.]

About 20 species, natives of western and central North America. Type species: *Ptiloria pauciflora* (Torr.) Raf.

Involucre about 5″ high; pappus brownish, plumose to below the middle.   1. *P. pauciflora*.
Involucre about 4″ high; pappus white, plumose almost to the base.   2. *P. ramosa*.

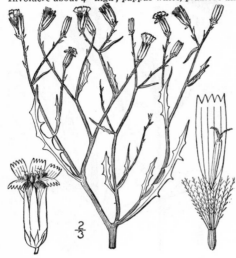

$\frac{2}{3}$

1. Ptiloria pauciflòra (Torr.) Raf.
Brown-plumed Ptiloria.   Fig. 4056.

Prenanthes? pauciflora Torr. Ann. Lyc. N. Y.
2: 210. 1827.

Ptiloria pauciflora Raf. Atl. Journ. 145. 1832.

Stephanomeria runcinata Nutt. Trans. Am.
Phil. Soc. (II.) 7: 427. 1841.

Perennial; stem rather stout, striate, rigid, divergently branched, 1°–2° high. Basal and lower leaves runcinate-pinnatifid, 1′–2½′ long, 3″–6″ wide, the upper all short and narrowly linear or reduced to scales; heads somewhat racemose-paniculate along the branches, usually about 5-flowered; involucre 4″–5″ high; rays 1″–2″ long; pappus brownish, plumose to below the middle.

Plains, Nebraska, Kansas to Wyoming, Texas and California. Summer.

2. Ptiloria ramòsa Rydb. White-plumed Ptiloria. Fig. 4057.

Ptiloria ramosa Rydb. Mem. N. Y. Bot. Gard.
1: 453. 1900.

Similar to the preceding species, but commonly lower, bushy-branched, the branches ascending. Basal leaves runcinate-pinnatifid, those of the stem linear or filiform, entire, or sometimes runcinate-dentate, the uppermost reduced to small scales; heads numerous, usually solitary at the ends of the branchlets; involucre about 4″ high; pappus bright white, very plumose to near the base.

Plains and dry, rocky soil, western Nebraska to Wyoming, Montana and Colorado. In first edition of this work not distinguished from the western *Ptiloria tenuifolia* (Torr.) Raf. May–Aug.

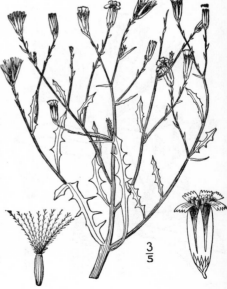

$\frac{3}{5}$

12. TRAGOPÒGON [Tourn.] L. Sp. Pl. 789. 1753.

Biennial or perennial, erect usually branched, somewhat succulent herbs, with slender fleshy tap-roots, alternate entire linear-lanceolate long-acuminate leaves, clasping at the base, and long-peduncled large heads of yellow or purple flowers, opening in the early morning, usually closed by noon. Involucre cylindric or narrowly campanulate, its bracts in 1 series, nearly equal, acuminate, united at the base. Rays truncate and 5-toothed at the apex. Anthers sagittate at the base. Style-branches slender. Achenes linear, terete, or 5-angled, 5-10-ribbed, terminated by slender beaks, or the outermost beakless. Pappus-bristles in 1 series, plumose, connate at the base, the plume-branches interwebbed. [Greek, goats-beard.]

About 35 species, natives of the Old World. Type species : *Tragopogon pratensis* L.

Flowers yellow ; involucral bracts equalling or shorter than the rays.	1. *T. pratensis.*
Flowers purple ; involucral bracts much longer than the rays.	2. *T. porrifolius.*

1. Tragopogon praténsis L.   Yellow Goat's-beard.   Meadow Salsify.
Fig. 4058.

Tragopogon pratensis L. Sp. Pl. 789. 1753.

Stem branched, 1¼°–3° high. Leaves keeled, tapering from the broad, more or less clasping base to a very long acuminate tip, the lower sometimes 10′ long and 1′ wide ; peduncles thickened at the top ; heads 1′–2½′ broad ; bracts of the involucre about 8, lanceolate, acuminate, shorter than or equalling (rarely exceeding) the yellow rays ; marginal achenes striate, smooth or roughened.

In fields and waste places, Nova Scotia to New Jersey, Ontario, Ohio, Manitoba and Colorado. Naturalized from Europe. June–Oct. Called also buck's-beard, noon-flower, star-of-jerusalem, noon-tide, joseph's flower, go-to-bed-at-noon.

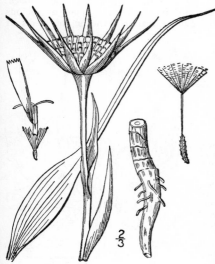

2. Tragopogon porrifòlius L.   Oyster Plant.   Salsify.   Purple Goat's-beard.
Fig. 4059.

Tragopogon porrifolius L. Sp. Pl. 789. 1753.

Taller, sometimes 4½° high. Peduncles very much thickened and hollow for 1 to 3 inches below the heads ; heads 2′–4′ broad, very showy ; bracts of the involucre linear-lanceolate, acuminate, usually much longer than the purple rays ; achenes sometimes 2′ long, the outer ones covered with scale-like tubercles, especially on the ribs below.

In fields and waste places, Ontario to New Jersey, Virginia, Minnesota, British Columbia, Nebraska and California, mostly escaped from gardens, where it is common. Native of Europe. Called also vegetable oyster, jerusalem-star, nap-at-noon, oyster-root. Naturalized as a weed on the Pacific Coast. The root is the familiar vegetable known as oyster-plant. An apparent hybrid between this and the preceding species has been noticed at New Brunswick, N. J. June–Oct.

13. MALÁCOTHRIX DC. Prodr. 7: 192. 1838.

Annual or perennial, branching or scapose herbs, with alternate or basal, mostly pinnatifid leaves, and long-peduncled panicled or solitary heads of yellow or rarely white flowers. Involucre campanulate, its principal bracts in 1 or 2 series, equal or nearly so, with several series of shorter exterior ones. Receptacle flat, naked or bristly. Rays truncate and 5-toothed at the apex. Anthers sagittate at the base. Style-branches slender. Achenes oblong or linear, glabrous, 10–15-ribbed, 4 or 5 of the ribs usually more prominent than the others, truncate, or margined and 4–5-toothed at the summit. Pappus-bristles in 2 series, the inner naked or minutely serrulate, slender, coherent at the base and deciduous in a ring, the outer few (1–8), more persistent, or all deciduous in our species. [Greek, soft-hair, in allusion to the soft pappus.]

About 15 species, natives of the western and southwestern United States and lower California. Type species: *Malacothrix californica* DC.

1. Malacothrix sonchoìdes (Nutt.) T. & G.
Malacothrix. Fig. 4060.

Leptoseris sonchoides Nutt. Trans. Am. Phil. Soc. (II.) **7**: 439. 1841.

Malacothrix sonchoides T. & G. Fl. N. A. **2**: 486. 1843.

Annual, glabrous throughout, or slightly glandular; stem branched, 6′–12′ high. Leaves somewhat fleshy, oblong or linear-oblong in outline, pinnatifid and the lobes dentate with mucronate-pointed teeth, the basal ones 1½′–3′ long, narrowed into short broad petioles, those of the stem smaller, sessile; heads several or numerous, 8″–13″ broad; principal bracts of the involucre linear, acute, scarious-margined, the outer short, oblong, obtuse, or acutish; achenes linear-oblong, margined at the summit by a 15-denticulate white border; pappus-bristles all deciduous.

On dry plains, western Nebraska and Kansas to California and Arizona. May–Aug.

14. CHONDRÍLLA [Tourn.] L. Sp. Pl. 796. 1753.

Perennial herbs, with stiff divaricately branched stems, the basal leaves large and mostly pinnatifid, those of the stem small, narrow, alternate, and few middle sized heads of yellow flowers mostly solitary at the ends of the branches. Involucre cylindric, several-flowered, its inner bracts in 1 or 2 series, nearly equal, with several series of small or minute outer ones. Receptacle flat, naked. Rays truncate and 5-toothed at the summit. Anthers sagittate at the base. Style-branches slender. Achenes oblong or linear, 4–5-angled, many-ribbed, more or less spiny near the summit, abruptly contracted into a beak. Pappus of copious soft white simple bristles. [Greek, lump, from the gummy matter borne on the stems of some species.]

About 18 species, natives of the Old World, the following typical.

1. Chondrilla júncea L. Gum Succory. Fig. 4061.

Chondrilla juncea L. Sp. Pl. 796. 1753.

Stem rush-like, hirsute at the base, glabrous above, much branched, 1°–3° high. Basal leaves runcinate-pinnatifid, those of the stem linear or linear-lanceolate, acute, dentate or entire, sessile, ½′–1½′ long, ½″–1½″ wide; heads terminal and lateral on the branches, short-peduncled or sessile, 4″–6″ broad; involucre glabrous or nearly so, about 4″ high, its inner bracts narrowly linear; achenes muricate and spiny near the summit, slightly shorter than the filiform beak.

In dry fields and waste places, Delaware to Maryland and Virginia. Naturalized from Europe. July–Aug. Naked-weed. Skeleton-weed. Devil's-grass. Hog-bite.

15. LEÓNTODON L. Sp. Pl. 798. 1753.

[TARÁXACUM (Hall.) Ludwig, Def. 175. 1760.]

Perennial acaulescent herbs, with basal tufted pinnatifid or sinuate-dentate leaves, and large heads of yellow flowers, solitary, or very rarely 2 or 3 together at the ends of naked hollow scapes. Involucre oblong or campanulate, its inner bracts in 1 series, nearly equal, slightly united at the base, the outer of several series of shorter somewhat spreading ones, often reflexed at maturity. Receptacle flat, naked. Rays truncate and 5-toothed at the summit. Anthers sagittate at the base. Style-branches slender, obtusish. Achenes oblong or linear-fusiform, 4–5-angled, 5–10-nerved, roughened or spinulose, at least above, tapering into a very slender beak. Pappus of numerous filiform unequal simple persistent bristles. [Greek, lion's-tooth.]

About 20 species, natives of the northern hemisphere and southern South America. Type species: *Leontodon Taraxacum* L.

Outer involucral bracts reflexed; achenes greenish brown, the beak 2–3 times their length.
1. *L. Taraxacum.*
Outer involucral bracts spreading or ascending.
Achenes greenish, the beak 2–3 times their length.
2. *L. latilobum.*
Achenes red, the beak not more than twice their length.
3. *L. erythrospermum.*

1. Leontodon Taráxacum L. Dandelion. Blowball. Fig. 4062.

Leontodon Taraxacum L. Sp. Pl. 798. 1753.
T. officinale Weber, Prim. Pl. Holst. 56. 1780.
T. Dens-leonis Desf. Fl. Atlant. 2: 228. 1800.
T. Taraxacum Karst. Deutsch. Fl. 1138. 1880–83.

Root thick, deep, often 10' long, bitter. Leaves oblong to spatulate in outline, usually pubescent, at least when young, acute or obtuse, pinnatifid, sinuate-dentate or rarely nearly entire, rather succulent, 3'–10' long, $\frac{1}{2}$'–2$\frac{1}{2}$' wide, narrowed into petioles; scape erect, 2'–18' high; head 1'–2' broad; containing very numerous golden-yellow flowers (150–200), inner bracts of the involucre linear or linear-lanceolate, the outer similar, shorter, not glaucous, reflexed, all acute; achenes greenish-brown, fusiform, spinulose above, narrowed into a filiform beak 2–3 times their length, which support the copious white pappus, the fruiting mass of which becomes globose when ripe.

In fields and waste places, naturalized as a weed from Europe. Also in Asia and distributed as a weed in all civilized parts of the world. Jan.–Dec. Called also lion's-tooth, cankerwort, milk-witch- or yellow-gowan, Irish daisy, monk's-head, priest's-crown, puff-ball. Arnica.

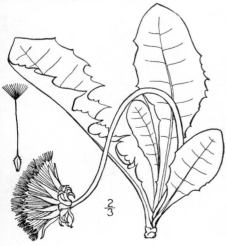

2. Leontodon latilòbum (DC.) Britton. Mountain Dandelion. Fig. 4063.

Taraxacum latilobum DC. Prodr. 7: 146. 1838.

T. Taraxacum alpinum Porter, Mem. Torr. Club 5: 349. 1894.

Similar to the preceding but scape lower, 2'–7' high. Leaves sinuately lobed with broadly triangular lobes, but less deeply so than in *L. Taraxacum* and rarely as far as half way to the midrib, or often merely dentate or subentire; heads smaller, or about 1' wide; bracts fewer, the outer broadly ovate, appressed or merely spreading.

In moist places, Newfoundland and Labrador to Quebec. Europe. Summer. Rocky Mountain and northwestern plants formerly regarded as included in this species are now considered distinct.

3. Leontodon erythrospérmum (Andrz.) Britton. Red-seeded Dandelion. Fig. 4064.

Taraxacum erythrospermum Andrz. in Bess. Enum. Pl. Vilh. 75. 1821.

Similar to the two preceding species, the leaves glabrous, very deeply runcinate-pinnatifid or pinnately divided into narrower triangular-lanceolate usually long-pointed segments; scapes glabrous, or pubescent above; heads rarely more than 1' broad, 70–90-flowered; bracts of the involucre glaucous, the outer lanceolate, spreading or ascending, the inner linear, longer, each usually with an appendage just below the tip; flowers sulphur-yellow, the outer rays purplish without; achenes narrower, bright red, or red-brown, spinulose above, the filiform beak not more than twice their length; pappus dirty white.

In fields and woods, Maine and Vermont to New York, Pennsylvania, Ontario, Alberta, Illinois and Wyoming. Naturalized from Europe. April–June. April–June.

16. SÓNCHUS [Tourn.] L. Sp. Pl. 793. 1753.

Annual or perennial succulent herbs, with alternate, mostly auriculate-clasping, entire dentate lobed or pinnatifid, prickly-margined leaves, and large or middle-sized, peduncled corymbose or paniculate heads of yellow flowers. Involucre ovoid or campanulate, usually becoming thickened and more or less conic at the base when old, its bracts herbaceous or membranous, imbricated in several series, the outer successively smaller. Receptacle flat, naked. Rays truncate and 5-toothed at the apex. Anthers sagittate at the base. Style-branches slender. Achenes oval, oblong, or linear, more or less flattened, 10–20-ribbed, somewhat narrowed at the base, truncate. Pappus of very copious soft white simple capillary bristles, usually falling away connected, sometimes with 1 or 2 stouter ones which fall separately. [The Greek name of the Sow-thistle.]

About 45 species, natives of the Old World. Besides the following, another occurs on the Pacific Coast. Type species: *Sonchus oleraceus* L.

Involucre glandular-pubescent; heads nearly 1' high.	1. *S. arvensis.*
Involucre glabrous; heads 6"–8" high.	
Auricles of the leaves acute; achenes striate and transversely wrinkled.	2. *S. oleraceus.*
Auricles rounded; achenes ribbed, not transversely wrinkled.	3. *S. asper.*

1. Sonchus arvénsis L. Corn Sow-Thistle. Milk Thistle. Fig. 4065.

Sonchus arvensis L. Sp. Pl. 793. 1753.

Perennial by deep roots and creeping rootstocks, glabrous; stem leafy below, paniculately branched and nearly naked above, 2°–4° high. Lower and basal leaves runcinate-pinnatifid, often 12' long, spinulose-dentate, narrowed into short petioles, the upper pinnatifid or entire, lanceolate, clasping; heads several, or numerous, corymbose-paniculate, 1'–2' broad, bright yellow, very showy; involucre nearly 1' high, its bracts as also the peduncles glandular-bristly; achenes oblong, compressed, with about 10 rugose longitudinal ribs.

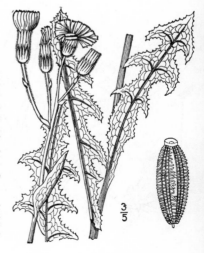

In low grounds, fields and waste places, New Jersey to Quebec, Newfoundland,. Minnesota, Colorado and British Columbia. Naturalized from Europe. Native also of Asia. July–Oct. Dindle. Gutweed. Swine-thistle. Tree sow-thistle.

2. Sonchus oleràceus L. Annual Sow-Thistle. Hare's Lettuce. Fig. 4066.

Sonchus oleraceus L. Sp. Pl. 794. 1753.

Annual, with fibrous roots; stem leafy below, nearly simple, 1°–10° high. Basal and lower leaves petioled, lyrate-pinnatifid, 4′–10′ long, the terminal segment commonly large and triangular, the margins denticulate with mucronate or scarcely spiny teeth; upper leaves pinnatifid, clasping by an auricled or sagittate base, the auricles acute or acuminate; uppermost leaves often lanceolate and entire; heads several or numerous, pale yellow, 9″–15″ broad; involucre glabrous, 6″–8″ high; achenes flat, longitudinally ribbed and transversely rugose.

In fields and waste places, a common weed in most cultivated parts of the globe except the extreme north. Also in Central and South America. Naturalized from Europe. Hare's-colewort or -thistle. Milk-weed. Milk-thistle. Milky tassel. Swinies. The leaves used as a salad and as a pot herb. May–Nov.

3. Sonchus ásper (L.) Hill. Spiny or Sharp-fringed Sow-Thistle. Fig. 4067.

Sonchus oleraceus var. *asper* L. Sp. Pl. 794. 1753.

Sonchus asper Hill, Herb. Brit. 47. 1769.

Annual, similar to the preceding species; leaves undivided, lobed or sometimes pinnatifid, spinulose-dentate to spinulose-denticulate, the lower and basal ones obovate or spatulate, petioled, the upper oblong or lanceolate, clasping by an auricled base, the auricles rounded; heads several or numerous, 1′ broad or less; flowers pale yellow; involucre glabrous, about 6″ high; achenes flat, longitudinally ribbed.

In waste places throughout most of our area and in tropical and South America. Widely distributed as a weed in nearly all cultivated parts of the earth. Naturalized from Europe. May–Nov.

17. LACTÙCA [Tourn.] L. Sp. Pl. 795. 1753.

Tall leafy herbs, with small panicled heads of yellow, white, pink, or blue flowers, and alternate leaves. Involucre cylindric, its bracts imbricated in several series, the outer shorter, or of 1 or 2 series of principal nearly equal inner bracts, and several rows of short outer ones. Receptacle flat, naked. Rays truncate and 5-toothed at the summit. Anthers sagittate at the base. Style-branches mostly slender. Achenes oval, oblong or linear, flat, 3–5-ribbed on each face, narrowed above or tipped by a filiform beak, which is somewhat expanded at the summit into a small disk bearing the copious soft capillary white or brown pappus-bristles. [The Ancient Latin name, from *lac*, milk, referring to the milky juice.]

About 100 species, natives of the northern hemisphere. Type species: *Lactuca sativa* L.

      A. **Achenes filiform-beaked; rays mainly yellow.**

Introduced European species; heads few-flowered.

Panicle widely branching; achene about as long as its beak.	1. *L. virosa.*
Panicle-branches nearly erect; achene shorter than its beak.	2. *L. saligna.*

Native species; heads several- to many-flowered.

Leaves, or their lobes, spinulose-denticulate; stem leafy below.	3. *L. ludoviciana.*
Leaves entire to pinnatifid, the teeth not spinulose; stem leafy to the inflorescence.	
Leaves hirsute or bristly on the veins beneath.	
Plant hirsute below; leaves pinnatifid or the upper entire.	4. *L. hirsuta.*
Plant hirsute up to the inflorescence; leaves merely dentate or denticulate.	
	5. *L. Steelei.*

Leaves glabrous.
   Leaves entire to pinnatifid, not sagittate at base.   6. *L. canadensis.*
   Leaves entire or denticulate, sagittate at base.   7. *L. sagittifolia.*

    B. **Achenes truncate, or narrowed into stout beaks; rays blue to white.**

Perennial; achenes flattened.   8. *L. pulchella.*
Annuals or biennials; achenes swollen.
   Pappus bright white.
      Achenes beakless; leaves dentate, acuminate.   9. *L. villosa.*
      Achenes beaked; leaves pinnatifid.   10. *L. floridana.*
   Pappus brown; achenes short-beaked.   11. *L. spicata.*

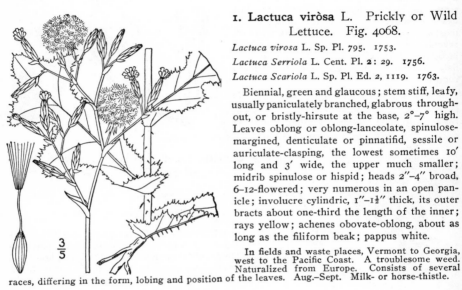

1. **Lactuca viròsa** L.   Prickly or Wild Lettuce.   Fig. 4068.

Lactuca virosa L. Sp. Pl. 795. 1753.

Lactuca Serriola L. Cent. Pl. 2: 29. 1756.

Lactuca Scariola L. Sp. Pl. Ed. 2, 1119. 1763.

Biennial, green and glaucous; stem stiff, leafy, usually paniculately branched, glabrous throughout, or bristly-hirsute at the base, 2°–7° high. Leaves oblong or oblong-lanceolate, spinulose-margined, denticulate or pinnatifid, sessile or auriculate-clasping, the lowest sometimes 10′ long and 3′ wide, the upper much smaller; midrib spinulose or hispid; heads 2″–4″ broad, 6–12-flowered; very numerous in an open panicle; involucre cylindric, 1″–1½″ thick, its outer bracts about one-third the length of the inner; rays yellow; achenes obovate-oblong, about as long as the filiform beak; pappus white.

   In fields and waste places, Vermont to Georgia, west to the Pacific Coast. A troublesome weed. Naturalized from Europe. Consists of several races, differing in the form, lobing and position of the leaves. Aug.–Sept. Milk- or horse-thistle.

2. **Lactuca salígna** L.   Willow Lettuce.   Fig. 4069.

Lactuca saligna L. Sp. Pl. 796. 1753.

Biennial, light green; stem slender, smooth or nearly so, 2° high or more, the very slender branches erect-ascending. Leaves glabrous, linear to oblong, 6′ long or less, runcinate-pinnatifid with pointed divergent lobes, or entire, sagittate at the base, the midvein sometimes spinulose; panicles narrow; heads about 3″ broad, few-flowered; involucre subcylindric; outer bracts much shorter than the inner; achenes oblong, about one-half as long as the filiform beak; pappus white.

   Waste and cultivated grounds, Ohio. Naturalized from Europe. July–Aug.

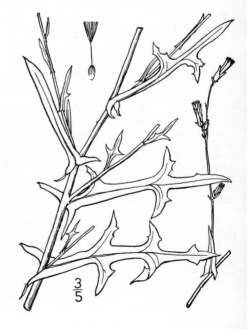

3. Lactuca ludoviciàna (Nutt.) DC. Western Lettuce. Fig. 4070.

Sonchus ludovicianus Nutt. Gen. **2**: 125. 1818.

Lactuca ludoviciana DC. Prodr. **7**: 141. 1838.

Biennial, glabrous throughout, leafy up to inflorescence, paniculately branched, 2°–5° high. Leaves oblong to ovate-oblong, acute or acutish, 2'–4' long, auriculate-clasping, spinulose-denticulate, sinuate-lobed, or pinnatifid with spinulose segments; heads 3''–5'' broad, numerous in an open panicle, their peduncles bracteolate; involucre cylindric or ovoid-cylindric, glabrous, 8''–9'' high, its bracts successively shorter and broader, the lower ones ovate; rays yellow; achenes oval to obovate, flat, about the length of their filiform beak; pappus white.

Plains and banks, Iowa, Minnesota and South Dakota to Kansas and Texas. July–Sept.

$\frac{3}{5}$

4. Lactuca hirsùta Muhl. Hairy or Red Wood-Lettuce. Fig. 4071.

Lactuca hirsuta Muhl. Cat. 69. 1813.
Lactuca sanguinea Bigel. Fl. Bost. Ed. 2, 287. 1824.
Lactuca elongata var. *sanguinea* T. & G. Fl. N. A. **2**: 496. 1843.

Stem 1°–6° high, naked and paniculately branched above, usually hirsute, at least below. Leaves, or most of them, sinuate-pinnatifid, pubescent on both sides, or on the midrib beneath, those of the stem mostly sessile or auriculate-clasping, 3'–7' long, the uppermost sometimes lanceolate and entire, the basal petioled; heads numerous, 2''–3'' broad; involucre glabrous, 5''–9'' high; outer bracts shorter than the inner; rays reddish-yellow or paler; achenes oblong-oval, flat, about the length of the beak; pappus white.

In dry soil, Quebec to Ontario, Minnesota, Alabama and Texas. Stem, peduncles and involucre often red or purple. July–Sept.

$\frac{1}{2}$

5. Lactuca Steèlei Britton. Steele's Wild Lettuce. Fig. 4072.

L. Steelei Britton, Man. 899. 1901.

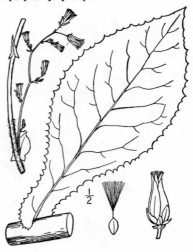

Stem and branches hirsute. Leaves obovate to oval, the larger about 8' long and 4' wide, sessile, irregularly dentate, not lobed, hispid on the veins beneath, short-acuminate; involucre about 6'' high; rays not seen; achenes oval, 3-ribbed, about 1½'' long, the filiform beak slightly shorter; pappus bright white.

Near Washington, D. C., and, apparently, in Delaware. July–Aug.

$\frac{1}{2}$

6. Lactuca canadénsis L.   Wild or Tall Lettuce.   Wild Opium.   Fig. 4073.

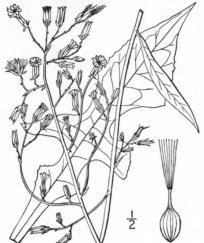

Lactuca canadensis L. Sp. Pl. 796. 1753.
Lactuca elongata Muhl.; Willd. Sp. Pl. 1525. 1804.
Lactuca canadensis montana Britton, in Britton and Brown, Ill. Fl. 3: 274. 1898.

Biennial or annual, glabrous throughout, somewhat glaucous; stem leafy up to the inflorescence, 3°–10° high, branching above into a narrow panicle. Leaves mostly sinuate-pinnatifid, those of the stem sessile or auriculate-clasping, 2′–8′ long, the uppermost smaller, often lanceolate, acuminate and entire, sometimes all lanceolate and entire, the basal often 12′ long, narrowed into petioles; heads numerous, 2″–3″ broad; involucre cylindric, 4″–6″ high, its outer bracts shorter than the inner; rays yellow; achenes oval, flat, about as long as the filiform beak; pappus white.

In moist, open places, Nova Scotia to British Columbia, Georgia, Alabama, Louisiana, Arkansas and Colorado. Santo Domingo. Trumpetweed, trumpet-milkweed, trumpets, and fire-weed. Butter- or horse-weed. Devil's-weed. Devil's-iron-weed. June–Nov.

Lactuca Mórssii Robinson, with blue rays, and achenes 3 or 4 times longer than their filiform beaks, may be a hybrid with *L. spicata.*

7. Lactuca sagittifòlia Ell.   Arrow-leaved Lettuce.   Fig. 4074.

L. sagittifolia Ell. Bot. S. C. & Ga. 2: 253. 1821–24.
Lactuca integrifolia Bigel. Fl. Bost. Ed. 2, 287. 1824. Not Nutt. 1818.
L. elongata var. integ. T. & G. Fl. N. A. 2: 496. 1843.

Biennial; stem glabrous throughout, or hirsute below, leafy nearly up to the usually paniculate inflorescence, 2°–6° high. Leaves oblong, oblong-lanceolate or lanceolate, acuminate or acute, entire or denticulate, the lower rarely pinnatifid, sometimes spinulose on the margins, those of the stem sessile or sagittate-clasping, 3′–6′ long, ½′–1½′ wide, the basal and lower ones petioled; heads commonly very numerous, 2″–4″ broad; involucre cylindric, 5″–7″ high, the outer bracts shorter than the inner; rays yellow or reddish; achenes oval, flat, rather longer than their filiform beaks; pappus white.

In dry, open soil, New Brunswick and Ontario to Georgia, Idaho and Kansas. Called also devil's-iron-weed. July–Sept.

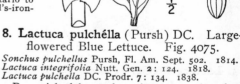

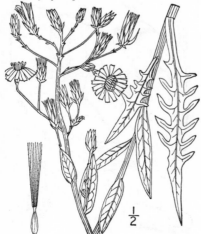

8. Lactuca pulchélla (Pursh) DC.   Large-flowered Blue Lettuce.   Fig. 4075.

Sonchus pulchellus Pursh, Fl. Am. Sept. 502. 1814.
Lactuca integrifolia Nutt. Gen. 2: 124. 1818.
Lactuca pulchella DC. Prodr. 7: 134. 1838.

Perennial, glabrous throughout, somewhat glaucous; stem rather slender, leafy up to the corymbose-paniculate inflorescence, 1°–3° high. Leaves linear-lanceolate, lanceolate or oblong, acute, entire, dentate, lobed or pinnatifid, those of the stem sessile or partly clasping, 2′–8′ long; 2″–18″ wide, the lowest and basal ones sometimes petioled; heads mostly numerous, 6″–10″ broad; branches and peduncles scaly; involucre well imbricated, 8″–10″ high, its outer bracts successively shorter, ovate-lanceolate; rays bright blue or violet; achenes oblong-lanceolate, flat, twice as long as their tapering beaks; pappus white.

In moist soil, western Ontario to British Columbia, Michigan, Iowa, Kansas, New Mexico and California. June–Sept.

Lactuca campéstris Greene, of the prairie region, is described as differing from this by yellow rays.

9. Lactuca villòsa Jacq. Hairy-veined Blue Lettuce. Fig. 4076.

L. villosa Jacq. Hort. Schoen. **3**: 62. *pl. 367.* 1798.
Sonchus acuminatus Willd. Sp. Pl. **3**: 1521. 1804.
Mulgedium acuminatum DC. Prodr. **7**: 249. 1838.
L. acuminata A. Gray, Proc. Am. Acad. **19**: 73. 1883.

Annual or biennial; stem glabrous, leafy up to the paniculate inflorescence, 2°–6° high. Leaves oblong, ovate or lanceolate, acuminate, acutely dentate or the teeth mucronate-tipped, glabrous above, pubescent with short stiff hairs on the veins beneath, sessile and slightly clasping at the base, or petioled, 4′–6′ long, 1′–2½′ wide, the lowest sometimes lobed at the base; heads numerous, 3″–5″ broad; peduncles usually minutely scaly; rays blue; involucre about 5″ high, its outer bracts much shorter than the inner, some or all of them obtuse; achenes thick, oblong, little flattened, narrowed above; pappus white.

In thickets, New York to Illinois, Nebraska, south to Florida, Georgia and Kentucky. July–Sept. False lettuce.

$\frac{3}{5}$

$\frac{1}{2}$

10. Lactuca floridàna (L.) Gaertn. False or Florida Lettuce. Fig. 4077.

Sonchus floridanus L. Sp. Pl. 794. 1753.
Lactuca floridana Gaertn. Fruct. & Sem. **2**: 362. 1791.
Mulgedium floridanum DC. Prodr. **7**: 349. 1791.

Annual or biennial; stem glabrous, rather stout, leafy up to the large, paniculate inflorescence, 3°–7° high. Leaves deeply lyrate-pinnatifid, or sometimes cordate-ovate, sessile or petioled, 4′–12′ long, glabrous above, pubescent on the veins beneath, the terminal segment usually broad, triangular, acute or acuminate, the lateral ones lanceolate to oval, acute, all usually dentate, or the leaves irregularly lobed; heads numerous, 3″–5″ broad; peduncles commonly scaly; rays blue; involucre about 6″ high, its outer bracts much smaller than the inner; achenes thick, somewhat compressed, narrowed above into a short beak; pappus white.

In moist, open places, southern New York and Pennsylvania to Illinois, Kansas, Florida and Texas. Porto Rico. July–Sept.

11. Lactuca spicàta (Lam.) Hitchc. Tall Blue Lettuce. Fig. 4078.

Sonchus spicatus Lam. Encycl. **3**: 401. 1789.
Mulgedium leucophaeum DC. Prodr. **7**: 250. 1838.
Lactuca leucophaea var. *integrifolia* A. Gray, Syn. Fl. **1**: Part 2, 444. 1884.
Lactuca spicata integrifolia Britton, Mem. Torr. Club **5**: 350. 1894.
L. spicata Hitchc.; Britt. & Brown, Ill. Fl. **3**: 276. 1898.

Annual or biennial; stem usually stout, glabrous, 3°–12° high, leafy up to the large, rather dense panicle. Leaves deeply pinnatifid or lobed to entire, sharply dentate with mucronate-pointed teeth, sessile, or the lower narrowed into margined petioles, glabrous on both sides, or pubescent on the veins beneath, 5′–12′ long, 2′–6′ wide; heads very numerous, about 2″ broad; peduncles minutely scaly; rays blue to white; achenes oblong, compressed, narrowed above into a short neck; pappus brown.

In moist soil, Newfoundland to Manitoba, North Carolina, Tennessee, Iowa, South Dakota and Colorado. Ascends to 2000 ft. in North Carolina. Races differ in leaf-form and in color of the flowers. Milk-weed. July–Oct.

$\frac{3}{5}$

18. LYGODÉSMIA D. Don, Edinb. Phil. Journ. 6: 311. 1829.

Perennial or annual glabrous rigid branching herbs, with linear leaves, or the basal and lower ones sometimes broader and pinnatifid, those of the stem very narrow and entire or reduced to linear scales, and middle-sized 3–12-flowered heads of pink or purple flowers, solitary and erect at the ends of the stem and branches, or sometimes racemose. Involucre cylindric, its principal bracts 5–8, linear, scarious-margined, equal, slightly united at the base, with several very short outer ones. Receptacle flat, naked. Rays truncate and 5-toothed at the apex. Anthers sagittate at the base. Style-branches slender. Achenes linear, smooth or striate. Pappus of copious somewhat unequal simple bristles. [Greek, twig-bundle, from the numerous branches.]

About 6 species, natives of western and southern North America. Type species: *Prenanthes juncea* Pursh.

Heads solitary at the ends of the branches; leaves linear or subulate.   1. *L. juncea.*
Heads racemose along the branches; leaves elongated-linear.   2. *L. rostrata.*

3/5

1. Lygodesmia júnceá (Pursh) D. Don. Rush-like Lygodesmia.
Fig. 4079.

Prenanthes juncea Pursh, Fl. Am. Sept. 498. 1814.

Lygodesmia juncea D. Don; Hook. Fl. Bor. Am. 1: 295. 1833.

Perennial by a thick woody root; stems stiff, striate, much branched, 8′–18′ high, the branches erect. Lower leaves linear-lanceolate, rigid, entire, acute or acuminate, ¼′–2′ long, ½″–1½″ wide, the upper similar but smaller, or reduced to subulate scales; heads 6″–8″ broad, mostly 5-flowered, solitary at the ends of the branches; involucre 6″–8″ high; achenes narrowly columnar or slightly tapering, truncate at the summit, about 8-nerved or ribbed, 2½″–3½″ long; pappus light brown.

Plains, Minnesota to Saskatchewan, Montana, Wisconsin, Missouri, Nebraska, Kansas and Arizona. Often infested by a globose gall 2″–5″ in diameter. June–Aug.

2. Lygodesmia rostràta A. Gray. Beaked Lygodesmia. Fig. 4080.

L. juncea var. *rostrata* A. Gray, Proc. Phil. Acad. 1863: 69. 1863.

Lygodesmia rostrata A. Gray, Proc. Am. Acad. 9: 217. 1874.

Annual, less rigid; stem striate, leafy, paniculately branched, 1°–3° high. Leaves elongated-linear, acuminate, entire, 3-nerved, the lower 3′–7′ long, 1″–1½″ wide, the uppermost very small and subulate; heads numerous, 7–10-flowered, about ½′ broad, racemose along the branches on scaly short erect peduncles; involucre 5″–7″ high; achenes narrowly fusiform, narrowed or somewhat beaked at the summit, 5–8-ribbed or -striate, 4″–5″ long, longer than the whitish pappus.

Plains and canyons, South Dakota to Saskatchewan, Nebraska, Kansas, Colorado and Wyoming. Aug.–Sept.

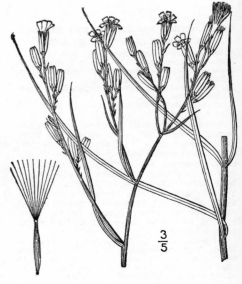

3/5

19. AGOSERIS Raf.; D. Dietr. Syn. Pl. **4**: 1332. 1847.

[Troximon Nutt. Fras. Cat. 1813. Not Gaertn. 1791.]

Perennial or annual herbs, mostly acaulescent, with tufted usually sessile basal leaves, and solitary heads of yellow or rarely purple flowers at the end of a naked or bracted scape. Involucre campanulate or oblong, its bracts imbricated in several series, appressed, or with spreading tips, membranous or herbaceous, not thickened after flowering, the outer ones gradually shorter and broader. Receptacle flat, naked or foveolate. Rays truncate and 5-toothed at the apex. Anthers sagittate at the base. Style-branches slender. Achenes oblong, obovate, or linear, 10-ribbed, not flattened, beaked or beakless. Pappus of copious slender simple white bristles or soft unequal narrow scales. [Greek, head- or chief-succory.]

About 30 species, natives of western and southern North America and southern South America. Besides the following, some 20 others occur in the western parts of the United States. Type species: *Agoseris cuspidata* (Pursh) D. Dietr.

Achenes beaked.
    Head 1′–2′ broad, achenes 5″–6″ long.   1. *A. glauca.*
    Head 1′–2′ broad; achenes 5″–6″ long.   2. *A. parviflora.*
Achenes beakless.   3. *A. cuspidata.*

1. Agoseris glàuca (Pursh) D. Dietr. Large-flowered Agoseris. Fig. 4081.

T. glaucum Pursh, Fl. Am. Sept. 505. 1814.

Agoseris glauca D. Dietr. Syn. Pl. **4**: 1332. 1847.

Perennial, pale or glaucous, glabrous throughout or a little woolly below. Leaves linear, lanceolate, or oblong, entire, dentate or pinnatifid, 2′–10′ long, 2″–10″ wide, acuminate at the apex, narrowed at the base, sometimes into margined petioles; scapes stout, glabrous or slightly pubescent, longer than the leaves, often 1½° high; head 1′–2′ broad; involucre oblong-campanulate, or broader in fruit, commonly quite glabrous, its bracts lanceolate, acuminate, often hyaline-margined; achenes conspicuously beaked, 5″–6″ long, when mature longer than the copious pappus of rather rigid scabrous or denticulate bristles.

Minnesota to South Dakota, Saskatchewan, south to Kansas (according to Smyth), Colorado and Utah. May–July.

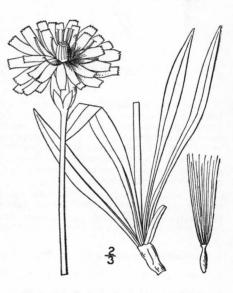

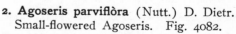

2. Agoseris parviflòra (Nutt.) D. Dietr. Small-flowered Agoseris. Fig. 4082.

Troximon parviflorum Nutt. Trans. Am. Phil. Soc. **7**: 434. 1841.
Troximon glaucum var. *parviflorum* A. Gray, Syn. Fl. **1**: Part 2, 437. 1884.
Agoseris parviflora D. Dietr. Syn. Pl. **4**: 1332. 1847.

Perennial, glabrous throughout; scape slender, much longer than the leaves, 5′–15′ high. Leaves narrowly linear, acuminate, entire, 3′–8′ long, 1″–2½″ wide; head 1′ broad or less; involucre oblong-ovoid, becoming nearly hemispheric in fruit, 6″–8″ high, glabrous, its bracts lanceolate, acuminate; achenes conspicuously beaked, about 4″ long; pappus of numerous unequal very slender bristles.

Plains, North Dakota to Nebraska, Manitoba, Alberta, Idaho and New Mexico. Called also false dandelion. May–July.

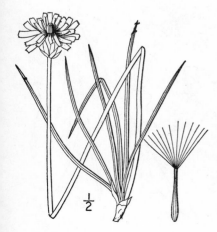

3. Agoseris cuspidàta (Pursh) D. Dietr. Prairie False Dandelion. Fig. 4083.

Troximon cuspidatum Pursh, Fl. Am. Sept. 742.   1814.

Troximon marginatum Nutt. Gen. **2**: 128.   1818.

Agoseris cuspidata D. Dietr. Syn. Pl. **4**: 1332.   1847.

Nothocalais cuspidata Greene, Bull. Cal. Acad. (II.) **2** : 55.   1886.

Leaves linear, long-acuminate, thick, pubescent or glabrate, 4′–8′ long, 2″–5″ wide, somewhat conduplicate, their margins conspicuously white-to-mentose and crisped, or entire.   Scape stout, to-mentose, at least above, shorter than or equalling the leaves; head 1′–2′ broad; involucre usually quite glabrous, nearly 1′ high; achenes slightly contracted at the summit; about 3″ long, beakless; pappus of 40–50 unequal bristles.

In dry soil, on prairies and rocky hills, Illinois to North Dakota, Nebraska, Kansas and Colorado. Called also false dandelion.   April–June.

20. SITÍLIAS Raf. New Fl. N. A. 4: 85.   1836.

[PYRRHOPAPPUS DC. Prodr. 7: 144.   1838.]

Annual or perennial herbs, with alternate or basal leaves, and mostly large, solitary or few heads of yellow flowers, borne on long, usually bracted peduncles. Involucre oblong or campanulate, its principal bracts in 1 series, nearly equal, slightly united at the base, with several series of smaller outer ones. Rays truncate and 5-toothed at the summit. Anthers sagittate at the base. Style-branches short, obtusish. Achenes oblong or fusiform, mostly 5-ribbed, roughened or hirsute, abruptly narrowed into a long filiform beak. Pappus of numerous soft simple brownish somewhat unequal bristles, surrounded at the base by a villous white ring. [Name unexplained.]

Six known species, natives of North America and Mexico. Besides the following, 3 others occur in the southwestern United States. Type species: *Sitilias caroliniana* (Walt.) Raf.

Stem leafy, usually branched; plant glabrous, or nearly so.   1. *S. caroliniana.*
Scape naked, monocephalous; plant hirsute, or pubescent.   2. *S. grandiflora.*

1. Sitilias caroliniàna (Walt.) Raf.   Leafy-stemmed False Dandelion.   Fig. 4084.

Leontodon carolinianum Walt. Fl. Car. 192.   1788.

S. caroliniana Raf. New Fl. N. A. Part 4, 85.   1836.

Pyrrhopappus carolinianus DC. Prodr. 7: 144.   1838.

Annual or biennial, glabrous or nearly so; stem leafy, usually branched, 2°–5° high. Basal leaves oblong or oblong-lanceolate, pinnatifid, lobed, coarsely dentate or some of them entire, acute, acuminate, or obtusish, 3′–8′ long, ½′–1½′ wide, narrowed into margined petioles; stem leaves sessile or partly clasping, the upper usually lanceolate, entire and acuminate; peduncles usually puberulent; heads 1 or several, 1′–1½′ broad; involucre commonly puberulent or pubescent, about 1′ high, its outer bracts setaceous or subulate, spreading, the inner corniculate at the apex; achenes 2″–3″ long, tipped with a filiform beak of about three times their length.

In dry fields, Delaware to Florida, Kentucky, Missouri, Louisiana and Texas.   April–July.

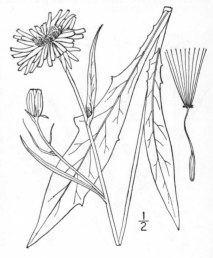

2. Sitilias grandiflòra (Nutt.) Greene.
Rough False Dandelion. Fig. 4085.

Barkhausia grandiflora Nutt. Journ. Phila. Acad. **7**:
69. 1834.

Pyrrhopappus scaposus DC. Prodr. **7**: 144. 1838.

Sitilias grandiflora Greene, Pittonia **2**: 180. 1891.

Hirsute or pubescent; root tuberous-thickened.
Leaves all basal, oblong or spatulate in outline,
deeply pinnatifid, 3'–7' long, 1'–1½' wide, narrowed
into margined petioles; scape naked or sometimes
with a small leaf near its base; head solitary,
1'–2' broad; outer bracts of the involucre small,
short, subulate, the inner ones obscurely corniculate at the tip.

On prairies, Kansas to Texas. April–June.

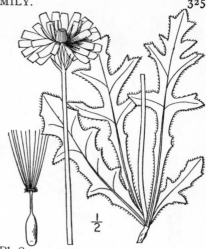

21. CRÉPIS L. Sp. Pl. 805. 1753.

Perennial, biennial or annual herbs, with alternate or basal, mostly toothed or pinnatifid
leaves, and small or middle-sized heads, usually paniculate or corymbose, of yellow or orange
flowers. Involucre cylindric, campanulate, or swollen at the base, its principal bracts in 1
series, equal, with a number of exterior smaller ones. Receptacle mostly flat, naked or short-
fimbrillate. Rays truncate and 5-toothed at the apex. Anthers sagittate at the base. Style-
branches slender. Achenes linear-oblong, 10–20-ribbed or nerved, not transversely rugose,
narrowed at the base and apex, beakless in our species. Pappus copious, of very slender
white bristles. [Greek, sandal; application not explained.]

About 200 species, natives of the northern hemisphere. Besides the following, about 20 others
occur in western North America. Type species: *Crepis tectòrum* L.

Involucre glabrous.
 Involucre cylindric; plant pubescent below; introduced.   1. *C. pulchra.*
 Involucre campanulate; plant glaucous; native, western.   2. *C. glauca.*
Involucre pubescent, glandular, or canescent.
 Foliage not canescent nor scurfy, sometimes hirsute.
  Stems naked, or with 1 or 2 small leaves; western, native. 3. *C. runcinata.*
  Stems leafy; introduced annuals or biennials.
   Stem leaves narrow, revolute-margined, sessile.   4. *C. tectorum.*
   Stem leaves lanceolate, clasping, not revolute-margined.
    Involucre 3"–4" high; achenes 10-striate.   5. *C. capillaris.*
    Involucre 4"–6" high; achenes 13-striate.   6. *C. biennis.*
 Foliage cinereous, canescent, or scurfy, sometimes also hirsute.
  Inner bracts of the involucre 5–8; flowers 5–8.   7. *C. intermedia.*
  Inner bracts of the involucre 9–24; flowers 10–30. 8. *C. occidentalis.*

1. Crepis púlchra L. Small-flowered Hawks-
beard. Fig. 4086.

Crepis pulchra L. Sp. Pl. 806. 1753.

Annual; stem 2°–4½° high, leafy and pubescent
below, mostly glabrous, naked, and paniculately
branched above. Stem leaves oblong or oblong-
lanceolate, dentate, clasping by an auricled base or
truncate, pubescent, 3'–4' long, acute or obtuse, or
the uppermost lanceolate, acuminate and entire;
basal leaves runcinate, narrowed into petioles; heads
very numerous, about 3" broad, in a large naked
panicle; involucre narrow, cylindric, glabrous, about
5" high, its principal bracts 12–15, linear, acuminate,
the outer ones very short, ovate, appressed; achenes
linear, faintly 10-nerved, slightly narrowed above.

Along railroad near Culpepper, Virginia, very abundant in 1890. Naturalized or adventive from Continental Europe. May–July.

2. Crepis glàuca (Nutt.) T. & G. Glaucous Hawksbeard. Fig. 4087.

Crepidium glaucum Nutt. Trans. Am. Phil. Soc. (II.) **7** : 436. 1841.

Crepis glauca T. & G. Fl. N. A. **2** : 488. 1843.

Perennial; scapose, or rarely with 1 or 2 leaves on the stem, 1°–2½° high, glabrous and glaucous throughout. Basal leaves spatulate, oblanceolate, or obovate, acute or obtuse at the apex, gradually narrowed into margined petioles, entire, dentate, or pinnatifid, 2′–6′ long, ½′–1′ wide; heads not numerous, long-peduncled, 6″–12″ broad; peduncles glabrous; involucre campanulate, its principal bracts lanceolate, acute, the outer ones very short, ovate, appressed; achenes oblong-cylindric, slightly narrowed above, strongly 10-ribbed.

In moist and saline soil, Nebraska to Saskatchewan, Utah and Nevada. July–Aug.

3. Crepis runcinàta (James) T. & G. Naked Stemmed Hawksbeard. Fig. 4088.

Hieracium runcinatum James in Long's Exp. **1** : 453. 1823.
Crepis runcinata T. & G. Fl. N. A. **2** : 487. 1843.
Crepis perplexans Rydb. Bull. Torr. Club **32** : 134. 1906.

Perennial, similar to the preceding species, but not glaucous or scarcely so, often pubescent below; stem leafless or with 1 or 2 small leaves, 1°–3° high. Basal leaves spatulate, obovate, or oblong, obtuse or acute, entire, repand, dentate, or rarely runcinate-pinnatifid, 2′–6′ long, ½′–1½′ wide; heads several, long-peduncled, nearly 1′ broad; peduncles glabrous or glandular-pubescent; involucre campanulate, pubescent or glandular, its principal bracts linear-lanceolate, acute, outer ones short, lanceolate, appressed; achenes linear-oblong, somewhat narrowed above, 10-ribbed.

In moist soil, Iowa to North Dakota, Manitoba, Alberta, Utah and Montana. June–July.

Crepis ripària A. Nelson, with broadly obovate basal leaves and larger flowers, is recorded from Nebraska.

4. Crepis tectòrum L. Narrow-leaved Hawksbeard. Fig. 4089.

Crepis tectorum L. Sp. Pl. 807. 1753.

Annual; stem slender, puberulent or pubescent, leafy, branched, 1°–2° high. Basal leaves lanceolate, dentate, or runcinate-pinnatifid, 4′–6′ long; stem leaves sessile, sometimes slightly sagittate at the base, linear, entire, dentate, or lobed, their margins revolute; heads numerous, corymbose, 6″–10″ broad; involucre narrowly campanulate, canescent or pubescent, 3″–5″ high, its principal bracts lanceolate, acuminate, downy within, the exterior ones linear, spreading; peduncles usually canescent; achenes 10-ribbed, narrowed above into a short beak, the ribs minutely scabrous.

In waste places and on ballast, New York and New Jersey to Connecticut, Ontario, Michigan and Nebraska. Naturalized from Europe. June–July.

5. Crepis capillàris (L.) Wallr.　Smooth Hawksbeard.　Fig. 4090.

Lapsana capillaris L. Sp. Pl. 812.　1753.
Crepis virens L. Sp. Pl. Ed. 2, 1134.　1763.
Crepis polymorpha Wallr. Sched. Crit. 426.　1822.
Crepis capillaris Wallr. Fl. Hereyn. 287.　1840.

Annual; stem stout or slender, leafy, corymbosely branched above, glabrous or somewhat
hirsute below, 1°–2½° high.　Basal leaves spatulate, pinnatifid, or dentate, sometimes 8' long and
2' wide, narrowed into petioles; stem leaves lanceolate or oblong, clasping by a sagittate base,
flat, the upper mostly very small and usually entire; heads numerous, 5″–8″ broad, slender-peduncled; peduncles glabrous or glandular; involucre
oblong, more or less pubescent or glandular, 3″–4″
high, its principal bracts lanceolate, glabrous
within, the outer mostly appressed; achenes 10-
ribbed, smooth, slightly narrowed at both ends.

In fields and waste places, Connecticut, New York,
New Jersey and Pennsylvania, and in ballast about
the seaports.　Also on the Pacific Coast.　Adventive
from Europe.　July–Sept.

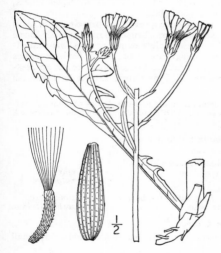

6. Crepis biénnis L.　Rough Hawksbeard. Fig. 4091.

Crepis biennis L. Sp. Pl. 807.　1753.

Biennial, or sometimes annual; stem pubescent
or hirsute, leafy, at least below, branched above,
2°–3° high.　Leaves runcinate-pinnatifid, usually
hirsute, 2'–6' long, oblong or spatulate, the lower
and basal ones narrowed into petioles and sometimes merely dentate, the uppermost lanceolate,
clasping, their margins not revolute; heads several, subcorymbose, 1'–1½' broad; involucre canescent or pubescent, 4″–6″ high, its principal bracts
linear-lanceolate, downy within, the outer ones
linear-oblong or lanceolate, spreading; achenes
oblong, slightly narrowed above, 13-striate, glabrous.

In waste places, Vermont to Pennsylvania, Michigan, and in ballast about the seaports.　Naturalized
or adventive from Europe.　June–Aug.

7. Crepis intermèdia A. Gray.　Small-flowered Gray Hawksbeard.　Fig. 4092.

Crepis intermedia A. Gray, Syn. Fl. 1: Part 2, 432.　1884.

Perennial, cinerous-puberulent or scurfy; stem rather
slender, 1–3-leaved, 1°–2° high, naked and branched
above.　Basal leaves lanceolate or oblong-lanceolate,
laciniate-pinnatifid, long-acuminate at the apex, narrowed at the base, 4'–6' long; stem leaves lanceolate,
sessile, less divided; heads several, 6″–9″ broad, 5–8-
flowered; involucre oblong-campanulate, its principal
bracts 5–8, lanceolate or linear-oblong, canescent, acutish, somewhat keeled by the thickened midrib when old,
the outer ones few and short; achenes oblong, narrowed
above, not beaked, strongly 10-ribbed.

In dry soil, western Nebraska (according to Williams);
Colorado to California, Montana and British Columbia.
May–Sept.

½

8. Crepis occidentàlis Nutt. Large-flowered Gray Hawksbeard. Fig. 4093.

Crepis occidentalis Nutt. Journ. Acad. Phila. **7** : 29. 1834.

Perennial, scurfy and canescent, sometimes also hirsute; stem rather stout, leafy, branched, 6′–18′ high. Leaves oblong-lanceolate or oblong in outline, laciniate-pinnatifid, acute or acuminate, the lower and basal ones 6′–10′ long, narrowed into petioles, the upper sessile and slightly clasping; heads few or several, corymbose, stout-peduncled, about 1′ broad, 10–30-flowered; involucre oblong-campanulate, canescent, its principal bracts 9–24, linear, acute; achenes oblong, sharply 10-ribbed, glabrous.

Plains, western Nebraska (according to Gray) ; Colorado to California, north to Montana and British Columbia. May–July.

22. HIERÀCIUM [Tourn.] L. Sp. Pl. 799. 1753.

Perennial hispid scabrous glandular or glabrous herbs, with alternate or basal leaves, and small middle-sized or large, solitary corymbose or paniculate heads of yellow orange or red flowers. Involucre cylindric, campanulate, or nearly hemispheric, its principal bracts in 1–3 series, the outer either regularly and gradually smaller or abruptly much smaller, none of them much thickened at the base after flowering. Receptacle flat, naked or short-fimbrillate. Rays truncate and 5-toothed at the apex. Anthers sagittate at the base. Style-branches slender. Achenes oblong, columnar, or fusiform, not beaked, 10–15-ribbed, terete or 4–5-angled. Pappus copious, or 1–2 rows of simple rather stiff persistent brown or brownish bristles. [Greek, hawk.]

Not fewer than 300 species, natives of the north temperate zone and the Andes of South America. Besides the following, some 20 others occur in western North America. Known as Hawkweed, Hawk-bit, or Speerhawk. Type species: *Hieracium muròrum* L.

A. Rootstock short, erect; plants without stolons.
a. Heads 1′–2′ broad.

Stem with 1–5 leaves only ; basal leaves tufted ; introduced species.
    Leaves cordate or subcordate ; scape naked or 1-leaved.   1. *H. murorum.*
    Leaves narrowed at the base ; stem several-leaved.   2. *H. vulgatum.*
Stem very leafy up to the inflorescence ; native species.
    Leaves short, ovate to ovate-lanceolate, rounded or clasping at the base.   3. *H. canadense.*
    Leaves elongated, linear-lanceolate, narrowed at the base.   4. *H. scabriusculum.*

b. Heads less than 1′ broad.

* Stem leaf-bearing nearly or quite up to the inflorescence, the upper leaves sometimes very small and distant.
    Pubescence of abundant brownish or whitish hairs ½′–1′ long; inflorescence elongated.
                                                       5. *H. longipilum.*
    Pubescence of short hairs, or nearly or quite wanting.
        Leaves lanceolate to oblong-lanceolate, acute or acuminate, glabrous.   6. *H. paniculatum.*
        Leaves elliptic to obovate, mostly obtuse.
            Stem hispid-pubescent, densely glandular-hispid above; the peduncles stout, spreading.
                                                       7. *H. scabrum.*
            Stem loosely pubescent ; pedicels slender.
                Inflorescence elongated ; achenes spindle-shaped.   8. *H. Gronovii.*
                Inflorescence corymbiform ; achenes columnar, truncate.   9. *H. marianum.*
** Leaves all basal or 1 or 2 borne on the stem above.
    Basal leaves elliptic to obovate or oblong-spatulate ; native species.
        Pedicels and involucres glabrous or nearly so ; leaves glabrous or loosely pubescent.
                                                  10. *H. venosum.*
        Pedicels and involucres glandular-pubescent ; leaves villous.   11. *H. Greenii.*
    Basal leaves narrowly oblanceolate, introduced.   12. *H. florentinum.*

B. Rootstock elongated, slender; plants mostly stoloniferous, scapose.

Scape bearing a single head, rarely 2 to 4.   13. *H. Pilosella.*
Heads several or many, corymbose.
    Flowers yellow.
        Plant glaucous ; leaves glabrous or nearly so above.   14. *H. floribundum.*
        Plant not glaucous ; leaves hirsute on both sides.   15. *H. pratense.*
    Flowers orange.   16. *H. aurantiacum.*

1. Hieracium muròrum L.　Wall Hawk-weed.　Fig. 4094.

Hieracium murorum L. Sp. Pl. 802. 1753.

Stem pubescent or glabrate, simple, or with 1 or 2 branches, 1°–2½° high.　Basal leaves thin, ovate or oblong, obtuse or acute, cordate or truncate at the base, or abruptly narrowed into petioles, coarsely dentate or laciniate, at least near the base, 2′–4′ long, 1′–2′ wide, the petioles villous; stem leaves 1 or 2, short-petioled or sessile, sometimes none; heads 2–several, corymbose, about 1′ broad; peduncles ascending, usually glandular; involucre 4″–5″ high, its bracts linear-lanceolate, acute, glandular-pubescent, imbricated in 2 or 3 series; achenes columnar, truncate; pappus of slender nearly white bristles.

Woodlands near Brooklyn, N. Y., Northampton, Mass., and about Quebec.　Adventive or fugitive from Europe.　French or golden lungwort.　June–Aug.

2. Hieracium vulgàtum Fries.　Hawkweed.　Fig. 4095.

H. molle Pursh, Fl. Am. Sept. 503.　1814.　Not Jacq. 1774.
H. vulgatum Fries, Fl. Hall. 128.　1817–18.

Similar to the preceding species, sometimes taller and slightly glaucous; stem 2–5-leaved, pubescent or glabrate.　Basal leaves oblong or lanceolate, acute at both ends, or some of them obtuse at the apex, coarsely dentate or denticulate, petioled, 2′–5′ long, ½′–1½′ wide, often mottled; stem leaves similar, short-petioled or sessile; petioles more or less pubescent; heads several, corymbose, smaller than those of *H. murorum* or as large; peduncles mostly glandular, straight; bracts of the involucre imbricated in 2 or 3 series, linear, acuminate, mostly glandular; achenes columnar, truncate; pappus copious.

Labrador and Newfoundland to Quebec, and in southern New York and New Jersey.　Naturalized from Europe. Also in Greenland, northern Europe and Asia.　July–Sept.

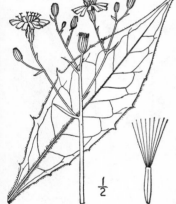

3. Hieracium canadénse Michx.　Canada Hawkweed.　Fig. 4096.

H. canadense Michx. Fl. Bor. Am. **2**: 86.　1803.

Stem erect, firm, glabrate or pubescent, leafy, 1°–5° high.　Leaves numerous, ovate, oblong-lanceolate, ovate-oblong, or lanceolate, acute or acuminate at the apex, rounded, sessile, and, at least the upper ones, clasping at the base, 1′–3′ long, 3″–12″ wide, serrate or incised, the margins sometimes ciliolate, glabrous or pubescent beneath, the lowest somewhat spatulate and petioled; no tuft of basal leaves at flowering time; heads usually numerous, corymbose-paniculate, about 1′ broad; involucre about 6″ high, pubescent or puberulent, its bracts imbricated in 2–3 series, the outer spreading; flowers yellow; achenes columnar, truncate; pappus copious, brown.

In dry woods and thickets, Newfoundland and Nova Scotia to Ontario, British Columbia, New Jersey, Indiana, Michigan, South Dakota and Oregon.　July–Sept.　High dandelion.

3/5

4. Hieracium scabriúsculum Schwein. Narrow-leaved Hawkweed. Fig. 4097.

H. scabriusculum Schwein. in Long's Exp. **2**: 394. 1824.

Stem rather slender, glabrous or puberulent, sometimes hispid below, usually very leafy nearly or quite up to the inflorescence, usually simple, 1°–2½° high. Leaves lanceolate or linear-lanceolate or the lowest spatulate, entire, denticulate or sometimes laciniate-dentate, acute or acuminate, narrowed to a sessile base, 1′–3′ long, 2″–7″ wide, glabrous above, mostly somewhat pubescent beneath, the margins commonly ciliolate; no tuft of basal leaves at flowering time; heads few or several, corymbose, about 1′ broad; peduncles rather stout, canescent; involucres 5″–7″ high, glabrous or somewhat pubescent, its bracts imbricated in 2–3 series, the outer spreading; flowers bright yellow; achenes columnar, truncate; pappus copious, brownish.

Minnesota to Nebraska, Saskatchewan, British Columbia and Oregon. Included in our first edition in the similar Old World *H. umbellatum* L. Apparently erroneously recorded from the St. Lawrence River. June–Aug.

5. Hieracium longípilum Torr. Long-bearded Hawkweed. Fig. 4098.

H. longipilum Torr.; Hook. Fl. Bor. Am. **1**: 298. 1833. *Hieracium barbatum* Nutt. Journ. Phil. Acad. **7**: 70. 1834. Not Tausch. 1828.

Stem, at least its lower portion, and leaves densely covered with long brown rather rigid bristly hairs ½′–1′ long, arising from papillae; stem very leafy below, stiff, simple, 2°–3½° high. Basal and lower leaves spatulate or oblong, obtuse, mostly entire, 4′–8′ long, narrowed into margined petioles, the upper lanceolate or spatulate, mostly sessile, the uppermost small and bract-like; heads not very numerous, racemose or racemose-paniculate, 8″–10″ broad; peduncles short, stout, tomentose and glandular; involucre 4″–5″ high, its principal bracts in 1 series, linear-lanceolate, acuminate, with several short subulate outer ones; flowers yellow; achenes fusiform with a slightly tapering summit; pappus brown.

Prairies and dry woods, Ontario to Minnesota, Illinois, Kansas and Texas. July–Sept.

1/2

1/2

6. Hieracium paniculàtum L. Panicled Hawkweed. Fig. 4099.

Hieracium paniculatum L. Sp. Pl. 802. 1753.

Glabrous throughout, or somewhat pilose-pubescent below, stem paniculately branched above, leafy, slender, 1°–3° high. Leaves thin, lanceolate or oblong-lanceolate, acute or acuminate at the apex, narrowed to a sessile base, or the lowest into petioles, denticulate or dentate, 2′–6′ long, 3″–12″ wide; no tuft of basal leaves at flowering time; heads 5″–7″ broad, commonly numerous, corymbose-paniculate, 12–20-flowered; peduncles slender, often drooping, quite glabrous or sometimes glandular; involucre about 3″ high, glabrous or nearly so, its principal bracts in 1 series, linear, acute with a few very small outer ones at the base; flowers yellow; achenes columnar, truncate; pappus brown, not very copious.

In dry woods, Nova Scotia to Ontario, Michigan, Georgia, Alabama and Tennessee. Ascends to 4600 ft. in Virginia. July–Sept.

7. Hieracium scàbrum Michx. Rough Hawkweed. Fig. 4100.

H. scabrum Michx. Fl. Bor. Am. **2** : 86. 1803.

Stem stout, leafy, mostly hirsute or hispid below and glandular-pubescent above, strict, 1°–4° high. Leaves hirsute, obovate, oblong, or broadly spatulate, 2′–4′ long, 1′–2′ wide, obtuse at the apex, narrowed to the sessile base or the lowest into margined petioles, denticulate; no tuft of basal leaves at flowering time; heads usually numerous, 6″–8″ broad, corymbose- or racemose-paniculate; peduncles stout, densely glandular; involucre 4″–5″ high, glandular, its principal bracts in 1 series, linear, acute with a few very small outer ones; flowers yellow; achenes columnar, truncate; pappus brown.

In dry woods and clearings, Nova Scotia to Minnesota, Georgia, Iowa, and recorded from Nebraska and Kansas. July–Sept.

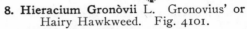

8. Hieracium Gronòvii L. Gronovius' or Hairy Hawkweed. Fig. 4101.

Hieracium Gronovii L. Sp. Pl. 802. 1753.

Stem stiff, mostly slender, leafy and villous or hirsute, at least below, sometimes nearly leafless, 1°–3° high. Leaves villous or hirsute, the basal and lower ones obovate or spatulate, denticulate or entire, obtuse, 2′–6′ long, usually narrowed into petioles; stem leaves mostly sessile, oblong or oval, obtuse or acute, narrowed or broad at the base, the upper gradually smaller; heads numerous, racemose-paniculate, 5″–8″ broad; peduncles glandular and canescent, slender; involucre about 4″ high, somewhat canescent, its principal bracts in 1 series, linear-lanceolate, acute, with several very small outer ones; flowers yellow; achenes spindle-shaped, with a tapering summit; pappus brown.

In dry soil, Massachusetts to Ontario, Illinois, Kansas, Florida, Louisiana and Texas. Santo Domingo. July–Sept. Cat's-ear.

9. Hieracium mariànum Willd. Maryland Hawkweed. Fig. 4102.

H. marianum Willd. Sp. Pl. **3** : 1572. 1804.

Stems usually solitary, slender, pilose-pubescent, at least below, paniculately branched above, 2°–3° high, bearing 2–7 leaves. Basal leaves obovate, oblanceolate or oblong, ascending or erect, obtuse, narrowed at the base, sessile or petioled, hirsute, at least on the veins beneath, entire or glandular-denticulate, 2′–8′ long, 1′–2′ wide, not purple-veined, mostly glabrous above, those of the stem similar, mostly smaller; heads commonly numerous, corymbose-paniculate, 6″–10″ broad, 15–40-flowered, slender-peduncled; peduncles more or less glandular and sometimes canescent; involucre about 4″ high, its principal bracts in 1 series, linear-lanceolate, acute, or acuminate, glabrous or nearly so, with a few short outer ones; achenes columnar, truncate; pappus brown, not copious.

In dry woods and thickets, New Hampshire to southern New York, Pennsylvania, Ohio, Kentucky, Alabama and Florida. May–July.

Hieracium alleghaniénse Britton, of the mountains of West Virginia, has the stem more abundantly leafy and no tuft of basal leaves at flowering time.

10. Hieracium venòsum L.   Rattlesnake-weed.   Poor Robin's Plantain. Fig. 4103.

Hieracium venosum L. Sp. Pl. 800. 1753.

Stems solitary or several from the same root, slender, glabrous, or with a few hispid hairs near the base, or also above, leafless or with 1–3 leaves, paniculately branched above, 1°–3° high. Basal leaves tufted, spreading on the ground, obovate, oval or oblong-spatulate, mostly obtuse, narrowed at the base, sessile or petioled, 1'–4' long, ½'–1½' wide, usually purple-veined, glabrous or more or less hirsute, pale beneath, some or all of them glandular-denticulate; heads commonly numerous, 5"–8" broad, 5–40-flowered, slender-peduncled; peduncles glabrous, or slightly glandular; involucre about 3" high, its principal bracts in 1 series, glabrous or nearly so, with a few short outer ones; achenes columnar, truncate; pappus brown, not copious.

Dry woods and thickets, Maine to Ontario and Manitoba, south to Georgia, Kentucky and Nebraska. Ascends to 4200 ft. in North Carolina. Early or vein-leaf hawkweed. Striped bloodwort. Snake-plantain. Hawkbit. Adder's-tongue. May–Oct.

11. Hieracium Greènii Porter and Britton. Green's Hawkweed. Fig. 4104.

Pilosella spathulata Sch. Bip. Flora **45**: 439. 1862. Not *Hieracium spathulatum* Scheele, 1863.
Hieracium marianum var. *spathulatum* A. Gray, Syn. Fl. **1**: Part 2, 446. 1886.
H. Greenii Porter and Britton, Bull. Torr. Club **20**: 120. 1893.

Stem entirely glabrous up to the branches, rather slender, leafless or rarely with 1 or 2 leaves. 1½°–2½° high. Basal leaves tufted, ascending, spatulate, oblong, or obovate, obtuse, narrowed at the base, mostly petioled, glandular-denticulate or entire, villous-pubescent or somewhat hispid, 4'–7' long, ½'–2' wide; heads corymbose-paniculate, several or numerous, 30–40-flowered, slender-peduncled, 8"–10" broad; peduncles and branches canescent-tomentose and glandular; involucre 5" high, its principal bracts in 1 series, linear, acute, densely pilose-glandular; flowers bright yellow; achenes columnar, truncate; pappus brownish, not copious.

In dry soil, mountains of Pennsylvania to Ohio, Virginia and West Virginia. May–June.

12. Hieracium florentìnum All.   King-devil.   Fig. 4105.

H. florentinum All. Fl. Ped. **1**: 213. 1785.

Stolons wanting; stem glabrous, or somewhat hispid, glaucous, slender, 1½°–3° high, bearing 1–3 leaves near the base. Basal leaves tufted, narrowly oblong, oblanceolate, or spatulate, entire, obtuse or acute at the apex, narrowed below into margined petioles, more or less hirsute with stiff hairs, or glabrous, 2'–4' long, 3"–7" wide; heads several or numerous, corymbose, 4"–6" broad; peduncles mostly short, pilose and glandular; involucre about 3" high, its bracts linear, acute or acuminate, pilose and somewhat glandular, imbricated in about 2 series; flowers yellow; achenes oblong, truncate; pappus a row of slender brownish bristles.

In fields, meadows and along roadsides, New York and Ontario to Quebec and Maine; a troublesome weed. Naturalized from Europe. Referred in our first edition to *H. praealtum*, also native of Europe, which differs in having long leafy branches from the base, and is recorded as established in a meadow at Andover, Massachusetts. June–Sept.

13. Hieracium Pilosélla L.   Mouse-ear Hawk-weed.   Fig. 4106.

Hieracium Pilosella L. Sp. Pl. 800. 1753.
Hieracium Pilosella peleterianum Mer. Nouv. Fl. Paris, Ed. 2, 230. 1821.

Stoloniferous, pilose-pubescent throughout; stolons leafy, rooting, slender, 3′–12′ long. Scape slender, erect, 4′–15′ high, leafless, with a single head, or sometimes 2–4; leaves oblong or spatulate, entire, obtuse or acutish at the apex, narrowed into petioles, often white-tomentose beneath, 1½′–3′ long, 4″–8″ wide; head 1′ broad or more; flowers yellow; principal bracts of the involucre in 1 or 2 series, linear, acuminate, pubescent, usually with 1 or 2 exterior ones; achenes oblong, truncate; pappus a single row of slender bristles.

Dooryards and fields, Prince Edward Island to Ontario, New York, Pennsylvania and Michigan. Adventive from Europe. Ling-gowans. Felon-herb. Mouse-bloodwort. May–Sept.

14. Hieracium floribúndum Wimm. & Grab.   Smoothish Hawkweed.   Fig. 4107.

Hieracum floribundum Wimm. & Grab. Fl. Siles. 2²: 204. 1829.

Stoloniferous; plant glaucous-green; scape loosely hirsute and more or less glandular-pubescent, slender, 1°–2½° high, the stolons sometimes 8′ long. Basal leaves tufted, narrowly oblanceolate, acutish or obtuse, 2′–6′ long, glabrous or very nearly so above, the margins and midvein beneath more or less hirsute; stem-leaves none, or rarely 1 or 2 near the base; flowers several, 1′ broad or less, corymbose; peduncles glandular; rays bright yellow; bracts of the involucre in about 2 series, hirsute.

In fields, New Brunswick and Maine to New York. Naturalized from Europe. June–Aug.

15. Hieracium praténse Tausch.   Field Hawkweed.   Fig. 4108.

H. pratense Tausch, Flora **11** : Part 1, Erg. 56. 1828.

Stoloniferous, hirsute or pilose with long hairs, those of the stem blackish. Stem scapose, simple, 1°–2° high, often bearing 1 or 2 leaves below the middle; basal leaves numerous, tufted, light green, oblanceolate to oblong, obtuse, 2′–5′ long, 5″–10″ wide, narrowed into margined petioles, or to a sessile base, entire, or with few distant minute glandular teeth, hirsute on both sides; heads several or numerous, corymbose-paniculate, 10″ wide, or less; flowers yellow; peduncles glandular and often tomentose; bracts of the involucre linear-lanceolate, acuminate, glandular and pilose; achenes columnar, truncate.

Fields and roadsides, Quebec to southern New York and Pennsylvania. Naturalized or adventive from Europe.

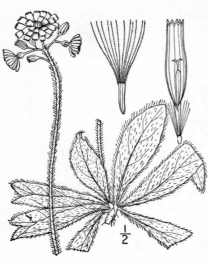

16. Hieracium aurantìacum L. Orange or Tawny Hawkweed. Golden Mouse-Ear Hawkweed. Fig. 4109.

Hieracium aurantiacum L. Sp. Pl. 801. 1753.

Stoloniferous; stem leafless or rarely with 1 or 2 small sessile leaves, hirsute, slender, 6′–20′ high. Basal leaves hirsute, tufted, spatulate or oblong, obtuse, narrowed at the base, entire, or sometimes slightly denticulate, 2′–5′ long, ¼′–1′ wide; heads several, short-peduncled, corymbose, 7″–12″ broad; peduncles glandular-pubescent; involucre 4″–5″ high, its bracts linear-lanceolate, acuminate, imbricated in 2 or 3 series, hirsute and sometimes glandular; flowers orange or red; achenes oblong, truncate; pappus a single row of slender brownish bristles.

In fields, woods and along roadsides, New Brunswick and Ontario to New York, New Jersey and Pennsylvania. Naturalized from Europe. Grim-the-collier. Devil's- or Flora's-paint-brush. Red daisy. Missionary-weed. June–Sept.

23. NÁBALUS Cass. Dict. Sci. Nat. 34: 94. 1825.

Perennial, herbs with alternate, mostly petioled, dentate lobed or pinnatifid leaves, or the upper auriculate and clasping, and numerous small heads of ligulate white yellowish or purplish flowers in open or spike-like terminal panicles, or also in axillary clusters, usually drooping. Involucre cylindric, usually narrow, its principal bracts in 1 or 2 series, nearly equal, with a few smaller exterior ones at the base. Receptacle flat, naked. Rays truncate and 5-toothed at the summit. Style-branches slender. Achenes oblong or narrowly columnar, truncate, terete or 4–5-angled, mostly 10-ribbed. Pappus of copious rather rigid simple white to reddish-brown bristles. [Modern Latin, from an Indian name for Rattlesnake-root.]

About 25 species, natives of America and Asia. Besides the following, two others occur in northwest America and one on the high mountains of North Carolina. Known by the general name of Rattlesnake-root or Drop-flower. Type species: *Nabalus trifoliatus* Cass. The European and African genus *Prenanthes* L. is here regarded as distinct from this.

* Bracts of the involucre glabrous, or with a few scattered hairs.

Heads 5–7-flowered; involucre very narrow, light green, 1″ thick; pappus light straw-color or brown.
　　　　　　　　　　　　　　　　　　　　　　　　　　　　　1. *N. altissimus.*
Heads 8–16-flowered; involucre broader, green, purple or glaucous, 1½″–3″ thick.
　Leaves, or some of them, lobed, divided, or pinnatifid (sometimes entire in No. 3); involucre
　　subcylindric, about 1½″ thick.
　　Pappus deep cinnamon-brown.   2. *N. albus.*
　　Pappus straw-color or light brown.
　　　Inflorescence paniculate.
　　　　Involucral bracts with some stiff hairs, obviously shorter than the pappus; panicle-
　　　　　branches divergent.   3. *N. serpentarius.*
　　　　Involucral bracts glabrous, equalling the pappus; panicle-branches ascending, or
　　　　　upcurved.   4. *N. trifoliolatus.*
　　　Inflorescence thyrsoid or glomerate.
　　　　Leaves palmately lobed or divided; northern.   5. *N. nanus.*
　　　　Leaves pinnately lobed or pinnatifid; southern.   6. *N. virgatus.*
　Leaves entire or denticulate; plant low, alpine; involucre narrowly campanulate, 2½″–3″ thick.
　　　　　　　　　　　　　　　　　　　　　　　　　　　　　7. *N. Boottii.*

** Bracts of the involucre hirsute-pubescent.

Inflorescence narrowly thyrsoid; heads 8–16-flowered.
　Leaves and stem rough-puberulent or scabrous.   8. *N. asper.*
　Leaves and stem glabrous, glaucous.   9. *N. racemosus.*
Inflorescence corymbose-paniculate; heads 20–25-flowered.   10. *N. crepidineus.*

1. Nabalus altíssimus (L.) Hook.   Tall White Lettuce.   Fig. 4110.

Prenanthes altissima L. Sp. Pl. 797. 1753.
N. altissimus Hook. Fl. Bor. Am. 1 : 294. 1833.

Glabrous, or sometimes hispidulous, not glaucous; stem slender, 3°–7° high, green, or sometimes purplish. Leaves thin, hastate, cordate, ovate, or the uppermost lanceolate, entire, denticulate, dentate or palmately lobed or divided, most of them long-petioled, the larger sometimes 6′ long; heads very numerous, in a narrow panicle, and often in axillary clusters, 5–7-flowered, pendulous, about 2″ broad; inflorescence often narrow; involucre narrowly cylindric, 5″–6″ long, about 1″ thick, green, glabrous, its principal bracts about 5; flowers greenish or yellowish white; pappus light straw-color, or cinnamon-brown.

In woods and thickets, Newfoundland to Manitoba, Missouri, Georgia and Louisiana. Lion's-foot. Rattlesnake-root. Ascends to 2500 ft. in the Catskills. Wild lettuce. Joy-leaf. Milk-weed. Bird-bell. Races differ in leaf-form, pubescence and in color of the pappus. July–Oct.

$\frac{2}{3}$

2. Nabalus álbus (L.) Hook.   Rattlesnake-root. White Lettuce. Fig. 4111.

Prenanthes alba L. Sp. Pl. 798. 1753.
Nabalus albus Hook. Fl. Bor. Am. 1 : 294. 1833.

Glabrous and glaucous; stem commonly purple, 2°–5° high. Leaves hastate, ovate, cordate, denticulate, dentate. lobed, or palmately divided, or the upper lanceolate, entire, thicker than those of the preceding species, the larger sometimes 8′ long; heads numerous, pendulous, 8–15-flowered, about 3″ broad, paniculate, or thyrsoid, and often in axillary clusters; involucre glabrous, or with a few scattered hairs, glaucous, 5″–7″ high, about 1½″ thick, its principal bracts about 8, purplish, with minute outer ones; flowers greenish or yellowish white, fragrant; pappus cinnamon-brown.

In woods, Maine and Ontario to Manitoba, Saskatchewan, Georgia, Kentucky, Wisconsin and North Dakota. Lion's-foot. White cankerweed. Wild lettuce. Milk-weed. Joy-leaf. Cancer-weed. Aug.–Sept.

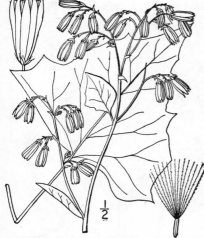

$\frac{1}{2}$

3. Nabalus serpentàrius (Pursh) Hook.   Lion's-foot. Gall-of-the-Earth. Fig. 4112.

Prenanthes serpentaria Pursh, Fl. Am. Sept. 499. *pl. 24.* 1814.
Nabalus integrifolius Cass. Dict. Sci. Nat. 34 : 95. 1825.
Nabalus serpentarius Hook. Fl. Bor. Am. 1 : 294. 1833.

Glabrous or sparingly pubescent, green; stem stout or slender, not glaucous, 1°–4° high. Leaves rather firm, similar to those of the preceding species, equally variable in outline, pinnatifid or pinnately lobed, palmately divided, or merely dentate, or entire; inflorescence paniculate, the branches divaricate, upcurved; heads numerous, about 3″ broad, 8–12-flowered, pendulous, paniculate, and commonly also in axillary clusters; involucre more or less bristly-hispid, rarely glabrous, green or purplish, about 1½″ thick, 5″–7″ long, its principal bracts about 8, shorter than the pappus, with several minute lanceolate outer ones; flowers whitish or cream-color, rarely yellow; achenes about 3″ long; pappus light brown or straw-color.

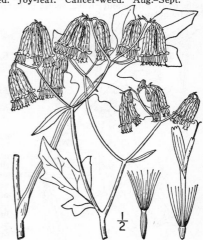

$\frac{1}{2}$

In fields and thickets, Massachusetts to southern New York, Florida, Alabama, Kentucky and Mississippi. Canker-weed. Rattlesnake-root. White lettuce. Snake-gentian. July–Oct.

Nabalus cylindrìcus Small, of the southern mountains, ranging north to Kentucky, differs by an elongated narrow panicle.

4. **Nabalus trifoliolàtus** Cass.   Tall Rattlesnake-root.   Fig. 4113.

Nabalus trifolilatus Cass. Dict. Sci. Nat. **34**: 95. 1825.
Prenanthes trifoliata Fernald; Brainerd, Jones & Eggleston, Fl. Verm. 89. 1900.

Glabrous throughout; stem usually stout, 3°–9° high. Leaves thinnish, the lower long-petioled, usually 3-divided with the divisions stalked or sessile, the upper short-petioled or sessile, all commonly lobed or dentate, but the upper sometimes lanceolate, acuminate and entire; inflorescence paniculate, the panicle-branches ascending, or nearly erect; heads few in the clusters, drooping, 7–12-flowered; involucre pale green or purplish, glabrous, about 1½″ thick and 6″ long, its principal bracts 6–8, equalling the pappus, the short outer ones ovate to lanceolate; flowers whitish or pale yellow; achenes 2″–3″ long; pappus light brown.

In woods and thickets, Newfoundland to Pennsylvania, Indiana, Delaware and Tennessee. Aug.–Oct.

5. **Nabalus nànus** (Bigel.) DC.   Low Rattlesnake-root, or Lion's-foot.   Fig. 4114.

Prenanthes alba var. *nana* Bigel. Fl. Bost. Ed. *2*, 286. 1824.
Nabalus nanus DC. Prodr. **7** : 241. 1838.
Prenanthes nana Torr.; Robinson & Fernald in A. Gray, Man. Ed. 7, 871. 1908.
Prenanthes serpentaria var. *nana* A. Gray, Syn. Fl. **1** : Part 2, 434. 1884.

Glabrous throughout; stem simple, erect, 4′–16′ high. Basal and lower leaves slender-petioled, 3-divided, or sometimes broadly hastate, the divisions variously lobed, toothed, or entire, usually sessile, occasionally stalked; upper leaves much smaller, entire, toothed, or lobed, sessile, or short-petioled; inflorescence thyrsoid, glomerate-spicate or racemose, rarely with 1 or 2 short ascending branches; involucre dark purple-brown or nearly black, glabrous, 4″–6″ long, its inner bracts 6–8, slightly ciliate at the apex, about as long as the usually bright brownish pappus; outer bracts lanceolate to ovate-lanceolate.

Alpine summits of the Adirondacks and the mountains of New England; Nova Scotia to Labrador and Newfoundland. Aug.–Sept.

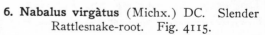

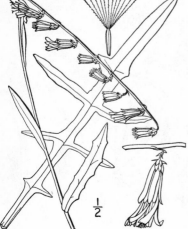

6. **Nabalus virgàtus** (Michx.) DC.   Slender Rattlesnake-root.   Fig. 4115.

Prenanthes virgata Michx. Fl. Bor. Am. **2** : 84. 1803.

Nabalus virgatus DC. Prodr. **7** : 242. 1838.

Glabrous, somewhat glaucous; stem strict, mostly simple, 2°–4° high. Leaves lanceolate or oblong-lanceolate, the basal and lower ones sinuate-pinnatifid or pinnately parted, petioled, often 10′ long, the lobes entire or dentate, distant; upper leaves all sessile, pinnately lobed, or the uppermost entire, very small and bract-like; heads very numerous, pendulous, about 2″ broad, in a narrow, simple or branched, terminal thyrsus, often unilateral; involucre purplish, about 1½″ thick and 5″ long, its principal bracts about 8, with several minute outer ones; flowers white or pinkish; pappus straw-color.

In moist sandy soil, New Jersey to Florida, near the coast. Called also De Witt's-snakeroot. Sept.–Oct.

7. Nabalus Boòttii DC.   Boott's Rattle-snake-root.   Fig. 4116.

Nabalus Boottii DC. Prodr. **7**: 241. 1838.
Prenanthes Boottii A. Gray, Syn. Fl. **1**: Part 2, 435. 1884.

Glabrous below, commonly pubescent above; stem simple, 4′–12′ high. Leaves thickish, the basal and lower ones ovate, hastate, or deltoid, petioled, mostly obtuse, entire, or denticulate, 1′–2′ long, the upper ovate or oblong, usually entire, short-petioled or sessile, much smaller; heads several or numerous, 10–18-flowered, erect, spreading, or some of them pendulous, racemose or somewhat thyrsoid, 4″–5″ broad; involucre campanulate-oblong, 2½″–3″ thick, 4″–7″ long, dark purplish-green, its principal bracts 8–10, obtuse or obtusish, with several shorter outer ones; flowers whitish, odorous; pappus brownish.

Alpine summits of the mountains of northern New England and New York. July–Aug.

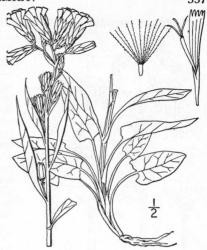

8. Nabalus ásper (Michx.) T. & G.   Rough White-lettuce.   Fig. 4117.

Prenanthes aspera Michx. Fl. Bor. Am. **2**: 83. 1803.
Nabalus asper T. & G. Fl. N. A. **2**: 483. 1843.

Scabrous or rough-puberulent; stem virgate, simple, 2°–4° high. Leaves firm, oval, oblong, or oblanceolate, those of the stem all closely sessile, acute or acutish, dentate, denticulate, or the uppermost entire, the lower sometimes clasping, 2′–3′ long, ½′–1′ wide, the basal tapering into winged petioles, commonly obtuse; heads very numerous, erect, spreading, or slightly drooping, 3″–4″ broad, 12–16-flowered, in a long narrow thyrsus; involucre oblong, 2″–2½″ thick, 5″–7″ high, very hirsute, its principal bracts 8–10, with several short outer ones; flowers light yellow; pappus straw-color.

On dry prairies, Ohio to South Dakota, Nebraska, Kentucky, Louisiana and Kansas. Rattlesnake-root. Aug.–Sept.

9. Nabalus racemòsus (Michx.) DC.   Glaucous White-lettuce.   Fig. 4118.

Prenanthes racemosa Michx. Fl. Bor. Am. **2**: 83. 1803.
Nabalus racemosus DC. Prodr. **7**: 242. 1838.
Nabalus racemosus pinnatifidus Britton; Britt. & Brown, Ill. Fl. **3**: 291. 1898.
Prenanthes racemosus var. *pinnatifida* A. Gray, Syn. Fl. **1**: Part 2, 433. 1884.

Stem virgate, rather stout, glabrous and somewhat glaucous; stem striate, 2°–6° high. Leaves thickish, glabrous and glaucous, the lower and basal ones oval, oblong, oblanceolate, or obovate, dentate, denticulate, pinnatifid or pinnately lobed, 4′–8′ long, mostly obtuse, tapering into long margined petioles; upper leaves sessile, smaller and partly clasping, lanceolate to ovate-lanceolate, denticulate, entire, or pinnatifid, mostly acute; heads very numerous, erect, spreading, or slightly drooping, 12–16-flowered, 2″–3″ broad, in a long narrow thyrsus; involucre oblong-cylindric, hirsute, 5″–6″ long, 1½″–2½″ thick, longer than the hirsute peduncle, its principal bracts 8–10, with several small outer ones; flowers purplish; pappus straw-color.

In moist open places, New Brunswick and Quebec to Maine, Manitoba, Alberta, southern New York, New Jersey, Iowa, Missouri and Colorado. Aug.–Sept.

Prenanthes mainénsis A. Gray, from northern Maine and New Brunswick, is probably a hybrid between *N. racemosus* and *N. trifoliolatus.*

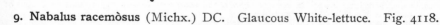

10. **Nabalus crepidíneus** (Michx.) DC.　Corymbed Rattlesnake-root.　Fig. 4119.

Prenanthes crepidinea Michx. Fl. Bor. Am. **2** : 84.　1803.
Nabalus crepidineus DC. Prodr. **7** : 242.　1838.

Stem glabrous or very nearly so below, corymbosely branched and sometimes puberulent above, stout, 5°–9° high. Leaves thin, the basal and lower ones hastate, ovate, oblong, or deltoid, sharply dentate, lobed, or incised, sometimes 10′ long, usually with broadly winged petioles; the upper much smaller, sessile or short-petioled, narrowed at the base, not clasping, ovate, deltoid, or lanceolate, acute; heads numerous, pendulous, short-peduncled, corymbose-paniculate, 4″–6″ broad, 20–35-flowered; involucre oblong or oblong-campanulate, hirsute, 5″–8″ long, about 3″ thick, dark green or purplish, its principal bracts 12–15, with several very short outer ones; flowers cream color; pappus brown.

In fields and thickets, western Pennsylvania and New York to Kentucky, west to Minnesota and Kansas. Aug.–Oct.

Family 45.　**AMBROSIÀCEAE** Reichenb. Consp. 112.　1828.

RAGWEED FAMILY.

Annual or perennial herbs, monoecious, or sometimes dioecious, many of them weeds, some shrubby, with alternate leaves, or the lower opposite, and small heads of greenish or white flowers subtended by an involucre of few, separate or united bracts, the pistillate heads sometimes larger and nut-like or bur-like. Staminate and pistillate flowers in the same heads, or in separate heads. Receptacle chaffy. Pistillate flowers with no corolla, or this reduced to a short tube or ring; calyx adnate to the 1-celled ovary, its limb none, or a mere border; style 2-cleft. Staminate flowers with a funnelform tubular or obconic 4–5-lobed corolla; stamens mostly 5, separate, or their anthers merely connivent, not truly syngenesious, with short inflexed appendages; ovary rudimentary; summit of the style often hairy or penicillate.

Eight genera and about 60 species, mostly natives of America, a few only of the Old World.

Staminate and pistillate flowers in the same heads; involucre of a few rounded bracts.　1. *Iva.*
Staminate and pistillate flowers in separate heads, the staminate mostly uppermost; involucre of the pistillate heads bur-like or nut-like.
　Involucral bracts of the staminate heads united.
　　Involucre of the pistillate heads with several tubercles or prickles in a single series.
　　　　　　　　　　　　　　　　　　　　　　　　　　　　　　2. *Ambrosia.*
　　Involucre of the pistillate heads with numerous prickles in several series.　　3. *Gaertneria.*
　Involucral bracts of the staminate heads separate; involucre of pistillate heads an oblong bur.
　　　　　　　　　　　　　　　　　　　　　　　　　　　　　　4. *Xanthium.*

1.　**ÌVA** L. Sp. Pl. 988.　1753.

Puberulent or scabrous herbs, with thick opposite leaves, or the upper alternate, and small nodding, axillary and solitary, spicate racemose or paniculate heads of greenish flowers. Involucre hemispheric or cup-shaped, its bracts few, rounded. Receptacle chaffy, the linear or spatulate chaff enveloping the flowers. Marginal flowers 1–6, pistillate, fertile, their corollas short, tubular or none. Disk-flowers perfect, sterile, their corollas funnelform, 5-lobed, their styles undivided, dilated at the apex. Anthers entire at the base, yellow, scarcely coherent with each other, tipped with mucronate appendages. Achenes compressed, obovoid, glabrous. Pappus none. [Named after *Ajuga Iva,* from its similar smell.]

About 15 species, natives of America. Besides the following, 7 others occur in the southern and western United States. Type species : *Iva annua* L.

Heads spicate or racemose, each subtended by a linear or oblong leaf.
　Heads solitary, pedicelled.
　　Bracts of the involucre 4–5 ; heads 1½″–2″ high.
　　　Leaves serrate, oval or oblong ; eastern.　　　　　　　　　　　　　　1. *I. frutescens.*
　　　Leaves entire or nearly so, obovate or oblong ; western.　　　　　　　2. *I. axillaris.*
　　Bracts of the involucre 6–9 ; heads 3″–4″ high ; southeastern.　　　　　3. *I. imbricata.*
　Heads spicate-paniculate ; leaves dentate.　　　　　　　　　　　　　　　4. *I. ciliata.*
Heads spicate-paniculate, not subtended by leaves.　　　　　　　　　　　　5. *I. xanthiifolia.*

1. Iva frutéscens L. Marsh Elder. High-water Shrub. Fig. 4120.

Iva frutescens L. Sp. Pl. 989. 1753.
Iva oraria Bartlett, Rhodora **8**: 26. 1906.

Perennial, shrubby or herbaceous, somewhat fleshy; stem paniculately branched above, minutely pubescent, or sometimes glabrous below, 3°–12° high. Leaves oval, oblong, or oblong-lanceolate, all the lower ones opposite, short-petioled, 3-nerved, acute or obtusish, serrate, narrowed at the base, the lower 4′–6′ long, 1′–2′ wide, the upper smaller and narrower, passing gradually into those of the racemose inflorescence which are much longer than the short-pedicelled heads; involucre depressed-hemispheric, its bracts about 5, orbicular-obovate, separate; fertile flowers about 5, their corollas tubular.

Along salt marshes and on muddy sea-shores, Massachusetts to Florida and Texas, the northern plant (*I. oraria*) mainly broader-leaved and less shrubby than the southern. Jesuits'- or false Jesuits'-bark. July–Sept.

2. Iva axillàris Pursh. Small-flowered Marsh Elder. Fig. 4121.

Iva axillaris Pursh, Fl. Am. Sept. 743. 1814.

Perennial by woody roots; stems herbaceous, ascending, glabrous or sparingly pubescent, simple or branched, 1°–2° high. Leaves sessile, entire or very nearly so, obtuse, faintly 3-nerved, obovate, oblong, or linear-oblong, ½′–1½′ long, thick, somewhat fleshy, glabrous or pubescent, the lower opposite, the upper alternate and smaller, passing gradually into those of the inflorescence; heads mostly solitary in the axils of the leaves, 2″–3″ broad, short-peduncled; involucre hemispheric, about 1½″ high; its bracts about 5, connate at the base, or united nearly to the summit; pistillate flowers 4 or 5, their corollas tubular.

In saline or alkaline soil, Manitoba and North Dakota to western Nebraska, New Mexico, British Columbia and California. May–Sept.

3. Iva imbricàta Walt. Sea-coast Marsh Elder. Fig. 4122.

Iva imbricata Walt. Fl. Car. 232. 1788.

Perennial by woody roots, glabrous or nearly so throughout, fleshy; stem 1°–2° high, simple, or sparingly branched. Leaves all but the lowest alternate, sessile, oblong-spatulate, or lanceolate, obtusish, mucronulate, entire, or rarely serrate, obscurely 3-nerved, the larger 1′–2′ long, 3″–5″ wide; heads about 4″ broad, short-peduncled or nearly sessile, the upper often longer than their subtending leaves; involucre broadly campanulate, its bracts 6–9, not united, somewhat imbricated in 2 series; fertile flowers 2–4, their corollas tubular, the staminate ones much more numerous; chaff of the receptacle spatulate.

On sandy sea-shores, southeastern Virginia to Florida and Louisiana. Bahamas; Cuba. July–Oct.

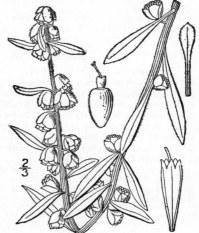

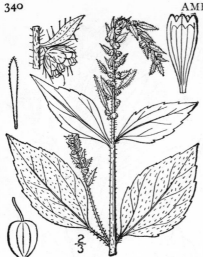

2/3

4. Iva ciliàta Willd. Rough Marsh Elder. Fig. 4123.

Iva annua Michx. Fl. Bor. Am. **2** : 184. 1803. Not L.

Iva ciliata Willd. Sp. Pl. **3** : 2386. 1804.

Annual, hispid-pubescent; stem simple, or branched, 2°–6° high. Leaves nearly all opposite, ovate, petioled, scarcely fleshy, puberulent beneath, acuminate at the apex, abruptly or gradually narrowed at the base, coarsely and irregularly dentate, 3-nerved, the lower 4′–5′ long; heads spicate-paniculate, about 1″ broad; spikes dense or interrupted, erect, 2′–8′ long; upper leaves linear-lanceolate, hispid, squarrose, much longer than the heads; bracts of the involucre 3–5, distinct or united below, hispid; fertile flowers 3–5, their corollas slender; staminate flowers 10–15.

In moist soil, Illinois to Nebraska, south to Louisiana and New Mexico. Plant with the aspect of an *Ambrosia*. Aug.–Oct.

5. Iva xanthiifòlia Nutt. Burweed Marsh Elder. Fig. 4124.

Iva xanthiifolia Nutt. Gen. **2** : 185. 1818.

Cyclachaena xanthiifolia Fresen. Ind. Sem. Hort. Franc. 4. 1836.

Annual; stem much branched, pubescent or puberulent above, glabrous below, 3°–6° high. Leaves nearly all opposite, broadly ovate, long-petioled, acuminate at the apex, abruptly or gradually contracted at the base, coarsely and irregularly dentate, pale and canescent or puberulent beneath, roughish above, 3-ribbed, the lower often 6′ long and wide; inflorescence spicate-paniculate, terminal and axillary, ample, naked; heads sessile or minutely peduncled, 1″ broad or less; bracts of the involucre in 2 series of 5, the outer ovate, the inner obovate or truncate, concave, subtending the usually 5 achenes; corollas of the fertile flowers rudimentary or none; staminate flowers 10–15, their corollas obconic.

In moist soil, or sometimes in waste places, Ontario to Assiniboia, south to Wisconsin, Nebraska, New Mexico and Utah. In waste grounds farther east. Plant with the aspect of a *Chenopodium*. July–Sept.

3/4

2. AMBRÒSIA [Tourn.] L. Sp. Pl. 987. 1753.

Monoecious (rarely dioecious) branching herbs, with alternate or opposite, mostly lobed or divided leaves, and small heads of green flowers, the staminate spicate or racemose, the pistillate solitary or clustered in the upper axils. Involucre of the pistillate heads globose-ovoid or top-shaped, closed, 1-flowered, usually armed with 4–8 tubercles or spines; corolla none; stamens none; style-branches filiform; achenes ovoid or obovoid; pappus none. Involucre of the staminate heads mostly hemispheric or saucer-shaped, 5–12-lobed, open, many-flowered; receptacle nearly flat, naked, or with filiform chaff; corolla funnelform, 5-toothed; anthers scarcely coherent, mucronate-tipped; style undivided, penicillate at the summit. [The ancient classical name.]

About 15 species, mostly natives of America. Besides the following, some 5 others occur in the southern and western United States. Type species: *Ambrosia maritima* L.

Sterile heads sessile; a lanceolate hispid lobe on inner border of involucre.	1. *A. bidentata.*
Sterile heads short-pedicelled, involucre depressed-hemispheric.	
Leaves opposite, palmately 3–5-lobed, or undivided; receptacle naked.	2. *A. trifida.*
Leaves opposite and alternate, 1–2-pinnatifid; receptacle chaffy.	
Annual; leaves thin; fruiting involucre spiny.	3. *A. elatior.*
Perennial; leaves thick; fruiting involucre naked or tubercled.	4. *A. psilostachya.*

1. Ambrosia bidentàta Michx. Lance-leaved Ragweed. Fig. 4125.

Ambrosia bidentata Michx. Fl. Bor. Am. **2** : 182. 1803.

Annual, hirsute, usually much branched, very leafy, 1°–3° high. Leaves lanceolate, mainly alternate, sessile and somewhat cordate-clasping at the base, acuminate at the apex, 1-nerved, 1'–3' long, 2''–4'' wide, usually with 1 or 2 sharp lobes at the base and a few minute sharp teeth above, or the upper ones quite entire, rough and hirsute or ciliate; spikes of staminate heads dense, 3'–7' long, their involucres turbinate, bearing a long lanceolate hispid reflexed lobe appearing like a bract on the inner border, their receptacles chaffy; fertile heads solitary, or clustered, oblong, 4-angled, 3''–4'' long, bearing 4 sharp spines.

Prairies, Illinois to Missouri, Kansas, Louisiana and Texas. July–Sept.

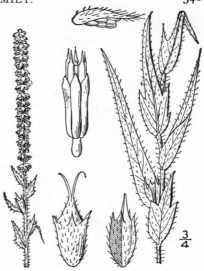

2. Ambrosia trífida L. Horse-cane. Bitter-weed. Great Ragweed. Fig. 4126.

Ambrosia trifida L. Sp. Pl. 987. 1753.
A. integrifolia Muhl.; Willd. Sp. Pl. 4 : 375. 1805.
Ambrosia trifida integrifolia T. & G. Fl. N. A. **2** : 290. 1841.

Annual, scabrous or hispid, or nearly glabrous, branched, 3°–17° high. Leaves all opposite, petioled, 3-nerved, deeply 3–5-lobed, or undivided, the lobes lanceolate or ovate, serrate, acute or acuminate; lower leaves often 1° wide; racemes of sterile heads 3'–10' long, their involucres saucer-shaped, 3-ribbed on the outer side, crenate-margined or truncate, their receptacles naked; fertile heads usually clustered in the axils of the upper bract-like leaves, turbinate to obovoid, 5–7-ribbed, conic-beaked, 3''–4'' long, each rib bearing a tubercle near the summit.

In moist soil, Quebec to Florida, west to Manitoba, Nebraska, Colorado and New Mexico. Tall ambrosia. Richweed. Wild hemp. Horse-weed. Buffalo-weed. Hay-fever weed. July–Oct.

3. Ambrosia elátior L. Ragweed. Roman Wormwood. Hogweed. Wild Tansy. Fig. 4127.

Ambrosia elatior L. Sp. Pl. 987. 1753.
Ambrosia artemisiaefolia L. Sp. Pl. 988. 1753.

Annual, pubescent, puberulent or hirsute, panicu-lately branched, 1°–6° high. Leaves thin, 1–2-pin-natifid, petioled, 2'–4' long, the upper alternate, the lower mostly opposite, pale or canescent beneath, the lobes oblong or lanceolate, obtuse or acute; uppermost leaves of the branches sometimes linear-lanceolate and entire; racemes of sterile heads very numerous, 1'–6' long, the involucres hemispheric, crenate, the receptacle chaffy; fertile heads obovoid or subglobose, mostly clustered, 1½''–2'' long, short-beaked, 4–6-spined near the summit, sparingly pu-bescent.

In dry soil, often a pernicious weed in cultivated fields, Nova Scotia to Florida, west to British Colum-bia and Mexico. Bermuda. Introduced into Europe as a weed. Consists of several slightly differing races. Also called bitterweed, stickweed, stammerwort, carrot-weed, black, or tassel-weed, hay-fever weed. July–Oct.

4. Ambrosia psilostàchya DC. Western Ragweed. Fig. 4128.

Ambrosia psilostachya DC. Prodr. **5**: 526. 1836.

Similar to the preceding species, but perennial by long rootstocks, the leaves thick, the pubescence strigose or hispid. Stems usually much branched, 2°–6° high, rather stout; leaves 1–2-pinnatifid, 2′–5′ long, the lobes acutish; racemes of sterile heads several or numerous, 2′–6′ long, the involucres campanulate, the receptacles chaffy; fertile heads mostly solitary, ovoid or obovoid, reticulated, short-pointed, unarmed, or with about 4 short tubercles, pubescent, 1½″–2″ long.

In moist open soil, Illinois to Saskatchewan, Texas, Mexico and California. July–Oct.

$\frac{2}{3}$

3. GAERTNÈRIA Med. Act. Pal. 3: 244. 1785. Phil. Bot. 45. 1789.

[FRANSERIA Cav. Icon. **2**: 78. *pl. 200.* 1793.]

Hispid or tomentose branching herbs, with the aspect of *Ambrosias,* sometimes woody at the base, with mostly alternate lobed or divided leaves, and small monoecious greenish heads of discoidal flowers, the staminate in terminal spikes or racemes, the pistillate solitary or clustered in the upper axils. Involucre of the pistillate heads ovoid or globose, closed, 1–4-celled, 1–4-beaked, armed with several rows of spines and forming a bur in fruit; corolla none or rudimentary; style deeply bifid, its branches exserted; stamens none; achenes obovoid, thick, solitary in the cells; pappus none. Staminate heads sessile, or short-peduncled, their involucres broadly hemispheric, open, 5–12-lobed; receptacle chaffy; corolla regular, the tube short, the limb 5-lobed; style undivided; anthers scarcely coherent; mucronate-tipped. [In honor of Joseph Gaertner, 1732–1791, German botanist.]

About 25 species, natives of America. In addition to the following, some 12 others occur in the western and southwestern United States. Type species: *Xanthium fruticosum* L. f.

Plant hirsute; annual; spines of the fruiting involucre long, flat.	1. *G. acanthicarpa.*
Leaves densely white-tomentose beneath; spines short, conic; perennials.	
Leaves bipinnatifid.	2. *G. discolor.*
Leaves pinnately divided, the terminal segment large.	3. *G. tomentosa.*

1. Gaertneria acanthicàrpa (Hook.) Britton. Hooker's Gaertneria. Fig. 4129.

Ambrosia acanthicarpa Hook. Fl. Bor. Am. **1**: 309. 1833.
Franseria Hookeriana Nutt. Trans. Am. Phil. Soc. (II.) **7**: 345. 1841.
Gaertneria acanthicarpa Britton, Mem. Torr. Club **5**: 332. 1894.
Franseria acanthicarpa Coville, Contr. Nat. Herb. **4**: 129. 1893.

Annual, erect or diffuse, paniculately branched, 1°–2° high; stem hirsute or hispid. Lower and basal leaves slender-petioled, bipinnatifid, 2′–4′ long, the upper short-petioled or sessile, once-pinnatifid, or merely lobed; racemes of sterile heads usually numerous, 1′–3′ long; fruiting involucres clustered in the axils, 3″–4″ long, commonly 1-flowered, armed with numerous long flat straight spines.

In moist soil, Saskatchewan to western Nebraska and Texas, west to British Columbia and California. Sand-bur. Recorded from Minnesota. July–Sept.

$\frac{3}{4}$

Gaertneria tenuifòlia (A. Gray) Kuntze, a southwestern perennial species with pinnately dissected leaves, the fruiting involucre only about 1″ long, is recorded as extending eastward into Kansas.

2. Gaertneria díscolor (Nutt.) Kuntze.
White-leaved Gaertneria. Fig. 4130.

Ambrosia tomentosa Nutt. Gen. **2**: 186. 1818. Not
Gaertneria tomentosa (A. Gray) Kuntze.

Franseria discolor Nutt. Trans. Am. Phil. Soc. (II.)
7: 345. 1841.

Gaertneria discolor Kuntze, Rev. Gen. Pl. 339. 1891.

Erect or ascending from perennial rootstocks,
branched, about 1° high. Leaves nearly all bipinnatifid, petioled, densely white-tomentose beneath,
green and pubescent or glabrate above, 2′–5′ long;
sterile racemes narrow, commonly solitary, 1′–2′
long; fruiting involucres clustered in the axils,
finely canescent, about 3″ long, mostly 2-flowered,
armed with short sharp conic spines.

In dry soil, South Dakota, Nebraska, Wyoming,
Kansas, Colorado and New Mexico. Aug.–Sept.

3. Gaertneria tomentòsa (A. Gray)
Kuntze. Woolly Gaertneria.
Fig. 4131.

Franseria tomentosa A. Gray, Mem. Am. Acad. **4**: 80.
1849.

G. tomentosa Kuntze, Rev. Gen. Pl. 339. 1891.

Gaertneria Grayi A. Nelson, Bot. Gaz. **34**: 35. 1902.

Erect from a deep perennial root, usually
branched at the base, 1°–3° high. Leaves pinnately lobed or divided, finely and densely tomentose on both sides, or ashy above, the terminal
segment lanceolate or oblong-lanceolate, acuminate, serrulate or entire, very much larger than
the 2–6 rather distinct narrow lateral ones; sterile
racemes solitary, 2′–4′ long; fruiting involucres
solitary, or 2–3 together in the upper axils, ovoid,
finely canescent or glabrate, 2-flowered, about 3″
long, armed with subulate-conic, very acute,
sometimes curved or hooked spines.

On rich prairies and along rivers, western Nebraska, Kansas and Colorado. Aug.–Sept.

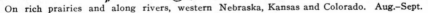

4. XÁNTHIUM [Tourn.] L. Sp. Pl. 987. 1753.

Monoecious annual branching coarse rough or spiny herbs, with alternate lobed or dentate
leaves, and rather small heads of greenish discoid flowers, the staminate ones capitate-
clustered at the ends of the branches, the pistillate axillary. Staminate heads with a short
involucre of 1 to 3 series of distinct bracts; receptacle cylindric, chaffy; corollas tubular,
5-toothed; anthers not coherent, mucronate at the apex; filaments monadelphous; style slender, undivided. Pistillate heads of an ovoid or oblong, closed involucre, covered with hooked
spines, 1–2-beaked, 2-celled, each cavity containing one obovoid or oblong achene; corolla
none; pappus none; style 2-cleft, its branches exserted. [Greek, yellow, from its yielding
a yellow hair-dye.]

About 15 species, of wide geographic distribution. Type species: *Xanthium strumarium* L.

Leaves lanceolate, acute at both ends; axils bearing 3-divided yellow spines.
                                                                            1. *X. spinosum.*
Leaves orbicular or broadly ovate, cordate to truncate at base; no axillary spines.
    Bur, or its prickles, or both, more or less hispid-pubescent; beaks incurved.
        Body of the bur ovoid to oval, twice as long as thick or shorter.
        Bur 1′ long or more, the prickles 4″–5″ long.   2. *X. speciosum.*

Bur 10″ long or less, the prickles 2″–3″ long.
  Bur densely prickly, its pubescence brown.   3. *X. echinatum.*
  Bur loosely prickly, its pubescence yellowish.   4. *X. glanduliferum.*
Body of the bur oblong, more than twice as long as thick.
  Prickles longer than the diameter of the body of the bur.   5. *X. inflexum.*
  Prickles shorter than the diameter of the body of the bur.
    Bur narrowly oblong.   6. *X. pennsylvanicum.*
    Bur broadly oblong.   7. *X. commune.*
Bur and its prickles glabrous, or merely puberulent; beaks nearly straight.   8. *X. americanum.*

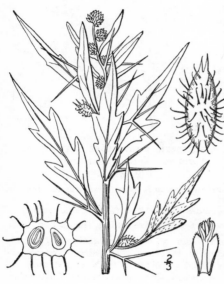

1. Xanthium spinòsum L.   Spiny or Thorny Clotbur, Clotweed or Burweed.   Fig. 4132.

Xanthium spinosum L. Sp. Pl. 987.   1753.

Stem pubescent or puberulent, much branched, ascending or erect, 1°–3° high.   Leaves lanceolate or ovate-lanceolate, acute or acuminate, lobed, or the upper entire, narrowed at the base, short-petioled, white-canescent beneath and on the whitish veins of the upper surface, 2′–5′ long; axils each with a short-stalked 3-pronged yellow spine nearly 1′ long; ripe fertile involucre (bur) oblong-cylindric, 4″–6″ long, about 2″ in diameter, pubescent, armed with short subulate rather inconspicuous beaks, and numerous glabrous spines about 1″ long.

In waste grounds, Maine to Ontario, Florida, Illinois, West Virginia, Missouri, Texas, New Mexico and California.   Widely distributed as a weed in tropical America.   Naturalized from Europe or Asia.   Cocklebur.   Dagger-cocklebur. Aug.–Nov.

2. Xanthium speciòsum Kearney.   Great Clotbur.   Fig. 4133.

Xanthium speciosum Kearney, Bull. Torr. Club **24**: 574.   1897.

Very stout, 3°–4½° high.   Stem sharply angled above; lower petioles 4′–6′ long; leafblades broadly triangular-ovate, the larger 6′–8′ wide, 3–5-lobed, dentate, scabrous on both surfaces; burs commonly clustered, oblong to ovoid-oblong, 1′ long or more, the stout beaks 5″–6″ long, somewhat incurved, strongly hooked at the apex, equalling or a little longer than the dense subulate uncinate prickles, which are hispid to above the middle, and 4″–5″ long.

Moist and waste grounds, North Dakota to Wisconsin, Tennessee, Montana, Nebraska and Texas, and locally in waste places eastward. Aug.–Sept.

3. Xanthium echinàtum Murr. Beach Clotbur.
Fig. 4134.

X. echinatum Murr. Comm. Goett. 6 : 32, *pl. 4.* 1783.

X. maculatum Raf. Am. Month. Mag. 344. 1818.

X. oviforme Wallr. Beitr. Bot. 1 : 240. 1842.

Stem rough, purplish or purple-blotched, 1°–2° high. Leaves firm, scabrous, with scattered short papillose hairs, obtusely toothed and lobed, somewhat resinous-glandular beneath; burs commonly clustered in the axils, ovoid to oval, 7″–11″ long, 4″–6″ thick, glandular; prickles very dense, densely hispid from the base to the middle or beyond, subulate, hooked, the longer about 2½″ long, and about equalling the stout hispid beaks.

Sea, lake and river beaches, occasionally in waste grounds, North Carolina to Nova Scotia, New York, Minnesota and North Dakota. Recorded west to Saskatchewan. Aug.–Sept.

4. Xanthium glandulìferum Greene.
Glandular Clotbur. Fig. 4135.

Xanthium glanduliferum Greene, Pittonia 4 : 61. 1899.

Similar to *X. echinatum.* Leaves very thick and scabrous with short stout papillae; burs oval, 5″–8″ long, 3″–4½″ thick, yellow, the prickles scattered, bristly-hispid nearly to the hooked apex, scarcely as long as the conic-subulate short-bristly beaks.

In dry soil, North Dakota to Assiniboia, British Columbia and Nebraska. Adventive in Missouri. June–Sept.

Xanthium Macoùnii Britton, known only from Lake Winnipeg, Manitoba, differs by a longer bur, 10″ long and 4″ thick.

5. Xanthium infléxum Mackenzie & Bush.
Missouri Clotbur. Fig. 4136.

Xanthium inflexum Mackenzie & Bush, Rep. Mo. Bot. Gard. 16 : 106. 1905.

Glabrate, or papillose-roughened above, 3°–4½° high. Leaves long-petioled, broadly ovate, more or less cordate, mostly 3-lobed, crenate-dentate; burs 1′ long or less, the body oblong, more than twice as long as thick, 3″–3½″ in diameter, glandular-pubescent; prickles hooked, stiff, longer than the diameter of the bur, glandular-pubescent below, glabrous above; beaks stout, about 5″ long, bent at the middle, strongly inflexed, hooked.

Sandy river-bottoms, Courtney, Missouri. Aug.–Sept.

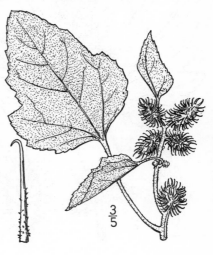

½

6. Xanthium pennsylvànicum Wallr. Pennsylvania Clotbur. Fig. 4137.

Xanthium pennsylvanicum Wallr. Beitr. Bot. 1: 236. 1842.

Stem comparatively slender, smooth below, roughish above, 1°–3° high. Leaves thin, long-petioled, sharply toothed and some of them 3–5-lobed, smoothish, or the upper surface scabrate, glandular; burs clustered in the axils, 7″–9″ long, about one-third as thick, puberulent and resinous-glandular, and commonly with a few longer hairs; prickles numerous, subulate, hooked, more or less hispid or glabrate, the longer ones nearly as long as the diameter of the bur; beaks slender, but stouter than the prickles, incurved and hooked.

Moist gorunds, Quebec to Minnesota, New York, Missouri and Colorado. Aug.–Oct. Referred in our first edition to *Xanthium canadense* Mill., a name which has been variously applied to different plants by authors.

7. Xanthium commùne Britton. Cocklebur or Clotbur. Fig. 4138.

Xanthium commune Britton, Manual 912. 1901.

Stem rather slender, 1°–2°, roughish. Leaves broadly ovate, more or less lobed, scabrous, especially above; burs commonly solitary in the axils, oblong, 7″–12″ long, half as thick, or less, the subulate-conic beaks slightly incurved, hooked at the apex, about as long as the subulate uncinate prickles, which are hispid to about the middle with brown hairs and shorter than the diameter of the bur.

Moist grounds, Quebec to Alberta, Maryland, Missouri, Utah and Arizona. Aug.–Oct.

⅗

½

8. Xanthium americànum Walt. American Cocklebur or Burweed. Fig. 4139.

X. americanum Walt. Fl. Car. 231. 1788.
X. macrocarpum glabratum DC. Prodr. 5: 523. 1836.
X. glabratum Britton, Manual 912. 1901.

Rough, 1°–6½° high. Leaves slender-petioled, broadly ovate to orbicular, 3-ribbed and cordate or cordate-reniform at the base, the lower often 10′ wide, the margins dentate, or more or less 3–5-lobed, both surfaces roughish and green; bur oblong, glabrous or merely puberulent, 6″–9″ long, about 3″ in diameter, its 2 sharp conical-subulate 2-toothed beaks straight or nearly so, equalling or slightly longer than the glabrous spines.

Moist grounds, Ontario to Florida, Michigan, Tennessee and Kansas. Referred, in our first edition, to the Old World *X. strumarium* L., from which it proves to be distinct. Sheep- or clot-bur. Button-bur. Small or lesser burdock. Aug.–Oct.

Xanthium orientàle L. (*X. canadense* Mill.), an Old World tropical species, is naturalized in the West Indies.

Family 46. COMPÓSITAE Adans. Fam. Pl. **2** : 103. 1763.

THISTLE FAMILY.

Herbs, rarely shrubs (some tropical forms trees), with watery or resinous (rarely milky) sap, and opposite alternate or basal exstipulate leaves. Flowers perfect, pistillate, or neutral, or sometimes monoecious or dioecious, borne on a common receptacle, forming heads, subtended by an involucre of few to many bracts arranged in one or more series. Receptacle naked, or with chaffy scales subtending the flowers, smooth, or variously pitted or honeycombed. Calyx-tube completely adnate to the ovary, the limb (pappus) of bristles, awns, teeth, scales, or crown-like, or cup-like, or wanting. Corolla tubular, usually 5-lobed or 5-cleft, the lobes valvate, or that of the marginal flowers of the head expanded into a ligule (ray) ; when the ray-flowers are absent the head is said to be discoid ; when present, radiate ; the tubular flowers form the disk. Stamens usually 5, borne on the corolla and alternate with its lobes, their anthers united into a tube (syngenesious), often appendaged at the apex, sometimes sagittate or tailed at the base ; pollen-grains globose, often rough or prickly. Ovary 1-celled ; ovule 1, anatropous ; style of fertile flowers 2-cleft, its branches variously appendaged, or unappendaged ; stigmas marginal ; style of sterile flowers commonly undivided. Fruit an achene. Seed erect ; endosperm none ; embryo straight ; hypocotyl inferior.

About 800 genera and not less than 10,000 species, of wide geographic distribution. The family is also known as CARDUACEAE, AGGREGATAE, and by the English name of *Asterworts*. In *Kuhnia,* the anthers are distinct, or nearly so.

* Anthers not tailed at the base ; stigmatic lines of the style-branches only at the base, or not extending beyond the middle ; flowers all tubular and perfect, never yellow ; rays none.

Style-branches filiform or subulate, hispidulous ; receptacle naked. Tribe 1. VERNONIEAE.
Style-branches thickened upward, obtuse, papillose. Tribe 2. EUPATORIEAE.

** Anthers tailed at the base, unappendaged at the tip ; heads small ; rays none (except in *Inula* where they are large, yellow). Tribe 4. INULEAE.

*** Anthers not tailed at the base ; stigmatic lines of the style-branches in the perfect flowers extending to the summit ; flowers tubular only, or tubular and radiate, often yellow.

    1. Receptacle naked (see also Nos. 85–88 of Tribe 7).
    a. Bracts of the involucre usually well imbricated.

Style-branches of the perfect flowers flattened, with terminal appendages. Tribe 3. ASTEREAE.
Style-branches truncate, or with hairy tips. Tribe 6. HELENIEAE.
    b. Bracts little imbricated, or not at all ; pappus of soft bristles.
        Tribe 8. SENECIONEAE.

    2. Receptacle chaffy (except in Nos. 85–88).
Bracts of the involucre herbaceous or foliaceous ; not scarious. Tribe 5. HELIANTHEAE.
Bracts of the involucre dry, scarious. Tribe 7. ANTHEMIDEAE.

**** Anthers long-tailed at the base, with elongated appendages at the tip ; heads large ; rays none (in *Centaurea* often with enlarged marginal flowers) ; bracts imbricated.
        Tribe 9. CYNAREAE.

Tribe 1. VERNONIEAE.

Pappus double, the inner of rough capillary bristles, the outer of scales, or short bristles ; heads not glomerate. 1. *Vernonia.*
Pappus a single series of bristles ; heads glomerate, subtended by sessile bracts. 2. *Elephantopus.*

Tribe 2. EUPATORIEAE.

* Achenes 3–5-angled, not ribbed.

Pappus of 5 broad obtuse scales ; aquatic herb with linear whorled leaves. 3. *Sclerolepis.*
Pappus of numerous capillary bristles.
    Involucre of more than 4 bracts ; erect herbs. 4. *Eupatorium.*
    Involucre of 4 bracts ; our species twining herbaceous vines. 5. *Mikania.*
    ** Achenes 8–10-ribbed, or 8–10-striate.

    1. Bracts of the involucre in several series, the outer successively shorter.
Bracts of the involucre strongly striate-nerved ; heads panicled or corymbed in our species.
    Pappus-bristles rough or serrate ; involucral bracts numerous. 6. *Coleosanthus.*
    Pappus-bristles plumose ; involucral bracts few. 7. *Kuhnia.*
Bracts of the involucre faintly striate, if at all ; heads spiked or racemed. 8. *Lacinaria.*
    2. Bracts of the involucre in only 2 or 3 series ; all nearly equal.
        9. *Trilisa.*

Tribe 3. ASTEREAE.

A. Ray-flowers yellow (white in one species of Solidago), or wanting ; plants not dioecious.
    * Pappus of scales, or awns, or wanting, never of numerous capillary bristles.

Heads small, not over 2″ high ; leaves narrowly linear, entire.
Perennial herbs, or shrubs ; all the flowers fertile. 10. *Gutierrezia.*

Annual herbs; disk-flowers sterile.   11. *Amphiachyris.*
Heads large, showy; leaves oblong to lanceolate, spinulose-dentate.   12. *Grindelia.*
** Pappus of either the radiate or tubular flowers, or both, of numerous capillary bristles, with or without
    an outer series of shorter ones, or of scales.

† *Pappus, at least of the disk-flowers, double, an inner series of capillary bristles, and an outer one*
    of scales or short bristles; heads large.

Achenes of the ray-flowers thick, the pappus obsolete, or of a few caducous bristles; achenes of the
    disk-flowers flat.   13. *Heterotheca.*
Achenes of both ray-flowers and disk-flowers flattened.   14. *Chrysopsis.*

†† *Pappus wholly of capillary bristles.*

    1. Heads wholly of disk-flowers (in our species); rays none; leaves narrowly linear.
        a. Perennial herbs; style-tips not exserted; eastern.   15. *Chondrophora.*
        b. Shrubs; style-tips mostly exserted; western.
Involucral bracts gradually narrowed to the tip, keeled, chartaceous.   16. *Chrysothamnus.*
Involucral bracts cuspidate, flat, herbaceous.   17. *Oonopsis.*
    2. Heads with both ray-flowers and disk-flowers (in our species).
        a. None of the leaves cordate; pappus mostly as long as the achene, or longer.
Leaves bristly-serrate or pinnatifid.
Achenes glabrous; pappus-bristles deciduous.   18. *Prionopsis.*
Achenes white-tomentose; pappus-bristles persistent.   19. *Sideranthus.*
Leaves entire, or toothed, not bristly-serrate.
Leaves narrow, coriaceous, evergreen, entire; low western undershrubs.   20. *Stenotus.*
Annual or perennial herbs; leaves not evergreen.
Pappus-bristles unequal; heads loosely panicled; annual.   21. *Isopappus.*
Pappus-bristles equal; heads variously clustered; perennial herbs.
Receptacle alveolate; rays mostly not more numerous than the disk-flowers.
                                22. *Solidago.*
Receptacle fimbrillate; rays more numerous than the disk-flowers; heads corymbose-
paniculate.   23. *Euthamia.*
        b. Lower and basal leaves cordate; pappus-bristles shorter than the achene.
                                24. *Brachychaeta.*

    B. **Ray-flowers present, not yellow in any of our species.**
* Pappus a mere crown, or of a few awns or bristles, or wanting, never of numerous capillary bristles.
                1. Receptacle conic.
Outer bracts of the involucre shorter than the inner; achenes prismatic.   25. *Aphanostephus.*
Bracts of the involucre all about equal; achenes flattened.   26. *Bellis.*
                2. Receptacle flat, or somewhat convex.
Achenes of the disk-flowers compressed, of the ray-flowers 3-angled; perennial herbs with large
    heads.   27. *Townsendia.*
Achenes fusiform, terete; annual herbs with small heads.   28. *Chaetopappa.*
Achenes obovate, flattened, with thickened or winged margins; perennials.   29. *Boltonia.*
        ** Pappus of numerous capillary bristles.

    1. Pappus a single series of capillary bristles; sometimes with an outer series of shorter ones.
        a. Bracts of the involucre in 2 to many series.
Bracts mostly in 2–5 series; teeth or lobes of the leaves not bristle-tipped.
Involucre narrow, its bracts firm; rays few, white.   30. *Sericocarpus.*
Involucre turbinate to hemispheric, bracts mostly thin; rays usually numerous, white to purple.
Disk-flowers turning red, purple, or brown.   31. *Aster.*
Disk-flowers permanently white; leaves subulate to spatulate, crowded, rigid.
                                32. *Leucelene.*
Involucre turbinate; rays not exceeding the mature pappus, or none.   33. *Brachyactis.*
Bracts in many series; teeth or lobes of the leaves bristle-tipped in our species. 34. *Machaeranthera.*
        b. Bracts of the involucre mostly in 1 or 2 series, very narrow; heads mostly long-peduncled.
Rays longer than the diameter of the disk.   35. *Erigeron.*
Rays not longer than the diameter of the disk; heads very small, panicled.   36. *Leptilon.*
        2. Pappus distinctly double, the inner bristles long, the outer shorter.
Leaves lanceolate, ovate, or obovate; rays white.   37. *Doellingeria.*
Leaves narrowly linear; rays violet.   38. *Ionactis.*
    C. **Ray-flowers none; dioecious shrubs; pappus capillary.**
                                39. *Baccharis.*

Tribe 4. INULEAE.

* **Heads small, rays none; flowers white, or whitish.**
                1. Receptacle chaffy.
Receptacle convex; pappus none.   40. *Filago.*
Receptacle subulate; pappus of the inner flowers of rough capillary bristles. 41. *Gifola.*
                2. Receptacle naked.
        a. Pappus, at least that of pistillate flowers, of capillary bristles.
Bracts of the involucre not scarious; plants pubescent or glabrous, not woolly. 42. *Pluchea.*
Bracts of the involucre scarious, mostly white or pink; plants woolly.
Plants dioecious, or polygamo-dioecious.
Pappus-bristles of staminate flowers thickened above.   43. *Antennaria.*
Pappus-bristles not thickened; stem leafy.   44. *Anaphalis.*
Plants not dioecious; flowers all fertile.   45. *Gnaphalium.*
        b. Pappus none; leaves broad, alternate, woolly beneath.
                                46. *Adenocaulon.*
      ** **Heads large; ray-flowers yellow.**   47. *Inula.*

Tribe 5. HELIANTHEAE.

A. Disk-flowers perfect, but sterile.

Achenes thick, short, not flattened ; pappus none.
  Achenes merely subtended by the inner involucral bracts.   48. *Polymnia.*
  Achenes embraced and enclosed by the inner involucral bracts.
    Involucral bracts unarmed.   49. *Melampodium.*
    Involucral bracts armed with hooked prickles.   50. *Acanthospermum.*
Achenes flattened.
  Ray-flowers in 2 or 3 series ; achenes falling away free.   51. *Silphium.*
  Ray-flowers in 1 series ; achenes adnate to 2 or 3 scales of the receptacle, falling away with them.
    Rays large, yellow.
      Leaves opposite and basal.   52. *Chrysogonum.*
      Leaves alternate.
        Pappus none, or of 2 caducous awns.   53. *Berlandiera.*
        Pappus a persistent irregularly cleft crown.   54. *Engelmannia.*
    Rays small, the head appearing discoid.   55. *Parthenium.*

B. Disk-flowers fertile.
* Ray-flowers persistent upon the achenes.

Achenes compressed, or 3-angled ; leaves entire.   56. *Crassina.*
Achenes short, thick ; leaves toothed.   57. *Heliopsis.*

** Ray-flowers deciduous, or none.

† *Pappus a cup, or crown, or of a few teeth, awns, or bristles.*

1. Achenes, at least those of disk-flowers, not compressed (except in *Ratibida* and *Phaethusa*) ;
scales of the receptacle mostly concave, or clasping.
Scales of the receptacle small, awn-like or bristle-like ; rays white, short.   58. *Verbesina.*
Scales of the receptacle broad, larger.
  Involucre of 4 large somewhat united bracts.   59. *Tetragonotheca.*
  Involucre of several or numerous separate bracts.
    Receptacle conic, or columnar.
      Ray-flowers fertile, or wanting ; leaves opposite.   60. *Spilanthes.*
      Ray-flowers sterile, or neutral ; leaves mostly alternate.
        Rays yellow.
          Achenes 4-angled or terete.
            Achenes 4-angled.   61. *Rudbeckia.*
            Achenes terete ; leaves cordate-clasping.   62. *Dracopis.*
          Achenes compressed, winged.   63. *Ratibida.*
        Rays rose-purple or yellow.   64. *Echinacea.*
    Receptacle flat, or convex (low-conic in species of Nos. 66 and 68).
      Low fleshy sea-coast shrubs.   65. *Borrichia.*
      Tall herbs, not fleshy.
        Achenes not much flattened, not winged, nor margined.   66. *Helianthus.*
        Achenes of disk-flowers flattened and margined, or winged.
          Involucre of a few deflexed bracts.   67. *Ridan.*
          Involucre of 2 series or more of appressed or spreading bracts.
            Perennials ; bracts erect or appressed.   68. *Phaethusa.*
            Annuals ; bracts spreading.   69. *Ximenesia.*
2. Achenes very flat ; scales of the receptacle flat, or but slightly concave.
a. Bracts of the involucre all separate.
Pappus of 2 short teeth or awns, or a mere border, or none.   70. *Coreopsis.*
Pappus of 2–6 awns or teeth, upwardly or downwardly barbed or hispid.
  Achenes flat, or angled.   71. *Bidens.*
  Achenes terete ; aquatic, the submerged leaves filiform-dissected.   72. *Megalodonta.*
b. Inner bracts of the involucre united to about the middle.   73. *Thelesperma.*
†† *Pappus of numerous scales.*
Leaves opposite, toothed ; ray-flowers fertile ; rays small.   74. *Galinsoga.*
Leaves alternate, entire.
  Rays large, neutral ; receptacle deeply honey-combed.   75. *Endorima.*
  Rays none ; scales of the receptacle narrow, rigid.   76. *Marshallia.*

Tribe 6. HELENIEAE.

A. Ray-flowers persistent on the achenes, falling away with them, papery.
                                                   77. *Psilostrophe.*

B. Ray-flowers deciduous, or wanting.
* Plants not dotted with oil-glands.
a. Pappus none.   78. *Flaveria.*
b. Pappus present (in all our species), of separate scales or bristles.
1. Bracts of the involucre petal-like, colored, their margins and apices scarious.
Leaves, at least the lower, pinnately parted, or pinnatifid ; rays none ; corolla-lobes of disk-flowers
  ovate.   79. *Hymenopappus.*
Leaves entire ; rays present, or none ; corolla-lobes of disk-flowers linear.   80. *Othake.*
2. Bracts of the involucre herbaceous, not scarious-tipped, nor petal-like, appressed, or spreading.
Receptacle naked.
  Bracts of the involucre appressed.
    Achenes 4-angled, linear or oblong.   81. *Picradeniopsis.*
    Achenes 5–10-ribbed or 5–10-angled, top-shaped.
      Involucral bracts separate to the base.   82. *Tetraneuris.*

Outer involucral bracts connate.   83. *Hymenoxys.*
Bracts of the involucre spreading, or reflexed at maturity.   84. *Helenium.*
Receptacle with bristle-like chaff.   85. *Galliardia.*
    ** Plants dotted with oil-glands, especially the leaves and involucre.
Involucral bracts more or less united; style-branches of the disk-flowers elongated.
  Involucral bracts united at the base.   86. *Boebera.*
  Involucral bracts united high up into a cup.   87. *Thymophylla.*
Involucral bracts separate; style-branches of the disk-flowers very short.   88. *Pectis.*

Tribe 7. ANTHEMIDEAE.
    * Receptacle chaffy.

Achenes flattened; involucre obovoid to campanulate; heads small.   89. *Achillea.*
Achenes terete; involucre hemispheric; heads large.   90. *Anthemis.*
    ** Receptacle not chaffy, naked, or sometimes hairy.
     1. Ray-flowers usually present, sometimes wanting.
Receptacle flat to hemispheric; bracts of the involucre in several series.   91. *Chrysanthemum.*
Receptacle conic to ovoid; bracts in few series.   92. *Matricaria.*
     2. Ray-flowers none; heads small.
Heads corymbed; pappus a short crown; flowers yellow.   93. *Tanacetum.*
Heads racemose, spicate or panicled; pappus none.   94. *Artemisia.*

Tribe 8. SENECIONEAE.

Leaves all basal; heads on scapes.
  Heads solitary; flowers yellow.   95. *Tussilago.*
  Heads corymbed; flowers white or purple.   96. *Petasites.*
Leaves opposite; rays yellow.
  Involucre of several thin herbaceous bracts.   97. *Arnica.* .
  Involucre of 4 or 5 broad fleshy bracts.   98. *Haploesthes.*
Leaves alternate.
  Flowers white, whitish or pinkish; rays none.
    Marginal flowers pistillate; disk-flowers perfect.   99. *Erechtites.*
    Flowers all perfect.
      Involucre of about 5 bracts; sap milky.   100. *Mesadenia.*
      Involucre of about 12 bracts and several smaller outer ones.   101. *Synosma.*
  Flowers yellow; ray-flowers mostly present.   102. *Senecio.*

Tribe 9. CYNAREAE.
    * Achenes inserted on the receptacle by their bases, not oblique.

Receptacle densely bristly.
  Filaments separate.
    Involucral bracts hooked at the tip; leaves not bristly.   103. *Arctium.*
    Involucral bracts not hooked; leaves bristly.
      Pappus-bristles plumose.   104. *Cirsium.*
      Pappus-bristles not plumose.   105. *Carduus.*
  Filaments united below.   106. *Mariana.*
Receptacle fleshy, not bristly.   107. *Onopordon.*
    ** Achenes obliquely inserted on the receptacle.
Heads not subtended by bristly leaves; involucral bracts often bristly.   108. *Centaurea.*
Heads sessile, subtended by bristly leaves.   109. *Cnicus.*

1. VERNONIA Schreb. Gen. Pl. 2 : 541. 1791.

Erect branching perennial herbs, or some tropical species shrubby, with alternate (very rarely oposite), in our species sessile leaves, and discoid cymose-paniculate heads of purple pink or white tubular flowers. Involucre hemispheric, campanulate or oblong-cylindric, its bracts imbricated in several or many series. Receptacle flat, naked. Corolla regular, 5-cleft. Anthers sagittate at the base, not caudate. Style-branches subulate, hispidulous their whole length. Achenes 8–10-ribbed, truncate. Pappus of our species in 2 series, the inner of numerous roughened capillary bristles, the outer of much shorter small scales or stout bristles. [Named after William Vernon, English botanist.]

More than 500 species, of wide distribution in warm-temperate regions, most abundant in South America. Besides the following, several others occur in the southern and southwestern United States. Type species: *Serratula noveboracensis* L.

Heads large, nearly 1' broad; involucral bracts with long filiform tips.   1. *V. crinita.*
Heads smaller, 6" broad or less.
  Involucral bracts with filiform tips.
    Leaves lanceolate, relatively narrow; pappus purple, rarely green.   2. *V. noveboracensis.*
    Leaves oval to lanceolate, relatively broad; pappus yellowish.   3. *V. glauca.*
  Involucral bracts acute or obtuse, not filiform-tipped.
    Leaves linear, 1-nerved.   4. *V. marginata.*
    Leaves lanceolate to oblong-lanceolate.
      Leaves glabrous or merely puberulent beneath.
        Heads loosely cymose.   5. *V. altissima.*
        Heads densely cymose.   6. *V. fasciculata.*
      Leaves tomentose beneath.
        Involucral bracts squarrose, acuminate.   7 *V. Baldwinii.*
        Involucral bracts obtuse or acute, appressed.   8. *V. missurica.*

1. Vernonia crinìta Raf. Great Iron-weed. Fig. 4140.

V. crinata Raf. New Flora N. A. **4** : 77. 1836.

Vernonia arkansana DC. Prodr. **7** : 264. 1838.

Cacalia arkansana Kuntze, Rev. Gen. Pl. 969. 1891.

Stout, glabrate or finely rough-pubescent, 8°–12° high, simple or little branched. Leaves narrowly lanceolate, finely denticulate, acuminate, 3′–12′ long, 3″–12″ wide; heads stout-peduncled, the peduncles thickened above; involucre hemispheric, 9″–12″ broad, 50–80-flowered; bracts green, or the upper reddish, very squarrose, all filiform-subulate from a broader base and equalling the head, the inner ones somewhat wider below; achenes glabrous or hispidulous on the ribs; pappus purplish.

On prairies and along streams, Missouri to Kansas and Texas. Aug.–Oct.

2. Vernonia noveboracénsis (L.) Willd. New York Iron-weed. Flat Top. Fig. 4141.

Serratula noveboracensis L. Sp. Pl. 818. 1753.
V. noveboracensis Willd. Sp. Pl. **3** : 1632. 1804.
C. noveboracensis Kuntze, Rev. Gen. Pl. 323. 1891.
Vernonia noveboracensis tomentosa Britton, Mem. Torr. Club **5** : 311. 1894.

Roughish-pubescent or glabrate, 3°–9° high. Leaves lanceolate or narrowly oblong, serrulate, 3′–10′ long, 5″–12″ wide, acuminate or acute; heads peduncled; involucre hemispheric, 20–40-flowered, 4″–5″ in diameter; bracts brownish-purple or greenish, ovate or ovate-lanceolate, with subulate spreading tips usually twice or three times their own length, or some of the lower linear-subulate, the upper sometimes merely acute; flowers deep purple, rarely white; achenes hispidulous on the ribs; pappus purple or purplish, rarely green.

In moist soil, Massachusetts to Pennsylvania, North Carolina, West Virginia, Mississippi and Missouri. Erroneously recorded west to Minnesota. July–Sept.

3. Vernonia glàuca (L.) Britton. Broad-leaved Iron-weed. Fig. 4142.

Serratula glauca L. Sp. Pl. 818. 1753.
Vernonia noveboracensis var. *latifolia* A. Gray, Syn. Fl. **1** : Part 2, 89. 1884.
Vernonia glauca Britton, Mem. Torr. Club **5** : 311. 1894.

Slender, glabrous or finely puberulent, 2°–5° high. Leaves thin, the lower broadly oval or slightly obovate, sharply serrate, acute or acuminate, 4′–7′ long, 1′–2½′ wide, the upper narrower and more finely toothed; inflorescence loosely branched; heads slender-peduncled, 10–20 flowered; involucre campanulate, 3″–4″ broad; bracts ovate, with filiform tips, appressed; achenes minutely hispidulous; pappus yellowish.

In woods, Pennsylvania and Maryland to Georgia and Alabama. Southern plants previously referred to this species prove to be distinct. Aug.–Sept.

5. Vernonia altissima Nutt. Tall Iron-weed. Fig. 4144.

Vernonia altissima Nutt. Gen. **2** : 134. 1818.
Vernonia maxima Small, Bull. Torr. Club **27** : 280. 1900.

Glabrous or nearly so, 5°–10° high. Leaves thin, lanceolate, sometimes broadly so, usually long-acuminate, finely serrate, 4′–12′ long, ½′–1½′ wide, glabrous on both surfaces, or puberulent beneath; inflorescence at length loosely branched and open; heads short-peduncled or some of them sessile; involucre campanulate or turbinate, 2″–3″ broad, 15–30-flowered; bracts obtuse or mucronate, more or less ciliate, appressed; achenes slightly hispidulous; pappus purplish.

In moist soil, New York to Florida, Illinois, Michigan, Missouri, Kentucky and Louisiana. July–Sept. Included in our first edition in *V. gigantea* (Walt.) Britton, of the Southern States.

4. Vernonia marginàta (Torr.) Raf. James' Iron-weed. Fig. 4143.

Vernonia altissima var. *marginata* Torr. Ann. Lyc. N. Y. **2** : 210. 1827.
Vernonia marginata Raf. Atl. Journ. **1** : 146. 1832.
Vernonia Jamesii T. & G. Fl. N. A. **2** : 58. 1841.
Cacalia marginata Kuntze, Rev. Gen. Pl. 968. 1891.

Glabrous or very nearly so, 1°–3° high. Leaves linear or linear-lanceolate, minutely denticulate, 1-nerved, firm, punctate, 2′–5′ long, 1½″–3″ wide, acuminate; inflorescence rather loose; heads slender-peduncled; involucre campanulate or turbinate, 15–30-flowered, 4″–6″ broad; bracts ovate or oval, acute, mucronate or obtusish, purplish, somewhat pubescent, appressed; achenes nearly glabrous, or somewhat pubescent; pappus brownish.

Prairies, Nebraska and Kansas to Texas and New Mexico. Autumn.

6. Vernonia fasciculàta Michx. Western Iron-weed. Fig. 4145.

Vernonia fasciculata Michx. Fl. Bor. Am. **2** : 94. 1803.
Cacalia fasciculata Kuntze, Rev. Gen. Pl. 970. 1891.

Glabrous, or puberulent above, 2°–6° high. Leaves firm, lanceolate or linear-lanceolate, long-acuminate, 3′–6′ long, 2″–4″ wide, glabrous or nearly so on both surfaces; inflorescence usually compact; heads short-peduncled, or some of them sessile; involucre campanulate, 2″–3″ broad, 20–30-flowered; bracts all appressed, ovate or oval, acute, ciliate, or sometimes pubescent; achenes glabrous, or a little pubescent; pappus purple.

In moist soil or on prairies, Ohio to Minnesota, Nebraska, Kansas and Oklahoma. Southern plants formerly referred to this species prove to be distinct. July–Sept.

Vernonia corymbòsa Schwein., ranging from Manitoba to western Nebraska, has broader leaves but is otherwise similar.

7. Vernonia Baldwínii Torr.   Baldwin's Iron-weed.   Fig. 4146.

V. Baldwinii Torr. Ann. Lyc. N. Y. **2**: 211.   1827.

Cacalia Baldwinii Kuntze, Rev. Gen. Pl. 969.   1891.

Vernonia interior Small, Bull. Torr. Club **27**: 279. 1900.

V. interior Baldwinii Mack. & Bush, Fl. Jackson Co. 190.   1903.

Stout, 2°–5½° high, finely and densely tomen tose-pubescent.   Leaves lanceolate or oblong-lanceolate, acuminate or acute at the apex, sharply serrate, 4′–8′ long, ½′–2′ wide, scabrate above, densely tomentulose beneath; heads stout-peduncled, 15–30-flowered; involucre hemispheric, 3″–4″ broad; bracts ovate, the acute tips recurved or spreading; pappus purple.

In dry soil, Iowa to Missouri, Nebraska, Kansas and Texas.   July–Sept.

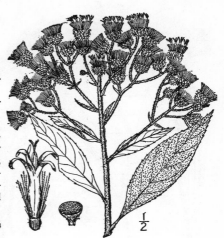

8. Vernonia missùrica Raf.   Drummond's Iron-weed.   Fig. 4147.

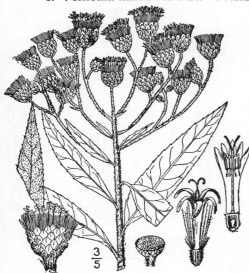

V. missurica Raf. Herb. Raf. 28.   1833.

Vernonia altissima var. *grandiflora* A. Gray, Syn. Fl. **1**: Part 2, 90.   1884.

Vernonia Drummondii Shuttlw.; Werner, Journ. Cinc. Soc. Nat. Hist. **16**: 171. 1894.

V. illinoensis Gleason, Bull. N. Y. Bot. Gard. **4**: 211.   1906.

Stout, densely tomentose, 3°–5° high. Leaves lanceolate to narrowly oblong, acuminate, finely serrate, 3′–6′ long, ½′–1½′ wide, scabrous above, densely pubescent beneath; inflorescence usually compact; heads short-peduncled; involucre hemispheric or short-cylindric, 2″–6″ in diameter, 30–60-flowered; bracts purplish, appressed, ovate, obtuse, acute, or mucronate, more or less floccose-pubescent or ciliate; achenes glabrous or a little pubescent; pappus purplish to tawny.

Prairies, Ontario and Ohio to Illinois, Kentucky, Missouri and Texas.   Autumn.

2. ELEPHÁNTOPUS [Vaill.] L. Sp. Pl. 814.   1753.

Perennial rigid pubescent herbs, with alternate or basal, simple pinnately-veined leaves, and in our species glomerate bracted heads of blue or purple flowers in branching corymbs. Heads discoid, 2–5-flowered.   Involucre compressed, oblong, its chaffy bracts imbricated in about 2 series, the 4 outer bracts shorter.   Bracts of the glomerules large, foliaceous.   Receptacle small, naked.   Corolla nearly regular, 5-lobed, but a little deeper cleft on the inner side.   Achenes 10-ribbed, truncate.   Pappus of rigid persistent awn-like scales or bristles in 1 or 2 rows.   [Greek, Elephant's-foot.]

About 14 species, natives of tropical or warm regions.   Besides the following, another occurs in the southern United States.   Type species: *Elephantopus scaber* L.

Stem and branches leafy.   1. *E. carolinianus.*
Stem scapiform, naked, or with 1 or 2 leaves.
    Leaves oblong or oblanceolate, 9″–2′ wide; heads 4″ long.   2. *E. nudatus.*
    Leaves ovate, oval, or obovate, 2′–4′ wide; heads 6″ long.   3. *E. tomentosus.*

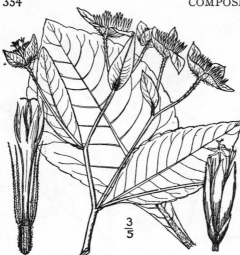

2. Elephantopus nudàtus A. Gray.
Smoothish Elephant's-foot.
Fig. 4149.

Elephantopus nudatus A. Gray, Proc. Am. Acad. **15**: 47. 1880.

Erect, rather stout, appressed-pubescent, or glabrate, 1°–2° high. Leaves oblong or oblanceolate, obtuse at the apex, gradually narrowed at the base, crenate or repand, 2'–10' long, 9''–2' wide, the basal ones usually much larger than those of the stem and branches, or these usually few and bract-like; glomerules 6''–9'' broad; heads about 4'' long; scales of the pappus ovate-triangular, abruptly narrowed into the awn.

In sandy woods, Delaware and Maryland to Florida, west to Arkansas and Louisiana. Aug.–Sept.

1. Elephantopus caroliniànus Willd.
Carolina Elephant's-foot.
Fig. 4148.

Elephantopus carolinianus Willd. Sp. Pl. **3**: 2390. 1804.

Erect, hirsute-pubescent, or glabrate above, corymbosely branched above, 1°–3° high. Leaves oval, ovate, or obovate, thin, the lower rather abruptly narrowed into margined petioles, obtuse, crenate-dentate, 3'–8' long, 2'–4' wide, the upper smaller, narrower and sessile; glomerules, including the bracts, nearly 1' broad; scales of the pappus lanceolate-subulate, gradually narrowed into a long slender awn.

In dry woods, southern New Jersey and Pennsylvania to Florida, Illinois, Kansas and Texas. Aug.–Sept.

3. Elephantopus tomentòsus L.
Woolly Elephant's-foot. Tobacco-weed. Fig. 4150.

Elephantopus tomentosus L. Sp. Pl. 814. 1753.

Erect, villous-pubescent, or sometimes glabrate, 1°–2° high, the stem branching above, leafless or with 1 or 2 leaves. Basal leaves ovate, oval, oblong, or obovate, obtuse, sessile, narrowed at the base, usually silky-pubescent beneath, crenate-dentate, 4'–9' long, 2'–4' wide; glomerules 1'–1½' broad; heads about 6'' long; scales of the pappus triangular-subulate, gradually narrowed into the awn.

In moist soil, Virginia to Florida, west to Kentucky, Arkansas and Louisiana. Called also devil's-grandmother. Aug.–Sept.

3. **SCLERÓLEPIS** Cass. Bull. Soc. Philom. **1816**: 198.   1816.

Slender aquatic herbs, with linear entire verticillate leaves, and solitary (rarely 2–4) discoid peduncled terminal heads of small purplish flowers. Involucre broadly campanulate or
hemispheric. Bracts imbricated in about 2 series. Receptacle conic, naked. Corolla regular,
its tube short, its limb campanulate, 5-lobed. Anthers obtuse at the base. Style-branches
slender, obtuse. Pappus of 5 broad cartilaginous obtuse scales. Achenes 5-angled. [Greek,
hard scale, referring to the pappus.]

A monotypic genus of eastern North America.

1. **Sclerolepis uniflòra** (Walt.) Porter.
   Sclerolepis. Fig. 4151.

Aethulia uniflora Walt. Fl. Car. 195.   1788.
Sparganophorus verticillatus Michx. Fl. Bor. Am. **2**: 98.
   1803.
Sclerolepis verticillata Cass. Dict. **48**: 155.   1827.
Sclerolepis uniflora Porter, Mem. Torr. Club **5**: 311.
   1894.

Perennial; stem simple, decumbent at the base,
erect or ascending, glabrous or slightly pubescent,
1°–2° long, leafy. Leaves sessile, verticillate in
4's–6's, linear, 1-nerved, 4″–12″ long, ½″–1″ wide, or
the submerged ones filiform; head about 5″ broad;
bracts of the involucre linear-oblong, acutish, usually
puberulent.

In shallow ponds and streams, New Hampshire to
Florida. July–Sept.

4. **EUPATÒRIUM** [Tourn.] L. Sp. Pl.
   836.   1753.

Erect, mostly branching, perennial herbs, with opposite or verticillate, or sometimes alternate, often punctate leaves, and in our species cymose-paniculate discoid heads of white, blue
or purple flowers. Involucre oblong, ovoid, campanulate, or hemispheric, the bracts imbricated in 2–several series. Receptacle flat, convex, or conic, naked. Corolla regular, its tube
slender, its limb 5-lobed or 5-toothed. Anthers obtuse and entire at the base, appendiculate
at the apex. Style-branches elongated, flattened, or thickened above, stigmatic at the base.
Achenes 5-angled, truncate. Pappus of numerous capillary usually scabrous bristles arranged
in 1 row. [Named for Mithridates *Eupator, i. e.,* of a noble father.]

Over 500 species, mostly of warm or tropical regions. Besides the following, some 35 others
occur in the southern and western parts of North America. Type species: *Eupatorium canná-
binum* L.

　　　　　* Leaves alternate, pinnatifid into filiform segments.   1. *E. capillifolium.*
** **Leaves petioled, verticillate in 3's–6's, or the upper opposite; involucral bracts in several series.**
Leaves thick, rugose, pubescent; inflorescence depressed.
　　Leaves ovate, acute.　　　　　　　　　　　　　　　　　　　　　2. *E. maculatum.*
　　Leaves lanceolate to ovate-lanceolate, acuminate.　　　　　　　3. *E. Bruneri.*
Leaves thin, nearly glabrous; inflorescence pyramidal.　　　　　　4. *E. purpureum.*
　　　*** **Leaves opposite (rarely in 3's), or the uppermost alternate.**
　　　† *Involucral bracts imbricated in 2 or more series, the outer shorter.*
　　　　　1. Leaves, at least the lower, slender-petioled.　　　　　5. *E. serotinum.*
　　　2. Leaves all sessile, short-petioled or connate-perfoliate.
　　　　a. Leaves not clasping nor connate-perfoliate.
　　　　　§ Leaves narrowed at the base.
Bracts of the involucre acute or cuspidate, scarious-tipped, white.
　　Leaves linear-lanceolate, sparingly toothed, 2″–6″ wide.　　　6. *E. leucolepis.*
　　Leaves oblong or lanceolate, coarsely toothed, ½′–1½′ wide.　7. *E. album.*
Bracts of the involucre obtuse, not scarious, or but slightly so.
　　Leaves linear, crowded, usually entire, obtuse.　　　　　　　　8. *E. hyssopifolium.*
　　Leaves lanceolate, oval, or oblong, usually dentate.
　　　Leaves linear-lanceolate, acute, dentate.　　　　　　　　　　9. *E. Torreyanum.*
　　　Leaves oblong to oval, sharply dentate, obtusish or acute.　10. *E. semiserratum.*
　　　Leaves lanceolate, sparingly dentate, long-acuminate.　　　11. *E. altissimum.*
　　　　　§§ Leaves rounded, obtuse or truncate at the base.
Plant glabrous; leaves lanceolate, long-acuminate.　　　　　　　12. *E. sessilifolium.*
Plants pubescent; leaves ovate or oblong, acute or obtuse.
　　Leaves ovate-oblong, rounded or narrowed at the base, usually obtuse.　13. *E. verbenaefolium.*
　　Leaves broadly ovate, crenate-dentate, mostly truncate at the base, obtusish. 14. *E. rotundifolium.*
　　Leaves ovate, dentate, acute.　　　　　　　　　　　　　　　　15. *E. pubescens.*
　　　　b. Leaves clasping or connate-perfoliate at the base.
Leaves connate-perfoliate; involucral bracts acute.　　　　　　　16. *E. perfoliatum.*
Leaves merely clasping; involucral bracts obtuse.　　　　　　　　17. *E. resinosum.*
　　†† *Involucral bracts in 1 or 2 series, all equal or nearly so.*

Receptacle flat.
  Flowers white; leaves ovate.
    Leaves thin, 2′–5′ long, sharply dentate, acuminate.   18. *E. urticaefolium.*
    Leaves firm, 1′–2′ long, obtusely dentate, acute or obtusish.   19. *E. aromaticum.*
  Flowers pink to purple; leaves deltoid-ovate.   20. *E. incarnatum.*
Receptacle conic; flowers blue or violet; leaves petioled.   21. *E. coelestinum.*

$\frac{2}{3}$

1. Eupatorium capillifòlium (Lam.) Small. Dog-fennel. Hog-weed. Fig. 4152.

Artemisia capillifolia Lam. Encycl. 1 : 267. 1783.
Eupatorium foeniculoides Walt. Fl. Car. 199. 1788.
E. foeniculaceum Willd. Sp. Pl. 3 : 1750. 1804.
E. capillifolium Small, Mem. Torr. Club 5 : 311. 1894.

Erect, paniculately much branched, with the aspect of an *Artemisia*, the stem finely pubescent, 4°–10° high. Leaves crowded, glabrous or nearly so, alternate, pinnatifid into filiform segments, the lower petioled, the upper sessile; heads very numerous, about 1½″ high, short-pedicelled, racemose-paniculate, 3–6-flowered; bracts of the involucre in about 2 series, linear, cuspidate, narrowly scarious-margined, glabrous; flowers greenish-white.

In fields, Virginia to Florida. In ballast, at Philadelphia. Also in the West Indies. Sept.

2. Eupatorium maculàtum L. Spotted Joe-Pye Weed. Fig. 4153.

E. maculatum L. Amoen. Acad. 4 : 288. 1755.
Eupatorium purpureum var. *maculatum* Darl. Fl. Cest. 453. 1837.
Eupatorium maculatum amoenum Britton, Mem. Torr. Club 5 : 312. 1894.

Similar to the two following species, scabrous or pubescent, often densely so, 2°–6° high. Stem usually striate, often rough and spotted with purple; leaves thick, ovate or ovate-lanceolate, coarsely dentate, verticillate in 3′s–5′s, or the upper ones opposite; inflorescence depressed, cymose-paniculate; pedicels and outer scales of the involucre pubescent; flowers pink or purple.

In moist soil, Newfoundland to New York, Kentucky, British Columbia, Kansas and New Mexico. Spotted boneset. Perhaps to be regarded as a race of *E. purpureum.* Aug.–Sept.

$\frac{3}{5}$

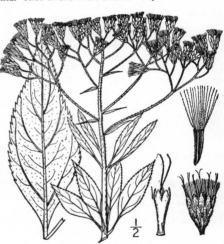

$\frac{1}{2}$

3. Eupatorium Brùneri A. Gray. Bruner's Trumpet-weed. Fig. 4154.

Eupatorium Bruneri A. Gray, Syn. Fl. 1² : 96. 1884.

Eupatorium Rydbergi Britton, Manual 921. 1901.

Stem tall, pubescent, often densely so, at least above. Leaves verticillate in 3′s–5′s, rather slender-petioled, lanceolate, serrate, acuminate at the apex, narrowed at the base, scabrous above, finely densely pubescent and reticulate-veined beneath, 4′–6′ long, ¾′–2′ wide; inflorescence depressed or subpyramidal; outer bracts of the cylindric involucre pubescent; flowers pink or purple.

In moist soil, South Dakota to Wyoming, Nebraska and Colorado. Apparently erroneously recorded from Iowa. July–Sept.

4. Eupatorium purpùreum L. Joe-Pye or Trumpet-weed. Gravel-root. Tall or Purple Boneset. Fig. 4155.

Eupatorium purpureum L. Sp. Pl. 838. 1753.
E. falcatum Michx. Fl. Bor. Am. **2**: 99. 1803.
Eupatorium purpureum var. *angustifolium* T. & G. Fl. N. A. **2**: 82. 1841.
Eupatorium purpureum falcatum Britton, Mem. Torr. Club **5**: 312. 1894.

Glabrous or sparingly pubescent, simple or branched at the summit, 3°–10° high. Stem green or purple, terete or striate, usually smooth; leaves thin, verticillate in 3's–6's, ovate, oval, ovate-lanceolate or narrowly lanceolate, petioled, acuminate, serrate, 4′–12′ long, 6″–3′ wide, glabrous or slightly pubescent along the veins on the lower surface; inflorescence usually elongated; heads very numerous; involucre cylindric, its bracts pink, oblong, obtuse, imbricated in 4 or 5 series, the outer shorter; flowers pink or purple, occasionally white.

In moist soil, New Brunswick to Manitoba, Florida and Texas. Kidney-root. Skunk-weed. Indian gravel-root. Marsh-milk weed. Nigger-weed. Quillwort. Motherwort. King- or queen-of-the-meadow. Aug.–Sept.

Eupatorium trifoliàtum L. has the teeth of the leaves bluntly apiculate, but otherwise closely resembles *E. purpureum* and may not be specifically distinct.

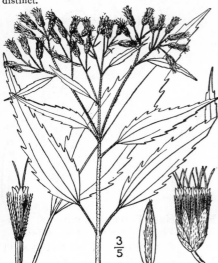

6. Eupatorium leucólepis T. & G. White-bracted Thoroughwort. Justice-weed. Fig. 4157.

E. leucolepis T. & G. Fl. N. A. **2**: 84. 1841.

Slender, puberulent, branched above, 1°–2° high. Leaves opposite, sessile, linear, oblong or oblong-lanceolate, glaucous green, rough on both sides, thick, blunt-pointed, sparingly serrate, or the upper entire, 1′–3′ long, 2″–5″ wide, obscurely 3-nerved and narrowed at the base; inflorescence cymose-paniculate; heads 3″–4″ high, about 5-flowered; bracts of the narrow involucre imbricated in about 3 series, white, lanceolate, acute, densely canescent, the outer shorter; flowers white.

In moist places, Massachusetts and Long Island to Florida, Georgia and Louisiana. Aug.–Sept.

5. Eupatorium serótinum Michx. Late-flowering Thoroughwort. Fig. 4156.

E. serotinum Michx. Fl. Bor. Am. **2**: 100. 1803.

Much branched, finely and densely pubescent, or glabrate below, 4°–8° high. Leaves all slender-petioled, lanceolate or ovate-lanceolate, acuminate, sharply serrate, 3′–6′ long, ½′–2′ wide, 5-nerved at the base, the lower opposite, the upper alternate; heads very numerous, the inflorescence broadly cymose; heads 7–15-flowered, 2″–3″ high; involucre campanulate, its bracts pubescent, linear-oblong, obtuse or truncate, imbricated in 2 or 3 series, the outer shorter; flowers white.

In moist soil, Delaware to Florida, Minnesota, Iowa, Kansas and Texas. Sept.–Nov.

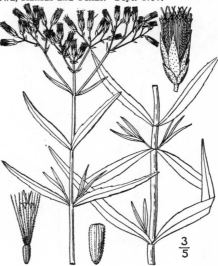

$\frac{3}{5}$

7. Eupatorium álbum L. White Thoroughwort. Fig. 4158.

Eupatorium album L. Mant. 111. 1767.
Eupatorium glandulosum Michx. Fl. Bor. Am. 2: 98. 1803.
Eupatorium album subvenosum A. Gray, Syn. Fl. 1: Part 2, 98. 1884.

Pubescent with spreading hairs, branched above, 1°–3° high. Leaves opposite, sessile or nearly so, oblong or oblong-lanceolate, obtuse, obtusish or the upper acute at the apex, narrowed at the base, coarsely or rather finely serrate, 1'–4' long, ½'–1' wide, rather thick, minutely scabrous above, more or less pubescent beneath; inflorescence cymose-paniculate; heads numerous, 4''–5'' long, 5–7-flowered; involucre narrow, its bracts bright white, linear, cuspidate, imbricated in 3–4 series, the outer short and usually pubescent, the inner much longer, glabrous and shining; flowers white.

In sandy soil, Long Island to Florida, Tennessee, Arkansas and Louisiana. Ascends to 3000 ft. in Virginia. Aug.–Sept.

8. Eupatorium hyssopifòlium L. Hyssopleaved Thoroughwort. Fig. 4159.

Eupatorium hyssopifolium L. Sp. Pl. 836. 1753.
E. linearifolium Walt. Fl. Car. 199. 1788.

Roughish-puberulent, densely corymbosely branched above, bushy, 1°–2° high. Leaves linear, opposite, and fascicled in the axils of the stem, or on short axillary branches, entire or very nearly so, ½'–2' long, 1''–2'' wide, firm, obtuse at the apex, narrowed at the base; inflorescence densely cymose-paniculate; heads 3''–4'' long, about 5-flowered; involucre campanulate, its bracts linear or linear-oblong, obtuse or truncate, sometime apiculate, puberulent, imbricated in about 3 series, the outer shorter; flowers white.

In dry fields, Massachusetts to Florida and Texas. Justice-weed. Leaf-margins usually revolute. Aug.–Sept. A plant from the coast of Maryland with very narrow leaves, closely approaches *Eupatorium lecheaefolium* Greene, from Florida.

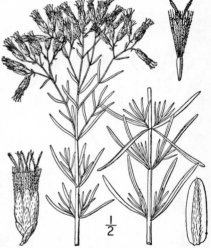

$\frac{1}{2}$

9. Eupatorium Torreyànum Short. Torrey's Thoroughwort. Fig. 4160.

Eupatorium Torreyanum Short, 2nd Suppl. Cat. Pl. Ky. 5. 1836.
Eupatorium hyssopifolium var. *laciniatum* A. Gray, Syn. Fl. 1: Part 2, 98. 1884.

Rootstock tuberous-thickened; stem usually densely puberulent, mostly branched above, 1½°–3° high. Leaves opposite, or sometimes in 3's, commonly with short leafy branches in their axils, often drooping, lanceolate, acute at the apex, narrowed at the base, sessile, 1'–3' long, 2''–6'' wide, usually distinctly 3-nerved, distantly serrate, the upper and those of the branches much smaller, linear, entire; inflorescence mostly loose; heads about 3'' high, generally 5-flowered; bracts of the campanulate involucre linear-oblong, obtuse, pubescent, or puberulent, imbricated in about 3 series, the outer shorter; flowers white.

In dry soil, Pennsylvania to North Carolina, Kentucky, Florida and Texas. July–Sept.

$\frac{1}{2}$

10. Eupatorium semiserràtum DC.
Small-flowered Thoroughwort.
Fig. 4161.

E. semiserratum DC. Prodr. **5**: 177. 1836.

Pubescent or puberulent, loosely branched above, 2°–3° high. Leaves rather thin, short-petioled, oblong-lanceolate to spatulate, acute or obtusish at the apex, narrowed at the base, sharply serrate, at least above the middle, 2′–4′ long, 5″–12″ wide, 3-nerved, usually with short branches in their axils; inflorescence cymose-paniculate; heads 2″–3″ high, about 5-flowered; involucre campanulate, its bracts linear-oblong, obtuse, canescent, imbricated in 2 or 3 series, the outer shorter, flowers white.

In dry soil, Virginia to Florida, Missouri, Arkansas and Texas. Aug.–Sept.

½

11. Eupatorium altíssimum L.   Tall Thoroughwort.   Fig. 4162.

Eupatorium altissimum L. Sp. Pl. 837. 1753.

Densely and finely pubescent, corymbosely much-branched above, 4°–8° high. Leaves lanceolate, acuminate at the apex, tapering below into a short petiole, roughish, rather thick, sparingly dentate above the middle, or some of them entire, strongly 3-ribbed, 2′–5′ long, 5″–12″ wide; inflorescence densely cymose-paniculate; heads about 5-flowered, 3″–4″ high; involucre campanulate, its bracts oblong, obtuse or truncate, densely pubescent, imbricated in about 3 series, the outer shorter; flowers white.

In dry open places, Pennsylvania to North Carolina, Alabama, Illinois, Minnesota, Nebraska and Texas. Sept.–Oct.

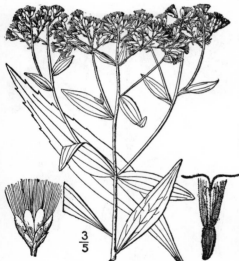

3/5

12. Eupatorium sessilifòlium L.   Upland or Bastard Boneset.   Fig. 4163.

Eupatorium sessilifolium L. Sp. Pl. 837. 1753.

Glabrous throughout, or pubescent above, branched near the summit, 2°–6° high. Leaves opposite, or the lower rarely in 3′s, closely sessile, lanceolate or ovate-lanceolate, long-acuminate, thin, sharply serrate, 3′–6′ long, ½′–1½′ wide, inflorescence cymose-paniculate; heads 4″–5″ high, about 5-flowered; involucre campanulate, its bracts linear-oblong, imbricated in about 3 series, the inner obtuse, the outer acutish, shorter; flowers white.

In dry woods, Vermont to Massachusetts, Pennsylvania, Georgia, Alabama and Missouri. Aug.–Oct. A related plant, described under the name *Eupatorium sessilifòlium Brittoniànum* Porter, has leaves ovate or oblong-ovate, sparingly and finely serrate, firm, rounded at the base, acute, the upper 1′–2′ long, 6″–9″ wide, pinnately veined, dark green; cymes dense, their branches puberulent.

3/5

It is known only from Budd's Lake, N. J.

13. Eupatorium verbenaefòlium Michx.   Rough or Vervain Thoroughwort.
Fig. 4164.

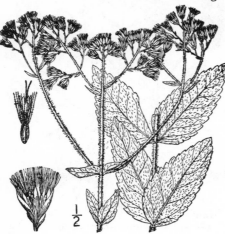

$\frac{1}{2}$

Eupatorium pilosum Walt. Fl. Car. 199. 1788. ?
Eupatorium verbenaefolium Michx. Fl. Bor.
Am. **2**: 98. 1803.
E. *teucriifolium* Willd. Sp. Pl. **3**: 1753. 1804.
E. *verbenaefolium Saundersi* Porter; Britton,
Man. 923. 1901.

Rough-pubescent, slender, 3°–8° high,
branched at the summit. Leaves opposite,
or occasionally in 3's, ovate-oblong, usually
obtuse or blunt-pointed, closely sessile or
rarely short-petioled, rounded or narrowed
at the base, crenate-dentate, or incised,
2′–4′ long, ½′–1′ wide, the upper pairs distant and small; inflorescence cymose-
paniculate; heads 5-flowered, about 3″ high;
involucre campanulate, its bracts linear-
lanceolate, acute, imbricated in about 3
series, densely pubescent, the outer shorter;
flowers white.

In moist soil, Massachusetts to Pennsylvania, West Virginia, Florida and Louisiana,
mostly near the coast. Called also wild hoarhound. July–Sept.

14. Eupatorium rotundifòlium L.
Round-leaved Thoroughwort.
Wild Hoarhound.   Fig. 4165.

E. *rotundifolium* L. Sp. Pl. 837. 1753.

Pubescent, branched at the summit, 1°–3°
high. Leaves opposite, sessile, broadly ovate,
often as wide as long, acutish or obtuse, truncate to subcordate at the base, coarsely dentate-
crenate, 1′–2′ long, ascending; inflorescence
cymose-paniculate; heads about 5-flowered,
2″–3″ high; involucre campanulate, its bracts
linear-oblong, acutish, densely pubescent, imbricated in about 3 series, the outer shorter;
flowers white.

In dry soil, Rhode Island to Pennsylvania,
Florida, Kentucky, Arkansas and Texas. Erroneously reported from Canada. July–Sept.

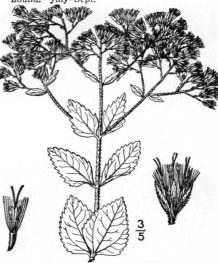

$\frac{3}{5}$

15. Eupatorium pubéscens Muhl.   Hairy
Thoroughwort.   Fig. 4166.

E. *pubescens* Muhl.; Willd. Sp. Pl. **3**: 1755. 1804.

Eupatorium rotundifolium var. *ovatum* Torr.; DC.
Prodr. **5**: 178. 1836.

Puberulent or pubescent, branched above,
usually taller than the preceding species. Leaves
opposite, ovate, sessile, often twice as long as
wide, acute at the apex, rounded or subtruncate
at the base, coarsely dentate with acute or acutish teeth, or the lower incised; inflorescence
cymose-paniculate, sometimes 10′ broad; heads
5–8-flowered, 2″–3″ high; involucre campanulate, its linear-oblong bracts in 2 or 3 series,
the outer shorter; flowers white.

In dry soil, Maine to Pennsylvania, South Carolina, West Virginia and Tennessee. Recorded
from Florida. Ascends to 3000 ft. in Virginia.
July–Sept.

$\frac{3}{5}$

16. Eupatorium perfoliàtum L.
Common Thoroughwort. Boneset. Indian Sage. Fig. 4167.

E. perfoliatum L. Sp. Pl. 838. 1753.
Eupatorium truncatum Muhl.; Willd. Sp.
Pl. **3**: 1751. 1804.
Eupatorium perfoliatum truncatum A. Gray,
Syn. Fl. 1 : Part 2, 99. 1804.

Pubescent, stout, branched above, 2°–5°
high. Leaves opposite, or rarely in 3's,
connate-perfoliate, or the upper, rarely
all, truncate and separated at the base,
divaricate, lanceolate, long-acuminate
with a slender apex, finely crenate-serrate, rugose and pubescent beneath, 4′–8′
long, 1′–1½′ wide; heads crowded, 10–16-
flowered, 2″–3″ high; involucre campanulate, its bracts lanceolate, acutish,
in 2 or 3 series, pubescent, the outer
shorter; flowers white, rarely blue.

In wet places, Nova Scotia and New
Brunswick to Manitoba, Florida, Nebraska
and Texas. Called also ague-weed, cross-
wort, wild sage, thorough-wax, thorough-
grow, thorough-stem. July–Sept.

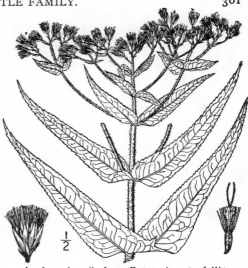

A plant described as *Eupatorium perfoliàtum
cuneàtum* Engelm., with leaves smaller, narrowed
and separated at the base, from Arkansas and Missouri, is probably a hybrid with *E. serotinum*.

17. Eupatorium resinòsum Torr. Resin
Boneset. Fig. 4168.

E. resinosum Torr. DC. Prodr. **5**: 176. 1836.

Slender, finely pubescent and resinous, 2°–3°
high, branched at the summit, the inflorescence
comparatively small, 3′–4′ broad. Leaves opposite, closely sessile, clasping, or slightly connate-perfoliate at the base, linear-lanceolate,
long-acuminate, sharply serrate, 3′–6′ long,
3″–6″ wide, roughish above, canescent beneath;
heads 10–15-flowered; involucre campanulate,
about 2″ high, its bracts oblong, obtuse, in 2
or 3 series, the outer shorter; flowers white.

In wet pine-barrens of New Jersey. Aug.–Sept.

18. Eupatorium urticaefòlium Reichard.
White Snake-root. Fig. 4169.

E. urticaefolium Reichard, Syst. **3**: 719.
1780.
E. ageratoides L. f. Suppl. 355. 1781.
E. boreale Greene, Rhodora **3**: 83. 1901.

Glabrous or villous and somewhat viscid, much branched, 1°–4° high. Leaves
opposite, ovate, thin, acuminate at apex,
rounded, truncate or cordate at the base,
or abruptly narrowed into the slender
petiole, coarsely and usually sharply dentate-serrate, sometimes crenate, 3′–6′
long, 1′–3′ wide, 3-nerved and veiny;
petioles ½–2½′ long; inflorescence rather
loose, ample; heads 10–30-flowered; receptacle flat; involucre narrowly campanulate, about 2″ high, its bracts linear,
acute or acuminate, in 1 or 2 series,
equal or nearly so; flowers bright white.

In rich woods, New Brunswick to Florida, Ontario, Nebraska and Louisiana. Indian sanicle. Richweed. Stevia (Wis.).
White sanicle. Deerwort boneset. July–Nov.

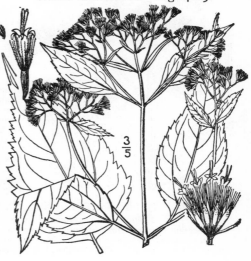

$\frac{3}{5}$

20. Eupatorium incarnàtum Walt. Pink Thoroughwort. Fig. 4171.

E. incarnatum Walt. Fl. Car. 200. 1788.

Minutely pubescent, 2°–4° high, sometimes reclining, often diffusely branched. Leaves opposite, deltoid-ovate, $\frac{3}{4}'$–$2\frac{1}{4}'$ long, long-pointed, rather coarsely blunt-serrate or crenate-serrate, truncate or cordate at the base, the petioles slender, often as long as the blades; heads about 20-flowered; involucres campanulate, about 2″ high, its bracts few, narrowly linear, slightly ribbed when dry, acute; corolla pink or pale purple.

In woods and sandy soil, Virginia to Missouri, Florida and Mexico. Aug.–Oct.

19. Eupatorium aromáticum L. Smaller White Snake-root. Fig. 4170.

Eupatorium aromaticum L. Sp. Pl. 839. 1753.

Puberulent or glabrate, slender, branched at the summit, 1°–2° high. Leaves opposite, petioled, firm, obtuse or acutish at the apex, rounded, cordate or sometimes narrowed at the base, crenate-dentate, $1\frac{1}{2}'$–3′ long, 9″–18″ wide, 3-nerved and veiny; petioles 2″–8″ long; inflorescence usually compact; heads 10–25-flowered; receptacle flat; involucre campanulate, $1\frac{1}{2}$″–2″ high, its bracts linear, generally obtuse, in 1 or 2 series; flowers white.

In dry soil, Massachusetts to Florida, Pennsylvania, West Virginia, Tennessee and Mississippi. Blooms rather later than the preceding species where the two grow together, near New York. Called also poolroot, poolwort, and wild hoarhound. Aug.–Oct.

$\frac{3}{5}$

21. Eupatorium coelestìnum L. Mist-flower. Blue Boneset. Fig. 4172.

Eupatorium coelestinum L. Sp. Pl. 838. 1753.
Conoclinium coelestinum DC. Prodr. 5 : 135. 1836.

Pubescent or puberulent, branched, 1°–3° high. Leaves opposite, petioled, ovate, obtuse or acute at the apex, truncate at the base, or abruptly contracted into the petiole, crenate-dentate, $1\frac{1}{2}'$–3′ long, 9″–18″ wide; inflorescence cymose-corymbose, rather compact; heads 2″–3″ high; involucre broadly campanulate, its bracts linear-lanceolate, acuminate, equal or nearly so, in 1 or 2 series; receptacle conic; flowers blue or violet.

In moist soil, New Jersey to Florida, Illinois, Michigan, Kansas, Arkansas and Texas. Also in Cuba. Aug.–Oct.

$\frac{1}{2}$

5. MIKÀNIA Willd. Sp. Pl. **3**: 1742. 1804.

[WILLUGHBAEA Neck. Elem. 1 : 82. Hyponym. 1790.]

Herbaceous twining vines, or some tropical species erect shrubs, with opposite, petioled leaves, and discoid, mostly cymose-paniculate heads of whitish flowers. Heads 4-flowered. Involucre oblong, of 4 slightly unequal narrow bracts. Receptacle small, naked. Corolla regular, its tube slender, the limb campanulate, 5-cleft. Anthers entire and obtuse at the base. Style-branches elongated, acutish. Achenes truncate, 5-angled. Pappus of numerous capillary bristles. [In honor of Joseph Gottfried Mikan, 1743–1814, professor at Prague.]

About 150 species, natives of America. Besides the following, two others occur in the southern United States. Type species: *Mikania hastata* (L.) Willd.

1. Mikania scándens (L.) Willd. Climbing Hempweed or Boneset. Fig. 4173.

Eupatorium scandens L. Sp. Pl. 836. 1753.
M. scandens Willd. Sp. Pl. **3**: 1743. 1804.
Willoughbya scandens Kuntze, Rev. Gen. Pl. 371. 1891.

Glabrous or nearly so, twining over bushes, 5°–15° long. Leaves ovate or hastate, deeply cordate at the base with the lobes rounded or truncate, acuminate at the apex, repand or obtusely dentate, 2'–4' long, 1'–2' wide; petioles slender, shorter than the blades; heads in compound clusters borne at the ends of the branches; involucre about 3" long, its bracts acuminate or apiculate; flowers white or pink; achenes resinous.

In swamps and moist soil, Maine to western Ontario, Florida and Texas. West Indies and South America. July–Sept.

6. COLEOSÁNTHUS Cass. Dict. 10: 36. 1817.

[Brickellia Ell. Bot. S. C. & Ga. **2**: 290. 1824.]

Herbs or shrubs, with opposite or alternate leaves, and discoid heads of white yellowish or pink flowers, in panicles or cymes, or rarely solitary. Involucre campanulate or oblong, its bracts striate, imbricated in several series, the exterior ones successively smaller. Receptacle flat or convex, naked. Corolla regular, 5-lobed. Anthers obtuse at the base. Style-branches long, obtuse. Achenes 10-striate or 10-ribbed. Pappus a single row of numerous rough or serrate bristles. [Greek, sheath-flower.]

A genus of about 70 species. Besides the following, some 45 others occur in the southern and western United States. Type species: *Coleosanthus Cavanillèsii* Cass.

1. Coleosanthus grandiflòrus (Hook.) Kuntze. Large-flowered Thoroughwort. Fig. 4174.

Eupatorium grandiflorum Hook. Fl. Bor. Am. **2**: 26. 1834.
Brickellia grandiflora Nutt. Trans. Am. Phil. Soc. (II.) **7**: 287. 1841.
Coleosanthus grandiflorus Kuntze, Rev. Gen. Pl. 328. 1891.
C. umbellatus Greene, Pittonia **4**: 238. 1901.

Erect, glabrous or puberulent, much branched, 2°–3° high. Leaves slender-petioled, deltoid-ovate, cordate at the base, acuminate at the apex, coarsely crenate-dentate, 2'–4' long, 1'–2' wide; petioles shorter than the leaves; inflorescence cymose-paniculate; heads short-peduncled, 6"–7" long; peduncles pubescent; involucre campanulate, 30–45-flowered, the outer bracts ovate, pubescent, usually abruptly acuminate or subulate-tipped, the inner linear, glabrous, striate, obtuse or acute; pappus bristles scabrous.

Montana to Wyoming, Missouri, New Mexico, Washington and Arizona. Aug.–Sept. Tassel-flower.

7. KÙHNIA L. Sp. Pl. Ed. 2, 1662. 1763.

Perennial herbs, with alternate punctate resinous-dotted leaves, and discoid heads of white or purplish flowers in terminal cymose corymbs. Involucre turbinate-campanulate, its bracts striate, imbricated in several series, the outer shorter. Corolla regular, the tube slender, the limb 5-lobed. Anthers obtuse and entire at the base, nearly or quite separate. Style-branches slender, obtusish. Achenes 10-striate. Pappus a single row of numerous very plumose bristles. [Named for Dr. Adam Kuhn, of Philadelphia, a pupil of Linnaeus.]

About 8 species, native of North America and Mexico. Type species: *Kuhnia eupatorioides* L.

Puberulent; leaves sparingly dentate, or entire; heads loosely clustered, 4″–5″ high.
1. *K. eupatorioides.*
Pubescent or tomentulose; leaves sharply serrate; heads densely clustered, 6″–8″ high.
2. *K. glutinosa.*

1. Kuhnia eupatorioìdes L. False Boneset. Fig. 4175.

Kuhnia eupatorioides L. Sp. Pl. Ed. 2, 1662. 1763.

Eupatorium alternifolium Ard. Spec. Bot. **2**: 40. *pl. 20.* 1764.

Erect, puberulent and resinous, 1°–3° high, branched above. Leaves lanceolate or linear-lanceolate, acute or obtusish at the apex, narrowed at the base, sparingly dentate, or entire, the upper sessile, the lower usually short-petioled; heads several or numerous, peduncled, 4″–5″ high, loosely clustered; outer bracts of the involucre lanceolate or ovate-lanceolate, acuminate, the inner much longer, linear, cuspidate; pappus tawny, or sometimes nearly white.

In dry soil, New Jersey to Georgia, Ohio, Minnesota, West Virginia and Texas. Ascends to 3300 ft. in West Virginia. Aug.–Sept.

2. Kuhnia glutinòsa Ell. Prairie False Boneset. Fig. 4176.

Kuhnia glutinosa Ell. Bot. S. C. & Ga. **2**: 292. 1821–24.
Kuhnia suaveolens Fresen. Ind. Sem. Francf. 1838.
Kuhnia eupatorioides var. *corymbulosa* T. & G. Fl. N. A. **2**: 78. 1841.

Stouter and often taller than the preceding species, corymbosely or paniculately branched, pubescent or tomentulose, somewhat viscid. Leaves all sessile, lanceolate to ovate-lanceolate, usually sharply serrate with distinct teeth, veiny, 1′–3′ long, 3″–10″ wide, or those of the branches linear-lanceolate and entire; heads numerous, 6″–8″ high, densely clustered in the cymes, their peduncles mostly short; inner bracts of the involucre lanceolate, acuminate; pappus tawny or brown.

In dry soil, Illinois to North Dakota, Colorado, Alabama and Texas. Perhaps a race of the preceding species. Aug.–Oct.

Kuhnia Hitchcóckii A. Nelson, a little known species of Kansas, differs by having very small linear leaves, at least on the upper part of the plant.

8. LACINÀRIA Hill, Veg. Syst. 4: 49. *pl. 46.* 1762.

[LIATRIS Schreb. Gen. Pl. 542. 1791.]

Erect perennial herbs, usually from a globular tuber, simple or little branched, with alternate, entire, narrow 1–5-nerved leaves, and spicate or racemose discoid heads of rose-purple or white flowers. Involucre oblong, ovoid or subhemispheric, its bracts imbricated in several series, the outer shorter. Receptacle flat, or slightly convex, naked. Corolla regular, its tube slender, its limb 5-lobed or 5-cleft. Anthers obtuse at the base. Style-branches elongated, obtuse or flattened at the apex. Achenes 10-ribbed, slender, tapering to the base. Pappus of 1 or 2 series of slender barbellate or plumose bristles. [Latin, fringed, from the appearance of the heads.]

About 35 species, natives of eastern and central North America, known as Blazing Star, or Button Snakeroot from the globular tubers. Type species: *Serratula squarrosa* L.

* Bracts of the involucre acute, acuminate or mucronate.

Involucre cylindric, or turbinate, 15–60-flowered, its base rounded.

Bracts with lanceolate spreading rigid tips.	1. *L. squarrosa.*
Bracts mucronate, closely appressed.	2. *L. cylindrica.*

Involucre oblong, or narrowly campanulate, 3–6-flowered.

Inner bracts with prolonged petaloid tips.	3. *L. elegans.*

Bracts all acute, mucronate or acuminate.

Bracts appressed; pappus-bristles very plumose.

Leaves 1″–2″ wide; spike usually leafy below.	4. *L. punctata.*
Leaves less than 1″ wide; spike mostly naked.	5. *L. acidota.*
Tips of the bracts spreading; pappus-bristles barbellate.	6. *L. pycnostachya.*

** Bracts of the involucre rounded, obtuse or acutish.

Involucre hemispheric, ½′–1′ broad, 15–45-flowered; heads peduncled.   7. *L. scariosa.*

Involucre oblong, 2″–4″ broad, 5–15-flowered.

Bracts obtuse, rounded.

Involucre rounded at base; bracts usually not punctate; heads mostly sessile.

        8. *L. spicata.*

Involucre narrowed at base; bracts usually punctate; heads peduncled.

Leaves, even the lower, narrowly linear; involucre narrowly obovoid, 5″–6″ high.

        9. *L. graminifolia.*

Lower leaves linear-oblong; involucre broadly obovoid, 6″–7″ high.   10. *L. pilosa.*

Bracts acutish, punctate.   11. *L. Smallii.*

1. Lacinaria squarròsa (L.) Hill. Scaly Blazing Star. Colic-root. Fig. 4177.

Serratula squarrosa L. Sp. Pl. 818. 1753.
Lacinaria squarrosa Hill, Hort. Kew. 70. 1769.
Liatris squarrosa Willd. Sp. Pl. 3: 1634. 1804.
Liatris intermedia Lindl. Bot. Reg. *pl. 948.* 1825.
Lacinaria squarrosa intermedia Porter, Mem. Torr. Club 5: 314. 1894.
Liatris squarrosa var. *intermedia* DC. Prodr. 5: 129. 1836.

Usually stout, ½°–2° high, pubescent or glabrous. Leaves narrowly linear, rigid, sparingly punctate, 3′–6′ long, 1″–2½″ wide; heads sessile or short-peduncled, 15–60-flowered, usually few, or sometimes solitary. ½′–1½′ long, 4″–8″ thick; bracts of the involucre imbricated in 5–7 series, lanceolate, rigid, acuminate, glabrous or pubescent, their tips more or less spreading when old; flowers purple; pappus very plumose.

In dry soil, western Ontario to Pennsylvania, Virginia, Florida, South Dakota, Nebraska and Texas. Called also rattlesnake-master. Races differ in pubescence and in size of the heads. June–Sept.

2. Lacinaria cylindràcea (Michx.) Kuntze. Cylindric Blazing Star. Fig. 4178.

Liatris cylindracea Michx. Fl. Bor. Am. 2: 93. 1803.
Liatris graminifolia Willd. Sp. Pl. 3: 1636. 1804.
Lacinaria cylindracea Kuntze, Rev. Gen. Pl. 349. 1891.

Glabrous or nearly so, stout, 1°–1½° high, sometimes branched above. Leaves narrowly linear, rigid, scarcely punctate, 3′–7′ long, 1″–2″ wide; heads several or numerous (rarely solitary), peduncled, or the lower sessile, turbinate-cylindric, ½′–1′ high, 4″–6″ thick, 15–60-flowered; bracts of the involucre imbricated in 5 or 6 series, broadly oval, appressed, abruptly acuminate at the apex; flowers purple; pappus very plumose.

In dry soil, western Ontario to Minnesota, south to Illinois and Missouri. July–Sept.

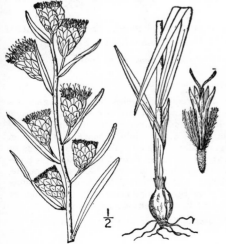

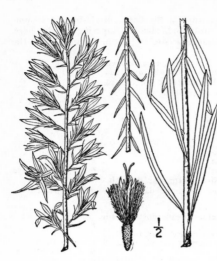

3. Lacinaria élegans (Walt.) Kuntze.
Handsome Blazing Star. Fig. 4179.

Stoepelina elegans Walt. Fl. Car. 202. 1788.

Liatris elegans Willd. Sp. Pl. **3**: 1635. 1804.

Lacinaria elegans Kuntze, Rev. Gen. Pl. 349. 1891.

Densely and finely pubescent, rarely glabrate, 2°–3° high. Leaves linear, very punctate, 1'–5' long, 1"–3" wide, the upper much smaller than the lower and soon reflexed; heads 4–5-flowered, 6"–7" high, narrowly campanulate, very numerous in a dense spike or raceme sometimes a foot long; bracts of the involucre in 2 or 3 series, the inner ones linear, dilated above into oblong or lanceolate acuminate rose-colored petaloid tips, or sometimes white; pappus very plumose; flowers purple.

In dry soil, Virginia to Florida, Alabama, Arkansas and Texas. Aug.–Oct.

4. Lacinaria punctàta (Hook.) Kuntze.
Dotted Button-Snakeroot. Fig. 4180.

Liatris punctata Hook. Fl. Bor. Am. **1**: 306. *pl. 55.* 1833.

Lacinaria punctata Kuntze, Rev. Gen. Pl. 349. 1891.

Glabrous, or sparingly pubescent, 6'–30' high; rootstock stout, branching, or globose. Leaves linear, rigid, very punctate, 2'–6' long, about 1" wide, or the lower 2", the upper gradually shorter, all erect or ascending; heads 3–6-flowered, 6"–8" long, sessile, crowded into a densè spike; spike commonly leafy below; involucre narrowly campanulate, acute or acutish at the base, its bracts oblong, cuspidate or acuminate, often ciliate on the margins, sometimes woolly, imbricated in 4 or 5 series; flowers purple; pappus very plumose.

In dry soil, Minnesota to Manitoba, Saskatchewan, Montana, Missouri, Texas, New Mexico and Sonora. Recorded from Ohio. Aug.–Oct.

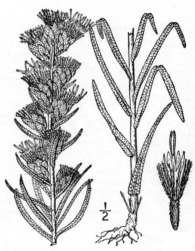

5. Lacinaria acidòta (Engelm. & Gray) Kuntze.
Slender Button-Snakeroot. Fig. 4181.

Liatris acidota Engelm. & Gray, Bost. Journ. Nat. Hist. **5**: 218. 1847.

Lacinaria acidota Kuntze, Rev. Gen. Pl. 349. 1891.

Similar to the preceding species, but usually glabrous throughout, the rootstock globose, or elongating. Stem 2° high, or less, slender. Leaves narrowly linear, ¼"–½" wide, or the lower wider, slightly punctate, 1'–4' long, the upper gradually shorter; spike slender, naked, or sparingly leafy at the base, 4'–10' long; heads 3–5-flowered, 6"–9" long; involucre narrowly oblong-campanulate, its bracts oblong to ovate-lanceolate, more or less punctate, abruptly or gradually acuminate; glabrous or sparingly ciliate; flowers purple; pappus plumose.

Prairies, Kansas to Arkansas and Texas. Aug.–Oct.

6. Lacinaria pycnostàchya (Michx.) Kuntze.
Prairie or Hairy Button-Snakeroot.
Fig. 4182.

Liatris pycnostachya Michx. Fl. Bor. Am. **2**: 91. 1803.
Lacinaria pycnostachya Kuntze, Rev. Gen. Pl. 349. 1891.

Hirsute-pubescent above, usually glabrous below, 2°–5° high, very leafy. Lower leaves linear-lanceolate, narrowed below the middle into a slender margined petiole, acuminate at the apex, often 1° long and ½′ wide, the upper much smaller, linear-subulate, rigid, punctate; spikes very dense, 6′–18′ long; heads 4″–6″ long, 3–6-flowered; involucre oblong or cylindric, its bracts oblong, pubescent and ciliate with acute, spreading, colored tips; flowers purple; pappus barbellate or roughened, scarcely plumose.

On prairies, Indiana to Minnesota, Nebraska, Kentucky, Louisiana and Texas. Aug.–Sept.

7. Lacinaria scariòsa (L.) Hill. Large
Button-Snakeroot. Fig. 4183.

$\frac{1}{2}$

Serratula scariosa L. Sp. Pl. 818. 1753.
Lacinaria scariosa Hill, Hort. Kew. 70. 1769.
Liatris scariosa Willd. Sp. Pl. **3**: 1635. 1804.
Liatris squarrulosa Michx. Fl. Bor. Am. **2**: 92. 1803.

Finely pubescent, at least above, 1°–6° high. Lower leaves oblanceolate, spatulate or oblong-lanceolate, narrowed into a margined petiole, acute or obtusish at the apex, often 1° long and 1½′ wide; upper leaves linear or linear-lanceolate, acute, or sometimes very obtuse, much smaller, all densely punctate; heads hemispheric, ½′–1′ broad, 15–45-flowered, on stout peduncles 2″–2′ long, or sometimes sessile; bracts of the involucre imbricated in 5 or 6 series, spatulate-linear, oblanceolate or obovate, rounded at the apex, appressed, their tips dry and scarious, often colored; flowers bluish purple; pappus barbellate.

In dry soil, Maine to Florida, western Ontario, Manitoba, Nebraska and Texas. Consists of many races, differing in leaf-form and size of heads. Called also blue blazing-star, gray feather, rattlesnake-master, devil's-bite. Aug.–Sept.

8. Lacinaria spicàta (L.) Kuntze.
Dense Button-Snakeroot. Gay Feather. Devil's Bit. Fig. 4184.

Serrulata spicata L. Sp. Pl. 819. 1753.
Liatris spicata Willd. Sp. Pl. **3**: 1636. 1804.
?Liatris pumila Lodd. Bot. Cab. *pl. 147.* 1821.
L. spicata Kuntze, Rev. Gen. Pl. 349. 1891.

Glabrous or nearly so, 2°–6° high. Lower leaves linear-lanceolate or linear-oblong, usually blunt-pointed, sometimes 1° long and 5″ wide, the upper linear or even subulate, somewhat or obscurely punctate; spike generally dense, 4′–15′ long; heads short-oblong or cylindric, 5–13-flowered, 2″–4″ broad, mostly sessile; involucre subcampanulate, rounded or obtuse at the base, its bracts appressed, oblong, obtuse and scarious-margined at the apex, obscurely punctate, imbricated in 4–6 series; flowers blue-purple, occasionally white; pappus roughened or barbellate.

In moist soil, Massachusetts to Florida, Ontario, Wisconsin, Kentucky, Louisiana and Arizona. Called also rough- or backache-root, throat-wort, prairie-pine, colic-root. Aug.–Oct.

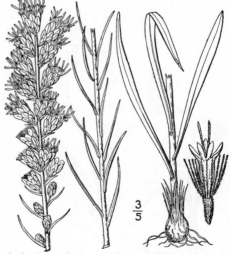

$\frac{3}{5}$

Lacinaria kansàna Britton, of Kansas, differs in having a short, very dense spike, a more leafy stem, the involucral bracts purple and the pappus plumose.

10. **Lacinaria pilòsa** (Ait.) Heller. Mountain Button-Snakeroot. Fig. 4186.

Serratula pilosa Ait. Hort. Kew. **3**: 138. 1789.
L. pilosa Heller, Muhlenbergia **1**: 6. 1900.

Glabrous, except the ciliate leaf-bases and involucral bracts, 5′–3¾° tall, simple. Leaves rather numerous, the basal often fully half as long as the stem, with broadly linear or linear-oblong blades, the upper stem leaves much smaller, with linear blades, all more or less ciliate near the base; involucres turbinate, 2½″–3″ broad, sessile or short-peduncled, the bracts purple, the outer ovate or oval, the middle ones with oval or ovate blades and short claws, the inner linear or linear-spatulate, all obtuse and rather broadly margined; pappus barbellate; achenes 2½″–3″ long, hairy.

In dry or rocky soil, Virginia and West Virginia to Georgia and Alabama. July–Sept.

9. **Lacinaria graminifòlia** (Walt.) Kuntze. Loose-flowered Button-Snakeroot. Fig. 4185.

Anonymos graminifolius Walt. Fl. Car. 197. 1788.
Liatris graminifolia Pursh, Fl. Am. Sept. **2**: 508. 1814. Not Willd. 1804.
L. graminif. Kuntze, Rev. Gen. Pl. 349. 1891.
Liatris graminifolia var. *dubia* A. Gray, Man. Ed. 2, 185. 1856.

Glabrous, or sparingly pubescent, 1°–3° high. Leaves similar to those of the preceding species, but more conspicuously punctate, usually somewhat ciliate, at least near the base and acute or acutish at the apex; heads spicate, racemose or sometimes panicled, mostly peduncled; involucre obovoid, narrowed or acute at the base, 2″–4″ broad, its bracts distinctly punctate, appressed, rounded, more or less scarious-margined, thick; flowers purple; pappus barbellate; achenes hairy.

In dry soil, New Jersey to Florida. Fine-leaved blazing-star. Aug.–Sept.

11. **Lacinaria Smàllii** Britton. Small's Button-Snakeroot. Fig. 4187.

Lacinaria Smallii Britton, Manual 927. 1901.

Similar to *L. graminifolia.* Stem very slender, 1°–2° high. Leaves scattered, linear, 2″–5″ wide, acute, pubescent on the midvein beneath, spreading, very punctate; heads scattered, sessile in the axils of the upper small leaves, the uppermost leaves not larger than the involucral bracts; involucre oblong-campanulate, about 8-flowered, about 5″ high, its innermost bracts linear, acute or acutish, the outermost obtusish, all punctate; flowers purple; pappus plumose.

Iron Mountain, Smyth Co., Virginia. Aug.

Lacinaria Hélleri (Porter) Heller, of the North Carolina mountains, with narrower involucral bracts and erect or ascending leaves, is recorded from Virginia.

9. TRÍLISA Cass. Dict. Sci. Nat. 55: 310. 1828.

Erect perennial herbs, with fibrous roots and alternate simple leaves; those of the stem sessile or clasping, the basal ones narrowed into petioles. Heads small, discoid, of 5–10 purple, or white flowers, in terminal thyrsoid or corymbose panicles. Involucre campanulate, its bracts imbricated in 2 or 3 series, the outer scarcely shorter than the inner. Receptacle flat, naked. Corolla regular, its limb 5-lobed. Anthers obtuse and entire at the base. Achenes nearly terete, 10-ribbed. Style-branches slender, obtuse. Pappus of 1 or 2 series of elongated, barbed bristles. [Anagram of *Liatris*.]

Two knows species, natives of the southeastern United States. Type species: *Trilisia odoratissima* (Walt.) Cass.

Stem glabrous; heads ·corymbose-paniculate.   1. *T. odoratissima.*
Stem viscid-pubescent; heads thyrsoid-paniculate.   2. *T. paniculata.*

1. Trilisa odoratíssima (Walt.) Cass. Vanilla-leaf. Vanilla-plant. Dog's-tongue. Carolina Vanilla. Fig. 4188.

Anonymos odoratissima Walt. Fl. Car. 198. 1788.
Liatris odoratissima Michx. Fl. Bor. Am. 2: 93. 1803.
Trilisia odoratissima Cass. Dict. Sci. Nat. 55: 310. 1828.

Rather stout, glabrous, 2°–3° high. Leaves thick, entire, or sometimes dentate, the lower oblanceolate, oblong or spatulate, obtuse, 4′–10′ long, 1′–1½′ wide, those of the stem gradually smaller, oblong, ovate or oval, the uppermost bract-like; heads corymbose-paniculate, about 3″ high; bracts of the involucre oblong, obtusish; achenes glandular-pubescent.

In pine-barrens, North Carolina to Florida and Louisiana. Recorded from Virginia, and to be looked·for in the southeastern part of that state. Deer's- or hound's-togue. Aug.–Sept.

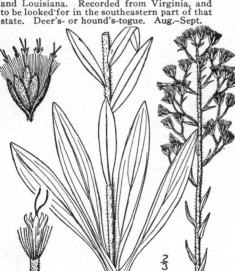

2. Trilisa paniculàta (Walt.) Cass. Hairy Trilisa. Fig. 4189.

Anonymos paniculatus Walt. Fl. Car. 198. 1788.
Liatris paniculata Michx. Fl. Bor. Am. 2: 93. 1803.
Trilisa paniculata Cass. Dict. Sci. Nat. 55: 310. 1828.

Stem viscid-pubescent, 1°–3° high. Leaves entire, the basal ones lanceolate or narrowly oblong, acute or obtusish, 3′–10′ long, ½′–1½′ wide, those of the stem much smaller, lanceolate or oblong-lanceolate; heads thyrsoid-paniculate, about 3″ high; bracts of the involucre oblong, obtusish; achenes finely pubescent.

In pine-barrens, Virginia to Florida. Sept.–Oct.

10. GUTIERRÈZIA Lag. Gen. & Sp. Nov. 30. 1816.

Erect branching, usually glabrous and often glutinous herbs, or shrubs, with linear entire alternate leaves, and small heads of both tubular and radiate yellow flowers, corymbosely paniculate in our species. Radiate flowers few, pistillate. Disk-flowers regular, mostly perfect, or some of them only staminate, the corolla-limb 5-lobed. Involucre ovoid, or narrowly campanulate, its bracts coriaceous, imbricated in few series. Receptacle flat, convex or conic, commonly foveolate. Anthers obtuse and entire at the base. Style-branches narrow, flattened, their appendages slender. Achenes terete, ribbed or 5-angled. Pappus of several scales, those of the ray-flowers shorter than or equalling those of the disk. [Named from Gutierrez, a noble Spanish family.]

About 25 species, natives of western North America, Mexico and western South America. Besides the following, several others occur in the western United States. Type species: *Gutierrezia linearifolia* Lag.

1. Gutierrezia Saròthrae (Pursh) Britton and Rusby. Broom-weed. Fig. 4190.

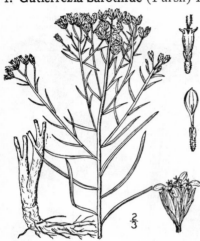

Solidago Sarothrae Pursh, Fl. Am. Sept. 540. 1814.
Gutierrezia Euthamiae T. & G. Fl. N. A. **2**: 193. 1841.
Gutierrezia Sarothrae Britton & Rusby, Trans. N. Y. Acad. Sci. **7**: 10. 1887.

Glabrous or minutely pubescent, bushy, woody at the base, ½°–1½° high, the branches rigid, ascending. Leaves linear, ½′–1½′ long, ½″–1″ wide, acute; heads oblong, 1½″–2″ high, few-flowered, usually in clusters of 2–5 at the ends of the numerous branches; rays 1–6, scarcely 1″ long; scales of the pappus linear-oblong, subulate or acuminate; achenes pubescent.

In dry or rocky soil, Manitoba and Minnesota to western Nebraska, Kansas, Texas and Sonora, west to Alberta and Utah. Adventive at Rochester, N. Y. Far western plants formerly referred to this species prove to be distinct. Rabbit-brush. Aug.–Sept.

Gutierrezia lineàris Rydb., of the Rocky Mountain region, ranging eastward into Kansas and Nebraska, differs in being more woody and has somewhat wider leaves.

11. AMPHIÁCHYRIS [DC.] Nutt. Trans. Am. Phil. Soc. II. 7: 313. 1841.

Erect, much branched, glabrous herbs, with small alternate entire leaves, and very numerous small heads of both tubular and radiate yellow flowers, solitary or clustered at the ends of the branches. Involucre ovoid or hemispheric, its bracts coriaceous, imbricated in few series. Receptacle naked. Ray-flowers pistillate. Disk-flowers perfect, but sterile, or staminate. Pappus of the ray-flowers obsolete or coroniform, that of the disk-flowers of 5–20 subulate scales or bristles somewhat united at the base. Achenes pubescent. [Greek, chaff-around.]

Two known species, natives of the south-central United States, the following typical.

1. Amphiachyris dracunculoìdes (DC.) Nutt. Amphiachyris. Fig. 4191.

Brachyris dracunculoides DC. Mem. Soc. Phys. Gen. **7**: Part 2, 265. *pl. 1.* 1836.
Amphiachyris dracunculoides Nutt. Trans. Am. Phil. Soc. (II.) **7**: 313. 1841.

Annual, slender, much branched, 6′–18′ high, the branches ascending. Leaves linear, 6″–18″ long, 1″–2″ wide, acutish, the uppermost almost filiform; heads solitary at the ends of short branches, 10–30-flowered, about 2″ high; involucre hemispheric, its bracts oval, obtuse; rays 5–10, about as long as the involucre; disk-flowers mostly staminate, their ovaries abortive, their pappus of 5–8 subulate aristate scales, united into a short cup at the base.

In dry soil, Missouri and Kansas to Texas and New Mexico. Found adventive at Easton, Penna. Sept.–Oct.

12. GRINDÈLIA Willd. Gesell. Nat. Fr. Berl. Mag. 1: 260.   1807.

Perennial or biennial herbs, sometimes woody at the base, with alternate sessile or clasping leaves, usually spinulose-dentate, and rather large heads of both discoid and radiate yellow flowers, solitary at the ends of the branches (rays rarely wanting). Involucre hemispheric or depressed, its bracts imbricated in several or many series, usually subulate-tipped. Receptacle flat or convex, naked, foveolate. Ray-flowers fertile. Disk-flowers perfect, or sometimes only staminate. Anthers obtuse and entire at the base. Style-branches narrow, flattened, their appendages linear or lanceolate. Achenes short, thick, sometimes compressed, glabrous, 4–5-ribbed. Pappus of 2–8 soon deciduous awns or bristles. [Named for Prof. H. Grindel, of Riga, 1776–1836.]

About 30 species, natives of western North America, Peru and Chile. Besides the following, some 15 others occur in the western and southwestern parts of North America. Known as Gumplant or Tar-weed. Type species: *Grindelia inuloides* Willd.

Leaves spatulate or oblong, obtuse or obtusish; achenes truncate; bracts squarrose.
1. *G. squarrosa.*
Leaves linear or linear-oblong, acute; achenes 1–2-toothed; bracts not squarrose.   2. *G. lanceolata.*

1. Grindelia squarròsa (Pursh) Dunal. Broad-leaved Gum-plant. Fig. 4192.

Donia squarrosa Pursh, Fl. Am. Sept. 559. 1814.
Grindelia squarrosa Dunal in DC. Prodr. 5: 315. 1836.
G. grandiflora Hook. Bot. Mag. *pl. 4628.* 1852.
Grindelia nuda Wood. Bot. Gaz. 3: 50. 1878.
Grindelia squarrosa nuda A. Gray, Syn. Fl. 1²: 118. 1884.

Glabrous, erect or ascending, branched, 10′–2° high. Leaves oblong or oblong-spatulate, obtuse, more or less clasping at the base, sharply spinulose-dentate, sometimes laciniate, ½′–1½′ long, 3″–6″ wide; heads 10″–15″ broad, very glutinous; bracts of the involucre linear-lanceolate, subulate-tipped, strongly squarrose; achenes truncate, those of the outer flowers usually thicker than those of the inner; rays up to 1′ long or wanting; pappus of 2 or 3 awns.

In dry soil, Illinois and Minnesota to Manitoba, Missouri, Texas, Arizona and Mexico. Adventive in southern New Jersey, Pennsylvania and New York. June–Sept.

2. Grindelia lanceolàta Nutt. Narrow-leaved Gum-plant. Fig. 4193.

Grindelia lanceolata Nutt. Journ. Acad. Phil. 7: 73. 1834.

Slender, erect, glabrous, branched, about 2° high. Leaves lanceolate or linear, acute at the apex, sessile or clasping at the base, spinulose-dentate, laciniate, or the upper entire, 1′–2′ long, 2″–5″ wide; heads nearly as large as those of the preceding species; bracts of the involucre linear-subulate, the inner erect, the outer spreading; achenes 1–2-toothed; pappus of 1 or 2 awns.

In dry soil, Tennessee to Missouri, Kansas, Louisiana and Texas. July–Sept.

13. HETEROTHÈCA Cass. Dict. Sci. Nat. **21**: 130. 1821.

Erect, hirsute or pubescent, branching herbs, with alternate, mostly dentate leaves, and rather large heads of both discoid and radiate yellow flowers, generally solitary at the ends of the branches. Involucre hemispheric or broadly campanulate, its bracts imbricated in several series, the outer shorter. Receptacle flat, alveolate. Ray-flowers pistillate. Disk-flowers perfect, or some of them only staminate. Style-branches flat, their appendages lanceolate or triangular. Achenes pubescent, obtuse, those of the ray-flowers thick, those of the disk-flowers flattened. Pappus of the ray-flowers obsolete or of a few caducous bristles, that of the disk-flowers of an inner row of numerous capillary rough bristles, and an outer row of shorter stouter bristles or scales. [Greek, different-case, from the dissimilar achenes.]

Five or 6 species, natives of the southern United States and Mexico, the following typical.

1. Heterotheca subaxillàris (Lam.) Britton & Rusby. Heterotheca. Fig. 4194.

Inula subaxillaris Lam. Encycl. **3**: 259. 1799.

Heterotheca Lamarckii Cass. Dict. Sci. Nat. **21**: 131. 1821.

Heterotheca subaxillaris Britton & Rusby, Trans. N. Y. Acad. Sci. **7**: 10. 1887.

Biennial or sometimes annual, 1°–3° high. Basal and lower leaves petioled, ovate or oblong, 2′–3′ long, the upper ones oblong, sessile or clasping, smaller, all acutish or obtuse, dentate; heads rather numerous, 6″–9″ broad; involucre nearly hemispheric, 3″–5″ high, its bracts linear, or slightly dilated above, the inner with scarious margins; rays 10–25; inner bristles of the pappus of the disk-flowers about 2″ long.

In dry soil, Delaware to Florida, Louisiana, Kansas, Arizona and Mexico. In ballast, at Philadelphia. July–Sept.

14. CHRYSÓPSIS [Nutt.] Ell. Bot. S. C. & Ga. **2**: 333. 1824.

[DIPLOGON Raf. Amer. Month. Mag. **2**: 268. 1818. Not Poiret, 1811.]

Perennial or biennial, rarely annual, branching herbs, with alternate sessile entire leaves, or the basal ones dentate, and large many-flowered heads of both tubular and radiate yellow flowers (rays wanting in some western species), loosely corymbose, or solitary at the ends of the branches. Involucre campanulate to hemispheric, its bracts narrow, imbricated in several series, the outer shorter. Receptacle usually flat, more or less foveolate. Ray-flowers pistillate. Disk-flowers mostly all perfect. Pappus double in both the disk- and ray-flowers, the inner of numerous rough capillary bristles, the outer of smaller or minute scales or bristles. Achenes flattened, oblong-linear or obovate, pubescent. Style-branches narrow, somewhat flattened, their appendages linear or subulate. [Greek, of golden aspect.]

About 20 species, natives of North America and Mexico. Besides the following, about 8 others occur in the southern and western United States. Type species: *Chrysopsis gossypina* (Michx.) Ell.

Leaves elongated-linear, entire, parallel-veined; achenes linear; involucre campanulate.
    Plants 1°–3° high, silvery-pubescent; leaves grass-like, 3′–12′ long.   1. *C. graminifolia.*
    Plants 4′–10′ high, woolly-pubescent; leaves rigid, 1′–4′ long.   2. *C. falcata.*
Leaves oblong, lanceolate, or linear, pinnately veined; achenes obovate, or oval; involucre hemispheric.
    Plant densely woolly-pubescent.   3. *G. gossypina.*
    Plants hirsute, or villous-pubescent.
        Heads numerous, corymbose-paniculate; pubescence of long deciduous hairs; eastern species.
             4. *C. mariana.*
        Heads fewer, corymbose, or terminating the branches; pubescence persistent; western species.
            Villous-pubescent, hirsute or hispid; perennials.
                Villous-pubescent and canescent with appressed hairs.   5. *C. villosa.*
                Hirsute or hispid-pubescent.
                    Leaves linear, acutish.   6. *C. stenophylla.*
                    Leaves mostly spatulate, obtuse.   7. *C. hispida.*
            Pilose-pubescent with soft spreading hairs; annual.   8. *C. pilosa.*

1. Chrysopsis graminifòlia (Michx.) Ell. Grass-leaved Golden Aster. Fig. 4195.

Inula graminifolia Michx. Fl. Bor. Am. **2**: 122. 1803.
Chrysopsis graminifolia Ell. Bot. S. C. & Ga. **2**: 334. 1824.

Slender, corymbosely branched above, very silvery-pubescent, 1°–3° high. Leaves linear, soft, grass-like, 3–5-nerved, shining, the basal ones 4′–12′ long, 2″–5″ wide, the upper much smaller, and the uppermost subulate and erect; heads several or numerous, about ½′ broad, solitary at the ends of the branches; involucre campanulate, its bracts glabrate; achenes linear-fusiform.

In dry soil, Delaware to Florida, Ohio, Kentucky, Arkansas, Texas and Mexico. Great Bahama Island. Silver-grass. Scurvy-grass. Silk-grass. Aug.–Oct.

2. Chrysopsis falcàta (Pursh) Ell. Sickle-leaved Golden Aster. Fig. 4196.

Inula falcata Pursh, Fl. Am. Sept. 532. 1814.

Chrysopsis falcata Ell. Bot. S. C. & Ga. **2**: 336. 1824.

Corymbosely branched above, rather stiff, 4′–12′ high, leafy to the top, very woolly-pubescent, at least when young, or becoming glabrate. Leaves linear, rigid, spreading, sometimes curved, 1′–4′ long, 1″–3″ wide, obscurely parallel-nerved; heads rather few, corymbose, 3″–5″ broad, terminating the branches; involucre campanulate, its bracts slightly pubescent; achenes linear.

In sandy soil, eastern Massachusetts to New Jersey. Ground gold-flower. July–Aug.

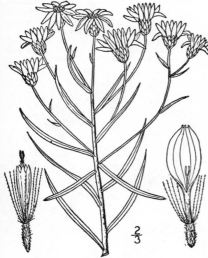

3. Chrysopsis gossýpina (Michx.) Ell. Cottony Golden Aster. Fig. 4197.

Erigeron pilosum Walt. Fl. Car. 206. 1788.
Inula gossypina Michx. Fl. Bor. Am. **2**: 122. 1803.
I. gossypina Nutt. Gen. **2**: 150. 1818.
C. gossypina Ell. Bot. S. C. & Ga. **2**: 337. 1824.
Chrysopsis pilosa Britton, Mem. Torr. Club **5**: 316. 1894. Not Nutt.

Stout, 1°–2° high, branched above, densely woolly-pubescent all over. Leaves spatulate or oblong, obtuse, the lower and basal ones 1′–3′ long, 3″–5″ wide, the uppermost much smaller; heads usually nearly 1′ broad, terminating the branches, bright yellow; involucre hemispheric, its bracts densely pubescent when young, becoming glabrate; achenes obovate.

In pine-barrens, Virginia to Florida and Alabama. Autumn.

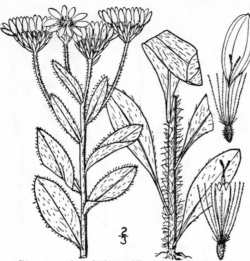

5. Chrysopsis villòsa (Pursh) Nutt.
Hairy Golden Aster.　Fig. 4199.

Amellus villosus Pursh, Fl. Am. Sept. 564. 1814.
Inula villosa Nutt. Gen. **2**: 151. 1818.
C. villosa Nutt. Trans. Am. Phil. Soc. II. **7**: 316. 1841.
C. foliosa Nutt. loc. cit. 316. 1841.
Chrysopsis camporum Greene, Pittonia **3**: 88. 1897.

Stem villous or strigose-pubescent, 1°–2° high. Leaves oblong, lanceolate, or oblanceolate, obtuse or acutish, 1′–2′ long, 2″–5″ wide, the upper sessile, the lower narrowed into a petiole, pale, persistently canescent with appressed hairs; heads rather few, 1′ broad or more, terminating the short branches; rays oblong-linear, golden yellow; involucre hemispheric, its bracts 4″–5″ high, linear-subulate, pubescent and often ciliate; achenes obovate, 3–5-nerved.

In dry soil, Minnesota and Illinois to Alabama, Manitoba, British Columbia, Nebraska and New Mexico. July–Aug. Rosin-wood.

4. Chrysopsis mariàna (L.)
Ell.　Maryland Golden Aster.　Fig. 4198.

Inula mariana L. Sp. Pl. Ed. 2, 1240. 1763.
I. mariana Nutt. Gen. **2**: 151. 1818.
C. mariana Ell. Bot. S. C. & Ga. **2**: 335. 1824.

Stout, 1°–2½° high, loosely villous-pubescent with silky hairs when young, at length nearly glabrous, corymbosely branched at the summit. Upper leaves oblong or lanceolate, acutish or obtuse, sessile, 1′–2′ long, the lower oblanceolate or spatulate and narrowed into a petiole, generally obtuse, 2′–4′ long and sometimes 1′ wide; heads commonly numerous, 9″–12″ broad, on glandular peduncles; involucre hemispheric, its bracts glandular, acute, viscid-pubescent; achenes obovate.

In dry soil, southern New York and Pennsylvania to Tennessee, Florida and Louisiana. Aug.–Sept. Golden-star.

6. Chrysopsis stenophýlla (A. Gray)
Greene.　Stiff-leaved Golden Aster.　Fig. 4200.

Chrysopsis villosa var. *stenophylla* A. Gray, Syn. Fl. **1**: Part 2, 123. 1884.
C. stenophylla Greene, Erythea **2**: 96. 1894.
C. angustifolia Rydb. Bull. Torr. Club **37**: 128. 1910.

Low, slender, hirsute or rough-pubescent, 6′–10′ high. Leaves linear or slightly broadened above, densely canescent and ciliate, acutish, 9″–15″ long, 1″–2″ wide, the margins revolute in drying; involucre hemispheric or broadly campanulate, its bracts pubescent or the outer densely ciliate; heads few, 6″–10″ broad.

In dry soil, Missouri and Nebraska to Arkansas and Texas. Aug.–Sept.

7. Chrysopsis híspida (Hook.) Nutt.   Hispid Golden Aster.   Fig. 4201.

Diplopappus hispidus Hook. Fl. Bor. Am. **2**: 22. 1834.
Chrysopsis hispidus Nutt. Trans. Am. Phil. Soc. (II.) **7**: 316. 1841.
Chrysopsis villosa var. *hispida* A. Gray, Syn. Fl. **1**: Part 2, 123. 1884.

Lower than *C. villosa,* stem rarely over 1° high, with spreading, sparse or copious, hirsute or hispid pubescence, sometimes viscid. Leaves spatulate to oblong, entire, spreading, 9″–18″ long, obtuse at the apex, narrowed at the base, often into petioles half as long as the blade or more; heads smaller, often more numerous; involucre not over 4″ high, its bracts lanceolate, hirsute; achenes 3–5-nerved.

In dry soil, Manitoba to Idaho, Kansas, Texas and Arizona. Consists of several races, differing in size, pubescence and leaf-form. July–Sept.

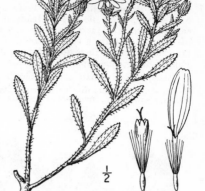

8. Chrysopsis pilòsa Nutt.   Nuttall's Golden Aster.   Fig. 4202.

Chrysopsis pilosa Nutt. Journ. Acad. Nat. Sci. Phila. **7**: 66. 1834.
C. Nuttallii Britton, Mem. Torr. Club **5**: 316. 1894.

Pilose-pubescent with soft hairs and minutely glandular, 1°–2° high, branched above. Leaves oblong-lanceolate, mostly acute, 1′–2½′ long, 2″–5″ wide, entire, or the lower serrate or even incised; heads few, 8″–12″ broad, terminating the branches; involucre hemispheric, its bracts glandular-viscid; achenes obovate, 10-nerved.

In dry soil, Missouri and Kansas to Louisiana, Arkansas and Texas. July–Sept.

15. CHONDRÓPHORA Raf. New Fl. N. A. 4: 79. 1836.

[BIGELOWIA DC. Mem. Comp. *pl. 5.* 1833. Not *Bigelovia* Spreng. 1821.]

Perennial stiff herbs, with alternate entire leaves, and corymbose-paniculate discoid heads of yellow flowers. Involucre oblong or campanulate, its bracts rigid and glutinous, imbricated in several series. Receptacle flat, generally foveolate, naked. Corolla regular, tubular, the limb 5-cleft. Anthers obtuse at the base. Style-branches flattened, their appendages lanceolate. Achenes oblong, somewhat compressed, 1–2-ribbed on each side. Pappus of 1 or 2 series of numerous capillary unequal bristles. [Greek, cartilage-bearing.]

Two species, native of eastern North America, the following typical.

1. Chondrophora nudàta (Michx.) Britton.   Rayless Golden-rod.   Fig. 4203.

Chrysocoma nudata Michx. Fl. Bor. Am. **2**: 101. 1803.
Bigelowia nudata DC. Prodr. **5**: 329. 1836.
C. nudata Britton, Mem. Torr. Club **5**: 317. 1894.

Glabrous, erect, simple, 1°–2½° high. Lower and basal leaves spatulate, obtuse, or obtusish, 3′–6′ long, 3″–6″ wide, attenuate into a margined petiole; stem leaves distant, small, linear or subulate; heads numerous, 2″–3″ high, crowded in a compound terminal corymbose cyme; involucre narrowly campanulate, acute at the base, its bracts coriaceous, appressed, linear-oblong, imbricated in 3 or 4 series, the outer much smaller than the inner; achenes short-turbinate; pappus-bristles rigid; edges of the depressions in the receptacle prolonged into subulate teeth.

In moist pine-barrens, New Jersey (?) to Florida and Texas. Aug.–Oct.

Chondrophora virgàta (Nutt.) Greene, with narrowly linear basal leaves, or some of them linear-spatulate, though originally cited by Nuttall as from New Jersey, is not definitely known from north of North Carolina.

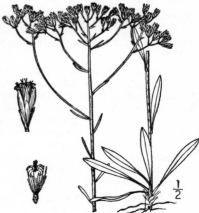

16. CHRYSOTHÁMNUS Nutt. Trans. Am. Phil. Soc. (II) 7 : 323. 1840.

Low shrubs, with equably leafy branches, hard wood, linear leaves, and discoid heads of yellow perfect flowers. Heads narrow, 5–7-flowered. Involucre oblong to narrowly campanulate, its bracts more or less keeled, thin, or papery, impricated in several series, often so as to form 5 vertical rows. Corolla 5-toothed. Anthers obtuse at the base. Style-branches exserted, their appendages subulate to filiform. Achenes narrow, mostly pubescent. Pappus of copious capillary roughened bristles. [Greek, golden-bush.]

About 30 species, natives of western North America. Type species : Chrysothamnus pumilus Nutt.

Heads about 6″ high ; involucral bracts obtuse or mucronulate.   1. C. graveolens.
Heads 7″–10″ high ; involucral bracts subulate-tipped.   2. C. Howardi.

1. Chrysothamnus gravèolens (Nutt.) Greene. Fetid Rayless Golden-rod. Fig. 4204:

Chrysocoma graveolens Nutt. Gen. 2 : 136.
  1818.
Bigelovia graveolens A. Gray, Proc. Am. Acad.
  8 : 644. 1873.
Chrysothamnus graveolens Greene, Erythea 3 :
  108. 1895.

Sparingly tomentose above, or glabrate, much branched, very leafy, 1°–4½° high, odorous. Leaves linear, 1′–3′ long, 1″–2″ wide ; heads 5″–6″ high, very numerous and crowded in terminal compound corymbose cymes ; rays none ; involucre narrowly campanulate, acute at the base, its bracts oblong or linear-oblong, acute or acutish, imbricated in about 4 series ; achenes linear ; pappus-bristles soft, copious.

In sterile, especially alkaline soil, Montana to western Nebraska, Utah and New Mexico. Included in our first edition in the C. nauseosa (Pursh) Britton, a densely tomentose far western species. Rabbit-brush. Aug.–Oct.

2. Chrysothamnus Hówardi (Parry) Greene. Howard's Rayless Golden-rod. Fig. 4205.

Linosyris Howardi Parry ; A. Gray, Proc. Am.
  Acad. 6 : 541. 1865.

Bigelovia Howardi A. Gray, Proc. Am. Acad. 8 :
  641. 1873.

Chrysothamnus Howardi Greene, Erythea 3 : 113.
  1895.

Similar to the preceding species, white-tomentose when young, glabrous or nearly so when old, tufted, much branched, 6′–10′ high. Leaves narrowly linear, entire, 1′–2′ long, about 1″ wide ; heads 7″–10″ long, narrow, 5-flowered, more or less glomerate, usually surpassed by the upper leaves ; rays none ; involucre slightly arachnoid-pubescent, its bracts lanceolate, thin, acuminate or subulate-tipped ; achenes nearly linear, pubescent.

In dry soil, western Nebraska to Wyoming, Colorado, Utah and New Mexico. July–Sept.

17.　OONÓPSIS [Nutt.] Greene, Pittonia 3: 45.　1896.

Shrubs or herbs, the following species glabrous, mostly tufted, with woody roots, the stems leafy to the top. Leaves entire, linear to lanceolate. Heads cymose at the end of the stem or branches. Involucre campanulate to ovoid, its bracts flat, imbricated in several series, herbaceous, cuspidate, appressed and erect, or the outer spreading. Disk-flowers with a nearly cylindric 5-toothed corolla. Stamens and style included or scarcely exserted. Ray-flowers present or wanting, pistillate. Style-appendages ovate to subulate. Achenes glabrous, or somewhat pubescent. Pappus-bristles coarse, rigid. [Greek, resembling an egg, referring to the ovoid involucre.]

Four known species, natives of central North America. Type species: *Oonopsis multicaulis* (Nutt.) Greene.

1. Oonopsis Engelmánni (A. Gray) Greene.　　Engelmann's Oonopsis.　Fig. 4206.

Bigelovia Engelmanni A. Gray, Proc. Am. Acad. 11: 75. 1876.

Oonopsis Engelmanni Greene, Pittonia 3: 45. 1896.

Perennial by a deep woody root, glabrous throughout; stems stiff, about 8′ high, densely leafy. Leaves narrowly linear, sessile, 9″–2′ long, less than 1″ wide, brittle when dry; heads clustered, or sometimes solitary at the ends of the branches, ½′ wide or less, sessile among the upper leaves; involucre oblong-campanulate, its bracts in about 4 series, oblong to spatulate, short-acuminate or mucronate, appressed; ray-flowers none; disk-flowers about as long as the rather rigid capillary pappus-bristles; achenes linear-oblong, narrowed at the base, many-striate.

Western Kansas, Nebraska and Colorado. Sept.–Oct.

18.　PRIONÓPSIS Nutt. Trans. Am. Phil. Soc. (II) 7: 329.　1841.

A glabrous annual or biennial herb, leafy to the top, with sessile spinulose-dentate leaves, and large heads of yellow radiate and tubular flowers. Involucre broadly hemispheric, its bracts imbricated in several series, lanceolate, acuminate, the outer more or less spreading. Receptacle naked. Disk-flowers perfect, their corollas 5-toothed. Ray-flowers very numerous, pistillate. Achenes glabrous, those of the ray-flowers broader than those of the disk; pappus of a few deciduous, rigid, unequal bristles, the outer very short. [Greek, resembling a saw, referring to the leaf-margins.]

A monotypic genus of south-central United States.

1. Prionopsis ciliàta Nutt.　Prionopsis. Fig. 4207.

Donia ciliata Nutt. Journ. Acad. Nat. Sci. Phila. 2: 118. 1821.
Aplopappus ciliatus DC. Prodr. 5: 346. 1836.
Prionopsis ciliata Nutt. Trans. Am. Phil. Soc. (II.) 7: 329. 1841.

Stem erect, stout, branched, very leafy, 2°–5° high. Leaves sessile, oval or the lower obovate, obtuse, conspicuously veined, 1′–3′ long, ½′–1½′ wide, sharply serrate with bristle-pointed teeth; heads few, clustered, stalked or nearly sessile, 1′–1½′ broad; involucre depressed-hemispheric, its bracts glabrous; achenes of the ray-flowers ellipsoid, those of the disk-flowers oblong, the central sterile; pappus-bristles rigid, the inner ones rough or ciliate.

On hillsides and river-banks, Missouri and Kansas to Texas. Aug.–Sept.

19. SIDERÁNTHUS Fraser, Cat. 1813; Sweet, Hort. Brit. 227. 1826.

[ERIOCARPUM Nutt. Trans. Am. Phil. Soc. (II) 7: 320. 1841.]

Perennial or annual herbs or shrubs with alternate spinulose-dentate or lobed leaves and many-flowered heads of tubular or of both tubular and radiate yellow flowers (heads rarely without rays). Involucre hemispheric to campanulate, its bracts imbricated in several series, the outer ones gradually smaller. Receptacle flat or convex, generally foveolate, naked. Ray-flowers fertile. Disk-flowers usually perfect. Anthers obtuse and entire at the base. Style-branches flattened, their appendages short, lanceolate. Achenes oblong or obovoid, obtuse, white-tomentose, or canescent, usually 8–10-nerved. Pappus of 1–3 series of numerous capillary persistent more or less unequal bristles. [Greek, iron-flower.]

About 15 species, natives of America. Besides the following, about 10 others occur in the western parts of the United States. Type species: *Sideranthus spinulosus* (Nutt.) Sweet.

Rays none; leaves dentate.   1. *S. grindelioides.*
Rays present.
    Leaves dentate; annual.   2. *S. annuus.*
    Leaves pinnatifid; perennial.   3. *S. spinulosus.*

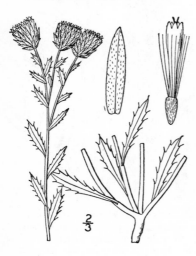

1. Sideranthus grindelioìdes (Nutt.) Britton. Rayless Sideranthus. Fig. 4208.

Eriocarpum grindelioides Nutt. Trans. Am. Phil. Soc. (II.) **7**: 321. 1841.

Aplopappus Nuttallii T. & G. Fl. N. A. **2**: 240. 1842.

Sideranthus grindelioides Britton, Manual 932. 1901.

Perennial by a deep woody root, finely pubescent; stems tufted, simple, erect, 4′–12′ high. Leaves oblong-lanceolate to spatulate, sessile, or the lower petioled, ½′–1′ long, firm, acute or obtusish, spinulose-dentate; heads several or solitary, terminating the stem or branches; peduncles 1′ long, or less; involucre campanulate, its bracts linear, acute, puberulent, their tips somewhat spreading, the outer shorter than the inner; achenes densely silky tomentose.

In dry soil, South Dakota to Assiniboia, Nebraska, New Mexico and Arizona. July–Aug.

2. Sideranthus ánnuus Rydb. Viscid Sideranthus. Fig. 4209.

Sideranthus annuus Rydb. Bull. Torr. Club **31**: 653. 1904.
Aplopappus rubiginosus A. Gray, Syn. Fl. 1²: 130. 1884. Not T. & G.

Viscid, glandular-pubescent, erect, annual, branched near the summit, 1°–3° high. Leaves sessile, or the lowest narrowed into short petioles, oblong, lanceolate, or oblanceolate, conspicuously dentate with distant awn-pointed teeth, acute or obtusish at the apex, 1½′–2½′ long, 2″–6″ wide; heads several, cymose-paniculate, 8″–15″ broad; involucre hemispheric, its bracts linear-subulate with spreading tips; rays large; pappus bristles rigid, very unequal; achenes villous-canescent, turbinate, not compressed.

On plains and in canons, Nebraska, Kansas and Colorado. Erroneously referred in our first edition to *Eriocarpum rubiginosum.* Aug.–Sept.

3. Sideranthus spinulòsus (Nutt.) Sweet.
Cut-leaved Sideranthus. Sapo.
Fig. 4210.

Amellus spinulosus Pursh, Fl. Am. Sept. **2**: 564. 1814.
Sideranthus spinulosus Sweet, Hort. Brit. 227. 1826.
Aplopappus spinulosus DC. Prodr. **5**: 347. 1836.
Eriocarpum spinulosum Greene, Erythea **2**: 108. 1894.
S. glaberrimus Rydb. Bull. Torr. Club **27**: 621. 1900.

Canescent or glabrate, much branched at the base, perennial by thick woody roots, 6'–15' high. Leaves pinnatifid, sessile, linear to ovate in outline, ½'–1½' long, 1''–2½'' wide, the lobes with bristle-pointed teeth; heads several or numerous (rarely solitary), 6''–12'' broad; involucre hemispheric, its bracts linear, acute, appressed; rays narrow; achenes pubescent, narrowed below; pappus soft and capillary.

In dry soil, Minnesota and North Dakota to Saskatchewan, Alberta, Colorado, Nebraska, Texas and Mexico. March–Sept.

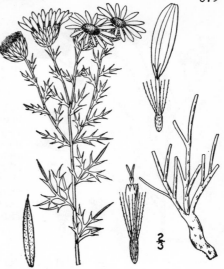

20. STENÒTUS Nutt. Trans. Am. Phil. Soc. (II) **7**: 334. 1841.

Low undershrubs, with coriaceous narrow entire evergreen leaves, scapose or leafy stems, and rather large heads of both radiate and tubular yellow flowers. Involucre mostly hemispheric, its bracts imbricated in several series, appressed, ovate to lanceolate. Receptacle alveolate. Disk-flowers perfect, their corollas tubular, usually somewhat enlarged upward, deeply 5-toothed. Ray-flowers fertile. Anthers obtuse at the base. Appendages of the style-branches short, lanceolate. Achenes white-villous. Pappus of soft white capillary bristles. [Greek, narrow, referring to the leaves.]

About 18 species, natives of western North America. Type species: *Stenotus acaulis* Nutt.

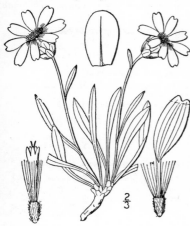

1. Stenotus armerioìdes Nutt. Narrow-leaved
Stenotus. Fig. 4211.

Stenotus armerioides Nutt. Trans. Am. Phil. Soc. (II.) **7**: 335. 1841.
Aplopappus armerioides A. Gray, Syn. Fl. **1**: Part. 2, 132. 1884.

Perennial, tufted from a branched woody caudex, glabrous throughout; flowering stems slender, naked above, or quite leafless, 4'–8' high. Basal leaves numerous, narrowly spatulate or linear, acute or acutish, firm, 1'–3' long, 1''–2'' wide, entire, narrowed below; stem leaves usually 1–3, sessile, linear, sometimes none; head commonly solitary, about 1' broad; involucre campanulate, 4''–6'' high, its bracts broadly oval, green, appressed, obtuse or retuse, scarious-margined, or the inner ovate-oblong and acutish; rays 8–10; achenes canescent or villous; pappus bristles soft, white.

In dry, mostly rocky soil, western Nebraska to Wyoming, Utah and New Mexico. June–July.

21. ISOPÁPPUS T. & G. Fl. N. A. **2**: 239. 1841.

Rough-hairy annual or biennial herbs, loosely paniculately branched, with alternate linear to lanceolate, 1-nerved, entire or somewhat toothed leaves, and small slender-peduncled heads of radiate and tubular yellow flowers. Involucre campanulate-cylindric, its appressed lanceolate or subulate bracts in 2 or 3 series. Receptacle alveolate. Ray-flowers 5–12, pistillate. Disk-flowers 10–20, perfect. Anthers not sagittate. Style-appendages narrow, hirsute. Achenes terete, narrowed below, silky-villous. Pappus a single series of rough capillary bristles, nearly equal in length. [Greek, equal-pappus.]

Two known species, natives of the southern United States, the following typical.

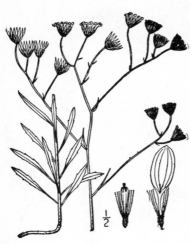

1. **Isopappus divaricàtus** (Nutt.) T. & G.
Isopappus.   Fig. 4212.

Inula divaricata Nutt. Gen. **2**: 152. 1818.

Aplopappus divaricatus A. Gray, Syn. Fl. **1**: Part 2, 130. 1884.

Isopappus divaricatus T. & G. Fl. N. A. **2**: 239. 1841.

Annual or biennial, erect, paniculately much branched, slender, rough-pubescent or glandular, 1°–3° high. Leaves linear, linear-lanceolate, or the lowest linear-spatulate, acute or cuspidate, dentate with distant teeth, or sometimes entire, 1′–3′ long, 1″–3″ wide, the uppermost much smaller, subulate or bract-like; heads numerous, 3″–5″ broad; involucre campanulate, its bracts linear-lanceolate, subulate-tipped, pubescent, peduncles very slender, or filiform.

In dry soil, Kansas to Texas, east to Georgia and Florida. Aug.–Oct.

22. SOLIDÀGO L. Sp. Pl. 878. 1753.

Perennial erect herbs, sometimes woody at the base, simple, or little branched, with alternate simple toothed or entire leaves, and small heads of both tubular and radiate, yellow or rarely white flowers, in terminal or axillary panicles, thyrsi, or cymose-corymbose or capitate clusters. Involucre oblong or narrowly campanulate, its bracts imbricated in several series, the outer successively shorter. Receptacle small, flat, or somewhat convex, generally alveolate. Ray-flowers in one series, pistillate. Disk-flowers mostly all perfect, their corollas tubular or narrowly campanulate, 5-cleft or 5-lobed. Anthers obtuse and entire at the base. Style-branches flattened, their appendages lanceolate. Achenes terete or angled, usually ribbed. Pappus of numerous capillary rough nearly equal bristles in 1 or 2 series. [Greek, to make whole.] Golden-rod; also called yellow-top, yellow-weed or flower-of-gold.

About 125 species, mostly of North America, 2 or 3 in Europe, a few in Mexico and South America. Type species: *Solidago Virgaurea* L.

A. Tips of the involucral bracts, or some of them spreading or recurved.

Leaves glabrous or glabrate, 4′–10′ long.   1. *S. squarrosa.*
Leaves rough-ciliate, 1′–2′ long.   2. *S. petiolaris.*

B. Tips of the involucral bracts all erect and appressed.

* Heads in axillary clusters, or also in a terminal spike-like, sometimes branched thyrsus.

1. Heads 2″–3″ high, chiefly in axillary clusters; achenes pubescent.

Stem and branches terete; leaves lanceolate to oblong.   3. *S. caesia.*
Stem and branches grooved or angled.
  Leaves broadly oval, contracted into margined petioles.   4. *S. flexicaulis.*
  Leaves lanceolate, sessile.   5. *S. Curtisii.*

2. Heads 2″–3″ high, chiefly in a terminal spike-like thyrsus; achenes glabrous, or nearly so.

Rays white; stem pubescent.   6. *S. bicolor.*
Rays yellow; stem densely pubescent.   7. *S. hispida.*
Rays yellow; stem glabrous, or sparingly pubescent.
  Leaves thick, not acuminate, dentate, or the upper entire.   8. *S. erec'a.*
  Leaves thin, acuminate, sharply serrate.   9. *S. monticola.*

3. Heads 5″–6″ high; bracts elongated, acute; leaves ovate. 10. *S. macrophylla.*

** Heads in a terminal simple or branched thyrsus, not at all secund on its branches, or scarcely so.

Plant rough-pubescent or scabrous; leaves sessile, entire; bracts acute.   11. *S. Lindheimeriana.*
Plants glabrous, puberulent, or sparingly pubescent above.
  Low arctic-alpine species, seldom over 10′ high.
    Heads with 30 flowers or more.   12. *S. Cutleri.*
    Heads with fewer than 30 flowers.   13. *S. multiradiata.*
  Taller species, not arctic-alpine.
    Bracts of the involucre linear-subulate, very acute; stem puberulent. 14. *S. puberula.*
    Bracts of the involucre obtuse or merely acutish; stem glabrous, or sparingly pubescent above.
      Glabrous throughout; upper leaves very small, appressed. 15. *S. stricta.*
      At least the involucre or peduncles pubescent or puberulent.

Bog species; inflorescence wand-like.   16. *S. uliginosa.*
Upland species, the inflorescence various.
    Heads very short-peduncled.
       Leaves thick, firm in texture, little toothed or entire; tall species.
         Lower leaves ovate to broadly oval, serrate.   17. *S. speciosa.*
         Leaves lanceolate to oblong-lanceolate, entire.   18. *S. rigidiuscula.*
       Leaves thin in texture, at least the lower serrate; low species.
                                         19. *S. Randii.*
    Heads distinctly slender-peduncled.
       Basal leaves narrowly oblanceolate, 4″ wide or less.   20. *S. racemosa.*
       Basal leaves broadly oblanceolate or obovate, 6″–18″ wide.
         Heads 4″ high; basal leaves sharply serrate.   21. *S. Gillmani.*
         Heads 3″ high; leaves nearly entire.   22. *S. sciaphila.*

*** **Heads in a terminal, usually large panicle, secund on its spreading or recurved branches.**

      ‡ Plant maritime; leaves thick, fleshy, entire.   23.   *S. sempervirens.*

        ‡‡ Plants not maritime; leaves not fleshy.

          1.   Leaves pinnately-veined, not triple-nerved.

            (a) Leaves all entire, thin and glabrous.   24.   *S. odora.*

        (b) Leaves, at least the lower, more or less dentate or serrate.

          † *Leaves linear or linear-oblong, 1′–2′ long, scabrous.*   25.   *S. tortifolia.*

        †† *Leaves broader, lanceolate, oblong, or ovate, 2′–10′ long.*

         ‡ Stem densely pubescent; leaves more or less so.

Leaves not rugose, sparingly dentate or entire.   26.   *S. fistulosa.*
Leaves rugose-veiny beneath, sharply serrate.   27.   *S. rugosa.*

       ‡‡ Stem glabrous, or merely puberulent above.

Leaves very rough on the upper surface, serrulate.   28.   *S. patula.*
Leaves smooth, or minutely roughened on the upper surface.
    Racemes few, widely divergent, very slender.
       Lower leaves oblong, coarsely serrate, thin.   29.   *S. ulmifolia.*
       Lower leaves ovate or lanceolate, rather finely serrate, firm.   30.   *S. Boottii.*
    Racemes numerous, spreading, recurved or ascending.
       Leaves all oblong or oblong-lanceolate, sessile.   31.   *S. Elliottii.*
       At least the lower leaves petioled, lanceolate or ovate-lanceolate.
         Leaves firm, ovate-lanceolate or oblong-lanceolate; heads about 2″ high; racemes short;
           rays several.   32.   *S. neglecta.*
         Leaves firm, narrowly lanceolate; heads about 2″ high; racemes few, short; rays 1–5.
                                         33.   *S. uniligulata.*
         Leaves firm, lanceolate or oval-lanceolate; heads 1½″–2½″ high; racemes numerous,
           slender.   34.   *S. juncea.*
         Leaves thin, the lower broadly ovate, short-acuminate; heads 2½″–3½″ high; racemes
           numerous.   35.   *S. arguta.*

   2.   Leaves triple-nerved, *i. e.,* with a pair of lateral veins much stronger than the others.

Heads small, the involucre only 1¼″ high or less; stem glabrous, or pubescent.   36.   *S. canadensis.*
Heads larger, the involucre 1½″–3″ high.
    Stem glabrous.
       Leaves, and involucral bracts thin.   37.   *S. serotina.*
       Leaves, and involucral bracts firm, somewhat rigid.
         Leaves linear-lanceolate; achenes glabrous.   38.   *S. glaberrima.*
         Leaves oblong-lanceolate; achenes silky-pubescent.   39.   *S. Shortii.*
         Basal leaves oblanceolate, upper bract-like.   40.   *S. Gattingeri.*
    Stem pubescent or scabrous.
       Leaves lanceolate, sharply serrate or entire, rough above.   41.   *S. altissima.*
       Leaves oblanceolate, spatulate, oblong, or ovate, the lower crenate.
         Minutely rough-pubescent, grayish; lower leaves oblanceolate; heads 2″–3″ high.
                                         42.   *S. nemoralis.*
         Canescent and pale; leaves oblong or ovate; heads 3″ high.   43.   *S. mollis.*
         Very scabrous, green, not grayish, nor canescent.   44.   *S. radula.*
         Leaves broadly ovate-oval, sharply serrate, finely pubescent.   45.   *S. Drummondii.*

**** **Heads in a terminal, corymbiform, sometimes thyrsoid cyme, forming a flat-topped inflorescence.**

                 (Genus Oligoneuron Small.)

Leaves ovate, oblong, or oval, mostly rough on both sides.   46.   *S. rigida.*
Leaves lanceolate, linear, oblong, or oblanceolate, glabrous or nearly so.
    Lower leaves oblong-lanceolate, serrulate.   47.   *S. ohioensis.*
    Leaves all lanceolate or linear, entire.
       Stout; leaves lanceolate, the basal 8′–12′ long.   48.   *S. Riddellii.*
       Slender; leaves linear, the basal 4′–5′ long.   49.   *S. Houghtoni.*

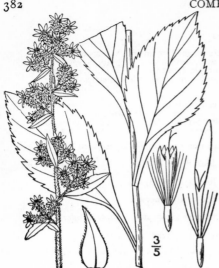

1. Solidago squarròsa Muhl. Stout Ragged Golden-rod. Fig. 4213.

Solidago squarrosa Muhl. Cat. 76. 1813.

Stem stout, simple, or rarely branched above, glabrous or puberulent, 2°–5° high. Upper leaves oblong, acute, entire or nearly so, sessile; lower and basal leaves obovate, oval, or broadly spatulate, acute or obtuse, 4′–10′ long, 1′–3′ wide, sharply dentate, often narrowed into a margined petiole, all glabrous, or sometimes slightly pubescent; heads 15–25-flowered, 4″–5″ high, numerous in a terminal narrow often leafy thyrsus sometimes 12′ in length; rays 10–16, showy, 2″–3″ long; tips of the involucral bracts green, acute or obtuse, rarely some of them erect, all usually strongly recurved, giving the heads a ragged appearance; achenes glabrous.

In rocky soil, New Brunswick to Ontario, North Carolina and Ohio. Ascends to 2000 ft. in the Catskills. Lower branches of the inflorescence sometimes elongated. Aug.–Oct.

2. Solidago petiolàris Ait. Downy Ragged Golden-rod. Fig. 4214.

Solidago petiolaris Ait. Hort. Kew. **3**: 216. 1789.
Solidago Wardii Britton, Man. 935. 1901.

Stem rather slender, pubescent or puberulent, 1°–3° high, simple, or branched above. Leaves sessile, or very short-petioled, oblong to oval, acute, obtuse or mucronate, entire or dentate, ½′–3′ long, rough or ciliate on the margins, often silvery-pubescent; heads 3″–4″ high, in a terminal narrow more or less compound thyrsus; involucral bracts pubescent, with green acute to acuminate tips, the outer spreading, the inner appressed; achenes glabrous or nearly so.

In dry soil, Illinois to Kansas and Texas, east to North Carolina and Florida. Races differ in pubescence and in leaf-form and leaf-serration. Sept.–Oct.

3. Solidago caèsia L. Blue-stemmed or Wreath Golden-rod. Fig. 4215.

Solidago caesia L. Sp. Pl. 879. 1753.
S. gracilis Poir. in Lam. Encycl. **8** : 476. 1808.
S. axillaris Pursh, Fl. Am. Sept. 542. 1814.
S. caesia axillaris A. Gray, Proc. Am. Acad. **17** : 189. 1882.

Stem glabrous, slender, often glaucous, usually bluish or purple, branched or simple, terete, 1°–3° high. Leaves lanceolate or oblong-lanceolate, sessile, acuminate at the apex, narrowed at the base, glabrous, sharply serrate, 2′–5′ long, 3″–15″ wide; heads 2″–3″ high, in axillary clusters or racemes, or occasionally with some in a short terminal thyrsus; bracts of the involucre obtuse, appressed; achenes pubescent.

Woods and thickets, Nova Scotia to Ontario, Minnesota, Florida, Arkansas and Texas. Consists of several slightly differing races. Woodland golden-rod. Aug.–Oct.

4. Solidago flexicàulis L. Zig-zag or Broad-leaved Golden-rod. Fig. 4216.

Solidago flexicaulis L. Sp. Pl. 879. 1753.
Solidago latifolia L. loc. cit. 1753.

Stem glabrous, angled, usually simple, zig-zag, 1°–3° high. Leaves thin, ovate, acuminate at the apex, abruptly narrowed at the base into margined petioles, somewhat pubescent, or glabrous beneath, sharply serrate, 2′–7′ long, 1′–4′ wide, the uppermost sometimes lanceolate and entire or nearly so; heads about 3″ high, in short axillary racemose clusters, and rarely also in a narrow terminal thyrsus; bracts of the involucre obtuse to acutish, appressed; achenes hirsute-pubescent.

In rich woods, Nova Scotia and New Brunswick to Georgia, Tennessee, Minnesota and Missouri. Ascends to 2300 ft. in the Catskills. July–Sept.

5. Solidago Curtísii T. & G. Curtis' Golden-rod. Fig. 4217.

Solidago Curtisii T. & G. Fl. N. A. **2**: 200. 1841.

Stem glabrous or sparingly pubescent, simple or branched, slender, 1½°–3° high, angled and grooved. Leaves thin, sessile, elongated-lanceolate or sometimes broader above the middle, long-acuminate, narrowed below into an entire base, sharply serrate, 3′–6′ long, 4″–12″ wide, glabrous or nearly so; heads 2″–3″ high, in rather loose axillary clusters and sometimes also in a narrow terminal thyrsus; bracts of the involucre few, obtuse.

In mountain woods, Virginia and West Virginia to Kentucky and Georgia. Aug.–Sept.

Solidago pùbens M. A. Curtis, of nearly the same range, differs in being quite densely pubescent.

6. Solidago bícolor L. White or Pale Golden-rod. Silver-rod. Fig. 4218.

Solidago bicolor L. Mant. 114. 1767.

Stem rather stout, hirsute-pubescent, or nearly glabrous, 6′–4° high, simple or branched. Basal and lower leaves obovate or broadly oblong, mostly obtuse, 2′–4′ long, 1′–2′ wide, narrowed into long margined petioles, dentate or crenate-dentate, more or less pubescent; upper leaves smaller and narrower, oblong or sometimes lanceolate, obtusish or acute, sessile or nearly so, often entire; heads 2″–3″ high, crowded in a terminal narrow thyrsus 2′–7′ long, and sometimes also clustered in the upper axils; rays white; bracts of the involucre whitish, obtuse, the midvein broadened above; achenes glabrous.

In dry soil, Prince Edward Island to Georgia, west to Ontario, Minnesota and Tennessee. Ascends to 6300 ft. in North Carolina. Belly-ache-weed. Silver-weed. July–Sept.

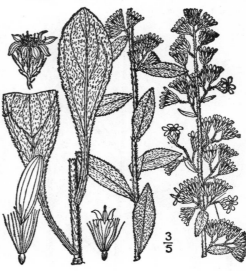

8. Solidago erécta Pursh. Slender Golden-rod. Fig. 4220.

Solidago erecta Pursh, Fl. Am. Sept. 542. 1814.
Solidago speciosa angustata T. & G. Fl. N. A. 2: 205. 1841.

Stem slender, glabrous, or puberulent above, 2°–3° high, simple or rarely branched. Leaves firm, nearly glabrous on both sides, ciliolate on the margins, the lower and basal ones broadly oblong or oval, obtuse or obtusish, crenate-dentate, the upper lanceolate or oblong-lanceolate, acute, usually quite entire; heads 2″–3″ high in a very narrow terminal thyrsus, rarely also with a few clustered in the upper axils; bracts of the involucre obtuse; rays light yellow; achenes glabrous.

In dry soil, southeastern New York, New Jersey and Pennsylvania to Kentucky, Georgia and Alabama. Aug.–Sept.

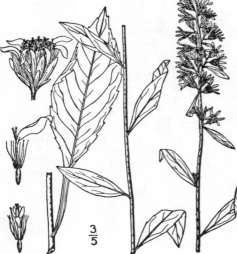

7. Solidago híspida Muhl. Hairy Golden-rod. Fig. 4219.

S. hispida Muhl.; Willd. Sp. Pl. 3: 2063. 1804.
Solidago hirsuta Nutt. Journ. Phil. Acad. 7: 103. 1834.
Solidago bicolor concolor T. & G. Fl. N. A. 2: 197. 1841.

Stout, stem densely pubescent or hirsute, simple or sometimes branched, 1½°–3° high. Lower leaves oval, acute, or obtuse, petioled, pubescent on both sides, usually dentate, 2′–5′ long, 1′–2′ wide; upper leaves oblong, sessile, acute, dentate or entire, smaller, sessile, heads about 3″ high, crowded in a dense narrow terminal thyrsus and also often in racemose clusters in the upper axils; rays yellow; involucral bracts yellowish, obtuse, the midvein narrow; achenes with a few appressed hairs, or glabrous.

In dry soil, Newfoundland to western Ontario, Manitoba, New Jersey, Pennsylvania, Georgia and Missouri. Ascends to 2000 ft. in the Catskills. Aug.–Oct.

9. Solidago montícola T. & G. Mountain Golden-rod. Fig. 4221.

Solidago Curtisii var. *monticola* T. & G. Fl. N. A. 2: 200. 1841.
Solidago monticola T. & G.; Chapm. Fl. S. States 209. 1860.

Slender, glabrous or nearly so, 1°–3° high. Stem leaves ovate-oblong, or oblong-lanceolate, thin, acuminate at the apex, narrowed at the base, sharply and sparingly serrate, or the upper entire, 1′–6′ long, 4″–1½′ wide, the upper sessile, the lower petioled; basal leaves broadly oblong, obtuse, with slender petioles; heads about 2″ high; in a terminal spike-like, simple or branched thyrsus; bracts of the involucre acutish or obtuse; achenes glabrous.

In mountain woods, Pennsylvania and Maryland to Georgia and Alabama.

10. Solidago macrophýlla Pursh. Large-leaved Golden-rod. Fig. 4222.

Solidago macrophylla Pursh, Fl. Am. Sept. 542. 1814.
Solidago thyrsoidea E. Meyer, Pl. Lab. 63. 1830.

Stem striate, glabrous or sparingly pubescent, stout, 6'–4° high. Leaves thin, ovate, acuminate, or the basal ones obtuse, sharply serrate, glabrous or sparingly pubescent beneath, 3'–5' long, 1'–2½' wide, abruptly contracted into margined petioles, or the uppermost lanceolate, entire, sessile; heads 4''–6'' high, in a terminal compact or loose thyrsus and usually also in axillary clusters; bracts of the involucre linear, acute; rays 8–10, linear-oblong, conspicuous; achenes glabrous or nearly so.

In rocky woods, Catskill Mountains, N. Y., and Greylock Mt., Mass., to Newfoundland, Labrador, Hudson Bay and Lake Superior. Ascends to 4000 ft. in the Adirondacks. July–Sept.

Solidago calcícola Fernald, a related plant found in Maine and Quebec, has smaller heads, 3''–4'' high, and pubescent achenes.

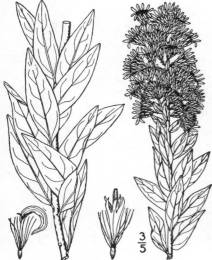

11. Solidago Lindheimeriàna Scheele. Lindheimer's Golden-rod. Fig. 4223.

Solidago Lindheimeriana Scheele, Linnaea **21**: 599. 1848.

Scabrous, simple, 1½°–3° high, leafy, rather stout and rigid. Leaves oblong, oblong-lanceolate or oval, acute or acutish at the apex, narrowed or rounded at the base, all entire, thickish, rough on both surfaces, sessile or the lowest petioled; heads about 3'' high, in a terminal, often short thyrsus; bracts of the involucre acute or the outer obtuse, puberulent; achenes nearly glabrous.

Southern Kansas to Texas and northern Mexico. Aug.–Nov.

Solidago Bigelòvii A. Gray, another southwestern species, which is rougher than this, with oval or oblong leaves obtuse or obtusish at both ends, is reported from Kansas, probably erroneously.

12. Solidago Cútleri Fernald. Cutler's Alpine Golden-rod. Fig. 4224.

Solidago Virgaurea alpina Bigel. Fl. Bost. Ed. 2, 307. 1824.
Solidago Cutleri Fernald, Rhodora **10**: 87. 1908.

Glabrous, or somewhat pubescent; stems simple, often tufted, 3'–12' high, ascending, or erect, angular. Basal leaves obovate, or broadly spatulate, serrate with low sharp or blunt teeth, at least above the middle, obtuse, or acute, 2'–4' long, narrowed into petioles; stem leaves few, oblanceolate, spatulate, or oblong, sessile, or the lower petioled, mostly distant; inflorescence a short raceme or thyrsus, and often with clusters of heads in the axils of the leaves; heads 3''–4'' high, over 30-flowered; bracts of the involucre obtuse to acute; achenes hirsute.

Alpine summits of the mountains of Maine, New Hampshire, Vermont and northern New York, mostly above timber line. Referred, in our first edition, to the European *S. alpestris* Waldst. & Kit., which it resembles. Aug.–Sept.

½

13. Solidago multiradiàta Ait. Northern Golden-rod. Fig. 4225.

Solidago multiradiata Ait. Hort. Kew. **3**: 218. 1789.

Stem glabrous or somewhat pubescent above, rather slender, 6′–15′ high. Leaves firm, glabrous or very nearly so, the basal and lower ones spatulate or oblanceolate, entire, or sparingly serrate, obtuse, finely reticulate-veined, 3′–5′ long, 3″–9″ wide, the upper smaller, narrower, sessile, entire; heads about 4″ high, usually few in a terminal rather compact, corymbose cyme, but the inflorescence sometimes elongated and thyrsoid; bracts of the involucre thin, linear-lanceolate, acute or acutish; glabrous; rays 8–15, prominent, linear, achenes pubescent.

Labrador and Hudson Bay to British Columbia, the Rocky Mountains and Colorado. July–Aug.

Solidago decúmbens Greene, of the Rocky Mountains, with broader involucral bracts is recorded from Mt. Albert, Quebec.

14. Solidago pubérula Nutt. Downy Golden-rod. Fig. 4226.

Solidago puberula Nutt. Gen. **2**: 162. 1818.

Minutely puberulent, or glabrous, usually simple, rather slender, 1½°–3° high, leafy. Stem leaves oblong-lanceolate, acute, sparingly serrate or entire, 1′–2′ long, sessile, or the lower petioled, basal leaves and sometimes the lowest ones of the stem spatulate, obtuse, often sharply serrate, 2′–4′ long, narrowed into margined petioles; heads about 2½″ high, in a terminal, often leafy thyrsus, the branches of which are spreading or ascending; bracts of the involucre subulate, very. acute; achenes glabrous; heads rarely a little secund.

In sandy soil, Prince Edward Island to Florida and Mississippi, near the coast and on sandstone rocks in the Appalachian mountain system, west to Tennessee. Minaret-golden-rod. Aug.–Sept.

³⁄₅

15. Solidago strícta Ait. Wand-like or Willow-leaf Golden-rod. Fig. 4227.

Solidago stricta Ait. Hort. Kew. **3**: 216. 1789.

S. virgata Michx. Fl. Bor. Am. **2**: 117. 1803.

Glabrous throughout, slender, erect, simple, 2°–8° high. Basal and lowest stem leaves oblong, or somewhat spatulate, with few lateral veins, obtuse, entire, or very sparingly dentate, 3′–8′ long, ½′–1′ wide, narrowed into long petioles; upper stem leaves abruptly smaller, narrowly oblong, spatulate or linear, appressed, the uppermost very small and bract-like; heads about 3″ high, in a dense simple, or sometimes branched, naked thyrsus; bracts of the involucre oblong, obtuse, or the inner acutish; achenes glabrous, or sparingly pubescent.

In wet sandy pine-barrens, New Jersey to Florida and Louisiana. Also in western Cuba. Aug.–Oct.

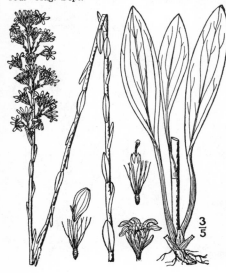

³⁄₅

16. Solidago uliginòsa Nutt. Bog or Swamp Golden-rod. Fig. 4228.

Solidago uliginosa Nutt. Journ. Phil. Acad. **7** : 101. 1834.

Stem glabrous, rather stout, simple, 2°–4° high, the branches of the inflorescence more or less pubescent.   Leaves oblong-lanceolate or lanceolate, glabrous, firm, more or less ciliolate or scabrous on the margins, few-veined, acute or acuminate, the lower and basal ones 4′–9′ long, ½′–1½′ wide, more or less serrate and narrowed into petioles, the upper smaller, mostly sessile and entire; heads 2″–3″ high, in a terminal oblong dense thyrsus, its branches appressed; bracts of the involucre linear-oblong, obtuse; achenes glabrous.

In swamps and bogs, Newfoundland to New Jersey, North Carolina, western Ontario, Minnesota and Wisconsin.   Aug.–Sept.

17. Solidago speciòsa Nutt. Showy or Noble Golden-rod. Fig. 4229.

Solidago speciosa Nutt. Gen. **2** : 160. 1818.

Stem stout, glabrous below, often rough above, usually simple, 3°–7° high.   Leaves glabrous, firm, the lower and basal ovate, or broadly oval, 4′–10′ long, 1′–4′ wide, dentate or crenate, pinnately veined, acute or obtuse at the apex, long-petioled; upper leaves smaller, oblong or oval, acute at each end, crenate-dentate, or entire, sessile or short-petioled, rough-margined; heads 3″–4″ high, in a large terminal thyrsus, the branches of which are ascending and often leafy; bracts of the involucre oblong, very obtuse; achenes glabrous or nearly so.

In rich soil, Massachusetts to North Carolina, west to Minnesota, Tennessee and Arkansas.   Apparently erroneously recorded from Canada and Nova Scotia. Aug.–Oct.

18. Solidago rigidiúscula (T. & G.) Porter. Slender Showy Golden-rod. Fig. 4230.

Solidago speciosa rigidiuscula T. & G. Fl. N. A. **2** : 205. 1841.
Solidago speciosa pallida Porter, Bull. Torr. Club **19** : 130. 1892.
Solidago rigidiuscula Porter, Mem. Torr. Club **5** : 319. 1894.
S. pallida Rydb. Bull. Torr. Club **33** : 153. 1906.

Stem rather slender, usually glabrous below, rough-pubescent above, simple, 2°–4° high.   Leaves lanceolate to ovate-lanceolate, entire, or the basal ones sometimes crenate, strongly ciliolate on the margins, 1′–5′ long, 3″–12″ wide, the upper sessile, the lower sometimes narrowed into petioles; heads similar to those of the preceding species; thyrsus generally narrow, dense, simple or branched.

In dry soil, mostly on prairies, Ohio to Alabama, Ontario, Minnesota, Colorado, Louisiana and Texas. Aug.–Oct.

19. Solidago Rándii (Porter) Britton. Rand's Golden-rod. Fig. 4231.

Solidago Virgaurea Randii Porter, Bull. Torr. Club **20**: 208. 1893.
Solidago Virgaurea Redfieldii Porter, Bull. Torr. Club **20**: 209. 1893.
Solidago Virgaura monticola Porter, Bull. Torr. Club **20**: 209. 1893.
Solidago Virgaurea Deanei Porter, Mem. Torr. Club **5**: 320. 1894.
Solidago Randii Britton, Manual 937. 1901.

Somewhat pubescent, at least above, often glutinous; stem usually simple, rather stout, 5′–2° high. Basal leaves oblanceolate, broadly spatulate, or obovate, 3′–8′ long, 1′ wide, or less, obtuse or acute, mostly dentate, narrowed into margined petioles; stem leaves few, sessile, or the lower petioled, oblong-lanceolate to spatulate, acute; heads 3″–4″ high, in a dense or interrupted, rarely branched thyrsus and often in axillary clusters; bracts of the involucre obtuse or acute; achenes more or less pubescent.

In dry, mostly rocky situations, Maine, New Hampshire, Vermont and northern New York to Michigan and on high mountains in Virginia. Referred, in our first edition, as by previous authors, to the Old World *Solidago Virgàurea* L. Aug.–Sept.

20. Solidago racemòsa Greene. River-bank Golden-rod. Fig. 4232.

S. racemosa Greene, Pittonia **3**: 160. 1897.

Nearly glabrous, sometimes glutinous; stems simple, usually somewhat glutinous, 6′–18′ high. Lower and basal leaves oblanceolate, obtuse or acutish, dentate, or crenate, 1½′–4′ long, 2½″–4″ wide, narrowed into slightly margined petioles; stem leaves sessile, lanceolate, oblong or linear, numerous, crenate or entire, mostly acute, smaller; heads about 3″ high, distinctly peduncled, in a terminal simple or branched thyrsus; bracts of the involucre linear-oblong, obtuse, or the inner acutish; achenes striate, pubescent.

On rocky river-banks, Newfoundland to northern New York, Vermont and Virginia. Described, in our first edition, under the name *S. Purshii* Porter, which proves to be untenable. July–Sept.

Solidago chrysólepis Fernald, of Quebec, has bright yellow acute involucral bracts.

21. Solidago Gíllmani (A. Gray) Steele. Gillman's Golden-rod. Fig. 4233.

S. humilis Gillmani A. Gray, Proc. Am. Acad. **17**: 191. 1882.
S. Virgaurea Gillmani Porter, Bull. Torr. Club **20**: 209. 1893.
S. Gillmani Steele, Contr. Nat. Herb. **13**: 367. 1911.
S. racemosa Gillmani Fernald, Rhodora **10**: 91. 1908.

Glabrous, except the puberulent inflorescence; stem erect or reclining, rather stout, sometimes 3° long. Lower and basal leaves spatulate or oblanceolate, dentate, 3′–12′ long, narrowed into long narrowly margined petioles; upper stem leaves lanceolate to linear-lanceolate, gradually smaller; inflorescence narrowly thyrsoid-paniculate, sometimes 16′ long; heads distinctly peduncled, about 4″ high; bracts of involucre oblong, scarcely glutinous; rays 6–10, deep yellow, 2″ long; achenes sparingly pubescent.

Sandy shores of Lakes Michigan and Superior. Aug.–Sept.

22. Solidago sciáphila Steele. Shadowy Golden-rod. Fig. 4234.

S. sciaphila Steele, Contr. Nat. Herb. **13**: 371. 1911.

Glabrous, except the ciliate leaf-margins and puberulent inflorescence, $2°-3\frac{1}{4}°$ high, light green. Leaves thin in texture, sparingly faintly veined, the lower spatulate to broadly oblanceolate, obtuse, $3'-5'$ long, narrowed into long petioles, the upper oblong to oblong-lanceolate, sessile, smaller; heads racemose on the slender branches of the narrow thyrsus, on peduncles as long as the involucre or longer; involucre campanulate, about $3''$ long, its bracts linear-oblong, obtuse; rays about 5, light yellow, short.

Shaded cliffs of the Wisconsin River, Sauk County, Wisconsin. Aug.–Sept.

23. Solidago sempérvirens L. Sea-side Golden-rod. Fig. 4235.

Solidago sempervirens L. Sp. Pl. 878. 1753.
S. angustifolia Ell. Bot. S. C. & Ga. **2** : 388. 1824.
Not Mill. 1768.

Stem stout, leafy, usually simple, $2°-8°$ high, glabrous or slightly puberulent above. Leaves thick, fleshy, entire, with 2–5 pairs of lateral veins, the lower and basal ones oblong, spatulate or lanceolate, mostly obtuse, sometimes $1°$ long, narrowed into long petioles; upper leaves sessile, lanceolate to oblong-lanceolate, acute; heads $3''-5''$ high, in secund racemes of a large terminal often leafy panicle; rays 8–10, showy; bracts of the involucre lanceolate, acute.

On salt marshes, sea-beaches, along tidal rivers and in sandy soil near the sea, Nova Scotia and New Brunswick to Florida and Mexico. Also in Bermuda. Salt-marsh or beach golden-rod. Aug.–Dec.

24. Solidago odòra Ait. Sweet or Anise-scented Golden-rod. Fig. 4236.

Solidago odora Ait. Hort. Kew. **3**: 214. 1789.
Solidago odora inodora A. Gray, Man. Ed. 5, 244. 1867.

Slender; stem simple, glabrous, or minutely pubescent above, $2°-4°$ high. Leaves usually punctate, anise-scented when bruised, or sometimes inodorous, lanceolate, quite entire, acute or acuminate, $2'-4'$ long, $3''-8''$ wide, sessile, or the lowermost petioled; heads $2''-2\frac{1}{2}''$ high, secund on the spreading racemes of the terminal, usually ample panicle; rays 3 or 4, $2''-3''$ long; bracts of the involucre oblong-lanceolate, acute, the inner much longer than the outer.

In dry soil, Nova Scotia (according to Sommers) ; New Hampshire to Florida, New York, Kentucky, Missouri and Texas. Blue mountain-tea. True golden-rod. July–Sept.

25. Solidago tortifòlia Ell. Twisted-leaf Golden-rod. Fig. 4237.

Solidago retrorsa Pursh, Fl. Am. Sept. 539. 1814. Not Michx. 1803.
S. tortifolia Ell. Bot. S. C. & Ga. **2**: 377. 1824.

Stem slender, rough-pubescent or puberulent, 2°–3° high, simple. Leaves linear or linear-oblong, often twisted, scabrous, sessile, acute, 1′–2′ long, 1½″–3″ wide, obscurely veined but with a distinct midrib, the lower serrate, the upper entire; heads about 2″ high, secund on the usually recurved branches of the terminal panicle; rays 3–5, short; bracts of the involucre linear, obtuse or obtusish.

In dry sandy soil, Virginia to Florida and Texas, mostly near the coast. Autumn.

26. Solidago fistulòsa Mill. Pine Barren Golden-rod. Fig. 4238.

Solidago fistulosa Mill. Gard. Dict. Ed. 8, No 19. 1768.
Solidago pilosa Walt. Fl. Car. 207. 1788. Not Mill. 1768.

Stem rather stout, simple or branched above, 3°–7° high, hirsute. Leaves numerous, sessile, ovate-oblong, oblong-lanceolate, or sometimes lanceolate, thick, rough or hirsute on the margins and midrib beneath, the upper small, obtuse or obtusish, entire, the lower sparingly serrate, obtuse or acute, 1′–4′ long with a broad base; heads about 2″ high, secund on the spreading or recurving branches of the panicle; rays 7–10, small; bracts of the involucre, at least the outer, acute.

In moist pine-barrens, New Jersey to Florida and Louisiana. Aug.–Oct.

27. Solidago rugòsa Mill. Wrinkle-leaved, Tall Hairy, or Pyramid Golden-rod. Bitter-weed. Fig. 4239.

S. rugosa Mill. Gard. Dict. Ed. 8, No. 25. 1768.
Solidago aspera Ait. Hort. Kew. **3**: 212. 1789.

Stem hirsute or scabrous, rarely glabrate, usually stout, 1°–7½° high, simple, or branched at the summit. Leaves more or less pubescent or scabrous, oval, oblong-lanceolate, or ovate-lanceolate, acute or acuminate, rarely obtusish at the apex, narrowed or obtuse at the base, rugosely veined on the lower surface, serrate, 1′–4′ long, 4″–18″ wide, sessile, or the lowest sometimes tapering into petioles; heads 1½″–2″ high, secund on the spreading or recurving, often leafy branches of the usually large and compound panicle; bracts of the involucre linear, obtuse or obtusish.

Usually in dry soil, in fields and along roadsides, Newfoundland to western Ontario, south to Florida and Texas. Consists of many races, differing in leaf-form, leaf-thickness and in pubescence. Supposed to hybridize with *S. sempervirens* L. Dyer's-weed. July–Nov.

28. Solidago pátula Muhl. Rough-leaved or Spreading Golden-rod. Fig. 4240.

Solidago patula Muhl.; Willd. Sp. Pl. 3: 2059. 1804.

Stem stout, rather rigid, usually simple, glabrous, or sometimes puberulent above, 2°–7° high. Leaves thick, glabrous beneath, exceedingly rough above, pinnately veined, the lower and basal ones very large, 3′–16′ long, 1½′–5′ wide, oval or elliptic, narrowed into margined petioles, the upper smaller, oval or oblong, sessile, acute, finely serrate or the uppermost entire; heads 2″–3½″ high, secund on the widely spreading and recurving branches of the loose panicle; rays small; bracts of the involucre linear-oblong, obtuse.

In swamps, Maine and Ontario to Minnesota, south to Georgia, Missouri and Texas. Ascends to 5000 ft. in North Carolina. Stem strongly angled, at least below. Aug.–Oct.

$\frac{3}{5}$

$\frac{1}{2}$

Solidago microphylla Engelm., ranging from Kansas to Texas, appears to be a race of this species with acutish inner bracts of the involucre.

30. Solidago Boòttii Hook. Boott's Golden-rod. Fig. 4242.

S. Boottii Hook. Comp. Bot. Mag. 1: 97. 1835.

Stem glabrous, or puberulent above, slender, usually branched near the summit, 2°–5° high. Leaves firm, pinnately and finely reticulate-veined, the lower and basal ones ovate or oblong-lanceolate, narrowed into margined, sometimes ciliate petioles, acuminate at the apex, sharply and rather coarsely serrate, 3′–6′ long, the upper smaller, entire, or finely serrate, sessile; heads 2″–3″ high, secund on the elongated, spreading or recurving branches of the usually ample panicle; rays few, small; bracts of the involucre linear-oblong, obtuse; achenes pubescent.

In dry woods, Virginia to Florida and Texas. Recorded from Missouri. Ascends to 3000 ft. in Virginia. July–Sept.

29. Solidago ulmifòlia Muhl. Elm-leaved Golden-rod. Fig. 4241.

Solidago ulmifolia Muhl.; Willd. Sp. Pl. 3: 2060. 1804.

Stem slender, glabrous, or puberulent at the summit, 2°–4° high, simple, or branched above, the arched branches puberulent or pubescent. Leaves thin, oblong to elliptic-lanceolate, acute or acuminate at the apex and base, coarsely and sharply serrate, pinnately veined, glabrous or sparingly pubescent, the lower and basal ones wider, 3′–5′ long, 1′–1½′ wide, narrowed into margined petioles, the upper smaller, sessile; heads 2″–3″ high, secund on the usually few and elongated, usually leafy branches of the panicle; rays few, small, deep yellow; bracts of the involucre oblong-lanceolate, obtusish.

In woods and copses, Nova Scotia to Georgia, west to Minnesota, Missouri and Texas. Ascends to 2100 ft. in Virginia. July–Sept.

$\frac{3}{5}$

$\frac{3}{5}$

31. Solidago Ellióttii T. & G. Elliott's Golden-rod. Fig. 4243.

Solidago Elliottii T. & G. Fl. N. A. **2**: 218. 1841.
Solidago elliptica Ell. Bot. S. C. & Ga. **2**: 376.
1824. Not Ait. 1789.

Stem glabrous, or minutely puberulent above, stout, 3°-6° high, simple, or branched at the inflorescence. Leaves firm, oblong or oblong-lanceolate, rarely ovate-oblong, acute or acuminate, sessile by a broad base, or sometimes narrowed below, finely serrate, crenate-serrate, or the upper entire, rough on the margins, pinnately veined, glabrous on both sides, or puberulent on the veins beneath, 1'-5' long, 4"-12" wide; heads about 3" high, more or less secund on the short, spreading or recurving branches of the narrow panicle; bracts of the involucre linear-oblong, obtuse; rays 6-12, short; achenes pubescent.

In swamps, Nova Scotia (?), Massachusetts to North Carolina and Georgia, mainly near the coast. Sept.–Oct.

32. Solidago neglécta T. & G. Swamp Golden-rod. Fig. 4244.

Solidago neglecta T. & G. Fl. N. A. **2**: 213. 1841.

Stem glabrous, or slightly rough above, simple, rather stout, 2°-4° high. Leaves firm, the basal and lower ones lanceolate or oblong-lanceolate, large, sometimes 12' long, acute or acutish, closely serrate or serrulate, tapering into margined petioles, rough on the margins; upper leaves smaller, lanceolate, acute, sessile, serrate or nearly entire; heads about 2½" high, more or less secund on the short branches of the thyrsoid panicle; rays 3-8, small; bracts of the involucre thin, linear-oblong, obtuse; achenes glabrous, or nearly so.

In swamps and bogs, Maine to Vermont, Michigan, Maryland, Illinois and Wisconsin. Recorded north to New Brunswick. Forms with the heads little secund resemble *S. uliginosa.* Pyramid-golden-rod. Aug.–Sept.

$\frac{1}{2}$

33. Solidago uniligulàta (DC.) Porter. Few-rayed Golden-rod. Fig. 4245.

Bigelovia (?) *uniligulata* DC. Prodr. **5**: 329. 1836.
Solidago linoides T. & G. Fl. N. A. **2**: 216. 1841.
Not Soland.
Solidago neglecta var. *linoides* A. Gray, Syn. Fl. **1**:
Part 2. 154. 1884.
S. uniligulata Porter, Mem. Torr. Club **5**: 320. 1894.

Stem simple, slender, 1½°-2½° high, glabrous, or slightly pubescent above. Leaves firm, obscurely pinnately veined, lanceolate or oblong-lanceolate, finely and sharply serrate, acute or acuminate, the lower long-petioled, 4'-9' long, 4"-9" wide, the upper sessile, the uppermost very small and erect; heads about 2" high, densely secund on the short spreading or recurving branches of the small naked panicle; rays 1-4; bracts of the involucre firm, linear-oblong, obtuse; achenes glabrous.

In bogs and swamps, Newfoundland to New York, New Jersey, Ontario and Illinois. Aug.–Sept.

$\frac{1}{2}$

34. Solidago júncea Ait. Early or Sharp-toothed Golden-rod. Fig. 4246.

Solidago juncea Ait. Hort. Kew. **3**: 213. 1789.
S. arguta scabrella T. & G. Fl. N. A. **2**: 214. 1841.
Solidago juncea scabrella A. Gray, Syn. Fl. **2**: Part 2, 155. 1884.
Solidago juncea ramosa Porter & Britton, Bull. Torr. Club **18**: 368. 1891.

Stem glabrous, or very nearly so throughout, rigid, rather stout, simple, or branched at the inflorescence, 1½°–4° high. Leaves firm, glabrous, sometimes rough, lanceolate or oval-lanceolate, acute or acuminate, serrate, serrulate, or nearly entire, the lower large, sometimes 12′ long and 2′ wide, long-petioled, the upper smaller, sessile; heads 1½″–2″ high, secund on the recurved or sometimes nearly erect branches of the usually ample spreading panicle; rays 7–12, small; bracts of the involucre oblong or ovate-oblong, obtuse or acute; achenes glabrous or sparingly pubescent.

In dry or rocky soil, New Brunswick to Hudson Bay, Saskatchewan, North Carolina and Missouri. One of the earliest flowering species. Yellow top. Plume or pyramid-golden-rod. June–Nov.

½

35. Solidago argùta Ait. Cut-leaved Golden-rod. Fig. 4247.

Solidago arguta Ait. Hort. Kew. **3**: 213. 1789.
S. Muhlenbergii T. & G. Fl. N. A. **2**: 214. 1841.
S. Vaseyi Heller, Muhlenbergia **1**: 7. 1900.
S. Harrisii Steele, Contr. Nat. Herb. **13**: 369. 1911.

Stem simple, rather stout, glabrous, or sparingly pubescent above, 2°–4° high. Leaves thin, pinnately veined, the lower and basal ones broadly ovate or oval, short-acuminate, 3′–16′ long, 1′–5′ wide, narrowed into margined petioles, or subcordate, sharply and coarsely serrate; upper leaves sessile, ovate to oblong, acute or acuminate, more or less serrate, smaller; heads 2½″–3″ high, secund on the lateral racemose branches of the terminal, often leafy panicle; rays 5–7, large; bracts of the involucre oblong, obtuse; achenes glabrous or nearly so.

In rich woods, Maine to Ontario, Virginia and Tennessee. Ascends to 2700 ft. in the Adirondacks. July–Oct.

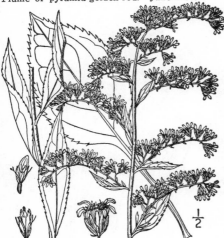

½

36. Solidago canadénsis L. Canada or Rock Golden-rod. Fig. 4248.

Solidago canadensis L. Sp. Pl. 878. 1753.
Solidago rupestris Raf. Ann. Nat. 14. 1820.
S. canadensis glabrata Porter, Bull. Torr. Club **21**: 310. 1894.

Stem slender, glabrous, or pubescent above, 1°–5° high. Leaves thin, triple-nerved, linear-lanceolate, 2′–5′ long, 3″–5″ wide, entire, or serrate with somewhat appressed teeth, acuminate at the apex, narrowed at the base, sessile, or the lowest petioled, glabrous, or pubescent; heads small, 1¼″ high or less, secund on the spreading branches of the often large panicle; rays 4–6, short; bracts of the involucre thin, linear, acutish; achenes small, nearly glabrous.

Hillsides, thickets and banks of streams, Newfoundland to Virginia, Ontario, Saskatchewan, Tennessee and South Dakota. Aug.–Oct.

Solidago gilvocanéscens Rydb. differs in being canescent; it ranges from Illinois to Manitoba, Saskatchewan, Nebraska and Utah.

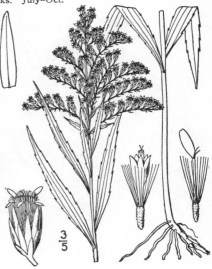

3/5

37. Solidago serótina Ait. Late Golden-rod. Fig. 4249.

Solidago serotina Ait. Hort. Kew. **3**: 211. 1789.
S. gigantea Ait. Hort. Kew. **3**: 211. 1789.
Solidago Pitcheri Nutt. Journ. Acad. Phil. **7**: 101. 1834.
S. serotina gigantea A. Gray, Proc. Am. Acad. **17**: 180. 1882.

Stem stout, 3°–8° high, glabrous, sometimes glaucous. Leaves lanceolate or oblong-lanceo-late, thin, triple-nerved, sharply serrate, or nearly entire, sessile, or the lowest petioled, glabrous on both sides or pubescent beneath, more or less rough-margined, 3′–6′ long, 2″–15″ wide, acumi-nate at the apex, narrowed at the base; heads 2½″–3″ high, crowded on the spreading or re-curving branches of the usually large and often leafy panicle, which are sometimes puberulent; bracts of the involucre oblong, thin, obtuse; rays 7–15, rather large; achenes finely pubescent.

In moist soil, New Brunswick to British Columbia and Oregon, south to Georgia, Texas and Utah. As-cends to 2300 ft. in Virginia. Aug.–Oct.

38. Solidago glabérrima Martens. Missouri Golden-rod. Fig. 4250.

Solidago glaberrima Martens, Bull. Acad. Brux. **8**: 68. 1841.
S. moritura Steele, Contr. Nat. Herb. **13**: 370. 1911.

Stem glabrous, rather slender, 3°–5° high. Leaves firm or thick, those of the stem linear-lanceolate to linear, sessile, acuminate at the apex, narrowed at the base, 2′–4′ long, rough-margined, triple-nerved, entire, or sparingly serrate with low sharp teeth, the basal and lowest ones mostly larger, spatulate, petioled; heads 2″–3″ high, se-cund on the spreading or recurving branches of the short and broad, usually naked panicle; bracts of the involucre oblong, greenish-tipped, obtuse, or the inner acute, thick; rays 6–13, short; achenes nearly glabrous.

On dry prairies, Manitoba and Minnesota to Ten-nessee, Alberta, Washington, Missouri and Texas. Autumn. Referred, in our first edition, to *S. mis-souriensis* Nutt. of the far west.

39. Solidago Shórtii T. & G. Short's Golden-rod. Fig. 4251.

Solidago Shortii T. & G. Fl. N. A. **2**: 222. 1841.

Stem slender, glabrous below, minutely pu-bescent above, 2°–4° high. Leaves firm, ob-long-lanceolate, sessile (the lowest petioled?), triple-nerved, acute or acuminate at the apex, glabrous on both sides, the larger 2′–4′ long, 5″–6″ wide, sharply serrate with rather small and distant teeth, the upper gradually smaller, sparingly serrate, or entire; heads about 3″ high, secund on the usually recurved branches of the commonly large puberulent panicle; in-volucre narrowly campanulate, its bracts linear-oblong, obtuse; rays 5–7, rather small; achenes silky-pubescent.

On rocks at the Falls of the Ohio river. Er-roneously recorded from northwestern Arkansas. July–Aug.

40. Solidago Gattíngeri Chapm. Gattin-ger's Golden-rod. Fig. 4252.

Solidago Gattingeri Chapm.; A. Gray, Syn. Fl. 1: Part 2, 156. 1884.

Stem slender, 2°–3° high, branched at the in-florescence, glabrous throughout. Leaves firm, glabrous beneath, rough above, ciliolate, the lower and basal ones oblanceolate or spatulate, acutish, 3′–6′ long, 6″–10″ wide, serrate with low distant teeth, narrowed into margined petioles; upper leaves abruptly smaller, linear-oblong or oblan-ceolate, bract-like, entire, sessile; heads 2″–2½″ high, somewhat secund on the spreading, often very slender and elongated branches of the pani-cle; bracts of the involucre oblong, very obtuse; rays 6–10; achenes puberulent, or glabrous below.

In dry soil, Tennessee and Missouri. July–Aug. A plant similar to this, but with larger serrate stem-leaves, occurs in central Nebraska.

½

41. Solidago altíssima L. Tall, High, or Double Golden-rod. Fig. 4253.

Solidago altissima L. Sp. Pl. 878. 1753.
S. procera Ait. Hort. Kew. **3**: 211. 1789.
Solidago canadensis procera T. & G. Fl. N. A. **2**: 224. 1841.
Solidago canadensis scabra T. & G. Fl. N. A. **2**: 224. 1841.

Stem stout, pubescent, or hirsute nearly through-out, 2°–8° high. Leaves lanceolate, triple-nerved, acute at each end, roughish above and pubescent beneath, the lower ones sharply serrate and petioled, 3′–6′ long, 4″–12″ wide, the upper smaller, often en-tire, sessile; heads 1½″–2½″ high, usually numerous, secund on the spreading or recurving branches of the usually large panicle; bracts of the involucre linear, obtuse or acutish; rays 9–15; achenes gla-brous or somewhat pubescent.

Usually in dry soil, Maine to Ontario, Nebraska, Geor-gia and Texas. Confused, in our first edition, with *S. canadensis* L. Yellow weed. Aug.–Nov.

½

42. Solidago nemoràlis Ait. Gray, Field, or Dwarf Golden-rod. Dyer's Weed. Fig. 4254.

Solidago nemoralis Ait. Hort. Kew. **3**: 213. 1789.
Solidago nemoralis arenicola Burgess; Britton & Brown, Ill. Fl. **3**: 344. 1898.

Slender, ashy-gray, 6′–2° high, erect, depressed or sometimes prostrate, finely and densely pubescent. Leaves thick, roughish, the basal and lower ones of the stem oblanceolate or spatulate, petioled, obscurely triple-nerved, obtuse or acutish, crenate-dentate, 3′–6′ long, 4″–12″ wide; upper leaves gradually smaller, oblanceolate or linear-oblong, acute or acutish, en-tire; heads 2″–3″ high, secund on the spreading or recurving branches of the terminal, usually one-sided panicle; bracts of the involucre linear-oblong; rays 5–9; achenes pubescent.

In dry soil, Nova Scotia and Quebec to Saskatchewan, Florida, Texas and Arizona. July–Nov.

Solidago pulchérrima A. Nelson (*S. longipetiolata* Mack. & Bush), of Central North America, ranging east-ward into Wisconsin and Missouri, differs mainly by somewhat larger heads, and is here regarded as one of the many races of *S. nemoràlis* Ait.

½

43. Solidago móllis Bartl. Velvety or Ground Golden-rod. Fig. 4255.

Solidago mollis Bartl. Ind. Sem. Goett. 5. 1836.

Solidago incana T. & G. Fl. N. A. **2** : 221. 1841.

Solidago nemoralis var. *incana* A. Gray, Proc. Am. Acad. **17** : 197. 1882.

Stem rigid, stout, low, canescent or slightly scabrous, 6′–12′ high. Leaves pale, canescent or rough, entire or dentate, strongly 3-nerved, oblong, ovate or oblanceolate, the lower petioled, 2′–3′ long, 3″–12″ wide, very obtuse, the upper sessile, smaller; heads 2″–3″ high, somewhat or scarcely secund on the short branches of the erect, scarcely one-sided, dense thyrsoid panicle; bracts of the involucre oblong, obtuse or acutish; rays 5–9; achenes pubescent.

Dry plains, Minnesota to Manitoba, Assiniboia, Kansas, Texas and Mexico. July–Sept.

44. Solidago rádula Nutt. Western Rough Golden-rod. Fig. 4256.

Solidago radula Nutt. Journ. Acad. Phila. **7** : 102. 1834.

Stem rough-pubescent, rather slender, 1°–3° high. Leaves thick, green, rough on both sides, the lower oblanceolate or spatulate, dentate-crenate, obtuse or acutish, petioled, 3–8′ long, 6″–20″ wide, obscurely 3-nerved, the upper smaller, oblanceolate, entire or nearly so, sessile, distinctly 3-nerved, the uppermost very small; heads 2″–3″ high, secund on the short, usually somewhat recurved branches of the dense, often one-sided panicle; bracts of the involucre oblong or linear-oblong, obtuse; rays 3–7, very short; achenes minutely pubescent.

In dry soil, Illinois to Missouri, Louisiana and Texas. Aug.–Sept.

45. Solidago Drummóndii T. & G. Drummond's Golden-rod. Fig. 4257.

Solidago Drummondii T. & G. Fl. N. A. **2** : 217. 1841.

Stem rather slender, 1°–3° high, finely soft-pubescent. Leaves rather thin, broadly ovate or oval, glabrous or nearly so above, finely, but sometimes sparingly, pubescent, or roughish beneath, sharply serrate, acute at the apex, narrowed at the base, 3-nerved and pinnately veined, petioled, or the upper sessile, the larger 3′–4′ long, 1½′–2′ wide; heads 2″–2½″ high, secund on the usually few, spreading or recurving branches of the panicle; bract-like upper leaves obtuse, or acute, entire; rays 4–6, conspicuous; bracts of the involucre oblong-lanceolate, obtuse; achenes pubescent.

In rocky soil, Illinois to Louisiana and Missouri. Sept.–Oct.

46. Solidago rígida L.   Stiff or Hard-leaved Golden-rod.   Fig. 4258.

Solidago rigida L. Sp. Pl. 880. 1753.
Oligoneuron canescens Rydb. Bull. Torr. Club **31**: 652. 1905.

Stem stout, simple, or branched above, densely and finely rough-pubescent, hoary, 1°–5° high. Leaves thick, flat, rigid, ovate to oblong, pinnately veined, often obtuse, rough on both sides, or smoothish beneath, the upper sessile, clasping, and rounded or sometimes narrowed at the base, 1′–2′ long, mostly entire; lower and basal leaves long-petioled, sometimes 1° long and 3′ wide, entire or serrulate; heads 4″–5″ high, many-flowered, in a terminal dense, compound corymb, the clusters sometimes slightly secund; involucre broadly campanulate, its bracts oblong, obtuse, the outer pubescent; rays 6–10, large; achenes glabrous, 10–15-nerved.

In dry sandy, gravelly or rocky soil, Ontario to Saskatchewan, Massachusetts, Georgia, Texas and Colorado. Aug.–Oct.

Solidago corymbòsa Ell., of the Southern States, differing by being less rough and with slightly smaller heads, probably a race of this species, ranges north into Ohio.

47. Solidago ohioénsis Riddell.   Ohio Golden-rod.   Fig. 4259.

Solidago ohioensis Riddell, Syn. Fl. West. States 57. 1835.

Very smooth throughout; stem rather slender, simple, 2°–3° high. Leaves firm, pinnately veined, flat, the basal and lower ones elongated-lanceolate or oblong-lanceolate, obtuse, long-petioled, serrulate toward the end, or entire, often 1° long; upper leaves sessile, lanceolate, entire, gradually smaller and those of the inflorescence bract-like; heads 2½″–3″ high, numerous in a terminal compound corymb, 15–25-flowered; rays 6–9, small; bracts of the narrowly campanulate involucre oblong, very obtuse, glabrous; achenes glabrous, 5-nerved.

In moist soil, western New York and southern Ontario to Illinois, Michigan and Wisconsin. Aug.–Sept.

48. Solidago Riddéllii Frank.   Riddell's Golden-rod.   Fig. 4260.

Solidago Riddellii Frank; Riddell, Syn. Fl. West. States 57. 1835.

Stem stout, glabrous, or slightly pubescent above, 1°–3° high. Leaves numerous, thick, glabrous on both sides, entire, acute at each end, the lower and basal ones long-petioled, elongated, lanceolate, somewhat triple-nerved and conduplicate, often 1° long, 4″–10″ wide, the upper smaller, similar, sessile and clasping at the base, conduplicate, somewhat falcate; heads 3″–4″ high, 20–30-flowered, very numerous in a dense corymb; involucre oblong-campanulate, its bracts broadly oblong, obtuse; rays 7–9, narrow; achenes 5-nerved, glabrous.

On moist prairies, Ontario to Minnesota, Ohio and Missouri. Also at Fortress Monroe, Va. A supposed hybrid with *S. rigida* L. occurs in Iowa. Aug.–Sept.

3/5

49. Solidago Hoùghtonii T. & G.
Houghton's Golden-rod.   Fig. 4261.

Solidago Houghtonii T. & G.; A. Gray, Man.
211.   1848.

Stem slender, glabrous below, sparingly
pubescent above, 1°–2° high.   Leaves linear,
the basal and lower ones petioled, 4'–5'
long, 2"–4" wide, 3-nerved, entire, acute at
each end, the upper smaller, sessile, slightly
conduplicate, otherwise similar, the uppermost small and bract-like; heads about 3"
high, few, in a small corymbose cyme, 20–
30-flowered; involucre broadly campanulate, its bracts oblong, obtuse; achenes glabrous, 4–5-nerved.

In swamps, north shores of Lakes Michigan
and Huron, and in Genesee Co., N. Y. Recorded from Lake Superior.   Autumn.

23. EUTHÀMIA Nutt. Gen. 2: 162.   1818.

Erect, paniculately-branched herbs, perennial by long rootstocks, with linear or linear-
lanceolate, entire, sessile, 1–5-nerved punctate leaves, and very numerous small heads of both
tubular and radiate yellow flowers, clustered in the large corymbose, convex or nearly flat-
topped inflorescence.   Bracts of the involucre obtuse, imbricated in several series, appressed,
somewhat glutinous.   Receptacle flattish, fimbrillate, or pilose.   Ray-flowers pistillate, usually
more numerous than the disk-flowers, the rays small.   Disk-flowers perfect.   Anthers obtuse
at the base.   Style-branches with lanceolate appendages.   Achenes top-shaped or oblong,
villous-pubescent.   [Greek, referring to the clustered heads.]

About 10 species, natives of North America.   Type species: *Euthamia graminifolia* (L.) Nutt.

Leaves distinctly 3–5-ribbed; heads 20–30-flowered.
   Involucre 2"–2½" high, the bracts yellowish.   1. *E. graminifolia.*
   Involucre less than 2" high, the bracts with appressed green tips.   2. *E. floribunda.*
Leaves 1-ribbed, or with a pair of indistinct lateral nerves; heads rarely more than 20-flowered.
   Involucre 2½"–3" high; southwestern species.
     Leaves 2"–4" wide; involucre scarcely viscid.   3. *E. leptocephala.*
     Leaves 1"–2½" wide; involucre very viscid.   4. *E. gymnospermoides.*
   Involucre 2" high, or less.
     Leaves 1"–2½" wide; involucre campanulate.   5. *E. tenuifolia.*
     Leaves less than 1" wide; involucre subcylindric.   6. *E. minor.*

1. Euthamia graminifòlia (L.) Nutt.
Bushy, Fragrant, or Flat-topped Golden-
rod.   Fig. 4262.

Chrysocoma graminifolia L. Sp. Pl. 841.   1753.
Solidago lanceolata L. Mant. 114.   1767.
S. graminifolia Salisb. Prodr. 109.   1796.
E. graminifolia Nutt. Gen. 2: 162.   1818.
E. Nuttallii Greene, Pittonia 5: 73.   1902.
E. camporum Greene, loc. cit. 74.   1902.

3/5

Stem paniculately much branched, or rarely
simple, glabrous or roughish-pubescent, 2°–4°
high.   Leaves numerous, linear-lanceolate,
acuminate or acute at each end, 1'–5' long,
2"–4" wide, 3–5-nerved, minutely rough-pubescent on the margins and nerves of the lower
surface; resinous dots few; heads 2"–2½" high,
sessile in capitate clusters arranged in a flat-
topped compound corymb; involucre ovoid-
campanulate to subcylindric, its yellowish
bracts oblong or oblong-lanceolate, slightly
viscid; rays 12–20; disk-flowers 8–12.

In moist soil, fields and roadsides, New Brunswick to Saskatchewan, Alberta, Florida, Nebraska
and Wyoming.   Fragrant.   July–Sept.

2. Euthamia floribunda Greene. Small-headed
Bushy Golden-rod. Fig. 4263.

E. floribunda Greene, Pittonia **5**: 74. 1902.

Solidago polycephala Fernald, Rhodora **10**: 93. 1908.

Finely roughish-pubescent, at least above, paniculately branched, 2°–3° high. Leaves linear-lanceolate, the larger 2′–3′ long, 2½″–3″ wide, 3-nerved,
those of the branches much smaller, spreading or
deflexed; heads numerous, small, 1½″–2″ high, sessile or very nearly so in small corymbed clusters;
involucre turbinate, its glutinous bracts puberulent,
their triangular-lanceolate green tips appressed.

Fields and borders of marshes, southern New Jersey,
and recorded from eastern Pennsylvania. Aug–Oct.

3. Euthamia leptocéphala (T. & G.) Greene.
Western Bushy Golden-rod. Fig. 4264.

Solidago leptocephala T. & G. Fl. N. A. **2**: 226. 1841.

Euthamia leptocephala Greene, Mem. Torr. Club **5**: 321.
1894.

Stem smooth, 1½°–2½° high, branched above. Leaves
linear-lanceolate, acuminate or acute at each end,
1-nerved, or with a pair of indistinct lateral nerves,
rough-margined, those of the stem usually 2′–3′ long,
2″–4″ wide; heads about 3″ high, rather narrow, sessile in the clusters of the flat-topped inflorescence;
bracts of the subturbinate involucre linear-oblong,
scarcely viscid; disk-flowers 3 or 4; ray-flowers 7–10.

In moist soil, Missouri to Louisiana and Texas. Aug.–
Oct.

4. Euthamia gymnospermoìdes
Greene. Viscid Bushy Golden-
rod. Fig. 4265.

Euthamia gymnospermoides Greene, Pittonia
5: 75. 1902.

Solidago gymnospermoides Fernald, Rhodora
10: 93. 1908.

Usually branched from the base or from
below the middle, glabrous, resinous, 1½°–2°
high, the branches strict, ascending. Leaves
narrowly linear, 1″–2″ wide, 3′ long or less,
light green, 1-nerved, or the larger 3-nerved;
heads numerous, sessile in the clusters of
the broad nearly flat-topped inflorescence;
involucre turbinate, about 2½″ high, its
bracts linear-oblong, blunt, very viscid;
disk-flowers 4–6; ray-flowers about 12.

Prairies, Nebraska to Kansas, Texas and
Louisiana. Aug.–Oct. Confused, in our first
edition, with the preceding species.

5. Euthamia tenuifòlia (Pursh) Greene. Slender Fragrant Golden-rod.
Quobsque-weed.   Fig. 4266.

?*Erigeron carolinianum* L. Sp. Pl. 863. 1753.

Solidago tenuifolia Pursh, Fl. Am. Sept. 540. 1814.

E. tenuifolia Greene, Pittonia 5 : 77. 1902.

E. remota Greene, loc. cit. 78. 1902.

?*S. Moseleyi* Fernald, Rhodora 10 : 93. 1908.

Glabrous and somewhat resinous, seldom over 1½° high, branched above. Leaves narrowly linear, entire, acuminate, sessile, narrowed at the base, 1-nerved or with an additional pair of faint lateral nerves, 1′–3′ long, 1″–2″ wide, punctate, often with smaller ones clustered in the axils, the resinous dots minute; heads about 1½″ high, very numerous and crowded in the dense nearly flat corymb; involucre oblong-campanulate, its bracts oblong; rays 6–12; disk-flowers 4–6.

In dry sandy soil, eastern Massachusetts to Illinois, Wisconsin, Florida and Louisiana. Referred, in our first edition, to *Euthamia caroliniana* (L.) Greene, but the identity of *Erigeron carolinianum* L. is doubtful. Aug.–Oct.

6. Euthamia minor (Michx.) Greene. Narrow-leaved Bushy Golden-rod. Fig. 4267.

Solidago lanceolata minor Michx. Fl. Bor. Am. 2 : 116. 1803.

Euthamia minor Greene, Pittonia 5 : 78. 1902.

Solidago minor Fernald, Rhodora 10 : 93. 1908.

Glabrous, bushy-branched above, 3° high or less. Leaves very narrowly linear, 1-nerved, the larger about 2½′ long, not over 1″ wide, often with tufts of smaller ones in the axils, the upper much smaller, often not more than ¼″ wide; heads very numerous, short-stalked, or sessile; involucre cylindraceous, about 2″ high, its yellowish oblong bracts appressed, viscid; ray-flowers about 10.

In dry sandy soil, Virginia to Florida and Mississippi. Sept.–Oct.

24. BRACHYCHAÈTA T. & G. Fl. N. A. 2 : 194. 1841.

An erect, perennial herb, with the aspect of a golden-rod. Leaves alternate, the lower and basal ones large, cordate, long-petioled, the upper ovate, short-petioled or sessile. Heads composed of both tubular and radiate flowers, sessile, in a terminal narrow spike-like thyrsus. Involucre narrowly campanulate, its bracts coriaceous, imbricated in few series, the outer successively smaller. Receptacle small, naked. Rays small, yellow, pistillate. Disk-flowers perfect, their corollas tubular, somewhat expanded above, 5-cleft. Anthers obtuse and entire at the base. Style-branches flattened, their appendages lanceolate. Achenes 8–10-ribbed. Pappus a single row of scale-like bristles, shorter than the achene. [Greek, short-bristle, referring to the pappus.]

A monotypic genus of eastern North America.

1. Brachychaeta sphacelàta (Raf.) Britton. False Golden-rod. Fig. 4268.

Solidago sphacelata Raf. Ann. Nat. 14. 1820.
S. cordata Short, Trans. Journ. Med. 7: 599. 1834.
Brachychaeta cordata T. & G. Fl. N. A. 2: 194. 1841.
B. sphacela.a Britton; Kearney, Bull. Torr. Club 20: 484. 1893.

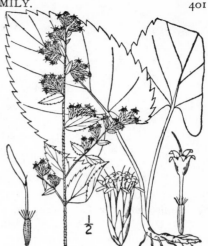

Stem erect, pubescent, simple or branched above, 2°–4° high. Basal and lower leaves broadly ovate, cordate or truncate at the base, acute at the apex, pinnately veined, sharply serrate, 3′–6′ long, the slender petioles 3′–9′ long, stem leaves gradually smaller and shorter-petioled, the uppermost very small and sessile; heads about 2½″ high, racemose-secund or densely clustered on the short branches of the narrow elongated terminal thyrsus; bracts of the involucre oblong or linear-oblong, obtuse or acutish; rays and disk-flowers each about 5.

In dry woods, Virginia to Indiana, western Kentucky, North Carolina, Georgia and Alabama. Aug.–Sept.

25. APHANÓSTEPHUS DC. Prodr. 5: 310. 1836

Erect or ascending canescent branching herbs, with alternate leaves, and rather large heads of both tubular and radiate flowers, solitary at the ends of the branches. Involucre hemispheric, its bracts lanceolate or linear, scarious-margined, imbricated in a few series, the outer smaller. Receptacle convex or conic, naked. Ray-flowers pistillate, white or purplish. Disk-flowers perfect, yellow, their corollas tubular, the limb expanded above, 5-dentate. Anthers obtuse and entire at the base. Style-branches flattened, their appendages short, obtuse. Achenes many-ribbed. Pappus a short dentate crown. [Greek, faint-crown.]

About 5 species, natives of the southwestern United States and Mexico. Type species: *Aphanostephus ramosissimus* DC.

1. Aphanostephus skirróbasis (DC.) Trelease. Aphanostephus. Fig. 4269.

Keerlia skirrobasis DC. Prodr. 5: 310. 1836.

Leucopsidium arkansanum DC. Prodr. 6: 43. 1837.

Aphanostephus arkansanus A. Gray, Pl. Wright. 1: 93. 1852.

Aphanostephus skirrobasis Trelease; Coville & Branner, Rep. Geol. Surv. Ark. 1884: Part 4, 191. 1891.

Erect, or diffusely branched, densely canescent, 6′–2° high. Basal and lower leaves spatulate, obtuse, 1′–4′ long, somewhat dentate, laciniate or entire, narrowed into margined petioles; upper leaves lanceolate, oblong or oblanceolate, obtuse or acute, mostly sessile, smaller; heads 8″–12″ broad, 3″–5″ high; rays numerous, narrow, entire; achenes ribbed and angled; pappus a lobed or dentate crown.

In dry soil, Kansas to Texas and Chihuahua, east to Florida. May–Aug.

26. BÉLLIS [Tourn.] L. Sp. Pl. 886. 1753.

Tufted herbs, with branching or scapose stems, alternate or basal leaves, and rather large heads of both tubular and radiate flowers, solitary at the ends of the branches, or of the monocephalous scape. Involucre hemispheric or broadly campanulate, its bracts herbaceous, imbricated in 1 or 2 series, nearly equal. Receptacle convex or conic, naked. Ray-flowers white or pink, pistillate. Disk-flowers yellow, perfect, their corollas tubular, the limb 4–5-toothed. Anthers obtuse and entire at the base. Style-branches flattened, their appendages short, triangular. Achenes flattened, obovate, nerved near the margins. Pappus none, or a ring of minute bristles. [Latin, pretty.]

About 9 species, natives of the northern hemisphere. Only the following are known to occur in the United States, but 2 others are found in Mexico. Type species: *Bellis perennis* L.

Stem branched, 6′–15′ high; involucral bracts acute.   1. *B. integrifolia.*
Scapes monocephalous, 1′–7′ high; involucral bracts obtuse.   2. *B. perennis.*

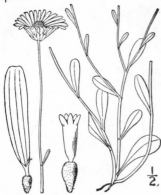

1. Bellis integrifòlia Michx. Western Daisy.
Fig. 4270.

Bellis integrifolia Michx. Fl. Bor. Am. **2**: 131. 1803.

Slender, diffusely branched, pubescent, 6'–15' high. Leaves thin, entire, obtuse, the lower and basal ones spatulate, 1'–3' long, narrowed into margined petioles, the upper smaller, oblong, oblanceolate or linear; heads 6''–15'' broad; bracts of the involucre acute or acuminate, scarious-margined, glabrous or nearly so; rays usually violet, oblong-linear; peduncles terminating the branches, 2'–7' long.

In moist soil, Kentucky and Tennessee to Arkansas and Texas. May–July.

2. Bellis perénnis L. European or Garden Daisy. Marguerite. Fig. 4271.

Bellis perennis L. Sp. Pl. 886. 1753.

Perennial, tufted. Leaves all basal, obovate, obtuse, slightly dentate, 1'–2' long, narrowed into margined petioles, pubescent and ciliate; scapes naked, 1'–7' high, usually several from the same root, pubescent; heads 6''–12'' broad; rays numerous, linear, white, pink, or purple; bracts of the involucre oblong, obtuse, usually purple.

In waste places, or occasionally spontaneous on lawns, southern New York and eastern Pennsylvania to Nova Scotia and Ontario. Fugitive from Europe. Native also of Asia. Naturalized in California and British Columbia. Herb-margaret. Ewe- or may-gowan. Childing-daisy. Bone- or bruise-wort. Bone-flower. Hen-and-chickens. Ban-wort. Bennert. March daisy. Bairn-wort. April–Nov.

27. TOWNSÉNDIA Hook. Fl. Bor. Am. **2**: 16. 1834.

Tufted scapose or branching herbs, with alternate, entire, linear or spatulate leaves, and large heads of both tubular and radiate flowers. Involucre hemispheric or broadly campanulate; bracts imbricated in several series, the outer shorter. Receptacle nearly flat, naked or fimbrillate. Ray-flowers pink or white, pistillate. Disk-flowers tubular, mostly perfect, their corollas regular, 5-lobed. Anthers obtuse and entire at the base. Style-branches flattened, their appendages lanceolate. Achenes of the disk-flowers compressed, those of the rays commonly 3-angled. Pappus a single series of rigid bristles or short scales. [Named for David Townsend, botanist, of Philadelphia.]

About 25 species, natives of western North America. Type species: *Townsendia sericea* Hook.

Branching from the base; heads terminal.   1. *T. grandiflora.*
Acaulescent, or nearly so; heads sessile among the leaves.   2. *T. exscapa.*

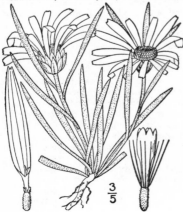

1. Townsendia grandiflòra Nutt. Large-flowered Townsendia. Fig. 4272.

Townsendia grandiflora Nutt. Trans. Am. Phil. Soc. (II) **7**: 306. 1841.

Perennial from a long woody root, branching at the base and sometimes also above, pubescent, or at length glabrate, 2'–8' high. Leaves linear or linear-spatulate, 1'–3' long, 1½''–3'' wide, canescent heads 1'–1½' broad, solitary at the ends of the branches; involucre hemispheric, its bracts scarious-margined, lanceolate, conspicuously acuminate; rays violet or purple; pappus of the ray-flowers a crown of short scales, that of the disk-flowers of rigid bristles longer than the achene, which is pubescent with 2-toothed hairs.

In dry soil, South Dakota to western Nebraska, Wyoming, Texas and New Mexico. May–Aug.

2. Townsendia exscàpa (Richards) Porter.
Silky or Low Townsendia. Fig. 4273.

Aster (?) *exscapus* Richards. App. Frank. Journ. 32. 1823.
Townsendia sericea Hook. Fl. Bor. Am. **2**: 16. *pl. 119.*
1834.
Townsendia exscapa Porter, Mem. Torr. Club **5**: 321.
1894.

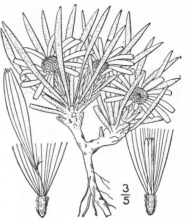

Acaulescent, or nearly so, from a deep woody root,
tufted, 1'–3' high. Leaves all clustered at the base,
narrowly linear or slightly spatulate, 1'–2' long, 1''–2½''
wide; heads closely sessile, 1'–1½' broad, equalled or
surpassed by the leaves; involucre broadly campanulate, its bracts lanceolate, acute, the inner scarious-
margined; rays white or purplish; pappus of both
ray- and disk-flowers a row of rigid bristles, those of
the rays shorter and smaller.

In dry soil, Saskatchewan to Montana, Nebraska,
Texas and New Mexico. Often silky-pubescent. April–
July.

Townsendia intermedia Rydb., ranging from Nebraska and Colorado to Texas, appears to be a race of
this species with leaves wider than in the type.

28. CHAETOPÁPPA DC. Prodr. 5: 301. 1836.

Annual slender much branched herbs, with small alternate entire leaves, and small long-
peduncled heads of both tubular and white or purple radiate flowers. Involucre narrowly
campanulate, its bracts lanceolate, acute or acuminate, imbricated in few series, the outer
slightly shorter. Receptacle small, naked. Ray-flowers in 1 row, pistillate. Disk-flowers
perfect, or the central ones staminate, their corollas 5-lobed. Anthers obtuse and entire at
the base. Style-branches narrow, flattened, their appendages short, obtuse. Achenes nearly
terete, fusiform, or linear, 5-ribbed. Pappus usually of 5 rigid awn-like scabrous bristles,
alternating with as many short scales or more. [Greek, bristle-pappus.]

Two known species, natives of the southwestern United States, the following typical. The genus
Distasis DC. (*Chaetopappa modesta* A. Gray) is here regarded as distinct.

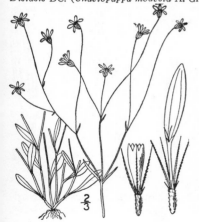

1. **Chaetopappa asteroìdes** DC. Chaetopappa.
Fig. 4274.

Chaetopappa asteroides DC. Prodr. **5**: 301. 1836.

Pubescent, erect, 2'–12' high, the branches filiform.
Lower and basal leaves spatulate, obtuse, ¾–1' long,
petioled, the upper linear, much smaller and bract-
like, sessile; heads about 3'' broad; involucre about
2'' high, its bracts scarious-margined, glabrous or
nearly so; rays 5–12, raised on slender tubes; central sterile flowers usually awnless; shorter scales
of the pappus of the fertile flowers hyaline, sometimes lacerate.

In dry soil, Missouri to Texas and northern Mexico.
April–July.

29. BOLTÒNIA L'Her. Sert. Angl. 16. 1788.

Perennial glabrous branching herbs, with striate or angled stems, alternate, entire, sessile
or clasping leaves, and numerous rather large heads of both tubular and radiate flowers,
paniculate, or solitary at the ends of the branches. Involucre hemispheric or broadly campanulate, its bracts scarious-margined, imbricated in few series, the outer slightly shorter.
Receptacle convex or conic, foveolate. Ray-flowers pistillate. Disk-flowers perfect, their
corollas elongated, 5-lobed. Anthers obtuse and entire at the base. Style-branches flattened,
their appendages short, lanceolate. Achenes flattened, obovate, their margins thickened or
narrowly winged, glabrous or nearly so. Pappus a series of short scales, usually with 2–4
slender rigid bristles. [Named for James Bolton, an English botanist of the 18th century.]

As here limited, the genus consists of the 4 following species, with perhaps a fifth in Oregon.
Type species: *Boltonia glastifolia* (Hill) L'Her.

Disk about 2″ broad ; leaves linear.
Disk 3″–6″ broad ; leaves lanceolate to oblanceolate.
    Leaves narrowed at the base, sessile, not decurrent on the stem.
        Involucre-bracts lanceolate, acute.
        Involucre-bracts spatulate, obtuse, or mucronate.
    Stem leaves, and sometimes those of the branches decurrent, sagittate.

1. *B. diffusa.*

2. *B. asteroides.*
3. *B. latisquama.*
4. *B. decurrens.*

1. Boltonia diffùsa Ell. Panicled Boltonia. Fig. 4275.

Boltonia diffusa Ell. Bot. S. C. & Ga. **2**: 400. 1824.

Paniculately much branched, 2°–7° high, the branches
very slender or filiform. Leaves linear, or the lower
linear-lanceolate, acutish, the larger 1′–2′ long, 1½″–2″
wide, those of the branches very small and subulate;
heads about 2″ high; disk about 2″ broad; rays usually
white, 1″–2″ long; involucre broadly campanulate, its
bracts oblong or oblong-lanceolate, acutish or obtuse;
achenes obovate, narrowly winged; pappus of several
short scales and 2 subulate bristles shorter than the
achene.

In dry soil, southern Illinois to Texas, east to South Carolina and Florida. Aug.–Oct.

2. Boltonia asteroìdes (L.) L'Her. Aster-like Boltonia. Fig. 4276.

Matricaria asteroides L. Mant. 116. 1767.
Matricaria glastifolia Hill, Hort. Kew. **19**: *pl. 3.* 1769.
Boltonia glastifolia L'Her. Sert. Angl. 16. 1788.
Boltonia asteroides L'Her. Sert. Angl. 16. 1788.

Rather stout, 2°–8° high, somewhat cymosely
paniculate. Leaves lanceolate, to oblanceolate, sessile, 2′–5′ long, 3″–12″ wide, the upper linear-lanceolate, smaller; heads 2″–4″ high; disk 3″–6″ wide;
rays white, pink or purple, 3″–6″ long; involucre
hemispheric, its bracts lanceolate or oblong-lanceolate, acute or acuminate; pappus of setose scales,
with or without 2–4 slender bristles nearly as long
as the obovate or oval achene.

In moist soil, Connecticut to Florida, west to Minnesota, Nebraska and Louisiana. July–Sept.

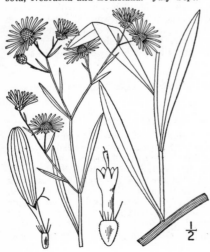

3. Boltonia latisquàma A. Gray. Broad-scaled Boltonia. Fig. 4277.

Boltonia latisquama A. Gray, Am. Journ. Sci. (II)
33: 238. 1862.

Similar to the preceding species and perhaps a
race of it. Leaves lanceolate, acute, sessile; heads
rather larger; rays violet-blue; bracts of the involucre oblong-spatulate, obtuse or mucronate;
pappus of numerous small short broad scales and
2 long bristles.

Western Missouri, Kansas, Arkansas and Oklahoma. Autumn.

4. Boltonia decúrrens (T. & G.) Wood.
Clasping-leaved Boltonia.   Fig. 4278.

Bol:onia glastifolia var. (?) *decurrens* T. & G. Fl. N.
A. 2 : 188. 1841.

Boltonia decurrens Wood, Bot. & Flor. 166. 1870.

Boltonia asteroides var. *decurrens* Engelm.; A. Gray,
Syn. Fl. 1 : Part 2, 166. 1884.

Stout, 3°–6° high, branched above. Leaves oblong-
lanceolate or elongated-lanceolate, mucronate at
the apex, those of the stem decurrent and sagittate
at the base, 3′–6′ long, 6″–8″ wide, those of the
branches smaller and merely sessile or some of them
also decurrent; heads 2½″–3″ high; involucre hemispheric; rays about 3″ long, violet or purple; pappus of several or numerous short scales and 2 very
slender bristles.

In wet prairies, Illinois and Missouri. Aug.–Sept.
Perhaps a race of *B. asteroides*.

30. SERICOCÀRPUS Nees, Gen. & Sp. Ast. 148. 1833.

Erect perennial herbs, with alternate leaves, and middle-sized heads of both tubular and
radiate flowers, in terminal cymose panicles. Involucre ovoid, oblong, or campanulate, its
bracts coriaceous, with herbaceous or squarrose tips, imbricated in several series, the outer
shorter. Receptacle small, foveolate. Ray-flowers white, pistillate. Disk-flowers mostly
perfect, their corollas tubular, narrow, yellowish or purplish, 5-lobed. Anthers obtuse and
entire at the base. Style-branches with lanceolate-subulate appendages. Achenes slightly
compressed, linear-oblong, 1-nerved on each side, pubescent. Pappus of numerous capillary
scabrous bristles, the outer usually shorter. [Greek, silky fruit, referring to the achenes.]

Four known species, natives of North America. Besides the following, another occurs on the
northwestern coast of North America. Type species: *Sericocarpus solidagineus* (Michx.) Nees.

Leaves entire, linear, spatulate, or obovate, rigid.
    Glabrous, or nearly so ; leaves linear or linear-spatulate.   1. *S. linifolius.*
    Puberulent or scabrous ; leaves obovate.   2. *S. bifoliatus.*
Leaves dentate, oblong, or obovate, thin.   3. *S. asteroides.*

1. Sericocarpus linifòlius (L.) B.S.P. Narrow-leaved White-topped Aster.
Fig. 4279.

Conyza linifolia L. Sp. Pl. 861. 1753.
Aster solidagineus Michx. Fl. Bor. Am. 2 : 108.
    1803.
Sericocarpus solidagineus Nees, Gen. & Sp. Ast.
    149. 1832.
Sericocarpus linifolius B.S.P. Prel. Cat. N. Y. 26.
    1888.

Glabrous or very nearly so throughout; stem
rather slender, striate, 1°–2½° high. Leaves
linear or linear-spatulate, spreading, faintly
3-nerved, thick, entire, obtuse at the apex, narrowed at the base, 1′–2′ long, 1½″–3″ wide, sessile, or the lowest on short margined petioles,
their margins scabrous; heads about 3″–4″
high, clustered in 2′s–6′s at the ends of the
cymose branches; involucre oblong-campanulate, 2″–3½″ long, its bracts oblong, obtuse, the
outer with somewhat spreading or reflexed
green tips, the inner scarious and often lacerate or ciliate at the apex; rays 4 or 5, about
5″ long; pappus white.

In dry, usually sandy soil, Maine to Ohio, Georgia and Louisiana. Erroneously recorded from
Canada. June–Sept.

$\frac{1}{2}$

In dry soil, Virginia to Florida and Louisiana, mainly near the coast. Rattlesnake-master. July–Sept.

2. Sericocarpus bifoliàtus (Walt.) Porter.
Rough White-topped or Silk-seed-Aster. Fig. 4280.

Conyza bifoliata Walt. Fl. Car. 204. 1788.
Aster tortifolius Michx. Fl. Bor. Am. **2**: 109. 1803.
S. tortifolius Nees, Gen. & Sp. Ast. 151. 1832.
S. bifoliatus Porter, Mem. Torr. Club **5**: 322. 1894.

Densely puberulent or scabrous, about 2° high, the stem terete, or slightly angled. Leaves sessile, obscurely veined, obovate or broadly spatulate, thick, obtuse, ascending or erect by a twist at the base, $\frac{1}{2}'$–$1'$ long, $4''$–$8''$ wide, entire, mucronulate, densely appressed-puberulent on both sides, the upper mostly oblong, much smaller; heads $4''$–$5''$ high, cymose-paniculate; involucre narrowly campanulate, its bracts oblong or the inner linear, pubescent, their tips somewhat spreading; rays short; pappus white.

3. Sericocarpus asteroìdes (L.) B.S.P.
Toothed White-topped Aster. Fig 4281.
Conyza asteroides L. Sp. Pl. 861. 1753.
S. conyzoides Nees, Gen. & Sp. Ast. 150. 1832.
S. asteroides B.S.P. Prel. Cat. N. Y. 26. 1888.

Stem pubescent, or glabrate, slightly angled, $1°$–$2°$ high. Leaves thin, pubescent or glabrous, ciliate, pinnately veined and faintly 3-nerved, the basal and lower ones obovate or spatulate, dentate or rarely entire, $2'$–$4'$ long, $1'$–$1\frac{1}{2}'$ wide, narrowed into margined petioles; upper leaves oblong or oblong-lanceolate, entire or dentate, acute or obtuse, smaller; heads $5''$–$6''$ high, densely clustered; involucre campanulate, its bracts oblong, ciliate or pubescent, the outer with green reflexed tips; pappus brown or white.

In dry woods, Maine to Florida, west to Ohio, Kentucky and Alabama. July–Sept.

$\frac{1}{2}$

31. ÁSTER L. Sp. Pl. 872. 1753.*

Perennial or rarely annual, mostly branching herbs, with alternate leaves, and corymbose or paniculate (rarely racemose or solitary) heads of both tubular and radiate flowers. Involucre hemispheric, campanulate or turbinate, its bracts various, imbricated in several series, the exterior ones usually smaller and shorter. Receptacle flat or convex, generally foveolate. Ray-flowers white, pink, purple, blue, or violet (very rarely yellowish), pistillate. Disk-flowers tubular, perfect, their corollas 5-lobed, usually yellow and changing to red, brown, or purple. Anthers obtuse and entire at the base. Style-branches flattened, their appendages subulate, lanceolate or ovate, acute. Pappus-bristles slender, numerous, scabrous or denticulate, usually in 1 series, sometimes in 2 series. Achenes mostly flattened and nerved. [Greek, star.]

A genus of not less than 250 species, most abundant in North America, where, in addition to the following, many others occur beyond our limits. A large number of the species apparently consist of many slightly differing races, and hybridism is also suspected. Known as Asters or Starworts. Type species: *Aster amellus* L.

A. Basal and lower leaves, or some of them, cordate and slender-petioled. (See No. 50.)

    1. *None of the stem leaves cordate-clasping.*
        * Rays white, violet, or rose.

        § Rays white, or rarely rose, usually 2-toothed; plants not glandular.
† Involucre ovoid, campanulate or turbinate, its bracts mostly obtuse or rounded; basal leaves few and small, or commonly none (except in No. 5).

        (a) Leaves membranous or thin, smooth, or nearly so.
Heads short-peduncled, $9''$ broad or less, the disk turning crimson; leaves acute or short-acuminate.
                          1. *A. carmesinus.*

* Text prepared for our first edition with the assistance of Prof. EDWARD S. BURGESS, here somewhat revised.

Heads long-peduncled, 9″ broad or more, the disk turning brown or reddish brown ; leaves long-
　　acuminate.
Heads 1′ broad or more ; leaves of the branches large, long, lanceolate, acuminate.
　　　　　　　　　　　　　　　　　　　　　　　　　　　　　　　　2. *A. tenebrosus.*
　　Heads 9″–12″ broad ; leaves of the branches small, obtuse, or acute.　3. *A. divaricatus.*
　　　　　　　　　　(b) Leaves thick, firm, rough.
Heads 6″–8″ high ; inflorescence forking.　　　　　　　　　　　　4. *A. furcatus.*
Heads 4″–5″ high ; inflorescence paniculate or glomerate.
　　Leaves acute, or short-acuminate, pilose beneath ; inflorescence glomerate.　5. *A. glomeratus.*
　　Leaves long-acuminate, not pilose beneath ; inflorescence open-paniculate.　6. *A. Claytoni.*
　　　†† Involucre cylindric, its bracts tapering to an obtuse apex ; basal leaves large, tufted.
Bracts of the involucre pale, scarious, usually without herbaceous tips.　7. *A. curvescens.*
Bracts of the involucre broader, with herbaceous tips.　　　　　　　8. *A. Schreberi.*
　　　　　§§ Rays violet, usually 3-toothed ; plants glandular.
　　　† Predominant glands large, capitate ; leaves thick, coarse, heavy.
Sinus broad ; glands chiefly confined to the inflorescence ; plant usually harsh.　9. *A. macrophyllus.*
Sinus narrow ; glands abundant on the leaves and stem ; growing plant clammy.　10. *A. roscidus.*
　　　†† Predominant glands minute, scarcely capitate ; leaves usually thin.
　　(a) Inflorescence rather regular, flat, or convex-topped ; plants usually less than 2½° tall.
Sinus broad, shallow.
　　Broader leaves orbicular-cordate, their teeth and the inflorescence-leaves inconspicuous.
　　　　　　　　　　　　　　　　　　　　　　　　　　　　　　　　11. *A. ianthinus.*
　　Broader leaves reniform, sharply incised ; some inflorescence-leaves conspicuous.
　　　　　　　　　　　　　　　　　　　　　　　　　　　　　　　　12. *A. violaris.*
Sinus rather deep and narrow ; broader leaves ovate-cordate, sharply serrate.　13. *A. multiformis.*
(b) Inflorescence very irregular, paniculate-corymbose ; plants often 4°–5° high ; broader leaves
　　　　　　　　　　　　　　　　　　　　　　large, cordate, acute.　14. *A. nobilis.*
　　　　** Rays blue or purple ; plants not glandular.
　　　† Bracts of the involucre spreading or recurved ; rays 30–45.　15. *A. anomalus.*
　　　†† Bracts of the involucre appressed, or erect ; rays 8–20.
　　　(a) Leaves all entire, or nearly so, thick, or firm.
Leaves nearly or quite glabrous above.　　　　　　　　　　　　　16. *A. Shortii.*
Leaves rough-puberulent on both sides, the upper bract-like.　　　17. *A. azureus.*
　　　　　(b) Leaves nearly all sharply serrate, thin.
Heads 2″–3″ high, numerous ; bracts obtuse or obtusish.
　　Leaves rough ; petioles not wing-margined ; bracts appressed.　　18. *A. cordifolius.*
　　Leaves smooth, or nearly so ; petioles, or some of them, wing-margined.　19. *A. Lowrieanus.*
Heads 4″–5″ high, usually few ; bracts acute or acuminate.　　　20. *A. Lindleyanus.*
Heads 3″–5″ high, numerous ; bracts acute or acuminate.
　　Stem densely and finely pubescent.　　　　　　　　　　　　　21. *A. Drummondii.*
　　Stem glabrous or nearly so ; bract-tips spreading.　　　　　　22. *A. sagittifolius.*
　　2. *Stem leaves, or some of them, cordate-clasping ; plant rough when dry.*
　　　　　　　　　　　　　　　　　　　　　　　　　　　　　　　　23. *A. undulatus.*

**B. No cordate and petioled leaves ; those of the stem, or some of them, with more or less cordate
　or auricled clasping bases (only slightly auricled in *A. tardiflorus*, and sometimes in *A. laevis*).**
　　　　　　1. *Stem rough, or hirsute-pubescent.*
　　　　　　　* Leaves entire, oblong, linear, or lanceolate.
　　　　§ Heads 1′–2′ broad ; leaves sessile, strongly cordate-clasping.
　　　　† Stem rough ; leaves oblong to lanceolate ; involucre turbinate.
Leaves thick, firm, very rough, oblong to oval.　　　　　　　　　24. *A. patens.*
Leaves thin, roughish, oblong-lanceolate.　　　　　　　　　　　25. *A. phlogifolius.*
　　　†† Stem hirsute ; leaves lanceolate ; involucre hemispheric ; bracts viscid.
　　　　　　　　　　　　　　　　　　　　　　　　　　　　　　　　26. *A. novae-angliae.*
　　　　§§ Heads ½′–1′ broad ; leaves but slightly clasping.
Involucre hemispheric, its bracts glandular.　　　　　　　　　　27. *A. oblongifolius.*
Involucral bracts hispid or ciliate.
　　Leaves lanceolate to oblong-lanceolate, 4″–6″ wide.　　　　　28. *A. nebraskensis.*
　　Leaves linear to linear-lanceolate, 2″–3″ wide.　　　　　　　29. *A. amethystinus.*
　　　　** Leaves, at least the lower, serrate.
Stem usually pilose ; bracts very glandular.　　　　　　　　　　30. *A. modestus.*
Stem hispid-pubescent ; bracts glabrous, or ciliate.　　　　　　31. *A. puniceus.*
　　　　2. *Stem glabrous, or only sparingly pubescent above.*
　　　　　　　* Leaves sharply serrate.
　　　　　　§ Leaves tapering to the base.
Leaves narrowed to the base, the lower into winged petioles.　　32. *A. tardiflorus.*
Leaves scarcely or gradually narrowed to the base.　　　　　　　31. *A. puniceus.*
　　§§ Leaves abruptly contracted into margined petioles, often enlarged near the base.
　　　　　　　　　　　　　　　　　　　　　　　　　　　　　　　　33. *A. prenanthoides.*
　　　§§§ Leaves usually strongly cordate-clasping ; bracts green-tipped.　34. *A. laevis.*
　　　　　** Leaves entire, or very nearly so.
　　　　§ Involucre campanulate, its bracts appressed, green-tipped.
　　　　　† Bracts of the involucre with rhomboid green tips.
Stem leaves oblong, lanceolate, or oval-lanceolate.　　　　　　　34. *A. laevis.*
Stem leaves elongated-lanceolate.　　　　　　　　　　　　　　　35. *A. concinnus.*
　†† Bracts of the involucre linear, the tips narrower, lanceolate ; stem leaves linear or narrowly
　　　　　　　　　　　　　lanceolate.　36. *A. purpuratus.*
　　　　　§§ Involucre hemispheric.
　　　　† Bracts of the involucre narrow, not foliaceous.

(a) Bracts in several series, unequal.

Bracts linear-subulate; leaves narrowly linear.   37. *A. junceus.*
Bracts lanceolate, linear, or spatulate; leaves lanceolate to linear.
    Western dry soil plant, 1°–2° tall; leaves 1′–3′ long.   38. *A. adscendens.*
    Eastern swamp plant, 2°–5° tall; leaves 2′–6′ long.   39. *A. novi-belgii.*
        (b) Bracts in only 1 or 2 series; leaves linear to lanceolate. 40. *A. longifolius.*
      †† Bracts of the involucre green, foliaceous; western.
Leaves lanceolate, firm; heads few.   41. *A. foliaceus.*
Leaves oblong-lanceolate, thin; heads many.   42. *A. phyllodes.*
    C. **Leaves sessile or petioled, not at all clasping, or scarcely so.**
      1. *Leaves silky, silvery or canescent on both sides, entire.*
Heads corymbose-paniculate; bracts oblong; achenes glabrous.   43. *A. sericeus.*
Heads in a narrow raceme; bracts linear; achenes silky.   44. *A. concolor.*
      2. *Leaves neither silvery, silky nor canescent, entire or toothed.*
        * Bracts of the involucre with herbaceous tips.
        † Bracts thin or coriaceous, relatively small.
        ‡ Leaves bristly-ciliate, linear, rigid; western species. 45. *A. Fendleri.*
        ‡‡ Leaves not bristly-ciliate.

○ Tips of the involucral bracts spreading (little spreading in *A. Radula,* erect or spreading in
    A. Herveyi); heads large and showy; rays violet to purple.
Leaves linear to linear-oblong, rigid, obtuse, entire; heads 2′ broad.   46. *A. grandiflorus.*
Leaves lanceolate to oblong, the lower sparingly dentate.
    Basal leaves with margined petioles.
      Involucre hemispheric to campanulate; heads 1′ or more.
        Bracts of the involucre glandular.   47. *A. spectabilis.*
        Bracts of the involucre ciliate, or glabrous.   48. *A. surculosus.*
      Involucre turbinate; heads 6″–9″ broad.   49. *A. gracilis.*
    Basal leaves with unmargined petioles.   51. *A. Herveyi.*
Leaves oblong-lanceolate, sharply serrate, rugose, the basal usually wanting. 50. *A. Radula.*
○○ Involucral bracts all appressed (except in *A. multiflorus* and *A. commutatus,* small-headed species).
        Δ Involucre top-shaped; rays violet.   52. *A. turbinellus.*
    ΔΔ Involucre hemispheric to campanulate; rays mostly white, sometimes purple.
Heads unilaterally racemose.
    Stem leaves oval, oblong, or lanceolate, serrate, or chiefly so.
      Stem pubescent or glabrate.   53. *A. lateriflorus.*
      Stem villous; leaves narrowly lanceolate, thin.   54. *A. hirsuticaulis.*
    Stem leaves linear-lanceolate to linear, nearly entire; stem glabrate. 55. *A. vimineus.*
Heads not unilaterally racemose, mostly paniculate.
    Involucral bracts spatulate, mostly ciliate, somewhat spreading, at least the outer obtuse; plants
      roughish-puberulent.
      Heads 3″–4″ broad, numerous; rays 10–20.   56. *A. multiflorus.*
      Heads 5″–8″ broad, relatively few; rays 20–30.   57. *A. commutatus.*
    Involucral bracts appressed, acute.
      Heads solitary at the ends of very small-leaved branchlets.   58. *A. dumosus.*
      Heads paniculate.
        Stem leaves lanceolate, serrate or entire.
          Heads 8″–10″ broad.
            Plants glabrous, or sparingly pubescent above.
              Leaves firm, roughish or rough; rays often purplish; involucral bracts acute.
                      59. *A. salicifolius.*
              Leaves thin, smoothish; rays chiefly white; involucral bracts acuminate.
                      60. *A. paniculatus.*
            Plant puberulent all over.   61. *A. missouriensis.*
            Heads 6″–8″ broad; stem leaves narrowly lanceolate.   62. *A. Tradescanti.*
        Stem leaves linear-lanceolate to subulate, mostly entire.
          Heads scattered, 6″–9″ broad; upper leaves linear.   63. *A. Faxoni.*
          Heads numerous, 4″–7″ broad; upper leaves subulate.
            Involucre subhemispheric, 2½″–3″ high.
              Rays usually white; heads 4″–7″ broad.
                Paniculately branched, bushy.   64. *A. ericoides.*
                Simple, or with slender ascending branches.   65. *A. Pringlei.*
             Rays purple; heads 8″–12″ broad.   66. *A. Priceae.*
            Involucre top-shoped, 2½″ high or less.
              Plant pilose-pubescent, 1½°–2° high.   67. *A. parviceps.*
              Plant glabrous, very slender, 1° high or less. 68. *A. depauperatus.*
†† Bracts stiff, relatively large; leaves narrow, rigid, entire. [Genus HELEASTRUM DC.]
                            69. *A. paludosus.*
        ** Bracts of the involucre without herbaceous tips.
Bracts linear-subulate, acuminate.
    Leaves firm, 3′ long or less, entire or sparingly serrate.   70. *A. nemoralis.*
    Leaves thin, 6′ long or less, sharply serrate.   71. *A. acuminatus.*
Bracts oblong or oblong-lanceolate, obtuse or acutish; leaves narrow, entire. [Genus UNAMIA
    Greene.]   72. *A. ptarmicoides.*

D. **Leaves fleshy, narrow, entire; plants of salt marshes or saline soil (No. 74 sometimes in
    non-saline situations).**

Perennial; heads 6″–12″ broad; involucral bracts lanceolate, acuminate.   73. *A. tenuifolius.*
Annuals; head 3″–5″ broad; involucral bracts linear-subulate. [Genus TRIPOLIUM Nees.]
    Involucre campanulate; disk-flowers more numerous than the rays; rays about 2″ long.
                            74. *A. exilis.*
    Involucre cylindraceous; disk-flowers fewer than the very short rays. 75. *A. subulatus.*

1. Aster carmesìnus Burgess.   Crimson-disk Aster.   Fig. 4282.

Stems erect, delicate, closely tufted, 1°–2° high, glabrous, reddish brown, terete. Leaves all petioled, glabrate, very thin, but firm and crisp, the lower and basal ones oval, rounded, or with a small deep and rounded sinus at the base, bluntly acute or short-acuminate at the apex, crenate-serrate, the upper ones sometimes ovate-lanceolate, the uppermost short-elliptic; petioles slender, the uppermost sometimes winged; inflorescence 5′ broad, or less, usually of about 5 convex glomerules, each often of 10–15 short-peduncled heads, its branches spreading, 3′ long, or less; rays chiefly 6, white; disk at first golden yellow, finally deep purplish crimson; florets broadly bell-shaped; outer bracts obtuse, ciliate, pale, with a green tip; achenes glabrous.

On shaded rocks, near Yonkers, N. Y. Peculiar in its dense glomerules subtended by large short-elliptic leaves, but probably a race of *A. divaricatus* L. September.

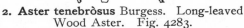

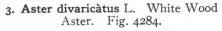

2. Aster tenebròsus Burgess.   Long-leaved Wood Aster.   Fig. 4283.

Stems solitary or scattered, glabrate, striate, about 3° high. Leaves very thin and smooth, slender-petioled, broadly oblong, coarsely toothed with remote acuminate teeth, abruptly long-acuminate at the apex, the basal sinus broad, rounded, shallow, except in the lowest ones; leaves of the inflorescence lanceolate, subentire, sessile, sometimes 4′ long; inflorescence broadly corymbose, heads about 4″ high, often 1¼′ broad; rays usually 9–12; disk pale yellow, becoming purplish brown, the florets funnelform with a long slender tube; outer bracts chiefly elongated-triangular, acute, green, the others linear, obtus-ish, the green tip lance-linear; achenes generally glabrous.

In moist dark woodlands, New York to Virginia. Peculiar in its large dark leaves with coarser teeth than in the next species. Aug.–Sept.

3. Aster divaricàtus L.   White Wood Aster.   Fig. 4284.

Aster divaricatus L. Sp. Pl. 873. 1753.

Aster corymbosus Ait. Hort. Kew. 3: 207. 1789.

Stems tufted, assurgent, flexuous, brittle, terete, 1½°–2½° high, glabrate at maturity. Leaves thin, smoothish, slender-petioled, ovate-lanceolate, dentate with sharp teeth, or the small basal ones coarsely serrate, acute to acuminate, the basal sinus broad or narrow; leaves of the inflorescence small, ovate and acute to orbicular; corymb broad, flattish, repeatedly forked, the slender branches long, divergent; heads 9″–12″ broad; rays chiefly 6–9, linear, white; disk turning brown; bracts of the involucre broad, ciliate, the rounded tip with an inconspicuous green spot.

In open woodlands and thickets, in rather dry soil, Quebec to Manitoba, Georgia and Tennessee. Here regarded as consisting of many slightly dif-fering races, a number of which have been con-sidered species and varieties. Sept.–Oct.

Aster virìdis Nees, remarkable for its coarse rough basal leaves, and large oval rhomboid rameal ones, occasionally from New York and Pennsylvania to Virginia, may prove to be a hybrid between the preceding and *A. macrophyllus* L.

4. **Aster furcàtus** Burgess.   Forking Aster. Fig. 4285.

Aster furcatus Burgess; Britt. & Brown, Ill. Fl. **3**: 358. 1898.

Stem leafy, 1½° high, or less, loosely forked above. Leaves hispid above, hispidulous beneath, firm, saliently cut-toothed, the lower ovate, short-petioled, with a small or shallow sinus, the upper sessile, with broad laciniate winged bases, often 5′ long by 2½′ wide, the uppermost elliptic-oblong, often 3½′ long; teeth long and low, sharp; heads few (5–20), slender-peduncled; involucre turbinate to campanulate, with a truncate or rounded base; rays 3-toothed; disk turning brown, the florets funnelform with rather broad lobes; pappus long, straight; achenes pubescent, subangular, not constricted at the summit.

In woods, especially on shaded cliffs, Indiana, Illinois and Missouri. Aug.–Oct.

5. **Aster glomeràtus** (Nees) Bernh.   Bernhardi's Aster.   Fig. 4286.

Eurybia glomerata Nees, Gen. & Sp. Ast. 139. 1832.
Aster glomeratus Bernh.; Burgess in Britt. & Brown, Ill. Fl. **3**: 358. 1898.

Loosely clustered, dull green. Leaves not large, mostly short-pilose beneath, thickish, rough above; basal leaves present, these and the lower stem leaves cordate with a deep, narrow sinus, the teeth sharp, rather close and small; petioles slender, ciliate; upper leaves much smaller, ovate, truncate with a short broadly winged base, or the uppermost ovate to lanceolate, sessile, entire; inflorescence compact, of many glomerate clusters, round-topped; heads about 4″ high; bracts pubescent, obtuse, green, the inner twice as long as the outer; rays about 6, cream-white, short, soon deciduous; disk turning brown.

In moist thickets or swamps, especially in ravines, Maine to New York and Virginia. July.

6. **Aster Clàytoni** Burgess.   Clayton's Aster.   Fig. 4287.

Aster Claytoni Burgess; Britt. & Brown, Ill. Fl. **3**: 358. 1898.

Similar to *A. divaricatus*, stems red, tough. Leaves chiefly ovate-lanceolate, not large, rough, thick, slender-petioled, coarsely serrate, pale, or dull, the apex incurved-acuminate, the upper spreading or deflexed, sessile by a broad base, lanceolate-triangular, serrulate; inflorescence high, with percurrent axis, the long suberect branches each bearing a small umbelliform cluster of heads; peduncles filiform, as long as the heads, 3″–4″ long; bracts pale; rays short, narrow, chiefly 6, snow-white; disks at first golden-yellow, finally sienna-brown; florets about 20, achenes densely short-hairy.

In sunny or slightly shaded rocky places, Maine to New York and the mountains of Virginia. Sept.

7. Aster curvéscens Burgess. Dome-topped Aster. Fig. 4288.

Aster curvescens Burgess; Britt. & Brown, Ill. Fl. **3**: 359. 1898.

Dark green, chiefly glabrous; rootstocks often 10′ long; stem smooth, striate, delicate, 1½°–3½° high. Basal leaves tufted, conspicuous, these and the lowest stem leaves with a broad sinus tapering into a petiole 1–2 times as long as the blade, abruptly incurved-acuminate; middle leaves ovate, short-petioled, rounded at the base, the upper lanceolate, slenderly acuminate, often falcate; leaves firm, smoothish, the teeth broad, curved; inflorescence mostly convex, 3′–5′ broad, its short filiform naked branches widely ascending; heads 4″–5″ high; lower bracts short, obtuse, the others longer, nearly uniform, scarious, shining, linear, often acute, usually glabrous; rays about 8, cream-white, about 5″ long; disk becoming purple-brown; pappus early reddening; achenes slender, glabrous.

In loose moist shaded soil, New England and New York to Virginia. Aug.–Sept.

8. Aster Schréberi Nees. Schreber's Aster. Fig. 4289.

Aster Schreberi Nees, Syn. Ast. 16. 1818.

Stem stout, 2°–3° high, with long internodes. Basal leaves often in extensive colonies, thin, dull green, firm, rough above, with scattered slender appressed bristles, pubescent beneath on the veins, reniform-cordate or cordate-triangular, often 7′ long by 5′ wide, the basal sinus when well developed rectangular, 2′ across and 1′ deep; upper leaves ovate-oblong to lanceolate, with a short broad basal wing, or sessile; petioles of the lower leaves long, conspicuously ciliate when young; inflorescence decompound, flattish, or irregularly convex, 6′–12′ broad; heads about 5″ high; bracts greenish, mostly obtuse, ribs and midrib dark green, ciliate; rays usually 10.

In borders of woods, and along fence rows in partial shade, New York to Michigan and Virginia. July–Aug.

9. Aster macróphýllus L. Large-leaved Aster. Fig. 4290.

Aster macrophyllus L. Sp. Pl. Ed. 2, 1232. 1763.

Rough; rootstocks long, thick; stem reddish, angular, 2°–3° high. Basal leaves forming large colonies, 3 or 4 to each stem, broad, cordate with a large irregular sinus, rough above, harsh, thick, the teeth broad, curved, pointed, the petioles long, narrow; upper stem leaves oblong with short broadly winged petioles, the uppermost sessile, acute; inflorescence strigose and glandular, broadly corymbose, irregular; heads 5″–6″ high; peduncles rigid, thickish; rays about 16, 5″–7″ long, chiefly lavender, sometimes violet, rarely pale; bracts conspicuously green-tipped, the lower acute, the inner oblong, obtuse; disk turning reddish brown; florets short-lobed.

In moderately dry soil, in shaded places, Canada to Minnesota and North Carolina. Here regarded as consisting of numerous slightly differing races, perhaps including the five following described as species. Aug.

10. Aster róscidus Burgess. Dewy-leaf Aster. Fig. 4291.

Aster roscidus Burgess; Britt. & Brown, Ill. Fl. **3**: 360. 1898.

Clammy-hairy, odorous, copiously glandular when young, somewhat so at maturity; stem 3° high, or less. Basal leaves in close colonies, coriaceous, the earlier ones cordate-quadrate, low-serrate, the sinus deep, narrow, the later, or winter leaves, elliptic, long-petioled, often prostrate, often 5′ long; stem leaves chiefly orbicular and not cordate, with short broadly winged petioles, rarely slender-petioled; inflorescence convex, sometimes irregular; involucre hemispheric, its bracts chiefly with rounded ciliate tips; rays 14–16, broad, clear violet; disks at first golden yellow, soon turning red; pappus long, white, copious.

In slight shade and rich cleared woodlands, Maine to Pennsylvania and Michigan. Aug.–Sept.

11. Aster iánthinus Burgess. Violet Wood Aster. Fig. 4292.

Aster ianthinus Burgess; Britt. & Brown, Ill. Fl. **3**: 360. 1898.

Glandular, dark green, slightly strigose-pubescent; stem erect, or decumbent, 2°–3° tall. Leaves thinnish, rough, the lower and basal ones orbicular to oblong, 5′ long, or less, abruptly acuminate, low-serrate or crenate; the sinus broad, open, shallow, upper leaves sessile by a narrowed base, crenate-serrate; inflorescence open, nearly naked, peduncles slender, divergent; heads large; rays 10–13, long, very deep violet or sometimes pale, 4″–6″ long; bracts green-tipped, little pubescent.

On shaded banks and along woodland paths, Maine to Lake Erie and West Virginia. July–Oct.

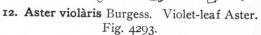

12. Aster violàris Burgess. Violet-leaf Aster. Fig. 4293.

Aster violaris Burgess; Britt. & Brown, Ill. Fl. **3**: 361. 1898.

Caudex thick, fleshy; plant glabrate, bluish green, minutely glandular; stem slender, erect, or assurgent, 2° high, or less. Basal and lower leaves broadly reniform, abruptly acuminate or apiculate, often 3′ long by 4′ wide, their slender petioles 6′–8′ long, the sinus very broad and shallow; middle stem leaves similar, not cordate; the upper numerous, long-elliptic, chiefly with narrowed bases, all thin, firm, rough above; inflorescence leafy, small, loose, rather narrow and high, paniculate-corymbose, nearly level-topped, its slender branches with nearly opposite, oblong leaves; heads 6″ high, or more; rays 12–15, pale violet, narrow.

In shaded moist places, sometimes in leaf-mold among rocks, New York from the Hudson to Lake Erie. Sept.–Oct.

13. Aster multifórmis Burgess. Various-
leaved Aster. Fig. 4294.

Aster multiformis Burgess; Britt. & Brown, Ill. Fl.
3: 361. 1898.

Deep green, minutely glandular; stem erect,
slender, 1°–2° high, angular-striate in drying.
Basal leaves usually 2, large, thick, cordate-
oblong, often accompanied by later smaller oblong ones; stem leaves very thin, sharply serrate, rough above, minutely puberulent beneath,
the lower ovate, acuminate, usually with a narrow sinus, the upper oval to ovate-lanceolate,
petioled, the uppermost elliptic-lanceolate, serrulate, sessile or nearly so; inflorescence small,
its branches upwardly curved; heads about 7″
high; rays about 13, rounded and retuse at the
apex; bracts green, glands few, almost hidden
by the minutely strigose pubescence of the peduncles.

In moist shaded places, Maine to western New
York, Pennsylvania and Maryland. July–Aug.

14. Aster nóbilis Burgess.   Stately Aster
Fig. 4295.

Aster nobilis Burgess; Britt. & Brown, Ill. Fl. **3**: 361.
1898.

Tall, minutely glandular above; stem shining,
bright green, 4°–5° high. Leaves thin, but firm,
smooth in growth, roughened in drying, minutely puberulent beneath, dark green, basal and
lower leaves large, the blade often 9′ long by 6′
wide, about as long as the stout petiole, sharply
toothed, the sinus deep, broad, or the lobes overlapping; stem leaves similar, the upper oblong-
lanceolate, sessile; inflorescence irregularly cymose-paniculate, with small subulate recurved
leaves; bracts long, acute, green; heads 6″ high,
or less; rays 13–15, violet-blue or pale violet;
disk-flowers not numerous, their corollas funnelform with a long capillary tube.

In leaf-mold, Lake Champlain to Lake Erie. Aug.

15. Aster anómalus Engelm. Many-
rayed Aster. Fig. 4296.

Aster anomalus Engelm.; T. & G. Fl. N. A. **2**: 503.
1843.

Stem rough, rather stout, branched above,
1°–3° high. Leaves thin, the lower and basal
ones deeply cordate, ovate, or ovate-lanceolate,
entire or slightly repand, rough-pubescent on
both surfaces, acute or acuminate at the apex,
3′–4′ long, 1′–2′ wide, on slender naked petioles; upper leaves short-petioled or sessile,
lanceolate, oblong, or linear, much smaller;
heads few, 4″–6″ high, 12″–15″ broad; receptacle hemispheric, its bracts lanceolate, acute,
or acuminate, hirsute, imbricated in several
series, their foliaceous tips spreading or reflexed; rays 30–45, 5″–6″ long, bright violet-
blue; pappus whitish.

On limestone cliffs, Illinois to Missouri and
Arkansas. Sept.

$\frac{1}{2}$

16. Aster Shórtii Hook.   Short's Aster.
Fig. 4297.

Aster Shortii Hook. Fl. Bor. Am. **2**: 9.   1834.

Stem roughish or smooth, slender, paniculately branched above, 2°–4° high.   Leaves
thick, glabrous or nearly so above, finely and
sparingly pubescent beneath, the lower and
basal ones ovate or ovate-lanceolate, cordate
at the base, acute or acuminate at the apex,
slightly dentate, or entire, 2′–6′ long, 1′–2′
wide, borne on slender naked petioles; upper
leaves lanceolate, entire, sessile or short-petioled, not cordate, those of the branches small
and bract-like; heads numerous, 5″–7″ high;
involucre broadly campanulate, its bracts linear, acute, puberulent, imbricated in several
series, their green tips appressed; rays 10–15,
linear, violet-blue, 5″–6″ long; pappus tawny.

On banks and along edges of woods, western
Pennsylvania to Virginia, Georgia, Illinois, Wisconsin and Tennessee.   Sept.–Oct.

17. Aster azùreus Lindl.   Sky-blue Aster.
Fig. 4298.

Aster azureus Lindl.; Hook. Comp. Bot. Mag. **1**:
98.   1835.
A. capillaceus Burgess; Small, Fl. S.E. U. S. 1215.
1903.

Stem slender, stiff, rough, branched above,
1°–4° high.   Leaves thick, usually all entire, scabrous on both sides, the lower and basal ones
cordate, ovate, ovate-lanceolate, or lanceolate,
acute, acuminate, or obtusish, 2′–6′ long, with
slender, naked, often pubescent petioles; upper
leaves short-petioled or sessile, lanceolate or
linear, those of the branches reduced to small
appressed bracts; heads numerous, 4″–5″ high;
involucre turbinate, its bracts glabrous, linear-
oblong, abruptly acute, imbricated in several
series, their green tips appressed; rays 10–20,
bright blue, 3″–4″ long; pappus tawny.

On prairies and along borders of woods, Ontario
and western New York to Georgia, Minnesota, Kansas, Alabama and Texas.   Aug.–Oct.

$\frac{1}{2}$

18. Aster cordifòlius L.   Common
Blue Wood Aster.   Fig. 4299.

Aster cordifolius L. Sp. Pl. 875.   1753.
Aster choralis Steele, Contr. Nat. Herb. **10**:
372.   1911.

Stem glabrous or nearly so, rarely pubescent, much branched, bushy, 1°–5° high.
Leaves thin, rough, more or less pubescent
with scattered hairs above and on the veins
beneath, sharply serrate, acuminate, the
lower and basal ones slender-petioled,
broadly ovate-cordate, 2′–5′ long, the upper
short-petioled or sessile, ovate or lanceolate, smaller; petioles scarcely margined;
heads usually very numerous, small, 2″–3″
high, 6″–9″ broad, handsome; involucre
turbinate to cylindric, its bracts oblong-
linear, obtuse or obtusish, green-tipped, appressed; rays 10–20, 3″–4″ long, violet or
blue, sometimes pale, rarely white.

Woods and thickets, Nova Scotia and New
Brunswick to Ontario, Minnesota, Georgia and
Missouri.   Consists of many races, differing
mainly in leaf-form and inflorescence.   Sept.–
Dec. Tongue.   Bee-weed.

$\frac{3}{5}$

19. Aster Lowrieànus Porter. Lowrie's or Fall Aster. Fig. 4300.

Aster cordifolius var. *laevigatus* Porter, Bull. Torr. Club **16**: 67. 1889. Not *A. laevigatus* Lam. 1783.
Aster Lowrieanus Porter, Bull. Torr. Club **21**: 121. 1894.

Glabrous, or very nearly so throughout; stem branched, 1°–4° high. Leaves thickish, firm, a little succulent, the basal slender-petioled, ovate to ovate-lanceolate, mostly cordate, acute or obtusish, serrate, or sometimes incised, 2′–6′ long, those of the stem ovate to oblong, often cordate, contracted into winged petioles, the uppermost lanceolate; heads usually not very numerous, 2½″–3″ high, loosely panicled; involucre turbinate, its bracts obtuse or obtusish, appressed; rays light blue, 3″–4″ long, but variable in length.

In woods, Connecticut and southern New York to Pennsylvania, Ontario, North Carolina and Kentucky. Races differ in leaf-form and serration. Sept.–Oct. Bee-weed. Blue-devil.

20. Aster Lindleyànus T. & G. Lindley's Aster. Fig. 4301.

As'er Lindleyanus T. & G. Fl. N. A. **2**: 122. 1841.
A. Wilsoni Rydb. Bull. Torr. Club **37**: 138. 1910.
Aster Lindleyanus eximius Burgess; Britt. & Brown, Ill. Fl. **3**: 364. 1898.

Stem usually stout, glabrous, or sometimes pubescent, 1°–6° high, branched above. Leaves rather thick, glabrous, or slightly pubescent, especially on the veins, the lower and basal ones cordate at the base, sharply serrate, ovate, acute or acuminate, 2′–4′ long, with slender naked petioles; upper leaves ovate, ovate-lanceolate, or lanceolate, less serrate, or entire, sessile, or with margined petioles, those of the branches lanceolate or linear-lanceolate, smaller; heads usually not numerous, 4″–5″ high; involucre broadly turbinate or nearly hemispheric, its bracts linear-lanceolate, acute, rather loosely imbricated, glabrous, or nearly so, their tips green; rays 10–20, blue or violet, 3″–5″ long; pappus nearly white.

In open places, Labrador to Mackenzie, Alberta, Maine, New York, Michigan and Montana. Aug.–Oct.

21. Aster Drummóndii Lindl. Drummond's Aster. Fig. 4302.

Aster Drummondii Lindl. in Hook. Comp. Bot. Mag. **1**: 97. 1835.
Aster hirtellus Lindl. in DC. Prodr. **5**: 233. 1836.

Stem usually stout, finely and densely canescent, branched above, 2°–5° high. Leaves mostly thin, ovate or ovate-lanceolate, acuminate, rough above, canescent beneath, the lower and basal ones cordate, with slender naked petioles, sharply toothed, 2′–4′ long, the upper cordate or rounded at the base, usually on margined petioles, those of the branches sessile and entire or nearly so, much smaller; heads 3″–4″ high, rather numerous on the racemose branches; involucre turbinate, its bracts linear, slightly pubescent, acute or acuminate, their green tips appressed; rays 8–15, blue, 3″–4″ long; pappus whitish.

In dry soil, borders of woods and on prairies, Ohio to Minnesota, Kentucky, Arkansas and Texas. Perhaps not specifically distinct from the following. Sept.–Oct.

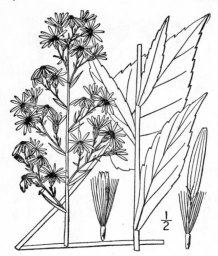

22. Aster sagittifòlius Willd. Arrow-leaved Aster. Fig. 4303.

Aster sagittifolius Willd. Sp. Pl. 3 : 2035. 1804.

Stem stout, or slender, strict, glabrous, or sparingly pubescent above, 2°–5° high, paniculately branched at the inflorescence, the branches ascending. Leaves thin, slightly roughened, or glabrous above, glabrate or pubescent beneath, the lower and basal ones cordate or sagittate, ovate-lanceolate, sharply serrate, acuminate, 3′–6′ long, with slender naked or narrowly margined petioles; upper leaves lanceolate, sessile, or on short and usually margined petioles, serrate or entire, those of the branches very small; heads 2½″–4″ high, 8″–10″ broad, numerous, crowded, racemose; involucre turbinate, its bracts linear-subulate, glabrous or nearly so, their tips green and slightly spreading; rays 10–15, light blue or purplish, 3′–4′ long; pappus whitish.

In dry soil, New Brunswick to Ontario, North Dakota, New Jersey, Georgia and Missouri. Aug.–Oct.

Aster Saundersii Burgess, of the Dakotas and Iowa, differs by a corymbose inflorescence.

23. Aster undulàtus L. Wavy-leaf Aster. Fig. 4304.

Aster undulatus L. Sp. Pl. 875. 1753.

Stem stiff, very rough and pubescent, divaricately branched above, 1°–3½° high. Leaves usually thick, rough on both sides, pubescent beneath, dentate, undulate or entire, acute or acuminate, the lowest and basal ones ovate, cordate, 2′–6′ long, with naked or margined petioles; middle ones ovate, lanceolate or oblong, with margined petioles dilated and clasping at the base, the upper sessile or clasping, those of the branches small and subulate; heads numerous, racemose and often secund on the spreading branches, about 4″ high, 8″–10″ broad; involucre broadly turbinate, its bracts linear-oblong, pubescent, acute or acutish, their green tips appressed; rays 8–15, pale blue to violet, 3″–5″ long; pappus whitish.

In dry soil, New Brunswick and Ontario to Minnesota, Florida, Alabama, Louisiana and Arkansas. Various-leaved aster. Races differ in leaf-form and inflorescence. Sept.–Oct.

24. Aster pàtens Ait. Late Purple Aster. Purple Daisy. Fig. 4305.

Aster patens Ait. Hort. Kew. 3 : 201. 1789.
A. patens gracilis Hook. Comp. Bot. Mag. 1 : 97. 1835.

Stem slender, rough, 1°–3° high, divergently branched. Leaves ovate-oblong to oblong-lanceolate, rough or pubescent, thick and somewhat rigid, strongly cordate or auriculate-clasping at the broad base, entire, acute, or the lowest obtuse, 1′–3′ long, those of the branches much smaller and bract-like, the margins rough-ciliate; heads 1′ broad or more, solitary at the ends of the branches; involucre broadly turbinate, its bracts linear-oblong, finely pubescent or scabrous and somewhat glandular, imbricated in several series, their green acute tips spreading; rays 20–30, purplish-blue, or deep violet, 4″–6″ long; pappus tawny; achenes pubescent.

In dry, open places, Maine to northern New York, Minnesota, Florida, Louisiana and Texas. Reported from Canada. Races differ in leaf-form and pubescence. Aug.–Oct.

25. Aster phlogifòlius Muhl.   Thin-leaved Purple Aster.   Fig. 4306.

A. phlogifolius Muhl.; Willd. Sp. Pl. 3: 2034. 1804.

Aster patens var. *phlogifolius* Nees, Gen. & Sp. Ast. 49. 1832.

Similar to the preceding species, usually taller. Leaves larger, lanceolate to oblong-lanceolate, entire, thin, or membranous, acuminate at the apex, strongly auriculate-clasping at the base, roughish above, pubescent beneath, usually narrowed below the middle, sometimes 6′ long; heads usually numerous, 1′–2′ broad, panicled, or somewhat racemose on the branches; bracts of the involucre lanceolate, glabrate, rather loose, with herbaceous tips; rays numerous, purple-blue.

In woods and thickets, New York to Ohio, North Carolina and Tennessee. Perhaps a sylvan race of the preceding species. Aug.–Sept.

26. Aster nòvae-angliae L.   New England Aster.   Fig. 4307.

Aster novae-angliae L. Sp. Pl. 875. 1753.

A. roseus Desf. Cat. Hort. Paris, Ed. 3, 401. 1812.

Stem stout, hispid pubescent, corymbosely branched above, 2°–8° high, very leafy. Leaves lanceolate, entire, rather thin, acute, pubescent, 2′–5′ long, 6″–12″ wide, clasping the stem by an auriculate or broadly cordate base; heads numerous, 1′–2′ broad, clustered at the ends of the branches; involucre hemispheric, its bracts linear-subulate, somewhat unequal, green, spreading, pubescent and more or less glandular, viscid; rays 40–50, linear, 5″–8″ long, violet-purple, rarely pink or red, or white; achenes pubescent; pappus reddish-white.

In fields and along swamps, Quebec to Saskatchewan, South Carolina, Alabama, Kansas and Colorado. One of the most beautiful of the genus. Aug.–Oct.

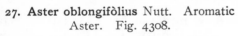

27. Aster oblongifòlius Nutt.   Aromatic Aster.   Fig. 4308.

Aster oblongifolius Nutt. Gen. **2**: 156. 1818.
Aster oblongifolius var. *rigidulus* A. Gray, Syn. Fl. **1**: Part 2, 179. 1884.
Aster Kumleinii Fries; Rydb. Fl. Colo. 354. 1906.

Stem much branched, hirsute-pubescent, 1°–2½° high, the branches divaricate or ascending. Leaves crowded, oblong, or oblong-lanceolate, sessile by a broad, partly clasping base, usually rigid, entire, acute or mucronulate at the apex, rough or hispidulous on both sides, rough-margined, those of the stem 1′–2′ long, 2″–4″ wide, those of the branches gradually smaller; heads corymbose, nearly 1′ broad; involucre hemispheric, its bracts much imbricated, glandular, aromatic, linear or linear-oblong, the acute green tips spreading; rays 20–30, violet-purple, rarely rose-pink, 3″–5″ long; pappus light brown; achenes canescent.

On prairies and bluffs, central Pennsylvania to Minnesota, North Dakota, Nebraska, Colorado, Virginia, Tennessee and Texas. Races differ in leaf-form and pubescence. Plant odorous. Aug.–Oct.

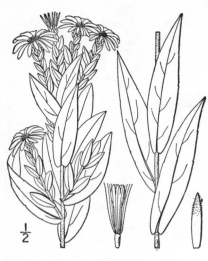

1/2

28. Aster nebraskénsis Britton. Nebraska Aster. Fig. 4309.

Aster nebraskensis Britton, in Britt. & Brown, Ill. Fl. 3 : 375. 1898.

Stem strictly erect, slender, stiff, rough to the base, simple, or with a few short nearly erect branches, very leafy, 1½°–2½° tall. Leaves thick, rather rigid, ascending, lanceolate to oblong-lanceolate, entire, sessile by a subcordate base, acute or acuminate at the apex, 1′–3′ long, 4″–6″ wide, very rough on both sides, the midvein prominent beneath, the lateral veins obscure; heads few, terminating short leafy branchlets, 1′–1¼′ broad; involucre broadly campanulate or hemispheric, about 3″ high, its bracts green, oblong, acute, imbricated in several series, the outer quite foliaceous; rays purple, about 6″ long.

Lake shores, central Nebraska. Sept. Lower and basal leaves not seen.

29. Aster amethýstinus Nutt. Amethyst Aster. Fig. 4310.

Aster amethystinus Nutt. Trans. Am. Phil. Soc. (II) 7 : 294. 1841.

Resembles *Aster novae-angliae*, but is often taller, sometimes 5° high. Leaves often crowded, linear-lanceolate, entire, rough or hispidulous on both sides, partly clasping, though sometimes slightly so, at the sessile base, acute at the apex, those of the stem 1′–2′ long, 2″–3″ wide; heads rather numerous, racemose or corymbose, ½′–1′ broad; involucre broadly turbinate, its bracts much imbricated, linear, hispid, not glandular, the acutish green tips spreading; rays 20–30, blue or violet, about 3″ long; pappus brown; achenes canescent.

In moist soil, Vermont and Massachusetts to New York, Pennsylvania, Illinois, Iowa and Nebraska. Sept.–Oct. Specimens have intermediate characters between *Aster novae-angliae* L. and *Aster multiflorus* L., and hybridism is suspected.

3/5

30. Aster modéstus Lindl. Great Northern Aster. Fig. 4311.

Aster unalaschensis var. *major* Hook. Fl. Bor. Am. 2 : 7 1834.
Aster modestus Lindl.; Hook. loc. cit. 8. 1834.
Aster majus Porter, Mem. Torr. Club 5 : 325. 1894.

Stem stout, leafy to the summit, usually densely pilose-pubescent with many-celled hairs, rarely glabrate, branched above, 4°–6° high. Leaves membranous, lanceolate, partly clasping by a narrowed base, acuminate at the apex, sharply serrate with low, distant teeth, dark green and slightly pubescent above, villous-pubescent on the veins beneath, 3′–5′ long, 5″–10″ wide; heads mostly solitary at the ends of short branches, 1½′ broad; involucre hemispheric, its bracts little imbricated, green, linear-subulate, densely glandular; rays 35–45, purple to violet, 5″–7″ long; achenes appressed-pubescent; pappus tawny.

In moist soil, western Ontario to Minnesota, Oregon and British Columbia. Sept.–Oct.

1/2

31. Aster puníceus L. Red-stalk or Purple-stem Aster. Fig. 4312.

Aster puniceus L. Sp. Pl. 875. 1753.

Stem usually stout, reddish, corymbosely or racemosely branched above, hispid with rigid hairs to glabrous, 3°–8° high. Leaves lanceolate to oblong-lanceolate, acuminate, sessile and clasping by a broad or narrowed base, sharply serrate, or entire, usually very rough above, pubescent on the midrib or glabrous beneath, 3′–6′ long, ½′–1½′ wide; heads generally numerous, 1′–1½′ broad; involucre nearly hemispheric, its bracts linear or oblong, attenuate, imbricated in about 2 series, glabrous or ciliate, green, loose, spreading, nearly equal, sometimes broadened; rays 20–40, violet-purple or pale (rarely white), 5″–7″ long, showy; pappus nearly white; achenes pubescent.

In swamps, Newfoundland to Ontario, Manitoba, Minnesota, Georgia, Tennessee, Ohio and Michigan. Races differ in pubescence, leaf-form and leaf-serration. Early purple aster. Swan-weed. Cocash. Meadow-scabish. July–Nov.

32. Aster tardiflòrus L. Northeastern Aster. Fig. 4313.

Aster ardiflorus L. Sp. Pl. Ed. 2, 1231. 1763.

Aster patulus Lam. Encycl. 1 : 308. 1783.

Stem glabrous, slightly pubescent, or villous, corymbosely branched near the summit, 1°–3° high. Leaves lanceolate, oblong-lanceolate, or ovate-lanceolate, serrate with low teeth, or some of them entire, acuminate at the apex, narrowed into a slightly clasping base, or the lower into winged petioles, glabrous or nearly so on both sides, roughish-margined, 3′–6′ long, 4″–10″ wide; heads about 1′ broad, not very numerous, involucre hemispheric, its bracts often 6″ long, acute, somewhat unequal; rays 20–30, violet; pappus nearly white; achenes pubescent.

Along streams, New Brunswick to Pennsylvania. Aug.–Oct.

33. Aster prenanthoìdes Muhl. Crooked-stem Aster. Fig. 4314.

A. prenanthoides Muhl.; Willd. Sp. Pl. 3 : 2046. 1804.

Aster prenanthoides porrectifolius Porter, Mem. Torr. Club 5 : 326. 1894.

Stem glabrous, or pubescent in lines above, flexuous, much branched, 1°–2° high. Leaves thin, oblong to ovate-lanceolate, or lanceolate, sharply and coarsely serrate, scabrous above, glabrous or nearly so beneath, 3′–8′ long, 9″–18″ wide, acuminate at the apex, abruptly narrowed below into a broad margined entire petiole, the base auriculate-clasping; heads usually numerous, 1′ broad or more; involucre hemispheric, its bracts linear, acute, green, spreading, imbricated in 3 or 4 series, the outer shorter; rays 20–30, violet, 4″–6″ long; pappus tawny; achenes pubescent.

In moist soil, Massachusetts to Minnesota, Virginia, Kentucky and Iowa. Aug.–Oct.

Aster schistòsus Steele, of West Virginia, is intermediate in characters between this species and *A. Lowrieanus* Porter, and may be a hybrid.

$\frac{1}{2}$

34. Aster laèvis L.   Smooth Aster.
Fig. 4315.

Aster laevis L. Sp. Pl. 876. 1753.
Aster laevis amplifolius Porter, Mem. Torr. Club **5** : 324. 1894.
Aster laevis potomacensis Burgess; Britt. & Brown, Ill. Fl. **3** : 369. 1898.

Stem usually stout, glabrous, often glaucous, 2°–4° high, branched or simple. Leaves thick, entire, or serrate, glabrous, slightly rough-margined, the upper all sessile and usually cordate-clasping, lanceolate, oblong-lanceolate, oblanceolate or ovate, acute or obtusish, 1′–4′ long, 4″–2′ wide, the basal and lower gradually narrowed into winged petioles, those of the branches often small and bract-like; heads usually numerous, about 1′ broad; involucre campanulate, its bracts rigid, acute, appressed, green-tipped, imbricated in several series; rays 15–30, blue or violet; pappus tawny; achenes glabrous or nearly so.

Usually in dry soil, Maine to Ontario, Virginia, Alabama, Louisiana, Saskatchewan, Missouri and Colorado. Races differ in leaf-form. Sept.–Oct.

35. Aster concínnus Willd. Narrow-leaved Smooth Aster.   Fig. 4316.

Aster concinnus Willd. Enum. 884. 1809.

Similar to narrow-leaved forms of *Aster laevis,* and perhaps a race of that species, glabrous, or sparingly pubescent above, not glaucous; stem paniculately branched, 1°–3° high. Leaves light green, lanceolate to linear, entire, or sometimes serrulate, the upper sessile, somewhat clasping, 1′–3′ long, the lower and basal ones spatulate, or oblong, narrowed into margined petioles, sometimes coarsely toothed; heads usually numerous, about 10″ broad; bracts of the involucre with rhomboid acute herbaceous tips; rays violet to purple.

Woodlands, Connecticut to Pennsylvania, Virginia, North Carolina, Missouri and Arkansas. Sept.–Oct.

$\frac{1}{2}$

$\frac{1}{2}$

36. Aster purpuràtus Nees.   Southern Smooth Aster.   Fig. 4317.

Aster virgatus Ell. Bot. S. C. & Ga. **2** : 353. 1824. Not Moench, 1802.
A. purpuratus Nees, Gen. & Sp. Ast. 118. 1832.

Stem slender, glabrous, simple, or branched above, 1½°–3° high, the branches sometimes puberulent. Leaves firm, glabrous, dark green, entire, the upper sessile and clasping at the base, elongated-lanceolate or linear-lanceolate, acuminate, 2′–6′ long, 2″–4″ wide, the lower and basal ones petioled, oblong-lanceolate, obtusish, those of the branches very small; heads rather few, loosely paniculate, 8″–12″ broad; involucre campanulate to turbinate, its bracts coriaceous, linear, appressed, green-tipped, acute, imbricated in several series, the outer shorter; rays 5–10, blue or violet, 3″–5″ long, pappus tawny; achenes glabrous.

Virginia and West Virginia to Georgia and Texas. Aug.–Sept.

37. Aster júnceus Ait. Rush Aster.
Fig. 4318.

Aster junceus Ait. Hort. Kew. **3**: 204. 1789.
Aster longulus Sheldon, Bull. Geol. Surv. Minn. **9**: 18,
pl. 2. 1894.
Aster junciformis Rydb. Bull. Torr. Club **37**: 142. 1910.

Stem very slender, glabrous, or pubescent above,
simple or little branched, 1°–3° high. Leaves firm,
glabrous, roughish-margined, narrowly linear, entire
or sometimes with a few distant teeth, acute or
acuminate at the apex, sessile by a broad clasping
and often slightly cordate base, 3′–6′ long, 1½″–4″
wide; heads paniculate, rather distant, about 1′
broad; involucre hemispheric, about 3″ high, its
bracts glabrous, linear-subulate, very acute, imbricated in 3 or 4 series, the outer shorter; rays violet
to white, 4″–5″ long; pappus pale.

In swamps and bogs, Nova Scotia to British Columbia,
New Jersey, Ohio, Wisconsin and Colorado. July–Sept.

38. Aster adscéndens Lindl. Western Aster. Fig. 4319.

A. adscendens Lindl.; Hook. Fl. Bor. Am. **2**: 8. 1834.

Stem slender, rigid, glabrous, or sparingly
hirsute-pubescent, branched or simple, 6′–2° high.
Leaves firm, entire, rough-margined, sometimes
ciliolate, those of the stem linear-lanceolate or
linear-oblong, acute or obtusish, 1′–3′ long, 2″–5″
wide, sessile by a more or less clasping base;
basal leaves spatulate, narrowed into short petioles; heads not numerous, about 1′ broad; involucre hemispheric, its bracts imbricated in 3–5
series, oblong-linear or spatulate, their tips obtuse
or obtusish, slightly spreading, the inner often
mucronulate; pappus nearly white; achenes pubescent.

On prairies and moist banks, western Nebraska to
Wyoming, Montana, Assiniboia, Colorado, New Mexico and Nevada. July–Sept.

39. Aster nòvi-bélgii L. New York Aster. Fig. 4320.

Aster novi-belgii L. Sp. Pl. 877. 1753.
Aster novi-belgii elodes A. Gray, Syn. Fl. 1²: 190. 1884.
Aster novi-belgii litoreus A. Gray, loc. cit. 189. 1884.
Aster novi-belgii atlanticus Burgess; Britt. & Brown,
Ill. Fl. **3**: 370. 1898.
Aster novi-belgii Brittonii Burgess, loc. cit. 371. 1898.

Stem slender, usually much branched, glabrous, or
slightly pubescent above, 1°–3° high. Leaves lanceolate, oblong-lanceolate, or linear-lanceolate, firm,
often somewhat fleshy, entire, or slightly serrate,
glabrous, or very nearly so, acuminate at the apex,
narrowed, sessile and more or less clasping at the
base, 2′–6′ long, 3″–8″ wide, the lowest petioled;
heads corymbose-paniculate, usually numerous, 10″–
15″ broad; involucre hemispheric to campanulate,
its bracts linear, acute, or obtusish, green, somewhat
spreading, in 3–5 series, the outer shorter; rays 15–
25, violet, 4″–5″ long; pappus whitish; achenes glabrous or nearly so.

In swamps, Newfoundland to Maine and Georgia, mainly near the coast. Races differ in leaf-form and in the involucral bracts. Aug.–Oct.

40. **Aster longifòlius** Lam.   Long-leaved Aster.   Fig. 4321.

Aster longifolius Lam. Encycl. 1 : 306. 1783.

Aster longifolius villicaulis A. Gray, Syn. Fl. 1 : Part 2, 189. 1884.

Stem glabrous, or pubescent, leafy, paniculately branched, 1°–3° high. Leaves lanceolate to linear-lanceolate, entire, or nearly so, acuminate at the apex, narrowed into a sessile clasping usually slightly cordate base, 3′–8′ long, 2″–6″ wide; heads rather numerous, about 1′ broad; involucre hemispheric, 4″–5″ high, its bracts glabrous, narrow, green, acute, imbricated in few series, nearly equal; rays numerous, 2½″–7″ long, violet or pale purple; pappus pale.

In swamps and moist ground, Labrador to Saskatchewan, northern New England, Ontario and Montana. Summer.

41. **Aster foliàceus** Lindl.   Leafy-bracted Aster.   Fig. 4322.

Aster foliaceus Lindl. in DC. Prodr. 5 : 228. 1835.

Stem usually stout, sparingly pubescent in lines, 2°–3° high, branched above, the branches ascending. Leaves lanceolate, acute at the apex, entire, or with a few small distant teeth, glabrous on both sides, very rough-margined, 2′–4′ long, ½′–1′ wide, the upper clasping at the base, the lowest petioled; heads few, 1′ broad or more; involucre hemispheric, its bracts green, foliaceous, oblong, the inner narrower and acute; rays about 30, violet, 4″–5″ long; pappus nearly white; achenes pubescent.

Mountains of Quebec; Oregon to Alaska. July–Sept. The figure of this species in our first edition was erroneously stated to have been drawn from specimens collected in western Nebraska.

42. **Aster phyllòdes** Rydb.   Large-bracted Aster.   Fig. 4323.

Aster phyllodes Rydb. Bull. Torr. Club 37 : 145. 1910.

Stem rather slender, somewhat branched, loosely pubescent, 2°–3° high. Leaves oblong-lanceolate, entire, or sparingly low-dentate, the larger 4′ long or less, ½′–1′ wide, acute or acutish at the apex, narrowed to a subcordate and slightly clasping base, rough-margined, glabrous on both sides; heads leafy-paniculate, about 1′ broad; involucre hemispheric, its bracts narrowly oblong, foliaceous, acute; rays purplish; achenes pubescent.

Wet grounds, western Nebraska and eastern Colorado. Aug.–Sept. Included in *A. foliaceus* Lindl. in our first edition.

43. Aster seríceus Vent. Western Silvery or Silky Aster. Fig. 4324.

Aster sericeus Vent. Hort. Cels, *pl. 33.* 1800.

Aster argenteus Michx. Fl. Bor. Am. **2**: 111. 1803.

Stem slender, paniculately or corymbosely branched, stiff, glabrous, leafy, 1°–2° high. Stem leaves sessile, with a broad base, oblong, entire, mucronate, $\frac{1}{2}'$–$1\frac{1}{2}'$ long, 2″–5″ wide, erect or ascending, with a dense silvery-white silky pubescence on both sides; basal and lowest leaves oblanceolate, narrowed into margined petioles; heads numerous, about $1\frac{1}{2}'$ broad; involucre turbinate, its bracts oblong, or the inner lanceolate, canescent, imbricated in 3 or 4 series, their tips green, acute, spreading; rays 15–25, violet-blue, 6″–8″ long; pappus tawny; achenes glabrous.

In dry open soil, Illinois to Minnesota, Manitoba, South Dakota, Tennessee, Missouri and Texas. Aug.–Sept.

44. Aster cóncolor L. Eastern Silvery Aster. Fig. 4325.

Aster concolor L. Sp. Pl. Ed. 2, 1228. 1763.

Stem slender, glabrous, or pubescent above, 1°–2½° high, leafy, simple, or with few erect branches. Leaves oblong or linear-oblong, finely and densely canescent on both sides, or the lower glabrate, sessile, obtuse or mucronate, $1\frac{1}{2}'$–2′ long; heads numerous in an elongated narrow raceme resembling *Lacinaria;* involucre broadly turbinate, its bracts linear or linear-oblong, appressed, canescent, imbricated in 4 or 5 series, their tips green, acute, the outer shorter; rays 10–15, lilac, 3″–4″ long; pappus tawny; achenes villous.

In dry sandy soil, eastern Massachusetts and Rhode Island to Florida, Tennessee and Louisiana, mostly near the coast. Lilac-flowered aster. Aug.–Oct.

45. Aster Féndleri A. Gray. Fendler's Aster. Fig. 4326.

Aster Fendleri A. Gray, Mem. Am. Acad. (II) **4**: 66. 1849.

Aster Nuttallii var. *Fendleri* A. Gray, Pac. R. R. Rep. **4**: 97. 1856.

Stems several or solitary from thick woody roots, rigid, hirsute, 6′–12′ high. Leaves linear, rigid, 1-nerved, acute or acuminate, 8″–15″ long, 1″–1½″ wide, glabrous on both sides, but the margins bristly-ciliate; heads usually few and racemose, $\frac{1}{2}'$–1′ broad; involucre turbinate, its bracts glandular, linear-oblong, imbricated in about 4 series, the inner acute, the outer shorter and obtuse; rays 10–15, violet, 3″–5″ long.

In dry soil on the plains, Nebraska and Kansas to Colorado and New Mexico. Aug.–Sept.

½

47. Aster spectábilis Ait. Low showy Aster. Seaside Purple Aster. Fig. 4328.

Aster spectabilis Ait. Hort. Kew. **3**: 209. 1789.

Stem stiff, simple, or corymbosely branched above, puberulent, or rough below, more or less glandular above, 1°–2° high. Leaves firm, thickish, the basal and lower ones oval, acute or acutish, 3′–5′ long, 1′–1½′ wide, sparingly dentate with low teeth, narrowed at the base into slender petioles; upper leaves sessile, entire or very nearly so, acute, linear-oblong; heads several or numerous, about 1½′ broad, corymbose, very showy; involucre nearly hemispheric, its bracts linear-oblong or slightly spatulate, glandular, viscid, imbricated in about 5 series, their green obtusish tips spreading; rays 15–30, bright violet, 6″–10″ long; pappus whitish; achenes slightly pubescent.

In dry sandy soil, Massachusetts to Delaware, mostly near the coast. Aug.–Oct.

46. Aster grandiflòrus L. Large-flowered Aster. Fig. 4327.

Aster grandiflorus L. Sp. Pl. 877. 1753.

Stem rather stiff, divaricately much branched, hispid with short hairs, 1°–2½° high. Leaves oblong, linear, or somewhat spatulate, rigid, sessile by a broad, sometimes slightly clasping base, reflexed, entire, obtusish, hispid, the larger 2′ long and 4″ wide, those of the branches very numerous, 2′–5″ long; heads about 2′ broad, terminating the branches; involucre hemispheric, its bracts very squarrose and foliaceous, imbricated in 5–7 series, linear, or linear-oblong, glandular, the outer obtusish, the inner acute; rays very numerous, deep violet, nearly 1′ long, 1½″ wide; pappus brownish; achenes ribbed, canescent.

In dry soil, Virginia, east of the mountains, to Florida. Sept.

½

48. Aster surculòsus Michx. Creeping Aster. Fig. 4329.

Aster surculosus Michx. Fl. Bor. Am. **2**: 112. 1803.

Stem slender, from elongated-filiform rootstocks, minutely scabrous-pubescent, 10′–18′ high, corymbosely branched above. Leaves firm, lanceolate or linear, the lower petioled, 2′–3′ long, 4″–8″ wide, rough-margined, slightly scabrous above, sparingly dentate, the upper narrower, sessile, entire; heads few, or sometimes solitary, about 15″ broad; involucre turbinate-hemispheric, its bracts coriaceous, imbricated in about 5 series, ciliate, but scarcely glandular, their green tips spreading; rays 15–30, violet; pappus whitish; achenes nearly glabrous.

In sandy or gravelly soil, Kentucky, North Carolina and Georgia. Erroneously reported from New Jersey. Sept.–Oct.

½

49. Aster grácilis Nutt.   Slender or Tuber Aster.   Fig. 4330.

Aster gracilis Nutt. Gen. **2**: 158. 1818.

Stem slender, finely puberulent and scabrous, corymbosely branched above, 1°–1½°
high. Leaves minutely scabrous, the basal
and lower ones oval, acute or obtusish, 2′–3′
long, 4″–8″ wide, dentate, narrowed into
slender petioles; upper leaves linear, linear-
oblong, or slightly oblanceolate, acute, entire, sessile or a little clasping; heads usually numerous, 6″–10″ broad; involucre
narrowly turbinate, its bracts coriaceous,
glabrous or very nearly so, imbricated in
about 5 series, their tips green and spreading, obtusish; rays 9–15, violet, 3″–4½″ long;
pappus nearly white; achenes minutely pubescent.

In dry sandy soil, New Jersey to Kentucky,
Tennessee and South Carolina.   Rootstock
tuberous-thickened.   July–Sept.

50. Aster Rádula Ait.   Low Rough Aster. File-blade Aster.   Fig. 4331.

Aster Radula Ait. Hort. Kew. **3**: 210. 1789.
Aster biflorus Michx. Fl. Bor. Am. **2**: 114. 1803.
Aster strictus Pursh, Fl. Am. Sept. 556. 1814.
Aster Radula var. *strictus* A. Gray, Syn. Fl. **1**: Part
2, 176. 1884.
Aster Radula biflorus Porter, Mem. Torr. Club **5**:
326. 1894.

Stem glabrous, or puberulent above, slender,
corymbosely branched near the summit, or simple, 4′–2° high. Leaves sessile, rough above,
more or less pubescent beneath, lanceolate to
oblong-lanceolate, acute, sharply serrate, strongly pinnately veined, 2′–3′ long, 3″–12″ wide;
heads several, or sometimes numerous, 1′–1½′
broad; involucre hemispheric, its bracts oblong
or oblong-spatulate, coriaceous, appressed-pubescent, conspicuously ciliolate, their green obtuse or acutish tips very little spreading; rays
20–30, violet, 4″–6″ long; achenes glabrous,
striate; pappus nearly white.

In swamps, Newfoundland to Delaware, Pennsylvania, Maryland and West Virginia.   July–Sept.

51. Aster Hérveyi A. Gray.   Hervey's Aster.   Fig. 4332.

Aster Herveyi A. Gray, Man. Ed. 5, 229. 1867.

Stem roughish, at least above, slender, simple
or corymbosely branched, rarely paniculate, 1°–3°
high, the branches glandular-puberulent. Leaves
firm, rough above, pubescent on the veins beneath,
the basal and lower ones on slender naked petioles, ovate, dentate with low usually distant teeth,
acute at the apex, narrowed, rounded or rarely
cordate at the base, 2′–6′ long, 1′–3′ wide; upper
leaves sessile, or narrowed into winged petioles,
smaller, entire or nearly so; heads 1′–1½′ broad;
involucre turbinate or campanulate, its bracts appressed, or sometimes spreading, densely glandular, oblong or spatulate, obtuse or mucronulate;
rays 15–25, violet, 5″–7″ long; achenes minutely
pubescent, striate; pappus nearly white.

In dry soil, eastern Massachusetts, Rhode Island,
Connecticut and Long Island.   Aug.–Oct.

52. Aster turbinéllus Lindl. Prairie Aster. Fig. 4333.

A. turbinellus Lindl. Comp. Bot. Mag. **1** : 98. 1835.

Stem slender, paniculately branched, glabrous below, puberulent above, 2°–3° high. Leaves firm, lanceolate, or oblong-lanceolate, entire, ciliate, acute or acuminate, 2′–3′ long, the lower and basal ones petioled, the upper sessile, those of the branches much smaller; heads about 1′ broad, mostly solitary at the ends of the branches; involucre turbinate, its bracts oblong, coriaceous, obtuse, appressed, imbricated in 5 or 6 series, their tips green only at the apex; rays 10–20, 3″–5″ long, violet; pappus tawny; achenes finely pubescent.

In dry soil, especially on prairies, Illinois to Missouri, Nebraska, Kansas, Louisiana and Arkansas. Sept.–Oct.

53. Aster lateriflòrus (L.) Britton. Starved Aster. Calico Aster. Fig. 4334.

Solidago lateriflora L. Sp. Pl. 879. 1753.
Aster diffusus Ait. Hort. Kew. **3** : 205. 1789.
Aster miser Nutt. Gen. **2** : 158. 1818.
A. lateriflorus Britton, Trans. N. Y. Acad. Sci. **9** : 10. 1889.

Stem puberulent, or nearly glabrous, slender, divergently branched, 1°–5° high. Basal leaves ovate, slender-petioled; stem leaves broadly lanceolate or oblong-lanceolate, mostly acuminate, serrate, 2′–5′ long, 6″–12″ wide, those of the branches smaller, oblong or linear-oblong; heads 3″–5″ broad, racemosely unilateral on the branches, short-peduncled or sessile, usually numerous and crowded; involucre turbinate, its bracts linear-oblong, obtuse or acutish, imbricated in about 4 series, their short green tips appressed or slightly spreading; rays numerous, short, white or pale purple; disk-flowers purple; pappus white; achenes minutely pubescent.

In dry or moist soil, Nova Scotia to western Ontario, south to North Carolina, Louisiana and Texas. Consists of many races, differing in leaf-form, inflorescence and pubescence. Called in Maryland rosemary. Aug.–Oct.

54. Aster hirsuticaùlis Lindl. Hairy-stemmed Aster. Fig. 4335.

A. hirsuticaulis Lindl.; DC. Prodr. **5** : 242. 1836.

Aster lateriflorus hirsuticaulis Porter, Mem. Torr. Club **5** : 324. 1894.

Stem slender, erect, 1½°–3° high, pubescent, often nearly or quite to the base, the usually short branches spreading or ascending. Leaves thin, glabrous above, usually pubescent on the midvein beneath, serrate with a few appressed teeth, or entire, linear-lanceolate to lanceolate, sometimes 6′ long, 2″–7″ wide, sessile, or the basal ones spatulate and petioled; heads more or less unilateral on the branches, densely or loosely clustered, often also solitary or few in the lower axils; bracts of the involucre in 3 or 4 series, linear-lanceolate, acuminate or acute; rays white, about 2″ long.

In woods and thickets, New Brunswick to Pennsylvania, Kentucky and Michigan. Perhaps a race of the preceding species. Aug.–Oct. Wiseweed. Old-field sweet. Farewell-summer. Old-Virginia. Stickweed. White-devil. Nail-rod.

55. Aster vimíneus Lam. Small White Aster. Fig. 4336.

Aster vimineus Lam. Encycl. **1**: 306. 1783.
Aster foliolosus Ait. Hórt. Kew. **3**: 203. 1789.
Aster Tradescanti T. & G. Fl. N. A. **2**: 129. 1841. Not L. 1753.
Aster vimineus foliolosus A. Gray, Syn. Fl. 1: Part 2, 186. 1884.

Glabrous or nearly so throughout; stem slender, divergently branched, 2°–5° high. Stem leaves linear-lanceolate, entire, or with a few low teeth, 3′–5′ long, 2″–4″ wide, acuminate at the apex, narrowed to a sessile base, those of the branches much smaller; heads very numerous, small, 3″–5″ broad, 2″–3″ high, generally densely racemose-secund, sometimes subpaniculate, short-peduncled; involucre broadly turbinate, its bracts linear, acute or acutish, imbricated in about 3 series, green-tipped, appressed; rays numerous, about 2″ long, white to purplish; pappus white; achenes minutely pubescent.

In moist soil, Ontario to Florida, west to Minnesota and Arkansas. Aug.–Sept.

Aster racemòsus Ell., with smaller heads, the leaves all very small, of the Southeastern States, apparently occurs in extreme southern Illinois.

½

56. Aster multiflòrus Ait. Dense-flowered Aster. White Wreath Aster. Fig. 4337.

Aster multiflorus Ait. Hort. Kew. **3**: 203. 1789.
A. multiflorus stricticaulis T. & G. Fl. N. A. **2**: 125. 1841.
A. multiflorus exiguus Fernald, Rhodora **1**: 187. 1899.
A. exiguus Rydb. Bull. Torr. Club **28**: 505. 1901.
A. polycephalus Rydb. Bull. Torr. Club **33**: 153. 1906.

Stem strict, much branched and bushy, rough-pubescent or scabrous, 1°–7° high, the branches ascending or spreading. Leaves rigid, linear or linear-oblong, entire, mostly obtuse, sessile or slightly clasping at the base, rough and ciliate, those of the stem ½–1½′ long, those of the branches very small and crowded; heads 3″–4″ broad, densely crowded, nearly sessile, sometimes slightly secund on the branches; involucre 2″–3″ high, its bracts coriaceous, mostly ciliate or pubescent, in 3 or 4 series, their short green tips obtuse or mucronate, spreading; rays 10–20, white, 1½″–2″ long; pappus brownish white; achenes puberulent.

In dry open places, Maine and Ontario to Alberta, Georgia, Texas and Arizona. Consists of many slightly differing races. Fall-flower. Aug.–Nov.

½

57. Aster commutàtus (T. & G.) A. Gray. White Prairie Aster. Fig. 4338.

Aster ramulosus var. *incanopilosus* Lindl. in DC. Prodr. **5**: 243. 1836.
Aster multiflorus var. *commutatus* T. & G. Fl. N. A. **2**: 125. 1841.
Aster commutatus A. Gray, Syn. Fl. 1: Part 2, 185. 1884.
A. incanopilosus Sheldon, Bull. Torr. Club **20**: 286. 1893.

Similar to the preceding species, except in the inflorescence, the stem rough-pubescent, or sometimes nearly glabrous, 1½°–2½° high, with ascending or divergent branches. Leaves linear or linear-oblong, obtuse, entire, sessile or slightly clasping at the base, those of the stem 1′ 5 long; heads larger than those of *A. multiflorus*, 6″–8″ broad; involucre 3″–4″ high, its ciliate bracts squarrose-tipped and sometimes foliaceous; rays 20–30, about 3″ long.

On prairies and along rivers, Minnesota to Saskatchewan, Nebraska, Texas and New Mexico. Aug.–Oct.

½

58. **Aster dumòsus** L. Bushy Aster. Rice-button Aster. Fig. 4339.

Aster dumosus L. Sp. Pl. 873. 1753.
Aster coridifolius Michx.; Willd. Sp. Pl. 3: 2028. 1804.
As·er dumosus strictior T. & G. Fl. N. A. 2: 128. 1841.
Aster dumosus coridifolius T. & G. Fl. N. A. 2: 128. 1841.

Glabrous or nearly so throughout, rather stiff and viscid, usually paniculately branched, 1°–3° high. Leaves firm, those of the stem linear or linear-lanceolate, entire, acute, or obtusish, 1′–3′ long, 1½″–3″ wide, roughish-margined, often reflexed, those of the branches very numerous, small and bract-like, the basal ones spatulate, dentate; heads 4″–7″ broad, terminating the slender minutely leafy branches and branchlets, usually numerous; involucre broadly campanulate, its bracts linear-subulate, obtuse or acutish, appressed, imbricated in about 4 series, green-tipped; rays 15–30, blue to pale violet or white, 2″–4″ long; pappus white; achenes pubescent.

Sandy soil, Massachusetts to western New York, Ontario, Florida, Louisiana and Missouri. Aug.–Oct.

Aster Grávesii Burgess, known only from Waterford, Conn., has oblong-lanceolate leaves, acuminate at both ends, the larger 8″ wide, the rays bright purple.

59. **Aster salicifòlius** Lam. Willow Aster. Fig. 4340.

Aster salicifolius Lam. Encycl. 1: 306. 1783.
Aster subasper Lindl. Comp. Bot. Mag. 1: 97. 1835.
Aster stenophyllus Lindl. DC. Prodr. 5: 242. 1836.
Aster salicifolius subasper A. Gray, Syn. Fl. 1: Part 2, 188. 1884.

Stem rather slender, paniculately much branched, usually very leafy, 2°–5° high, glabrous, roughish or somewhat pubescent above. Leaves firm, lanceolate or linear-lanceolate, roughish or rough-margined, acute or acuminate at the apex, narrowed and sessile or slightly clasping at the base, entire or sparingly dentate with low teeth, glabrous or nearly so, 2′–4′ long, 2″–6″ wide, the lowest sometimes petioled, those of the branches gradually smaller; heads numerous, 8″–12″ broad; involucre 3″–4″ high, its bracts linear-oblong, appressed, imbricated in 4 or 5 series, their green tips acute or obtusish; rays numerous, violet, or violet-purple, or sometimes white, 3″–4″ long; pappus white; achenes pubescent.

In moist soil, Maine to Massachusetts, Florida, Ontario, Assiniboia, Texas and Colorado. Aug.–Oct.

60. **Aster paniculàtus** Lam. Tall White or Panicled Aster. Fig. 4341.

Aster paniculatus Lam. Encycl. 1: 306. 1783.
Aster bellidiflorus Willd. Enum. 886. 1809.
Aster tenuifolius var. bellidiflorus T. & G. Fl. N. A. 2: 132. 1841.

Stem glabrous or nearly so, or sometimes pubescent, paniculately much branched, 2°–8° high. Leaves lanceolate to oblong-lanceolate or linear, acuminate at the apex, narrowed to a sessile or slightly clasping base, glabrous, usually thin, roughish-margined, those of the stem sparingly serrate, or sometimes entire, 3′–6′ long, 3″–6″ wide, the upper and those of the branches gradually smaller; heads numerous, 8″–10″ broad; involucre 3″–4″ high, its bracts narrowly linear-lanceolate, acuminate, appressed, green-tipped, imbricated in 4 or 5 series; rays numerous, white, or faintly tinged with violet, 3″–4″ long; pappus white or nearly so; achenes minutely pubescent.

In moist soil, New Brunswick to western Ontario and Montana, south to New Jersey, Virginia, Kentucky, Louisiana and Missouri. Consists of many races, differing in leaf-form, leaf-serration, size of heads, color of rays, and pubescence. Aug.–Oct.

61. Aster missouriénsis Britton.　Missouri Aster.　Fig. 4342.

Aster missouriensis Britton, in Britt. & Brown, Ill. Fl. 3: 378.　1898.

Stem densely puberulent or pubescent, at least above, much branched, 2° high or more.　Leaves thin, oblong-lanceolate to oblanceolate, acute or acuminate at the apex, sharply serrate above the middle, gradually tapering to an entire sessile or slightly clasping base, or the lower petioled, puberulent above, finely pubescent beneath, the larger 3′–4′ long, the upper much smaller, entire; heads 6″–8″ broad, panicled, short-peduncled, or terminating short leafy branchlets, sometimes somewhat secund; involucre 2″–3″ high, its linear acute bracts well imbricated, ciliate or pubescent; rays white.

In moist soil, Kansas, Missouri and Iowa.　Sept.–Oct.

62. Aster Tradescánti L.　Tradescant's Aster.　Michaelmas Daisy.　Fig. 4343.

Aster Tradescanti L. Sp. Pl. 876.　1753.

Stem slender, paniculately branched, 2°–5° high, the branches usually ascending and often pubescent in lines.　Stem leaves linear-lanceolate or lanceolate, acuminate at the apex, narrowed to a sessile base, 3′–6′ long, 1½″–6″ wide, glabrous or nearly so on both sides, commonly thin, sharply serrate in the middle with low teeth, or sometimes entire; heads very numerous, racemose but not secund on the branches, 5″–8″ broad; involucre hemispheric to broadly turbinate, 2″–3″ high, its bracts linear, acute, appressed, green-tipped, imbricated in 4 or 5 series; rays white or nearly so, numerous, 2″–3″ long; pappus white; achenes minutely pubescent.

In fields and swamps, Maine to Virginia, Ontario, Illinois and Missouri.　Aug.–Oct.

Aster saxátilis (Fernald) Blanchard, of rocky situations in New England, appears to be a low race of this species, rather than of *A. vimineus.*

63. Aster Fáxoni Porter.　Faxon's Aster.　Fig. 4344.

Aster polyphyllus Willd. Enum. 888.　1809.　Not Moench, 1802.
Aster Faxoni Porter, Mem. Torr. Club 5: 323.　1894.
A. ericoides Randi Britton, in Britt. & Brown, Ill. Fl. 3: 379.　1898.

Glabrous throughout; stem paniculately or corymbosely branched; rather stout, ½°–5° high.　Stem leaves lanceolate or linear-lanceolate, acute or acuminate, narrowed to a sessile base, or the lower into margined petioles, entire or nearly so, firm, 2′–5′ long, 2″–4″ wide, those of the branches gradually smaller; basal leaves oblong to spatulate, obtuse, dentate, or entire; heads not very numerous, 6″–9″ broad; involucre hemispheric, nearly 4″ high, its bracts linear-lanceolate, acute or subulate, greentipped or green on the back, imbricated in about 3 series, the outer shorter; rays bright white, rarely purplish, 3″–4″ long, numerous; pappus white; achenes minutely pubescent.

On moist cliffs, Maine and Vermont to Pennsylvania, Wisconsin and North Carolina.　Aug.–Sept.

3/5

64. Aster ericoìdes L. White Heath Aster. Frost-weed Aster. Fig. 4345.

Aster ericoides L. Sp. Pl. 875. 1753.
Aster villosus Michx. Fl. Bor. Am. **2**: 113. 1803. Not Thunb. 1800.
Aster ericoides var. *villosus* T. & G. Fl. N. A. **2**: 124. 1841.
Aster ericoides pilosus Porter, Mem. Torr. Club **5**: 323. 1894.

Stem glabrous, villous, or hirsute, paniculately branched, usually bushy, $1°$–$3°$ high, the branches racemose, and the branchlets often somewhat secund. Leaves firm or rigid, the basal ones spatulate, obtuse, dentate, narrowed into margined petioles, glabrous or ciliate; stem leaves narrowly linear to linear-lanceolate, acute, entire, $1'$–$3'$ long, $1''$–$3''$ wide, those of the branches linear-subulate, numerous; heads usually very numerous, $4''$–$7''$ broad; involucre campanulate to hemispheric, its bracts coriaceous, lanceolate or linear-lanceolate, abruptly acute or acuminate, green-tipped, imbricated in about 3 series; rays 15–25, white, or tinged with rose; pappus white; achenes finely pubescent.

In dry soil, Maine to Ontario, Florida, Minnesota and Missouri. Frost-weed. Michaelmas daisy. Farewell-summer. White rosemary. Dog-fennel. Mare's-tail. Scrub-bush. Steel-weed. Sept.–Dec.

A densely villous, broad-leaved relative or race of this species, ranging from Ohio to North Carolina and Michigan, is known as *Aster ericoides platyphyllus* T. & G.

This species apparently hybridizes with *A. paniculatus* Lam. where the two grow together.

65. Aster Prínglei (A. Gray) Britton. Pringle's Aster. Fig. 4346.

Aster ericoides var. *Pringlei* A. Gray, Syn. Fl. **1**: Part 2, 184. 1884.
Aster Pringlei Britton, in Britt. & Brown, Ill. Fl. **3**: 379. 1898.

Stem very slender, glabrous, simple, or with few or numerous slender ascending branches, not bushy, $6'$–$2°$ high. Basal leaves lanceolate, oblong or oblanceolate, $2'$–$6'$ long, $2''$–$6''$ wide, entire, or slightly toothed, ciliate and sometimes a little pubescent, at least on the slender petioles which are often as long as the blades; stem leaves narrowly linear, those of the branches small and subulate; heads as large as those of *A. ericoides,* or commonly smaller, usually fewer, solitary at the ends of the branches and branchlets; bracts of the involucre with short green tips; rays white.

On banks, especially in rocky places, Massachusetts and Vermont to Wisconsin. Aug.–Oct.

3/5

3/5

66. Aster Priceae Britton. Miss Price's Aster. Fig. 4347.

Aster Priceae Britton, Manual 960. 1901.

Stem pubescent, widely branched, $1°$–$2\frac{1}{2}°$ high. Basal leaves oblanceolate, obtuse or acutish, entire, petioled, $1'$–$3'$ long, the petioles ciliate, broad; stem leaves linear-lanceolate, sessile, ciliate, acuminate, $\frac{3}{4}'$–$2'$ long, those of the branches similar but smaller; involucre nearly hemispheric, about $3''$ high, its bracts linear, the outer gradually acuminate, green, the inner a little broader, abruptly acuminate, acute or obtusish; heads about $1'$ broad; rays bright purple or pink.

In dry soil, Kentucky and North Carolina. October.

Aster kentuckiénsis Britton, also of Kentucky and North Carolina, differs in being glabrous, its involucral bracts all gradually acuminate.

67. Aster parviceps (Burgess) Mackenzie & Bush. Small-headed Aster. Fig. 4348.

Aster ericoides parviceps Burgess; Britt. & Brown, Ill. Fl. **3**: 379. 1898.

Aster parviceps Mackenzie & Bush, Fl. Jackson Co. 196. 1902.

A. depauperatus parviceps Fernald, Rhodora **10**: 94. 1908.

Pilose, sometimes slightly so, much branched above, 1°–2½° high, the branches ascending. Basal leaves spatulate; stem leaves linear to linear-lanceolate, 1′–3′ long, often with short leafy branches in their axils, those of the branches very small; heads many, paniculate; involucre about 2″ high, turbinate, its linear-subulate bracts imbricated in several series; rays 10–12, white, about 2″ long.

Dry prairies and open woodlands, Illinois, Iowa and Missouri. Aug.–Sept.

$\frac{2}{3}$

$\frac{3}{4}$

68. Aster depauperàtus (Porter) Fernald. Serpentine Aster. Fig. 4349.

Aster ericoides pusillus A. Gray, Syn. Fl. 1²: 184. 1884. Not *A. pusillus* Horn.
Aster ericoides depauperatus Porter, Mem. Torr. Club **5**: 323. 1894.
Aster depauperatus Fernald, Rhodora **10**: 94. 1908.
Aster parviceps pusillus Fernald, Rhodora **11**: 59. 1909.

Glabrous, slender, widely branched, 4′–15′ high. Basal leaves spatulate or oblanceolate, obtusish, ½′–1½′ long, narrowed into petioles; stem leaves linear to linear-subulate, 1′ long or less, ½″–1″ wide, those of the branches minute; heads many, terminating short branchlets; involucre turbinate, about 2″ high, its bracts linear-subulate, acute, rays white, about 2″ long.

On serpentine barrens, southern Pennsylvania and West Virginia. Aug.–Oct.

69. Aster paludòsus Ait. Southern Swamp Aster. Fig. 4350.

Aster paludosus Ait. Hort. Kew. **3**: 310. 1789.

Heleastrum paludosum DC. Prodr. **5**: 264. 1836.

Stem roughish, or rough-pubescent, slender, simple, or somewhat branched above, 1°–2½° high. Leaves linear, entire, glabrous, but margins rough or ciliate, rigid, 2′–6′ long, 2″–4″ wide, mostly 1-nerved, acute, the lower part commonly sheathing the stem; heads few or several, racemose or paniculate, 1½′–2′ broad; involucre broadly campanulate or hemispheric, its bracts imbricated in about 5 series, foliaceous, ciliate, the outer lanceolate, acute, the inner oblong or spatulate; rays 20–30, deep violet, 5″–7″ long, pappus tawny; achenes 8–10-nerved, glabrous, or nearly so.

In swamps, Kansas and Missouri to Texas, east to North Carolina and Florida. Aug.–Oct.

$\frac{1}{2}$

70. **Aster nemoràlis** Ait.   Bog Aster.
Fig. 4351.

Aster nemoralis Ait. Hort. Kew. **3**: 198.   1789.
Aster nemoralis Blakei Porter, Bull. Torr. Club **21**: 311.   1894.

Stem puberulent, slender, simple, or corymbosely branched above, 6'–2° high. Leaves sessile, oblong-lanceolate or linear-oblong, acute at each end, pubescent or puberulent on both sides, dentate or entire, 1'–3' long, 1½''–10'' wide, margins often revolute; heads several, or solitary, 1'–1½' broad, the peduncles slender; involucre broadly obconic to hemispheric, its bracts appressed, linear-subulate, acute or acuminate, imbricated in about 3 series; rays 15–25, light violet-purple to rose-pink; achenes glandular-pubescent; pappus white.

In sandy bogs, New Jersey to northern New York, Ontario, Newfoundland and Hudson Bay. Races differ in leaf-form and serration. Aug.–Sept.

71. **Aster acuminàtus** Michx.   Whorled or Mountain Aster.   Fig. 4352.

Aster divaricatus Lam. Encycl. **1**: 305.   1783.   Not L. 1753.
Aster acuminatus Michx. Fl. Bor. Am. **2**: 109.   1803.

Stem pubescent or puberulent, zigzag, corymbosely branched, often leafless below, 1°–3° high. Leaves thin, broadly oblong, acuminate at the apex, narrowed to a somewhat cuneate sessile base, sharply and coarsely dentate, pinnately veined, glabrous or pubescent above, pubescent at least on the veins beneath, 3'–6' long, ½'–1½' wide, often approximate above, and appearing whorled; heads several or numerous, 1'–1½' broad; involucre nearly hemispheric, its bracts subulate-linear, acuminate, the outer much shorter; rays 12–18, narrow, 6''–8'' long, white or purplish; pappus soft, fine, nearly white; achenes pubescent.

Moist woods, Labrador to Ontario, western New York, and in the mountains to Tennessee and Georgia. July–Oct.

72. **Aster ptarmicoìdes** (Nees) T. & G.   Upland White Aster.   Fig. 4353.

Chrysopsis alba Nutt. Gen. **2**: 152.   1818.   Not *A. albus* Willd.
Doellingeria ptarmicoides Nees, Gen. & Sp. Ast. 183.   1832.
A. ptarmicoides T. & G. Fl. N. A. **2**: 160. 1841.
Unamia alba Rydb. Bull. Torr. Club **37**: 146.   1910.

Stems tufted, slender, rigid, usually rough above, corymbosely branched near the summit, 1°–2° high. Leaves linear-lanceolate, 1–3-ribbed, entire, or with a few distant teeth, firm, shining, rough-margined or ciliate, sometimes scabrous, acute, narrowed to a sessile base, or the lower petioled, the lowest and basal ones 3'–6' long, 2''–4'' wide, the upper smaller, those of the branches linear-subulate; heads 8''–12'' broad; involucre nearly hemispheric, 2''–3'' high, its bracts linear-oblong, obtuse, or the outer acutish, appressed, nearly glabrous, green, imbricated in about 4 series; rays 10–20, white, 3''–4'' long; pappus white; achenes glabrous.

In dry or rocky soil, Massachusetts, Vermont and Ontario to Saskatchewan, Illinois, Missouri and Colorado. July–Sept.

Aster lutéscens (Lindl.) T. & G. is a very interesting race with light yellow rays, known from Illinois, Wisconsin and Saskatchewan. Yellow rays are otherwise almost or quite unknown in the genus *Aster* as here limited.

73. Aster tenuifòlius L. Perennial Salt-marsh Aster. Fig. 4354.

Aster tenuifolius L. Sp. Pl. 873. 1753.
Aster flexuosus Nutt. Gen. **2**: 154. 1818.

Perennial, glabrous and fleshy; stem flexuous, striate, at least when dry, sparingly and loosely branched, 1°–2° high. Stem leaves linear, entire, acute, sessile or partly clasping at the base, the lowest lanceolate-linear, 2′–6′ long, 2″–3″ wide, those of the branches minute, bract-like, appressed; heads rather few, 6″–12″ broad, terminating the branches; involucre turbinate, about 4″ high, its bracts lanceolate, acuminate or mucronate, glabrous, green on the back or tip, appressed, imbricated in about 5 series, the outer shorter; rays numerous, longer than the pappus, pale purple or nearly white; pappus tawny; achenes hispid-pubescent, 5-nerved.

In salt marshes, coast of Massachusetts to Florida. Aug.–Oct.

74. Aster exìlis Ell. Slim Aster. Fig. 4355.

Aster exilis Ell. Bot. S. C. & Ga. **2**: 344. 1824.

Aster divaricatus T. & G. Fl. N. A. **2**: 163. 1841. Not L. 1753.

Annual, glabrous, fleshy; stem slender, usually much branched, the branches usually divergent. Leaves linear to linear-lanceolate, 1′–4′ long, 1″–2½″ wide, entire, sessile, acute or acuminate, or the lowest narrowly oblong, 3″–4″ wide and petioled, those of the branches subulate; heads numerous, panicled, about 5″ broad; involucre campanulate, about 3″ high, its bracts linear-subulate, appressed, imbricated in 3 or 4 series; rays purplish, about 2″ long, mostly fewer than the disk-flowers, longer than the pappus; achenes somewhat pubescent.

In moist or wet soil, especially in saline situations, Kansas to Texas, South Carolina and Florida. Bahamas; Cuba. Aug.–Oct.

75. Aster subulàtus Michx. Annual Salt-marsh Aster. Fig. 4356.

Aster subulatus Michx. Fl. Bor. Am. **2**: 111. 1803.
Aster linifolius T. & G. Fl. N. A. **2**: 162. 1841. Not L. 1753.

Annual, glabrous and fleshy; stem paniculately branched, flexuous above, 1°–6° high, slightly angled, sometimes 1′ in diameter at the base, but usually smaller. Stem leaves linear-lanceolate, acute, entire, sessile by a broad or slightly clasping base, 2′–10′ long, 1″–8″ wide, those of the branches very small and subulate; heads numerous, 3″–5″ broad; involucre campanulate, or at length hemispheric, 2″–3″ high, its bracts linear-subulate, green, imbricated in 3 or 4 series, the outer shorter; rays 20–30, purplish, scarcely exceeding the nearly white pappus, more numerous than the disk-flowers; achenes compressed, minutely pubescent.

In salt marshes, coast of New Brunswick to Florida. Also on salt lands Onondaga Lake, N. Y. Aug.–Nov.

32. LEUCELÈNE Greene, Pittonia 3: 147. 1896.

Low perennial herbs, with much branched leafy stems, sessile, rather rigid, narrow, entire leaves and small heads of both tubular and radiate white flowers, solitary at the ends of the numerous slender branchlets, involucre turbinate, its bracts well imbricated. Disk-flowers perfect, their corollas white, tubular-funnelform, 5-toothed. Ray-flowers numerous, white, or drying red to rose, pistillate. Style appendages acutish. Achenes elongated, flattened, hispidulous. Pappus a single series of slender rough white bristles. [Greek, referring to the white disk.]

Two or three species, natives of the central and southwestern States and Mexico, the following typical.

1. Leucelene ericoìdes (Torr.) Greene. Rose Heath Aster. Fig. 4357.

Inula (?) *ericoides* Torr. Ann. Lyc. N. Y. 2: 212. 1828.

Aster ericaefolius Rothrock, Bot. Gaz. 2: 70. 1877.

Leucelene ericoides Greene, Pittonia 3: 148. 1896.

Stems tufted from deep woody roots, corymbosely much branched, 3'–12' high, hispid or scabrous, the branches erect or diffuse. Leaves hispid-ciliate, erect, or slightly spreading, obtusish or mucronulate, the lower and basal ones spatulate, 3''–6'' long, tapering into short petioles, the upper sessile, linear or linear-spatulate; heads terminating the branches, 5''–8'' broad; involucre broadly turbinate, its bracts lanceolate, appressed, scarious-margined, imbricated in 3 or 4 series; rays 12–15, white to rose, 2''–4'' long.

In dry soil, western Nebraska to Kansas, Texas and New Mexico. May–Aug.

33. BRACHYÁCTIS Ledeb. Fl. Ross. 2: 495. 1846.

Annual, nearly glabrous, somewhat fleshy herbs, with narrow chiefly entire leaves, and small racemose or racemose-paniculate heads of tubular, or also radiate purplish flowers. Involucre campanulate. Central flowers of the head few, perfect, their narrow corollas 4–5-toothed; outer flowers pistillate, usually in 2 series or more, and more numerous than the perfect ones; style-appendages lanceolate; rays very short, or none. Achenes 2–3-nerved, slender, appressed-pubescent. Pappus a single series of nearly white bristles. [Greek, short rays.]

About 5 species, natives of western North America and northern Asia. Type species: *Brachyactis ciliata* Ledeb.

1. Brachyactis angústa (Lindl.) Britton. Rayless Aster. Fig. 4358.

Tripolium angustum Lindl.; Hook. Fl. Bor. Am. 2: 15. 1834.

Aster angustus T. & G. Fl. N. A. 2: 162. 1841.

Brachyactis angusta Britton, in Britt. & Brown; Ill. Fl. 3: 383. 1898.

Stem usually sparsely pubescent, at least above, racemosely or rarely paniculately branched, 6'–24' high, striate, at least when dry. Leaves linear, fleshy, ciliate on the margins, acutish, entire, sessile by a rather broad base, the basal (when present) spatulate; heads 4''–6'' broad, racemose on the ascending branches, or terminating them; involucre campanulate or nearly hemispheric, 2''–3'' high, its bracts linear or linear-oblong, somewhat foliaceous, green, acute or acutish, imbricated in 2 or 3 series, glabrous or slightly ciliate, nearly equal; rays none, or rudimentary; pappus soft and copious.

In wet saline soil, or sometimes in waste places, Minnesota to Saskatchewan, Utah and Colorado, and along the St. Lawrence River in Quebec. Found also about Chicago. July–Sept.

Brachyactis frondòsa (Nutt.) A. Gray, of the Rocky Mountain region, differing by bluntly pointed leaves and oblong or oblanceolate involucral bracts, has been found on Prince Edward Island.

34. MACHAERANTHÈRA Nees, Gen. & Sp. Ast. 224. 1832.

Annual, biennial or perennial branched herbs, with leafy stems, alternate, mostly serrate or pinnatifid leaves, the teeth or lobes usually bristle-tipped, and large heads of both tubular and radiate flowers. Involucre of numerous series of imbricated canescent or glandular bracts with herbaceous or foliaceous spreading or appressed tips. Receptacle alveolate, the alveoli usually toothed or lacerate. Ray-flowers numerous, violet to red or purple, pistillate. Disk-flowers perfect, their corollas tubular, 5-lobed, yellow, changing to red or brown; anthers exserted, appendaged at the tip, rounded at the base; style-appendages subulate to lanceolate. Achenes turbinate, narrowed below, pubescent. Pappus of numrous stiff, rough unequal bristles. [Greek, sickle-anther.]

About 15 species, natives of western North America. Type species: *Machaeranthera tanacetifolia* (H.B.K.) Nees.

Annual or biennial; leaves pinnatifid.   1. *M. tanacetifolia.*
Perennial or biennial; leaves sharply serrate.   2. *M. sessiliflora.*

1. Machaeranthera tanacetifòlia (H.B.K.) Nees. Tansy Aster. Dagger-flower. Fig. 4359.

Aster tanacetifolius H.B.K. Nov. Gen. Sp. 4: 95. 1820.

M. tanacetifolia Nees, Gen. & Sp. Ast. 225. 1832.

Annual or biennial; stem glandular-pubescent, often viscid, densely leafy, much branched and bushy, 1°–2° high. Leaves sessile or short-petioled, pubescent, the lowest 1′–3′ long, 2–3-pinnatifid, their lobes linear or oblong, acute or mucronate, the upper pinnatifid, those of the branches sometimes entire; heads numerous, corymbose-paniculate, 1′–2′ broad; involucre hemispheric, 4″–6″ high, its bracts linear, glandular, imbricated in 5–7 series, their green tips very squarrose; rays 15–25, violet-purple, 5″–8″ long, pappus copious, tawny; achenes villous.

In dry soil, South Dakota to Nebraska, Texas, Mexico, Montana and California. June–Aug.

2. Machaeranthera sessiliflòra (Nutt.) Greene. Viscid Aster. Fig. 4360.

Dieteria sessiliflora Nutt. Trans. Am. Phil. Soc. 7: 301. 1840.

M. sessiliflora Greene, Pittonia 3: 60. 1896.

Stem usually stout, finely rough-pubescent or canescent, branched, and viscid-glandular above, 1°–2° high. Leaves lanceolate, linear, or the lowest spatulate, sessile, somewhat viscid, sharply incised-dentate, the larger 1′–3′ long, the teeth bristle-tipped; heads numerous, racemose, or corymbose above, 1′–1½′ broad, the lower often nearly sessile; involucre broadly turbinate or hemispheric, 4″–6″ high, its bracts acute, imbricated in 6–10 series, their tips strongly squarrose; rays numerous, violet, 4″–6″ long; pappus copious; achenes narrow, appressed-pubescent.

In dry soil, central and western Nebraska and Colorado. July–Oct.

A Kansas plant differs from this species by having acute appressed tips to the involucral bracts.

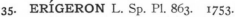

35. ERÍGERON L. Sp. Pl. 863. 1753.

Branching or scapose herbs, with alternate or basal leaves, and corymbose, paniculate or solitary, peduncled heads of both tubular and radiate (rarely all tubular) flowers. Involucre hemispheric, cylindric or campanulate, its bracts narrow, nearly equal, mostly imbricated in but 1 or 2 series. Receptacle nearly flat, usually naked. Ray-flowers, in our species, white, violet or purple, pistillate. Disk-flowers yellow, tubular, perfect, their corollas mostly 5-lobed. Anthers obtuse and entire at the base. Style-branches more or less flattened, their appendages short, mostly rounded or obtuse. Achenes flattened, usually 2-nerved. Pappus-bristles fragile, slender, scabrous or denticulate, in 1 series, or often an additional outer short series. [Greek, early-old, alluding to the early hoary pappus.]

A genus of 130 species or more, of wide geographic distribution, most abundant in the New World. In addition to the following, many others occur in the southern and western parts of North America. Type species: *Erigeron acris* L.

* **Rays long, narrow, usually equalling or longer than the diameter of the disk.**

　† Stem 1′–10′ high, simple, leafy; head solitary; involucre woolly.　　 1. *E. uniflorus.*
　†† Stem 4′–3° high, leafy, usually branched; involucre hirsute or glabrous.

　　　　1. Roots perennial, thick and woody.

Heads 1′–2′ broad; leaves lanceolate, ovate, oblong, or spatulate.
　Rays violet or purple; stem leaves ovate, lanceolate, or oblong.
　　Stem leaves linear-lanceolate, the upper much smaller.　　　　2. *E. asper.*
　　Stem leaves ovate-lanceolate, the upper little smaller.　　　　3. *E. subtrinervis.*
　Rays mostly white; stem leaves linear or linear-oblong.　　　　4. *E. caespitosus.*
Heads ½′–1′ broad; leaves linear.
　Plants hirsute or canescent; pappus double; western species.
　　Stem hirsute; achenes pubescent; flowers white.　　　　5. *E. pumilus.*
　　Stem appressed-canescent; achenes glabrous; flowers purple or white.　6. *E. canus.*
　Plant nearly glabrous; pappus simple; northern.　·　　　　7. *E. hyssopifolius.*
　　　　2. Perennial by decumbent, rooting stems or stolons.　8. *E. flagellaris.*
　　　3. Roots annual or biennial, fibrous; plants often perennial by offsets.
Heads 1′–1½′ broad, few; stem simple; eastern.　　　　9. *E. pulchellus.*
Heads ½′–1′ broad, numerous; stem branched.
　Rays 100–150, narrow, mostly purple or violet.
　　Pappus simple; plant erect, corymbosely branched.　　　10. *E. philadelphicus.*
　　Pappus double; plant diffusely branched, western.　　　11. *E. divergens.*
　Rays much less numerous, purplish or white.
　　Plants 6′–12′ high, diffuse, western; pappus simple.　　　12. *E. Bellidiastrum.*
　　Plants 6′–3° high, erect, branched; pappus double.
　　　Pappus of the ray-flowers and disk-flowers unlike.
　　　　Stem leaves lanceolate, nearly all sharply serrate.　　13. *E. annuus.*
　　　　Stem leaves linear-lanceolate or oblong, nearly all entire.　14. *E. ramosus.*
　　　Pappus of the ray-flowers and disk-flowers alike; plant hirsute.　15. *E. tenuis.*
　　　††† Stem leafless or nearly so; heads ½′ broad, corymbose. 16. *E. vernus.*
** **Rays inconspicuous or short; a row of tubular pistillate flowers inside the row of rays.**
　　　　　　　　　　　　　　　　　　　　　　　　17. *E. acris.*

1. Erigeron uniflòrus L.　Arctic Erigeron.　Fig. 4361.

Erigeron uniflorus L. Sp. Pl. 864.　1753.

Perennial by short branching rootstocks; stems slender, single or tufted, more or less pubescent, simple, erect, 1′–10′ high. Basal leaves petioled, spatulate, obtuse, entire, 1′–2′ long; stem leaves sessile, lanceolate or linear-lanceolate, entire, acute or obtuse; head solitary, peduncled, ½′–1′ broad; rays about 100, purple or purplish, 2″–4″ long; involucre hemispheric, its bracts linear-lanceolate, acute, copiously woolly; pappus simple.

Labrador and Arctic America to Alaska, south in the Rocky Mountains to Colorado and in the Sierra Nevada to California. Also in Europe. Summer.

$\frac{3}{4}$

2. Erigeron ásper Nutt.　Rough Erigeron.　Fig. 4362.

Erigeron asper Nutt. Gen. **2**: 147.　1818.
Erigeron glabellus Nutt. loc. cit.　1818.

Perennial by a woody root; stem simple, or branched above, more or less pubescent, sometimes hirsute, 6′–24′ high. Leaves glabrous, pubescent or ciliate, entire, the basal ones spatulate, obtuse, 2′–4′ long, 3″–1′ wide, narrowed into margined petioles; stem leaves oblong-lanceolate or linear-lanceolate, obtuse or acute, the upper smaller; heads several or solitary, slender-peduncled, 1′–2′ broad; involucre hemispheric, its bracts linear, acute, hirsute or pubescent; rays 100–150, very narrow, violet, purple, or nearly white, 4″–7″ long; pappus double, the outer row of bristles much shorter than the inner.

$\frac{1}{2}$

　In dry soil, Minnesota to Wisconsin, Nebraska, Manitoba, Utah and New Mexico. Races differ in pubescence. June–Sept.

3. Erigeron subtrinérvis Rydberg. Three-nerved Fleabane. Fig. 4363.

Erigeron glabellus var. *mollis* A. Gray, Proc. Acad. Phila. **1863**: 64. 1864. Not *E. mollis* D. Don.

Erigeron subtrinervis Rydberg, Mem. Torr. Club **5**: 328. 1894.

Similar to the preceding species, perennial by a woody root, finely pubescent all over; stems leafy to the inflorescence. Leaves entire, thin, the basal and lower ones oblanceolate to oblong, obtuse or acute, petioled, the upper lanceolate or ovate-lanceolate, sessile or somewhat clasping, acute, rather distinctly 3-nerved; heads 1′–1½′ broad, corymbose, or rarely solitary; involucre hemispheric, hirsute; rays numerous, blue to pink; pappus double, the outer bristles very short.

In dry soil, South Dakota to Wyoming, Nebraska, Utah and New Mexico. July–Sept.

4. Erigeron caespitòsus Nutt. Tufted Erigeron. Fig. 4364.

Diplopappus canescens Hook. Fl. Bor. Am. **2**: 21. 1834. Not *E. canescens* Willd. 1804.
Erigeron caespitosus Nutt. Trans. Am. Phil. Soc. (II) **7**: 307. 1841.

Perennial by a deep root; stems tufted, canescent, simple, or branched above, 6′–12′ high. Leaves canescent or pubescent, entire, the lower and basal ones petioled, narrowly oblanceolate or spatulate, obtuse or acutish, 1′–3′ long; stem leaves linear or linear-oblong, acute or obtuse, sessile, the upper gradually shorter, heads solitary or several, short-peduncled, 1′–1½′ broad; involucre hemispheric, its bracts lanceolate or linear-oblong, acute, canescent; rays 40–60, 3″–6″ long, white or pinkish; pappus double, the outer series of bristles very short.

In dry soil, Manitoba to Yukon, Nebraska (according to Webber), British Columbia and Colorado. June–Aug.

5. Erigeron pùmilus Nutt. Low Erigeron. Daisy. Fig. 4365.

Erigeron pumilis Nutt. Gen. **2**: 147. 1818.

Perennial by a deep root; stems tufted, hirsute, slender, simple, or branched, 4′–10′ high. Leaves entire, hirsute, the lower and basal ones narrowly spatulate or linear, petioled, obtuse or acutish, 1′–4′ long, 1″–2″ wide; stem leaves linear, sessile, ½–2½′ long, acute; heads solitary or several, 6″–10″ broad, short-peduncled; involucre hemispheric, its bracts linear, acute, hirsute; rays 50–80, white, 3″–4″ long, at length deflexed; pappus double, the outer row of bristles short and more or less intermixed with the inner; achenes pubescent.

Dry plains, North Dakota to western Nebraska, Kansas, British Columbia, Colorado and Utah. May–Sept.

6. Erigeron cànus A. Gray. Hoary Erigeron.
Fig. 4366.

Erigeron canus A. Gray, Mem. Am. Acad. (II) **4**: 67. 1849.

Perennial by a deep woody root, resembling the preceding species; stems slender, erect, tufted, appressed-canescent, 6′–10′ high, simple, or branched above. Leaves narrow, entire, canescent, the basal and lower ones narrowly spatulate, petioled, 2′–4′ long, the upper linear, sessile, acute, gradually smaller; heads solitary, or 2–4, peduncled, 6″–8″ broad; involucre hemispheric, its bracts linear, acute, densely canescent; rays 40–50, purple or white, 2″–3″ long; achenes glabrous, 8–10-nerved; pappus double, the outer row of bristles rather conspicuous.

In dry soil, South Dakota to western Nebraska, Wyoming, Colorado and New Mexico. June–Aug.

7. Erigeron hyssopifòlius Michx. Hyssop-leaved Erigeron. Fig. 4367.

Erigèron hyssopifolius Michx. Fl. Bor. Am. **2**: 123. 1803.
Aster graminifolius Pursh, Fl. Am. Sept. 545. 1814.

Perennial by slender rootstocks; stems tufted or single, very slender, simple or branched, glabrous or very nearly so, 4′–15′ high. Leaves narrow, thin, the basal and lower ones oblong or spatulate, short-petioled, 1′–1½′ long, 1½″–2″ wide, the upper linear or linear-oblong, acute, usually numerous; heads solitary or several, slender-peduncled, 5″–8″ broad; peduncles appressed-pubescent; involucre nearly cylindric at flowering time, its bracts linear-lanceolate, sparingly pubescent; rays 12–30, white or purplish, 3″–6″ long; pappus simple.

On moist cliffs, Newfoundland to northern Vermont, Mackenzie and Lake Superior. July–Aug.

8. Erigeron flagellàris A. Gray. Running Fleabane. Fig. 4368.

Erigeron flagellaris A. Gray, Mem. Am. Acad. (II) **4**: 68. 1849.

Appressed-pubescent, sometimes densely so, perennial by decumbent rooting stems or stolons; root slender; stem slender, branched, the branches elongated. Leaves entire, the basal and lower ones spatulate or oblong, obtuse or acute, 1′–2′ long, narrowed into long petioles, the upper sessile, linear or linear-spatulate, much smaller; peduncles solitary, elongated; heads about 1′ broad and ¾′ high; involucre hemispheric, its narrow bracts pubescent; rays very numerous, white to pink; pappus double, the outer series of subulate bristles.

In moist soil, South Dakota to Wyoming, Utah, western Texas and New Mexico. May–July.

9. Erigeron pulchéllus Michx. Robin's or Poor Robin's Plantain. Fig. 4369.

E. pulchellus Michx. Fl. Bor. Am. **2** : *124.* 1803.
E. bellidifolius Muhl. ; Willd. Sp. Pl. **3** : 1958. 1804.

Perennial by stolons and offsets, villous-pubescent; stems simple, slender, 10′–24′ high. Basal leaves tufted, spatulate or obovate, somewhat cuneate at the base, narrowed into short margined petioles, obtuse at the apex, 1′–3′ long, ½′–2′ wide, dentate or serrate; stem leaves sessile, partly clasping, oblong, lanceolate or ovate, mostly acute, entire, or sparingly serrate; heads 1–6, slender-peduncled, 1′–1½′ broad; involucre depressed-hemispheric, its bracts linear, acuminate, villous; rays numerous, violet or purplish, 4″–7″ long; achenes nearly glabrous; pappus simple.

On hills and banks, Maine to Ontario and Minnesota, Kansas, Florida and Louisiana. Recorded from Quebec and Nova Scotia. Rose-petty. Robert's-plantain. Blue spring-daisy. April–June.

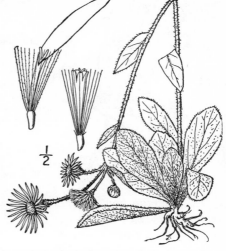

10. Erigeron philadélphicus L. Philadelphia Fleabane. Skevish. Fig. 4370.

Erigeron philadelphicus L. Sp. Pl. 863. 1753.

Perennial by stolons and offsets, soft-pubescent or sometimes nearly glabrous; stems slender, mostly branched above, 1°–3° high. Basal and lower leaves spatulate or obovate, obtuse, dentate, 1′–3′ long, narrowed into short petioles; upper stem leaves clasping and often cordate at the base, obtuse or acute, dentate or entire; heads several or numerous, corymbose-paniculate, 5″–12″ broad, slender-peduncled; peduncles thickened at the summit; involucre depressed-hemispheric, its bracts linear, usually scarious-margined; rays 100–150, 2″–4″ long, light rose-purple to pinkish; pappus simple; achenes puberulent.

In fields and woods, Labrador to British Columbia, Florida and California, but locally rare. Races differ in leaf-form and pubescence. Buds drooping. Sweet scabious. Daisy-fleabane. April–Aug.

11. Erigeron divérgens T. & G. Spreading Fleabane. Fig. 4371.

Erigeron divergens T. & G. Fl. N. A. **2** : 175. 1841.

Annual or biennial, diffusely branched, 6′–15′ high, densely cinereous-pubescent or hirsute. Basal and lower leaves spatulate or oblanceolate, acute or obtuse, mostly petioled, entire, dentate or lobed, 1′–2′ long, 2″–5″ wide, the upper sessile, linear or narrowly spatulate, usually acute, gradually smaller; heads slender-peduncled, 8″–12″ broad, usually numerous; involucre hemispheric, about 2″ high, its bracts linear, acute, hirsute or canescent; rays about 100, purplish, violet or nearly white, 2″–4″ long; pappus double, the shorter outer row of bristles subulate.

In moist soil, Montana to Nebraska, Texas, Mexico, Washington and California. April–Sept.

$\frac{2}{3}$

12. Erigeron Bellidiástrum Nutt. Western Daisy Fleabane. Fig. 4372.

Erigeron Bellidiastrum Nutt. Trans. Am. Phil. Soc. (II) **7**: 307. 1841.

Annual, much branched, 6′–15′ high, cinereous-pubescent throughout. Leaves entire, linear-spatulate, the lower and basal petioled, 1′–1½′ long, the upper sessile and smaller; heads several or numerous, 5″–8″ broad; involucre hemispheric, about 2″ high, its bracts linear, hirsute, acute; rays 30–65, purplish, about 2″ long; pappus a single series of capillary bristles.

In moist soil, South Dakota to Nebraska, Kansas, Texas, Wyoming and Arizona. May–July.

13. Erigeron ánnuus (L.) Pers. Sweet Scabious. White-top. Fig. 4373.

Aster annuus L. Sp. Pl. 875. 1753.

Erigeron annuus Pers. Syn. **2**: 431. 1807.

Annual, sparingly pubescent with spreading hairs; stem erect, corymbosely branched, 1°–4° high. Leaves thin, the lower and basal ones ovate or ovate-lanceolate, mostly obtuse, petioled, usually coarsely dentate, 2′–6′ long, 1′–3′ wide, the upper sessile or short-petioled, lanceolate, oblong, or linear-lanceolate, acute or acuminate, mostly sharply dentate in the middle, those of the branches narrower and often entire; heads rather numerous, 5″–7″ broad, mostly short-peduncled; bracts of the hemispheric involucre somewhat hispid; rays 40–70, linear, white, or commonly tinged with purple, 2″–4″ long; pappus double, the inner a series of slender fragile deciduous bristles, often wanting in the ray-flowers, the outer a persistent series of short, partly united, slender scales.

$\frac{1}{2}$

In fields, Nova Scotia to Manitoba, Georgia, Kentucky and Missouri. Naturalized in Bermuda and in continental Europe. Daisy-fleabane. Lace-buttons. May–Nov.

$\frac{1}{2}$

14. Erigeron ramòsus (Walt.) B.S.P. Daisy Fleabane. Fig. 4374.

Doronicum ramosum Walt. Fl. Car. 205. 1788.
E. strigosus Muhl.; Willd. Sp. Pl. **3**: 1956. 1804.
Stenactis Beyrichii F. & M. Index Sem. Hort. Petrop. **5**: 1838.
Erigeron ramosus B.S.P. Prel. Cat. N. Y. 27. 1888.
Erigeron ramosus Beyrichii Smith & Pond, Bot. Surv. Neb. **2**: 11. 1893.

Resembles the preceding species, but is usually lower and the pubescence more appressed. Stem leaves linear-oblong or linear-lanceolate, nearly all of them entire; basal and lowest spatulate or oblong, usually serrate; bracts of the involucre glabrous or nearly so; pappus similar to that of the preceding; rays white, or sometimes purplish, occasionally minute or wanting.

In fields, Nova Scotia to British Columbia, Florida, Louisiana, Texas and California. Naturalized in Europe. White-top. May–Nov.

15. Erigeron tenuis T. & G.   Slender Rough Fleabane.   Fig. 4375.

Erigeron tenuis T. & G. Fl. N. A. **2**: 175. 1841.

Annual or biennial, branched from the base and some-times also above, strigose-pubescent; stems slender, erect or ascending, 1° high or less. Basal leaves obovate to spatulate, 1'–2½' long, 3''–6'' wide, usually toothed; stem leaves linear or linear-oblong, toothed or entire; heads several or solitary, slender-peduncled, 7''–10'' broad; involucre 2''–3'' high, its linear bracts glabrous or spar-ingly pubescent; rays white or purplish, numerous; pappus of ray-flowers and disk-flowers alike, of few long bristles and short small scales.

Moist prairies and plains, Missouri to Texas and Louis-iana. April–May.

16. Erigeron vérnus (L.) T. & G.   Early Fleabane.   Fig. 4376.

Aster vernus L. Sp. Pl. 876. 1753.
E. nudicaulis Michx. Fl. Bor. Am. **2**: 124. 1803.
Erigeron vernus T. & G. Fl. N. A. **2**: 176. 1841.

Perennial by stolons and offsets; stem slender, simple or branched above, glabrous, or the branches pubescent, 1°–2½° high. Leaves mainly in a basal rosette, glabrous, obovate, oval or spat-ulate, obtuse, repand-denticulate or entire, 2'–4' long, narrowed into margined petioles; stem leaves mostly reduced to subulate-lanceolate scales, the lowest sometimes spatulate or oblong; heads not numerous, corymbose, peduncled, about 5'' broad; involucre hemispheric, its bracts linear-subulate; rays 20–30, white or pink, 2''–3'' long; pappus simple; achenes usually 4-nerved.

In marshes and moist soil, Virginia to Florida and Louisiana. April–May.

17. Erigeron àcris L.   Blue or Bitter Flea-bane.   Fig. 4377.

Erigeron acris L. Sp. Pl. 863. 1753.
Erigeron Droebachianus O. F. Mueller, Fl. Dan. *pl. 874.* 1782.
Erigeron acris Droebachianus Blytt, Norg. Fl. **1**: 562. 1861.
Erigeron acris debilis A. Gray, Syn. Fl. **1**: Part 2, 220. 1884.

Biennial or perennial; stem hirsute-pubescent or glabrate, slender, simple, or branched above, 6'–2° high. Leaves pubescent or glabrous, entire, the basal and lower ones spatulate, mostly obtuse, 1'–3' long, petioled, those of the stem mostly oblong or oblanceolate, obtuse or acutish, sessile, shorter; heads several or numerous, racemose or paniculate, peduncled, 5''–6'' broad; involucre hemispheric, its bracts linear, hirsute to glabrous; rays numerous, purple, equalling or slightly exceeding the brownish pappus; tubular pistillate flowers filiform, numer-ous; pappus simple or nearly so, copious.

Labrador to Alaska, Maine, Ontario, south in the Rocky Mountains to Colorado and Utah. Also in Eu-rope and Asia. Races differ in size, pubescence and length of rays. July–Aug.

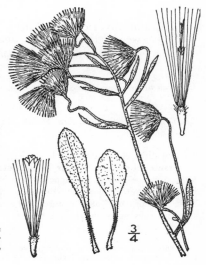

36. LÉPTILON Raf. Am. Month. Mag. 2: 268. 1818.

[CAENOTUS Raf. Fl. Tell. 2: 50. 1836.]

Annual or biennial herbs, with small racemose thyrsoid or panicled heads of white flowers, the rays small, usually shorter than the diameter of the disk, or none. Involucre mostly campanulate, its narrow bracts in 2 or 3 series. Receptacle naked. Ray-flowers pistillate; disk-flowers perfect, their corollas usually 4-lobed or 4-toothed, the anthers obtuse at the base; style-branches somewhat flattened, their appendages short. Achenes flattened. Pappus of numerous simple fragile bristles in 1 series. [Greek, referring to the small heads.]

About 20 species, natives of America and Asia. Besides the following, 2 or 3 others occur in the southwestern United States and one is naturalized from Asia on the southern Atlantic and Pacific coast. Type species: *Leptilon divaricatum* (Michx.) Raf.

1. Leptilon canadénse (L.) Britton. Horse-weed. Canada Fleabane. Fig. 4378.

Erigeron canadensis L. Sp. Pl. 863. 1753.

Leptilon canadense Britton, in Britt. & Brown, Ill. Fl. 3: 391. 1898.

Stem hispid-pubescent or glabrate, 3'–10° high, the larger plants paniculately much branched. Leaves usually pubescent or ciliate, the basal and lower spatulate, petioled, incised, dentate or entire, 1'–4' long, obtuse or acutish, those of the stem linear and mainly entire; heads usually very numerous; about 2'' broad; involucre campanulate, 1''–1½'' high, its bracts linear, acute, glabrate, the outer shorter; rays numerous, white, shorter than the pappus and mostly shorter than their tubes.

In fields and waste places, a common weed throughout North America except the extreme north. Widely distributed as a weed in the Old World, the West Indies and South America. Butter-weed. Prideweed. Fireweed. Blood-staunch. Cow's-, mare's- or colt's-tail. Hogweed. Bitter-weed. June–Nov.

2. Leptilon divaricàtum (Michx.) Raf. Low Horse-weed. Purple Horse-weed. Dwarf Fleabane. Fig. 4379.

Erigeron divaricatus Michx. Fl. Bor. Am. 2: 123. 1803.

Leptilon divaricatum Raf. Am. Month. Mag. 2: 268. 1818.

Stem diffusely much branched, 3'–12' high, pubescent or hirsute. Leaves all linear or subulate, entire, 4''–12'' long, about ½'' wide, the uppermost minute; heads numerous, about 2'' broad; involucre campanulate, 1'' high, its bracts linear, acute, pubescent, the outer shorter; rays purplish, shorter than their tubes.

In sandy soil, especially along rivers, Indiana to Minnesota, Tennessee, Louisiana, Nebraska and Texas. June–Oct.

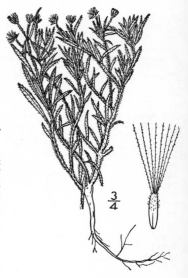

37. DOELLINGÈRIA Nees, Gen. & Sp. Ast. 177. 1832.

Perennial leafy herbs, the lower leaves reduced to scales or sheaths, the upper large, mostly entire, acute or acuminate in our species. Basal leaves none. Heads corymbose, usually numerous; involucre campanulate to hemispheric, its bracts imbricated in several series, appressed, thin, sometimes scarious-margined, their tips not herbaceous nor foliaceous. Receptacle foveolate. Ray-flowers white, pistillate, not very numerous. Disk-flowers perfect, the corolla with a slender tube abruptly expanded into a campanulate 5-lobed limb, white to greenish in our species. Anthers obtuse at the base; style-appendages ovate to subulate (rarely obtuse). Achenes obovoid, glabrous, or pubescent. Pappus double, the outer series of numerous short bristles or scales, the inner series of long capillary bristles, some or all of which have thickened tips. [In honor of Th. Dollinger, botanical explorer.]

About 5 species, natives of eastern North America. Type species: *Doellingeria umbellata* (Mill.) Nees.

Leaves lanceolate to ovate; heads mostly numerous.
  Leaves lanceolate to oblong-lanceolate, acuminate.   1. *D. umbellata.*
  Leaves ovate to ovate-lanceolate, acute.   2. *D. humilis.*
Leaves, at least the lower, obovate; heads commonly few.   3. *D. infirma.*

1. Doellingeria umbellàta (Mill.) Nees. Tall Flat-top White Aster. Fig. 4380.

Aster umbellatus Mill. Gard. Dict. Ed. 8, No. 22. 1768.
Diplopappus umbellatus Hook. Fl. Bor. Am. **2**: 22. 1834.
D. umbellata Nees, Gen. & Sp. Ast. 178. 1832.
Aster umbellatus var. *pubens* A. Gray, Syn. Fl. **1**: Part 2, 197. 1884.
D. pubens Rydb. Bull. Torr. Club **37**: 147. 1910.

Stem glabrous or pubescent above, striate, corymbosely branched at the summit, 1°–8° high. Leaves lanceolate to oblong-lanceolate, ascending, glabrous above, usually pubescent at least on the veins beneath, membranous, acuminate at the apex, narrowed at the base into short petioles, or the uppermost sessile, hispid-margined, those of the stem 5′–6′ long, ½′–1′ wide; heads numerous, 6″–10″ broad, in terminal compound corymbs; involucre broadly campanulate or hemispheric, 1½″–2″ high, its bracts lanceolate, usually pubescent or ciliate, acutish or obtuse, imbricated in 3 or 4 series, the outer shorter; rays 10–15, white; style-appendages ovate, acute; pappus nearly white; achenes nerved, slightly pubescent.

In moist soil, Newfoundland to Georgia, Saskatchewan, Iowa and Michigan. July–Oct.

2. Doellingeria hùmilis (Willd.) Britton. Broad-leaved Flat-top White Aster. Fig. 4381.

Aster humilis Willd. Sp. Pl. **3**: 2038. 1804.
D. amygdalina Nees, Gen. & Sp. Ast. 179. 1832.
Aster umbellatus var. *latifolius* A. Gray, Syn. Fl. **1**: Part 2, 197. 1884.
Doellingeria humilis Britton, in Britt. & Brown, Ill. Fl. **3**: 392. 1898.

Similar to the preceding species, usually lower, seldom over 4° high; stem striate, corymbosely branched above, glabrous, or somewhat pubescent. Leaves ovate to ovate-lanceolate, rather firm, sessile, or the lower very short-petioled, acute or short-acuminate at the apex, narrowed, or sometimes rounded at the base, rough-margined, those of the stem 1′–3′ long, ½′–2′ wide, inflorescence as in *D. umbellatus;* pappus dirty white; achenes somewhat pubescent.

In moist soil, eastern Massachusetts to New Jersey, Pennsylvania, Florida and Texas. Perhaps a broad-leaved race of the preceding species. July–Sept.

½

3. Doellingeria infirma (Michx.) Greene. Cornel-leaved Aster. Fig. 4382.

Aster infirmus Michx. Fl. Bor. Am. **2**: 109. 1803.
Diplopappus cornifolius Less.; Darl. Fl. Cestr. 474. 1837.
D. infirma Greene, Pittonia **3**: 52. 1896.

Stem slender, glabrous, or roughish above, sparingly branched at the summit, terete, $1\frac{1}{2}°-3°$ high. Leaves entire, hispid-margined, glabrous above, sparingly hispid on the veins beneath, the lower obovate, small, obtuse, usually sessile, the upper larger, oblong-lanceolate, acute, $2'-5'$ long, $1'-1\frac{1}{2}'$ wide; heads few, about $1'$ broad, in a divergently branched terminal corymbose cluster; involucre broadly campanulate, $2''-3''$ high, its bracts oblong-lanceolate, obtuse, sparingly pubescent, imbricated in about 4 series, the outer much shorter; rays 8–15, white; style-appendages subulate; pappus tawny; achenes nerved, glabrous.

In dry, usually rocky soil, Massachusetts to New York, Georgia, Tennessee and Alabama. Aug.–Sept.

38. IONÁCTIS Greene, Pittonia **3**: 245. 1897.

Low, mostly branching, perennial herbs with numerous narrow, entire, 1-nerved leaves, and rather large and showy heads of tubular and radiate flowers terminating the stem and branches. Basal leaves none. Involucral bracts coriaceous, imbricated in several series, appressed, their tips not herbaceous. Ray-flowers normally violet, pistillate. Disk-flowers perfect, the corolla with a campanulate limb. Achenes villous. Pappus double, the inner series of long capillary bristles, the outer much shorter. [Greek, violet rays.]

Three known species, natives of North America, the following typical.

1. Ionactis linariifòlius (L.) Greene. Stiff or Savory-leaved Aster. Fig. 4383.

Aster linariifolius L. Sp. Pl. 874. 1753.
Diplopappus linariifolius Hook. Fl. Bor. Am. **2**: 21. 1834.
Ionactis linariifolius Greene, Pittonia **3**: 245. 1897.

Stems tufted, stiff, puberulent or scabrous, very leafy, corymbosely branched above, $6'-2°$ high. Leaves linear or spatulate, spreading, 1-nerved, rigid, entire, rough, usually ciliolate on the margins, mucronulate, $9''-18''$ long, $1''-2''$ wide, sessile, those of the branches much smaller and somewhat appressed; heads several, terminating the branchlets, about $1'$ broad; involucre broadly turbinate, its bracts linear-lanceolate, keeled, green on the back, appressed, imbricated in 4 or 5 series, the inner obtuse, the outer usually acute; rays 10–15, violet, rarely white, $4''-5''$ long, entire, or their tips dentate, or even laciniate; pappus bristles in 2 series, tawny, the outer ones setose; achenes silky.

In dry or rocky soil, Maine to Florida, west to Minnesota, Mississippi and Texas. Recorded from Newfoundland, New Brunswick and Quebec, apparently erroneously. Sandpaper- or pine-starwort. July–Oct.

3/5

39. BÁCCHARIS L. Sp. Pl. 860. 1753.

Dioecious shrubs or herbs, with alternate leaves, and paniculate or corymbose heads of tubular flowers. Involucre campanulate in our species, its bracts imbricated in several series, the outer shorter. Receptacle flat, naked, commonly foveolate. Corolla of the pistillate flowers slender, that of the staminate tubular, 5-lobed. Anthers obtuse and entire at the base. Style-branches narrow or subulate, those of the fertile flowers smooth, exserted, those of the sterile flowers rudimentary, tipped with an ovate pubescent appendage. Achenes more or less compressed, ribbed. Pappus of the fertile flowers copious, capillary, that of the sterile flowers short. [Named for Bacchus; originally applied to some different shrubs.]

About 300 species, all American, most abundant in South America. Besides the following, some 18 others occur in the southern and western United States. Type species: *Baccharis ivifolia* L.

Shrubs; pappus of fertile flowers in 1 or 2 series.
　Leaves oblong, or lance-oblong, mostly obtuse, sparingly dentate.　　　　　　　　1. *B. salicina.*
　Leaves deltoid-obovate, or oblong, the lower coarsely dentate.　　　　　　　　　2. *B. halimifolia.*
　Leaves linear or linear-lanceolate, mostly acute, sparingly dentate.　　　　　　　3. *B. neglecta.*
Herbaceous, from a woody base; pappus of fertile flowers in several series.　　　　4. *B. Wrightii.*

1. Baccharis salícina T. & G. Willow Bac-charis. Fig. 4384.

Baccharis salicina T. & G. Fl. N. A. **2**: 258. 1841.

A glabrous glutinous much-branched shrub, 3°–6° high, the branches ascending. Leaves firm, oblong-lanceolate or somewhat oblanceolate, more or less conspicuously 3-nerved, mostly obtuse at the apex, narrowed into a cuneate subsessile base, 1′–1½′ long, 2″–6″ wide, sparingly repand-dentate, or entire; heads in peduncled clusters of 1–7, the involucre of both sterile and fertile ones campanulate, 2½″–3″ high, its bracts ovate or ovate-lanceolate, acute or subacute; pappus usually but a single series of nearly white capillary bristles.

Western Kansas and eastern Colorado to Texas and New Mexico. May–July.

2. Baccharis halimifòlia L. Groundsel-tree or -bush. Pencil-tree. Fig. 4385.

Baccharis halimifolia L. Sp. Pl. 860. 1753.

A branching glabrous shrub, 3°–10° high, the branch-lets angled, sometimes minutely scurfy. Leaves thick, those of the stem and larger branches obovate or del-toid-obovate, obtuse, petioled, coarsely angular-dentate, 1′–3′ long, ½′–2′ wide, those of the branchlets oblanceo-late, short-petioled or sessile, entire, or few-toothed toward the apex; heads in peduncled clusters of 1–5, those of the sterile plant nearly globose when young, the bracts of the involucre oblong-ovate, obtuse, gluti-nous, appressed, the inner ones of the pistillate heads lanceolate, acute or acutish; fertile pappus bright white, 3″–4″ long, of 1–2 series of capillary bristles, much ex-ceeding the involucre.

Along salt marshes and tidal rivers, extending beyond saline influence, Massachusetts to Florida and Texas. Ba-hamas; Cuba. The white pappus is very conspicuous in autumn. Cotton-seed tree. Ploughman's-spikenard. Sept.–Nov.

Baccharis glomeruliflòra Pers., which has larger heads glomerate in the axils of the upper leaves, is doubtfully re-ported from southern Virginia, but occurs along the coast from North Carolina to Florida, and in Bermuda.

3. Baccharis neglécta Britton. Linear-leaved Baccharis. Fig. 4386.

Baccharis neglecta Britton, in Britt. & Brown, Ill. Fl. **3**: 394. 1898.

A much-branched, glabrous or slightly glutinous shrub, 3° high or more, the branches paniculate, slender, ascending. Leaves narrowly linear to linear-lanceolate, faintly 3-nerved, acute, or the lower suboctuse at the apex, gradually attenuate into a nearly sessile base, 1′–3′ long, 1″–3″ wide, entire, or remotely dentate or denticulate, green in drying; heads in short-peduncled clusters; in-volucre of both kinds of heads campanulate, 2″ high, its outer bracts ovate, acute or somewhat obtuse, the inner lanceolate, acuminate; pappus of the fertile flowers a single series of capillary dull-white bristles.

Nebraska to Texas and North Mexico. July–Sept.

4. Baccharis Wrìghtii A. Gray. Wright's Baccharis. Fig. 4387.

Baccharis Wrightii A. Gray, Pl. Wright. **1**: 101. 1852.

Herbaceous from a thick woody base, much branched, glabrous, not glutinous, 1°–3° high, the branches straight, nearly erect, slender, striate. Leaves linear, sessile, 1-nerved, entire, 3″–12″ long, ½″–1″ wide; heads solitary at the ends of the branches, 5″–6″ broad; involucre of the sterile heads hemispheric, about 3″ high, that of the fertile ones somewhat campanulate and longer; bracts of both involucres lanceolate, acuminate, with scarious margins and a green back; pappus of the fertile flowers of several series of tawny or purplish capillary bristles.

Western Kansas and Colorado to Texas, Arizona and Chihuahua. April–July.

40. FILÀGO Loefl.; L. Sp. Pl. 927. 1753.

[Evax Gaertn. Fr. & Sem. **2**: 393. *pl. 165. f. 3.* 1791.]

White-woolly annual herbs, with alternate entire leaves, and small discoid clustered heads, usually subtended by leafy bracts. Bracts of the involucre few and scarious. Receptacle convex or elongated, chaffy, each chaffy scale subtending an achene. Outer flowers of the heads in several series, pistillate, fertile, their corollas filiform, minutely 2–4-dentate. Central flowers few, perfect, mainly sterile, their corollas tubular, 4–5-toothed. Anthers sagittate at the base, the auricles acuminate. Achenes compressed or terete. Pappus none. [Latin *filum*, a thread.]

About 12 species, natives of temperate or warm regions of both the New World and the Old. In addition to the following, 3 others occur in the western and southwestern United States. Type species: *Filago pygmaea* L.

1. Filago prolífera (Nutt.) Britton. Filago. Fig. 4388.

Evax prolifera Nutt.; DC. Prodr. **5**: 459. 1836.
Diaperia prolifera Nutt. Trans. Am. Phil. Soc. (II) **7**: 338. 1841.
Filago prolifera Britton, Mem. Torr. Club **5**: 329. 1894.

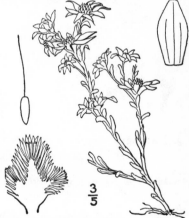

Stem simple, or branched at the base, very leafy, 2′–6′ high. Leaves spatulate, obtuse, sessile, ascending or appressed, 4″–8″ long, 1″–2½″ wide; heads in a sessile leafy-bracted cluster, usually subtended by 1 or several slender, nearly leafless branches, each terminated by a similar cluster, or these again proliferous; heads oblong or fusiform; receptacle convex; chaff of the central sterile flowers woolly-tipped, that of the fertile flowers scarious, mostly glabrous.

In dry soil, Texas to western Kansas and South Dakota, west to Colorado and New Mexico. April–July.

41. GÍFOLA Cass. Bull. Soc. Philom. 1819: 143. 1819.

[Filago L. Gen. Pl. Ed. 5. 1758. Not Sp. Pl. 927. 1753.]

White-woolly herbs, closely resembling those of the preceding genus, with alternate, entire leaves, and small discoid glomerate heads, often subtended by leafy bracts, the clusters proliferous in our species. Involucre small, its bracts scarious, imbricated in several series, the outer usually tomentose. Receptacle subulate, cylindric or obconic, chaffy, each chaffy scale near its base enclosing an achene. Outermost flowers pistillate, fertile, with filiform corollas and no pappus, or the pappus a few rudimentary bristles; inner flowers also pistillate and fertile, but with a pappus of capillary scabrous bristles; central (uppermost) flowers few, perfect, their corollas tubular, their pappus capillary and scabrous. Anthers sagittate at the base. Achenes terete or slightly compressed. [Anagram of *Filago*.]

About 10 species, natives of warm and temperate regions, the following typical. In addition to the following, 3 Californian species are probably to be referred to this genus.

1. Gifola germánica (L.) Dumort. Cudweed. Cotton Rose. Herb Impius. Fig. 4389.

Gnaphalium germanicum L. Sp. Pl. 857. 1753.

Filago germanica L. Sp. Pl. Ed. 2, 1311. 1763.

Gifola germanica Dumort. Fl. Belg. 68. 1827.

Annual, erect, cottony, 4'–18' high, simple, or branched at the base, very leafy. Leaves sessile, lanceolate, linear, or slightly spatulate, erect or ascending, obtuse or acutish, 3"–12" long; stem terminated by a sessile dense cluster of heads, usually subtended by several leafy branches terminated by similar clusters and these often again proliferous; heads 12–30 in each cluster, many-flowered; involucre ovoid, light yellow, its bracts mainly acute. $\frac{2}{3}$

In dry fields, southern New York and New Jersey to Pennsylvania, West Virginia and North Carolina. Old names downweed, hoarwort, owl's-crown, chafeweed, childing cudweed. May–Sept.

42. PLÙCHEA Cass. Bull. Soc. Philom. 1817: 31. 1817.

Pubescent or glabrous herbs, or some tropical species shrubby, with alternate dentate leaves, and small heads of tubular flowers in terminal corymbose cymes. Involucre ovoid, campanulate, or nearly hemispheric, its bracts appressed, herbaceous, imbricated in several series. Receptacle flat, naked. Outer flowers of the head pistillate, their corollas filiform, 3-cleft or dentate at the apex. Central flowers perfect, but mainly sterile, their corollas 5-cleft. Anthers sagittate at the base, the auricles caudate. Style of the perfect flowers 2-cleft or undivided. Achenes 4–5-angled. Pappus a single series of capillary scabrous bristles. [Named for the Abbé N. A. Pluche, of Paris.]

About 35 species, widely distributed in warm and temperate regions. In addition to the following, 2 or 3 other indigenous species occurs in the southern United States, and two introduced ones have been found in waste places in Florida. Type species: *Conyza marilandica* Michx.

Perennial; leaves sessile, cordate, or clasping at the base.   1. *P. foetida.*
Annual; leaves, at least those of the stem, petioled.
  Leaves short-petioled; heads about 3" high; involucral bracts densely puberulent.
            2. *P. camphorata.*
  Leaves slender-petioled; heads 2"–2½" high; involucral bracts granulose, ciliate.
            3. *P. petiolata.*

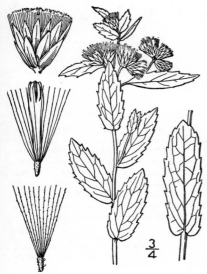

1. Pluchea foètida (L.) DC. Viscid Marsh Fleabane. Fig. 4390.

Baccharis foetida L. Sp. Pl. 861. 1753.

Baccharis viscosa Walt. Fl. Car. 202. 1788.

Pluchea bifrons DC. Prodr. 5: 451. 1836.

Pluchea foetida DC. Prodr. 5: 452. 1836.

Root perennial; stem simple or sparingly branched at the summit, puberulent and slightly viscid, 1½°–3° high. Leaves oblong, ovate or ovate-lanceolate, closely sessile and more or less cordate-clasping at the base, obtuse or acute at the apex, sharply denticulate, pubescent or puberulent, 2'–4' long, ¾'–1½' wide, reticulate-veiny; clusters of heads sessile, or stalked, compact, leafy-bracted; involucre 2½"–3" high, its bracts lanceolate, acute, viscid-puberulent.

$\frac{3}{4}$

In swamps, southern New Jersey to Florida and Texas, mainly near the coast. Also in the West Indies. Foetid marsh-fleabane. July–Sept.

2/3

2. Pluchea camphoràta (L.) DC. Spicy or Salt-marsh Fleabane. Fig. 4391.

Erigeron camphoratum L. Sp. Pl. Ed. 2, 1212. 1763.
Conyza marilandica Michx. Fl. Bor. Am. 2 : 126. 1803.
Pluchea camphorata DC. Prodr. 5 : 451. 1836.

Root annual; stem usually branched, finely viscid-puberulent, or nearly glabrous, 2°–3° high, somewhat channeled. Leaves ovate, oblong or lanceolate, puberulent or glabrous, acute or acuminate at the apex, narrowed at the base, short-petioled, or the upper sessile but not clasping, 3'–8' long, 1'–3' wide, serrate or denticulate, not conspicuously reticulate-veined; heads about 3'' high, rarely leafy-bracted, mostly in naked open corymbiform cymes; bracts of the involucre ovate-lanceolate, or lanceolate, acute, puberulent; flowers purplish; achenes pubescent.

In salt marshes, coast of Massachusetts to Florida, Texas and Mexico. Called also ploughman's-wort. Aug.–Oct.

3. Pluchea petiolàta Cass. Inland Marsh Flea-bane. Fig. 4392.

Pluchea petiolata Cass. Dict. Sci. Nat. 42 : 2. 1826.

Pluchea foetida DC. Prodr. 5 : 452. 1830.

Similar to the preceding species, but glabrate, darker green, usually taller, 2½°–4° high, the stem stout, rather strongly channelled. Leaves ovate-lanceolate to oval, thin, 4'–10' long, 1½'–3' wide, mostly acuminate at the apex, cuneate-narrowed at the base, long-petioled, irregularly serrate; petioles of the larger leaves 8''–12'' long; heads 2''–2½'' high, in terminal and often also axillary clusters; bracts of the involucre granular, ciliate; achenes short-pubescent.

In moist soil, often in woods, Maryland to Florida, Illinois, Missouri and Oklahoma. Aug.–Oct.

½

43. ANTENNÀRIA Gaertn. Fruct. & Sem. 2 : 410. *pl. 167.* 1791.

Perennial woolly dioecious or polygamo-dioecious herbs, with alternate and basal leaves, and small discoid many-flowered heads usually capitate, glomerate or corymbose. Involucre oblong, ovoid or campanulate, its bracts scarious, imbricated in several series, the outer shorter, usually woolly. Receptacle convex, or nearly flat, foveolate, not chaffy. Staminate flowers with a truncate or minutely dentate corolla, usually undivided style and scanty pappus of club-shaped smooth or minutely barbed bristles. Pistillate or perfect flowers with tubular mostly 5-toothed corollas, 2-cleft style, and copious pappus of capillary naked bristles, slightly united at the base, the style often crimson. Achenes oblong, terete, or slightly compressed. [Latin, in allusion to the fancied resemblance of the sterile pappus to insect antennae.]

About 50 species, natives of the north temperate zone and southern South America. In addition to the following, many others occur in the western parts of North America. The patches of fertile and sterile plants are usually quite distinct, and the sterile plants of some species are much less abundant than the pistillate. Perfect achenes are, however, abundant, parthenogenesis being a known feature of this genus. Type species: *Antennaria dioica* (L.) Gaertn.

1. Antennaria carpáthica (Wahl.) Hook.   Carpathian Everlasting.   Fig. 4393.

Gnaphalium carpathicum Wahl. Fl. Carp. 258. *pl. 3.* 1814.

Antennaria carpathica Hook. Fl. Bor. Am. 1: 329. 1833.

Plant floccose-woolly throughout, not stoloniferous; stem 2′–10′ high, simple. Basal leaves oblanceolate or oblong, obtuse or acutish at the apex, narrowed at the base into short petioles, 1′–2′ long, 2″–4″ wide; stem leaves linear, acute or acutish, erect, the upper gradually smaller; heads in a terminal subcapitate cluster, or rarely solitary, 2½″–3″ broad; involucre 2″–3″ high, woolly at the base, its bracts brownish purple, the inner ones of the fertile heads mostly acutish, those of the sterile heads mainly obtuse.

$\frac{3}{5}$

In dry soil, Labrador and Anticosti to Hudson Bay.   Also in Europe and northern Asia.   May–Sept.

2. Antennaria alpìna (L.) Gaertn.   Alpine Everlasting.   Fig. 4394.

Gnaphalium alpinum L. Sp. Pl. 856. 1753.

Antennaria alpina Gaertn. Fr. & Sem. 2: 410. 1791.

?*A. labradorica* Nutt. Trans. Am. Phil. Soc. (II) 7: 406. 1841.

A. angustata Greene, Pittonia 3: 284. 1898.

Surculose by short stolons; stems floccose-woolly, 1′–4′ high. Basal leaves usually numerous, tufted, spatulate or linear-oblong, obtuse, silvery-woolly on both sides, or glabrate and green above, 4″–12″ long; stem leaves linear and small, few, sessile, woolly; heads in a terminal capitate, or seldom somewhat corymbose cluster, rarely solitary, 2″–2½″ broad; involucre about 2½″ high, the bracts of the fertile heads dark brownish-green, the inner ones acute or acuminate, those of the sterile heads lighter, brownish, broader and obtuse; achenes glandular.

Labrador and Arctic America to Alaska and British Columbia.   Western plants previously referred to this species prove to be distinct.   April–Aug.

3. Antennaria canadénsis Greene. Canadian Cat's-foot. Fig. 4395.

Antennaria canadensis Greene, Pittonia 3 : 275. 1898.

Stems 8'–12' high, slender. Basal leaves and those of the ends of stolons spatulate to oblanceolate, obtuse or apiculate, 1'–1½' long, 6" wide or less, gradually tapering from above the middle to a long narrow base, 1-nerved or with two faint lateral nerves, bright green and glabrous above, lanate beneath; stolons short, leafy, assurgent; stem-leaves linear-lanceolate, distant, about 1" wide; inflorescence capitate to racemose-corymbose; fertile involucre about 4½" high, its outer bracts oblong, obtuse, the inner lanceolate, acute to attenuate; bracts of the staminate involucre white-tipped.

In dry soil, Newfoundland to Connecticut, Manitoba and Michigan. May–July.

The Old World **Antennaria dioìca** (L.) Gaertn., with stem leaves close together and rose-colored involucral bracts, is recorded as long ago found at Providence, R. I.

4. Antennaria Parlìnii Fernald. Parlin's Cat's-foot. Fig. 4396.

Antennaria Parlinii Fernald, Gard. & For. **10** : 284. 1897.
A. arnoglossa Greene, Pittonia **3** : 318. 1898.
A. Parlinii arnoglossa Fernald, Proc. Bost. Soc. Nat. Hist. **28** : 243. 1898.
A. propinqua Greene, Pittonia **4** : 83. 1899.

More or less glandular-pubescent; stems of fertile plant 1°–1½° tall. Leaves bright green and devoid of tomentum on the upper surface from the time of unfolding, or very slightly floccose when very young, the basal ones obovate or spatulate to elliptic, obtuse or acutish, gradually contracted into a narrow base about as long as the expanded part, 2'–3½' long, ¾'–1½' wide; stem-leaves lanceolate or the lower narrowly oblong; heads corymbose; involucre 3½"–5" high, its bracts all lanceolate-acuminate or the outer ones linear-oblong and obtusish.

Fields, hillsides and woodlands, Maine to Ontario, Virginia and Iowa. May–July.

5. Antennaria solitària Rydb. Single-headed Cat's-foot. Fig. 4397.

Antennaria plantaginifolia monocephala T. & G. Fl. N. A. **2** : 431. 1843.

Antennaria monocephala Greene, Pittonia **3** : 176. 1896. Not DC. 1836.

Antennaria solitaria Rydb. Bull. Torr. Club **24** : 304. 1897.

Stem slender, weak, floccose-woolly, 2'–10' long, bearing a solitary head. Basal leaves obovate to oblong-obovate or broadly spatulate, 3½' long or less, 8"–16" wide, obtuse or apiculate, densely floccose beneath, loosely floccose, becoming glabrate above, 3–5-nerved; stem-leaves linear, few and distant; stolons procumbent, leafy at the ends; involucre 4"–6" high, its linear white-tipped bracts very woolly.

Woodlands, Pennsylvania to Georgia, Ohio, Alabama and Louisiana. March–May.

6. Antennaria plantaginifòlia (L.) Richards. Plantain-leaf Everlasting. Fig. 4398.

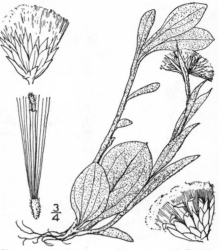

Gnaphalium plantaginifolium L. Sp. Pl. 850. 1753.
Antennaria plantaginifolia Richards. App. Frank.
   Journ. Ed. 2, 30. 1823.

Floccose-woolly, stoloniferous, forming broad
patches; flowering stems of fertile plant 4´-20´
high, slender or stout, sometimes with glandular hairs. Basal leaves obovate, spatulate, or
broadly oval, obtuse or acutish, distinctly 3-
ribbed, petioled, dull dark green and arachnoid
above, silvery beneath, 1½´-3´ long, 5´´-18´´
wide; stem-leaves sessile, oblong or lanceolate,
the upper usually small and distant; heads in
corymbose or often subcapitate clusters, 4´´-5´´
broad; involucre 3´´-4½´´ high, its bracts greenish-white, linear to lanceolate, acute or acutish;
achenes minutely glandular; sterile plant
smaller, 3´-8´ high; basal leaves somewhat
smaller; heads smaller, 3´´-4´´ broad; bracts
oblong, obtuse.

In dry soil, especially in open woods, Quebec to
Florida, Minnesota, Nebraska and Texas. Spring-
or early everlasting. White plantain. Pussy-toes.
Ladies´-tobacco. Dog-toes. Four-toes. Love's-test. Indian- or woman's-tobacco. Poverty-weed. Pearly
mouse-ear everlasting. Consists of many races differing in size, leaf-form, leaf-size, size of heads
and shape of involucral bracts; these have been variously regarded by authors as species and as
varieties. April–June.

7. Antennaria microphỳlla Rydb. Small-leaved Cat's-foot. Fig. 4399.

A. microphylla Rydb. Bull. Torr. Club **24**: 303. 1897.

?*A. parvifolia* Nutt. Trans. Am. Phil. Soc. (II) **7**: 406. 1841.

Stem slender, 8´-12´ high; stolons short, not over 2½´
long. Basal leaves and those of the ends of the stolons
spatulate, obtuse, or apiculate, 2´´-8´´ long, 1´´-2´´ wide,
narrowed from above the middle; stem-leaves linear-
oblong, or the leaves linear-spatulate, often longer than
the basal ones; heads corymbose, rather numerous;
involucre 2½´´-3½´´ high; bracts of the fertile heads
linear-oblong, acute or acutish, those of the sterile obtuse.

Dry plains and hills, Saskatchewan to Nebraska, British
Columbia and New Mexico. July–Aug.

8. Antennaria neodioìca Greene. Smaller Cat's-foot. Fig. 4400.

Antennaria neodioica Greene, Pittonia **3**: 184. 1897.
A. alsinoides Greene, Pittonia **4**: 83. 1899.
A. rupicola Fernald, Rhodora **1**: 74. 1899.

Floccose-woolly, with numerous stolons which
are leafy throughout; stem of fertile plants slender, about 1° high. Basal leaves about 1´ long,
3´´-5´´ wide, broadly obovate to spatulate, 1-nerved,
or indistinctly 3-nerved, white-tomentose beneath,
becoming glabrate above, usually narrowed into
distinct petioles; stem-leaves linear, acute; heads
loosely corymbose, 3´´-4´´ broad; outermost bracts
of the involucre obtuse, the rest lanceolate, acute,
or acuminate, all greenish or brownish below,
with white scarious tips; achenes obtusely 4-an-
gled, granular-papillose; sterile plant lower, 3´-8´
high; heads more densely clustered, the bracts of
the involucre oblong, obtuse.

In dry places, Newfoundland to Virginia, Quebec, Michigan and South Dakota. April–July.

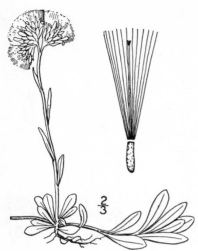

$\frac{2}{3}$

9. Antennaria áprica Greene. Rocky Mountain Cudweed. Fig. 4401.

Antennaria aprica Greene, Pittonia 3: 282. 1898.

Floccose-woolly or canescent, surculose, forming broad patches; flowering stems 2′–12′ high. Basal leaves spatulate or obovate, obtuse, narrowed into short petioles, white-canescent on both sides, 4″–15″ long, 1″–4″ wide; stem-leaves linear, sessile; heads 3″–4″ broad in a terminal capitate or corymbose cluster; involucre 2″–3″ high, the bracts of the fertile heads oblong, white or pink, all obtuse, or the inner ones acute; those of the sterile heads oval or elliptical, obtuse; achenes glabrous, or minutely glandular.

In dry soil, South Dakota to Nebraska, Alberta, Utah and New Mexico. June–Sept. Erroneously referred, in our first edition, as by previous authors, to the Old World *Antennaria dioica* (L.) Gaertn.

10. Antennaria campéstris Rydberg. Prairie Cat's-foot. Fig. 4402.

Antennaria campestris Rydberg, Bull. Torr. Club 24: 304. 1897.

Stolons short, leafy; flowering stems of both fertile and sterile plants 2′–6′ high. Basal leaves obovate-cuneate, without a distinct petiole, white-tomentose beneath, glabrate above, 1-nerved, or indistinctly 3-nerved; stem-leaves small, linear; heads 3″–4″ broad in subcapitate clusters; bracts of the fertile heads lanceolate, greenish below, brownish at the middle, the apex white, acute or acuminate; bracts of sterile heads elliptic, obtuse.

On dry prairies, Nebraska and Kansas to Saskatchewan. May–June.

$\frac{3}{4}$

$\frac{2}{3}$

11. Antennaria neglécta Greene. Field Cat's-foot. Fig. 4403.

Antennaria neglecta Greene, Pittonia 3: 173. 1897.
An ennaria petaloidea Fernald, Rhodora 1: 73. 1899.
A. neglecta simplex Peck, Bull. N. Y. State Mus. 67: Bot. 6: 33. 1903.

Stoloniferous, the stolons long and slender, bearing small leaves, except at the ends, where they are normally developed. Basal leaves oblanceolate or cuneate-spatulate, gradually tapering to a sessile base, without a distinct petiole, white-tomentose beneath, glabrate above, 1-nerved; stem-leaves linear; fertile plant nearly 1° high; heads 3″–4″ broad, corymbose or sometimes only 1 or 2; bracts brownish, with white tips, lanceolate, acute; sterile plant 4′–8′ high, the heads densely clustered, the bracts oblong, obtuse.

In fields and pastures, Maine to New York, Virginia and Wisconsin. April–June.

12. Antennaria dimórpha (Nutt.) T. & G.
   Low Everlasting. Fig. 4404.

Gnaphalium dimorphum Nutt. Trans. Am. Phil. Soc.
   (II) **7**: 405. 1841.
A. dimorpha T. & G. Fl. N. A. **2**: 431. 1843.

Tufted from a thick woody often branched caudex, 1′–1½′ high. Leaves all in a basal cluster, spatulate, white-canescent or tomentose on both sides,
obtuse or acutish, ½′–1′ long, 1″–2″ wide, narrowed
into short petioles; heads of staminate flowers
about 3″ broad and high, solitary and sessile
among the leaves, or raised on a very short sparingly leafy stem, with obtuse or obtusish brownish involucral bracts, those of pistillate flowers
longer, their inner bracts linear-lanceolate, acuminate; pappus of the staminate flowers slender,
scarcely thickened, but barbellate at the tips, that
of the pistillate flowers of fine and smooth bristles.

Dry soil, Nebraska to Utah and California, north
to Montana and British Columbia. April–June.

44. ANÁPHALIS DC. Prodr. **6**: 271. 1837.

Perennial white-tomentose or woolly herbs, with leafy erect stems, in our species alternate entire leaves, and small corymbose discoid heads of dioecious flowers. Involucre oblong
to campanulate, its bracts scarious, imbricated in several series, mostly white, the outer
shorter. Receptacle mostly convex, not chaffy. Staminate flowers with a slender or filiform
corolla, an undivided style, and a pappus of slender bristles, not thickened at the summit, or
scarcely so; anthers tailed at the base. Pistillate flowers with a tubular 5-toothed corolla,
2-cleft style, and a pappus of capillary bristles. [Greek name of some similar plant.]

About 35 species, natives of the north temperate zone. Besides the following, 2 or 3 other
species occur in western North America. Type species: *Anaphalis nubigena* (Wall.) DC.

1. Anaphalis margaritàcea (L.) Benth. &
   Hook. Pearly or Large-flowered Ever-
      lasting. Fig. 4405

Gnaphalium margaritaceum L. Sp. Pl. 850. 1753.
Antennaria margaritacea Hook. Fl. Bor. Am. **1**: 329. 1833.
A. margaritacea Benth. & Hook. Gen. Pl. **2**: 303. 1873.

Stem floccose-woolly, corymbosely branched at the
summit, leafy, 1°–3° high. Leaves linear-lanceolate,
narrowed to a sessile base, revolute, green, but mostly
more or less pubescent above, woolly beneath, 3′–5′
long, 2″–4″ wide, the lowest shorter, spatulate, usually
obtuse; corymb compound, 2′–8′ broad; heads very numerous, short-peduncled or sessile, about 3″ high, 4″
broad when expanded; involucre campanulate, its
bracts ovate-lanceolate, obtuse, finely striate, pearly
white, mostly glabrous; pappus-bristles of the fertile
flowers distinct and falling away separately.

Dry soil, Newfoundland to Alaska, Pennsylvania, Kansas,
Oregon and northern Asia. Recorded south to North Caro-
lina. Adventive in Europe. Silver-leaf. Life-everlasting.
Moonshine. Cotton-weed. None-so-pretty. Lady-never-fade.
Indian-posy. Ladies'-tobacco. Poverty-weed. Silver-button.
July–Sept.

Anaphalis occidentàlis (Greene) Heller, occurring from Alaska to California and in Newfoundland and Quebec, differs by its mostly broader leaves being green and glabrous.

45. GNAPHÀLIUM L. Sp. Pl. 850. 1753.

Woolly erect or diffusely branched herbs, with alternate leaves, and discoid heads of
pistillate and perfect flowers arranged in corymbs, spikes, racemes, or capitate. Receptacle
flat, concex or conic, not chaffy, usually foveolate. Pistillate flowers in several series, their
corollas filiform, minutely dentate or 3-4-lobed. Central flowers perfect, tubular, few, their
corollas 5-toothed or 5-lobed. Anthers sagittate at the base, the auricles tailed. Achenes
oblong or obovate, terete or slightly compressed, not ribbed. Pappus a single series of capillary bristles, sometimes thickened above. [Greek, referring to the wool.]

About 120 species, widely distributed. Type species: *Gnaphalium luteo-album* L.

1. Pappus-bristles distinct.

Tall, erect; inflorescence corymbose, or paniculate.
   Leaves sessile; plant not viscid.   1. *G. obtusifolium.*
   Leaves sessile; plant glandular-viscid.   2. *G. Helleri.*
   Leaves decurrent; plant glandular-viscid.   3. *G. decurrens.*

Low, diffuse; inflorescence mostly capitate; pappus-bristles distinct.
  Floccose-woolly; involucral bracts yellowish, or white.   4. *G. palustre.*
  Appressed-woolly; involucral bracts becoming dark brown.   5. *G. uliginosum.*
  Tufted low mountain herbs; heads few; bracts brown; pappus-bristles distinct.   6. *G. supinum.*

2. Slender, simple; heads spicate; pappus-bristles united at base.

Leaves linear or lanceolate-spatulate, acute; heads about 3″ high; northeastern.
  Bracts dark brown; stem leaves lanceolate-spatulate.   7. *G. norvegicum.*
  Bracts brownish tipped; stem leaves linear.   8. *G. sylvaticum.*
Leaves spatulate, obtuse or obtusish; heads 2″–2½″ high; eastern and southern.   9. *G. purpureum.*

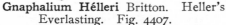

1. Gnaphalium obtusifòlium L. Sweet or White Balsam. Sweet or Fragrant Life Everlasting. Fig. 4406.

Gnaphalium obtusifolium L. Sp. Pl. 851: 1753.
G. polycephalum Michx. Fl. Bor. Am. 2: 127. 1803.

Annual or winter-annual, fragrant; stem erect, simple, or branched above, tomentose, 1°–3° high. Leaves lanceolate or linear-lanceolate, sessile, acute or acutish, or the lower obtuse at the apex, narrowed at the base, densely white-woolly beneath, glabrate and commonly dark green above, 1′–3′ long, 2″–4″ wide, the margins undulate; heads in numerous corymbose or paniculate clusters of 1–5, about 3″ high; bracts of the involucre white, or tinged with brown, oblong, thin and scarious, obtuse, the outer woolly at the base; pappus-bristles distinct, separately deciduous; achenes glabrous.

In dry, mostly open places, Nova Scotia to Florida, Manitoba, Kansas and Texas. Jamaica. Poverty-, chafe- or balsam-weed. Old-field balsam. Indian-posy. Feather-weed. Fussy-gussy. Rabbit-tobacco. Life-of-man. Moonshine. Leaves of rosettes oblong. Aug.–Sept.

2. Gnaphalium Hélleri Britton. Heller's Everlasting. Fig. 4407.

G. Helleri Britton, Bull. Torr. Club 20: 280. 1893.
G. polycephalum Helleri Fernald, Rhodora 10: 94. 1908.

Similar to the preceding species, corymbosely or somewhat paniculately branched above, 1½°–2° high, the stem and branches densely glandular-pubescent, not tomentose. Leaves oblong-lanceolate, sessile, acuminate at both ends, green and hispidulous above, white-tomentose beneath, the larger about 2′ long and 5″ wide, the uppermost much smaller and narrower; heads very numerous, corymbose or corymbose-paniculate, sessile or short-peduncled in the clusters, about 2½″ broad; involucre oblong, or becoming campanulate, 3″ high, its bracts bright white, tomentose, the outer oblong, the inner linear-oblong, all obtuse; pappus-bristles distinct; achenes glabrous.

In fields and woods, New York and New Jersey to Virginia, Kentucky and Georgia. Sept.–Oct.

3. Gnaphalium decúrrens Ives. Clammy Everlasting. Winged Cudweed. Fig. 4408.

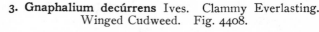

Gnaphalium decurrens Ives, Am. Journ. Sci. 1: 380. *pl. 1.* 1819.

Annual or biennial, similar to the two preceding species, fragrant; stem very leafy, glandular-viscid, corymbosely branched above, 2°–3° high. Leaves lanceolate or broadly linear, acutish at the apex, densely white-woolly beneath, glabrate or loosely woolly above, sessile and decurrent on the stem at the base, 1′–3′ long, 2″–3″ wide, or the lowest shorter and slightly spatulate; heads in several or numerous corymbose glomerules of 2–6, about 3″ high; bracts of the involucre yellowish-white or brownish, ovate, acutish or the inner lanceolate and acute, the outer woolly at base; pappus-bristles distinct; achenes glabrous.

In open, moist or dry places, Nova Scotia to Pennsylvania, West Virginia ?, Ontario, British Columbia, Michigan, south in the Rocky Mountains to Arizona. Sweet balsam. Balsam-weed. July–Sept.

4. Gnaphalium palústre Nutt. Western Marsh Cudweed. Fig. 4409.

Gnaphalium palustre Nutt. Trans. Am. Phil. Soc. (II) **7**: 403. 1841.

Annual; diffusely branched from the base, densely but loosely floccose-woolly all over, 2′–8′ high. Leaves sessile, oblong, linear-oblong, or the lower spatulate, obtuse or acutish, 6″–12″ long, 1½″–3″ wide; heads about 2″ high, several together in leafy-bracted glomerules; involucre more or less woolly, its bracts linear or lanceolate, acute, white or pale yellow; pappus-bristles distinct, separately deciduous.

In moist wet soil, Saskatchewan to Montana, western Nebraska and New Mexico, west to British Columbia and California. May–Aug.

5. Gnaphalium uliginòsum L. Low or Marsh Cudweed. Wartwort. Mouse-ear. Fig. 4410.

Gnaphalium uliginosum L. Sp. Pl. 856. 1753.

Annual; diffusely branched from the base, or the stems sometimes erect or ascending, appressed-woolly all over, 2′–8′ high. Leaves sessile, spatulate-linear, linear, or the lower oblanceolate or spatulate and narrowed into petioles, all obtuse or obtusish, generally mucronulate, 1′–1½′ long; heads about 2″ high, numerous in dense leafy-bracted terminal glomerules; bracts of the involucre oblong or oblong-lanceolate, brown, the outer obtuse or obtusish and more or less woolly, the inner acute; pappus-bristles distinct, separately deciduous.

In damp soil, Newfoundland to Virginia, west to western Ontario, Saskatchewan, British Columbia and Oregon, and Indiana. Also in Europe. July–Sept.

6. Gnaphalium supìnum L. Dwarf Cudweed. Fig. 4411.

Gnaphalium supinum L. Syst. Ed. 2, 234. 1767.

Perennial, white-woolly, much tufted; stems simple, 1′–3½′ high. Leaves mainly basal, linear, acute, narrowed at the base, sessile, 6″–12″ long, 1″–2″ wide; heads few or several, capitate or short-spicate, about 3″ high; flowers yellowish; bracts of the involucre brown, glabrous, lanceolate or oblong-lanceolate, acute; pappus-bristles distinct, separately deciduous.

Alpine summit of the White Mountains of New Hampshire, and of Mt. Katahdin, Maine; Labrador and Greenland, and on high mountains in Europe and Asia. Called also mountain-cudweed. July–Aug.

7. Gnaphalium norvégicum Gunner. Norwegian Cudweed. Fig. 4412.

G. norvegicum Gunner, Fl. Norveg. **2**: 105. 1772.

Perennial; stem simple, 6′–18′ high. Leaves lanceolate to spatulate, elongated, acute, narrowed at the base, woolly on both sides, or green and glabrate above, 3′–6′ long, 2″–5″ wide, the lower and basal ones petioled; heads about 3″ high, numerous in a more or less leafy spike, the lowest often distant, solitary or glomerate in the upper axils; bracts of the involucre ovate-oblong, dark brown, or brown-tipped, glabrous or slightly woolly, obtuse; pappus-bristles united at the base, falling away in a ring; achenes hispidulous.

Mt. Albert, Gaspé, Quebec, north to Greenland and Arctic America. Also in Europe. July–Aug.

8. Gnaphalium sylváticum L. Wood Cudweed. Chafweed. Owl's Crown. Golden Motherwort. Fig. 4413.

Gnaphalium sylvaticum L. Sp. Pl. 856. 1753.

Perennial; stem slender, simple, 6′–18′ high. Leaves linear, acute, 1′–2′ long, 1″–2½″ wide, or the lowest linear-spatulate, woolly beneath, glabrous or glabrate above; heads about 3″ high, numerous in a more or less leafy spike, or the lowest solitary or glomerate in the upper axils; bracts of the involucre linear-oblong, obtuse, mostly glabrous, yellowish or greenish with a brown spot at or just below the apex; pappus-bristles united at the base; achenes hispidulous.

New Brunswick and Cape Breton Island to Quebec and northern Maine and New Hampshire. Widely distributed in Europe and northern Asia. June–Aug.

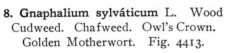

9. Gnaphalium purpùreum L. Purplish Cudweed. Fig. 4414.

Gnaphalium purpureum L. Sp. Pl. 854. 1753.

Annual or biennial, simple and erect or branched from the base and the branches ascending, 2′–2° high. Leaves spatulate, or the uppermost linear, mostly obtuse, mucronulate, woolly beneath, usually green and glabrous or nearly so above when old, sessile, or the lowest narrowed into petioles, 1′–2′ long, 2″–6″ wide; heads 2″–2½″ high in a terminal, sometimes leafy, often interrupted spike, or the lowest ones distant and axillary; bracts of the involucre yellowish brown or purplish, lanceolate-oblong, acute or acutish, the outer woolly at the base; pappus-bristles united below; achenes roughish.

In dry sandy soil, eastern Maine to Florida, Pennsylvania, West Virginia, Kentucky, Kansas and Texas. Bermuda; Jamaica; Mexico. Far western plants formerly referred to this species prove to be distinct. May–Sept.

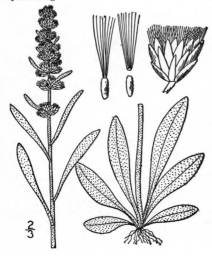

46. ADENOCAÙLON Hook. Bot. Misc. 1: 19. 1830.

Perennial herbs, with broad alternate slender-petioled leaves, woolly beneath, and pani-cled small 5–10-flowered heads of tubular flowers. Involucre campanulate, composed of a few herbaceous bracts. Receptacle nearly flat, naked. Corollas all tubular, 4–5-lobed. Mar-ginal flowers pistillate, fertile. Central flowers perfect, sterile, the style undivided; anthers slightly sagittate at the base. Pappus none. Achenes obovoid or clavate, very obtuse, faintly nerved, glandular above, longer than the bracts of the involucre. [Greek, gland-stem.]

Two species, natives of North America, Japan and the Himalayas. Only the following typical one is known in North America.

1. Adenocaulon bícolor Hook. Adenocaulon.
Fig. 4415.

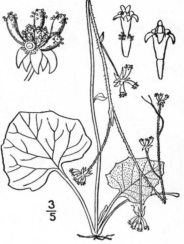

Adenocaulon bicolor Hook. Bot. Misc. 1: 19. *pl. 15.* 1830.

Stem floccose-woolly, or at length glabrous, 1°–3° high, leafless and mostly paniculately branched above. Leaves all basal or nearly so, deltoid-ovate, obtuse or acute at the apex, deeply cordate at the base, coarsely repand-toothed or lobed, thin, green and glabrous above, densely and persistently white-woolly beneath, 2'–6' long and broad, with slender narrowly margined peti-oles; heads numerous, very slender-peduncled, small; bracts of the involucre 4 or 5, ovate to lanceolate, re-flexed in fruit, at length deciduous; achenes 3"–4" long, ½" thick, the upper part beset with nail-shaped glands.

3/5

In moist woods, northern Michigan and Lake Superior to British Columbia, Montana and California. May–July.

47. ÍNULA L. Sp. Pl. 881. 1753.

Perennial, mostly tomentose or woolly herbs, with alternate and basal leaves, and large heads of both tubular and radiate yellow flowers. Involucre hemispheric or campanulate, its bracts imbricated in several series, the outer often foliaceous. Receptacle flat or convex, areolate or foveolate, not chaffy. Ray-flowers pistillate, their ligules 3-toothed. Disk-flowers perfect, their corollas tubular, 5-toothed. Anthers sagittate at the base, the auricles caudate. Style-branches of the disk-flowers linear, obtuse. Achenes 4–5-ribbed; pappus of capillary rough bristles in our species. [The ancient Latin name.]

About 90 species, natives of Europe, Asia and Africa, the following typical.

1. Inula Helènium L. Elecampane.
Horseheal. Fig. 4416.

Inula Helenium L. Sp. Pl. 881. 1753.

Stems tufted from large thick roots, simple or rarely somewhat branched, 2°–6° high, densely pubescent above. Leaves large, broadly oblong, rough above, densely pubescent be-neath, denticulate, the basal ones acute at each end, long-petioled, 10'–20' long, 4'–8' wide; stem leaves sessile, or cordate-clasping at the base, acute at the apex, smaller; heads solitary or few, terminal, stout-peduncled, 2'–4' broad; involucre hemispheric, nearly 1' high, its outer bracts ovate, foliaceous, pubescent; rays nu-merous, linear; achenes glabrous, 4-sided.

Along roadsides and in fields, Nova Scotia to Ontario and Minnesota, south to North Carolina and Missouri. Naturalized from Europe. Native also in Asia. Scabwort. Horse-elder. Yellow star-wort. Elfdock. Elfwort. Wild sunflower.

48. POLÝMNIA L. Sp. Pl. 926. 1753.

Perennial herbs (some tropical species woody), with opposite membranous lobed or angled leaves, or the lower alternate, and mostly large corymbose-paniculate heads of both tubular and radiate yellow or whitish flowers, or rays sometimes wanting. Involucre hemispheric or broader, of about 5 large outer bracts, and more numerous smaller inner ones. Receptacle chaffy. Ray-flowers pistillate, fertile, subtended by the inner involucral bracts, the ligules elongated, minute, or none. Disk-flowers subtended by the chaffy scales of the receptacle, perfect, sterile, their corollas tubular, 5-toothed. Anthers 2-toothed at the base. Pappus none. Achenes thick, short, turgid, glabrous. [From the Muse Polhymnia.]

About 10 species, natives of America. Only the following are known in North America. Type species: *Polymnia canadensis* L.

Rays commonly 6″ long or more, yellow; achenes strongly striate.   1. *P. Uvedalia.*
Rays commonly minute or up to 6″ long, whitish, or none; achenes 3-ribbed.   2. *P. canadensis.*

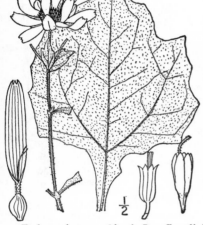

1. Polymnia Uvedàlia L.   Yellow or Large-flowered Leaf-cup.   Fig. 4417.

Polymnia Uvedalia L. Sp. Pl. Ed. 2, 1303. 1763.

Rough-pubescent, stout, branched, 3°–10° high. Leaves broadly ovate or deltoid, 3-nerved, abruptly contracted above the base, minutely ciliate, more or less pubescent on both sides, angulate-lobed, the lower often 1° long and broad, petioled, the upper sessile, somewhat clasping; heads few in terminal clusters, peduncled, 1½′–3′ broad; rays 10–15, commonly 6″–12″ long, linear-oblong, bright yellow, 3-toothed or entire; exterior bracts of the cup-like involucre ovate-oblong, obtuse, ciliate, 4″–10″ long; achenes slightly oblique and laterally compressed, strongly striate, nearly 3″ long.

In rich woods, New York to Indiana, Florida, Missouri, Oklahoma and Texas. Bermuda. Yellow bearsfoot. July–Aug.

2. Polymnia canadénsis L.   Small-flowered Leaf-cup.   Fig. 4418.

Polymnia canadensis L. Sp. Pl. 926. 1753.
Polymnia canadensis radiata A. Gray, Syn. Fl. N. A.
  1: Part 2, 238. 1884.
P. radiata Small, Fl. S.E. U. S. 1239. 1903.

Rather slender, viscid-pubescent, at least above, simple or branched, 2°–5° high. Leaves deltoid-ovate to hastate, usually very thin, all petioled, deeply angulàte-lobed and the lobes dentate, or the lower lyrate-pinnatifid, 4′–10′ long, the uppermost sometimes ovate and entire or merely denticulate; heads few in terminal clusters, short-peduncled or sessile, 4″–6″ broad; outer bracts of the involucre ovate to lanceolate, obtuse or acutish, 2″–3″ long; rays small, minute or none, or sometimes up to 6″ long and 3-lobed, whitish or yellowish; achenes 3-angled, obovoid, obcompressed, 3-ribbed, not striate.

In damp, rich shaded places, Vermont and Ontario to Minnesota, Pennsylvania, North Carolina, Tennessee and Arkansas. June–Sept.

49. MELAMPÒDIUM L. Sp. Pl. 921. 1753.

Herbs, some species woody, with opposite entire or dentate leaves, and terminal peduncled heads of both tubular and radiate, white or yellow flowers in our species. Involucre hemispheric, its bracts in 2 series, the 4 or 5 outer ones broad, often connate at the base, the inner hooded, embracing or permanently surrounding the pistillate fertile ray-flowers. Receptacle convex or conic, chaffy. Ray-flowers in 1 series, the rays spreading, 2–3-lobed or entire. Disk-flowers perfect, sterile, their corollas with a narrowly campanulate 5-toothed limb, the anthers entire at the base, the style undivided. Achenes obovoid, more or less incurved. Pappus none. [Greek, black-foot, of doubtful significance, but the stem bases of the typical species are dark-colored.]

About 35 species, natives of the warmer parts of America. Besides the following, 2 or 3 others occur in the southwestern United States. Type species: *Melampodium americanum* L.

1. Melampodium leucanthum T. & G.
Plains Melampodium.   Fig. 4419.

Melampodium leucanthum T. & G. Fl. N. A. **2**: 271. 1842.

Perennial, woody at the base, branched, canescent, 4′–12′ high, the branches slender. Leaves linear, lanceolate, or the lower spatulate, sessile, entire, or nearly so, canescent, 1′–2′ long, 1½″–3″ wide, obtuse or obtusish at the apex; heads ½′–¾′ broad, terminating the branches; peduncles slender, 1′–3′ long; outer bracts of the involucre ovate or oval, obtuse, united below; rays 5–9, cuneate-oblong, white, 2–3-lobed, firm in texture, veiny, persistent; inner bracts turbinate or terete, hooded, muricate, the hood wider than the body.

In dry soil, Kansas to Colorado, Arizona, Texas and Mexico. June–Oct. Not distinguished, in our first edition, from *M. cinereum* DC. of Texas and northern Mexico.

50. ACANTHOSPÉRMUM Schrank, Pl. Rar. Hort. Monac. *pl. 53*. 1819.

Annual rather coarse herbs, with pubescent foliage and erect or creeping stems, opposite, broad, often leathery, toothed leaves, and radiate but inconspicuous heads, axillary to leaf-like bracts. Involucre double, an outer one of flat herbaceous bracts, and an inner one of several smaller bracts which become bur-like and fall away enclosing an achene at maturity. Receptacle concave or convex. Ray-flowers few, in 1 series, the rays very small, yellowish, concave or hooded. Disk-flowers perfect, sterile. Anthers entire at the base. Achenes broadest above the middle, slightly curved. Pappus wanting. [Greek, thorn-seed, from the prickly, bur-like fruits.]

About 3 species, natives of tropical America. Type species: *Acanthospermum brasilium* Schrank.

1. Acanthospermum austràle (Loefl.) Kuntze.
Spiny-bur.   Fig. 4420.

Melampodium australe Loefl. Iter. Hisp. 268. 1758.

A. xanthoides DC. Prodr. **5**: 521. 1836.

A. australe Kuntze, Rev. Gen. Pl. 303. 1891.

Stems branching at the base, the branches prostrate or creeping, ⅓°–2° long. Leaves ovate, oval or rhombic, ⅓′–1′ long, usually acute, serrate-dentate above the middle, cuneate at the base; peduncles shorter than the subtending leaves; involucre campanulate, the bracts broadly ovate, obtuse, 1″–1½″ long, uniformly prickly; disk-flowers with corollas 1″ long, puberulent; mature inner bracts of the involucre forming a starfish-like bur, each lobe 4″–5″ long, densely beset with uniform blunt weak prickles.

In waste places and dry soil, Virginia to Florida and Louisiana. June–Oct. Widely distributed as a weed in tropical regions.

51. SÍLPHIUM L. Sp. Pl. 919. 1753.

Tall perennial herbs, with resinous juice, opposite whorled or alternate leaves, and large corymbose or paniculate (rarely solitary) peduncled heads of both tubular and radiate yellow flowers. Involucre hemispheric or campanulate, its bracts imbricated in few series. Receptacle flat or nearly so, chaffy, the chaff subtending the disk-flowers. Ray-flowers in 2 or 3 series, pistillate, fertile, the ligules numerous, linear. Disk-flowers perfect but sterile, their corollas tubular, 5-toothed, the style undivided. Anthers minutely 2-toothed or entire at the base. Achenes broad, dorsally flattened, 2-winged, notched at the apex. Pappus none, or of 2 awns confluent with the wings of the achene. [Greek, from the resinous juice.]

About 12 species, natives of North America, known as Rosin-weed or Rosin-plant. Type species: *Silphium Asteriscus* L.

Stem leafy, the leaves opposite, alternate, or verticillate.
  Leaves, or their petiole-bases, connate-perfoliate; stem square.   1. *S. perfoliatum.*
  Leaves not connate-perfoliate, sessile or petioled.
    Leaves opposite, or the uppermost alternate; cauline sessile.   2. *S. integrifolium.*
    Leaves, or some of them, verticillate in 3's or 4's, petioled.   3. *S. trifoliatum.*
    Most or all of the leaves alternate, entire or dentate.   4. *S. Asteriscus.*
    Leaves all alternate, pinnatifid or bipinnatifid, large.   5. *S. laciniatum.*
Stem leafless or nearly so, scaly above; leaves basal, large.
  Leaves sharply serrate to pinnatifid; achenes obovate.   6. *S. terebinthinaceum.*
  Leaves coarsely dentate; achenes suborbicular.   7. *S. reniforme.*

3/5

1. Silphium perfoliàtum L. Cup-plant. Indian-cup. Fig. 4421.

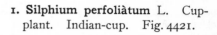
S. perfoliatum L. Sp. Pl. Ed. 2, 1301. 1763.

Stem square, glabrous, or rarely somewhat hispid, branched above, or sometimes simple, 4°–8° high. Leaves ovate or deltoid-ovate, opposite, the upper connate-perfoliate, the lower abruptly contracted into margined petioles, all thin, usually scabrous on both sides, or pubescent beneath, coarsely angulate-dentate, or the upper entire, the larger 6′–12′ long, 4′–8′ wide; heads commonly numerous, 2′–3′ broad; rays 20–30, about 1′ long and 2″ wide; involucre depressed-hemispheric, its outer bracts broad, ovate, ciliolate, spreading or erect; achenes obovate, emarginate, sometimes 2-toothed.

In moist soil, southern Ontario to Minnesota, South Dakota, New Jersey, Georgia, Nebraska and Louisiana. Naturalized near New York City, and elsewhere escaped from cultivation. Called also ragged-cup. July–Sept.

2. Silphium integrifòlium Michx. Entire-leaved Rosin-wood. Fig. 4422.

S. integrifolium Michx. Fl. Bor. Am. 2: 146. 1803.

S. speciosum Nutt. Trans. Am. Phil. Soc. (II) 7: 341. 1841.

Stem glabrous, rough or sometimes hirsute, corymbosely branched above, 2°–5° high. Leaves ovate to ovate-lanceolate, acute or acuminate, entire, denticulate or remotely dentate, rough above, pubescent or glabrous beneath, those of the stem all closely sessile, often half-clasping but not connate-perfoliate at the rounded base, 3′–5′ long, 1′–2′ wide; heads usually numerous, 1′–2′ broad; involucre nearly hemispheric, its outer bracts ovate or ovate-lanceolate, acute, spreading, ciliolate or pubescent; rays 15–25; achenes oval or obovate, 4″–5″ long, deeply emarginate.

On prairies, Ohio to Minnesota, south to Louisiana, Nebraska, Arkansas and Texas. Aug.–Sept.

1/2

3. Silphium trifoliàtum L. Whorled Rosin-weed. Fig. 4423.

Silphium trifoliatum L. Sp. Pl. 920. 1753.

Stem glabrous, sometimes glaucous, corymbosely branched at the summit, 4°–7° high. Leaves lanceolate or oblong-lanceolate, the middle ones almost always whorled in 3's or 4's, acuminate at the apex, narrowed at the base and usually somewhat petioled, rough or roughish above, pubescent or nearly glabrous beneath, entire or denticulate, 3′–7′ long, ½′–1½′ wide; heads several or numerous, 1½′–2′ broad; involucre hemispheric, its outer bracts ovate or oval, acute or obtuse, glabrous or slightly pubescent, ciliolate; rays 15–20; achenes oval·or obovate, narrowly winged, emarginate, sharply 2-toothed.

In woods, Pennsylvania to Ohio, Ontario, Virginia and Alabama. July–Oct.

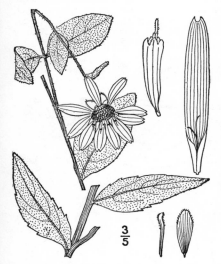

4. Silphium Asteríscus L. Starry Rosin-weed. Fig. 4424.

Silphium Asteriscus L. Sp. Pl. 920. 1753.

Stem hispid-pubescent, simple or branched above, 2°–4° high, usually purple. Leaves nearly all alternate, ovate, ovate-oblong, or lanceolate, acute or obtusish, sessile, somewhat clasping, or the lower narrowed into short petioles, dentate, or the upper entire, 2′–5′ long, ½′–1′ wide; heads commonly few, 1′–2′ broad; rays 12–15; involucre hemispheric, its bracts mostly hispid, ovate to oblong, acute or obtuse, squarrose; achenes oval or obovate, narrowly winged, 2-toothed.

In dry soil, Maryland to Tennessee and Missouri, south to Florida and Louisiana. June–Sept.

5. Silphium laciniàtum L. Compass-plant. Pilot-weed. Fig. 4425.

Silphium laciniatum L. Sp. Pl. 919. 1753.

Rough or hispid, very resinous; stem 6°–12° high; basal leaves pinnatifid or bipinnatifid, long-petioled, 1° long or more, the lobes oblong or lanceolate; stem leaves alternate, vertical, their edges tending to point north and south, sessile, or the lower short-petioled, the upper cordate-clasping at the base, gradually smaller and less divided; heads several or numerous, sessile or short-peduncled, 2′–5′ broad, the peduncles bracted at the base; rays 20–30, 1′–2′ long; involucre nearly hemispheric, its bracts large, rigid, lanceolate or ovate, very squarrose; achenes oval, about 6″ long, the wing broader above than below, notched at the apex, awnless.

On prairies, Ohio to South Dakota, south to Alabama, Louisiana and Texas. Turpentine-weed, polar-plant, rosin-weed. July–Sept.

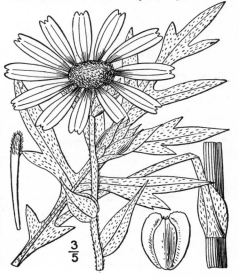

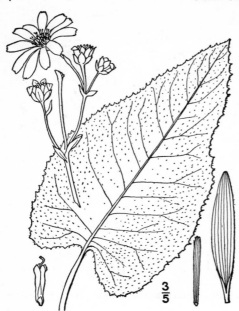

3/5

6. Silphium terebinthinàceum Jacq.
Prairie Dock. Prairie Burdock.
Fig. 4426.

Silphium terebinthinaceum Jacq. Hort. Vind. 1: *pl. 43.* 1770.

S. pinnatifidum Ell. Bot. S. C. & Ga. 2: 462. 1824.

Silphium terebinthinaceum pinnatifidum A. Gray, Man. 220. 1848.

Stem glabrous or nearly so, branched and scaly above, 4°–10° high. Leaves all basal or nearly so, coriaceous, ovate, mostly long-petioled, acute at the apex, cordate at the base, rough on both sides, often 12′ long and 6′ wide, sharply dentate, laciniate or pinnatifid; heads numerous, 1½′–3′ broad, borne on glabrous peduncles; involucre hemispheric, its bracts ovate-oblong, erect, glabrous or minutely pubescent; rays 12–20; achenes obovate, narrowly winged, slightly 2-toothed and emarginate at the apex.

On prairies and in dry woods, southern Ontario and Ohio to Minnesota, south to Georgia, Iowa and Louisiana. Rosin-plant. July–Sept.

7. Silphium reniforme Raf. Kidney-leaved Silphium. Fig. 4427.

Silphium reniforme Raf.; Nutt. Trans. Am. Phil. Soc. (II) 7: 342. 1840.

S. compositum reniforme T. & G. Fl. N. A. 2: 276. 1842.

Stem scape-like, 3°–6° tall, glabrous or nearly so, corymbosely branched above. Leaves mostly basal, broadly ovate to reniform, 4′–15′ long, obtuse or abruptly pointed, coarsely dentate, deeply cordate at the base, long-petioled; heads numerous, about 1′ broad, the peduncles glabrous; involucre campanulate, the bracts ovate or broadly oblong to narrowly oblong or oblong-spatulate, obtuse; rays several, yellow; achenes suborbicular or orbicular-obovate, 3½″ long, the wings prolonged enough to form a shallow apical sinus.

In dry or stony soil, mountains of Virginia and North Carolina. July–Aug.

Silphium compositum Michx., a related southern species with pedately parted leaves, is reported as observed in southern Virginia.

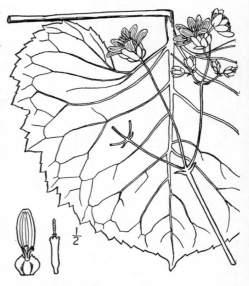

½

52. CHRYSÓGONUM L. Sp. Pl. 920. 1753.

Pubescent perennial herbs, with opposite and basal petioled leaves, and slender-peduncled-axillary and terminal, rather large heads of both tubular and radiate, yellow flowers. Involucre hemispheric, its bracts in 2 series of 5, the outer large, obovate or spatulate, foliaceous, the inner oval, firm, each subtending a pistillate ray-flower. Receptacle chaffy, each scale subtending and partly enclosing a perfect but sterile tubular flower with a 5-toothed corolla. Anthers nearly entire at the base. Achenes obovate, compressed, their margins acute, not winged, 1-nerved on the back, 1-2-ribbed on the inner side. Pappus a short half-cup-shaped crown. [Greek, golden-knee.]

A monotypic genus of eastern North America.

1. Chrysogonum virginiànum L. Chryso-gonum. Fig. 4428.

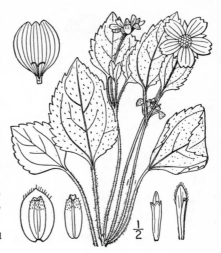

Chrysogonum virginianum L. Sp. Pl. 920. 1753.

Chrysogonum virginianum dentatum A. Gray, Bot. Gaz. **7**: 31. 1882.

Perennial by rootstocks or runners, pubescent or hirsute throughout, branched from the base, or at first acaulescent, 3′–12′ high. Leaves ovate or oblong, obtuse or acutish at the apex, the upper sometimes subcordate at the base, dentate or crenate-dentate, rather thin, 1′–3′ long, ½′–2′ wide, the basal ones with petioles as long as the blade or longer, those of the upper one shorter; pedun-cles 1′–4′ long; heads 1′–1½′ broad; outer bracts of the involucre obtuse or acute; rays about 5, 4″–7″ long.

In dry soil, southern Pennsylvania to Florida and Alabama. April–July.

53. BERLANDIÈRA DC. Prodr. 5: 517. 1836.

Perennial canescent or pubescent herbs, with alternate leaves and rather large, pedunclled solitary or corymbose heads of both tubular and radiate yellow-flowers. Involucre depressed-hemispheric, its bracts imbricated in about 3 series, the outermost small, mostly oblong, the second series broader, oval or obovate, the inner membranous, similar, reticulated when mature, subtending the ray-flowers and exceeding the disk. Receptacle nearly flat, chaffy, the chaff subtending the disk-flowers. Ray-flowers 5–12, pistillate, fertile. Disk-flowers perfect, tubular, sterile, their corollas 5-toothed. Anthers entire, or minutely 2-toothed at the base. Style of the tubular flowers undivided, hirsute. Achenes obovate, compressed, not winged, 1-ribbed on the inner side, the pappus obsolete, early deciduous or of 2 caducous awns. [Named after J. L. Berlandier, a Swiss botanical collector in Texas and Mexico.]

About 8 species, natives of the southern United States and Mexico. Type species: *Berlandiera texana* DC.

Stem leafy; leaves ovate to oblong, crenate.   1. *B. texana.*
Plant acaulescent, or nearly so; leaves lyrate-pinnatifid.   2. *B. lyrata.*

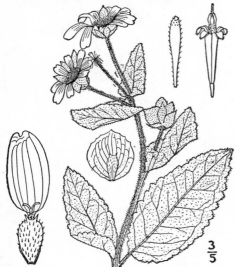

1. Berlandiera texàna DC. Texan Berlandiera. Fig. 4429.

Berlandiera texana DC. Prodr. **5**: 517. 1836.

Hirsute-pubescent throughout; stem erect, branched above, or simple, 2°–3° high, leafy. Leaves ovate, or the basal oblong, crenate, acutish or obtuse at the apex, rounded or cordate at the base, 2′–4′ long, 1′–2′ wide, the upper sessile, the lower petioled; heads few or several, 1′–1½′ broad, in a terminal corym-bose-cymose cluster; peduncles ¼′–1½′ long; inner bracts of the involucre twice as large as the outer.

In dry soil, Missouri and Kansas to Arkansas and Louisiana. July–Aug.

2. Berlandiera lyràta Benth. Lyre-leaved Berlandiera. Fig. 4430.

Silphium Nuttallianum Torr. Ann. Lyc. N. Y. **2**: 216. Name only. 1827.

Berlandiera lyrata Benth. Pl. Hartw. 17. 1839.

Finely whitish-canescent, acaulescent or short-stemmed; scapes or peduncles slender, 3′–8′ long, bearing a solitary head, or rarely 2. Leaves lyrate-pinnatifid, obtuse, petioled, the terminal segment usually larger than the lateral ones, the lower ones very small, all obtuse, mostly crenate, sometimes becoming green and glabrate above; head about 1′ broad; inner bracts of the involucre much broader than the outer, orbicular, or wider than long; achenes obovate, keeled on the inner face.

In dry soil, Kansas to Texas, Arizona and Mexico.

54. ENGELMÁNNIA T. & G. Fl. N. A. **2**: 283. 1841.

Perennial hirsute herbs, with alternate pinnatifid leaves, and corymbose slender-peduncled rather large heads of both tubular and radiate yellow flowers. Involucre hemispheric, its bracts imbricated in 2 or 3 series, the outer linear, loose, hirsute, ciliate, the inner oval or obovate, concave, appressed, subtending the ray-flowers. Receptacle flat, chaffy, the chaff subtending and partly enclosing the disk-flowers. Rays 8–10, pistillate, fertile. Disk-flowers about as many, tubular, perfect, sterile, the corolla 5-toothed. Anthers minutely 2-dentate at the base. Style of the tubular flowers undivided. Achenes obovate, compressed, not winged, 1-ribbed on each face. Pappus a persistent irregularly cleft crown. [Named for Dr. Geo. Engelmann, 1809–1884, botanist, of St. Louis.]

A monotypic genus of the south-central United States.

1. Engelmannia pinnatífida T. & G. Engelmannia. Fig. 4431.

E. pinnatifida T. & G. Fl. N. A. **2**: 283. 1841.

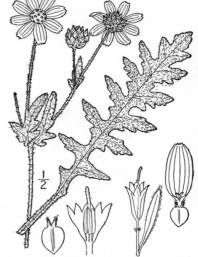

Stem usually branched above, 1°–3° high. Basal leaves slender-petioled, 4′–8′ long, their lobes lanceolate or oblong, dentate or entire, obtuse or acutish; upper leaves smaller, sessile, less divided, the uppermost sometimes entire, or with a pair of basal lobes; heads usually numerous, about 1′ broad; peduncles 1′–5′ long; outer bracts of the involucre somewhat in 2 series, the first linear, the second broadened at the base.

In dry soil, Kansas to Colorado, Louisiana, Arizona and North Mexico. Reported as found along railroads in western Missouri. May–Aug.

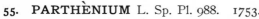

55. PARTHÈNIUM L. Sp. Pl. 988. 1753.

Perennial, mostly pubescent or canescent herbs, or shrubs, with alternate leaves, and small corymbose or paniculate heads of both tubular and radiate white or yellow flowers. Involucre broadly campanulate or hemispheric, its bracts imbricated in 2 or 3 series, obtuse, appressed, nearly equal. Receptacle convex or conic, chaffy, the chaff membranous, surrounding the disk-flowers. Ray-flowers about 5, pistillate, fertile, their ligules short, broad, 2-toothed or obcordate. Disk-flowers perfect, sterile, their corollas 5-toothed, the style undivided. Anthers entire at the base. Achenes compressed, keeled on the inner face, margined, bearing the persistent rays on the summit. Pappus of 2–3 scales or awns. [Greek, virgin.]

About 12 species, natives of North America, Mexico, the West Indies and northern South America. Besides the following, some 3 others occur in the southern and southwestern United States. Type species: *Parthenium Hysterophorus* L.

Leaves 1–2-pinnatifid ; annual weed.
Leaves crenate-dentate, or somewhat lyrate ; perennials.
   Stem glabrous, or pubescent or puberulent above ; rootstock
     tuberous-thickened.
   Stem hirsute or villous.
     Stem leaves auricled, clasping at base ; rootstock thick.
     Stem leaves sessile, not auricled ; rootstock long, slender.

1. *P. Hysterophorus.*

2. *P. integrifolium.*

3. *P. auriculatum.*
4. *P. hispidum.*

1. Parthenium Hysteróphorus L.   Santa Maria.
Fig. 4432.

P. Hysterophorus L. Sp. Pl. 988. 1753.

   Annual, strigose-pubescent or somewhat villous, usually much branched, 1°–2½° high. Leaves ovate to oblong in outline, 1–2-pinnately parted into linear or lanceolate toothed or pinnatifid segments, thin and flaccid ;
heads numerous, 2½″–3″ broad ; involucre saucer-shaped,
its bracts concave, the outer ones rhombic, the inner
broader ; ray-flowers few ; rays reniform, white, about
½″ wide ; achenes obovate, about ½″ long.

   Waste and cultivated grounds, southern Pennsylvania to
Illinois, Missouri, Florida and Texas. Throughout tropical
America. July–Sept.

2. Parthenium integrifòlium L.   Amer-
ican Fever-few. Prairie Dock.
Fig. 4433.

P. integrifolium L. Sp. Pl. 988. 1753.

   Stem stout, striate, finely pubescent with
short hairs, or glabrous below, corymbosely
branched above, 1°–4° high. Rootstocks tuberous thickened ; leaves firm, ovate or ovate-
oblong, acute or acuminate, crenate-dentate
or somewhat lyrate at the base, hispidulous
and roughish on both sides, the lower and
basal ones petioled, often 12′ long and 5′ wide,
the upper smaller, sessile ; heads numerous
in a dense terminal corymb ; involucre nearly
hemispheric, about 3″ high, its bracts firm,
the outer oblong, densely appressed-pubescent, the inner broader, glabrous, or ciliate
on the margins ; rays white or whitish.

   In dry soil, Maryland to Minnesota, south to
Georgia, Missouri and Arkansas. Cutting-almond.
Wild quinine. May–Sept.

3. Parthenium auriculàtum Britton.   Auri-
cled Parthenium. Fig. 4434.

P. auriculatum Britton, in Britt. & Brown, Ill. Fl. **3** :
521. 1898.

   Rootstock an oval erect tuber twice as long as
thick ; stem villous-pubescent, 1½°–2½° high. Leaves
rough above, villous, especially on the veins beneath, oval, ovate or oblong, irregularly crenate-
dentate, some or all of them laciniate or pinnatifid
at the base, the basal and lower slender-petioled
with petiole as long as the blade, or longer, the
upper with a sessile clasping auricled base, or
with margined clasping petioles ; inflorescence
densely corymbose, its branches villous-tomen-
tose ; bracts of the involucre densely canescent.

   Mountains of Virginia.

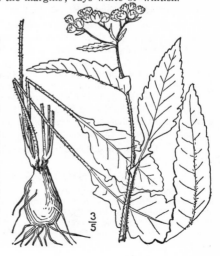

4. Parthenium híspidum Raf. Creeping or Hairy Parthenium. Fig. 4435.

Parthenium híspidum Raf. New Fl. N. A. **2**: 35. 1836.

P. repens Eggert, Cat. Pl. St. Louis 16. 1891.

Similar to the two preceding species, but lower, seldom over 2° high. Rootstocks slender, forming runners; stem pilose or hispid with spreading hairs; leaves hispid on both sides, irregularly crenate, sometimes lyrate at the base, the teeth rounded and obtuse; heads fewer, slightly larger, in a small loose corymb; outer bracts of the involucre proportionately broader.

Barren rocky soil, Missouri to Kansas and Texas. April–July.

56. CRASSÌNA Scepin, Sched. Acido Veg. 42. 1758.
[ZINNIA L. Syst. Ed. 10, 1221. 1759.]

Annual or perennial herbs, some species shrubby, with opposite, entire, or sparingly serrate, mostly narrow and sessile leaves, and large or middle-sized heads of both tubular and radiate flowers. Ray-flowers pistillate, yellow, or variegated, persistent on the achene. Disk-flowers perfect, fertile; corolla cylindraceous, its lobes villous. Involucre campanulate to nearly cylindric, its bracts obtuse, dry, firm, appressed, imbricated in 3 series or more, the outer gradually shorter. Receptacle conic or cylindric, chaffy, the chaff subtending and enwrapping the disk-flowers. Style-branches elongated, not appendages. Achenes of the ray-flowers somewhat 3-angled, those of the disk flattened. Pappus of few awns or teeth. [In honor of Paul Crassus, an Italian botanist of the sixteenth century.]

About 12 species, natives of the United States and Mexico. Type species: *Chrysogonum peruviànum* L.

1. Crassina grandiflòra (Nutt.) Kuntze. Prairie Zinnia. Fig. 4436.

Zinnia grandiflora Nutt. Trans. Am. Phil. Soc. (II) **7**: 348. 1841.

Crassina grandiflora Kuntze, Rev. Gen. Pl. 331. 1891.

Perennial, woody at the base, tufted, much branched, rough, 4′–6′ high. Leaves rather rigid, linear to linear-lanceolate, entire, 6″–15″ long, 1½″ wide, or less, connate at the base, acute or acutish, crowded; heads numerous, peduncled, terminating the branches, 10″–18″ broad; rays 4 or 5, broad, yellow, rounded, or emarginate, their achenes with a pappus of 2 or 4 awns; involucre campanulate-cylindric, 3″–4″ high; style-branches of the disk flowers subulate.

In dry soil, Kansas and Colorado to Texas, Mexico and Arizona. June–Sept.

57. HELIÓPSIS Pers. Syn. **2**: 473. 1807.

Perennial herbs (a tropical species annual), with opposite petioled 3-ribbed leaves, and large peduncled terminal and axillary heads of tubular and radiate yellow flowers. Involucre hemispheric or broadly campanulate, its bracts oblong or lanceolate, imbricated in 2 or 3 series. Receptacle convex or conic, chaffy, the chaff enveloping the disk-flowers. Ray-flowers pistillate, fertile, the rays spreading, the tube very short, commonly persistent on the achene. Disk-flowers perfect, the tube short, the limb elongated, 5-toothed. Anthers

entire, or minutely 2-toothed at the base. Style-branches tipped with small hirsute appendages. Achenes thick, obtusely 3-4-angled, the summit truncate. Pappus none, or of 2-4
teeth, or a coroniform border. [Greek, sun-like.]

About 6 species, natives of America. Besides the following, two others occur in the southern
and southwestern United States. Type species: *Heliopsis laevis* Pers.

Leaves mostly smooth, or nearly so; pappus none, or of 2-4 stout teeth.   1. *H. helianthoides*
Leaves rough; pappus crown-like, or of 1-3 sharp teeth.   2. *H. scabra.*

1. Heliopsis helianthòides (L.) Sweet. Ox-eye. False Sunflower. Fig. 4437.

Buphthalmum helianthoides L. Sp. Pl. 904. 1753.
Heliopsis laevis Pers. Syn. 2: 473. 1807.
Heliopsis helianthoides Sweet, Hort. Brit. 487.
1827.

Stem glabrous, branched above, 3°-5° high.
Leaves opposite, or rarely in 3's, ovate or
ovate-lanceolate, rather thin, acuminate at the
apex, usually abruptly narrowed at the base,
sharply and nearly equally dentate, smooth on
both sides, or roughish, 3'-6' long, 1'-2½' wide;
heads long-peduncled, somewhat corymbose,
1¾'-2½' broad; rays 9''-12'' long, persistent, or
at length decaying away from the achenes;
bracts of the involucre oblong or linear-oblong,
obtuse or acutish, the outer commonly longer
than the inner; achenes glabrous, the summit
truncate; pappus none, or of 2-4 short teeth.

In open places, Ontario to New York, Florida,
North Dakota, Illinois and Tennessee. July-Sept.

2. Heliopsis scàbra Dunal. Rough Ox-eye. Fig. 4438.

Heliopsis scabra Dunal, Mem. Mus. Paris 5: 56. *pl. 4.*
1819.
Heliopsis laevis var. *scabra* T. & G. Fl. N. A. 2: 303.
1842.

Similar to the preceding species, but stem
rough, at least above, simple or branched, 2°-4°
high. Leaves ovate or ovate-lanceolate, acute or
sometimes acuminate, sharply dentate, rough on
both sides, firm, 2'-5' long, 1½'-2' wide, abruptly
narrowed at the base, short-petioled; heads few,
or sometimes solitary, long-peduncled, 2'-2½'
broad; rays usually 1' long, or more; bracts of
the involucre canescent, oblong or linear-oblong;
achenes pubescent on the margins when young;
pappus a short laciniate crown, or 1-3 sharp teeth.

Usually in dry soil, Maine to New York, New Jersey, Manitoba, British Columbia, Arkansas and New
Mexico. June-Sept. False sunflower.

58. VERBESINA L. Sp. Pl. 901. 1753.

Erect or diffuse branching pubescent or hirsute herbs, with opposite leaves, and small
peduncled terminal and axillary heads of tubular and radiate whitish flowers. Involucre
hemispheric or broadly campanulate, its bracts imbricated in about 2 series, nearly equal, or
the outer longer. Receptacle flat or convex, chaffy, the chaff awn-like, subtending the achenes.
Ray-flowers pistillate, fertile. Disk-flowers perfect, mostly fertile, their corollas tubular,
4-toothed or rarely 5-toothed. Anthers entire or minutely 2-toothed at the base. Style-
branches of the disk-flowers with obtuse or triangular tips. Achenes thick, those of the rays
3-sided or 4-sided, those of the disk compressed. Pappus none, or of a few short teeth.
[Name changed from *Verbena.*]

About 4 species, mostly of tropical distribution, the following typical.

1. Verbesina álba L.   Yerba de tajo.   Fig. 4439.

Verbesina alba L. Sp. Pl. 902. 1753.

Eclipta erecta L. Mant. **2** : 286. 1771.

Eclipta procumbens Michx. Fl. Bor. Am. **2** : 129. 1803.

Eclipta alba Hassk. Pl. Jav. Rar. 528. 1848.

Annual, rough with appressed pubescence, erect or diffuse, 6′–3° high. Leaves lanceolate, oblong-lanceolate or linear-lanceolate, acute or acuminate, denticulate or entire, narrowed to a sessile base, or the lower petioled, 1′–5′ long, 2″–10″ wide; heads commonly numerous, 3″–6″ broad, nearly sessile, or slender-peduncled; rays short, nearly white; anthers brown; achenes 4-toothed, or at length truncate.

Along streams, and in waste places, Massachusetts to Illinois, Nebraska, Florida, Texas and Mexico. Naturalized from the south in its northeastern range and widely distributed in warm regions as a weed. July–Oct.

59. TETRAGONOTHÈCA (Dill.) L. Sp. Pl. 903. 1753.

Erect perennial mostly branched herbs, with opposite, sessile or connate-perfoliate, broad dentate leaves, and large peduncled heads of tubular and radiate yellow flowers. Involucre depressed-hemispheric, its principal bracts 4, large and foliaceous, inserted in 1 series; inner bracts 6–15, small, subtending the pistillate ray-flowers. Receptacle conic, chaffy, the chaff concave, enwrapping the perfect fertile disk-flowers, the corollas of which are slender and 5-toothed. Anthers entire or minutely 2-toothed at the base. Style-branches of the disk-flowers hispid, tipped with elongated appendages. Achenes thick, 4-sided, truncate at the summit. Pappus none, or of several short scales. [Greek, 4-angled-case, referring to the involucre.]

Four known species, natives of the southern United States and northern Mexico, the following typical.

1. Tetragonotheca helianthoìdes L.
Tetragonotheca.   Fig. 4440.

Tetragonotheca helianthoides L. Sp. Pl. 903. 1753.

Viscidly pubescent; stem branched or simple, 1°–2½° high   Leaves ovate, ovate-oblong, or somewhat rhomboid, thin, coarsely and unequally dentate, pinnately veined, acute at the apex, narrowed at the sessile or somewhat clasping base, 2′–6′ long, 1′–3′ wide; heads usually few, 1½′–3′ broad; involucre 4-angled in the bud, its principal bracts broadly ovate, acute; rays 6-10, strongly parallel-nerved, 2–3-toothed; corolla-tube villous below; achenes 4-sided, or nearly terete; pappus none.

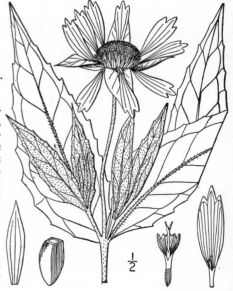

In dry soil, Virginia to Tennessee, Florida and Alabama. May–June. Sometimes flowering again in autumn.

60. SPILÁNTHES Jacq. Stirp. Am. 214. *pl. 126.* 1763.

Annual or perennial branching herbs, with opposite, usually toothed leaves and rather small, long-peduncled discoid and radiate heads, terminal, or in the upper axils, or rays wanting in some species. Involucre campanulate, its bracts in about 2 series, herbaceous, loosely appressed. Receptacle convex or elongated, chaffy, its chaff embracing the disk-achenes and at length falling away with them. Ray-flowers yellow, or white, pistillate, sometimes wanting. Disk-flowers yellow, perfect, their corollas tubular with an expanded 4–5-cleft limb. Anthers truncate at the base. Style-branches of the disk-flowers long, sometimes penicillate at the summit. Ray-achenes 3-sided, or compressed, those of the disk-flowers compressed, margined. Pappus of 1–3 awns, or more. [Greek, spot- or stain-flower, not significant.]

About 30 species, natives of warm and tropical regions. Type species: *Spilanthes urens* Jacq.

1. Spilanthes rèpens (Walt.) Michx.
Spilanthes. Fig. 4441.

Anthemis repens Walt. Fl. Car. 211. 1788.
Spilanthes repens Michx. Fl. Bor. Am. 2: 131. 1803.
S. americana repens A. H. Moore, Proc. Am. Acad. 42: 547. 1907.

Perennial, usually rooting at the lower nodes; stem slender, simple or branched, spreading or ascending, 8′–2° long, pubescent, or nearly glabrous. Leaves ovate to lanceolate, petioled, acute or acuminate at the apex, or the lower obtuse, coarsely toothed, or nearly entire, 1′–3′ long; heads long-peduncled, solitary at the end of the stem and branches, 6″–10″ broad; bracts of the involucre oblong to oblong-lanceolate, obtuse or acute; rays 8–12, yellow; receptacle narrowly conic; achenes oblong, most of them roughened when mature and hispidulous; pappus of 1 or 2 very short awns, or none.

In moist or wet soil, Missouri to Texas, east to South Carolina and Florida. June–Sept.

61. RUDBÉCKIA L. Sp. Pl. 906. 1753.

Perennial or biennial (rarely annual), mostly rigid, usually rough or hispid herbs, with alternate undivided lobed or pinnatifid leaves, and large long-peduncled heads of tubular (mostly purple) and radiate (yellow) flowers. Involucre hemispheric, its bracts imbricated in 2–4 series. Receptacle conic or convex, with chaffy concave scales subtending or enveloping the disk-flowers. Ray-flowers neutral, the rays entire or toothed. Disk-flowers perfect, fertile, their corollas 5-lobed. Anthers entire or minutely 2-mucronate at the base. Style-branches tipped with hirsute appendages. Achenes 4-angled, obtuse or truncate at the apex. Pappus coroniform, sometimes of 2–4 short teeth, or none. [In honor of Claus Rudbeck, 1630–1702, Swedish anatomist and botanist.]

About 30 species, natives of North America and Mexico. In addition to the following, some 20 others occur in the southern and western United States. Type species: *Rudbeckia hirta* L.

Disk globose or ovoid and purple or dark brown in fruit; lower leaves entire or lobed.
  Lower leaves deeply 3-lobed or 3-divided.
    Plant more or less hirsute; leaves thin; chaff awned.   1. *R. triloba.*
    Plant scabrous; leaves thick; chaff blunt, pubescent at apex.   2. *R. subtomentosa.*
  Leaves neither 3-lobed nor 3-divided.
    Plants hispid; style-branches subulate.
      Stem leaves lanceolate to oblong; involucre shorter than the rays.   3. *R. hir'a.*
      Stem leaves oval to obovate; involucral bracts foliaceous, nearly as long as the rays.
                                             4. *R. Brittonii.*
    Plants pubescent or glabrate; style-branches obtuse.
      Chaff merely ciliate.
        Leaves denticulate or entire; rays 9″–12″ long.
          Basal leaves narrowed at base.   5. *R. fulgida.*
          Basal leaves cordate at base.   6. *R. umbrosa.*
        Leaves dentate or laciniate; rays about 18″ long.   7. *R. speciosa.*
      Chaff canescent.   8. *R. grandiflora.*
Disk elongated or cylindric in fruit, yellowish or gray.
  Leaves very thick, shallowly toothed.   9. *R. maxima.*
  Leaves thin, pinnately divided or pinnatifid.   10. *R. laciniata.*

$\frac{3}{5}$

1. Rudbeckia tríloba L. Thin-leaved Cone-flower. Fig. 4442.

Rudbeckia triloba L. Sp. Pl. 907. 1753.

Stem somewhat pubescent and rough, rarely glabrate; branched, 2°–5° high. Leaves thin, rough on both sides, bright green, the basal and lower ones petioled, some or all of them 3-lobed or 3-parted, the lobes lanceolate or oblong, acuminate, sharply serrate; upper leaves ovate, ovate-lanceolate, or lanceolate, acuminate or acute, narrowed to a sessile base or into short margined petioles, serrate or entire, 2′–4′ long, ½′–1′ wide; heads nearly 2′ broad, corymbed; bracts of the involucre linear, acute; pubescent, soon reflexed; rays 8–12, yellow, or the base orange or brownish-purple; disk dark purple, ovoid, about 6″ broad; chaff of the receptacle awn-pointed; pappus a minute crown.

In moist soil, New Jersey to Georgia, west to Michigan, Minnesota, Missouri, Kansas and Louisiana. Sometimes escaped from gardens to roadsides. Brown-eyed susan. June–Oct.

2. Rudbeckia subtomentòsa Pursh. Sweet Cone-flower. Fig. 4443.

Rudbeckia subtomentosa Pursh, Fl. Am. Sept. 575. 1814.

Densely and finely cinereous-pubescent and scabrous; stem branched above, 2°–6° high. Leaves thick, some or all of the lower ones deeply 3-lobed or 3-parted, petioled, 3′–5′ long, the lobes oblong or lanceolate, acute or acuminate, dentate; upper leaves, or some of them, lanceolate or ovate, acuminate, sessile or nearly so; heads numerous, 2′–3′ broad; rays 15–20, yellow, or with a darker base; disc subglobose, rounded, purple or brown, 6″–8″ broad; bracts of the involucre linear-lanceolate, acuminate, squarrose, sweet-scented; chaff of the receptacle linear, obtuse or obtusish, pubescent, or somewhat glandular at the apex; pappus a short crenate crown.

On prairies and along rivers, Illinois to Louisiana, Kansas and Texas. July–Sept.

$\frac{1}{2}$

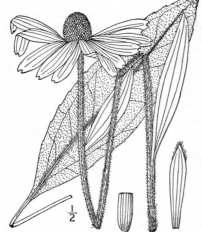

$\frac{1}{2}$

3. Rudbeckia hírta L. Black Eyed Susan. Yellow Daisy. Fig. 4444.

Rudbeckia hirta L. Sp. Pl. 907. 1753.

Hirsute or hispid throughout, biennial or sometimes annual; stems simple or sparingly branched, often tufted, 1°–3° high. Leaves thick, sparingly serrate with low teeth, or entire, lanceolate or oblong, the lower and basal ones petioled, mostly obtuse, 3–5-nerved, 2′–7′ long, ½′–2′ wide, the upper sessile, narrower, acute or acutish; heads commonly few or solitary, 2′–4′ broad; rays 10–20, orange or orange-yellow, rarely darker at the base; bracts of the involucre very hirsute, spreading or reflexed, much shorter than the rays; disk globose-ovoid, purple-brown; chaff of the receptacle linear, acute or acutish, hirsute at the apex; style-tips acute; pappus none.

Prairies and plains, Ontario to Manitoba, Florida, Colorado and Texas. Widely distributed in the east as a weed, north to Quebec. Races differ in pubescence and in length and color of the rays. Nigger- or darkey-head. Nigger- or poor-land daisy. Golden-jerusalem. Yellow ox-eye-daisy. English bull's-eye. Brown daisy or betty. Brown-eyed susan. May–Sept.

Rudbeckia montícola Small, of the southern Alleghanies, with broader, ovate, acute or acuminate stem leaves, is recorded as extending northward into Pennsylvania.

4. Rudbeckia Brittònii Small. Britton's Cone-flower. Fig. 4445.

R. Brittonii Small, Mem. Torr. Club 4: 130. 1894.

Stem stout, hispid, erect, $1\frac{1}{2}°$–$2\frac{1}{2}°$ high, simple, grooved, leafy, at least below. Leaves serrate or crenate-serrate, strigose-pubescent, the basal ones ovate to ovate-lanceolate, 3'–4' long, obtuse, long-petioled; stem leaves obovate to oval, often with a lateral lobe, the petioles wing-margined; uppermost leaves often ovate-lanceolate, sessile, cordate; bracts of the involucre foliaceous, often 1' long or more; head 2'–3' broad; rays about 12, 2-lobed; outer chaff oblanceolate, the inner linear, acute, purple-tipped, fringed with jointed hairs; style-tips slender, acute.

In woods, Pennsylvania to Virginia and Tennessee. May–July.

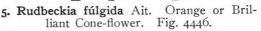

5. Rudbeckia fúlgida Ait. Orange or Brilliant Cone-flower. Fig. 4446.

Rudbeckia fulgida Ait. Hort. Kew. 3: 251. 1789.
R. spathulata Michx. Fl. Bor. Am. 2: 144. 1803.
Rudbeckia missouriensis Engelm.; Boynton & Beadle, Biltmore Bot. Studies 1: 17. 1901.
R. palustris Eggert; Boynton & Beadle, loc. cit. 16. 1901.

Perennial; stem hirsute, or strigose-pubescent, slender, sparingly branched or simple, $1°$–$3°$ high. Leaves entire or sparingly serrate with distant teeth, more or less hirsute or pubescent on both sides, the basal and lower ones oblong or spatulate, obtuse, 2'–4' long, 3-nerved, narrowed into margined petioles, the upper lanceolate, oblong-lanceolate, ovate or obovate, sessile, or slightly clasping at the base; heads few, 1'–$1\frac{1}{2}'$ broad; bracts of the involucre oblong or lanceolate, 3''–8'' long; rays 8–15, linear, bright yellow or with an orange base; disk globose or globose-ovoid, brown-purple, 5''–7'' broad; chaff of the receptacle linear-oblong, glabrous, or ciliate at the summit; pappus a minute crown.

In dry soil, New Jersey and Pennsylvania to Florida, west to Missouri and Texas. Consists of races differing in pubescence and leaf-form. Aug.–Oct.

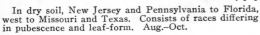

6. Rudbeckia umbròsa Boynton & Beadle. Woodland Cone-flower. Fig. 4447.

Rudbeckia umbrosa Boynton & Beadle, Biltmore Bot. Studies 1: 16. 1901.

Pubescent, perennial; stems $1\frac{3}{4}°$–$3\frac{1}{2}°$ tall, striate, simple or branched. Leaves thin, the basal and lower cauline ones ovate, 2'–$4\frac{1}{2}'$ long, coarsely serrate, rounded, truncate or cordate at the base, acute at the apex, the upper stem leaves diminishing in size, short-petioled or subsessile, narrower and less toothed than the lower; heads mostly several, showy; rays 8–12, yellow or orange-yellow, 7''–10'' long; bracts of the involucre oblong to linear-oblong, 5''–$7\frac{1}{2}''$ long, pointed; disk somewhat depressed, 5''–$7\frac{1}{2}''$ wide, dark purple; chaff broad, densely ciliate at the apex; pappus coroniform.

In moist soil and woodlands, Kentucky, Tennessee and northwestern Georgia. Aug.–Sept.

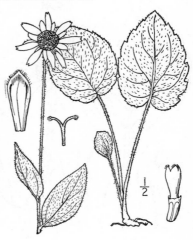

8. Rudbeckia grandiflòra Gmelin.
Large-flowered Cone-flower.
Fig. 4449.

Rudbeckia grandiflora Gmelin; DC. Prodr. **5**: 556. 1836.

Perennial; stem 1¾°–3° tall, scabrous or hispid throughout. Leaves mainly on the lower part of the stem, very rough on both surfaces, ovate-elliptic to lanceolate, 2½'–6' long, acute or acuminate at the apex, cuneate or more abruptly contracted at the base, shallowly serrate or denticulate, the petioles of the lower cauline and basal leaves as long as the blades or longer; heads large, solitary or few, showy; rays several, yellow, 1¼'–1½' long, drooping; bracts of the involucre linear, acuminate; disk ovoid or ovoid-globose, ⅜'–1' thick; chaff obtuse, canescent; pappus conspicuous, crenate or toothed.

On dry prairies, Oklahoma to Louisiana and Texas; introduced into Missouri. June–Aug.

7. Rudbeckia speciòsa Wenderoth. Showy Cone-flower. Fig. 4448.

Rudbeckia aspera Pers. Syn. **2**: 477. 1807?
R. speciosa Wendler. Ind. Sem. Hort. Marb. 1828.

Perennial, more or less hirsute or hispid; stem branched above, 1°–4° high. Leaves firm, slender-petioled, 2'–5' long, 1'–2' wide, dentate with low teeth, acute or sometimes acuminate, 3–5-nerved; stem leaves sessile or partly clasping, or narrowed into broad margined petioles, laciniate or serrate, lanceolate to ovate, acuminate, often 6' long, the uppermost smaller and sometimes entire; heads several, 2'–3' broad; bracts of the involucre linear-lanceolate, acute; rays 12–20, 1'–1½' long, bright yellow, usually orange at the base; disk depressed-globose, 5''–8'' broad, brown-purple; chaff of the receptacle obtusish or acute, ciliate or naked; pappus a short crown.

In moist soil, New Jersey to Michigan, south to Alabama and Arkansas. Aug.–Oct.

Rudbeckia Sullivántii Boynton & Beadle has been separated from *R. speciosa* on account of its broader leaves, larger disk-flowers and larger achenes.

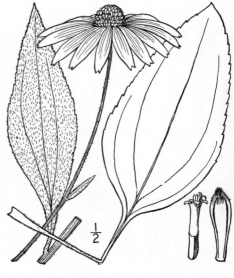

9. Rudbeckia máxima Nutt. Great Cone-flower. Fig. 4450.

Rudbeckia maxima Nutt. Trans. Am. Phil. Soc. (II) **7**: 354. 1841.

Perennial; stem 3°–9° tall, simple or branched above, smooth, glaucous. Leaves oblong, oval, or ovate, or pandurate, 2½'–8' long, mostly obtuse, undulate, repand-denticulate or entire, the upper sessile and partly clasping; heads large, showy; rays several, yellow, 7''–20'' long; bracts of the involucres linear or linear-lanceolate, acute, short; disk cylindric to conic-cylindric, 1'–2½' long; chaff abruptly short-pointed, pubescent at the summit; pappus conspicuous, denticulate, accentuated at the angles.

In moist soil, Missouri to Louisiana and Texas. June–Aug.

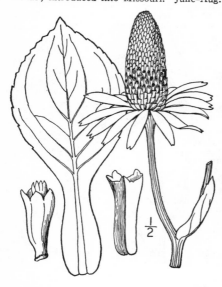

10. Rudbeckia laciniàta L.   Tall, or Green-
headed Cone-flower.   Fig. 4451.

Rudbeckia laciniata L. Sp. Pl. 906.   1753.

Perennial; stem much branched, glabrous, or nearly
so, 3°–12° high.   Leaves rather thin, minutely pubescent on the margins and upper surface, broad, the
basal and lower ones long-petioled, often 1° wide,
pinnately 3–7-divided, the segments variously toothed
and lobed; stem leaves shorter-petioled, 3–5-parted
or divided, the uppermost much smaller, 3-lobed,
dentate or entire; heads several or numerous, 2½′–4′
broad; rays 6–10, bright yellow, drooping; bracts of
the involucre unequal; chaff of the receptacle truncate and canescent at the apex; disk greenish-yellow,
at length oblong and twice as long as thick or longer;
pappus a short crown.

In moist thickets, Quebec to Manitoba, Idaho, Colorado, Florida and Arizona.   Thimble-weed.   A double-
flowered form in cultivation is called golden-glow.   July–
Sept.
A southern mountain race, lower, often only 1° high,
with smaller heads, has been described as *Rudbeckia
laciniata humilis* A. Gray.

62. **DRACOPIS** Cass. Dict. Sci. Nat. **35**: 273. 1825. **46**: 400. 1827.

An annual caulescent herb, with smooth and glaucous foliage, and alternate broad thickish.
entire or slightly serrate, 1-ribbed, clasping leaves.   Heads radiate, showy.   Involucre flat,
many-flowered, of few narrow, somewhat foliaceous bracts.   Receptacle slender, with early
deciduous chaffy scales.   Ray-flowers few, neutral, the rays yellow or often brownish-purple
at the base.   Disk-flowers perfect, fertile, brownish, their corollas 5-lobed.   Style-branches
with small pubescent appendages.   Achenes terete or nearly so, not angled, striate and
minutely transversely wrinkled.   Pappus wanting.   [Greek, dragon-like, referring to the
appendaged style-branches.]

A monotypic genus of the southeastern United States.

1. Dracopis amplexicaùlis (Vahl) Cass.
Clasping-leaved Cone-flower.   Fig. 4452.

Rudbeckia amplexicaulis Vahl, Act. Havn. **2**: 29. *pl. 4.*
1783.
Dracopis amplexicaulis Cass.; DC. Prodr. **5**: 558. 1836.

Annual; glabrous throughout, somewhat glaucous;
stem branched, grooved, 1°–2° high, the branches
ascending.   Leaves entire or sparingly toothed, 1-ribbed,
reticulate-veined, the lower oblong to spatulate, sessile, the upper ovate, ovate-oblong, or lanceolate,
acute, cordate-clasping; heads solitary at the ends of
the branches, long-peduncled, about 2′ broad; bracts
of the involucre few, lanceolate, acuminate; rays
yellow, or sometimes brown at the base; disk ovoid-
oblong, often becoming 1′ high; achenes not angled,
obliquely attached to the elongated receptacle; chaff
at length deciduous.

In wet soil, Missouri to Oklahoma, Louisiana and
Texas.   June–Aug.

63. **RATÍBIDA** Raf. Am. Month. Mag. **2**: 268.   1818.

[LÉPACHYS Raf Journ. Phys. **89**: 100. 1819.]

Perennial herbs, with alternate pinnately divided or parted leaves, and long-peduncled
terminal heads of tubular and radiate flowers, the disk-flowers gray or yellow, becoming
brown, the rays yellow, or with brown bases, drooping or spreading.   Involucral bracts in
2 or 3 series.   Disk globose, oblong or cylindric.   Receptacle columnar to subulate, the concave chaff subtending or enveloping the disk-flowers, truncate, the tips inflexed, canescent.
Ray-flowers neutral.   Disk-flowers perfect, fertile, their corollas with scarcely any tube.

Achenes short, flattened, sharp-margined, or winged, at length deciduous with the chaff. Pappus with 1 or 2 teeth, or none. [Name unexplained.]

About 4 species, natives of North America. Type species: *Rudbeckia columnaris* Sims.

Style-tips lanceolate-subulate; leaf-segments lanceolate; rays 1′–3′ long.   1. *R. pinnata*.
Style-tips short, blunt; leaf-segments linear, rays 3″–15″ long.
    Disk cylindric, at length 1′ long or more; rays mostly as long, or longer.   2. *R. columnaris*.
    Disk globose to short-oblong, about ½′ high; rays mostly short.   3. *R. Tagetes*.

1. Ratibida pinnàta (Vent.) Barnhart.   Gray-headed Cone-flower.   Fig. 4453.

Rudbeckia pinnata Vent. Hort. Cels. *pl. 71.* 1800.
Lepachys pinnata T. & G. Fl. N. A. 2 : 314. 1842.
Ratibida pinnata Barnhart, Bull. Torr. Club 24: 410. 1897.

Rough and strigose-pubescent throughout; stem branched or simple, 3°–5° high. Leaves pinnately 3–7-divided, the basal ones sometimes 10′ long, petioled, the segments lanceolate, dentate, cleft or entire, acute or acuminate; upper leaves sessile or nearly so, the uppermost commonly small and entire; bracts of the involucre linear or linear-oblong, short, reflexed; rays 4–10, yellow, 1′–3′ long, 3″–9″ wide, drooping; style-tips lance-subulate; disk oblong, gray or becoming brown, rounded, at length twice as long as thick; chaff of the receptacle canescent at the summit; achenes compressed, acutely margined, the inner margin produced into a short tooth.

On dry prairies, Ontario and western New York to Florida, South Dakota, Nebraska, Kansas and Louisiana. Adventive eastward to Massachusetts. June–Sept.

2. Ratibida columnàris (Sims) D. Don.   Long-headed or Prairie Cone-flower. Fig. 4454.

Rudbeckia columnaris Sims, Bot. Mag. *pl. 1601.* 1813.
Ratibida columnaris D. Don; Sweet, Brit. Fl. Gard. 2 : 361. 1838.
Lepachys columnaris T. & G. Fl. N. A. 2 : 313. 1842.
Lepachys columnaris var. *pulcherrima* T. & G. loc. cit. 1842.

Strigose-pubescent and scabrous; stem slender, usually branched, 1°–2½° high. Leaves thick, pinnately divided into linear or linear-oblong, acute or obtuse, entire dentate or cleft segments, the cauline short-petioled or sessile, 2′–4′ long, the basal ones sometimes oblong, obtuse and undivided, slender-petioled; bracts of the involucre short, linear-lanceolate or subulate, reflexed; rays 4–10, yellow, brown at the base, or brown all over, 4″–15″ long, drooping; disk gray, elongated-conic or cylindric, blunt, at length 3 or 4 times as long as thick; chaff of the receptacle canescent at the apex; achenes scarious-margined or narrowly winged on the inner side; pappus of 1 or 2 subulate teeth usually with several short intermediate scales.

On dry prairies, Minnesota to Assiniboia, British Columbia, Montana, Nebraska, Texas, Mexico and Arizona. Also in Tennessee. Brush. May–Aug.

3. Ratibida Tagètes (James) Barnhart.
Short-rayed Cone-flower.   Fig. 4455.

Rudbeckia Tagetes James in Long's Exp. **2** : 68.   1823.

Lepachys Tagetes A. Gray, Pac. R. R. Rep. **4**: 103. 1856.

Ratibida Tagetes Barnhart, Bull. Torr. Club **24**: 100. 1897.

Rough-canescent; stem $1°–1\frac{1}{2}°$ high, usually much branched, leafy. Leaves firm, pinnately divided into 3–7 narrowly linear, mostly entire segments; peduncles terminal, $\frac{1}{2}'–2'$ long; heads $1'$ broad, or less; bracts of the receptacle narrow, deflexed; rays few, mostly shorter than the globose to short-oval disk; style-tips obtuse; achenes scarious-margined; pappus of 1 or 2 subulate deciduous teeth, with no short intermediate teeth.

On dry plains and rocky hills, Kansas to Texas, Colorado, Chihuahua, New Mexico and Arizona. July–Sept.

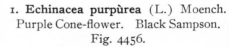

64. ECHINÀCEA Moench, Meth. 591.   1794.
[Brauneria Neck. Elem. **1** : 17. Hyponym. 1790.]

Perennial erect branched or simple herbs, with thick black roots, thick rough alternate or opposite, 3–5-nerved entire or dentate, undivided leaves, and large long-peduncled heads of tubular and radiate flowers, the rays purple, purplish, crimson or yellow, the disk green or purple, at length ovoid or conic. Involucre depressed-hemispheric, its bracts lanceolate, spreading or appressed, imbricated in 2–4 series. Receptacle conic, chaffy, the chaff carinate and cuspidate. Ray-flowers neutral, or with a rudimentary pistil. Disk-flowers perfect, the corolla cylindric, 5-toothed. Achenes 4-sided, obpyramidal, thick. Pappus a short dentate crown. [Greek, referring to the spiny chaff of the receptacle.]

Four species, natives of eastern and central North America. Type species: *Echinacea purpurea* (L.) Moench.

Rays purple, crimson, pink or white.
  Leaves broad, ovate to ovate-lanceolate, often toothed.   1. *E. purpurea.*
  Leaves narrow, linear to lanceolate, entire.
    Rays about $1'$ long, spreading.   2. *E. angustifolia.*
    Rays $1\frac{1}{2}'–3'$ long, drooping.   3. *E. pallida.*
Rays bright yellow, drooping.   4. *E. paradoxa.*

1. Echinacea purpùrea (L.) Moench.
Purple Cone-flower.   Black Sampson.
Fig. 4456.

Rudbeckia purpurea L. Sp. Pl. 907. 1753.

Echinacea purpurea Moench, Meth. 591. 1794.

Brauneria purpurea Britton, Mem. Torr. Club **5** : 334. 1894.

Stem glabrous, or sparingly hispid, usually stout, $2°–5°$ high. Lower and basal leaves slender-petioled, ovate, mostly 5-nerved, acute or acuminate at the apex, abruptly narrowed or rarely cordate at the base, commonly sharply dentate, $3'–8'$ long, $1'–3'$ wide; petioles mostly winged at the summit; upper leaves lanceolate or ovate-lanceolate, 3-nerved, sessile or nearly so, often entire; rays 12–20, purple, crimson, or rarely pale, $1\frac{1}{2}'–3'$ long, spreading or drooping.

In moist, rich soil, Pennsylvania to Alabama, Georgia, Michigan, Kentucky, Louisiana and Arkansas. Called also Red sunflower. July–Oct.

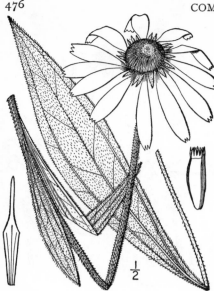

2. Echinacea angustifòlia DC. Narrow-leaved Purple Cone-flower. Fig. 4457.

Echinacea angustifolia DC. Prodr. **5**: 554. 1836.
Brauneria angustifolia Heller, Muhlenbergia **1**: 5. 1900.

Stem hispid or hirsute, slender, often simple, 1°–2° high. Leaves lanceolate, oblong-lanceolate, or linear-lanceolate, hirsute, acute and about equally narrowed at each end, strongly 3-nerved and sometimes with an additional pair of marginal less distinct nerves, entire, 3′–8′ long, 4″–12″ wide, the lower and basal ones slender-petioled, the upper short-petioled or sessile; heads and flowers similar to those of the preceding species, but the rays usually shorter, spreading.

In dry soil, especially on prairies, Minnesota to Saskatchewan, Nebraska and Texas. Confused with the following species in our first edition. June–Oct. Comb.

3. Echinacea pállida (Nutt.) Britton. Pale Purple Cone-flower. Fig. 4458.

Rudbeckia pallida Nutt. Journ. Acad. Phila. **7**: 77. 1834.
Brauneria pallida Britton, Mem. Torr. Club **5**: 333. 1894.

Similar to the preceding species, but often taller, sometimes 3° high. Leaves elongated-lanceolate or linear-lanceolate, entire; rays narrow, linear, elongated, drooping, 1½′–3′ long, 1½″–3″ wide, rose-purple or nearly white.

In dry soil on prairies, Illinois to Michigan, Alabama and Texas. May–July.

4. Echinacea paradóxa (Norton) Britton. Bush's Cone-flower. Fig. 4459.

Brauneria atrorubens Boynton & Beadle, Biltmore Bot. Stud. **1**: 11. 1901. Not *Rudbeckia atrorubens* Nutt.
Brauneria paradoxa Norton, Trans. Acad. St. Louis **12**: 40. 1902.

Stem glabrous to somewhat hispid, 1°–2½° high. Leaves smooth or sparingly rough-hairy, lanceolate to linear-lanceolate, the lower ones petioled, sometimes 1° long, ½′–1′ wide, 3–5-nerved, the upper smaller and nearly sessile; involucre about 1′ high and broad, the disk-flowers brown; rays bright yellow, drooping or somewhat spreading, 1½′–2½′ long.

Prairies and barren soil, Missouri to Texas. June.

65. BORRÍCHIA Adans. Fam. Pl. 2: 130. 1763.

Fleshy, more or less canescent, branching shrubs of the sea-coast, with opposite entire or denticulate, cuneate oblong spatulate or obovate, 1–3-nerved leaves, and terminal large long-peduncled heads of both tubular and radiate yellow flowers. Involucre hemispheric, its bracts slightly unequal, imbricated in 2 or 3 series, the inner ones coriaceous. Receptacle convex, chaffy, the chaff rigid, concave, subtending or enwrapping the disk-flowers. Ray-flowers pistillate, fertile. Disk-flowers perfect, the corolla tubular, 5-toothed, the style-branches elongated, hispid. Anthers dark-colored, entire at the base, or minutely sagittate. Achenes of the ray-flowers 3-sided, those of the disk-flowers 4-sided. Pappus a short dentate crown. [Named for Olaf Borrick, a Danish botanist.]

About 5 species, natives of America. In addition to the following typical one, another occurs in South Florida.

1. Borrichia frutéscens (L.) DC. Sea Ox-eye. Fig. 4460.

Buphthalmum frutescens L. Sp. Pl. 903. 1753.
Borrichia frutescens DC. Prodr. 5: 488. 1836.

Finely canescent, even when old; stems terete, sparingly branched, 1°–4° high. Leaves mostly erect or ascending, lanceolate, spatulate or obovate, obtuse or acutish and mucronulate at the apex, fleshy, tapering to the sessile base, somewhat connate, 1′–3′ long, 2″–7″ wide; heads solitary or few, about 1′ broad; rays 15–25, rather short; exterior bracts of the involucre ovate and somewhat spreading, the inner ones and the chaff of the receptacle cuspidate.

Sea-coast, Virginia to Florida and Texas. Also on the coasts of Mexico and in Bermuda. April–Oct.

66. HELIÁNTHUS [Vaill.] L. Sp. Pl. 904. 1753.

Erect, annual or perennial, mostly branched herbs, with opposite or alternate, simple leaves, and large peduncled corymbose or solitary heads of both tubular and radiate flowers, the rays yellow, the disk yellow, brown, or purple. Involucre hemispheric, or depressed, its bracts imbricated in several series. Receptacle flat, convex or conic, chaffy, the chaff subtending the disk-flowers. Ray-flowers neutral (in our species), the rays spreading, mostly entire. Disk-flowers perfect, fertile, the corolla tubular, the tube short, the limb 5-lobed. Anthers entire, or minutely 2-toothed at the base. Style-branches tipped with hirsute appendages. Achenes thick, oblong or obovate, compressed, or somewhat 4-angled. Pappus of 2 scales or awns, or sometimes with 2–4 additional shorter ones, deciduous. [Greek, sunflower.]

About 70 species, natives of the New World. Besides the following, about 30 others occur in the southern and western parts of North America, and hybrids may exist. Type species: *Helianthus annuus* L.

A. Annual species; disk brown or purple; heads large.

Leaves dentate; bracts ovate to ovate-lanceolate, strongly ciliate.   1. *H. annuus.*
Leaves mostly entire; bracts lanceolate to ovate-lanceolate, canescent, sometimes ciliate.
         2. *H. petiolaris.*

B. Perennial species.
* Disk purple or purple-brown.

Leaves narrowly linear or the lower broader, many of them alternate.
   Stem rough; leaves linear or linear-lanceolate.   3. *H. angustifolius.*
   Stem smooth; leaves elongated-lanceolate.   4. *H. orgyalis.*
Leaves lanceolate to ovate, mainly opposite.
   Leaves hispid, rather thin, abruptly contracted into winged petioles.   5. *H. atrorubens.*
   Leaves firm in texture, gradually narrowed into petioles.
     Leaves lanceolate; bracts acute or obtusish.   6. *H. scaberrimus.*
     Leaves rhombic-ovate to rhombic-lanceolate; bracts acute or acuminate.
         7. *H. subrhomboideus.*

** Disk yellow or yellowish; receptacle convex or conic.

   † *Leaves nearly all basal or near the base; upper bract-like.*   8. *H. occidentalis.*
     †† *Stem leafy; leaves alternate or opposite.*
   1. Leaves prevailingly lanceolate and 3–8 times as long as wide.
     a. Leaves glabrous on both sides; heads 1′–1½′ broad.   9. *H. laevigatus.*
     b. Leaves scabrous, at least on the upper surface.
Heads 1′–1½′ broad; leaves thin; stem glabrous.   10. *H. microcephalus.*
Heads 1½′–3½′ broad; leaves firm, many of them often alternate.
   Stem scabrous, scabrate or hispid; leaves sessile or nearly so.
     Leaves lanceolate, scabrous above, hirsute beneath, flat.   11. *H. giganteus.*

Leaves very scabrous on both sides.
  Leaves elongated-lanceolate, conduplicate, pinnately-veined.
    Heads numerous; leaves mostly alternate.   12. *H. Maximiliani.*
    Heads only 1 or 2; leaves, all but the upper, opposite.   13. *H. Dalyi.*
    Leaves lanceolate to ovate-lanceolate, flat, 3-nerved.   14. *H. subtuberosus.*
  Stem glabrous; leaves sessile by a truncate base.   17. *H. divaricatus.*
  Stem glabrous; leaves petioled.
    Leaves lanceolate, firm, canescent and pale beneath.   15. *H. grosse-serratus.*
    Leaves linear-lanceolate, thin, green on both sides.   16. *H. Kellermani.*
      2. Leaves prevailingly ovate, ovate-lanceolate, or oblong.
        a. Leaves sessile, or very nearly so.
  Stem glabrous; leaves divaricate.   17. *H. divaricatus.*
  Stem hirsute or hispid; leaves ascending.
    Leaves cordate-clasping at the base.   18. *H. mollis.*
    Leaves narrowed from below the middle.   19. *H. doronicoides.*
        b. Leaves manifestly petioled.
  Stem puberulent or glabrous.
    Leaves membranous or thin, slender-petioled, sharply serrate.   20. *H. decapetalus.*
    Leaves firmer, shorter-petioled, less serrate or entire.
      Bracts of the involucre much longer than the disk.   21. *H. tracheliifolius.*
      Bracts of the involucre about equalling the disk.   22. *H. strumosus.*
  Stem hirsute, hispid, or scabrous.
    Leaves rounded or truncate at the base, short-petioled.   23. *H. hirsutus.*
    Leaves, at least the upper, narrowed at the base.
      Bracts of the involucre ovate-lanceolate, appressed.   24. *H. laetiflorus.*
      Bracts of the involucre lanceolate-acuminate, spreading.
        Leaves villous-pubescent beneath.   25. *H. tomentosus.*
        Leaves scabrous or puberulent beneath.   26. *H. tuberosus.*

1. Helianthus ánnuus L.   Common Sunflower.
Fig. 4461.

Helianthus annuus L. Sp. Pl. 904. 1753.
Helianthus lenticularis Dougl. Bot. Reg. *pl. 1265.* 1829.

Stem hispid or scabrous, stout, branched above, 3°–6°
high, or in cultivated races sometimes 15° high. Leaves
all but the lower alternate, broadly ovate, petioled,
3-nerved, dentate or denticulate, acute at the apex,
rough on both sides, sometimes pubescent beneath, the
lower cordate at the base, 3′–12′ long; heads in the wild
plant 3′–6′ broad; disk dark purple or brown, 10″–2′
broad; involucre depressed, its bracts ovate to ovate-
lanceolate, usually long-acuminate or aristate, hispid-
ciliate; chaff of the flat receptacle 3-cleft; achenes
obovate-oblong, appressed-pubescent, or nearly glabrous.

On prairies, etc., Minnesota to North Dakota, Idaho, Missouri, Texas and California. Recorded north to Saskatchewan. Much larger in cultivation; an occasional escape in
the east. Gold. Golden. Larea-bell. Comb-flower. Its flowers
yield honey and a yellow dye; its leaves fodder; its seeds,
an oil and food; and its stalks a textile fibre. July–Sept.

2. Helianthus petiolàris Nutt.   Prairie Sun-
flower. Fig. 4462.

H. petiolaris Nutt. Journ. Acad. Phila. **2**: 115. 1821.
H. aridus Rydb. Bull. Torr. Club **32**: 127. 1905.

Annual, similar to the preceding species, but smaller
and with smaller heads; stem strigose-hispid or hirsute, 1°–3° high. Leaves all but the lowest alternate,
petioled, oblong, ovate, or ovate-lanceolate, rough on
both sides, usually paler beneath than above, sometimes
canescent beneath, 1′–3′ long, entire, or denticulate, obtuse or acutish at the apex, mostly narrowed at the
base; heads 1½′–3′ broad; disk brown, mostly less than
10″ broad; involucre depressed-hemispheric, its bracts
lanceolate or oblong-lanceolate, densely canescent, sometimes hispid-ciliate, acute or short-acuminate; achenes
villous-pubescent, at least when young.

On dry prairies, Minnesota to Saskatchewan, Oregon,
Iowa, Missouri, Texas and California. Found rarely in
waste places farther east. Races differ in leaf-form, size
and pubescence. June–Sept.

3. Helianthus angustifòlius L.   Narrow-leaved or Swamp Sunflower.   Fig. 4463.

Helianthus angustifolius L. Sp. Pl. 906. 1753.

Perennial by slender rootstocks; stems branched above, or simple, slender, rough or roughish above, often hirsute below, 2°–7° high. Leaves firm, entire, sessile, linear, slightly scabrous, rarely somewhat canescent beneath, 2′–7′ long, 2″–3″ wide, the margins revolute when dry, the upper ones all alternate, the lower opposite; heads usually few, sometimes solitary, 2′–3′ broad; involucre hemispheric, its bracts linear-lanceolate, acute or acuminate, scarcely squarrose, pubescent; receptacle slightly convex; disk purple; chaff entire or 3-toothed; rays 12–20; achenes truncate, glabrous; pappus usually of 2 short awns.

In swamps, Long Island, N. Y., to Florida, Kentucky and Texas, mainly near the coast. Aug.–Oct.

4. Helianthus orgyàlis DC.   Linear-leaved Sunflower.   Fig. 4464.

H. giganteus var. *crinitus* Nutt. Gen. **2**: 177.   1818?
Helianthus orgyalis DC. Prodr. **5**: 586.   1836.

Perennial by slender rootstocks; stems glabrous, branched near the summit, very leafy to the top, 6°–10° high. Leaves sessile, entire, linear or nearly filiform, or the lowest lanceolate, remotely dentate and short-petioled, rough with mucronate-tipped papillae, especially on the lower surface, acuminate, 4′–16′ long, 1″–4″ wide, the upper all alternate and 1-nerved, the lower commonly opposite; heads numerous, about 2′ broad, terminating slender branches; involucre nearly hemispheric, its bracts linear-subulate to lanceolate, acuminate, squarrose, ciliate; disk purple or brown; receptacle convex, its chaff entire, or toothed, slightly ciliate; rays 10–20; achenes oblong-obovate, glabrous, 2½″–3″ long, 2–4-awned.

On dry plains, Missouri and Nebraska to Colorado and Texas. Sept.–Oct.

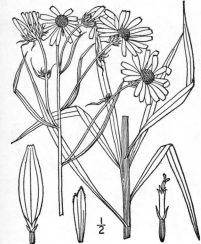

5. Helianthus atrórubens L.   Hairy Wood Sunflower.   Purple-disk Sunflower. Fig. 4465.

Helianthus atrorubens L. Sp. Pl. 906.   1753.

Perennial; stems hirsute below, often minutely pubescent above, branched at the summit, 2°–5° high. Leaves hirsute on both sides, or canescent beneath, mostly thin, ovate or ovate-lanceolate, acutish, contracted near the base into margined petioles, sometimes subcordate, dentate or crenate-dentate, 4′–10′ long, 1′–4′ wide, the lower opposite, the upper few, distant, small, mainly alternate; heads not numerous, slender-peduncled, about 2′ broad; involucre hemispheric, its bracts oblong to obovate, obtuse, ciliolate, appressed; disk purple; receptacle convex, its chaff acute, entire, or 3-toothed; rays 10–20; achenes obovate, truncate, finely pubescent, about 2″ long; pappus usually of 2 lanceolate awns.

In dry woods, Virginia to Florida, west to Ohio, Missouri, Arkansas and Louisiana. Aug.–Oct.

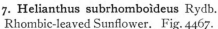

6. Helianthus scabérrimus Ell. Stiff Sunflower. Fig. 4466.

H. scaberrimus Ell. Bot. S. C. & Ga. **2**: 423. 1824.
H. rigidus Desf. Cat. Hort. Paris, Ed. 3, 184. 1829.

Perennial; stems simple or little branched, hispid or scabrate, 1°–8° high. Leaves thick, coriaceous, serrate or serrulate, very scabrous on both sides, 2′–7′ long, ½′–2′ wide, acute at the apex, narrowed at the base, the lower ovate or ovate-oblong, petioled, the upper lanceolate or oblong-lanceolate, sessile or short-petioled, all opposite, or the uppermost bract-like and alternate; heads solitary or few, 2′–3′ broad; involucre hemispheric, its bracts ovate, acute or obtusish, ciliate, appressed; disk purple or brown; receptacle convex, its chaff obtuse; rays 15–25, light yellow; achenes more or less pubescent, oblong-obovate; pappus of 2 broad scales or of 2–4 stout awns.

Prairies, Illinois to Minnesota, Iowa, Kansas, Georgia and Texas. Aug.–Sept.

7. Helianthus subrhomboìdeus Rydb. Rhombic-leaved Sunflower. Fig. 4467.

Helianthus subrhomboideus Rydb. Mem. N. Y. Bot. Gard. **1**: 419. 1900.

Stem simple, sparingly hirsute, usually tinged with red. Leaves opposite, firm, very scabrous, 3-nerved, slightly serrate, the basal ones broadly ovate to obovate-spatulate, those of the stem rhomboid-ovate to rhomboid-lanceolate, short-petioled, 2′–4′ long, the uppermost very small; heads 1–3, 1′–1¾′ in diameter; bracts of the involucre oblong, acutish, densely white-ciliate; disk purple.

Plains, Manitoba and Saskatchewan to South Dakota, Nebraska, Montana and New Mexico. Adventive, New Jersey to New Hampshire. July–Sept. Included in *H. scaberrimus* in our first edition.

8. Helianthus occidentàlis Riddell. Few-leaved Sunflower. Fig. 4468.

H. occidentalis Ridd. Suppl. Cat. Ohio Pl. 13. 1836.
H. illinoensis Gleason, Ohio Nat. **5**: 214. 1904.
H. occidentalis illinoensis Gates, Bull. Torr. Club **37**: 81. 1910.

Perennial; stems appressed-pubescent or sometimes nearly glabrous, slender, mostly simple, 1°–3° high. Leaves mostly basal, or below the middle of the stem, firm, ovate or oblong-lanceolate, obtuse or obtusish at the apex, narrowed at the base, 3–5-nerved, serrulate or entire, scabrous above, pubescent beneath, with slender petioles about as long as the blades; stem usually bearing 1 or 2 pairs of small distant leaves; heads several or solitary, 1½′–2½′ broad; involucre hemispheric, its bracts lanceolate or ovate-lanceolate, acute or acuminate, generally ciliate, appressed; receptacle convex, its chaff acute; disk yellow; rays 12–15; achenes truncate and pubescent at the summit; pappus of 2 lanceolate-subulate awns.

In dry soil, Ohio to Minnesota, south to Florida and Missouri. Aug.–Sept.

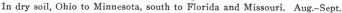

Helianthus Dowelliànus M. A. Curtis, which differs in being stouter, the stem leafy, the leaves merely puberulent, and ranges from the District of Columbia to Georgia, appears to be a race of this species.

9. Helianthus laevigàtus T. & G.   Smooth Sunflower. Fig. 4469.

Helianthus laevigatus T. & G. Fl. N. A. **2** : 330. 1842.

Stems slender, from a perennial root, simple or little branched, glabrous, or somewhat glaucous, 2°–6° high. Leaves nearly all opposite, firm, glabrous, lanceolate, short-petioled, or the upper sessile, serrulate or entire, pale beneath, acuminate, narrowed at the base, 3′–6′ long, ½′–1½′ wide, the margins sometimes roughish-ciliate; heads few or solitary, 1′–1½′ broad; involucre campanulate, its bracts lanceolate or ovate-lanceolate, firm, ciliolate, the tips of the outer ones spreading; rays 5–10; disk yellow; chaff linear; achenes slightly pubescent at the summit; pappus of 2 lanceolate or ovate owns, with or without 2 intermediate scales.

In dry soil, mountains of Virginia, West Virginia and North Carolina. Aug.–Oct.

10. Helianthus microcéphalus T. & G.   Small Wood Sunflower. Fig. 4470.

Helianhus parviflorus Bernh.; Spreng. Syst. **3** : 617. 1826. Not H.B.K. 1820.
H. microcephalus T. & G. Fl. N. A. **2** : 329. 1842.

Stems slender, glabrous, branched above, or rarely simple, 3°–6° high. Leaves thin or thinnish, petioled, most of them opposite, lanceolate, or the lower ovate-lanceolate, rough above, canescent or puberulent beneath, long-acuminate at the apex, narrowed at the base, serrulate, or the lower serrate, 3′–7′ long, ½′–1½′ wide; heads commonly several or numerous, 1′–1¼′ wide, borne on slender, sometimes roughish peduncles; involucre campanulate, 4″–5″ broad, its bracts lanceolate or ovate, acute or acuminate, ciliolate, the tips of the outer ones spreading; chaff of the receptacle oblong, entire, or 3-toothed; rays 5–10; disk yellow; achenes nearly glabrous; pappus usually of 2 subulate awns.

In moist woods and along streams, Pennsylvania to Georgia, west to Ohio, Missouri and Louisiana. July–Sept.

11. Helianthus gigantèus L.   Tall, Giant or Wild Sunflower. Fig. 4471.

Helianthus giganteus L. Sp. Pl. 905. 1753.

Perennial by fleshy roots and creeping rootstocks; stems hispid or scabrous, at least above, branched near the summit, or simple, 3°–12° high. Leaves sessile or short-petioled, firm, lanceolate, very rough above, rough-pubescent beneath, serrate or denticulate, acuminate at the apex, narrowed at the base, many or all of the upper ones alternate but sometimes all opposite, 2′–6′ long, ½′–1′ wide; heads usually several, mostly long-peduncled, 1½′–2½′ broad; involucre hemispheric, its bracts squarrose, lanceolate-subulate, hirsute or ciliate, commonly as long as the diameter of the yellowish disk; chaff of the receptacle oblong-linear, acute; achenes oblong, glabrous; rays 10–20; pappus of 2 subulate awns.

In swamps and wet meadows, Maine and Ontario to Saskatchewan, Florida, Louisiana and Colorado. Stem commonly purple. Aug.–Oct.

Helianthus ambíguus (A. Gray) Britton, differing in having all but the uppermost leaves opposite and rounded at the base, found on Long Island, New York, and recorded from Georgia, appears to be a hybrid, with this species one of its parents.

12. Helianthus Maximiliàni Schrad. Maximilian's Sunflower. Fig. 4472.

Helianthus Maximiliani Schrad. Ind. Sem. Hort. Goett. 1835.

Perennial by fleshy roots and thickened rootstocks; stems stout, scabrous or hispid below, 2°–12° high. Leaves sessile or short-petioled, long-lanceolate, folding in drying, alternate or the lower opposite, very rough on both sides, rigid, acuminate or acute at both ends, denticulate or entire, 3′–7′ long, ½′–1½′ wide; heads few or numerous, 2′–3′ broad on stout densely rough-pubescent peduncles; involucre hemispheric, its bracts lanceolate, acuminate, squarrose, densely strigose-pubescent, often 9″ long; disk yellowish; rays 15–30; chaff linear, acute, pubescent above; achenes linear-oblong, glabrous or nearly so; pappus commonly of 2 lanceolate awns.

On dry prairies, Minnesota and Manitoba to Saskatchewan, Missouri, Nebraska and Texas. Locally adventive eastward. Aug.–Oct.

13. Helianthus Dàlyi Britton. Judge Daly's Sunflower. Fig. 4473.

Helianthus Dalyi Britton, Journ. N. Y. Bot. Gard. **2**: 89. 1901.

Perennial by a fusiform tuber 1′–1¾′ long; stem slender, simple, roughish, appressed-pubescent, about 2° high. Leaves firm, conduplicate, drooping, finely rough-pubescent on both sides, narrowly lanceolate, acuminate at both ends, distantly serrate with low teeth, the larger about 4′ long and 7″ wide, all opposite or the uppermost alternate; heads usually solitary and long-peduncled, rarely 2, about 3½′ broad; involucral bracts narrowly lanceolate with long subulate tips; rays acute; pappus 2 subulate awns; chaff of the receptacle pubescent at the acute apex.

On a dry bank, Sag Harbor, New York. Sept.

14. Helianthus subtuberòsus Bourgeau. Indian Potato. Fig. 4474.

H. giganteus subtuberosus (Bourgeau) Britton, in Britt. & Brown, Ill. Fl. **3**: 425. 1898.
H. subtuberosus Bourgeau; Britton, Manual 993. 1901.

Fleshy roots, thick and edible. Leaves thick, more or less serrate, all or all but the upper distinctly petioled, sometimes all opposite, 2½′–5′ long, acuminate at the apex, mostly narrowed at the base, the petioles ciliate; heads not very numerous; bracts of the involucre lanceolate, acuminate, densely pubescent and white-ciliate, about 8″ long; rays 1′–1¾′ long.

Dry soil, Michigan and Minnesota to Saskatchewan, Montana and Wyoming. Aug.–Sept.

Helianthus Rýdbergi Britton, of western Nebraska, differs by broader, ovate-lanceolate leaves, rather abruptly narrowed at the base.

15. Helianthus grósse-serràtus Martens.
Saw-tooth Sunflower.   Fig. 4475.

Helianthus grösse-serratus Martens, Sel. Sem. Hort.
Loven. 1839.

Perennial by fleshy roots and slender rootstocks;
stems glabrous, glaucous, branched above, 6°–10°
high, the branches usually strigose-pubescent. Leaves
long-lanceolate, slender-petioled, the upper alternate,
the lower opposite, long-acuminate, narrowed at the
base, sharply serrate, or merely denticulate, the uppermost entire, rough above, densely puberulent or
canescent beneath, 4′–8′ long, ½′–1′ wide; heads several or numerous, 1½′–3′ broad; involucre hemispheric, its bracts squarrose, narrowly lanceolate,
acuminate, hirsute; chaff linear-oblong, pubescent at
the summit, acute, often 3-toothed; disk yellowish;
rays 10–20, deep yellow; achenes nearly glabrous;
pappus of 2 lanceolate awns.

In dry soil, Maine to Long Island, Pennsylvania, Ontario, South Dakota, Missouri, Kansas and Texas. Aug.–
Oct.

16. Helianthus Kellermáni Britton
Kellerman's Sunflower.   Fig. 4476.

Helianthus Kellermani Britton, Manual 994. 1901.

Stem 6°–10° high, very smooth, much branched above, the branches slender. Leaves narrowly elongated-lanceolate to linear-lanceolate,
drooping, rather thin, distantly serrate with
very small teeth, long-acuminate at the apex,
attenuate at the base into short petioles or the
upper sessile, scabrate and sparingly pubescent on both surfaces, pinnately veined, the
lower about 8′ long and 7″ wide; branches of
the inflorescence pubescent; bracts of the involucre linear-lanceolate, about 7″ long, and 1″
wide or less at the base, ciliate, long-acuminate; rays golden-yellow, 1′–1¾′ long; chaff of
the receptacle linear.

Dry soil, vicinity of Columbus, Ohio. Aug.–
Sept.

17. Helianthus divaricàtus L. Rough or
Woodland Sunflower. Fig. 4477.

Helianthus divaricatus L. Sp. Pl. 906. 1753.

Perennial by slender rootstocks; stem glabrous
throughout, or pubescent at the summit, slender,
2°–7° high. Leaves usually all opposite, divaricate,
lanceolate or ovate-lanceolate, firm, dentate or denticulate, 3-nerved, rough above, pubescent beneath,
sessile or nearly so by a truncate base, tapering
gradually to the long-acuminate apex, 3′–8′ long,
¼′–1½′ wide; heads few or solitary, about 2′ broad,
borne on strigose-pubescent peduncles; involucre
hemispheric, its bracts lanceolate or ovate-lanceolate, strigose or hirsute, the outer ones spreading;
disk yellow; rays 8–15; chaff of the receptacle
apiculate; achenes glabrous; pappus of 2 short
subulate awns.

In dry woodlands, Maine and Ontario to Manitoba,
Nebraska, Florida and Louisiana. July–Sept.

$\frac{1}{2}$

18. Helianthus móllis Lam.　Hairy Sunflower.　Fig. 4478.

Helianthus mollis Lam. Encycl. **3** : 85.　1789.

Perennial; stem stout, simple or sparingly branched above, densely hirsute, 2°–4° high. Leaves ovate or ovate-lanceolate, closely sessile and somewhat clasping by a broad cordate base, pinnately veined, 3-nerved above the base, all opposite, or the upper alternate, ascending, acute or acuminate, scabrous or cinereous-pubescent above, densely and finely pubescent beneath, serrulate, 2′–5′ long, 1′–2½′ wide; heads solitary or few, 2′–3′ broad; involucre hemispheric, its bracts lanceolate, acuminate, densely villous-canescent, somewhat spreading; disk yellow; rays 15–25; chaff canescent at the summit; mature achenes nearly glabrous; pappus of 2 lanceolate scales.

In dry barren soil, Massachusetts to New Jersey, Georgia, Iowa, Missouri, Kansas and Texas. Aug.–Sept.

19. Helianthus doronicoìdes Lam.　Oblong-leaved Sunflower.　Fig. 4479.

Helianthus doronicoides Lam. Encycl. **3** : 84.　1789.

Perennial by slender rootstocks; stems stout, rough, or finely rough-pubescent, branched above, 3°–7° high. Leaves oblong, ovate-oblong, or ovate, thick, ascending, serrate or serrulate, acute or acuminate at the apex, narrowed at or below the middle to a sessile or slightly clasping base, rough on both sides, or finely pubescent beneath, 3-nerved, 4′–8′ long, 1′–2½′ wide; heads commonly numerous, 2½′–4′ broad; involucre hemispheric, its bracts lanceolate, acuminate, pubescent or hirsute, somewhat spreading; disk yellow; rays 12–20, 4″–6″ wide; achenes glabrous; pappus of 2 subulate awns.

In dry soil, Ohio to Missouri and Arkansas. Recorded from Michigan. Aug.–Sept.

$\frac{1}{2}$

20. Helianthus decapétalus L.　Thin-leaved or Wild Sunflower.　Fig. 4480.

Hleianthus decapetalus L. Sp. Pl. 905.　1753.

Perennial by branched, sometimes thickened rootstocks; stem slender, glabrous or nearly so, branched above, 1°–5° high; branches puberulent. Leaves thin or membranous, 3′–8′ long, 1′–3′ wide, ovate or ovate-lanceolate, the lower all opposite and slender-petioled, the upper commonly alternate, all usually sharply serrate, roughish above, finely but often sparingly pubescent beneath, acuminate, the rounded or truncate base decurrent on the petiole; heads numerous, 2′–3′ broad; involucre hemispheric, its bracts linear-lanceolate, acuminate, hirsute, long-ciliate, spreading, often longer than the yellow disk; rays 8–15, light yellow; chaff entire or 3-toothed, pubescent at the apex; achenes glabrous; pappus of 2 subulate awns.

In moist woods and along streams, Quebec to Michigan, Georgia, Tennessee and Missouri. Aug.–Sept.

Helianthus scrophulariaefòlius Britton, from near Woodlawn, New York, differing by laciniate-serrate leaves, is probably a race of this species.

$\frac{1}{2}$

21. Helianthus tracheliifòlius Mill.   Throatwort Sunflower.   Fig. 4481.

Helianthus tracheliifolius Mill. Gard. Dict. Ed. 8, No. 7. 1768.

Similar to the following species, but the stem usually roughish-pubescent above. Leaves short-petioled, ovate-lanceolate, or lanceolate, 3-nerved, green both sides, but darker above, generally rougher on the upper surface, the lower ones sharply serrate; branches and peduncles scabrous; heads several, 2½′–3½′ broad; bracts of the hemispheric involucre linear-lanceolate, long-acuminate, ciliate and puberulent, longer than the yellow disk, sometimes foliaceous, and 3 times its length.

In dry soil, Connecticut to Pennsylvania, North Carolina, Ohio, Minnesota and Arkansas. Aug.–Sept.

22. Helianthus strumòsus L.   Pale-leaved Wood Sunflower.   Fig. 4482.

Helianthus strumosus L. Sp. Pl. 905. 1753.
H. mollis Willd. Sp. Pl. 3: 2240. 1804. Not Lam. 1789.
H. macrophyllus Willd. Hort. Berol. *pl. 70.* 1806.

Perennial by branched, sometimes tuberous-thickened rootstocks; stem glabrous below, sometimes glaucous, 3°–7° high, branched above, the branches usually pubescent. Leaves short-petioled, ovate or ovate-lanceolate, rarely lanceolate, not membranous, rough above, pale and somewhat puberulent or canescent beneath, serrate, serrulate, or nearly entire, acuminate, contracted much below the middle and decurrent on the petiole, 3-nerved above the base, 3′–8′ long, 1′–2½′ wide, mostly opposite, the upper often alternate; heads commonly several, 2½′–4′ broad; involucre hemispheric, its bracts lanceolate or ovate-lanceolate, acuminate, ciliate, equalling or a little longer than the diameter of the yellow disk; rays 5–15; chaff pubescent; achenes nearly glabrous.

In dry woods and on banks, Maine and Ontario to Minnesota, Georgia, Tennessee and Arkansas. July–Sept. Races differ in leaf-form and texture. A hybrid with *H. decapetalus* has been described.

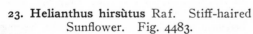

23. Helianthus hirsùtus Raf.   Stiff-haired Sunflower.   Fig. 4483.

Helianthus hirsutus Raf. Ann. Nat. 14. 1820.
Helianthus hirsutus trachyphyllus T. & G. Fl. N. A. 2: 329. 1842.

Perennial; stem densely hirsute, usually branched above, stout, rigid, 2°–4° high. Leaves ovate-lanceolate or the upper lanceolate, rounded, truncate or subcordate at the base, acuminate at the apex, thick, very rough above, rough-pubescent beneath, 3-nerved, serrate or serrulate, 3′–6′ long, ½′–2′ wide, short-petioled, or the uppermost sessile, nearly all opposite, the petioles of the lower ones ¼′–¾′ long; heads usually several, 2′–3½′ broad; involucre hemispheric, its bracts lanceolate, or ovate-lanceolate, acuminate, ciliate, mostly puberulent, erect or somewhat spreading, equalling or shorter than the yellow disk; rays 12–15; chaff obtusish, pubescent at the summit; achenes oval, rounded at the summit, glabrous; pappus of 1 or 2 subulate awns.

In dry soil, Pennsylvania and Ohio to Wisconsin, Nebraska, West Virginia, Georgia and Texas. July–Oct.

24. Helianthus laetiflòrus Pers. Showy Sunflower. Fig. 4484.

Helianthus laetiflorus Pers. Syn. **2** : 476. 1807.

Perennial; stem scabrous or hispid, leafy, 4°–8° high. Leaves oval-lanceolate or ovate-lanceolate, short-petioled, 3-nerved, rough on both sides, narrrowed at the base, acute or acuminate at the apex, serrate or serrulate, 4'–10' long, ½'–1½' wide, the upper often alternate; heads usually several, 2'–4' broad, mostly short-peduncled; bracts of the hemispheric involucre ovate-lanceolate, or oblong-lanceolate, imbricated in only 2 or 3 series, ciliate, otherwise nearly glabrous, appressed or but little spreading, shorter than or equalling the yellow disk; rays 15–25, showy; chaff of the receptacle entire, or sometimes 3-toothed.

On prairies and barrens, Pennsylvania to Minnesota. Recorded as adventive in Massachusetts. Aug.–Sept.

25. Helianthus tomentòsus Michx. Woolly Sunflower. Fig. 4485.

H. tomentosus Michx. Fl. Bor. Am. **2** : 141. 1803.

Perennial; stem stout, hirsute or hispid, especially above, branched, 4°–10° high. Leaves rather thin, ovate, or the lower oblong, mostly alternate, 3-ribbed above the base, gradually or abruptly contracted into margined petioles, rough above, softly villous-pubescent beneath, sparingly serrate, the lower often 1° long and 4' wide; heads commonly several or numerous, 3'–4' broad; involucre hemispheric, its bracts imbricated in many series, linear-lanceolate, long-acuminate, squarrose, densely hirsute and ciliate, usually longer than the broad yellowish disk; chaff of the receptacle and lobes of the disk corolla pubescent; pappus of 2 subulate awns.

In dry soil, Virginia to Georgia and Alabama. Reported from Illinois, probably erroneously. Aug.–Oct.

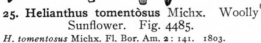

26. Helianthus tuberòsus L. Jerusalem Artichoke. Earth Apple. Fig. 4486.

Helianthus tuberosus L. Sp. Pl. 905. 1753.
Helianthus tuberosus subcanescens A. Gray, Syn. Fl. **1** : Part 2, 280. 1884.

Perennial by fleshy thickened rootstocks, bearing tubers; stems hirsute or pubescent, branched above, 6°–12° high. Leaves ovate or ovate-oblong, rarely ovate-lanceolate, firm, 3-nerved near the base, narrowed, or the lower rounded, truncate or subcordate at the base, acuminate at the apex, rather long-petioled, scabrous above, finely pubescent or canescent beneath, serrate, 4'–8' long, 1½'–3' wide, the upper alternate, the lower opposite; heads several or numerous, 2'–3½' broad; involucre hemispheric, its bracts lanceolate, acuminate, hirsute or ciliate, squarrose; disk yellow; rays 12–20; chaff of the receptacle acute and pubescent at the summit; achenes pubescent.

In moist soil, Nova Scotia and Ontario to Manitoba, Georgia and Arkansas and Kansas. Often occurs along roadsides in the east, a relic of cultivation by the aborigines. Now extensively grown for its edible tubers. Canada potato. Girasole. Topinambour. Sept.–Oct.

67. RÌDAN Adans. Fam. Pl. **2** : 130. 1763.

[ACTINOMERIS Nutt. Gen. **2** : 181. 1818.]

Rough-pubescent, rather coarse, herbs with alternate or opposite, simple, more or less toothed leaves, usually decurrent on the stem and branches, and corymbose, radiate or discoid heads. Involucre small, flattish, of few spreading or recurved narrow bracts. Receptacle convex or conic, becoming globose, chaffy. Ray-flowers neutral, yellow or white, sometimes wanting. Disk-flowers perfect, fertile, embraced by the chaff. Anthers entire at the base. Style-branches of the disk with acute appendages. Achenes flattened, more or less winged, spreading in all directions on the globose receptacle. Pappus of 2 or 3 finally spreading smooth awns, sometimes with 2–3 smaller awns or scales. [Name. unexplained.]

Two species, of eastern North America, the following typical.

1. Ridan alternifòlius (L.) Britton. Wing-stem. Yellow Iron weed. Fig. 4487.

Coreopsis alternifolia L. Sp. Pl. 909. 1753.
Actinomeris squarrosa Nutt. Gen. 2: 181. 1818.
Actinomeris alternifolia DC. Prodr. 5: 575. 1836.
Verbesina alternifolia Britton; Kearney, Bull. Torr. Club 20: 485. 1893.

Perennial; stem puberulent or glabrous, narrowly winged, or wingless, simple, or branched near the summit, leafy, 4°–9° high. Leaves oblong-lanceolate or lanceolate, acuminate, serrate, serrulate or entire, rough or roughish on both sides, gradually narrowed to the sessile base, or short-petioled, 4′–12′ long, ½′–2½′ wide, alternate, or the lower opposite and slender-petioled; heads numerous, 1′–2′ broad, corymbose-paniculate; rays 2–10, yellow; involucre of few lanceolate, at length deflexed bracts; disk globose, yellow; achenes broadly winged or nearly wingless, sparingly pubescent; pappus 2 divergent awns.

In rich soil, New Jersey to Iowa, Kansas, Florida and Louisiana. Winged ironweed. Aug.–Sept.

68. PHAETHÙSA Gaertn. Fr. & Sem. 2: 425. 1791.

Perennial, pubescent or scabrous herbs (some tropical species shrubby), with alternate or opposite dentate leaves often decurrent on the stem and branches, and corymbose or solitary heads of both tubular and radiate yellow or white flowers, or the rays sometimes wanting. Involucre campanulate or hemispheric, its bracts imbricated in few series. Receptacle convex or conic, chaffy, the chaff embracing the disk-flowers. Ray-flowers pistillate or neutral. Disk-flowers perfect, mostly fertile, their corollas with an expanded 5-lobed limb, usually longer than the tube. Style-branches of the disk-flowers with acute papillose appendages. Achenes flattened, or those of the ray-flowers 3-sided. Pappus of 2 (1–3) subulate awns, sometimes with 2 or 3 intermediate scales. [The daughter of Helios.]

A large genus, mainly natives of the New World. Besides the following, about 6 others occur in the southern and southwestern United States. Type species: *Phaethusa americana* Gaertn.

Involucre campanulate, 2″–3″ broad; heads small, numerous.
  Leaves alternate; rays white.   1. *P. virginica*.
  Leaves opposite; rays yellow.   2. *P. occidentalis*.
Involucre hemispheric, 7″–12″ broad; heads few, large.   3. *P. helianthoides*.

1. Phaethusa virgínica (L.) Britton. Small White or Virginia Crownbeard. Fig. 4488.

Verbesina virginica L. Sp. Pl. 901. 1753.

Perennial; stem densely puberulent, terete or winged, simple or branched, 3°–6° high. Leaves usually thin, alternate, ovate, roughish above, puberulent, canescent or glabrate beneath, acute or acuminate at the apex, 4′–10′ long, 1′–3′ wide, contracted at the base into winged petioles, the uppermost sessile, lanceolate, smaller, often entire; heads corymbose-paniculate at the ends of the stem and branches, numerous, 6″–10″ broad; involucre oblong-campanulate, 2″–3″ broad, its bracts narrowly lanceolate, erect, obtuse, pubescent; rays 3–5, obovate, white, pistillate; achenes minutely pubescent, winged or wingless; pappus of 2 slender awns, or sometimes none.

In dry soil, Pennsylvania to Virginia, Illinois, Missouri, Kansas, Florida and Texas. Aug.–Sept.

2. Phaethusa occidentàlis (L.) Britton. Small Yellow Crownbeard. Fig. 4489.

Siegesbeckia occidentalis L. Sp. Pl. 900. 1753.
Verbesina occidentalis Walt. Fl. Car. 213. 1788.
Phaethusa americana Gaertn. Fr. & Sem. **2** : 425. 1791.
V. Siegesbeckia Michx. Fl. Bor. Am. **2** : 134. 1803.

Perennial; stem glabrous, or puberulent above, usually much branched, narrowly 4-winged, 3°–7° high, the branches also winged and pubescent. Leaves thin, ovate, or the upper oblong, opposite, minutely rough-pubescent on both sides, or glabrate, acuminate at the apex, narrowed or contracted below into slender margined or naked petiole, serrate, 4'–10' long, 1'–3½' wide; heads numerous, 6''–12'' broad, corymbose at the ends of the stem and branches; involucre oblong-campanulate, 2''–3'' broad, its bracts lanceolate, obtuse, erect, or the tips slightly spreading, pubescent; rays 1–5, yellow, usually pistillate, rarely none; achenes wingless; pappus of 2 slender, at length divergent awns.

In dry thickets and on hillsides, Maryland and southern Pennsylvania to Illinois, Florida, Alabama and Texas. Aug.–Oct.

3. Phaethusa helianthoìdes (Michx.) Britton. Sunflower Crownbeard. Fig. 4490.

Verbesina helian:hoides Michx. Fl. Bor. Am. **2** : 135. 1803.
Actinomeris helianthoides Nutt. Gen. **2** : 181. 1818.

Perennial; stem hispid or hirsute, 4-winged, usually simple, 2°–4° high. Leaves ovate or oval, sessile, acute, acuminate or obtuse at the apex, narrowed at the base, serrate or serrulate, rough or appressed-hispid above, densely pubescent or canescent beneath, 2'–4' long, 1'–1½' wide, all alternate, or the lower opposite; heads solitary or few, 2'–3' broad; involucre hemispheric, about ½' high, its bracts lanceolate, acutish, canescent, appressed; rays 8–15, pistillate or neutral, linear-oblong, yellow; achenes scabrous or pubescent, broadly winged; pappus of 2 subulate awns.

On dry prairies and in thickets, Ohio to Georgia, west to Iowa, Missouri and Texas. June–July.

69. XIMENÈSIA Cav. Icones **2** : 60. 1793.

Pubescent caulescent herbs, with alternate or sometimes opposite, simple, toothed or somewhat laciniate leaves, and solitary or few, radiate, showy, peduncled heads. Involucre rather flat, of narrow spreading bracts. Receptacle convex, chaffy. Ray-flowers pistillate, fertile, numerous, the rays yellow. Disk-flowers numerous, perfect, fertile. Anthers somewhat sagittate at the base. Style-branches with slender pubescent appendages. Achenes flat, winged. Pappus of short and straight awns. [In honor of Joseph Ximenes, a Spanish apothecary and botanist.]

About 4 species, natives of America, the following typical.

1. **Ximenesia encelioìdes** Cav.   Golden Crownbeard.   Fig. 4491.

Ximenesia encelioides Cav. Icon. **2** : 60. *pl. 178*. 1793.
Verbesina encelioides A. Gray, Syn. Fl. **1** : Part 2, 288.
1884.

Annual; stem densely puberulent, much branched,
1°–2° high.   Leaves deltoid-ovate or deltoid-lanceo-
late, thin, 2′–4′ long, acuminate, acute or blunt at the
apex, coarsely dentate, or even laciniate, green and
minutely pubescent above, pale and densely canescent
beneath, all alternate, or the lowest opposite, narrowed at the base into naked or wing-margined petioles, which are often provided with dilated appendages at the base; heads several or numerous, 1′–2′
broad; involucre hemispheric, about ½′ high, its bracts
lanceolate, canescent; rays 12–15, bright golden yellow, 3-toothed; achenes of the disk-like flowers obovate, winged, pubescent, their pappus of 2 subulate
awns, those of the ray-flowers rugose, thickened,
often wingless.

In moist soil, Kansas to Texas, Arizona and Mexico,
and occasional in waste grounds farther east.   Also in
Florida and widely distributed in warm regions as a
weed.   Summer.

70. **COREÓPSIS** L. Sp. Pl. 907. 1753.

Anual or perennial, mostly erect herbs, with opposite leaves, or the upper alternate, and
large long-peduncled heads of both tubular and radiate flowers, the rays yellow, or brown
at the base, or brown throughout, or pink.   Involucre usually hemispheric, its bracts in 2
distinct series, all united at the base, those of the outer series commonly narrower and shorter
than the inner.   Receptacle flat or slightly convex, chaffy, the chaff flat or concave.   Ray-
flowers neutral.   Disk-flowers perfect, fertile, their corollas with slender tube and broader
5-toothed limb.   Anthers mostly entire at the base.   Style-tips truncate or subulate.   Achenes
flat, orbicular to oblong, winged or wingless.   Pappus of 2 short teeth, or a mere coroniform
border, or none.   [Greek, bug-like, referring to the achenes.]

About 55 species, natives of America, South Africa and Australia, known as Tickseed.   In
addition to the following, some 20 others occur in the southern and western United States.   Type
species: *Coreopsis lanceolata* L.

1. **Style-tips acute or cuspidate; outer involucral bracts about as long as the inner (except in**
     C. tripteris).
    Leaves simple, or pinnately lobed; achenes often with a callus on the incurved inner side.
        Achenes with thin flat broad wings.
            Leaves mostly near the base of the stem, the heads long-peduncled.
                Glabrous, or sparingly hairy below.   1. *C. lanceolata.*
                Pubescent or hirsute.   2. *C. crassifolia.*
            Stem leafy to near the shorter-peduncled heads.
                Pubescent or hirsute; leaves entire or with a few lateral lobes.   3. *C. pubescens.*
                Glabrous or very nearly so; leaves 1–2-pinnately parted into narrow segments.
                                                                                                 4. *C. grandiflora.*
        Achenes with thick, involute narrow wings.   5. *C. auricula!a.*

    Leaves palmately 3-cleft or divided; achenes without callus.
        Leaves petioled, 3-divided into lanceolate segments, or the upper entire.   6. *C. tripteris.*
        Leaves sessile.
            Leaves rigid, deeply 3-lobed above the base.   7. *C. palmata.*
            Leaves divided to the base.
                Leaf-segments entire; disk-flowers yellow.   8. *C. major.*
                Leaf-segments 1–2-parted.
                    Disk-flowers yellow; leaf-segments linear-filiform.   9. *C. verticillata.*
                    Disk-flowers purple-brown; leaf-segments 1″–3″ wide.   10. *C. delphinifolia.*
2. **Style-tips truncate or obtuse; outer involucral bracts much shorter than the inner.   [Genus**
     Calliopsis **Reichb.]**
    Achenes broadly winged; annual.   11. *C. cardaminefolia.*
    Achenes wingless.
        Rays yellow with brownish bases; annual with pinnately divided leaves.   12. *C. tinctoria.*
        Rays rose-pink, rarely white; perennial with linear entire leaves.   13. *C. rosea.*

1. Coreopsis lanceolàta L.　Lance-leaved Tickseed.　Fig. 4492.

Coreopsis lanceolata L. Sp. Pl. 908. 1753.
Coreopsis lanceolata var. *angustifolia* T. & G. Fl. N.
　A. 2 : 344. 1842.

Perennial; stem slender, glabrous, or sparingly hispid near the base, 1°–2° high. Leaves glabrous, the lower and basal ones slender-petioled, spatulate or narrowly oblong, very obtuse, 2′–6′ long, entire, or with 1–3 lateral obtuse entire lobes; stem leaves few, sessile or nearly so, lanceolate or oblong, obtuse or acutish, usually quite entire; heads few or solitary on elongated slender peduncles, 1½′–2½′ broad, showy; peduncles often 12′ long; involucre depressed-hemispheric, its bracts glabrous or ciliate, lanceolate or ovate-lanceolate, the outer narrower than the inner, but nearly as long; rays 6–10, bright yellow, cuneate, 3–7-lobed; achenes oblong, broadly winged, about 1″ long; pappus of 2 short teeth.

In dry or moist soil, Ontario to Virginia, Michigan, Illinois, Florida, Louisiana and Missouri. Escaped from cultivation eastward. May–Aug.

2. Coreopsis crassifòlia Ait.　Thick-leaved or Hairy Tickseed.　Fig. 4493.

Coreopsis crassifolia Ait. Hort. Kew. 3 : 253. 1789.
Coreopsis lanceolata var. *villosa* Michx. Fl. Bor. Am.
　2 : 137. 1803.

Perennial; stem hirsute or villous-pubescent below, ascending, little branched, 8′–20′ high. Leaves rather thick, hirsute or pubescent, all entire, the lower and basal ones petioled, oblong to obovate-spatulate, mostly very obtuse, 1′–3′ long, 4″–10″ wide; stem leaves few, short-petioled or sessile, obtuse or acutish, narrower; heads few, 1′–2′ broad, borne on slender, puberulent peduncles often 1° long, similar to those of *C. lanceolata*, as are the rays, involucral bracts and achenes.

Dry soil, Illinois and Missouri to Kansas, Louisiana, east to South Carolina and Florida. May–Aug.

3. Coreopsis pubéscens Ell.　Star Tickseed.　Fig. 4494.

C. pubescens Ell. Bot. S. C. & Ga. 2 : 441. 1824.

Perennial; stem pubescent or puberulent, leafy, little branched or simple, erect, 2°–4° high. Leaves firm in texture, pubescent or glabrous, entire, or some of them 3–5-lobed or divided, the basal and lower ones slender-petioled, obovate-oval, obtuse, the upper short-petioled or sessile, broadly lanceolate or oblong, acute or acutish, 2′–3′ long; heads few, 1′–1½′ broad, long-peduncled; involucre depressed-hemispheric, glabrous, star-like, its lanceolate obtuse or acute outer bracts nearly as long as but much narrower than the ovate inner ones; rays 8–10, yellow, cuneate, lobed at the apex; achenes similar to those of the preceding species or broader.

In dry woods, Virginia to Illinois and Missouri, south to Florida and Louisiana. June–Aug.

4. Coreopsis grandiflòra Hogg. Large-flowered Tickseed. Fig. 4495.

Coreopsis grandiflora Hogg; Sweet, Brit. Fl. Gard.
2: *pl. 175.* 1825–27.

Perennial; stem glabrous, usually branched
above, 1°–3° high. Leaves, or most of them,
1–2-pinnately parted, or the lower entire and
slender-petioled; segments of the lower stem
leaves oblong, obtuse, the terminal one larger
than the lateral; segments of most of the upper
leaves linear or even filiform, petioles ciliate;
heads commonly several, 1′–2′ broad, long-peduncled; outer bracts of the involucre lanceolate,
narrower and mostly somewhat shorter than the
oval or ovate-lanceolate inner ones; rays 6–10,
yellow, lobed; achenes oblong, broadly winged
when mature, the projections on the inner face
usually large; pappus of 2 short scales.

In moist soil, Kansas and Missouri to Texas, east
to Georgia. May–Aug.

5. Coreopsis auriculàta L. Running or Lobed Tickseed. Fig. 4496.

Coreopsis auriculata L. Sp. Pl. 908. 1753.

Perennial; stoloniferous; stems weak, very slender, decumbent or ascending, 6′–15′ long, little
branched, or simple, pubescent or hirsute, at least
below, mostly glabrous above. Leaves thin, entire or pinnately 3–5-lobed or 3–5-divided, the
terminal segment entire and much larger than the
lateral ones, the lower and basal more or less pubescent, broadly oblong or nearly orbicular, with
slender pubescent petioles; stem leaves few, petioled or sessile, mostly obtuse; heads 1′–1½′ broad,
slender-peduncled; outer bracts of the involucre
oblong to lanceolate, narrower than the mostly
ovate and acute inner ones; rays 6–10, cuneate,
about 4-toothed, yellow; achenes oval, narrowly
winged, the wings involute and thick.

In woods, Virginia to Illinois, Kentucky, Florida
and Louisiana. May–Aug.

6. Coreopsis trípteris L. Tall Tickseed. Fig. 4497.

Coreopsis tripteris L. Sp. Pl. 908. 1753.

Perennial; stem glabrous, much branched above,
4°–8° high. Leaves petioled, glabrous, or very
nearly so, firm, the lower all divided into lanceolate, entire, acute, rough-margined segments, which
are 2′–5′ long, ¼′–1′ wide and pinnately veined;
uppermost leaves lanceolate, entire; heads numerous, slender-peduncled, 1′–1½′ broad; outer bracts
of the involucre linear, obtusish, much narrower
and somewhat shorter than the ovate-oval or ovate-
lanceolate, acute inner ones; rays 6–10, yellow, obtuse, entire; achenes oblong to obovate, narrowly
winged, emarginate; pappus none.

In moist woods and thickets, southern Ontario to
Pennsylvania, Wisconsin, Arkansas, Virginia, Florida
and Louisiana. July–Oct.

$\frac{1}{2}$

Coreopsis major Walt. Fl. Car. 214. 1788.
Coreopsis senifolia Michx. Fl. Bor. Am. 2 : 138. 1803.
Coreopsis Oemleri Ell. Bot. S. C. & Ga. 2 : 435. 1824.
Coreopsis stella!a Nutt. Journ. Acad. Phil. 7 : 76. 1834.
Coreopsis senifolia var. stellata T. & G. Fl. N. A. 2 : 342. 1842.

Perennial; stem pubescent or glabrous, branched above, 2°–3° high. Leaves sessile, more or less pubescent, or glabrous, divided to the base into 3, lanceolate, linear, ovate-lanceolate or oblong, acute, entire segments 2′–4′ long, 2″–12″ wide, which appear as if in yerticils of 6; upper and lower leaves (rarely all of them), undivided and entire; heads several or numerous, slender-peduncled, 1′–2′ broad; bracts of the hemispheric involucre all united at the base, the outer ones linear-oblong, obtuse, equalling or shorter than the broader inner ones, all pubescent; rays 6–10, yellow, oblong, entire; disk yellow; achenes oblong to elliptic, winged, ½″–2″ long; pappus of 2 short deciduous teeth.

In dry sandy woods, Virginia to Kentucky, Florida and Alabama. July–Aug. Consists of several races, differing in pubescence and in shape and width of the leaf-segments.

7. Coreopsis palmàta Nutt. Stiff Tickseed. Fig. 4498.

Coreopsis palmata Nutt. Gen. 2 : 180. 1818.

Perennial; stems rigid, glabrous, simple, or little branched, very leafy, 1°–3° high. Leaves sessile, 2′–3′ long, palmately deeply 3-lobed at or below the middle, or the uppermost entire, thick, rigid, the lobes linear-oblong, obtusish, entire, or with 1–3 lateral lobes, their margins rough; heads few or solitary, short-peduncled, 1′–2′ broad; involucre hemispheric, its bracts somewhat united at the base, those of the outer series narrower and nearly as long as the inner ones; rays 6–10, bright yellow, oblong or obovate, mostly 3-toothed; achenes oblong, narrowly winged, slightly incurved; pappus of 2 short tips, or none.

On dry prairies and in thickets, Indiana to Missouri, Louisiana, Minnesota, Manitoba, Nebraska and Texas. June–July.

8. Coreopsis màjor Walt. Wood or Greater Tickseed. Fig. 4499.

$\frac{1}{2}$

9. Coreopsis verticillàta L. Whorled Tickseed. Fig. 4500.

Coreopsis verticillata L. Sp. Pl. 907. 1753.

Perennial; stem stiff, much branched, slender, leafy, 1°–2° high. Leaves sessile, glabrous, 2–3-ternately dissected into linear-filiform entire segments; heads numerous, 1′–1½′ broad; involucre hemispheric, or short-cylindric in fruit, glabrous, its outer bracts linear, obtuse, commonly somewhat shorter and much narrower than the ovate-oblong inner ones; rays 6–10, yellow, spatulate-oblong, obtuse; disk dull yellow; achenes oblong, narrowly winged, 2″ long; pappus of 2 short teeth.

In dry soil, Maryland to South Carolina, Kentucky, Nebraska and Arkansas. Apparently erroneously recorded from farther north. June–Sept.

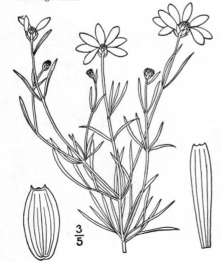

$\frac{3}{5}$

10. Coreopsis delphinifòlia Lam. Lark-spur Tickseed. Fig. 4501.

Coreopsis delphinifolia Lam. Encycl. **2** : 108. 1786.

Perennial; stem glabrous, branched above, rather slender, $1°-3°$ high. Leaves sessile, 1–2-ternately partly into linear or linear-lanceolate segments, which are $1'-2'$ long, $1''-3''$ wide; heads several or numerous, $1\frac{1}{2}'-2'$ broad; involucre hemispheric, its bracts glabrous, the outer linear-oblong, obtuse, shorter than or equalling the ovate-oblong inner ones; rays 6–10, yellow, entire; disk brown; achenes oblong to oval, narrowed at the base, narrowly winged; pappus of 2 short teeth.

In dry woods, Virginia (according to Torrey and Gray), North Carolina to Georgia and Alabama. Aug.–Sept.

11. Coreopsis cardaminefòlia (DC.) T. & G. Cress-leaved Tickseed. Fig. 4502.

Calliopsis cardaminefolia DC. Prodr. **5** : 568. 1836.
C. cardaminefolia T. & G. Fl. N. A. **2** : 346. 1842.

Annual; stem glabrous, branched, $1°-2\frac{1}{2}°$ high. Basal leaves petioled, $2'-4'$ long, 1–2-pinnately parted into oblong or oval obtuse segments, the petioles sometimes slightly ciliate; stem leaves distant, sessile, or nearly so, pinnately parted into linear segments, or the uppermost entire; heads $8''-12''$ broad, slender-peduncled; involucre hemispheric, its inner bracts brown, ovate or ovate-lanceolate, obtuse or obtusish, scarious-margined, much longer than the lanceolate obtusish outer ones; rays 4–8, yellow with a brown base, 3-toothed; achenes oval, $1''-1\frac{1}{2}''$ long, winged, smooth, or slightly papillose; pappus of 2 minute awns, or none.

In moist soil, Kansas to New Mexico, Louisiana and northern Mexico. May–Oct.

Coreopsis Atkinsoniàna Dougl., a northwestern species, with linear leaf-segments and very narrowly winged achenes, ranges eastward into North Dakota.

12. Coreopsis tinctòria Nutt. Golden Co-reopsis. Garden Tickseed. Fig. 4503.

C. tinctoria Nutt. Journ. Acad. Phila. **2** : 114. 1821.

Annual; stem glabrous, branched, $1°-3\frac{1}{2}°$ high. Leaves 1–2-pinnately divided into linear, obtusish, mostly entire segments, or the uppermost linear and entire, the lower petioled; heads slender-peduncled, $10''-12''$ broad, or in cultivation much broader; involucre hemispheric, its inner bracts brown, ovate or oblong, obtuse or acute, scarious-margined, 3–5 times as long as the obtuse outer ones; rays 6–10, cuneate, yellow with a brown base or brown all over; achenes linear or linear-oblong, about $1''$ long, wingless; pappus a mere border, or none.

In moist soil, Minnesota to Alberta, Nebraska, Louisiana and Arizona. Escaped from gardens to roadsides and waste places eastward. Wild flax. Nuttall's-weed. May–Sept.

13. Coreopsis ròsea Nutt.   Small Rose or Pink Tickseed.   Fig. 4504.

Coreopsis rosea Nutt. Gen. **2**: 179. 1818.

Perennial by slender rootstocks; stems at length much branched, slender, strict, glabrous, 6′–24′ high. Leaves opposite, linear, glabrous, entire, obscurely 1-nerved, 1′–2½′ long, ½″–1″ wide, sessile, or the lower petioled; heads slender-peduncled, several or numer-. ous, 6″–12″ broad; disk yellow; rays 4–8, pink or rose-colored (occasionally white), oblong to obovate, slight-ly 3-toothed or sometimes entire; inner bracts of the hemispheric involucre ovate-oblong, acutish or obtuse, glabrous, much longer than the lanceolate outer ones; achenes oblong or linear-oblong, thin, not winged, nearly straight, slightly ribbed on the inner face; pappus a . very short truncate crown.

In open swamps, eastern Massachusetts to Georgia, near the coast. July–Aug.

71. BÌDENS L. Sp. Pl. 831. 1753.

Annual or perennial herbs, with opposite serrate lobed divided or dissected leaves, or the uppermost alternate, and mostly large heads of both tubular and radiate flowers, or the rays none, or rudimentary. Involucre campanulate or hemispheric, its bracts in 2 series, distinct, or slightly united at the base; the outer often foliaceous and much larger than the inner. Receptacle flat or nearly so, chaffy, the chaff subtending the disk-flowers. Rays, when present, neutral, mostly entire, yellow in our species. Disk-flowers perfect, fertile, their corollas tubular, 5-toothed. Anthers entire, or minutely sagittate at the base. Style-branches with short or subulate tips. Achenes flat, or quadrangular, cuneate, oblong or linear, the outer ones often shorter than the inner. Pappus of 2–6 teeth or subulate awns, upwardly or downwardly barbed or hispid. [Latin, two-toothed, referring to the achenes.]

About 75 species of wide geographic distribution. Besides the following, about 12 others occur in the southern and southwestern United States. Type species: *Bidens tripartita* L.

Leaves lanceolate, serrate, undivided, rarely 3–5-lobed or incised.
  Rays present, large and conspicuous.
    Rays large, longer than the involucral bracts.   1. *B. laevis.*
    Rays short, rarely as long as the involucral bracts.   2. *B. cernua.*
  Rays rudimentary, or none.
    Heads nodding after flowering.   2. *B. cernua.*
    Heads persistently erect.
      Pappus awns downwardly barbed.
        Involucral bracts not foliaceous; stem purple; flowers orange.   3. *B. connata.*
        Involucral bracts foliaceous; stem straw-color; flowers greenish yellow.
                                          4. *B. comosa.*
      Pappus awns upwardly barbed; involucre narrow.   5. *B. bidentoides.*
Leaves, some or all of them, pinnately 1–3-parted or dissected.
  Rays rudimentary, or none, or very short.
    Achenes flat; leaves, some or all of them, 1–3-divided.
      Outer involucral bracts 4–8; achenes black or nearly black.
        Leaves membranous; heads 2″–3″ high; awns short.   6. *B. discoidea.*
        Leaves not membranous; heads 5″–7″ high; awns long.   7. *B. frondosa.*
      Outer involucral bracts 10–16; achenes brown.   8. *B. vulgata.*
    Achenes linear; leaves dissected.   9. *B. bipinnata.*
  Rays large and conspicuous.
    Achenes sparingly pubescent, not ciliate; pappus of 2 short teeth.   10. *B. coronata.*
    Achenes ciliate; pappus 2–4 subulate teeth or awns.
      Achenes cuneate, or linear-cuneate.   11. *B. trichosperma.*
      Achenes obovate, very flat.
        Bracts of the involucre glabrous, or ciliate, short.   12. *B. aristosa.*
        Outer bracts densely hispid, much longer than the inner.   13. *B. involucrata.*

1. Bidens laèvis (L.) B.S.P.   Larger or Smooth Bur-Marigold.   Brook Sunflower.   Fig. 4505.

Helianthus laevis L. Sp. Pl. 906. 1753.
Bidens chrysanthemoides Michx. Fl. Bor. Am. **2**: 136. 1803.
Bidens laevis B.S.P. Prel. Cat. N. Y. 29. 1888.
Bidens lugens Greene, Pittonia **4**: 254. 1901.

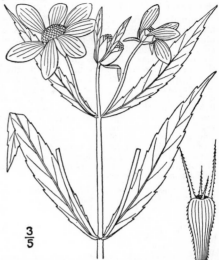

Annual; glabrous throughout; stems branched, erect or ascending, 1°–3° high. Leaves sessile, lanceolate, evenly serrate or serrulate, acuminate at the apex, narrowed to the sometimes connate-perfoliate base, 3′–8′ long, ¼′–1′ wide; heads numerous, short-peduncled, erect in flower, often declined in fruit, 1′–2½′ broad; rays very showy, golden yellow; involucre hemispheric, its outer bracts linear-oblong or spatulate, equalling or exceeding the broader, ovate or oblong, membranous inner ones, shorter than the rays; rays 8–10, obovate-oblong, obtuse; achenes cuneate, truncate, 2″–4″ long, retrorsely hispid on the margins; pappus of 2–4 (usually 2), rigid downwardly barbed awns, shorter than the achene.

$\frac{3}{5}$

In swamps and wet meadows, Massachusetts to Illinois, Kansas, Florida, Louisiana and Mexico. Northern records of this species apply to *Bidens cernua.* Aug.–Nov.

Bidens élegans Greene, of southeastern Virginia, appears to be a narrow-rayed race of this species.

2. Bidens cérnua L.   Smaller or Nodding Bur-Marigold.   Fig. 4506.

Bidens cernua L. Sp. Pl. 832. 1753.
Coreopsis Bidens L. loc. cit. 908. 1753.

Annual; stems glabrous, or hispid, usually erect, branched, 3′–3° high. Leaves sessile and commonly somewhat connate-perfoliate at the base, lanceolate to oblong-lanceolate, usually coarsely and sharply serrate, glabrous, acuminate, 3′–6′ long, ¼′–1′ wide; heads numerous, globose, short-peduncled, ½′–1′ broad, nodding after or during flowering; rays 6–10, short (3″–6″), or none; involucre depressed-hemispheric, its outer bracts commonly ciliate, often large, foliaceous and much exceeding the broad, yellowish-margined membranous inner ones; achenes cuneate, 2″ long, retrorsely hispid on the margins; pappus of 2–4 (usually 4), downwardly barbed awns, about half as long as the achene.

$\frac{1}{2}$

In wet soil, Nova Scotia to Hudson Bay and British Columbia, North Carolina, Missouri and California. Also in Europe and Asia. Consists of many races, differing in size. Water-agrimony. Double-tooth. Pitchforks. July–Oct.

Bidens Eatoni Fernald, known only from brackish soil, along the Merrimac River, Mass., has narrower heads and smaller achenes with awns either upwardly or downwardly barbed.

A plant, with all the leaves pinnately divided, growing in the vicinity of Minneapolis, Minn., described as *Bidens connata pinnata* S. Wats., may be a hybrid with *B. aristosa.*

3. Bidens connàta Muhl. Purple-stemmed Swamp Beggar-ticks. Fig. 4507.

B. connata Muhl.; Willd. Sp. Pl. **3**: 1718. 1804.

Annual; glabrous throughout; stem erect, usually branched, 6′–8° high, purple. Leaves petioled, lanceolate or oblong-lanceolate, sharply and coarsely serrate, thin, 2′–5′ long, ¼′–1′ wide, apex acuminate, base tapering, the uppermost sometimes sessile, nearly entire and acutish, the lower sometimes with a pair of basal lobes, decurrent on the petiole; heads several or numerous, peduncled, ½′–1½′ broad; involucre campanulate or hemispheric, the outer bracts somewhat exceeding the ovate-oblong, inner ones; rays none, or 1–5 and inconspicuous; disk-flowers orange; achenes cuneate or obovate, hairy and tubercled, or nearly glabrous, keeled, or angled, 2″–3″ long, the margins with either erect or retrorse hairs, or both, the 2–4 pappus awns downwardly barbed, half as long as the achene.

In swamps or moist soil, Rhode Island to Ontario, Minnesota, Delaware, Georgia, Kentucky, Missouri and Nebraska. Introduced into Europe. Cuckold. Harvest-lice. Pitchforks. Aug.–Oct.

½

4. Bidens comòsa (A. Gray) Wiegand. Leafy-bracted Tickseed. Fig. 4508

B. connata var. *comosa* A. Gray, Man. Ed. 5, 261. 1867.
B. comosa Wiegand, Bull. Torr. Club **24**: 436. 1897.
B. riparia Greene, Pittonia **4**: 261. 1901.
B. acuta (Wiegand) Britton, Man. 1001. 1901.

Annual, glabrous; stem erect, branched, 6′–4½° high, straw-colored. Leaves short-petioled, or subsessile, lanceolate to elliptic-lanceolate, coarsely serrate with mostly smaller teeth than in *B. connata,* tapering to each end, the petioles broadly margined; heads several or numerous, ½″–1½″ broad; outer bracts of the involucre linear, spatulate or lanceolate, foliaceous, erect or spreading, often toothed, 2–4 times as long as the head; rays none; corollas mostly 4-lobed, pale greenish yellow; stamens and style included; achenes larger, 3½″–5½″ long, evenly cuneate, very flat; pappus awns commonly 3, downwardly barbed, somewhat shorter than the achene.

In wet soil, Massachusetts to Illinois, North Dakota, New Jersey, West Virginia, Kentucky and Kansas. Leaves thicker and paler than in *B. connata.* Aug.–Oct.

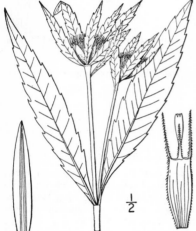

½

5. Bidens bidentoìdes (Nutt.) Britton. Swamp Beggar-ticks. Fig. 4509.

Diodonta bidentoides Nutt. Trans. Am. Phil. Soc. (II) **7**: 361. 1841.
Coreopsis bidentoides T. & G. Fl. N. A. **2**: 339. 1842.
B. bidentoides Britton, Bull. Torr. Club **20**: 281. 1893.

Closely resembles *B. connata,* glabrous throughout; stem branched, 1°–4° high. Leaves similar, lanceolate, sharply serrate, petioled, or the upper sessile and entire, acuminate at the apex, narrowed at the base; involucre narrowly or becoming somewhat broadly campanulate, its outer bracts linear, foliaceous, not ciliate, usually much exceeding the oblong inner ones; rays none, or rarely present and very short; achenes linear-cuneate, 3″–5″ long, their sides and the 2 slender pappus awns (rarely with 2 short intermediate awns) upwardly barbed or hispid.

Muddy shores of the Delaware River and Bay in Pennsylvania, New Jersey, Delaware and Maryland. Aug.–Oct.

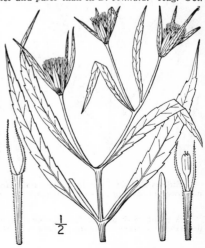

½

6. Bidens discoìdea (T. & G.) Britton. Small Beggar-ticks. Fig. 4510.

Coreopsis discoidea T. & G. Fl. N. A. **2**: 339. 1842.
B. discoidea Britton, Bull. Torr. Club **20**: 281. 1893.

Annual, glabrous, slender, branching, erect, 2′–6° high. Leaves membranous, very slender-petioled, all the lower ones divided into 3 lanceolate or oblong-lanceolate, dentate, acuminate segments which are 1′–3′ long; uppermost leaves commonly rhombic-lanceolate and undivided; heads usually numerous, slender-peduncled, 2″–4″ broad and about as high; involucre broadly campanulate or hemispheric, its outer bracts mostly 4, usually foliaceous and obtuse, usually much surpassing the inner ones; rays apparently always wanting; achenes flat, narrowly cuneate, upwardly strigose, about 2″ long; pappus of 2 short, upwardly hispid, rarely downwardly barbed awns.

In swamps and wet places, Massachusetts to Virginia, Ohio, Michigan, Louisiana and Texas. July–Sept.

7. Bidens frondòsa L. Beggar-ticks. Stick-tight. Fig. 4511.

Bidens frondosa L. Sp. Pl. 832. 1753.
B. melanocarpa Wiegand, Bull. Torr. Club **26**: 405. 1899.

Annual; stem erect, branched, glabrous, or nearly so, often purplish, 2°–3° high. Leaves thin, but not membranous, slender-petioled, pinnately 3–5-divided or the uppermost undivided, the segments lanceolate or oblong-lanceolate, sharply serrate, acuminate at the apex, narrowed at the base, usually slightly pubescent beneath, stalked, 2′–4′ long, ½′–1′ wide; heads usually numerous, long-peduncled, about 6″ high, 5″–10″ broad; involucre campanulate, becoming hemispheric, its outer bracts 4–8, more or less foliaceous, often much exceeding the ovate-lanceolate, scarious-margined inner ones; rays none or rudimentary and inconspicuous; disk-corollas orange; achenes flat, narrowly cuneate, nearly black, 3″–5″ long, ciliate, the two slender awns downwardly barbed, or sometimes upwardly hispid.

In moist soil, often a weed in fields, Nova Scotia to Florida, British Columbia, Texas, Colorado and California. Introduced as a weed into southern Europe. Rayless marigold. Beggar-lice. Devil's-pitchfork. Stick-seed. Common bur-marigold. Old-ladies clothes-pins. Cuckles. July–Oct.

8. Bidens vulgàta Greene. Tall Beggar-ticks. Fig. 4512.

Bidens vulgata Greene, Pittonia **4**: 72. 1899.
B. frondosa puberula Wiegand, Bull. Torr. Club **26**: 408. 1899.

Taller, sometimes 9° high, glabrous or nearly so, or crisp-pubescent above. Leaves pinnately 3–5-divided, the veins straight and prominent; heads larger, 7″–12″ broad, stout-peduncled; outer involucral bracts linear to linear-spatulate, ciliate, the inner mostly ovate or narrowly triangular, pubescent at apex; ray-flowers usually present, small, yellow; achenes very flat, 3″–4½″ long, 2″–2½″ wide, brown or greenish brown, the margins downwardly barbed above, upwardly hairy below; awns 2, half as long as the achene or more, downwardly barbed.

In moist soil, Quebec to British Columbia, New York, North Carolina, Missouri, Colorado and California. Aug.–Sept. Included in the preceding species in our first edition.

9. Bidens bipinnàta L. Spanish Needles. Cuckolds. Fig. 4513.

Bidens bipinnata L. Sp. Pl. 832. 1753.

Annual; stem quadrangular, erect, freely branched, rather slender, 1°–5° high. Leaves thin, acuminate, petioled, 1–3-pinnately dissected into ovate or oblong, toothed or lobed segments, the lower often 8′ long; heads usually numerous, long-peduncled, 2″–4″ broad; involucre narrow, its outer bracts linear, ascending, nearly as long as the broader erect inner ones; rays 3–4, yellow, short, sometimes none; achenes linear, 4-angled, slightly pubescent, narrowed upward into a beak, 5″–9″ long, the outer ones commonly shorter and thicker than the inner; pappus of 2–4 (usually 4), downwardly barbed, slightly unequal, spreading awns, much shorter than the achene.

In various situations, often a weed in cultivated fields, Rhode Island to Florida, Ohio, Nebraska, Kansas and Arizona. Introduced as a weed into southern Europe and Asia. July–Oct.

10. Bidens coronàta (L.) Fisch. Southern Tickseed-Sunflower. Fig. 4514.

Coreopsis coronata L. Sp. Pl. Ed. 2, 1281. 1763.
Coreopsis aurea Ait. Hort. Kew. 3: 252. 1789.
Bidens coronata Fisch.; Steudel, Nom. Ed. 2, 202. 1840.

Annual, glabrous or nearly so throughout; stem branched, 1°–3° high. Lower leaves petioled, 3′–5′ long, 3-divided, the terminal segment lanceolate, acute or acuminate, serrate, much larger than the serrate or entire lateral ones; upper leaves much smaller, 3-parted, 3-lobed or undivided, sessile or short-petioled, entire or serrate; leaves rarely all undivided; heads numerous, slender-peduncled, 1′–2′ broad; involucre hemispheric, its outer bracts linear-oblong, obtuse, equalling or slightly exceeding the broader inner ones; **rays 6–10, obtuse; achenes** broadly cuneate, slightly pubescent, 1″–2″ long; pappus of 2 chaffy blunt divergent somewhat laciniate teeth, rarely with 2 shorter intermediate ones.

In wet places, Virginia to Florida and Alabama. July–Sept.

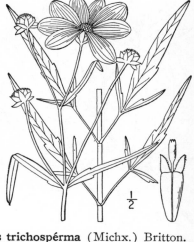

11. Bidens trichospérma (Michx.) Britton. Tall Tickseed-Sunflower. Fig. 4545.

C. trichosperma Michx. Fl. Bor. Am. 2: 139. 1803.
B. trichosperma Britton, Bull. Torr. Club 20: 281. 1893.
Coreopsis trichosperma var. *tenuiloba* A. Gray, Syn. Fl. 1: Part 2, 295. 1884.
Bidens trichosperma tenuiloba Britton, Bull. Torr. Club 20: 281. 1893.

Annual or biennial, glabrous; stem tall, obscurely quadrangular, much branched, 2°–5° high. Lower leaves petioled, 4′–8′ long, pinnately divided into 5–7 lanceolate or linear, acuminate, sharply serrate, incised or nearly entire, sessile or short-stalked segments; upper leaves sessile or nearly so, 3-lobed, 3-divided, or entire and linear-lanceolate; heads numerous, 1½′–2½′ broad, long-peduncled; involucre hemispheric, its outer bracts linear or spatulate, sometimes ciliate, about the length of the broader inner ones; rays 6–19, obtuse, golden yellow, 7″–12″ long; achenes oblong-cuneate or the inner ones narrower, hispid-pubescent and ciliate, 2″–4″ long; pappus of 2 short erect or divergent, upwardly hispid teeth or short awns.

In swamps and wet meadows, Massachusetts to Georgia; Illinois to Michigan and Kentucky. Recorded from Ontario and Minnesota. Aug.–Oct.

12. Bidens aristòsa (Michx.) Britton. Western Tickseed-Sunflower.
Fig. 4516.

Coreopsis aristosa Michx. Fl. Bor. Am. **2**: 140. 1803.
C. aristata Muhl.; Willd. Sp. Pl. **3**: 2253. 1804.
B. aristosa Britton, Bull. Torr. Club **20**: 281. 1893.

Annual or biennial; stem much branched, 1°–3°
high. Leaves thin, slender-petioled, pubescent
beneath, the lower ones pinnately 5–7-divided,
3′–6′ long, the segments lanceolate, serrate, incised or pinnatifid, acuminate, narrowed at the
base; upper leaves less divided, lanceolate, or
merely lobed, sessile or short-petioled; heads numerous, slender-peduncled, 1′–2′ broad; outer
bracts of the hemispheric involucre 8–10, linear
or spatulate, usually ciliate, not surpassing the
inner; rays 6–9, obtuse; achenes very flat, oblanceolate or obovate, upwardly ciliate and strigose-
pubescent; pappus of 2, rarely 4, slender upwardly or downwardly barbed awns, sometimes nearly
as long as the achene, rarely wanting.

In swamps and wet prairies, Ohio to Minnesota,
south to Louisiana and Missouri; southeastern Pennsylvania and Delaware. Aug.–Oct.

13. Bidens involucràta (Nutt.) Britton.
Long-bracted Tickseed-Sunflower.
Fig. 4517.

Coreopsis involucrata Nutt. Journ. Phil. Acad. **7**: 74.
1834.

Bidens involucra a Britton, Bull. Torr. Club **20**: 281.
1893.

Similar to the two preceding species, minutely
pubescent, 1°–3° high, much branched. Segments
of the leaves narrower, linear-lanceolate, incised
or pinnatifid, long-acuminate; heads numerous,
1′–2′ broad, on slender usually hispid peduncles;
outer bracts of the hemispheric involucre 10–20,
linear-lanceolate, acuminate, densely hispid and
ciliate, much exceeding the inner ones; rays
orange at the base; achenes flat, ciliate and strigose; pappus of 2 short teeth.

In swamps, Illinois to Kansas, Arkansas and Texas;
southeastern Pennsylvania and Delaware. July–Sept.

72. MEGALODÓNTA Greene, Pittonia **4**: 270. 1901.

Perennial aquatic herbs, with opposite or whorled leaves, those of the submersed ones
filiformly dissected, those of the emersed ones serrate to laciniate, and solitary peduncled
showy heads with both tubular and radiate flowers. Involucre hemispheric, its bracts in 2
series, rather broad, distinct or nearly so, the outer ones smaller than the inner, often lax.
Receptacle chaffy. Rays neutral, yellow. Disk-flowers perfect, fertile, their corollas tubular,
5-toothed. Anthers minutely sagittate at the base. Style-branches with subulate tips.
Achenes nearly terete, truncate at both ends. Pappus of 3–6 long-subulate awns which are
retrorsely barbed, except near the base. [Greek, large-toothed, referring to the pappus awns.]

Two known species, the following typical, the other occurring in Washington State.

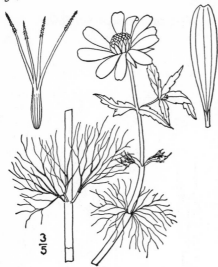

3/5

1. Megalodonta Béckii (Torr.) Greene. Water Marigold. Fig. 4518.

Bidens Beckii Torr.; Spreng. Neue Entdeck. **2**: 135. 1821.

M. Beckii Greene, Pittonia **4**: 271. 1901.

M. nudata Greene, loc. cit. 1901.

Stems simple, or little branched, 2°–8° long. Submersed leaves sessile, 1′–2′ long, repeatedly divided into numerous capillary segments; emersed leaves few, sessile, opposite, or sometimes in 3's, lanceolate or oblong, acute, serrate or laciniate, ½′–1½′ long; heads solitary or few, short-peduncled, 1′–1½′ broad; involucre hemispheric, its bracts oval or oblong, obtusish, glabrous, the outer somewhat shorter than the inner; rays 6–10, obovate, or oblong, notched, golden yellow; achenes nearly terete, 5″–7″ long; pappus of 3–6 slender awns, downwardly barbed above, smooth below, divergent, 6″–12″ long.

In ponds and streams, Quebec to New Jersey, west to Manitoba and Missouri. Aug.–Sept.

73. THELESPÉRMA Less. Linnaea 6: 511. 1831.

Glabrous annual or perennial herbs, with opposite linear and undivided, or finely dissected leaves, and long-peduncled heads of both tubular and radiate flowers, or the rays wanting. Involucre hemispheric or campanulate, of 2 distinct series of bracts, the outer short, narrow and somewhat spreading, the inner united nearly to or beyond the middle into a cup, their tips scarious-margined. Receptacle flat, chaffy, the 2-nerved broad white scarious chaff subtending the disk-flowers and achenes. Ray-flowers, when present, neutral, the rays yellow, entire or toothed. Disk-flowers perfect, fertile, their corolla with a slender tube and 5-toothed limb. Anthers obtuse and entire at the base. Style-tips acute. Achenes, some or all of them, papillose on the back, oblong or linear, slightly compressed or terete, wingless. Pappus of 2 retrorsely hispid awns or scales, or sometimes none. [Greek, nipple-seed.]

About 10 species, natives of the south central United States, Mexico and southern South America. Type species: *Thelesperma scabiousoides* Less.

Rays large; pappus-awns shorter than the width of the achene.
    Leaves not rigid, their segments filiform-linear; annual or biennial.   1. *T. trifidum.*
    Leaves rigid, their segments linear; perennial.   2. *T. intermedium.*
Rays inconspicuous, or none; awns longer than the width of the achene; perennial.
                                                      3. *T. gracile.*

1. Thelesperma trífidum (Poir.) Britton. Fine-leaved Thelesperma. Fig. 4519.

Coreopsis trifida Poir. in Lam. Encycl. Suppl. **2**: 353. 1811.

Thelesperma filifolium A. Gray, Kew. Journ. Bot. **1**: 252. 1849.

Thelesperma trifidum Britton, Trans. N. Y. Acad. Sci. **9**: 182. 1890.

Annual or biennial; stem branched, 1°–3° high. Leaves numerous, not rigid, 1½′–2′ long, bipinnately divided into filiform or linear-filiform segments; heads several or numerous, 12″–15″ broad; outer bracts of the involucre about 8, subulate-linear, equalling or more than half as long as the inner, which are united not higher than the middle; rays 6–10, somewhat spatulate, 3-lobed; disk purple or brown; achenes linear-oblong, straight, or slightly curved, the outer ones strongly papillose; awns of the pappus not longer than the width of the summit of the achene.

In dry soil, South Dakota, Missouri and Nebraska to Colorado, Texas, New Mexico and northern Mexico. June–Aug.

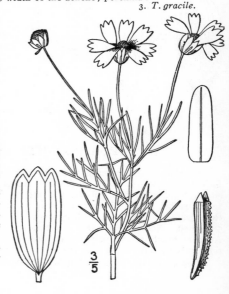

3/5

2. Thelesperma intermèdium Rydb.
Stiff Thelesperma. Fig. 4520.

Thelesperma intermedium Rydb. Bull. Torr. Club
27: 631. 1900.

Perennial from a deep woody root and slender rootstocks; stem rigid, usually much
branched, 1°–1½° high. Leaves usually numerous, 1½′–2′ long, bipinnately divided into entire,
rigid, linear segments, but less compound than
those of the preceding species; outer bracts of
the involucre lanceolate-subulate, usually much
shorter than the inner ones, which are united
to about the middle; rays and achenes similar
to those of the preceding.

In dry soil, on plains, Nebraska and Wyoming
to Colorado and New Mexico. In our first edition
included in *T. ambiguum* A. Gray, of the Southwest. June–Aug.

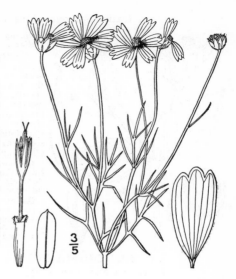

3. Thelesperma grácile (*Torr.*) A. Gray.
Rayless Thelesperma. Fig. 4521.

Bidens gracilis Torr. Ann. Lyc. N. Y. **2**: 215. 1827.
T. gracile A. Gray, Kew. Journ. Bot. **1**: 252. 1849.

Perennial from a deep root; stem rigid, branched,
1°–3° high, the branches nearly erect. Leaves rigid,
erect or ascending, 2′–3′ long, pinnately or bipinnately divided into linear segments, or the upper
linear and entire; heads 6″–10″ broad; rays usually
none, sometimes present and 2″–3″ long; outer bracts
of the involucre 4–6, oblong or ovate, mostly obtuse, very much shorter than the inner ones, which
are united to the middle or beyond; disk yellow or
brownish; outer achenes slightly papillose; pappus
awns longer than the width of the summit of the
achene.

On dry plains, Nebraska and Wyoming to Texas,
northern Mexico and Arizona. May–Aug.

74. GALINSÒGA Cav. Icon. 3: 41. 1794.

Annual branching herbs, with opposite, mostly petioled, dentate or entire leaves, and
small peduncled heads of both tubular and radiate flowers, terminal and in the upper axils.
Involucre hemispheric or broadly campanulate, its bracts in 2 series, ovate, obtuse, membranous, striate, nearly equal, or the outer shorter. Receptacle conic or elongated, its thin
chaff subtending the disk-flowers. Ray-flowers white or red, pistillate, fertile, the rays 4 or
5, short. Disk-flowers yellow, perfect, the corolla 5-toothed. Anthers minutely sagittate at
the base. Style-branches tipped with acute appendages. Achenes angled, or the outer ones
flat. Pappus of the disk-flowers of several short laciniate or fimbriate scales, that of the
ray-flowers of several or few short slender bristles, or none. [Named in honor of M. M.
Galinsoga, superintendent of the Botanic Gardens at Madrid.]

About 5 species, natives of tropical and warm temperate America, the following typical.

1. Galinsoga parviflòra Cav.   Galinsoga.   Fig. 4522.

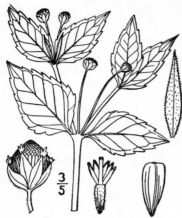

Galinsoga parviflora Cav. Icon. **3**: 41. *pl. 281.* 1794.
Galinsoga parviflora hispida DC. Prodr. **5**: 677. 1836.

Slightly appressed-pubescent or hirsute, 1°–3° high.
Leaves thin, ovate or deltoid-ovate, 3-nerved, 1′–3′ long,
acute at the apex, mostly obtuse at the base, dentate,
the lower slender-petioled, the upper short-petioled or
sessile, and sometimes nearly or quite entire; heads
usually numerous, 2″–3″ broad, slender-peduncled;
bracts of the involucre glabrous or nearly so, the outer
shorter; pappus of the disk-flowers 4–16, oblong to
spatulate, fimbriate obtusish or bristle-tipped scales,
somewhat shorter than or equalling the finely pubescent
obpyramidal achene.

In door-yards and waste places, Maine to Ontario, Oregon,
North Carolina, Missouri, Arizona, California and Mexico.
Bermuda; Jamaica.   Naturalized from tropical America.
Introduced into Europe as a weed. June–Nov.

Galinsoga caracasàna (DC.) Sch. Bip., similar to this
species but with reddish rays, and the pappus of the disk-
flowers only half as long as the achene, also tropical America, has been found in waste grounds in New Jersey and
Maryland.

75. ENDORÌMA Raf. Am. Month. Mag. **4**: 195. 1819.

[BALDUINA Nutt. Gen. **2**: 175. 1818. Not *Baldwinia* Raf. F. 1818.]

Perennial, caulescent, simple or branched, erect herbs, with alternate entire narrow punctate
leaves, and large terminal heads of both tubular and radiate, yellow flowers, or those of the
disk purple. Involucre hemispheric, its small bracts imbricated in several series, appressed,
or with spreading tips, the outer shorter. Receptacle convex, deeply honey-combed, chaffy,
the persistent chaff coriaceous or cartilaginous, laterally united, subtending the disk-flowers.
Rays large, neutral, toothed. Disk-flowers perfect, fertile, the corolla 5-toothed. Anthers
sagittate at the base. Style-branches with truncate subulate tips. Achenes turbinate, silky-
villous. Pappus of 7–12 scarious nearly equal scales. [Greek, perhaps referring to the
immersion of the achenes in the honeycombed receptacle.]

Two known species, natives of the southwestern United States. Type species: *Balduina uniflora* Nutt.

1. Endorima uniflòra (Nutt.) Barnhart.   One-headed Actinospermum.   Fig. 4523.

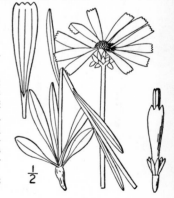

Balduina uniflora Nutt. Gen. **2**: 175. 1818.
Actinospermum uniflorum Barnhart, Bull. Torr. Club **24**: 411.
1897.
E. uniflora Barnhart; Small, Fl. SE. U. S. 1283. 1903.

Stem stout, puberulent, simple, or with a few erect
branches, 1°–3° high. Leaves thick, spatulate-linear or the
upper linear, sessile, erect or ascending, 1′–2′ long, the
lower 2″–3″ wide; heads long-peduncled, solitary, 2′–2½′
broad; bracts of the involucre ovate, acuminate, thick, their
tips at length spreading; rays 20–30, cuneate, 3–4-toothed
at the truncate apex; disk 8″–12″ broad; chaff of the receptacle cuneate, truncate, very cartilaginous, more or less
united laterally, the summit eroded; achenes obconic; pappus of 7–9 oblong scales about as long as the achene.

In wet pine-barrens, Virginia (according to Torrey and Gray);
North Carolina to Florida and Louisiana. July–Sept.

76. MARSHÁLLIA Schreb.; Gmelin, Syst. 1208. 1791.

Perennial, often tufted, simple or branched herbs, with basal or alternate, entire leaves,
and large long-peduncled discoid heads of purple, pink or white, glandular-pubescent flowers.
Involucre hemispheric or broadly campanulate, its bracts in 1 or 2 series, herbaceous, narrow,
nearly equal. Receptacle convex or at length conic, chaffy, the scales narrow, rigid, distinct.
Rays none. Flowers all perfect and fertile, their corollas with a deeply 5-lobed or 5-parted
campanulate limb and a slender tube. Anthers minutely sagittate at the base. Style-branches
long, truncate. Achenes turbinate, 5-ribbed and 5-angled. Pappus of 5 or 6 acute or acuminate, ovate or lanceolate-deltoid, nearly entire scales. [Named for Humphrey Marshall, of
Pennsylvania, botanical author.]

About 6 species, natives of the central United States. Type species: *Marshallia Schreberi* Gmel.

Leaves ovate, oval, or ovate-lanceolate, 3-nerved.
Leaves linear, lanceolate, or the basal spatulate, or obovate.
    Chaff of the receptacle linear; leaves linear; western.
    Chaff of the receptacle broader; leaves obovate to lanceolate; eastern.

1. *M. trinervia.*

2. *M. caespi·osa.*
3. *M. grandiflora.*

1. Marshallia trinérvia (Walt.) Porter. Broad-leaved Marshallia. Fig. 4524.

Athanasia trinervia Walt. Fl. Car. 201. 1788.
Marshallia Schreberi Gmelin, Syst. 1208. 1791.
Marshallia latifolia Pursh, Fl. Am. Sept. 519. 1814.
Marshallia trinervia Porter, Mem. Torr. Club 5 : 337. 1894.

Stem simple, or little branched, leafy to or beyond the middle, 1°–2° high. Leaves thin, those of the stem ovate, oval or ovate-lanceolate, 3-nerved, acute or acuminate at the apex, narrowed to a sessile base, 2′–3′ long, 9″–18″ wide; heads ½′–1′ broad, corolla purplish; bracts of the involucre linear-lanceolate, acute, rigid; chaff of the receptacle subulate-filiform; pappus scales lanceolate-acuminate from a triangular base; achenes glabrous when mature.

In dry soil, Virginia to Alabama and Mississippi. May–June.

$\frac{1}{2}$

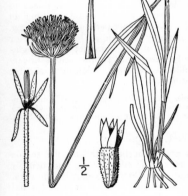

$\frac{1}{2}$

2. Marshallia caespitòsa Nutt. Narrow-leaved Marshallia. Fig. 4525.

Marshallia caespitosa Nutt.; DC. Prodr. 5 : 680. 1836.

Stems usually tufted and simple, sometimes sparingly branched, leafy either only near the base or to beyond the middle, 8′–15′ high. Leaves thick, mostly basal, faintly 3-nerved, linear or linear-spatulate, obtuse, sometimes 4′ long and 3″ wide, the upper ones linear, acutish, shorter; head about 1′ broad, borne on a peduncle often 10′ long; corollas pale rose or white; bracts of the involucre linear-lanceolate, acute or acutish; chaff of the receptacle linear, or slightly dilated above; achenes villous on the angles; scales of the pappus ovate, acutish, equalling or longer than the achene.

Prairies and hills, Missouri and Kansas to Texas. May–June.

3. Marshallia grandiflòra Beadle & Boynton. Large-flowered Marshallia. Fig. 4526.

Marshallia grandiflora Beadle & Boynton, Biltmore Bot. Stud. 1 : 7. 1901.

Stem simple, 1°–2° high, leafy to above the middle. Lower and basal leaves obovate to oblong-lanceolate, tapering into petioles often as long as the blade, obtuse or obtusish; upper leaves lanceolate, sessile, or more or less clasping; florets slightly larger than those of *M. trinervia*, 7″–10″ long; achenes larger, 2″–2½″ long, pubescent.

In moist soil, Pennsylvania to West Virginia and North Carolina. July–Aug.

Marshallia obovàta (Walt.) Beadle & Boynton, a lower plant of the Southern States, with obovate or spatulate leaves mainly basal, is recorded as extending northward to southwestern Pennsylvania.

$\frac{3}{5}$

77. PSILÓSTROPHE DC. Prodr. 7: 261. 1838.

[RIDDELLIA Nutt. Trans. Am. Phil. Soc. (II) 7: 271. 1841.]

Branched annual or perennial woolly herbs, often nearly glabrous when old, with alternate leaves, and middle-sized heads of both tubular and radiate yellow or orange flowers, corymbose, or clustered at the ends of the branches. Involucre cylindraceous, its bracts 4–10 in 1 series, narrow, equal, densely white-woolly, separate, but erect and connivent, commonly with 1–4 scarious ones within, and occasionally a narrow outer one. Rays broad, becoming papery and whitish, persistent, 5–7-nerved, 2–3-toothed, pistillate. Receptacle small, naked. Disk-flowers perfect, fertile, their corollas with a short proper tube and elongated cylindraceous limb, 5-toothed, the teeth glandular-bearded. Anthers obtuse and entire at the base. Style-branches of the disk-flowers capitellate. Achenes linear, striate. Pappus of 4–6 nerveless acute scales, glabrous or villous. [Greek, referring to the naked receptacle.]

About 7 species, natives of the southwestern United States and northern Mexico. Type species: *Psilostrophe gnaphalòdes* DC.

1. Psilostrophe villòsa Rydb. Plains Psilostrophe. Fig. 4527.

P. villosa Rydb.; Britton, Manual 1006. 1901.

Perennial, branched, 6'–2° high, loosely white-woolly. Basal and lower leaves spatulate, entire, dentate or rarely pinnatifid, mostly obtuse, 2'–4' long; upper leaves sessile, or nearly so, smaller, linear to spatulate, usually entire; heads several together in the clusters, 4"–6" broad, short-peduncled; rays few, lemon-yellow, commonly as wide as long, with 2 or 3 broad teeth or lobes at the summit; achenes glabrous, or sparingly pubescent; pappus scales linear-lanceolate to oblong-lanceolate, glabrous, shorter than the disk-corollas.

In dry sandy soil, Kansas to Texas and Arizona. Included, in our first edition, in *P. Tagetìnae* (Nutt.) Kuntze, which has much larger yellow rays. June–Sept.

78. FLAVÈRIA Juss.; Gmelin, Syst. 1269. 1791.

Glabrous or minutely puberulent, light-green, mostly annual herbs, with opposite sessile entire or serrate leaves, and small 1–several-flowered, usually sessile, oblong and densely cymose-capitate heads of tubular, or both tubular and radiate yellow or yellowish flowers. Involucre of 2–5 narrow, nearly equal, appressed bracts, sometimes with 1 or 2 additional small exterior ones. Receptacle small, naked or setose. Ray-flower commonly only 1, pistillate, fertile, sometimes wanting. Disk-flowers 1–15, perfect, fertile, their corollas 5-toothed. Anthers entire at the base. Style-branches of the disk-flowers truncate. Achenes oblong or linear-oblong, 8–10-ribbed. Pappus none. [Latin, *flavus*, yellow, from its dyeing properties.]

About 7 species, natives of the warmer parts of America. In addition to the following, 3 others occur in the southern United States. Type species: *Flaveria chilensis* Gmelin.

1. Flaveria campestris Johnston. Plains Flaveria. Fig. 4528.

Flaveria campestris Johnston, Proc. Am. Acad. 39: 287. 1903.

Annual, glabrous, erect, 1°–2° high, little branched. Leaves linear or lanceolate, serrulate or entire, 3-nerved, acuminate or acute at the apex, sessile by a broad and somewhat clasping base, 1'–2½' long, 2"–4" wide; heads about 3" high, closely sessile in terminal glomerules or these pedunculate from the upper axils; involucre of 3 oblong-lanceolate bracts, 2–5-flowered; ray equalling or longer than the breadth of the disk; achenes linear, glabrous, about 1½" long.

In alkaline soil, Missouri to Colorado, Texas and Mexico. Aug.–Oct. In our first edition included in the Mexican *F. angustifolia* (Cav.) Pers.

79. HYMENOPÁPPUS L'Her.; Michx. Fl. Bor. Am. 2: 103. 1803.

Perennial or biennial, erect herbs, with angled stems, alternate or basal, mostly pinnatifid or dissected leaves, and corymbose or solitary, small or rather large discoid heads, of white or yellow flowers. Involucre hemispheric or broadly campanulate, its bracts 6–12 in 1 or 2 series, nearly equal, mostly appressed, colored, petal-like, the margins and apices scarious. Receptacle small, naked. Rays none. Disk-flowers all perfect and fertile, their corollas with slender tubes and reflexed or spreading campanulate 5-lobed limbs, the lobes ovate. Anthers entire at the base. Style-branches with short conic appendages. Achenes obovoid or obpyramidal, 4–5-angled, the faces usually prominently 1–3-nerved. Pappus of 10–20 thin obtuse scales, sometimes very short or none. [Greek, membrane-pappus.]

About 7 species, natives of southern and central North America and Mexico. Type species: *Hymenopappus scabiosaèus* L'Her.

Bracts of the involucre broadly ovate or oval, bright white.   1. *H. carolinensis.*
Bracts obovate to oblong, green or with white tips.
  Heads numerous, 4″–6″ broad; biennials.
    Achenes puberulent; corolla white.   2. *H. corymbosus.*
    Achenes densely villous.
      Plant glabrate, or loosely woolly; corolla dull white.   3. *H. tenuifolius.*
      Plant densely white-woolly; corolla yellow.   4. *H. flavescens.*
  Heads few, 6″–12″ broad; corolla yellow; perennial.   5. *H. filifolius.*

1. Hymenopappus carolinénsis (Lam.) Porter. White-bracted Hymenopappus. Fig. 4529.

Rothia carolinensis Lam. Journ. Hist. Nat. 1: 16. *pl.*
  1. 1792.
Hymenopappus scabiosaeus L'Her.; Michx. Fl. Bor.
  Am. 2: 104. 1803.
Hymenopappus carolinensis Porter, Mem. Torr. Club
  5: 338. 1894.

Biennial; stem woolly-pubescent or glabrate, leafy below, corymbosely branched and nearly naked above, 2°–3° high. Basal and lower leaves petioled, 4′–6′ long, 1–2-pinnately parted or deeply pinnatifid into linear or oblong, obtuse or obtusish lobes, more or less white-tomentose beneath, green and glabrate above; upper leaves few, smaller, sessile, less divided; heads commonly numerous, corymbose, 6″–10″ broad; bracts of involucre oblong, ovate or oval, sometimes slightly obovate, thin, bright white, puberulent or glabrate; corolla-lobes about as long as the throat, white; achenes puberulent or pubescent; pappus of very small nerveless scales, shorter than the width of the top of the achene.

In dry sandy soil, Illinois to Texas, east to South Carolina and Florida. March–June.

$\frac{3}{5}$

2. Hymenopappus corymbòsus T. & G. Corymbed or Smooth White Hymenopappus. Fig. 4530.

Hymenopappus corymbosus T. & G. Fl. N. A. 2: 372. 1842.

Biennial; stem glabrous, or nearly so, corymbosely branched and nearly naked above, 1°–2° high. Lower and basal leaves petioled, 1–2-pinnately parted into linear or nearly filiform, acute or acutish, glabrous lobes, or somewhat tomentose beneath; upper leaves few, much smaller and less divided, or the uppermost reduced to linear scales; heads corymbose, numerous, 4″–6″ broad; bracts of the involucre obovate to oblong, puberulent, their tips greenish white; corolla white, its lobes about as long as the throat; achenes puberulent; pappus scales small, nerveless, shorter than the width of the top of the achene.

On dry prairies, Missouri and Nebraska to Texas. Summer.

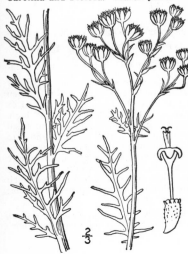

$\frac{2}{3}$

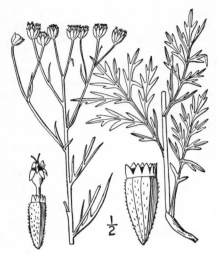

3. Hymenopappus tenuifòlius Pursh.
Woolly White Hymenopappus.
Fig. 4531.

Hymenopappus tenuifolius Pursh, Am. Sept. 742. 1814.

Biennial; stem lightly tomentose, or at length glabrate, 1°-2° high, slender, leafy below, corymbosely branched and nearly naked above. Lower and basal leaves petioled, 1-3-pinnately parted into linear or filiform lobes, woolly pubescent beneath, at least when young; upper leaves much smaller and less compound; heads numerous, corymbose, 4"-6" broad; bracts of the involucre obovate-oblong, usually densely tomentose; corolla white, its lobes slightly shorter than the throat; achenes densely villous-pubescent; pappus of several oblong to ovate, ribbed or nerved scales, which are about as long as the width of the top of the achene or shorter.

On dry prairies, South Dakota to Nebraska, Kansas and Texas. June-Sept.

4. Hymenopappus flavéscens A. Gray.
Woolly Yellow Hymenopappus.
Fig. 4532.

Hymenopappus flavescens A. Gray, Mem. Am. Acad. (II) 4: 97. 1849.

Biennial; stem densely white-woolly, at least when young, 1°-2½° high, leafy, branched above. Leaves 1-3-pinnately parted or divided into linear segments; heads numerous, usually larger than those of the preceding species; involucral bracts obovate to ovate with greenish white margins; corolla yellow or yellowish, the lobes about equalling the throat, achenes short-villous; pappus scales spatulate, shorter than the slender corolla-tube.

In sandy soil, Kansas to Texas, Arizona and northern Mexico.

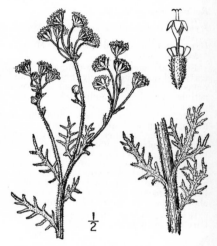

5. Hymenopappus filifòlius Hook. Low
Tufted Hymenopappus. Fig. 4533.

Hymenopappus filifolius Hook. Fl. Bor. Am. 1 : 317. 1833.

Perennial from a deep woody root; stems usually tufted, woolly when young, sometimes glabrate when old, densely leafy toward the base, usually naked or nearly so and sparingly branched above, 6'-18' high. Leaves tomentose when young, the lower and basal ones petioled, 1-3-pinnately parted or pinnatifid into narrowly linear, somewhat rigid lobes; heads commonly few, 6"-12" broad; bracts of the involucre obovate-oblong, usually densely woolly, their tips whitish; corolla yellow or yellowish, its lobes much shorter than the throat; achenes densely villous; pappus scales costate, short.

On prairies and in dry rocky soil, Saskatchewan to North Dakota, Montana, Nebraska and Colorado. June-Sept.

80. OTHÀKE Raf. New Fl. N. A. 4: 73. 1836.

Erect rough, glandular or cinereous, branching annual herbs, with alternate, mostly entire leaves, or the lower opposite, and corymbose or panicled heads of tubular or both tubular and radiate pink or purple flowers. Involucre campanulate or obconic, its bracts in 1 or 2 series, narrow, herbaceous, nearly equal, or with a few exterior shorter ones, appressed, usually colored. Receptacles small, flat, naked. Ray-flowers, when present, pistillate, fertile, the rays 3-cleft. Disk-flowers perfect, fertile, their corollas with slender tubes and deeply 5-parted campanulate limbs. Anthers entire or emarginate at the base. Style-branches filiform, acutish, glandular-pubescent throughout. Achenes linear or narrowly obpyramidal, quadrangular. Pappus of 6–12 lanceolate strongly costate scales, that of the outer achenes often much shorter. [Greek, warty apex, referring to the callous-tipped leaves of some species.]

About 6 species, natives of the south-central United States and Mexico; in our first edition referred to the genus *Polypteris* Nutt. Type species: *Othake tenuifolium* Raf.

Rays purple, deeply 3-lobed; leaves lanceolate.   1. *O. sphacelatum.*
Rays none; leaves linear.   2. *O. callosum.*

1. Othake sphacelàtum (Nutt.) Rydb. Hooker's Othake. Fig. 4534.

Stevia sphacelata Nutt.; Torr. Ann. Lyc. N. Y. **2**: 214. 1827.
Palafoxia Hookeriana T. & G. Fl. N. A. **2**: 368. 1842.
Polypteris Hookeriana A. Gray, Proc. Am. Acad. **19**: 31. 1883.
O. sphacelatum Rydb. Bull. Torr. Club **37**: 331. 1910.

Annual; stem rather stout, glandular-pubescent and viscid above, 1°–3° high. Leaves lanceolate, entire, acute or acuminate, narrowed at the base, rough on both sides, the upper alternate, the lower opposite and slender-petioled, 2′–4′ long, 3″–5″ wide; bracts of the involucre 10–16, linear-lanceolate or spatulate, glandular-hispid, the inner with purplish tips; ray-flowers 8–10; rays rose-purple, deeply 3-cleft, sometimes small, or none; achenes about 4″ long and ½″ thick; pappus scales of the disk-flowers 6–8, lanceolate, awned, more than half the length of the achene, those of the ray-flowers as many, spatulate, obtuse, shorter.

In dry soil, Nebraska to Colorado, Texas and Mexico. July–Sept.

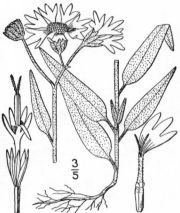

2. Othake callòsum (Nutt.) Bush. Rayless Othake. Fig. 4535.

S'evia callosa Nutt. Journ. Acad. Phila. **2**: 121. 1821.

O. tenuifolium Raf. New Fl. N. A. **4**: 74. 1836.

Polypteris callosa A. Gray, Proc. Am. Acad. **19**: 30. 1883.

O. callosum Bush, Trans. Acad. Sci. **14**: 174. 1904.

Annual, glandular, at least above; stem slender, paniculately branched, 1°–2° high. Leaves linear, or linear-lanceolate, 1′–2½′ long, 1″–2½″ wide, short-petioled, mostly alternate; heads discoid; bracts of the top-shaped involucre 8–10, linear or narrowly oblong, herbaceous, pubescent, about ¼′ long, their tips reddish; corollas purple, deeply 5-parted; achenes narrowly obpyramidal, pubescent or glabrous, nearly as long as the involucre; pappus scales obovate or nearly orbicular, rounded, or retuse, or sometimes minute, or none.

In dry soil, Missouri to Texas and New Mexico. June-Oct.

81. PICRADENIÓPSIS Rydb.; Britton, Man. 1008. 1901.

Herbs more or less woolly, with opposite leaves, and small corymbose heads, of both tubular and radiate yellow flowers. Involucre campanulate or hemispheric, its bracts in 1 or 2 series, herbaceous, obtuse, appressed, nearly equal. Receptacle small, nearly flat, naked, foveolate. Ray-flowers in 1 series, pistillate, fertile. Disk-flowers perfect, fertile, their corollas with campanulate or cylindric, 5-cleft limb. Anthers entire or emarginate at the base. Style-tips mostly truncate and obtuse. Achenes quadrangular, linear or oblong. Pappus of several scales, obtuse or truncate and scarious at the apex. [Named for its resemblance to *Picradenia*.]

Two known species, natives of western North America, the following typical.

1. Picradeniopsis oppositifòlia (Nutt.) Rydb. Picradeniopsis. Fig. 4536.

Trichophyllum oppositifolium Nutt. Gen. **2**: 167. 1818.

Bahia oppositifolium Nutt.; T. & G. Fl. N. A. **2**: 376. 1842.

P. oppositifolia Rydb.; Britton, Manual 1008. 1901.

Perennial, herbaceous; stem densely cinereous, much branched, 4'–12' high, very leafy. Leaves opposite, or the uppermost alternate, ½'–1½' long, palmately 2–5-parted into linear, obtuse or obtusish, entire segments, finely cinereous on both sides; heads short-peduncled, 6"–9" broad; involucre campanulate, or becoming hemispheric, its bracts oblong, obtuse, densely tomentose; rays 5–7, short; achenes linear-oblong, glandular-pubescent; pappus of 4–8 spatulate to lanceolate scales with thickened bases.

On plains, especially in alkaline soil, South Dakota to Montana, Nebraska, Texas, New Mexico. June–Sept.

82. TETRANEÙRIS Greene, Pittonia **3**: 265. 1898.

[ACTINELLA Nutt. (1818), not Pers. (1807), nor *Actinea* Juss. (1803).]

Branched or scapose, villous-pubescent or glabrous, bitter and aromatic punctate herbs, with alternate or basal, often punctate leaves, and small or rather large, peduncled heads of both tubular and radiate, yellow flowers, or rays rarely wanting. Involucre hemispheric, campanulate or depressed, its bracts imbricated in 2–3 series, appressed. Receptacle convex or conic, naked. Ray-flowers pistillate and fertile, the rays 3-toothed, 4-nerved. Disk-flowers perfect, fertile, their corollas with 4–5-toothed limbs. Anthers entire or minutely sagittate at the base. Style-branches truncate and penicillate at the summit. Achenes turbinate, 5–10-ribbed or angled, villous or pubescent. Pappus of 5–12 thin aristate, acuminate or truncate scales. [Greek, four-nerved.]

About 18 species, natives of western North America and Mexico. Besides the following, some 12 others occur in the western and southwestern parts of the United States. Type species: *Tetraneuris acaulis* (Pursh) Greene.

Stem leafy, branching; stem leaves linear; annual or biennial.   1. *T. linearifolia.*
Stems tufted, simple, scapose; leaves basal; perennials.
  Leaves narrowly linear; branches of the caudex slender.   2. *T. stenophylla.*
  Leaves broader, linear to spatulate; branches of the caudex short and thick.
    Bracts of the involucre acutish.   3. *T. acaulis.*
    Bracts of the involucre obtuse, rounded.   4. *T. herbacea.*

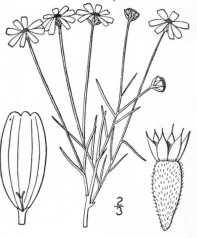

1. Tetraneuris linearifòlia (Hook.) Greene. Fine-leaved Tetraneuris. Fig. 4537.

Hymenoxys linearifolia Hook. Icon. *pl. 146.* 1837.

Actinella linearifolia T. & G. Fl. N. A. **2**: 383. 1842.

T. linearifolia Greene, Pittonia **3**: 369. 1898.

Annual or perhaps biennial; stem usually diffusely branched, finely hirsute, or glabrous, or woolly at the base, slender, 6'–15' high. Stem leaves narrowly linear, sessile, ½'–1½' long, ¼"–1" wide; basal leaves spatulate, often villous, much broader, obtuse, narrowed into margined petioles; heads numerous, slender-peduncled, 6"–8" broad; involucre broadly campanulate, its bracts oblong, obtuse, pubescent, distinct to the base, imbricated in about 2 series; receptacle conic; rays 6–10, oblong; achenes pubescent; pappus of 5 or 6 ovate awned scales.

In dry soil, Kansas to Louisiana, Texas and New Mexico. May–Sept.

2. Tetraneuris stenophýlla Rydb. Narrow-leaved Tetraneuris. Fig. 4538.

Tetraneuris stenophylla Rydb. Bull. Torr. Club **33**: 155. 1906.

Perennial with a branched caudex, the branches often 3½' long, the upper portion covered with the scarious bases of old leaves. Bases of the leaves dilated, sparingly silky-villous; blades narrowly linear, 1'–1¾' long, about 1" wide, glabrous, conspicuously punctate; scape 2'–4' high, minutely strigose; involucre 5"–7" wide; bracts oval or oblong-acutish; rays about 5" long and 2" wide.

In dry soil, Kansas to Colorado and Mexico. Formerly confused with *T. scaposa* (DC.) Greene.

Tetraneuris fastigiàta Greene, of Kansas, differs by the leaf-bases and involucre being more densely pubescent; it is known only from imperfect specimens, and may not be distinct from *T. stenophylla*, over which it has priority of publication.

3. Tetraneuris acaùlis (Pursh) Greene. Stemless Tetraneuris. Fig. 4539.

Gaillardia acaulis Pursh, Fl. Am. Sept. 743. 1814.
T. acaulis Greene, Pittonia **3**: 265. 1898.
Actinella acaulis Nutt. Gen. **2**: 173. 1818.
Picradenia acaulis Britton, Mem. Torr. Club **5**: 339. 1894.
T. simplex A. Nelson, Bot. Gaz. **28**: 127. 1899.

Perennial with thick roots and a stout branched caudex; scapes tufted, rather stout, or slender, densely silky or tomentose, 2'–8' high. Leaves all borne on the ends of the branches of the caudex, linear-spatulate, entire, obtuse or obtusish; 1'–3' long, 1½"–3" wide, densely silky or villous; heads 9"–18" broad; involucre hemispheric, its bracts densely villous, acutish; rays 10–15; pappus of 5 or 6, ovate or oblong, awned scales.

In dry or rocky soil, North Dakota to Assiniboia, Montana, Nebraska and New Mexico. May–Aug.

Tetraneuris scapòsa (DC.) Greene, admitted into our first edition as from Nebraska, is not definitely known north of Texas.

4. Tetraneuris herbacea Greene. Eastern Tetraneuris. Fig. 4540.

T. herbacea Greene, Pittonia **3**: 268. 1898.
Actinia herbacea Robinson, Rhodora **10**: 68. 1908.

Perennial by a stout thick caudex; scape stout, villous-pubescent, especially above, 6'–8' high. Leaves linear-spatulate, slightly fleshy, bluntish, strongly punctate, sparingly loosely long-hairy, at least toward the base, 2'–3' long, about 3" wide; heads nearly 2' broad; involucral bracts oblong, obtuse and rounded at the apex, densely villous; rays about 15; pappus scales ovate-oblong, obtuse, acute or apiculate.

Southern Ontario, Ohio and Illinois. May–June.

83. HYMENÓXYS Cass. Dict. Sci. Nat. 55: 278. 1828.

[PICRADENIA Hook. Fl. Bor. Am. **1**: 317. 1833.]

Pubescent herbs with branching stems and alternate entire or dissected leaves, the blades or divisions narrow, usually linear or filiform, and relatively small, peduncled, radiate or discoid yellow heads. Involucre turbinate to campanulate, its bracts in 2 series, appressed, rather broad, the outer connate at base. Receptacle flat, convex or conic. Ray-flowers fertile, the rays short and broad, 3-lobed. Disk-flowers perfect, fertile, with 5 short lobes. Anthers

notched at the base, with rounded auricles. Style-branches truncate and penicillate. Achenes turbinate, pubescent. Pappus of 5–8 acuminate or aristate hyaline scales. [Greek, referring to the thin and pointed pappus scales.]

Type species: *Hymenopappus anthemoides* Juss.

1. Hymenoxys odoràta DC. Limonillo.
Fig. 4541.

Hymenoxys odorata DC. Prodr. **5**: 661. 1836.
Actinella odorata A. Gray, Mem. Am. Acad. (II) **4**: 101. 1849.
Picradenia odorata Britton, in Britt. & Brown, Ill. Fl. **3**: 449. 1898.
Philozera multiflora Buckl. Proc. Acad. Phila. **1861**: 459. 1862.
H. multiflora Rydb. Bull. Torr. Club **33**: 157. 1906.

Annual; stem much branched, puberulent, sparingly hirsute or glabrous, 1°–2° high, leafy. Leaves 1′–2′ long, 1–3-parted into filiform, entire, somewhat pubescent segments about ½″ wide; heads commonly numerous, 6″–10″ broad; involucre campanulate, puberulent, its outer bracts 6–9, lanceolate, keeled, acute, united at the base; rays 7–10, cuneate.

In dry soil, Kansas and Colorado to Texas, Mexico and southern California. April–July.

84. HELÈNIUM L. Sp. Pl. 886. 1753.

Erect, mostly branching herbs, with alternate, mainly decurrent, punctate bitter entire or dentate leaves, and large peduncled heads of both tubular and radiate, yellow or brownish-yellow flowers, or rays sometimes wanting. Involucre broad and short, its bracts in 1 or 2 series, linear or subulate, reflexed or spreading. Receptacle convex, subglobose or oblong, naked. Ray-flowers pistillate and fertile, or neutral, the rays cuneate, 3–5-lobed. Disk-flowers perfect, fertile, their corollas 4–5-toothed, the teeth glandular-pubescent. Anthers 2-toothed or sagittate at the base. Style-branches of the disk-flowers dilated and truncate at the apex. Achenes turbinate, ribbed. Pappus of 5–8 entire, dentate or incised, acuminate or aristate scales. [The Greek name of some plant, from Helenus or Helena.]

About 24 species, natives of North and Central America. In addition to the following, some 18 others occur in the southern and southwestern United States. Type species: *Helenium autumnale* L.

Leaves oblong-lanceolate or ovate-lanceolate, dentate; rays fertile; disk yellow.	1. *H. autumnale.*
Leaves lanceolate or linear-lanceolate, mainly entire; rays neutral; disk purple.	2. *H. nudiflorum.*
Leaves all linear-filiform, entire; rays fertile.	3. *H. tenuifolium.*

1. Helenium autumnàle L. False or Swamp Sunflower. Fig. 4542.

Helenium autumnale L. Sp. Pl. 886. 1753.
Helenium pubescens Ait. Hort. Kew. **3**: 287. 1789.
Helenium autumnale pubescens Britton, Mem. Torr. Club **5**: 339. 1894.

Perennial; stem puberulent or glabrous, rather stout, narrowly winged by the decurrent bases of the leaves, corymbosely branched above, 2°–6° high. Leaves firm, oblong, lanceolate or ovate-lanceolate, acuminate or acute at apex, narrowed to the sessile base, pinnately few-veined, 2′–5′ long, ¼′–2′ wide, dentate, denticulate or entire, puberulent, glabrous or pubescent, bright green; heads numerous, 1′–2′ broad, borne on long puberulent peduncles; bracts of the flattish involucre densely canescent; rays 10–18, drooping, bright yellow, equalling or longer than the globose yellow disk, pistillate and fertile, 3-cleft; achenes pubescent on the angles; pappus scales ovate.

In swamps and wet meadows, Quebec to Florida, Manitoba, Oregon, Nevada and Arizona. Yellow-star. Ox-eye. Sneezeweed. Ascends to 2600 ft. in Virginia. Aug.–Oct.

2. Helenium nudiflòrum Nutt. Purple-head Sneezeweed. Fig. 4543.

Helenium nudiflorum Nutt. Trans. Am. Phil. Soc.
(II) **7**: 384. 1841.
Leptopoda brachypoda T. & G. Fl. N. A. **2**: 388.
1842.

Perennial; stem mostly slender, puberulent at
least above, corymbosely branched near the summit, 1°–3° high, narrowly winged by the decurrent leaf-bases. Stem leaves lanceolate or linear-
lanceolate, entire or sparingly denticulate, acute
or obtusish at the apex, 1½′–3′ long, 2″–6″ wide,
sessile; basal and lower leaves spatulate, obtuse,
more or less dentate, tapering into margined
petioles; heads several or numerous, 1′–1½′ broad,
on slender or short-puberulent peduncles; rays
10–15 (sometimes wanting), drooping, yellow,
yellow with a brown base, or brown throughout,
3-toothed, neutral, or with rudimentary pistils,
sterile, equalling or exceeding the brown or
purple globose disk; pappus scales ovate, aristate.

In moist soil, Missouri and Illinois to Texas, east
to North Carolina and Florida. Also locally naturalized from Pennsylvania to Connecticut. June–Oct.

3. Helenium tenuifòlium Nutt. Fine-leaved Sneezeweed. Fig. 4544.

H. tenuifolium Nutt. Journ. Phil. Acad. **7**: 66. 1834.

Annual; glabrous or minutely pubescent above;
stem slender, very leafy and usually much
branched, 8′–24′ high. Leaves all linear-filiform,
entire, sessile, often fascicled, ½′–1½′ long, ½″ or
less wide; heads several or numerous, corymbose, 9″–15″ broad, borne on slender or filiform
peduncles; bracts of the involucre few, linear or
subulate, sometimes pubescent, soon reflexed;
rays 4–8, fertile, 3–4-toothed, at length drooping,
longer than the globose disk; achenes villous;
pappus scales ovate, tipped with slender awns.

In moist soil, southeastern Virginia to Florida,
Missouri, Kansas and Texas. Naturalized in waste
places, northward to Massachusetts, and in Cuba and
Santo Domingo. Aug.–Oct.

85. GAILLARDIA Foug. Mem. Acad. Sci. Paris 1786: 5. *pl. 1, 2.* 1788.

Branching or scapose, more or less pubescent herbs, with alternate or basal leaves, and
large peduncled heads of both tubular and radiate flowers, or rays wanting. Involucre
depressed-hemispheric, or flatter, its bracts imbricated in 2 or 3 series, their tips spreading or
reflexed. Receptacle convex or globose, bristly, fimbrillate or nearly naked. Rays cuneate,
yellow, purple or parti-colored, neutral or rarèly pistillate, 3-toothed or 3-lobed. Disk-flowers
perfect, fertile, their corollas with slender tubes and 5-toothed limbs, the teeth pubescent with
jointed hairs. Anthers minutely sagittate or auricled at the base. Style-branches tipped
with filiform or short appendages. Achenes turbinate, 5-ribbed, densely villous, at least at
the base. Pappus of 6–12, 1-nerved awned scales, longer than the achene. [Named for M.
Gaillard de Marentonneau, a French botanist.]

About 15 species, natives of the south-central part of the United States, and Mexico, 1 in
southern South America. Type species: *Gaillardia pulchella* Foug. Called in Texas blanket-flower.

Stem leafy; style-tips with filiform hispid appendages.
  Fimbrillae of the receptacle obsolete, or short.   1. *G. lutea.*
  Fimbrillae subulate or bristle-like, mostly longer than the achenes.
    Rays yellow; fimbrillae exceeding the achenes.   2. *G. aristata.*
    Rays purple, or red at base; fimbrillae about equalling the achenes.   3. *G. pulchella.*
Leaves basal; style-tips with short naked appendages; rays none, or few.   4. *G. suavis.*

1. **Gaillardia lutea** Greene. Yellow Gaillardia.
Fig. 4545.

Gaillardia lutea Greene, Pittonia **5**: 57. 1902.

Stem roughish-puberulent or cinereous, usually branched, $1\frac{1}{2}°-2°$ high, the branches straight, nearly erect. Stem leaves sessile, oblong-lanceolate, serrate, roughish-puberulent, acute at the apex, narrowed to the base, $1'-2'$ long, $2''-5'''$ wide; heads about $2'$ broad, peduncled; bracts of the involucre about equalling the yellow disk; rays 8–12, yellow; style-tips with filiform hispid appendages; achenes villous at the base, or to beyond the middle; fimbrillae of the receptacle short or none; awns of the pappus slender.

In dry woods, Missouri to Texas. July–Sept. Included, in our first edition, in *G. lanceolata* Michx., of the Southern States.

2. **Gaillardia aristàta** Pursh. Great-flowered Gaillardia. Fig. 4546.

Gaillardia aristata Pursh, Fl. Am. Sept. 573. 1814.

Perennial; stem simple, or little branched, hirsute or densely pubescent with jointed hairs, $1°-3°$ high. Leaves firm, densely and finely pubescent, the lower and basal ones petioled, oblong or spatulate, laciniate, pinnatifid or entire, mostly obtuse, $2'-5'$ long; upper leaves sessile, lanceolate, or oblong, or slightly spatulate, smaller, entire or dentate, rarely pinnatifid; heads $1\frac{1}{2}'-4'$ broad, long-peduncled; bracts of the involucre lanceolate, acuminate, hirsute; rays 10–18, yellow; style-tips with filiform appendages; fimbrillae of the receptacle mostly longer than the achenes, which are villous at least at the base.

On plains and prairies, Minnesota to Saskatchewan, British Columbia, Colorado, New Mexico and Oregon. Adventive eastward. Leaves sometimes all basal. May–Sept.

3. **Gaillardia pulchélla** Foug. Showy Gaillardia. Fig. 4547.

Gaillardia pulchella Foug. Mem. Acad. Sci. Paris 1786: 5. 1786.

Annual; diffusely branched at the base, the branches ascending, $6'-15'$ high, or larger in cultivation, more or less hirsute or pubescent with jointed hairs. Leaves lanceolate, oblong, or the lower spatulate, $1'-3'$ long, entire, dentate or sinuate-pinnatifid, all but the lowest sessile; heads $1'-3'$ broad, long-peduncled, bracts of the involucre lanceolate, acuminate, hirsute or pubescent; rays 10–20, red or purple at the base, yellow toward the apex; style-tips with filiform hispid appendages; fimbrillae of the receptacle equalling or scarcely longer than the achenes, which are more or less villous, or glabrous.

In dry soil, Nebraska and Missouri to Louisiana, Mexico and Arizona. May–Sept.

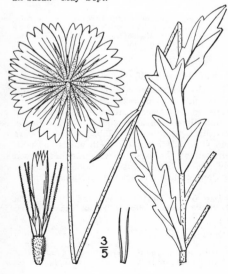

4. Gaillardia suàvis (A. Gray) Britt. & Rusby. Rayless Gaillardia. Fig. 4548.

Agassizia suavis A. Gray, Proc. Am. Acad. 1: 49. 1846.
Gaillardia simplex Scheele, Linnaea 22: 160. 1849.
Gaillardia suavis Britt. & Rusby, Trans. N. Y. Acad. Sci.
7: 11. 1887.

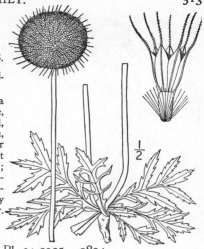

Annual or biennial. Leaves in a basal tuft, or a few near the base of the slender pubescent scape, spatulate or obovate in outline, 2'–6' long, pinnatifid, dentate, or some of them entire; scape 1°–2° high, monocephalous; head about 1' broad with the odor of heliotrope, globose in fruit; rays none, or short and pistillate, or a few of them longer and neutral; bracts of the involucre oblong or lanceolate, sparingly pubescent; fimbrillae of the receptacle obsolete; style-appendages short, naked; achenes densely villous; pappus scales broad, their awns very slender.

In dry rocky soil, Kansas to Texas. April–June.

86. BOÈBERA Willd. Sp. Pl. 3: 2125. 1804.

Erect or diffuse, branching, annual, or perennial, strong-scented, more or less glandular herbs, with opposite, mostly finely dissected leaves, and small peduncled heads of both tubular and radiate yellow flowers. Involucre campanulate or nearly hemispheric, its bracts in 1 series, united into a cup, with small additional outer ones. Receptacle flat, pubescent. Ray-flowers pistillate, the rays short. Disk-flowers perfect, their corollas 5-toothed. Anthers entire or minutely 2-toothed at the base. Style-branches of the disk-flowers hirsute, apiculate. Achenes narrowly obpyramidal, 3–5-angled, striate. Pappus of about 10 scales, parted to beyond the middle into numerous capillary, bristle-like segments. [In honor of J. von Boeber, a Russian botanist, died 1820.]

About 3 species, natives of the central United States and of Mexico, the following typical.

1. Boebera pappòsa (Vent.) Rydb.   Fetid Marigold. False Dog-fennel. Fig. 4549.

Tagetes papposa Vent. Hort. Cels. *pl. 36.* 1800.
Boebera chrysanthemoides Willd. Sp. Pl. 3: 2125. 1804.
Dysodia chrysanthemoides Lag. Gen. et Sp. Nov. 29. 1816.
D. papposa Hitchc. Trans. St. Louis Acad. 5: 503. 1891.
B. papposa Rydb.; Britton, Manual 1012. 1901.

Annual, very leafy, glabrous or finely pubescent, gland-dotted, much branched, 6'–18' high, the branches diffuse or erect. Leaves sessile, or short-petioled, ½'–1½' long, pinnately parted into linear or slightly spatulate, sharply serrate or incised segments; heads numerous, short-peduncled, 3''–5'' broad; involucre campanulate, of 8–10 appressed, oblong, obtuse, green or purplish, glabrous or ciliate bracts, with several narrow shorter outer ones; rays few, not longer than the width of the disk; receptacle and achenes pubescent.

Along streams and roadsides, Ohio to Minnesota, Montana, Louisiana, Mexico and Arizona. Occasionally found as a weed in waste places in the Eastern and Middle States, and in Ontario. Prairie-dogweed. July–Oct.

87. THYMOPHÝLLA Lag. Gen. et Sp. Nov. 25. 1816.

[HYMENATHERUM Cass. Bull. Soc. Philom. 1817: 12. 1817.]

Annual or perennial herbs, some species low undershrubs, with gland-dotted foliage and involucre, alternate or opposite leaves, and small heads of both tubular and radiate, mostly yellow flowers. Involucre campanulate, its principal bracts united into a cup, sometimes with smaller outer ones. Receptacle naked, or fimbrillate, not chaffy. Ray-flowers pistillate, fertile. Disk-flowers perfect, fertile. Style-branches truncate or blunt. Achenes striate. Pappus of several or numerous scales or bristles. [Greek, thyme-leaf, not applicable to the following species.]

About 15 species, natives of America. Besides the following, some 4 others occur in the western parts of the United States. Type species: *Thymophylla setifolia* Lag.

33

1. Thymophylla àurea (A. Gray) Greene.
Thyme-leaf.   Fig. 4550.

Lowellia aurea A. Gray, Mem. Am. Acad. (II) **4**: 91. 1849.
Hymena herum aureum A. Gray, Proc. Am. Acad. **19**: 42. 1883.
T. aurea Greene; Britt. & Brown, Ill. Fl. **3**: 453. 1898.

Annual, glabrous, 4′–12′ high, much branched; the leaves and involucre with large oval oil-glands. Leaves alternate, or the lower opposite, sessile or nearly so, very deeply parted into 5–9 linear-filiform, mostly entire, blunt segments; heads numerous, corymbose, 6″–10″ broad, terminating the branches; involucre about 3″ high, its bracts acute; rays about 12, 2½″–3″ long; pappus of 6–8 erose truncate scales, somewhat longer than the thickness of the achene.

Kansas and Colorado to Texas and New Mexico. June–Sept.

88. PÉCTIS L. Syst. Nat. Ed. 10, 1221. 1759.

Annual or perennial, diffuse prostrate or erect, mostly glabrous herbs, gland-dotted and strong-scented, with opposite narrow sometimes ciliate leaves, and small usually cymose heads of both tubular and radiate yellow flowers. Involucre cylindric, oblong or campanulate, its bracts in 1 series, narrow, keeled, distinct. Receptacle small, naked. Ray-flowers pistillate, the rays small, entire or 3-lobed. Disk-flowers perfect, their corollas with expanded, somewhat irregularly 5-cleft limbs. Anthers entire at the base. Style-branches of the disk-flowers very short, obtuse. Achenes linear, slightly angled, striate. Pappus of several or numerous scales, slender bristles or awns, sometimes with a few outer smaller additional ones. [Latin, *pecten*, comb, referring to the pappus.]

About 75 species, natives of the warmer parts of America. Besides the following, about 10 others occur in the southern and western parts of the United States. Type species: *Pectis ciliaris* L.

1. Pectis angustifòlia Torr.   Lemon-scented Pectis.   Fig. 4551.

Pectis angustifolia Torr. Ann. Lyc. N. Y. **2**: 214. 1827.

Annual; much branched, 4′–12′ high, the branches diffuse or ascending. Leaves narrowly linear, sessile, obtusish, ½′–2′ long, 1″ wide or less, often ciliate with a few bristles near the base; heads several or numerous, short-peduncled, about 3″ broad; involucre short-cylindric or narrowly campanulate, its bracts about 8, linear, acutish, partly enclosing the outer achenes; rays few, 3-toothed, or entire; pappus a crown of 4–6 somewhat united short scales, with or without 2 slender short awns.

In dry soil, Nebraska and Colorado to Mexico and Arizona. Plant with the odor of lemons. May–Oct.

89. ACHILLÈA [Vaill.] L. Sp. Pl. 898. 1753.

Herbs, mostly perennial, with erect leafy stems, finely dissected, pinnatifid or serrate alternate leaves, and small heads of both tubular and radiate flowers, corymbose at the ends of the stem and branches. Involucre obovoid, or campanulate, its bracts appressed, imbricated in few series, the outer shorter. Receptacle nearly flat or convex, chaffy, the membranous chaff subtending the disk-flowers. Ray-flowers pistillate, fertile, the rays white or pink. Disk-flowers perfect, fertile, their corollas yellow, 5-lobed. Anthers obtuse and entire at the base. Style-branches of the disk-flowers truncate. Achenes oblong or obovate, slightly compressed. Pappus none. [Named for Achilles.]

About 75 species, natives of the northern hemisphere, mostly of the Old World. Besides the following, another, or perhaps 2 others, occur in northwestern North America. Type species: *Achillea santolina* L.

Involucre broadly campanulate; leaves serrate.   1. *A. Ptarmica.*
Involucre ovoid to cylindric; leaves finely dissected.
  Involucral bracts greenish-yellow; rays small.
    Plant loosely woolly or nearly glabrous; inflorescence flat-topped.   2. *A. Millefolium.*
    Plant densely woolly; inflorescence convex.   3. *A. lanulosa.*
  Involucral bracts black-margined; rays large.   4. *A. borealis.*

1. Achillea Ptàrmica L. Sneezewort. White Tansy. Sneezewort-Yarrow. Fig. 4552.

Achillea Ptarmica L. Sp. Pl. 898. 1753.

Perennial from horizontal or creeping rootstocks; stem glabrous, or slightly pubescent, nearly or quite simple, 1°–2° high. Leaves linear or linear-lanceolate, sessile and slightly clasping at the base, acute at the apex, regularly and closely serrate, sometimes pubescent on the veins beneath, 1′–2½′ long, 1½″–3″ wide; heads not very numerous, 5″–9″ broad; peduncles puberulent; involucre broadly campanulate, its bracts ovate-oblong, obtuse or obtusish, slightly tomentose; rays 5–15, white, rather large.

In moist soil, Newfoundland, New Brunswick and Quebec to Massachusetts and Michigan. Naturalized from Europe. Native also of northern Asia. Goose-tongue. Wild, bastard- or european pellitory. Fair-maid-of-France. Sneezewort-tansy. July–Sept.

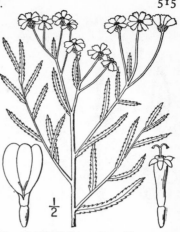

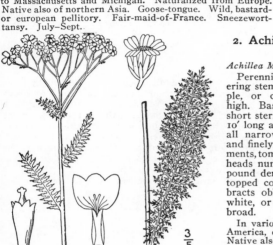

2. Achillea Millefòlium L. Yarrow. Milfoil. Fig. 4553.

Achillea Millefolium L. Sp. Pl. 899. 1753.

Perennial from horizontal rootstocks; flowering stems pubescent, or nearly glabrous, simple, or corymbosely branched above, 1°–2° high. Basal leaves, and those of the numerous short sterile shoots, mostly petioled, sometimes 10′ long and ½′ wide, those of the stem sessile, all narrowly oblong or lanceolate in outline and finely dissected into narrow pinnatifid segments, tomentose, pubescent or nearly glabrous; heads numerous, 2″–3″ broad, in terminal compound dense, somewhat convex or nearly flat-topped corymbs; involucre ovoid-cylindric, its bracts oblong, obtusish, pubescent; rays 4–6, white, or often pink or purple, less than 2″ broad.

In various situations throughout eastern North America, often occurring as a naturalized weed. Native also of Europe and Asia. Old names, sanguinary, thousand-leaf, nosebleed, old-man's-pepper, soldier's-woundwort, gordaldo. June–Nov.

3. Achillea lanulòsa Nutt. Woolly Yarrow. Fig. 4554.

Achillea lanulosa Nutt. Journ. Acad. Phila. **7**: 36. 1834.

Similar to the preceding species, perennial by rootstocks, 1°–2½° high, densely silky-woolly nearly all over. Leaves deeply bipinnatifid into narrow lobes and segments, those of the stem mostly sessile; inflorescence convex, 2′–4′ broad; involucre oblong-cylindric, its bracts greenish-yellow, with brownish margins; rays 1″–2½″ broad, white.

In dry soil, Quebec and Ontario to Michigan, Yukon, south to Oklahoma, Mexico and California. June–Sept. Locally naturalized eastward.

Achillea ligústica All., differing from *A. Millefolium* by being stouter with loosely corymbose heads, native of Europe, has been found in cultivated ground near Tannersville, New York.

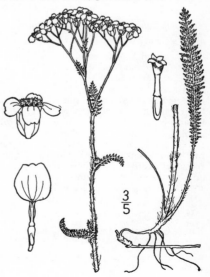

4. Achillea boreàlis Bongard. Northern Yarrow. Fig. 4555.

Achillea borealis Bongard, Veg. Sitch. 149. 1831.

More or less silky-woolly; stem erect, 16' high or less. Leaves deeply bipinnatifid into narrow crowded lobes and segments, those of the stem few, sessile or nearly so, the ultimate divisions very small; corymb dense, strongly convex, 2½' broad, or less; involucre about 3" high, its bracts with broad black or blackish margins; rays 10–20, white or pink, broadly oblong or suborbicular, often 2½" broad.

In wet places, on hillsides and rocks, Newfoundland to Quebec and Alaska. Summer. Rocky Mountain plants referred to this species appear to be distinct from it.

90. ÁNTHEMIS [Micheli] L. Sp. Pl. 893. 1753.

Annual or perennial herbs, with pinnatifid or dissected, alternate leaves, and usually large peduncled heads of both tubular and radiate flowers, terminating the branches, or heads rarely rayless. Involucre hemispheric, its bracts imbricated in several series, scarious-margined, appressed, the outer shorter. Receptacle convex, conic or oblong, chaffy at least toward the summit, the chaff subtending the disk-flowers. Ray-flowers pistillate and fertile, or neutral, the tube terete or 2-winged, the ray white or yellow, entire or 2–3-toothed. Disk-flowers perfect, fertile, yellow, their corollas with 5-cleft limbs. Anthers obtuse and entire at the base. Style-branches of the disk-flowers truncate. Achenes oblong, angled, ribbed or striate. Pappus none, or a short coroniform border. [Greek name of Camomile.]

About 60 species, natives of Europe, Asia and Africa. Type species: *Anthemis maritima* L.

Rays white.
  Rays neutral; plant glabrous, or nearly so, fetid.   1. *A. Cotula.*
  Rays pistillate; plants pubescent.
    Annual; chaff of the receptacle acute.   2. *A. arvensis.*
    Perennial; chaff of the receptacle obtuse.   3. *A. nobilis.*
Rays yellow; plant pubescent, or tomentose.   4. *A. tinctoria.*

1. Anthemis Cótula L. Mayweed. Dog's or Fetid Camomile. Dillweed. Fig. 4556.

Anthemis Cotula L. Sp. Pl. 894. 1753.

Maruta Co ula DC. Prodr. 6: 13. 1837.

Annual, glabrous, or sometimes pubescent above, glandular and with a fetid odor and acrid taste, much branched, 1°–2° high. Leaves mostly sessile, 1'–2' long, finely 1–3-pinnately dissected into narrow, or almost filiform, acute lobes; heads commonly numerous, about 1' broad; bracts of the involucre oblong, obtuse or obtusish, usually somewhat tomentose; rays 10–18, white, at length reflexed, neutral, or rarely with abortive pistils, mostly 3-toothed; receptacle convex, becoming oblong, its chaff bristly, subtending the central flowers; achenes 10-ribbed, rugose or glandular-tuberculate; pappus none.

In fields, waste places and along roadsides, all over North America except the extreme north. Naturalized from Europe, and widely distributed as a weed in Asia, Africa and Australia. Other names are mather, dog- or hog's-fennel, dog-finkle, morgan. Dog-daisy. Pig-sty-daisy. Maise. Chigger-weed. Balders. June–Nov.

2. Anthemis arvénsis L.   Corn or Field Camomile.   Fig. 4557.

Anthemis arvensis L. Sp. Pl. 894.   1753.

Annual or sometimes biennial, not fetid; stem finely pubescent, usually much branched, about 1° high, the branches decumbent or ascending. Leaves sessile, 1'-3' long, 1-2-pinnately parted into linear or lanceolate acute lobes, less divided than those of the preceding species and with broader segments; heads commonly numerous, 1'-1½' broad; bracts of the involucre oblong, obtuse, usually somewhat pubescent, with broad scarious margins; rays 10-18, white, pistillate, spreading, mostly 2-toothed; chaff of the obtuse receptacle lanceolate, acute or acuminate; achenes oblong, obtusely 4-angled; pappus a mere border.

In fields and waste places, Nova Scotia to Virginia, west to Michigan, Missouri, and on the Pacific coast. Naturalized from Europe.   May–Aug.

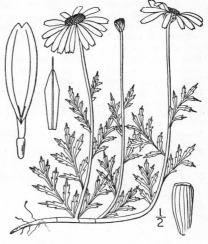

3. Anthemis nóbilis L.   Garden, Scotch, White or Low Camomile.   Fig. 4558.

Anthemis nobilis L. Sp. Pl. 894.   1753.

Perennial, pubescent, aromatic, much branched, 6'-18' high, the branches procumbent. Leaves numerous, 1'-2' long, finely and compactly dissected into nearly filiform lobes and segments; heads about 1' broad; bracts of the involucre obtuse, pubescent, their scarious margins broad; rays 12-18, white, spreading, pistillate, 2-3-toothed; chaff of the conic receptacle broad, membranous, obtuse; achenes oblong, obtusely 3-angled; pappus none.

Sparingly escaped from gardens, Rhode Island to Delaware, Michigan and Wyoming.   Adventive from Europe.   June–Aug.

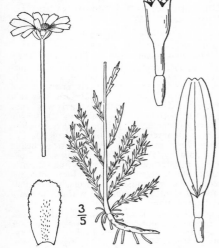

4. Anthemis tinctòria L.   Yellow or Oxeye Camomile.   Fig. 4559.

Anthemis tinctoria L. Sp. Pl. 896.   1753.

Perennial, pubescent or tomentose; stem erect, branched, 1°-3° high, with nearly erect branches. Leaves sessile, 1'-3' long, pinnately divided, the oblong segments pinnatifid into narrow acute lobes; heads few or several, 1'-1½' broad; bracts of the involucre oblong, obtuse, densely tomentose; rays 20-30, pistillate, usually 2-toothed, bright yellow or sometimes paler; chaff of the nearly hemispheric receptacle lanceolate, acuminate, rather rigid; achenes 4-angled, somewhat compressed; pappus a crown-like border.

In fields and waste places, New Brunswick to New Jersey, and locally escaped from gardens.   Adventive from Europe.   Native also of Asia.   June–Sept.

Anthemis aúrea (L.) DC., a species with small rayless heads, was found many years ago near St. Louis, Missouri.

91. CHRYSÁNTHEMUM [Tourn.] L. Sp. Pl. 888. 1753.

Perennial or annual, mostly erect and branching herbs, with alternate, dentate, incised or dissected leaves, and large, usually long-peduncled heads of both tubular and radiate flowers, or rays rarely wanting. Involucre hemispheric or depressed, its bracts appressed, imbricated in several series, the outer shorter. Receptacle flat, convex or hemispheric, naked. Ray-flowers pistillate, fertile, the rays white, yellow or rose-colored, entire or toothed. Disk-flowers perfect, fertile, their corollas with terete or 2-winged tubes and 4–5-cleft limbs. Anthers obtuse and entire at the base. Style-branches of the disk-flowers truncate, penicillate. Achenes angled or terete, 5–10-ribbed, those of the ray-flowers commonly 3-angled. Pappus none, or a scaly cup. [Greek, golden-flower.]

About 100 species, of wide geographic distribution in the northern hemisphere. Besides the following, 2 others occur in northwestern arctic America. Type species: *Chrysanthemum coronarium* L.

Heads large, few or solitary at the ends of the stem or branches.
     Rays white.
         Stem leaves linear-spatulate, pinnately incised ; weed.   1. *C. Leucanthemum.*
         Stem leaves cuneate-spatulate, toothed or lobed above ; arctic.   2. *C. arcticum.*
     Rays yellow.   3. *C. segetum.*
Heads numerous, small, corymbose ; plants ecsaped from gardens.
     Leaves pinnatifid, the segments incised.   4. *C. Parthenium.*
     Leaves oblong, serrate.   5. *C. Balsamita.*

$\frac{3}{5}$

1. Chrysanthemum Leucánthemum L. White-weed. White, Field or Ox-eye Daisy. Fig. 4560.

C. Leucanthemum L. Sp. Pl. 888. 1753.
Leucanthemum vulgare Lam. Fl. Fr. 2 : 137. 1778.

Perennial ; stems glabrous, or sparingly puberulent, simple or little branched, 1°–3° high, often tufted, the branches nearly erect. Basal leaves obovate, oblong, or spatulate, coarsely dentate, incised or pinnatifid, narrowed into long slender petioles ; stem leaves mostly sessile and partly clasping, 1′–3′ long, linear-spatulate or linear, pinnately incised or toothed, the uppermost very small and nearly entire ; heads few or solitary, 1′–2′ broad, on long naked peduncles ; rays 20–30, white, spreading, slightly 2–3-toothed ; bracts of the involucre oblong-lanceolate, obtuse, mostly glabrous, with scarious margins and a brown line within the margins ; pappus none.

In pastures, meadows and waste places, common throughout our area as a weed, but less abundant in the south and west. Bermuda. Naturalized from Europe. Native also of Asia. Other English names are dog-, bull-, butter-, big-, midsummer-, moon-, horse-, poorland- or maudlin-daisy ; dutch morgan, moon-flower, moon-penny, great white ox-eye, poverty-weed, white man's-weed, herb margaret ; bull's-eye daisy. Sheriff-pink. Dog-blow. Rays rarely short and tubular. Disk bright yellow. May–Nov.

2. Chrysanthemum àrcticum L. Arctic Daisy. Fig. 4561.

Chrysanthemum arcticum L. Sp. Pl. 889. 1753.
Leucanthemum arcticum DC. Prodr. 6 : 45. 1837.

Similar to the preceding species, but somewhat fleshy, lower, seldom over 1½° high. Leaves cuneate-spatulate, 1½′–3′ long, crenate or cleft at the apex, narrowed into a long tapering entire base, or the lower into slender petioles, slightly clasping at the base, the uppermost few, small, linear and nearly entire ; heads solitary or few, long-peduncled, 1′–2′ broad ; rays 20–30, white ; bracts of the involucre oblong, obtuse, brown, or with broad brown scarious margins, usually pubescent ; pappus none.

Coast of Hudson Bay to Alaska. Also in arctic Europe and Asia. Summer.

$\frac{1}{2}$

3. Chrysanthemum ségetum L.   Yellow Ox-eye.   Corn Marigold.   Fig. 4562.

Chrysanthemum segetum L. Sp. Pl. 889. 1753.

Annual, glabrous, $1\frac{1}{2}°$ high or less. Leaves oblong to oblanceolate, the upper auriculate-clasping, the lower petioled, dentate, incised, or nearly entire, $3'$ long or less; heads about $1\frac{1}{2}'$ broad; involucral bracts obtuse, scarious; rays obovate, yellow, emarginate; pappus a mere margin.

Waste grounds, New York, New Jersey, and in ballast about the seaports. Adventive from Europe.

Chrysanthemum coronàrium L., also European, with yellow rays and bipinnatifid leaves, has been found in Ontario.

4. Chrysanthemum Parthènium (L.) Pers.   Common Feverfew.   Feather-few.   Fig. 4563.

Matricaria Parthenium L. Sp. Pl. 890. 1753.
C. Parthenium Pers. Syn. 2 : 462. 1807.

Perennial; stem puberulent or glabrate, much branched, $1°$–$2\frac{1}{2}°$ high. Leaves thin, the lower often $6'$ long, petioled, or the upper sessile, pinnately parted into ovate or oblong, pinnatifid or incised segments; heads numerous, corymbose, slender-peduncled, $6''$–$10''$ broad; bracts of the depressed involucre lanceolate, rather rigid, keeled, pubescent, acute or acutish; rays 10–20, white, oval or obovate, spreading, mostly toothed, long-persistent; pappus a short toothed crown.

In waste places, New Brunswick and Ontario to New Jersey, Ohio, and in California, mostly escaped from gardens. Naturalized or adventive from Europe. Called also pellitory, wild camomile. Rays variable in length. Summer.

5. Chrysanthemum Balsámita L.   Costmary.   Mint Geranium.   Fig. 4564.

C. Balsamita L. Sp. Pl. Ed. 2, 1252. 1763.

Pyrethrum Balsamita Willd. Sp. Pl. 3 : 2153. 1804.

Perennial, puberulent or canescent; stem much branched, $2°$–$4°$ high. Leaves oblong, obtuse, crenate-dentate, $1'$–$2'$ long, those of the stem mostly sessile, and often with a pair of lateral lobes at the base; heads numerous, corymbose, slender-peduncled, $5''$–$8''$ broad, or when rayless only $3''$ broad; bracts of the involucre narrow, obtuse, pubescent; rays 10–15, white, spreading; pappus a short crown.

Sparingly escaped from gardens, Ohio to Michigan, Ontario and Nova Scotia. Native of the Old World. Other English names are cost, alecost, alecoast. Summer.

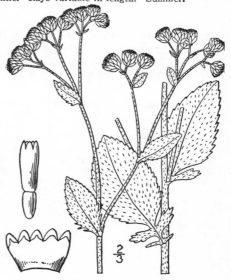

92. MATRICÀRIA L. Sp. Pl. 890. 1753.

Annual or perennial, mostly erect herbs, similar to some species of the preceding genus, with alternate leaves, dissected into filiform or narrowly linear segments and lobes, and peduncled heads of both tubular and radiate flowers, or rays wanting in some species. Involucre hemispheric, its bracts appressed, imbricated in few series, the outer shorter. Receptacle conic, elongated or hemispheric, naked. Rays, when present, white, pistillate and fertile. Disk-flowers yellow, perfect, fertile, their corollas 4–5-toothed. Anthers obtuse and entire at the base. Style-branches of the disk-flowers truncate, penicillate. Achenes 3–5-ribbed. Pappus none, or a coroniform border. [Latin, matrix, from its medicinal virtues.]

About 20 species, natives of the northern hemisphere and South Africa. The following are the only ones known to occur in North America. Type species: *Matricaria inodòra* L.

Rays present, white.
　Achenes obpyramidal, strongly 3-ribbed.
　　Plant tall, much branched; bracts of the involucre green.　　　　　1. *M. inodora.*
　　Plant low, nearly simple, arctic; bracts dark brown or black.　　　2. *M. grandiflora.*
　Achenes nearly terete, oblong, faintly 3–5-ribbed.　　　　　　　　3. *M. Chamomilla.*
Rays none; achenes oblong, faintly nerved.　　　　　　　　　　　4. *M. matricarioides.*

1. Matricaria inodòra L. Scentless Camomile. Corn Mayweed. Fig. 4565.

Matricaria inodora L. Fl. Suec. Ed. 2, 297. 1755.

Chrysanthemum inodorum L. Sp. Pl. Ed. 2, 1253. 1763.

Annual; stem usually much branched, glabrous, or very nearly so throughout, 1°–2° high. Leaves numerous, sessile, 2–3-pinnately dissected into filiform lobes, the rachis somewhat dilated at the base; heads several or numerous, terminating the branches, ¾'–1½' broad; bracts of the involucre lanceolate-oblong, obtuse, green with brown scarious margins; rays 20–30, white, spreading; receptacle hemispheric or ovoid; achenes obpyramidal with three prominent ribs; pappus a short entire or 4-toothed crown.

In fields and waste places, Newfoundland to New Jersey, Pennsylvania and Michigan. Naturalized or adventive from Europe. June–Sept.

2. Matricaria grandiflòra (Hook.) Britton. Arctic Camomile. Fig. 4566.

Chrysanthemum grandiflorum Hook. in Parry's 2d Voy. 398. 1825.

Pyrethrum inodorum var. *nanum* Hook. Fl. Bor. Am. 1: 320. 1833.

M. grandiflora Britton, Mem. Torr. Club 5: 340. 1894.

Perennial; stem usually simple and monocephalous, glabrous, 4'–12' high. Leaves sessile, or the lowest short-petioled, 1–2-pinnately dissected, 1'–2½' long; head not very long-peduncled, 1'–2' broad; bracts of the involucre ovate or ovate-oblong, obtuse, glabrous, brown or nearly black, or with broad, brown, scarious margins; rays 15–35, bright white, slightly 3–5-toothed at the summit; receptacle hemispheric when mature.

Coast of Hudson Bay to Alaska. Reported from Lake Huron. Summer.

3. Matricaria Chamomílla L. Wild or German Camomile. Fig. 4567.

Matricaria Chamomilla L. Sp. Pl. 891. 1753.

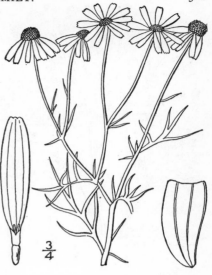

Annual, glabrous, much branched, 1°–2° high. Leaves aromatic, finely 2–3-pinnately dissected into numerous linear lobes; heads numerous, 8″–12″ broad, slender-peduncled at the ends of the branches; bracts of the involucre oblong, obtuse, green, or with brownish margins; rays 10–20, white, spreading; receptacle ovoid, becoming conic and hollow; achenes nearly oblong, or somewhat obovoid, faintly 3–5-ribbed; pappus none.

In waste places and on ballast, southern New York to Pennsylvania. Adventive or fugitive from Europe. Horse-gowan. Summer.

$\frac{3}{4}$

$\frac{3}{4}$

4. Matricaria matricarioìdes (Less.) Porter. Rayless Camomile. Wild Marigold. 4568.

Sántolina suaveolens Pursh, Fl. Am. Sept. 520. 1814. Not *M. suaveolens* L. 1755.
Artemisia matricarioides Less. Linnaea 6: 210. 1831.
Matricaria discoidea DC. Prodr. 6: 50. 1837.
Matricaria matricarioides Porter, Mem. Torr. Club 5: 341. 1894.
M. suaveolens Buchenau, Fl. Nord. Tief. 496. 1894.

Annual, glabrous; stem very leafy, at length much branched, 6′–18′ high. Leaves 2–3-pinnately dissected into linear acute lobes; heads numerous, 3″–4″ broad, peduncled; bracts of the involucre oval or oblong, green, with broad white scarious margins, much shorter than the ovoid yellow disk; rays none; receptacle conic; achenes oblong, slightly angular, faintly nerved; pappus an obscure crown, sometimes produced into 2 coriaceous oblique auricles.

In waste places, in ballast and along railroads, Missouri to Massachusetts and Maine. Adventive from the Pacific coast. Naturalized as a weed in northern Europe. May–Aug.

93. TANACÈTUM [Tourn.] L. Sp. Pl. 843. 1753.

Erect, strongly aromatic herbs, our species perennials, with alternate, 1–3-pinnately dissected or divided leaves, and numerous small corymbose heads of tubular flowers, or with rays sometimes present and imperfectly developed. Involucre hemispheric, depressed, or campanulate, its bracts appressed, imbricated in several series. Receptacle flat or convex, naked. Marginal flowers pistillate, fertile, their corollas 2–5-toothed or lobed, sometimes produced into short rays. Disk-flowers perfect, fertile, their corollas 5-toothed. Anthers obtuse and entire at the base, their tips broad. Style-branches truncate and penicillate at the summit. Achenes 5-angled or 5-ribbed, truncate or obtuse. Pappus none, or a short crown. [From tanasie, old French for tansy; Greek, athanasia, immortality.]

About 30 species, natives of the northern hemisphere. Besides the following, another occurs in California. Type species: *Tanacetum vulgare* L.

Glabrous, or nearly so; heads numerous, 3″–5″ broad.   1. *T. vulgare.*
Villous-pubescent; heads few, 6″–8″ broad.   2. *T. huronense.*

$\frac{2}{3}$

1. Tanacetum vulgàre L. Tansy.
Fig. 4569.

Tanacetum vulgare L. Sp. Pl. 844. 1753.
Tanacetum vulgare crispum DC. Prodr. **6**: 128. 1837.

Stem stout, usually simple up to the inflorescence, glabrous, or sparingly pubescent, $1\frac{1}{2}°-3°$ high. Leaves pinnately divided into linear-oblong, pinnatifid or incised, often crisped segments, the lobes acute, usually serrate; lower segments of the leaves often smaller than the others; basal leaves often 1° long; heads commonly numerous, 3″–5″ broad, rather short-peduncled; involucre depressed-hemispheric, its bracts oblong-lanceolate, obtuse, or the outer acute, slightly pubescent or ciliate; receptacle flat; flowers yellow; marginal corollas with short oblique 3-toothed limbs; pappus a short crown.

Along roadsides, mostly escaped from gardens, Nova Scotia and Ontario to Minnesota, Oregon, Nevada, North Carolina and Missouri. Naturalized or adventive from Europe. Bitter-buttons. Hindheal. Ginger-plant. July–Sept.

2. Tanacetum huronénse Nutt. Lake Huron Tansy. Fig. 4570.

Tanacetum huronense Nutt. Gen. **2**: 141. 1818.

Villous-pubescent throughout, at least when young, less so when mature, 1°–2° high. Leaves 2-pinnately divided, the lobes dentate or incised, acute, the lower segments commonly smaller than the others; heads 1–8, 6″–8″ broad, on very stout pubescent peduncles; involucre depressed-hemispheric; marginal flowers with 3–5-lobed limbs, often expanded into short rays; pappus a short crown.

In moist soil, especially along streams or lakes, New Brunswick to Hudson Bay, Maine, Lake Superior, Alaska and Oregon. July–Sept.

$\frac{3}{5}$

94. ARTEMÍSIA [Tourn.] L. Sp. Pl. 845. 1753.

Odorous herbs or shrubs, often canescent or tomentose, with alternate leaves, and small pendulous or erect, discoid racemose spicate glomerate or paniculate heads of greenish or yellowish flowers. Involucre ovoid, oblong, or broadly hemispheric, its bracts imbricated in few series, the outer gradually shorter. Receptacle flat, convex or hemispheric, naked or pubescent, not chaffy. Central flowers perfect, sometimes sterile, with abortive ovaries and undivided style, sometimes perfect and fertile, with truncate style-branches; marginal flowers usually pistillate and fertile, their corollas 2–3-toothed, or flowers all perfect and fertile in some species. Anthers obtuse and entire at the base, often tipped with subulate appendages. Achenes obovoid or oblong, 2-ribbed or striate, rounded at the summit, usually bearing a large epigynous disk. Pappus none. [Named for Artemisia, wife of Mausolus.]

About 225 species, natives of the northern hemisphere and southern South America. Besides the following, some 40 others occur in the western parts of North America. Type species: *Artemisia vulgaris* L.

* Marginal flowers pistillate; central flowers perfect, sterile.
         a. Biennial or perennial herbs.

Leaves pinnately dissected into narrowly linear lobes.	
Heads very numerous, 1″ broad; leaves mostly glabrous.	1. *A. caudata.*
Heads 2″ broad, in narrow panicles; leaves silky-pubescent.	
Heads few; involucre brown, mostly pubescent.	2. *A. borealis.*
Heads numerous; involucre green, mostly glabrous.	3. *A. canadensis.*
Leaves linear, the lower sometimes 3-cleft or pinnately divided.	
Leaves glabrous.	4. *A. dracunculoides.*
Leaves finely and densely pubescent.	5. *A. glauca.*
b. Shrubby, silvery-canescent; heads small and numerous.	6. *A. filifolia.*

** Marginal flowers pistillate; central flowers perfect, fertile.
　　　　a. Receptacle villous-pubescent.

Leaf-segments linear-filiform, short; native.　　　　　　　　　　7. *A. frigida.*
Leaf-segments oblong, or linear-oblong; introduced.　　　　　　　8. *A. Absinthium.*

　　　　b. Receptacle glabrous, or sparingly pubescent.
Leaves dissected, glabrous or pubescent, green, not tomentose.
　Heads about 2″ broad, numerous in panicled racemes; perennial.　　9. *A. Abrotanum.*
　Heads about 1″ broad, paniculate or spicate; annuals.
　　Leaves finely 2–3-pinnately divided; heads paniculate.　　　　10. *A. annua.*
　　Leaves pinnately divided; segments pinnatifid; heads in leafy spikes.　11. *A. biennis.*
Leaves densely white-canescent or tomentose, at least beneath.
　Leaves pinnatifid or dissected.
　　Heads 3″–4″ broad, racemose-glomerate; sea-beach plant.　　　12. *A. Stellariana.*
　　Heads 1″–2″ broad, spicate-paniculate or racemose.
　　　Leaves deeply pinnatifid, the segments mostly incised.　　　13. *A. vulgaris.*
　　　Leaves finely dissected into short linear lobes.　　　　　　14. *A. Pontica.*
　　　Leaves pinnately parted into 5–7 narrow entire segments.　　15. *A. kansana.*
　Leaves lanceolate or linear, serrate or entire, not pinnatifid.
　　Leaves lanceolate, sharply serrate, glabrous above.　　　　　16. *A. serrata.*
　　Leaves linear, oblong, lanceolate or obovate, entire or lobed.
　　　Leaves at length glabrous above.
　　　　Leaves linear, elongated, all entire.　　　　　　　　　17. *A. longifolia.*
　　　　Leaves various, at least the lower pinnately lobed or toothed.
　　　　　Involucre densely woolly; leaf-lobes broad.　　　　　18. *A. ludoviciana.*
　　　　　Involucre loosely woolly; leaf-lobes linear.　　　　　19. *A. mexicana.*
　　　Leaves shorter, oblong or lanceolate, tomentose both sides.　20. *A. gnaphalodes.*
　Leaves cuneate, ½′ long, 3-toothed at the apex.　　　　　　　　21. *A. Bigelovii.*

　　　*** Flowers all perfect and fertile; far western species.
Leaves cuneate, 3-toothed or 3-lobed.　　　　　　　　　　　　22. *A. tridentata.*
Leaves linear, entire.　　　　　　　　　　　　　　　　　　　23. *A. cana.*

1. Artemisia caudàta Michx. Tall or Wild Wormwood. Fig. 4571.

Artemisia caudata Michx. Fl. Bor. Am. **2**: 129. 1803.

Root biennial (sometimes perennial?); stems slender, glabrous, tufted, strict, very leafy, 2°–6° high, at length paniculately branched, the branches glàbrous, or rarely slightly pubescent, nearly erect. Lower and basal leaves and those of sterile shoots slender-petioled, sometimes a little pubescent, 3′–6′ long, 2–3-pinnately divided into narrowly linear, acute lobes, about ½″ wide; upper leaves sessile or nearly so, pinnately divided, or the uppermost entire and short; heads about 1″ broad, very short-peduncled, very numerous in a large somewhat leafy panicle, mostly nodding; bracts of the ovoid-campanulate involucre ovate, or the inner elliptic, glabrous; receptacle hemispheric, naked; central flowers sterile.

In dry sandy soil, abundant on sea-beaches, from Quebec to Florida, west to Ontario, Indiana, Manitoba, south to Nebraska and Texas. July–Sept.

2. Artemisia borèalis Pall. Northern Wormwood. Fig. 4572.

Artemisia borealis Pall. Iter. 129. *pl. hh, f. 1.* 1771.

Artemisia groenlandica Wormsk. Fl. Dan. *pl. 1585.* 1818.

Perennial, 5′–15′ high, densely silky-pubescent all over, resembling small forms of the following species. Leaves less divided, the basal and lower ones petioled, 1′–2½′ long, the upper sessile, linear and entire or merely 3-parted; heads about 2″ broad in a dense terminal rarely branched thyrsus; involucre nearly hemispheric, its bracts brown or brownish, pilose-pubescent or nearly glàbrous; receptacle convex, naked; disk-flowers sterile.

Quebec to Greenland, west through arctic America to Alaska, south in the Rocky Mountains to Colorado. Also in northern Asia. Apparently erroneously recorded from Maine. July–Aug.

3/4

3. Artemisia canadénsis Michx. Canada Worm-wood. Fig. 4573.

Artemisia canadensis Michx. Fl. Bor. Am. **2**: 129. 1803.

Root perennial (or sometimes biennial); stem pubescent or glabrous, strict, simple or branched, 1°–2° high, the branches appressed and erect. Leaves usually pubescent, but sometimes sparingly so, the basal and lower ones petioled, 2′–3′ long, 2-pinnately divided into linear, acute lobes which are shorter and broader than those of *Artemisia caudata;* upper leaves sessile, less divided; heads short-peduncled, about 2″ broad, commonly numerous in a narrow virgate panicle, mostly spreading or erect, in small forms the panicle reduced to a nearly or quite simple terminal raceme; involucre ovoid, its bracts ovate or oval, green, glabrous or pubescent; receptacle hemispheric; central flowers sterile.

In rocky soil, Newfoundland to Hudson Bay, Maine, Vermont, west along the Great Lakes to Minnesota and Manitoba and to the Canadian Pacific coast. Sea- or wild-wormwood. July–Aug.

Artemisia Forwoódii S. Wats., a taller plant of the Rocky Mountain region, with somewhat smaller heads, ranges eastward into Nebraska.

4. Artemisia dracunculoìdes Pursh. Linear-leaved Wormwood. Fig. 4574.

A. dracunculoides Pursh, Fl. Am. Sept. 742. 1814.

Perennial, glabrous; stem somewhat woody, usually much branched, 2°–4° high, the branches nearly erect. Leaves linear, 1′–3½′ long, 1″–2″ wide, acute, entire, or the lower and basal ones sometimes 3-cleft or even more divided; heads very numerous, 1″–1½″ broad, nodding, very short-peduncled, racemose-paniculate; involucre nearly hemispheric, its bracts ovate or oblong, green, scarious-margined; receptacle hemispheric, naked; central flowers sterile.

Dry plains and prairies, Manitoba to British Columbia, Illinois, Missouri, Nebraska, Texas, Chihuahua, New Mexico and California. July–Nov.

3/4

3/4

5. Artemisia glaùca Pall. Silky Wormwood. Fig. 4575.

A. glauca Pall.; Willd. Sp .Pl. **3**: 1831. 1804.
Artemisia dracunculoides var. *incana* T. & G. Fl. N. A. **2**: 416. 1843.

Perennial, similar to the preceding species; stems strict, leafy, usually simple or little branched, 1°–2° high, pubescent, tomentose or canescent, or glabrous below. Leaves linear, ½′–2½′ long, about 1″ wide, entire, finely and densely pubescent, obtuse or obtusish, or the lower or sometimes nearly all of them 3-cleft into linear lobes, 1′–1½′ long; panicle narrow, branched, its branches nearly erect; heads drooping, sessile, very numerous, scarcely more than 1½″ long; involucre hemispheric, its bracts scarious-margined, obtuse; receptacle naked; central flowers sterile.

Minnesota to North Dakota, Manitoba and Saskatchewan. June–Sept.

6. Artemisia filifòlia Torr.　Silvery Wormwood.　Fig. 4576.

Artemisia filifolia Torr. Ann. Lyc. N. Y. **2**: 211.　1827.

Shrubby, finely silvery-canescent throughout; stem branched, 1°–3° high, the rigid branches nearly erect. Leaves 1′–2′ long, nearly all 3-parted into filiform entire segments less than ½″ wide, or the uppermost undivided; heads exceedingly numerous, about ¼″ broad, racemose-paniculate, very short-peduncled, 3–5-flowered; involucre oblong, its bracts densely canescent; receptacle small, naked or slightly fimbrillate; central 1–3 flowers sterile.

On dry plains, Nebraska to Utah, Wyoming, Nevada, Texas, Mexico and Arizona. Wormwood-sage. July–Oct.

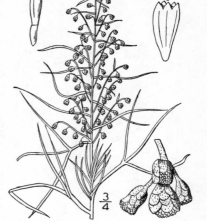

7. Artemisia frígida Willd.　Pasture Sage-Brush.　Wormwood Sage.　Fig. 4577.

Artemisia frigida Willd. Sp. Pl. **3**: 1838.　1804.

Perennial, woody at the base, densely silky-canescent all over; stem branched or simple, 10′–20′ high. Leaves ½′–1½′ long, ternately or 5-nately divided into numerous short acutish mostly entire lobes less than ½″ wide, the lower and basal ones petioled, and often with a pair of entire or 3-cleft divisions near the base of the petiole, the upper sessile and less divided; heads rather numerous, racemose or racemose-paniculate, short-peduncled, nodding, about 2″ broad; involucre hemispheric, its bracts oblong, canescent or tomentose; receptacle villous-pubescent; central flowers fertile.

On dry plains and in rocky soil, Minnesota to Saskatchewan, Yukon, Idaho, Nebraska, Texas and Arizona. Wild sage. July–Oct.

8. Artemisia Absínthium L.　Common Wormwood.　Absinth.　Fig. 4578.

Artemisia Absinthium L. Sp. Pl. 848.　1753.

Shrubby, finely canescent; stem much branched, 2°–4° high. Leaves 2′–5′ long, 1–3-pinnately divided into numerous linear to obovate, obtuse lobes, the lower long-petioled, the upper short-petioled or sessile, the uppermost commonly linear and entire; heads numerous, yellow, racemose-paniculate, drooping, short-peduncled, 2″–2½″ broad; involucre hemispheric, its outer bracts linear, the inner much broader, scarious-margined; receptacle pilose-pubescent; central flowers fertile, the marginal ones pistillate, fertile or sterile.

In waste places, Newfoundland and Hudson Bay to Massachusetts, Pennsylvania, North Carolina, western Ontario, New York, North Dakota and Montana. Naturalized or adventive from Europe, mostly escaped from gardens.　Old English names, madderwort, mugwort, mingwort, warmot. Boys'-love. July–Oct.

9. Artemisia Abrótanum L. Southernwood. Fig. 4579.

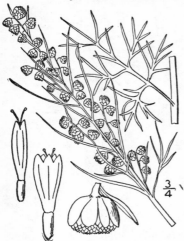

Artemisia Abrotanum L. Sp. Pl. 845. 1753.

Perennial, somewhat shrubby; stem puberulent or glabrous, much branched, 2°–4° high, the branches short, erect or ascending. Leaves glabrous or somewhat pubescent, 1′–3′ long, 1–3-pinnately parted into linear obtuse entire lobes about ½″ wide, or the uppermost linear and entire, the lowest petioled; heads several-flowered, yellow, very numerous, nodding, racemose-paniculate, 2″–2½″ broad; involucre nearly hemispheric, pubescent, its outer bracts lanceolate, acute, the inner ones obovate; receptacle glabrous; central flowers fertile.

In waste places, Massachusetts to western New York, southern Ontario, and Nebraska. Adventive from continental Europe. Old English names, lad's-love, boys'-love, slovenwood, old-man, sweet benjamin.

Artemisia prócera Willd., a similar species, but with glabrous involucre, is recorded as escaped from gardens at Buffalo, N. Y.

10. Artemisia ánnua L. Annual Wormwood. Fig. 4580.

Artemisia annua L. Sp. Pl. 847. 1753.

Annual, glabrous throughout, much branched, 2°–5° high. Leaves 2′–6′ long, finely 2–3-pinnately dissected into very narrow short, obtuse lobes, the lower and basal ones slender-petioled, the upper sessile and less divided, but none of them entire; heads very numerous, about 1″ broad, drooping, borne on very slender peduncles of about their own length or less; involucre hemispheric, glabrous, its bracts few, ovate to oblong; receptacle glabrous; flowers commonly all fertile.

In waste places, Ontario to New Hampshire, Virginia, West Virginia, Tennessee, Kansas and Arkansas, a bad weed in some places. Adventive or naturalized from Asia. Summer.

11. Artemisia biénnis Willd. Biennial Wormwood. Fig. 4581.

Artemisia biennis Willd. Phytogr. 11. 1794.

Annual or biennial, glabrous throughout; stem very leafy, usually branched, 1°–4° high, the branches nearly erect. Leaves 1′–3′ long, 1–2-pinnately divided into linear or linear-oblong, acutish, serrate or incised lobes, the lowest petioled, the uppermost less divided or rarely quite entire; heads about 1½″ broad, not drooping, sessile and exceedingly numerous in axillary glomerules which are crowded, forming a compound spicate inflorescence, the subtending leaves much exceeding the clusters; involucre nearly hemispheric, its bracts green, scarious-margined; receptacle naked; central flowers fertile.

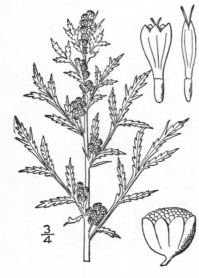

Native from Tennessee to Nebraska, Manitoba, British Columbia and California, now widely distributed as a weed from Manitoba to Nova Scotia, south to Missouri, Kentucky and Delaware. Aug.–Oct.

12. Artemisia Stelleriàna Bess. Beach Wormwood. Fig. 4582.

Artemisia Stelleriana Bess. Abrot. 79. *pl. 5.* 1829.

Perennial, densely white-tomentose; stem branched, 1°–2½° high, bushy, the branches ascending. Leaves obovate to spatulate, 1'–4' long, pinnatifid into oblong, obtuse, entire or few-toothed lobes, the lower petioled, the upper sessile, all densely tomentose beneath, but becoming green and glabrous above when old; heads racemose-spicate or racemose-glomerate, 3″–4″ broad, not drooping; involucre oblong-campanulate, its bracts tomentose, lanceolate or oblong-lanceolate; receptacle naked; central flowers fertile.

Sandy sea-beaches, Quebec to New Jersey; Oneida Lake, N. Y. Cultivated in gardens along the coast. Native of northeastern Asia. Occurs also on the coast of Sweden. Foliage similar to that of the dusty miller, *Cineraria maritima* L. July–Aug.

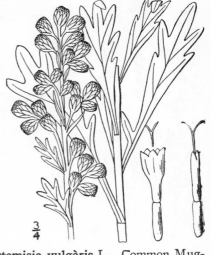

¾

13. Artemisia vulgàris L. Common Mugwort. Fig. 4583.

Artemisia vulgaris L. Sp. Pl. 848. 1753.

Perennial; stem glabrous or nearly so, much branched, 1°–3½° high. Leaves 1'–4½' long, deeply pinnatifid, into linear, oblong or somewhat spatulate, pinnatifid, toothed or entire lobes, densely white-tomentose beneath, dark green and glabrous above, the lower petioled and often with 1 or 2 pairs of small lateral divisions at or near the base of the petiole, the upper sessile, the uppermost sometimes linear and entire; heads numerous, erect, about 2″ broad, in panicled, simple or compound spikes; involucre oblong-campanulate, its bracts oblong, obtusish, scarious-margined, tomentose or glabrous; receptacle naked; central flowers fertile.

⅔

In waste places, Nova Scotia to Ontario, Michigan, New Jersey, Pennsylvania and Georgia. Naturalized from Europe. Native also of Asia. Reported as native of arctic America. Motherwort. Fellon-herb. Sailor's-tobacco. Wormwood. Bulwand. Green ginger. July–Oct.

Artemisia elatiòr (T. & G.) Rydb., a northwestern species, with elongated acuminate leaf-lobes, ranges eastward to Hudson Bay.

14. Artemisia póntica L. Roman or Hungarian Wormwood. Fig. 4584.

Artemisia pontica L. Sp. Pl. 847. 1753.

Perennial; stem branched, glabrous or canescent, 1°–3° high. Leaves 1½'–2½' long, 2–3-pinnately dissected into short narrow lobes less than 1″ wide, canescent on both sides, or tomentose beneath, the lower petioled and the petioles somewhat clasping or auricled at the base, the upper mostly linear and entire; heads numerous, 1″–2″ broad, drooping, slender-peduncled; involucre hemispheric, canescent, its bracts oblong or obovate, obtuse, the outer short, lanceolate; receptacle glabrous; central flowers fertile.

Waste grounds, Massachusetts, New Jersey, Pennsylvania, Ohio and Colorado. Fugitive or adventive from central Europe. July–Aug.

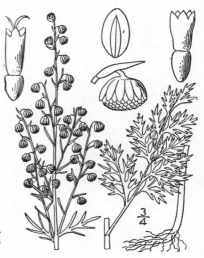

¾

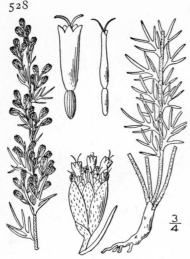

15. Artemisia kansàna Britton. Kansas Mug-wort. Fig. 4585.

?*A. Carruthii* A. Wood, Trans. Kans. Acad. Sci. **5**: 51. 1876.
A. kansana Britton, in Britt. & Brown, Ill. Fl. **3**: 466. 1898.

Densely white-woolly all over; stem erect, much branched, the branches strict, bearing very numerous small heads forming a narrow dense panicle. Leaves numerous, crowded, the lower pinnately divided into 3–7 narrowly linear revolute-margined segments ½″ wide or less, greenish above; upper leaves mostly narrowly linear and entire; heads oblong-oval, sessile, or very short-peduncled, erect, or somewhat spreading, 1½″ long; involucre very woolly, its bracts ovate-lanceolate to oblong-lanceolate, acute; receptacle naked.

Plains, Kansas to Colorado and New Mexico. Introduced in Missouri. July–Sept.

16. Artemisia serràta Nutt. Saw-leaf Mugwort. Fig. 4586.

Artemisia serra a Nutt. Gen. **2**: 142. 1818.

Perennial; stem stout, tomentose or becoming glabrous, much branched, 5°–10° high. Leaves lanceolate, 2′–6′ long, 3″–12″ wide, densely white-tomentose beneath, dark green and glabrous above, acuminate at the apex, narrowed to a sessile base, or the lowest petioled, sharply serrate or incised, or the upper entire; heads very numerous, greenish, erect, about 1½″ broad, sessile or short-peduncled in panicled spikes or racemes; involucre canescent, its bracts oblong, or the outer ones lanceolate; receptacle naked; central flowers fertile.

Prairies, Illinois to Minnesota and Dakota. Introduced on the Mohawk River, near Schenectady, N. Y. Aug.–Oct.

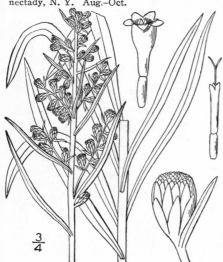

17. Artemisia longifòlia Nutt. Long-leaved Mugwort. Fig. 4587.

Artemisia longifolia Nutt. Gen. **2**: 142. 1818.

Perennial; stem densely white-tomentose, branched, 2°–5° high. Leaves linear or linear-lanceolate, elongated, entire, 2′–5′ long, 1″–5″ wide, acuminate, tapering to a sessile base, or the lower petioled, densely white-tomentose on both sides, or becoming green and glabrate above; heads numerous, erect, spicate-paniculate, about 2″ broad; involucre tomentose, its bracts narrowly oblong; receptacle naked; central flowers fertile.

In dry rocky soil, western Nebraska to Minnesota, Idaho, Oregon and Saskatchewan. Aug.-Sept.

18. Artemisia ludoviciàna Nutt. Dark-leaved Mugwort. Fig. 4588.

Artemisia ludoviciana Nutt. Gen. **2**: 143. 1818.

Perennial, 2°–4° high; stem woolly, branched above. Leaves linear to obovate, 3′ long or less, white-woolly beneath, at length dark green and glabrous, or very nearly so above, the base commonly narrowly cuneate, at least the lower ones pinnately lobed or toothed, their lobes lanceolate, acuminate; upper leaves often linear and entire; heads numerous, spicate-paniculate, 1″–1½″ broad; involucre oblong, tomentose; receptacle naked; central flowers fertile.

In dry soil, Missouri to Texas, Wyoming, Colorado and Arizona. Aug.–Nov.

19. Artemisia mexicàna Willd. Mexican Mugwort. Fig. 4589.

Artemisia mexicana Willd.; Spreng. Syst. **3**: 490. 1826.

Perennial, less densely pubescent than *A. ludoviciana,* 2°–3¾° high, often branched; stem finely pubescent and ultimately often floccose. Leaves ovate or orbicular in outline on the lower part of the stem, 2′–3½′ long, densely white-tomentulose beneath, green above, the lobes of the lower and the blades of the upper entire ones linear to narrowly linear or nearly so; heads small and numerous, usually inclined or nodding; involucre campanulate, loosely woolly, the pubescence sparse.

On prairies. hillsides and barrens, Missouri to Texas, Arkansas and Mexico. Sept.–Oct.

20. Artemisia gnaphalòdes Nutt. Prairie or Western Sage. Cud-weed Mugwort. Fig. 4590.

Artemisia gnaphalodes Nutt. Gen. **2**: 143. 1818.
Artemisia ludoviciana var. *gnaphalodes* T. & G. Fl. N. A. **2**: 420. 1843.

Perennial; stem white-tomentose, usually much branched, 1°–4° high. Leaves lanceolate or oblong, 1′–3′ long, 2″–6″ wide, entire, or the lower somewhat toothed, or rarely few-lobed, white-tomentose on both sides, acute or acuminate, sessile or the lower narrowed into short petioles; heads numerous, spicate-paniculate, about 1½″ broad; involucre oblong, tomentose; receptacle naked; central flowers fertile.

On prairies, plains, and dry banks, western Ontario and Illinois to Alberta, Missouri, Texas and Mexico. Locally established in waste grounds from New Hampshire to Delaware. Far western plants formerly referred to this species, which consists of many races, are, apparently, distinct.

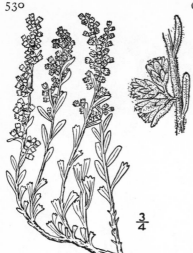

$\frac{3}{4}$

22. Artemisia tridentàta Nutt.   Common Sage-bush.   Sage-brush.   Sage-wood. Mountain Sage.   Fig. 4592.

Artemisia tridentata Nutt. Trans. Am. Phil. Soc. (II) **7**: 398. 1841.

Shrubby, silvery-canescent; stem much branch-ed, 1°–12° high. Leaves narrowly cuneate, ½′–1½′ long, 1″–3″ wide, sessile, 3–7-toothed at the truncate apex; heads very numerous, 5–8-flow-ered, about 1½″ broad, sessile, or very nearly so, in large dense panicles; involucre oblong, to-mentose, its inner bracts oblong, the outer short, ovate, all obtuse or obtusish; receptacle naked; flowers all perfect and fertile.

On dry plains and in rocky soil, western Ne-braska to Colorado, Utah and California, north to Montana and British Columbia. July–Sept.

$\frac{3}{4}$

21. Artemisia Bigelòvii A. Gray.   Bigelow's Sage-Bush.   Fig. 4591.

Artemisia Bigelovii A. Gray, Pac. R. R. Rep. **4**: 110. 1856.

Perennial, shrubby, silvery-canescent throughout, 8′–15′ high, much branched, the branches erect. Leaves narrowly cuneate, or oblong, obtuse, truncate, or 3–5-toothed at the apex, 5″–9″ long, about 1″ wide; heads very numerous, about 1″ broad, densely glom-erate-spicate in a narrow virgate panicle, 2–5-flow-ered, 1 or 2 of the marginal ones pistillate, the others perfect and fertile; involucre short-oblong, canescent or tomentose, its bracts obtuse; receptacle naked.

Kansas (according to Smyth) ; Colorado to Texas and Arizona. Aug.–Oct.

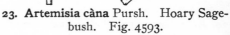

$\frac{3}{4}$

23. Artemisia càna Pursh.   Hoary Sage-bush.   Fig. 4593.

Artemisia cana Pursh, Fl. Am. Sept. **521**. 1814.

Shrubby, densely white-canescent; stem much branched, 1°–2½° high. Leaves linear, linear-oblong or narrowly lanceolate, sessile, acute at both ends, 1′–2′ long, 1½″–3″ wide, usually quite entire, rarely with 2 or 3 acute teeth or lobes; heads numerous, about 1½″ broad, glom-erate or sometimes solitary in the axils of the leaves, or crowded into a naked thyrsus at the summit, 5–9-flowered; involucre ob-long, canescent, its inner bracts oblong or lan-ceolate, obtuse, usually with 1–3 shorter outer ones; receptacle naked; flowers all perfect and fertile.

Plains, Nebraska and Colorado to North Da-kota, Montana and Saskatchewan. July–Sept.

95. TUSSILÀGO [Tourn.] L. Sp. Pl. 865. 1753.

An acaulescent herb, more or less white-tomentose, with slender perennial rootstocks, broad basal cordate, dentate or lobed, long-petioled leaves, and large solitary, monoecious heads of both tubular and radiate yellow flowers at the summit of a scaly scape, appearing before the leaves of the season. Involucre campanulate to cylindric, its principal bracts in a single series, equal, with or without a few shorter outer ones. Receptacle flat, naked. Ray-flowers in several series, pistillate, fertile. Disk-flowers perfect, sterile, the corolla 5-cleft, the style undivided and obtuse, lobed. Achenes of the ray-flowers linear, 5–10-ribbed. Pappus copious, of numerous slender roughish bristles, that of the sterile flowers shorter than that of the fertile. [Latin, *tussis,* cough, for which the plant was a reputed remedy.]

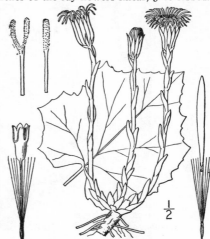

A monotypic genus of northern Europe and Asia.

1. Tussilago Fàrfara L. Coltsfoot. Coughwort. Fig. 4594.

Tussilago Farfara L. Sp. Pl. 865. 1753.

Scape slender, 3′–18′ high, bearing a solitary large head at the summit. Leaves nearly orbicular, or broadly ovate-reniform, angularly lobed and dentate, 3′–7′ broad, green and glabrous above, persistently white-tomentose beneath; head about 1′ broad; involucre campanulate; rays bright yellow, numerous, linear.

In moist soil, on banks and roadsides, Nova Scotia and New Brunswick to New Jersey, Pennsylvania and Minnesota. Naturalized from Europe. Horse-foot. Horse-hoof. Dove-dock. Sow-foot. Colt-herb. Hoofs. Cleats. Ass's-foot. Bull's-foot. Foal-foot. Ginger. Clay-weed. Butter-bur. Dummy-weed. April–June.

96. PETASÌTES [Tourn.] Mill. Gard. Dict. Abr. Ed. 4. 1754.

Herbs with perennial thick horizontal rootstocks, broad, basal, petioled leaves, and scaly scapes bearing racemose or corymbose heads of tubular or both tubular and radiate, white or purplish, often dioecious or subdioecious flowers. Involucre campanulate to cylindric, its bracts in 1 series, equal. Receptacle flat, or nearly so, not chaffy. Corolla of pistillate flowers very slender, 2–5-toothed, truncate or sometimes with a ray, marginal, or composing most of the head; perfect but sterile flowers with a tubular 5-cleft corolla, the style undivided. Anthers entire or minutely sagittate at the base. Fertile achenes linear, the pappus of numerous capillary rough or barbellate bristles. [Greek, a broad-rimmed hat, referring to the broad leaves of these plants.]

About 20 species, north temperate and subarctic. Type species: *Tussilago Petasites* L.

Flowers whitish, the pistillate radiate; natives; northern.
Leaves orbicular, 7–11-cleft nearly to the base.
Leaves deltoid-reniform, sinuate-lobed and toothed.
Leaves deltoid-ovate, repand-denticulate.
Flowers all rayless, purple; introduced.

1. *P. palmata.*
2. *P. trigonophylla.*
3. *P. sagittata.*
4. *P. Petasites.*

1. Petasites palmàta (Ait.) A. Gray. Palmate-leaf Sweet Coltsfoot. Fig. 4595.

Tussilago palmata Ait. Hort. Kew. 2: 188. *pl. 2.* 1789.
Nardosmia palmata Hook. Fl. Bor. Am. 1: 308. 1833.
P. palmata A. Gray, in Brew. & Wats. Bot. Cal. 1: 407. 1876.

Scape very scaly, stout, 6′–24′ high. Leaves nearly orbicular in outline, 3′–12′ broad, deeply 7–11-cleft to much beyond the middle, green and glabrous above, densely white-tomentose beneath, at least when young, sometimes becoming glabrate, the lobes oblong or obovate, acute, often somewhat cuneate, sharply dentate or incised; heads mostly dioecious, corymbose or racemose-corymbose, numerous, 4″–6″ broad; flowers nearly white, fragrant, the marginal ones of the pistillate heads radiate.

In swamps and along streams, Newfoundland to Massachusetts, New York, Wisconsin, Minnesota and Alberta. Far western plants, formerly included in this species, prove to be distinct. April–June. Butter-bur.

3. Petasites sagittàta (Pursh) A. Gray. Arrow-leaf Sweet Coltsfoot. Bitter-bur. Fig. 4597.

Tussilago sagittata Pursh, Fl. Am. Sept. 332. 1814.
Nardosmia sagittata Hook. Fl. Bor. Am. 1 : 307. 1833.
Petasites sagittata A. Gray, in Brew. & Wats. Bot. Cal. 1 : 407. 1876.

Scape and racemose-corymbose inflorescence similar to those of the two preceding species. Leaves deltoid-ovate to reniform-ovate, persistently white-tomentose beneath, glabrous or nearly so above, 4′–10′ long, their margins sinuate-denticulate, neither cleft nor lobed; involucre campanulate; flowers nearly white, the marginal ones of the pistillate heads radiate.

In wet grounds, Labrador to Hudson Bay, Manitoba and Minnesota, west to British Columbia, south in the Rocky Mountains to Colorado. May–June.

2. Petasites trigonophỳlla Greene. Arctic Sweet Coltsfoot. Fig. 4596.

Petasites trigonophylla Greene, Leaflets 1 : 180. 1906.

Scape very scaly, 3′–10′ high. Leaves deltoid-reniform to ovate-orbicular in outline, 2′–6′ long, irregularly lobed, green and glabrous above, persistently white-tomentose beneath, the lobes few-toothed; heads corymbose, the inflorescence about 4′ long; involucre campanulate; flowers nearly white, the marginal ones of the pistillate heads radiate.

Wet grounds, Quebec, Minnesota and Saskatchewan. June–Aug.

Petasites frigida (L.) Fries, admitted, in our first edition, as recorded from Lake Winnipeg, is a high boreal species, not known to occur within our area.

4. Petasites Petasìtes (L.) Karst. Butter-bur. Butterfly-dock. Fig. 4598.

Tussilago Petasites L. Sp. Pl. 866. 1753.
Petasites officinalis Moench. Meth. 568. 1794.
Petasites vulgaris Desf. Fl. Atlant. 2: 270. 1798.
P. Petasites Karst. Deutsch. Fl. 1062. 1880–83.

Scape very scaly, 6′–15′ high. Leaves orbicular or hastate-reniform, often 12′ broad when mature, rounded or pointed at the apex, repand-denticulate all around, persistently white-tomentose beneath, green and mostly glabrous above; heads 4″–6″ broad, mostly dioecious, in a dense raceme, the staminate ones smaller than the pistillate; flowers pink-purple, fragrant, none of them radiate.

In cultivated and waste ground, eastern Pennsylvania and Massachusetts. Naturalized from Europe. Native also of northern Asia. Batter- or flea-dock. Bog- or poison-rhubarb. Eldin. Gallon. Umbrella-leaves. Pestilence-wort. Ox-wort. April.

97. ÁRNICA L. Sp. Pl. 884. 1753.

Erect, simple or little branched herbs, with opposite leaves, or the upper rarely alternate, and large, long-peduncled heads of both tubular and radiate, yellow flowers, or rays wanting in some species. Involucre turbinate or campanulate, its bracts in 1 or 2 series, narrow, nearly equal. Receptacle flat, naked, fimbrillate or villous. Ray-flowers pistillate, fertile, the rays spreading, entire, or 2–3-toothed. Disk-flowers perfect, fertile, the corolla 5-lobed, the style with slender branches. Anthers entire or minutely 2-auriculate at the base. Achenes linear, 5–10-ribbed, more or less pubescent. Pappus a single series of rough or barbellate, rigid, slender bristles. [Derivation uncertain, perhaps from Ptarmica.]

About 45 species, natives of the northern hemisphere. Besides the following, many others occur in the western parts of North America. Type species: *Arnica montana* L.

Basal leaves ovate or oval, sessile; southern.   1. *A. acaulis.*
Basal leaves oblong, lanceolate, or cordate-ovate, petioled.
  Basal leaves cordate-ovate.   2. *A. cordifolia.*
  Basal leaves not cordate, tapering to the petiole.
    Leaves dentate.
      Pappus brownish, plumose.   3. *A. mollis.*
      Pappus white, barbellate.   4. *A. chionopappa.*
    Leaves entire or nearly so.   5. *A. alpina.*

1. Arnica acaùlis (Walt.) B.S.P.
Leopard's-bane. Fig. 4599.

Doronicum acaule Walt. Fl. Car. 205. 1788.
Arnica Claytoni Pursh, Fl. Am. Sept. 527. 1814.
Arnica nudicaulis Nutt. Gen. 2: 164. 1818.
Arnica acaulis B.S.P. Prel. Cat. N. Y. 30. 1888.

Glandular-hirsute; stem 1°–3° high, bearing several slender-peduncled heads at the summit. Basal leaves tufted, ovate or oval, obtuse, narrowed to a sessile base, denticulate or entire, 2′–5′ long, 1½′–3′ wide; stem leaves 1–3 pairs, and some alternate, very small ones above; heads 1′–1½′ broad; bracts of the involucre linear-lanceolate, acute or acutish; rays 12–15, commonly 3-toothed at the truncate apex; achenes pubescent when young, glabrous or nearly so when mature.

In low woods, Delaware and southern Pennsylvania to Florida. April–May.

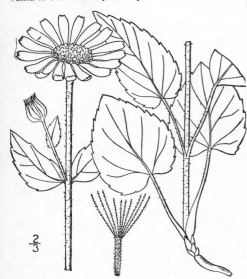

2. Arnica cordifòlia Hook. Heart-leaf Arnica. Fig. 4600.

Arnica cordifolia Hook. Fl. Bor. Am. 1: 331. 1833.

Villous or pubescent; stem simple or sparingly branched, glandular above, 1°–2° high. Basal and lower leaves ovate to nearly orbicular, obtuse or acute, deeply cordate at the base, dentate, 1′–3′ long, with slender sometimes margined petioles; stem leaves 1–3 pairs, ovate to oblong, sessile or short-petioled, much smaller; heads 1–8, 2′–3′ broad; bracts of the involucre acute or acuminate, villous, 6″–10″ long; rays 12–16, toothed at the apex; achenes hirsute-pubescent, or glabrous at the base; pappus barbellate, white.

Lake Superior to North Dakota, Yukon, Montana, New Mexico and California. Recorded from western Nebraska. May–July.

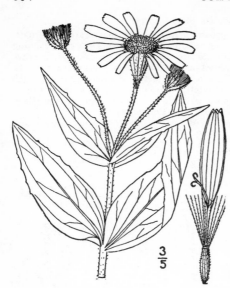

3. Arnica mollis Hook. Hairy Arnica. Fig. 4601.

Arnica mollis Hook. Fl. Bor. Am. **1** : 331. 1833.

Villous-pubescent; stem simple, or little branched, 1°–2½° high, bearing 1–6 heads at the summit. Leaves oblong or oblong-lanceolate, dentate or entire, acute or obtuse, 2'–5' long, 3''–9'' wide, the lower and basal ones narrowed into petioles, the upper sessile, and usually somewhat connate by a broad or narrowed base, those of the stem 3–5 pairs, usually with some alternate small ones on the branches; heads 1'–2' broad; bracts of the involucre acute; rays 10–15, 3-toothed; achenes hirsute-pubescent; pappus yellow-brown, plumose.

Mountains of northern New York, New Hampshire and Maine to New Brunswick, west to Lake Superior, British Columbia and California, south in the Rocky Mountains to Utah and Colorado. Included, in our first edition, in the northwestern *A. Chamissonis* Less. June–Aug.

4. Arnica chionopáppa Fernald. White-plumed Arnica. Fig. 4602.

Arnica chionopappa Fernald, Rhodora **7** : 148. 1905.

Stem 15' high or less, villous to the base. Leaves thin in texture, the lower and basal ones ovate to lanceolate, acute, narrowed at the base, sharply dentate with few teeth, petioled, 2½'–4½' long, the upper few pairs narrowly lanceolate, entire, sessile; heads solitary or few, about 1½' broad; involucre villous, 4''–5'' high, its bracts linear-lanceolate, acuminate; rays 10–15; pappus bright white, barbellate.

On wet cliffs, Quebec and New Brunswick. June–July.

Arnica gaspensis Fernald, from ledges in Gaspé County, Quebec, has creamy-white pappus, the involucre and peduncles glandular-pubescent.

5. Arnica alpìna (L.) Olin & Laden. Mountain Tobacco. Arctic Leopard's-bane. Arctic Arnica. Fig. 4603.

Arnica montana var. *alpina* L. Sp. Pl. 884. 1753.
Arnica alpina Olin & Laden, Diss. 11. 1799.
A. angustifolia Vahl, Fl. Dan. *pl. 1524.* 1814.
A. plantaginea Pursh, Fl. Am. Sept. 527. 1814.
A. Sornborgeri Fernald, Rhodora **7** : 147. 1905.

Stem slender, 6'–15' high, villous or pubescent, and glandular or glabrous below; stem simple, usually bearing but a single head, but sometimes with 1–3 additional ones from the axils of the upper leaves. Leaves lanceolate, linear-oblong, or the lowest spatulate, thickish, entire or denticulate, 3-nerved, the basal ones petioled, those of the stem 1–4 pairs, sessile or short-petioled, scarcely connate, the upper pair usually much smaller than the lower ones; heads about 2' broad; rays 10–15, 3-toothed; achenes hirsute; pappus brownish.

Labrador to Greenland and the Arctic Sea. Also in northern Europe. Races differ in pubescence. Rocky Mountain plants formerly referred to this species appear to be distinct. May–Sept.

98. HAPLOÉSTHES A. Gray, Mem. Am. Acad. (II) 4: 109. 1849.

Perennial caulescent, partly woody and partly fleshy plants, with opposite narrow entire leaves, the lower connate-sheathing, and corymbose heterogamous radiate heads. Involucre of few broad many-nerved bracts. Receptacle flat or slightly convex, naked. Ray-flowers pistillate, fertile, the rays yellow, spreading or recurved. Disk-flowers perfect, fertile. Anthers obtuse and entire at the base. Style-branches of the disk-flowers capitate-truncate. Achenes narrow, 10-ribbed. Pappus of 1 series of slender scabrous bristles. [Greek, simple garment, the involucre composed of few bracts.]

A monotypic genus of the south-central United States and Mexico.

1. Haploesthes Greggii A. Gray. Gregg's Haploesthes. Fig. 4604.

Haploesthes Greggii A. Gray, Mem. Am. Acad. (II) 4: 109. 1849.

Stems usually branched at the base, the branches 1°–2° tall, glabrous, striate, corymbose above. Leaves fleshy, the lower ones connate and sheathing the stem, narrowly linear or linear-filiform, ¾′–2′ long, entire; heads short-peduncled, few together in cymes; involucres 1½″–2″ high, the bracts oval to orbicular, rounded at the apex, thin-margined; rays yellow, 1″–2″ long; achenes 1″ long.

In saline and gypsum soil, Kansas and southern Colorado to Texas and Mexico. April–Sept.

99. ERECHTÌTES Raf. Fl. Ludov. 65. 1817.

Erect, usually branching herbs, with alternate leaves, and (in our species) rather large discoid many-flowered heads of whitish flowers, corymbose-paniculate at the ends of the stem and branches. Involucre cylindric, swollen at the base, its principal bracts in 1 series, linear, with or without some much smaller outer ones. Receptacle concave, naked. Marginal flowers in 2–several series, pistillate, fertile, their corollas filiform, the limb 2-4-toothed. Central flowers perfect, fertile; corolla narrowly tubular, the limb 4-5-toothed, the style-branches elongated, truncate or obtuse at the summit. Anthers obtuse and entire at the base. Achenes linear-oblong, angled or striate. Pappus of copious capillary soft smooth white bristles. [Ancient name of some groundsel.]

About 12 species, natives of America and Australasia. The following typical one is the only species known to occur in North America.

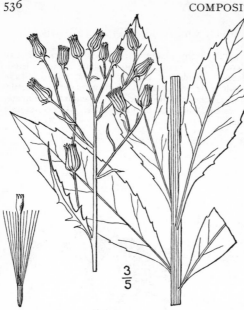

3
5

1. Erechtites hieracifòlia (L.) Raf. Fire-weed. Fig. 4605.

Senecio hieracifolius L. Sp. Pl. 866. 1753.
E. prealta Raf. Fl. Ludov. 65. 1817.
Erechtites hieracifolia Raf. DC. Prodr. **6**: 294. 1837.

Annual, glabrous, or somewhat hirsute; stem striate, succulent, usually branched, 1°–8° high. Leaves thin, lanceolate or ovate-lanceolate, dentate and often deeply incised, 2′–8′ long, the upper sessile or auriculate-clasping, mostly acuminate, the lower usually narrowed into petioles; heads 6″–10″ long, about 3″ in diameter, the involucre conspicuously swollen at the base before flowering, its bracts numerous, striate, green, with narrow scarious margins; pappus bright white.

In woodlands, thickets and waste places, very abundant after fires, Newfoundland to Florida, Ontario, Saskatchewan, Nebraska and Texas. Also in Mexico, the West Indies and South America. Pilewort. July–Sept.

100. MESADÈNIA Raf. New Fl. N. A. **4**: 78. 1836.

Tall perennial mostly glabrous herbs, with alternate petioled leaves and numerous, rather small, corymbose, discoid heads of white, yellowish or pinkish flowers, all tubular and perfect. Sap milky (at least in some species). Involucre cylindric or nearly so, its principal bracts 5, in 1 series, equal, usually with a few short outer ones. Receptacle flat, not chaffy, with a fleshy projection in the center. Corollas with somewhat spreading 5-cleft limbs, the lobes usually with a mid-nerve. Style-branches conic or obtuse at the apex. Achenes oblong, glabrous. Pappus of copious white scabrous bristles. [Greek, referring to the central projection of the receptacle.]

About 30 species, natives of North and Central America. Besides the following, 4 others occur in the southern and southwestern parts of the United States. Type species: *Mesadenia atriplicifolia* (L.) Raf.

Leaves reniform or fan-shaped, lobed, or angulate-dentate.
　　Leaves green both sides, angulate-dentate.　　　　　　　　　　　　　　　　　1. *M. reniformis.*
　　Leaves glaucous beneath, green above, angulate-lobed.　　　　　　　　　　　2. *M. atriplicifolia.*
Leaves thick, green both sides, ovate or oval, entire, or repand.　　　　　　　　3. *M. tuberosa.*

1. Mesadenia renifórmis (Muhl.) Raf. Great Indian Plantain. Wild Collard. Fig. 4606.

Cacalia reniformis Muhl.; Willd. Sp. Pl. **3**: 1753. 1804.
Mesadenia reniformis Raf. New Fl. **4**: 79. 1836.
M. rotundifolia Raf. New Fl. **4**: 79. 1836.

Glabrous; stem angled and grooved, 4°–10° high. Leaves thin, green both sides, coarsely angulate-dentate with mucronate-pointed teeth, the basal and lower reniform, long-petioled, sometimes 2° wide, the upper ovate or fan-shaped, mostly cuneate at the base, the uppermost small and oblong; heads numerous, mostly 5-flowered, about 2″ broad, in large compound corymbs; involucre 3″–4″ high, its bracts linear-oblong, obtuse or acutish, scarious-margined, with or without 1–3 minute outer ones.

In woods, New Jersey and Pennsylvania to Minnesota, south to North Carolina and Tennessee. July–Sept.

3
5

2. Mesadenia atriplicifòlia (L.) Raf.   Pale Indian Plantain.   Fig. 4607.

Cacalia atriplicifolia L. Sp. Pl. 835. 1753.

Senecio atriplicifolius Hook. Fl. Bor. Am. **1**: 332. 1833.

Mesadenia atriplicifolia Raf. New Fl. **4**: 79. 1836.

Stem terete, glabrous and glaucous, 3°–6° high. Leaves thin, angulate-lobed, palmately veined, glaucous beneath, the lower and basal ones slender-petioled, sometimes 6' wide, the upper reniform, fan-shaped, or triangular with a nearly truncate base, the uppermost commonly small, lanceolate or oblong and entire; heads very numerous, about 1½" broad, in large, loose compound corymbs; involucre 3"–4" high, its bracts linear-oblong, scarious-margined, with or without 1–3 minute outer ones.

In woods, New Jersey to Indiana, Minnesota, south to Florida, Tennessee, Missouri and Kansas. Recorded from Ontario. Called also wild caraway. July–Sept.

3. Mesadenia tuberòsa (Nutt.) Britton.   Tuberous Indian Plantain.   Fig. 4608.

Cacalia tuberosa Nutt. Gen. **2**: 138. 1818.
Mesadenia plantaginea Raf. New Fl. **4**: 79. 1836.
Senecio Nuttallii Sch. Bip. Flora **27**: 499. 1845.
M. tuberosa Britton in Britt. & Brown, Ill. Fl. **3**: 474. 1898.

Glabrous and green throughout; stem angled, stout, 2°–6° high. Leaves thick, strongly 5–9-nerved, the lower and basal ones oval, ovate, or ovate-lanceolate, obtuse or acutish, usually quite entire, but sometimes repand, long-petioled, narrowed at the base, or rarely subcordate, 4'–8' long, 1'–3' wide; upper leaves ovate to oblong or cuneate-obovate, sessile or short-petioled, much smaller, sometimes toothed toward the apex; heads very numerous in a compound corymb, about 2" broad, mostly 5-flowered; involucre 3"–4" high, its bracts linear-oblong, obtuse or obtusish, scarious-margined.

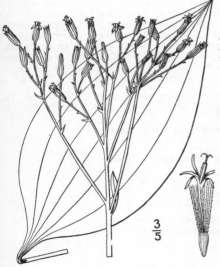

On wet prairies and in marshes, Ohio and western Ontario to Minnesota, Kansas, Alabama, Louisiana and Texas. June–Aug.

101. SYNÓSMA Raf.; Loud. Gard. Mag. **8**: 247. 1832.

A perennial leafy herb, glabrous or very nearly so, with triangular or hastate, alternate leaves, the lower petioled, the upper sessile, and several or numerous, corymbose or corymbose-paniculate, discoid heads of white or pinkish flowers. Involucre nearly cylindric, its principal bracts 12–15, linear, acute, usually with some subulate outer ones. Receptacle flat, naked. Flowers perfect; corolla 5-lobed. Style-branches not appendaged. Pappus of very numerous white soft capillary bristles. [Greek, perhaps signifying a fragrant composite.]

A monotypic genus of eastern North America.

1. Synosma suavèolens (L.) Raf. Sweet-scented Indian Plantain. Fig. 4609.

Cacalia suaveolens L. Sp. Pl. 835. 1753.

Senecio suaveolens Ell. Bot. S. C. & Ga. **2**: 328. 1821–24.

Synosma suaveolens Raf.; Loud. Gard. Mag. **8**: 247. 1832.

Glabrous or very nearly so throughout; stem striate, 3°–5° high, leafy to the inflorescence. Leaves triangular-lanceolate or hastate, sharply and irregularly serrate, acuminate, 4′-10′ long, 2′-6′ wide at the base, the auricles sometimes with 1 or 2 lobes on the lower side; petioles margined, or those of the basal leaves naked and slender; uppermost leaves sometimes merely lanceolate and sessile; heads 2″-3″ broad in a usually large and compound corymb; involucre 4″-6″ high, its principal bracts linear, acute; heads 20–30-flowered.

In woods, Rhode Island to Pennsylvania, New Jersey, Illinois, Minnesota, Florida, West Virginia and Kentucky. Called also wild caraway. Aug.–Oct.

102. SENÈCIO [Tourn.] L. Sp. Pl. 866. 1753.

Annual or perennial herbs (some tropical species shrubby or even arborescent), with alternate or basal leaves, and solitary corymbose or paniculate many-flowered heads, of both tubular and radiate, or only tubular flowers, in our species yellow. Involucre cylindric or campanulate, its principal bracts in 1 series, distinct, or united at the base, usually with some shorter outer ones. Receptacle flat or slightly convex, mostly naked, often honeycombed. Rays, when present, pistillate, fertile. Disk-flowers perfect, fertile, their corollas tubular, the limb 5-toothed or 5-lobed. Anthers obtuse and entire at the base, or rarely slightly sagittate. Style-branches of the disk-flowers usually recurving or spreading. Achenes terete, or those of the marginal flowers somewhat compressed, 5–10-ribbed, papillose or canescent, at least after wetting, and then usually emitting a pair of spiral threads. Pappus of numerous slender or capillary, smooth or rough, mostly white bristles. [Latin, *senex,* an old man, referring to the hoary character of some species, or to the white pappus.]

An immense genus of probably at least 1200 species, of very wide geographic distribution. In addition to the following, many others occur in the southern and western parts of North America. Our species known as Groundsel, Ragwort, or Squaw-weed. Type species: *Senecio vulgaris* L.

A. Annual or biennial species, with stems leafy throughout.

Rays none, or very short and inconspicuous; introduced.
The short outer involucral bracts black-tipped; rays none.   1. *S. vulgaris.*
The short outer involucral bracts not black-tipped, sometimes none; rays very short.
    Plant sparingly pubescent.   2. *S. sylvaticus.*
    Plant densely viscid-pubescent.   3. *S. viscosus.*
Rays large and conspicuous; native species.
    Leaves pinnately divided; heads 2″-3″ high.   4. *S. glabellus.*
    Leaves sinuate-dentate or the lower entire; heads 4″-5″ high. 5. *S. palustris.*

B. Perennial species, mostly with rootstocks.

    a. Stems woody at base; leaves linear, or pinnatifid with linear lobes; western species.
Leaves linear, entire or serrate, thin.   6. *S. spartioides.*
Leaves deeply pinnatifid into linear lobes, firm.   7. *S. Riddellii.*

    b. Stems wholly herbaceous; leaves various.

    * Heads very large; involucre 7″-10″ high; boreal species.
Stem stout, leafy above; heads several, 1½′-2′ broad.   8. *S. Pseudo-Arnica.*
Stem slender; upper leaves few and small; head mostly solitary, about 1′ broad.   9. *S. frigidus.*

    ** Heads smaller; involucre 4″-7″ high.

    † Leafy up to the inflorescence; stem leaves 2–3-pinnatifid; introduced species.
        10. *S. Jacobaea.*
    †† Leaves mostly borne on the lower part of the stem, the upper ones much smaller; native species.

    ‡ Leaves and stems more or less persistently woolly or tomentose.
Low species, seldom over 1° high, with small oval-oblong to spatulate basal leaves.
    Basal leaves angulate-dentate, oval.   11. *S. antennariifolius.*
    Basal leaves entire or sparingly toothed, oblong to spatulate. 12. *S. canus.*

Taller, up to 2½° high; basal leaves ovate to oblong-lanceolate.
  Densely persistently tomentose; stem-leaves mostly merely dentate.   13. *S. tomentosus.*
  Loosely tomentose, bcoming glabrate; stem leaves mostly pinnatifid.   14. *S. plattensis.*
  ‡‡ Plants glabrous, or nearly so, at least when mature, the stem sometimes tomentose at the base.
Basal leaves or some of them deeply cordate.
  Basal leaves lanceolate or oblong-lanceolate.   15. *S. Robbinsii.*
  Basal leaves orbicular to ovate.   16. *S. aureus.*
None of the leaves cordate.
  At least the stem-leaves lobed, laciniate, or pinnatifid; pubescence, if any, woolly; basal leaves
      dentate or crenate.
    Head rayless; boreal species.   17. *S. discoideus.*
    Heads radiate.
      Basal leaves obovate to suborbicular.   18. *S. obovatus.*
      Basal leaves oblong to spatulate.
        Basal leaves oblong to lanceolate, dentate or crenate.
          Involucre 4″ high; basal leaves sharply serrate.   19. *S. Crawfordii.*
          Involucre 2½″–3½″ high; basal leaves mostly crenate.
            Heads few; basal leaves mostly short.   20. *S. pauperculus.*
            Heads very numerous; basal leaves long.   21. *S. Smallii.*
        Basal leaves linear-cuneate, entire, or few-toothed at the apex. 22. *S. densus.*
  All leaves entire or very nearly so; pubescence, if any, of crisp hairs. 23. *S. integerrimus.*

1. Senecio vulgàris L.   Common Ground- sel.   Fig. 4610.

Senecio vulgaris L. Sp. Pl. 867. 1753.

   Annual, puberulent or glabrate; stem hollow,
usually much branched, 6′–15′ high. Leaves pinnatifid, 2′–6′ long, the lower spatulate in outline,
petioled, obtuse, the upper sessile or clasping at
the base, more deeply lobed or incised, their segments oblong, dentate; heads several or numerous in the corymbs, nearly 3″ broad, 4″–6″ high;
bracts of the involucre linear, with few or several subulate black-tipped outer ones; rays none;
achenes slightly canescent; pappus white.

   In cultivated ground and waste places, Newfoundland to Hudson Bay, North Carolina, Minnesota,
Michigan, and west to the Pacific Coast. Bermuda.
Naturalized from Europe. Other names are grinsel,
simson, birdseed, chickenweed. April–Oct.

$\frac{3}{4}$

2. Senecio sylváticus L.   Wood Groundsel. Fig. 4611.

Senecio sylvaticus L. Sp. Pl. 868. 1753.

   Annual, glabrous or puberulent; stem usually much branched, 1°–2½° high, leafy. Leaves pinnatifid, oblong or lanceolate in outline, the segments oblong or spatulate, obtuse,
dentate, lobed or entire, or the uppermost leaves linear and
merely dentate; heads several or numerous in the corymbs,
slender-peduncled, about 2″ broad, 3″–4″ high; involucre
usually quite naked and swollen at the base; rays very short
and recurved; achenes canescent; pappus white.

   In waste places Newfoundland to Nova Scotia, Quebec and
Maine. Also on the coasts of California and British Columbia.
Naturalized or adventive from Europe. April–Sept.

$\frac{2}{3}$

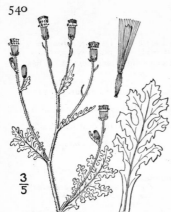

3/5

3. Senecio viscòsus L. Fetid or Viscous Groundsel. Fig. 4612.

Senecio viscosus L. Sp. Pl. 868. 1753.

Annual, viscid-pubescent, strong-scented; stem usually much branched, 1°-2° high. Leaves 1-2-pinnatifid, 1½'-3' long, oblong or somewhat spatulate in outline, the segments oblong or cuneate, dentate or incised; lower leaves petioled; heads few in the corymbs, 3"-4" broad, mostly slender-peduncled; involucre nearly cylindric, 4"-5" high, its bracts linear, acute, with 1-3 shorter outer ones; rays commonly about 20, very short, recurved and inconspicuous; achenes glabrous; pappus bright white, about one-third longer than the involucre.

In waste places and on ballast near the coast, Nova Scotia, New Brunswick and Quebec to North Carolina. July–Sept.

4. Senecio glabéllus Poir. Butterweed. Cress-leaved Groundsel. Fig. 4613.

S. lyratus Michx. Fl. Bor. Am. **2** : 120. 1803. Not L. 1753.

S. glabellus Poir. in Lam. Encycl. **7** : 102. 1806.

Senecio lobatus Pers. Syn. **2** : 436. 1807.

Annual, glabrous throughout, or slightly woolly when young, fleshy and tender; stem hollow, simple or branched, 1°-3° high. Leaves 2'-10' long, pinnately divided, the segments orbicular, oblong, obovate or cuneate, obtuse, sinuate-dentate, entire or lobed, the terminal segment usually larger than the others; lower and basal leaves slender-petioled; heads numerous, 7"-10" broad, slender-peduncled in terminal corymbs; involucre nearly cylindric, 2½" high, its bracts linear, acute, usually with no small outer ones; rays 6-12; achenes minutely hispidulous on some of the angles; pappus white, somewhat longer than the involucre.

In swamps, North Carolina to Kentucky, Illinois, Missouri, Arkansas, Florida, New Mexico and Mexico. April–Sept.

2/3

5. Senecio palústris (L.) Hook. Marsh Fleawort. Pale Ragwort. Marsh Groundsel. Fig. 4614.

1/2

Cineraria palustris L. Sp. Pl. Ed. 2, 243. 1763.

Senecio palustris Hook. Fl. Bor. Am. **1** : 334. 1833.

Annual or biennial, pubescent or glabrate; stem stout, simple, hollow, 6'-24' high. Leaves lanceolate, oblong or spatulate, entire, dentate, or laciniate, acute or obtuse, 2'-7' long, 3"-15" widé, or the upper linear-lanceolate and small, those of the stem sessile and somewhat auriculate-clasping, the basal petioled; heads numerous, 6"-12" broad, mostly short-peduncled in a large, rather dense, terminal corymb; involucre cylindric, becoming campanulate, 3"-4" high, its bracts linear, acute, more or less pubescent, with no shorter outer ones; rays 15-20 or more, pale-yellow; achenes glabrous; pappus white, elongated, at length twice the length of the involucre.

In swamps, Iowa and Wisconsin to Manitoba and arctic America, west to Alaska. Reported from Labrador. Also in Greenland, northern Europe and Asia. June–Aug.

6. Senecio spartioìdes T. & G.　Broom-like Senecio.　Fig. 4615.

Senecio spartioides T. & G. Fl. N. A. **2** : 438.　1843.

Woody at the base, usually branched, sometimes shrubby, glabrous or nearly so, leafy, 1°-6° high. Leaves sessile, or the lowest petioled, 1'-3' long, linear, entire, or more or less serrate, not lobed; heads corymbose at the ends of the branches, ½'-1' broad, slender-peduncled; involucre cylindric or becoming campanulate, 4"-5" high, its bracts linear, acute or acuminate, usually with some subulate exterior ones; rays 8-15; achenes canescent; pappus bright white.

Plains, in dry soil, Nebraska to Texas, Wyoming and Arizona.　June-Sept. This and the following species were included in the description of the far western *S. Douglasii* DC. in our first edition.

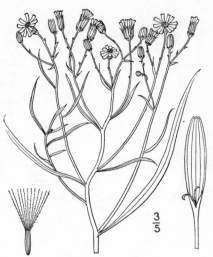

$\frac{3}{5}$

$\frac{2}{3}$

7. Senecio Riddellii T. & G.　Riddell's Senecio.　Fig. 4616.

S. Riddellii T. & G. Fl. N. A. **2** : 444.　1843.
S. Fremontii (T. & G.) Rydb.; Britton, Manual 1028.
1901. Not *S. Fremontii* T. & G.

Woody at the base, usually branched, sometimes shrubby, glabrous or nearly so, leafy, 1°-6° high. Leaves sessile, or the lowest petioled, thick, 1½'-3½' long, pinnately parted into 3-9 linear or filiform, entire segments, or the upper entire; heads corymbose at the ends of the branches, 5"-10" broad, slender-peduncled; involucre cylindric or becoming campanulate, 5"-8" high, its bracts linear, acute or acuminate, usually with some subulate exterior ones; rays 8-15; achenes canescent; pappus white.

Plains, in dry soil, Nebraska to Texas and Mexico. June-Sept.

8. Senecio Pseùdo-Arnica Less.　Sea-beach Senecio.　Fig. 4617.

Arnica maritima L. Sp. Pl. 884.　1753.　Not *S. maritimus* L.
Senecio Pseudo-Arnica Less. Linnaea **6** : 240.　1831.

Perennial, somewhat fleshy; stem stout, mostly simple, very leafy, 6'-3° high.　Leaves oblong-obovate, lanceolate, or the lower spatulate, acute or obtuse at the apex, 4'-8' long, ½'-2' wide, densely tomentose beneath, at least when young, repand-dentate or denticulate, narrowed to a sessile and partly clasping base, or the lowest into margined petioles; heads solitary, or several (2-7) and corymbose, stout-peduncled, 1½'-2' broad, 8"-10" high; involucre broadly campanulate, its bracts lanceolate, acuminate, mostly tomentose, commonly with several subulate spreading ones at the base; rays 12-25, linear, 3-toothed, conspicuous; disk-corollas 5-lobed; achenes glabrous; pappus dull.

On sea-beaches and rocks near the sea, Maine, New Brunswick and the lower St. Lawrence to Labrador and the Arctic Sea.　Also in Alaska.　July-Aug.

$\frac{1}{2}$

9. Senecio frígidus Less. Arctic Senecio. Fig. 4618.

Senecio frigidus Less. Linnaea 6: 239. 1831.

Perennial, more or less tomentose, or becoming glabrous when old; stem slender, 6'–12' high, bearing a solitary head (rarely 2 or 3) ½'–1' broad. Basal and lower leaves spatulate or obovate, 1'–2' long, petioled, obtuse, repand-dentate or entire; stem leaves oblong to linear-lanceolate, obtuse or acute, sessile, mostly entire, smaller; involucre broadly campanulate, about 7" high, its bracts lanceolate, acute, with no exterior smaller ones; rays 10–16, 6"–10" long, 3-toothed, linear-oblong, or cuneate at the base; achenes glabrous or sparingly pubescent; pappus white.

Labrador and arctic America to Alaska. Also in northeastern Asia. Summer.

10. Senecio Jacobaèa L. Tansy Ragwort. Staggerwort. Fig. 4619.

Senecio Jacobaea L. Sp. Pl. 870. 1753.

Perennial by short thick rootstocks, somewhat woolly, or glabrous; stems stout, simple, or branched above, 2°–4° high, very leafy. Stem leaves 2–3-pinnatifid, 2'–8' long, the lower petioled, the upper sessile, the lobes oblong-cuneate, dentate or incised; basal leaves lyrate-pinnatifid; heads very numerous, 6"–8" broad, shortpeduncled in large compact corymbs; involucre narrowly campanulate, about 2½" high, its bracts linear-lanceolate, acute, green, or tipped with black, usually with a few subulate outer ones; rays 12–15; achenes of the disk-flowers pubescent, those of the rays glabrous; pappus white.

In waste places, Newfoundland and Nova Scotia to Maine and Ontario, and in ballast about New York and Philadelphia. Adventive from Europe. Stavewort. Cankerweed. Kettle-dock. St. James'-wort. Felonweed. Fairies'-horse. Ragweed. Saracen's-compass. July–Sept.

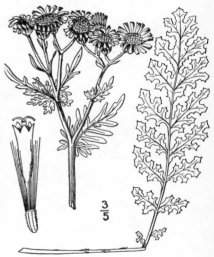

11. Senecio antennariifòlius Britton. Cat's-paw Ragwort. Fig. 4620.

Senecio antennariifolius Britton, in Britt. & Brown, Ill. Fl. 3: 478. 1898.

Perennial, tufted in mostly large clumps; stems slender, 8'–18' high, loosely white-woolly. Leaves nearly all basal, commonly numerous, oval to spatulate, angulately few-toothed or entire, mostly obtuse, narrowed into a petiole as long as the blade or longer, densely white-tomentose beneath, green and finally glabrous above, 1'–2½' long; stem leaves small, spatulate, laciniate, or the upper narrowly linear and entire; heads several, corymbose, slender-peduncled, rathed less than 1' broad; rays golden-yellow, showy; involucre 3" high, white-woolly; achenes glandular-pubescent.

Stony hillsides, mountains of Virginia and West Virginia. May–June.

12. Senecio cànus Hook. Silvery Ground-sel. Fig. 4621.

S. canus Hook. Fl. Bor. Am. 1: 333. *pl. 116.* 1833.

S. Purshianus Nutt. Trans. Am. Phil. Soc. (II) 7: 412. 1841.

Perennial, densely and persistently white-tomentose to the inflorescence; stems slender, usually tufted, 6'-18' high. Basal and lower leaves spatulate or oval, entire, or rarely some-what repand, very obtuse, 1'-2' long, narrowed into petioles; upper leaves oblong or spatulate, obtuse or acute, mostly sessile, smaller, entire or dentate; heads several or numerous, 8"-10" broad, usually slender-peduncled; involucre campanulate, or at first short-cylindric, about 5" high, its bracts linear-lanceolate, acute, spar-ingly tomentose, or glabrate, usually with no exterior smaller ones; rays 8-12; achenes gla-brous, at least below; pappus white.

In dry soil, Manitoba to North Dakota, Nebraska, west to British Columbia and California. Recorded from Minnesota. May-Aug.

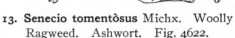

13. Senecio tomentòsus Michx. Woolly Ragweed. Ashwort. Fig. 4622.

S. tomentosus Michx. Fl. Bor. Am. 2: 119. 1803.

Perennial, more or less densely and persistently tomentose or woolly-canescent; stems rather stout, solitary, or sometimes tufted, 1°-2½° high. Basal and lower leaves ovate-lanceolate, oblong or rarely spatulate, long-petioled, erect, very obtuse, 2'-6' long, ½'-2' wide, crenate-dentate, narrowed or truncate at the base; stem leaves few and distant, small, linear-lanceolate or spatulate, crenate or rarely laciniate; heads several or numerous, co-rymbose, mostly long-peduncled, 8"-10" broad; involucre cylindric, or narrowly campanulate, its bracts linear-lanceolate, acute or acuminate, at length glabrate, with or without a few small outer ones; rays 10-15; achenes hispidulous, at least on the angles; pappus white.

In moist soil, southern New Jersey to Florida, Louisiana, Texas and Arkansas. April-June.

14. Senecio platténsis Nutt. Prairie Rag-wort. Fig. 4623.

S. plattensis Nutt. Trans. Am. Phil. Soc. (II) 7: 413. 1841.

Perennial, similar to the preceding species, usu-ally smaller-leaved, lower and less tomentose, or becoming glabrate in age, seldom over 1½° high. Basal leaves oval, ovate or oblong, some or all of them often more or less pinnatifid, with the termi-nal segment much larger than the lateral ones, crenulate or dentate, long-petioled; stem leaves mostly smaller than the basal ones, usually pin-natifid; heads several or numerous, compactly or loosely corymbose, conspicuously radiate.

Indiana and Illinois to Ontario, North Dakota, Colo-rado, Missouri and Texas. April-June.

Senecio pseudotomentòsus Mackenzie & Bush, of Missouri, differs in having the basal leaves mostly merely dentate.

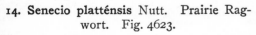

½

15. Senecio Robbínsii Oakes. Robbins' Squaw-weed. Fig. 4624.

Senecio Robbinsii Oakes; Rusby, Bull. Torr. Club **20**: 19. 1893.

Perennial, glabrous or very nearly so throughout; stems slender, 1°–2½° high. Basal leaves long-petioled, lanceolate, obtuse or acute at the apex, cordate, subcordate or truncate at the base, usually thin, 2'–4' long, ½'–1¼' wide, sharply dentate; stem leaves mostly pinnatifid or lobed, at least below the middle; heads several or numerous, slender-peduncled in an open corymb, 8''–10'' broad; rays 6–12; achenes glabrous or pubescent; pappus white.

In swamps and mountain meadows, Nova Scotia to New Hampshire, Vermont and New York. June–Sept.

16. Senecio aùreus L. Golden Ragwort. Life-root. Swamp Squaw-weed. Fig. 4625.

Senecio aureus L. Sp. Pl. 870. 1753.
Senecio pauciflorus Pursh, Fl. Am. Sept. **529**. 1814.
Senecio gracilis Pursh, Fl. Am. Sept. 529. 1814.

Perennial, glabrous or very nearly so throughout; stems rather slender, solitary or tufted, 6'–2½° high. Basal leaves cordate-ovate or cordate-orbicular or reniform, crenate-dentate, very obtuse and rounded, often purplish, 1'–6' long, with long slender petioles; lower stem leaves lanceolate or oblong, usually laciniate, pinnatifid or lyrate, the uppermost small, sessile, somewhat auriculate and clasping; heads usually several, 8''–10'' broad, 4''–5'' high, slender-peduncled in an open corymb; rays 8–12, golden-yellow; achenes glabrous; pappus white.

In swamps and wet meadows, Newfoundland to Florida, Ontario, Michigan, Missouri and Texas. Grundy-swallow. False-valerian. Root strong-scented. Races differ in size of plant, size of leaves and number of heads. May–July.

⅗

Senecio pseudaùreus Rydb. (*S. semi-cordatus* Mackenzie & Bush), of the Rocky Mountain region, differing by elliptic or broadly oblong basal leaves, ranges eastward into North Dakota and Missouri.

17. Senecio discòideus (Hook.) Britton. Northern Squaw-weed. Fig. 4626.

Senecio aureus var. *discoideus* Hook. Fl. Bor. Am. **1**: 333. 1833.
Senecio discoideus Britton, in Britt. & Brown, Ill. Fl. **3**: 479. 1898.

Perennial, glabrous except for small tufts of wool in the axils of the lower leaves; stem rather stout, 1°–2½° tall; basal leaves oval to ovate, obtuse, thin, sharply dentate, abruptly narrowed into petioles longer than the blade; stem leaves few, small, more or less laciniate; heads few or several, slender-peduncled, corymbose; principal bracts of the involucre narrowly linear, 3''–5'' long, the short outer ones few or none; rays none; achenes glabrous.

In moist places, Labrador to Yukon, Quebec, Michigan, Wyoming and British Columbia. June–Aug.

½

18. Senecio obovàtus Muhl. Round-leaf Squaw-weed. Fig. 4627.

Senecio obovatus Muhl.; Willd. Sp. Pl. 3: 1999. 1804.
Senecio Elliottii T. & G. Fl. N. A. 2: 443. 1843.
S. aureus var. *obovatus* T. & G. loc. cit. 442. 1843.
S. rotundus (Britton) Small, Fl. SE. U. S. 1304. 1903.

Perennial; stems glabrous, or a little woolly at the base, 9′–24′ high. Leaves glabrous, rather thick, the basal ones obovate with a cuneate base, suborbicular or broadly spatulate, very obtuse and rounded at the apex, 1′–3½′ long, ½′–2′ wide, crenate-dentate, often purplish; stem leaves commonly few and sessile, spatulate to oblong, often incised or pinnatifid; heads several, corymbose, 6″–8″ broad, about 3″ high, slender-peduncled; involucre nearly cylindric, its principal bracts linear-lanceolate, 2″–3″ long, acute, usually with 1–3 small exterior ones; rays 8–12, usually conspicuous, sometimes fewer and short; achenes glabrous; pappus white.

In moist soil on banks and in woods, Maine and Vermont to Florida, Ohio, Michigan, Alabama and Texas. Races differ in size, leaf-form, number and size or rays. Apparently erroneously recorded from Nova Scotia and Ontario. April–June.

19. Senecio Crawfòrdii Britton. Crawford's Squaw-weed. Fig. 4628.

Senecio Crawfordii Britton, Torreya 1: 21. 1901.
S. Balsamitae var. *Crawfordi* Greenman, Rhodora 10: 69. 1908.

Perennial, glabrous, or with sparse woolly pubescence below; stem slender, about 16′ high. Leaves thick, firm, the basal ones erect, the larger 8′–10′ long, the blades oval, oblong, or some of them narrowly obovate, mostly not more than one-half as long as the slender petioles, sharply and nearly equally serrate from the apex to the entire cuneate base; stem leaves lanceolate or narrower, mostly acuminate, incised-serrate, clasping, the upper sessile, the lower petioled, the uppermost very small; heads 3–7; peduncles slender, bracted, rarely forked; involucre 4″ high, its bracts linear-lanceolate, acuminate, shorter than the white barbellate pappus; rays 4″–5″ long; achenes linear, striate.

Wet meadows, southeastern Pennsylvania. May–June.

20. Senecio pauperculus Michx. Balsam Groundsel. Fig. 4629.

Senecio pauperculus Michx. Fl. Bor. Am. 2: 120. 1803.
Senecio Balsamitae Muhl.; Willd. Sp. Pl. 1999. 1804.
Senecio aureus var. *Balsamitae* T. & G. Fl. N. A. 2: 442. 1843.

Perennial, often tufted; stems slender, 1½′–20′ high, woolly at the base and in the axils of the lower leaves, or essentially glabrous. Basal leaves slender-petioled, oblong, rarely slightly spatulate, very obtuse, narrowed at the base, mostly thick, crenate, or rarely dentate, often purplish, 1′–5′ long, 3″–6″ wide, their petioles and sometimes their lower surfaces persistently tomentose or woolly, or glabrous throughout; lower stem leaves petioled, laciniate or pinnatifid, the upper sessile, very small; heads few or several, slender-peduncled, 6″–10″ broad; involucre about 3″ high; rays 8–12; achenes hispidulous or glabrous; pappus white.

In dry or rocky soil, Newfoundland to North Carolina, Ontario, British Columbia, Alabama, Tennessee and Nebraska. May–July.

21. Senecio Smàllii Britton. Small's Squaw-weed. Fig 4630.

S. *aureus* var. *angustifolius* Britton, Mem. Torr. Club
2 : 39. 1890. Not *S. angustifolius* Willd. 1804.

S. Smallii Britton, Mem. Torr. Club 4 : 132. 1893.

Similar to the preceding species but taller, grow-ing in large clumps; stem $1\frac{1}{2}°-2\frac{1}{2}°$ high, slender, densely and persistently floccose-woolly at the base and in the lower axils, or finally glabrate. Basal leaves elongated-oblong or linear-oblong, obtuse or acute, long-petioled, crenate-dentate, $3'-6'$ long, $3''-12''$ wide, at first tomentose, at length nearly glabrous; stem leaves several, deeply pinnatifid, or the lower lyrate, the uppermost very small; heads very numerous, $4''-5''$ broad, about $2\frac{1}{2}''$ high, slen-der-peduncled, forming large corymbs; rays 8–10; achenes hispidulous; pappus white.

In meadows and thickets, southeastern Pennsylva-nia to Florida and Alabama. May–June.

22. Senecio densus Greene. Western Squaw-weed. Fig. 4631.

Senecio aureus var. *compactus* A. Gray, Syn. Fl. 1 : Part 2,
391. 1884.
Senecio compactus Rydberg, Mem. Torr. Club 5 : 342. 1893.
Not T. Kirk.
Senecio densus Greene, Pittonia 4 : 226. 1900.

Perennial; stem usually tufted, low, rather stout, $6'-12'$ high, woolly at the base and in the lower axils, or glabrous. Basal leaves linear-cuneate, entire or 3-toothed at the apex, $1'-3'$ long, $2''-3''$ wide, thick, slender-petioled, the petioles commonly woolly-mar-gined; lower stem leaves often much larger and broader, usually laciniate or pinnatifid, but sometimes similar to the basal, the uppermost very small and sessile; heads several, $8''-10''$ broad, short-peduncled in a compact co-rymb; rays 10–15; achenes hispidulous; pappus white.

On dry plains, Manitoba to Nebraska, Colorado and Texas. May–June.

23. Senecio integérrimus Nutt. Entire-leaved Groundsel. Fig. 4632.

Senecio integerrimus Nutt. Gen. 2 : 165. 1818.

Perennial, more or less pubescent when young, glabrous or nearly so when old; stem stout, $1°-4°$ high. Leaves entire, or sparingly denticulate, somewhat fleshy, the lower and basal ones oval or oblong, obtuse or obtusish at the apex, $3'-8'$ long, $1'-1\frac{1}{2}'$ wide, petioled, the upper linear or lanceolate, acute, the uppermost very small; heads numerous, corymbose, long-peduncled, $6''-10''$ broad; involucre nearly cylindric, $4''-5''$ high, its principal bracts linear, acuminate, green, usually with a few subulate outer ones; rays 8–12, linear-oblong; achenes nearly glabrous; pappus white.

Iowa and Minnesota to Manitoba, Saskatchewan and Wyoming. June–July.

Senecio lùgens Richards., of northwestern North America, admitted in our first edition, is not defi-nitely known to occur within our area.

103. ÀRCTIUM L. Sp. Pl. 816. 1753.

Large coarse branching, rough or canescent, mostly biennial herbs, with broad alternate petioled leaves, and rather large heads of purple or white perfect tubular flowers, racemose, corymbose or paniculate at the ends of the branches. Involucre subglobose, its bracts rigid, lanceolate, tipped with spreading or erect hooked bristles, imbricated in many series. Receptacle flat, densely bristly. Corollas tubular with 5-cleft limbs. Filaments glabrous. Anthers sagittate at the base. Achenes oblong, somewhat compressed and 3-angled, ribbed, truncate. Pappus of numerous short serrulate scales. [Greek, bear, from the rough involucre.]

About 6 species, natives of Europe and Asia, readily distributed, their burs adhering to animals. Type species: *Arctium Lappa* L.

Bracts of the involucre densely cottony; heads corymbose.	1. *A. tomentosum.*
Bracts of the involucre glabrous, or slightly woolly.	
Involucre 1' broad or more; inner bracts equalling or exceeding the flowers.	2. *A. Lappa.*
Involucre 6"–9" broad; inner bracts not exceeding the flowers.	3. *A. minus.*

1. Arctium tomentòsum (Lam.) Schk. Woolly or Cottony Burdock. Fig. 4633.

Lappa tomentosa Lam. Encycl. 1: 377. 1783.

Arctium tomentosum Schk. Bot. Handb. 3: 49. 1803.

Arctium Lappa var. *tomentosum* A. Gray, Syn. Fl. 1: Part 2, 397. 1884.

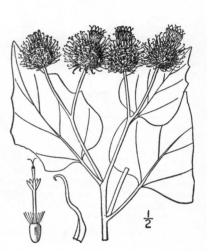

Similar to the following species; heads 8"–10" broad, corymbose at the ends of the branches, mostly long-peduncled; bracts of the involucre densely cottony, the inner ones erect and somewhat shorter than the flowers.

In waste places, Massachusetts to southern New York. Adventive from Europe. July–Aug.

2. Arctium Láppa L. Great Bur, Burdock or Clotbur. Fig. 4634.

Arctium Lappa L. Sp. Pl. 816. 1753.

Lappa major Gaertn. Fruct. & Sem. 2: 379. *pl. 162.* 1802.

Stem much branched, 4°–9° high. Leaves thin, broadly ovate, pale and tomentose-canescent beneath, obtuse, entire, repand or dentate, mostly cordate, the lower often 18' long; petioles solid, deeply furrowed; heads clustered or subcorymbose, sometimes long-peduncled, 1'–1½' broad; bracts of the involucre glabrous or nearly so, their spines all spreading, the inner ones equalling or exceeding the flowers; corolla-tube longer than the limb.

In waste places, New Brunswick and Ontario to southern New York, and locally in the interior. Not nearly as common as the next species in the Middle States. Naturalized from Europe. Other names are cockle-bur, cockle-button, cuckold-dock, hurr-bur, stick-button, hardock, bardane, beggar's-buttons. July–Oct.

3. Arctium mìnus Schk. Common Burdock.
Fig. 4635.

Arctium minus Schk. Bot. Handb. **3** : 49. 1803.

Lappa minor DC. Fl. Fran. **4** : 77. 1805.

Arctium Lappa var. *minus* A. Gray, Syn. Fl. **1** : Part 2, 397. 1884.

Smaller than the preceding species, seldom over 5° high. Leaves similar, the lower deeply cordate; petioles hollow, not deeply furrowed; heads numerous, racemose on the branches, short-peduncled or sessile, 6″–9″ broad; bracts of the involucre glabrous or slightly cottony, the spines of the outer ones spreading, those of the inner erect and shorter than the flowers; corolla-tube about as long as the limb.

In waste places, common nearly throughout our area, extending west to Colorado. Naturalized from Europe. Called also cuckoo-button. Leaves rarely laciniate or pinnatifid. July–Nov.

$\frac{1}{2}$

---

104. CIRSÌUM [Tourn.] Mill. Gard. Dict. Abr. Ed. 4. 1754.

Erect, branching or simple, prickly herbs, some species acaulescent, with alternate or basal, sinuate-dentate, lobed or pinnatifid, usually very spiny leaves, sometimes decurrent, and large, many-flowered, solitary or clustered, discoid heads of purple, yellow or white, tubular, perfect and fertile, or rarely dioecious flowers. Involucre ovoid or globose, its bracts prickle-tipped or unarmed, imbricated in many series. Receptacle flat or convex, bristly. Corolla-tube slender, the limb deeply 5-cleft. Filaments pilose, or rarely glabrous. Anthers sagittate at the base. Style-branches short or elongated, obtuse. Achenes obovate or oblong, compressed or obtusely 4-angled, glabrous, smooth or ribbed. Pappus of several series of slender, plumose bristles, connate at the base. [Greek, referring to the use of the thistle as a remedy for swollen veins.]

Over 200 species, widely distributed in the northern hemisphere. Besides the following, some 50 others occur in the southern and western parts of North America and many hybrids have been described. Type species: *Carduus heterophyllus* L.

† Outer involucral bracts, or all of them, strongly prickly-pointed.

1. Leaves glabrous or hispid above, tomentose beneath.

All the bracts of the involucre tipped with prickles; naturalized weed.	1. *C. lanceolatum.*
Outer bracts prickle-tipped, the inner merely acuminate; native species.	
Branches leafy up to the heads; involucral bracts firm or rigid.	
Leaves undivided, lobed or dentate, rarely pinnatifid.	2. *C. altissimum.*
Leaves deeply pinnatifid into lanceolate or linear segments.	3. *C. discolor.*
Heads naked-peduncled, 1′ high: involucral bracts thin.	4. *C. virginianum.*

2. Leaves tomentose on both sides, or becoming glabrous above; western.

Leaves pinnately parted; segments linear, entire or lobed.	5. *C. Pitcheri.*
Leaves pinnatifid into triangular or lanceolate dentate segments.	
Outer bracts with spines less than one-half their length.	
Leaf-lobes triangular; flowers pink or purple.	6. *C. undulatum.*
Leaf-lobes linear-lanceolate to oblong.	
Flowers yellow or cream-color.	7. *C. plattense.*
Flowers purple.	8. *C. Flodmani.*
Outer bracts with spines of nearly or quite their length.	9. *C. ochrocentrum.*
Leaves entire or undulate; outer pappus-bristles barbellate.	10. *C. nebraskense.*

3. Leaves green both sides, somewhat pubescent beneath.

Leaf-lobes acute; bracts of the involucre faintly nerved; roots solid.	11. *C. odoratum.*
Leaf-lobes blunt; bracts with prominent glutinous midnerve; root hollow.	12. *C. Hillii.*

†† Bracts of the involucre not at all prickly-pointed, or scarcely so.

Heads large, few, 1′–4′ broad; flowers all perfect and fertile.	
Heads involucrate by the upper very spiny leaves; flowers usually yellow.	13. *C. horridulum.*
Heads peduncled, naked, or with 1 or 2 bracts at the base; flowers purple.	14. *C. muticum.*
Heads small, numerous, 1′ or less broad.	
Heads partly dioecious; leaves not decurrent.	15. *C. arvense.*
Heads not dioecious; leaf-bases decurrent.	16. *C. palustre.*

1. Cirsium lanceolàtum (L.) Hill. Common Bur or Spear Thistle. Fig. 4636.

Carduus lanceolatus L. Sp. Pl. 821. 1753.
Cirsium lanceolatum Hill, Herb. Brit. 1: 80. 1769.
Cnicus lanceolatus Willd. Prodr. Fl. Berol. 259. 1787.

Biennial; stem stout, branched, more or less tomentose, 3°–5° high, leafy to the heads. Leaves dark green, lanceolate, acuminate, deeply pinnatifid, 3'–6' long, or the lowest larger, decurrent on the stem and branches, the lobes triangular-lanceolate, tipped with stout prickles, the margins and decurrent bases bristly, the upper surface strigose-pubescent or hispid, the lower brown-tomentose and midnerve pilose, especially when young; heads mostly solitary at the ends of the branches, 1½'–2' broad, 1½'–2' high; bracts of the involucre cottony, narrowly lanceolate, acuminate, all tipped with slender, erect or ascending prickles; flowers dark purple.

In fields and waste places, Newfoundland to Georgia, Minnesota, Nebraska, Oregon and California. Naturalized from Europe. Native also of Asia. Plume-, bank- or horse-thistle, bell-, bird-, blue-, button-, boar-, bull- or roadside-thistle. July–Nov.

2. Cirsium altíssimum (L.) Spreng. Tall or Roadside Thistle. Fig. 4637.

Carduus altissimus L. Sp. Pl. 824. 1753.
Cnicus altissimus Willd. Sp. Pl. 3: 1671. 1804.
Cirsium altissimum Spreng. Syst. 3: 373. 1826.

Biennial or perennial; roots often thickened; stem pubescent or tomentose, stout, branched, leafy to the heads, 3°–10° high. Leaves ovate-oblong or oblong-lanceolate, sessile or slightly clasping, sparingly pubescent above, densely white-tomentose beneath, scarcely or not at all decurrent, acute, spinulose-margined, entire, dentate with bristle-pointed teeth or lobed, sometimes pinnatifid into oblong or triangular-lanceolate segments, the lowest sometimes 8' long, narrowed into margined petioles, the uppermost linear or lanceolate, much smaller; heads about 2' broad, 1½'–2' high, mostly solitary at the ends of the branches; outer bracts of the involucre ovate or ovate-lanceolate, firm with a dark, slightly glandular spot or band on the beak, tipped with short prickles, the inner linear-lanceolate, acuminate, unarmed; flowers light purple.

In fields and thickets, Massachusetts to Ohio, Minnesota, Florida, Nebraska and Texas. Aug.–Sept.

Cirsium iowénse Pammel, with slightly larger heads and longer-tipped inner involucral bracts, appears to be a northwestern race of this species.

3. Cirsium díscolor (Muhl.) Spreng. Field Thistle. Fig. 4638.

Cnicus discolor Muhl.; Willd. Sp. Pl. 3: 1670. 1804.

Carduus discolor Nutt. Gen. 2: 130. 1818.

Cirsium discolor Spreng. Syst. 3: 373. 1826.

Similar to the preceding species, but lower and more leafy, seldom over 7° high. Leaves deeply pinnatifid into linear, linear-lanceolate or falcate, prickly toothed segments, white tomentose beneath, sessile. the basal ones sometimes 12' long; heads 1½'–2' broad, about 1½' high, usually involucrate by the upper leaves, mostly solitary at the ends of the branches; outer bracts of the involucre coriaceous, ovate, slightly woolly, tipped with slender bristles, which are longer than those of the preceding species; inner bracts lanceolate, acuminate, unarmed; flowers light purple or pink, rarely white.

In fields and along roadsides, New Brunswick to Ontario, Georgia, Minnesota, Nebraska and Missouri. July–Nov.

4. Cirsium virginiànum (L.) Michx. Virginia Thistle. Fig. 4639.

Carduus virginianus L. Sp. Pl. 824. 1753.
Cirsium virginianum Michx. Fl. Bor. Am. **2** : 90.   1803.
Cnicus virginianus Pursh, Fl. Am. Sept. 506.   1814.

Biennial; stem slender, naked or scaly above, pubescent or somewhat tomentose, simple or branched, 2°–3½° high. Leaves oblong, oblong-lanceolate, or the lowest slightly spatulate, sessile, or somewhat clasping, not decurrent, acute or acutish, spinulose-margined, entire, lobed or pinnatifid into triangular-lanceolate lobes, the lower sometimes 8' long and 2' wide, narrowed into margined petioles, all pubescent or glabrate above, and densely white-tomentose beneath; heads long-peduncled, 1'–1½' broad, about 1' high; outer bracts of the involucre not coriaceous, lanceolate or ovate-lanceolate, tipped with weak short bristles, the inner ones very narrow and merely acuminate; flowers purple.

In dry woods and thickets, Virginia to Kentucky, Ohio, Florida and Texas. April–Sept.

5. Cirsium Pítcheri (Torr.) T. & G.   Pitcher's Thistle.   Fig. 4640.

Cnicus Pitcheri Torr.; Eaton, Man. Ed. 5, 180.   1829.
Cirsium Pitcheri T. & G. Fl. N. A. **2** : 456.   1843.
Carduus Pitcheri Porter, Mem. Torr. Club **5** : 345.   1894.

Biennial, persistently white-tomentose throughout; stem stout, leafy up to the heads, usually branched, 1°–2° high. Leaves sessile, partly clasping or slightly decurrent, pinnately divided into narrowly linear, entire lobed or pinnatifid, acute sparingly prickly segments, 2''–3'' wide, with revolute margins; basal leaves often 12' long; heads solitary or several and racemose-spicate at the ends of the branches, about 1½' broad; outer bracts of the involucre ovate-lanceolate, sparingly pubescent and tomentose-ciliate, glutinous on the back, tipped with short spreading bristles, the inner narrowly lanceolate, acuminate or sometimes tipped with weak prickles; flowers cream color.

Shores of Lakes Michigan, Huron and Superior. June–Aug.

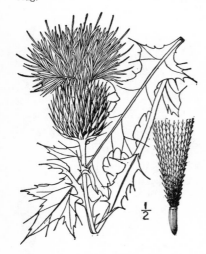

6. Cirsium undulàtum (Nutt.) Spreng. Wavy-leaved Thistle.   Fig. 4641.

Carduus undulatus Nutt. Gen. **2** : 130.   1818.
Cirsium undulatum Spreng. Syst. 3 : 374.   1826.
Cnicus undulatus A. Gray, Proc. Am. Acad. **10** : 42.   1874.
Cnicus undulatus var. megacephalus A. Gray, Proc. Am. Acad. **10** : 42.   1874.
Carduus undulatus megacephalus Porter, Mem. Torr. Club **5** : 345.   1894.

Biennial, persistently and densely white-tomentose throughout, or the upper surfaces of the leaves at length green and glabrous; stem stout, leafy, usually branched, 1°–3° high. Leaves lanceolate or oblong-lanceolate in outline, acute, sessile or decurrent, or the lowest petioled, undulate, lobed or pinnatifid, the lobes dentate, triangular, often very prickly; basal leaves often 8' long; heads 1½'–3' broad, and nearly as high, solitary at the ends of the branches; outer bracts of the involucre ovate or ovate-lanceolate, firm, glutinous on the back, tipped with short spreading prickles, the inner ones lanceolate, acuminate; flowers purple or pink.

On plains and prairies, Lake Huron to Assiniboia, Alberta, Kansas, New Mexico and Arizona. June–Sept.

7. Cirsium platténse (Rydb.) Britton.
Prairie Thistle. Fig. 4642.

Carduus plattensis Rydberg, Contr. Nat. Herb. **3** : 167.
pl. 2. 1895.

Perennial or biennial, the root thick and deep;
stem stout, simple, or little branched, $1\frac{1}{2}°-2\frac{1}{2}°$ tall,
densely white-felted. Leaves deeply pinnatifid,
white-tomentose beneath, green, loosely tomentose, or glabrate above, the lower 5'-7' long, the
lobes lanceolate to oblong, acute, prickly tipped
and margined; upper leaves smaller and less divided; heads few, about 2' high and broad; outer
bracts of the involucre lanceolate to ovate-lanceolate, firm, dark, tipped with a short weak spreading prickle, the inner linear-lanceolate, unarmed,
tipped with a scarious reflexed erose appendage;
corolla yellow, its lobes linear; pappus of outer
flowers merely barbellate.

Sand hills, Nebraska, Colorado and South Dakota.
May-July.

8. Cirsium Flódmani (Rydb.) Britton. Flod-
man's Thistle. Fig. 4643.

Carduus Flodmani Rydb. Mem. N. Y. Bot. Gard. **1** : 451.
1900.

Stem rather slender, $1\frac{1}{2}°-3°$ tall, loosely white-cottony,
usually more or less branched. Leaves deeply pinnatifid
into linear-oblong or lanceolate, acute or acuminate,
toothed or entire segments, floccose and green above,
densely white-cottony beneath, the lower 6' long or less;
heads $1\frac{1}{2}'-2'$ broad; involucre campanulate, its linear
bracts tipped with yellow prickles; flowers reddish-
purple to rose.

Meadows and river bottoms, Iowa and North Dakota to
Saskatchewan, Nebraska and Colorado. Has been referred
to the western *C. canescens.* July-Sept.

9. Cirsium ochrocéntrum A. Gray.
Yellow-spined Thistle. Fig. 4644.

Cirsium ochrocentrum A. Gray, Mem. Am. Acad.
1 : 110. 1849.
Cnicus ochrocentrus A. Gray, Proc. Am. Acad. **19** :
57. 1883.
Carduus ochrocentrus Greene, Proc. Phil. Acad.
1892 : 336. 1893.

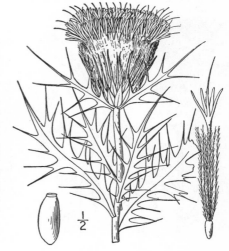

Similar to *Cirsium undulatum,* but commonly
taller and more leafy, often 6° high, equally
white-tomentose. Leaves oblong-lanceolate in
outline, usually very deeply pinnatifid into triangular-lanceolate, serrate or entire segments,
armed with numerous long yellow prickles;
lower leaves often 6'-8' long; heads about 2'
broad, $1\frac{1}{2}'-2'$ high, solitary at the ends of the
branches; outer bracts of the involucre lanceolate; tipped with stout yellow prickles of
nearly or quite their own length, the inner narrowly lanceolate, long-acuminate; flowers purple (rarely white?).

On plains, Nebraska to Texas, Nevada and Arizona. May-Sept.

10. Cirsium nebraskénse Britton. Nebraska Thistle. Fig. 4645.

Carduus nebraskensis Britton, in Britt. & Brown, Ill. Fl. 3 : 487. 1898.

Stem densely white-woolly, apparently over 1° high. Leaves linear-oblong to lanceolate, white-woolly beneath, green and sparingly loosely woolly above, irregularly slightly toothed or entire, the upper 3′–6′ long, $\frac{1}{4}$′–1′ wide, the margins prickly; heads solitary, or few, short-peduncled, about 1$\frac{1}{2}$′ high; outer bracts of the involucre lanceolate, prickle-tipped, the inner narrower with a reflexed acute scarious appendage; pappus bristles of inner flowers plumose, of the outer barbellate.

Western Nebraska and Wyoming. Summer.

11. Cirsium odoràtum (Muhl.) Britton. Pasture Thistle. Fragrant Thistle. Fig. 4646.

Cnicus odoratus Muhl. Cat. 70. 1813.
Carduus pumilus Nutt. Gen. **2**: 130. 1818.
Cnicus pumilus Torr. Compend. 282. 1826.
Carduus odoratus Porter, Mem. Torr. Club **5**: 345. 1894.

Biennial, more or less villous-pubescent; stem stout, simple or branched, leafy, 1°–3° high; roots thick, branched, solid; stem leaves green both sides, sessile and clasping, oblong or oblong-lanceolate in outline, acute, 3′–7′ long, 1′–2′ wide, pinnatifid into triangular, acute, dentate, prickly lobes. Basal leaves petioled; heads solitary, terminal, 2′–3′ broad, about 2′ high, often involucrate by the upper leaves; outer bracts lanceolate or ovate-lanceolate, with a slight glutinous strip on the back, glabrous or sparingly tomentose, tipped with slender prickles, the inner narrow, long-acuminate; flowers purple, rarely white, fragrant; tips of the pappus bristles usually spatulate.

In fields, Maine to Pennsylvania, Delaware and West Virginia. July–Sept.

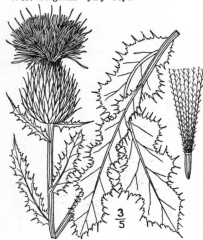

12. Cirsium Híllii (Canby) Fernald. Hill's Thistle. Fig. 4647.

Cnicus Hillii Canby, Gard. & For. **4**: 101. 1891.
Carduus Hillii Porter, Mem. Torr. Club **5**: 344. 1894.
Cirsium Hillii Fernald, Rhodora **10**: 95. 1908.

Perennial, low, villous-pubescent or somewhat woolly; stem leafy, simple or branched, 1°–2° high; root perpendicular, fusiform, slender and hollow above, enlarged below, 8′–12′ long; leaves green both sides, mostly obtuse, lobed or pinnatifid, the lobes mostly broad and rounded, dentate, spinulose or with some rather stout prickles, the upper oblong, sessile and clasping, the lower spatulate-oblong, narrowed at the base or the lowest ones petioled and 6′–8′ long; heads 2′–3′ broad, about 2′ high; outer bracts of the involucre ovate-lanceolate, tipped with short bristles, conspicuously glutinous on the back, the inner narrowly lanceolate, long-acuminate; flowers purple; pappus bristles slender-pointed or some of them slightly spatulate.

In fields, western Ontario to Minnesota, south to Pennsylvania, Illinois and Iowa. June–July.

13. Cirsium horridulum Michx. Yellow Thistle. Fig. 4648.

Carduus spinosissimus Walt. Fl. Car. 194. 1788. Not *Cirsium spinosissimum* (L.) Scop.
Cirsium horridulum Michx. Fl. Bor. Am. **2**: 90. 1803.
Cnicus horridulus Pursh, Fl. Am. Sept. 507. 1814.

Biennial or perennial, somewhat woolly when young, but becoming glabrate; stem branched, leafy, 2°–5° high. Leaves green both sides, lanceolate or oblong in outline, sessile and clasping or the basal ones short-petioled and somewhat spatulate, pinnatifid into triangular or broader, spinulose-margined and prickle-tipped, entire or dentate lobes; heads involucrate by the upper leaves, 2'–4' broad, 1½'–2½' high; bracts of the involucre narrowly lanceolate, roughish and ciliate, long-acuminate, unarmed; flowers pale yellow, yellowish, or occasionally purple.

In moist or dry sandy soil, Maine to Pennsylvania, Florida and Texas. Abundant along the edges of salt-meadows in New York and New Jersey. May–Aug., or earlier in the South.

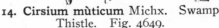

14. Cirsium mùticum Michx. Swamp Thistle. Fig. 4649.

Cirsium muticum Michx. Fl. Bor. Am. **2**: 89. 1803.
Carduus muticus Pers. Syn. **2**: 386. 1807.
Cnicus muticus Pursh, Fl. Am. Sept. 506. 1814.
Carduus muticus subpinnatifidus Britton, in Britt. & Brown, Ill. Fl. **3**: 489. 1898.
Cnicus muticus alpicola Fernald, Ott. Nat. **1905**: 166.

Biennial; stem woolly or villous when young, becoming glabrate, slender, striate, leafy, paniculately branched above, 3°–8° high. Leaves densely white-tomentose beneath when young, sometimes becoming glabrous on both sides, deeply pinnatifid into lanceolate or oblong, entire, lobed or dentate, spiny segments usually tipped with slender prickles, or sometimes merely lobed; basal leaves petioled, 4'–8' long, those of the stem sessile and smaller; heads about 1½' broad and high, solitary, terminal, naked-peduncled, or with a few small bract-like leaves near the base; outer bracts viscid, appressed, more or less cottony, ovate or ovate-lanceolate, the inner linear-lanceolate, acute, all unarmed; flowers purple.

In swamps and moist soil, Newfoundland to Florida, Saskatchewan and Texas. July–Oct.

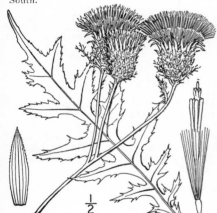

15. Cirsium arvénse (L.) Scop. Canada Thistle. Creeping Thistle. Fig. 4650.

Serratula arvensis L. Sp. Pl. 820. 1753.
Cirsium arvense Scop. Fl. Carn. Ed. 2, **2**: 126. 1772.
Carduus arvensis Robs. Brit. Fl. 163. 1777.
Cnicus arvensis Hoffm. Deutsch. Fl. Ed. 2, **1**: Part. 2, 130. 1804.

Perennial by horizontal rootstocks, forming patches, nearly glabrous, or the leaves sometimes woolly beneath; stems striate, 1°–3° high, branched above. Leaves sessile, slightly clasping, but not decurrent, lanceolate or oblong-lanceolate, deeply pinnatifid into very prickly, lobed or dentate segments, or sometimes nearly or quite entire; basal leaves sometimes petioled, 5'–8' long; heads numerous, corymbose, dioecious, 1' broad or less, nearly 1' high, purple or white, staminate heads globose, corollas projecting; pistillate heads oblong-campanulate, corollas shorter, the long pappus conspicuous; outer bracts ovate or ovate-lanceolate, appressed, tipped with short prickly points, inner bracts of the pistillate heads linear, elongated.

In fields and waste places, Newfoundland to Virginia, British Columbia, Nebraska and Utah. In many places a pernicious weed. Races differ in leaf-form and in pubescence. Naturalized from Europe. Way- or cursed thistle. Corn-, hard- or prickly-thistle. June–Sept.

16. Cirsium palústre (L.) Scop.   Marsh Thistle.
Fig. 4651.

Carduus palustris L. Sp. Pl. 822.   1753.

Cirsium palustre Scop. Fl. Carn. Ed. 2, **2** : 128.   1772.

Annual or biennial; stem little branched, 4°–5° high,
loosely floccose or glabrate and covered by the decurrent prickly margins of the leaves.   Leaves pinnatifid,
the lower often 6′–8′ long, linear-oblong in outline, the
segments lobed, loosely floccose beneath, spinulose;
heads usually many, rather less than 1′ broad, densely
clustered, short-peduncled, the involucre ovoid, its bracts
with very short, prickly tips.

Woodlands, East Andover, New Hampshire, recorded as
thoroughly naturalized.   Native of Europe and northern
Asia.   Summer.

Cirsium cànum (L.) Bieb., with larger, long-peduncled
heads, the decurrent leaf-bases merely ciliate, is recorded as
established in Massachusetts.   Adventive from Europe.

105.   CARDUUS [Vaill.] L. Sp. Pl. 820.   1753.

Herbs resembling *Cirsium* in habit, usually annual or biennial, the leaves decurrent on
the stem and branches as spiny wings, the heads often nodding.   Involucre ovoid to globose,
many-flowered, its bracts narrow, in many series.   Receptacle copiously bristly, flat or convex.
Corolla-tube slender, the limb deeply 5-cleft.   Filaments papillose-pubescent.   Anthers sagittate at the base and with slender auricular appendages.   Style-branches obtuse.   Achenes
mostly obovoid, sometimes angled or ribbed, glabrous.   Pappus of many naked or merely
roughened bristles.   [Ancient Latin name of these plants.]

About 80 species, natives of the Old World.   Type species : *Carduus nutans* L.

Heads solitary at end of stem or branches, nodding.   1. *C. nutans.*
Heads usually several, crowded at ends of winged branches.   2. *C. crispus.*

1. Carduus nùtans L.   Musk Thistle.   Plumeless Thistle.   Fig. 4652.

Carduus nutans L. Sp. Pl. 821.   1753.

Biennial, branched, sparingly tomentose, 2°–3°
high.   Leaves lanceolate in outline, deeply pinnatifid,
acuminate, 3′–6′ long, the lobes triangular, very
prickly; heads long-peduncled, solitary at the end
of the stem or branches, 1½′–2½′ broad, nodding, purple, rarely white, fragrant; involucre hemispheric, its
bracts in many series, lanceolate, long-acuminate, the
prominent mid-nerve prolonged into a prickle, or the
inner nerveless and awned; pappus bristles 10″–1′
long, white, very minutely barbed.

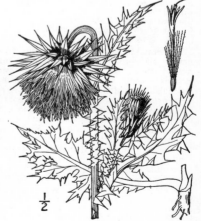

In waste places, District of Columbia, Pennsylvania
and New Jersey to New Brunswick, and in ballast about
the seaports.   Naturalized or adventive from Europe.
Native also of Asia.   Bank- or buck-thistle.   Queen Ann's-
thistle.   July–Oct.

2. Carduus críspus L. Curled Thistle. Welted Thistle. Fig. 4653.

Carduus crispus L. Sp. Pl. 821. 1753.

Biennial, somewhat tomentose; stem much branched, densely prickly, 2°–4° high. Leaves lanceolate in outline, with undulate and ciliate-spiny margins, all sinuate-pinnatifid into broad, 3-lobed, toothed segments, the teeth prickle-pointed; heads several, usually crowded at the ends of the winged branches, 1′ broad or less, purple or white, sessile or short-peduncled, or some of them rarely solitary and slender-peduncled; involucre ovoid, its bracts very numerous, linear, the outer prickle-tipped and rigid, the inner thinner and merely acuminate.

In fields and waste places, New Brunswick, Quebec, Nova Scotia and Pennsylvania, and in ballast about the seaports. Adventive from Europe. Native also of Asia. July–Sept.

Carduus acanthoïdes L., which resembles this species, but has larger hemispheric involucres, their outer bracts not rigid, has been collected in ballast on waste grounds about Atlantic seaports, and several other species of *Carduus* have been found in ballast deposits.

106. MARIÀNA Hill, Veg. Syst. 4: 19. 1762.

[SILYBUM Vaill.; Adans. Fam. Pl. 2: 116. 1763.]

Annual or biennial, simple or branched, nearly glabrous herbs, with large alternate clasping, sinuate-lobed or pinnatifid, white-blotched leaves, and large discoid heads of purple tubular flowers, solitary at the end of the stem or branches. Involucre broad, subglobose, its bracts rigid, imbricated in many series, the lower ones fimbriate-spinulose at the broad triangular summit, the middle ones similar but armed with huge spreading or recurved spines, the inner lanceolate, acuminate. Receptacle flat, densely bristly. Corolla-tube slender, the limb expanded and deeply 5-cleft. Filaments monadelphous below, glabrous. Anthers sagittate at the base. Style nearly entire. Achenes obovate-oblong, compressed, glabrous, surmounted by a papillose ring. Pappus bristles in several series, flattish, barbellate or scabrous. [St. Mary's thistle.]

A montypic genus of the Mediterranean region.

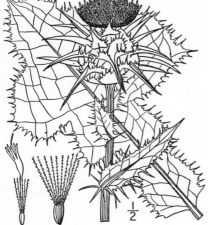

1. Mariana mariàna (L.) Hill. Milk Thistle. Fig. 4654.

Carduus marianus L. Sp. Pl. 823. 1753.
Mariana mariana Hill, Hort. Kew. 61. 1769.
Silybum marianum Gaertn. Fruct. & Sem. 2: 378. 1802.

Stem striate, glabrous or slightly woolly, little branched, 2°–4° high. Leaves oblong-lanceolate, prickly, strongly clasping, the lower often 12′ long and 6′ wide, the upper much smaller, scarcely lobed, acute; heads about 2½′ broad; spines of the middle involucral bracts often 1½′ long; pappus bristles white, barbellate.

Escaped from gardens near Kensington, Ontario (T. Walker, according to Macoun), in ballast and waste grounds about the eastern seaports, south to Alabama, and on the Pacific Coast from British Columbia to southern California, where it is naturalized. Virgin Mary's-thistle, lady's-milk, holy thistle. June–Aug.

107. ONOPÓRDON [Vaill.] L. Sp. Pl. 827. 1753.

Coarse, branching or rarely acaulescent, tomentose herbs, with stout stems winged by the decurrent bases of the alternate dentate or pinnatifid, prickly leaves, and large discoid heads of purple violet or white flowers, mostly solitary at the ends of the branches. Involucre nearly globular, its bracts imbricated in many series, all tipped with long spines in our species, the inner narrower than the outer. Receptacle flat, fleshy, honeycombed, not bristly.

Corolla-tube slender, the limb expanded and deeply 5-cleft. Filaments pilose. Anthers sagittate at the base. Achenes obovate or oblong, 4-angled or compressed, smooth or corrugated. Pappus bristles in several series, filiform, barbellate or plumose, united at the base. [Greek, Asses' thistle, the ancient name.]

About 12 species, natives of the Old World, the following typical.

1. Onopordon Acánthium L. Cotton Thistle. Scotch Thistle. Fig. 4655.

Onopordon Acanthium L. Sp. Pl. 827. 1753.

Biennial, white-tomentose all over; stem usually much branched, leafy, 3°–9° high. Leaves oblong, lobed and dentate, acute, very spiny, the lower often 12′ long; heads 1½′–2′ broad, about 1½′ high, solitary at the ends of the branches; outer bracts of the involucre ovate or oblong, minutely serrulate, tipped with long stout spreading spines; flowers pale purple; achenes slightly corrugated; pappus bristles brownish, longer than the achene.

In waste places, Nova Scotia and Ontario to New Jersey, Pennsylvania and Michigan. Naturalized from Europe. Native also of Asia. Argentine. Asses', oat or down-thistle. Queen Mary's-, silver- or musk-thistle. July–Sept.

108. CENTAURÈA L. Sp. Pl. 909. 1753.

Perennial or annual herbs, with alternate entire dentate incised or pinnatifid leaves, and large or middle-sized heads, of tubular purple violet white or rarely yellow flowers. Involucre ovoid or globose, its bracts imbricated in many series, appressed, fimbrillate, or dentate. Receptacle flat, densely bristly. Marginal flowers usually neutral and larger than the central ones, which are perfect and fertile, or flowers all perfect and fertile in some species. Corolla-tube slender, the limb regular or oblique, 5-cleft or 5-lobed, the segments sometimes appearing like rays. Anthers sagittate at the base. Style-branches short, somewhat connate, obtuse. Achenes oblong or obovoid, compressed or obtusely 4-angled, usually smooth and shining, obliquely or laterally attached to the receptacle, surmounted by a disk with an elevated margin. Pappus of several series of bristles or scales, rarely none. [Greek, of the Centaurs, who were said to use it in healing.]

About 350 species, mostly natives of the Old World. Type species: *Centaurea Centaurium* L.

Bracts of the involucre lacerate or fimbriate, not spiny.
　　Heads 2′ broad or less; achenes laterally attached; introduced species.
　　　　Annual; leaves entire. ....　1. *C. Cyanus.*
　　　　Perennials or biennials, or *C. maculosa* annual.
　　　　　　Bracts of the involucre laciniate or entire. ....　2. *C. Jacea.*
　　　　　　Bracts of the involucre, or their tips, pectinate-fringed.
　　　　　　　　Lower bracts of the involucre pectinate-fringed to below the middle.
　　　　　　　　　　Leaves entire or merely dentate. ....　3. *C. nigra.*
　　　　　　　　　　Leaves pinnatifid. ....　4. *C. Scabiosa.*
　　　　　　　　Lower bracts of the involucre pectinate-fringed only at the tips.
　　　　　　　　　　Leaves entire, toothed, or the lower lyrate. ....　5. *C. vochinensis.*
　　　　　　　　　　All but the upper leaves pinnatifid into linear segments.　6. *C. maculosa.*
　　Heads 2′–4′ broad; achenes obliquely attached; native western species.　7. *C. americana.*
Bracts of the involucre tipped with stout spines.
　　Flowers purple; stem wingless. ....　8. *C. Calcitrapa.*
　　Flowers yellow; stem winged by the decurrent leaf-bases.
　　　　Spines slender, purplish, 5″ long or less, branched below. ....　9. *C. melitensis.*
　　　　Spines stout, yellow, 6″–10″ long, with smaller ones at the base.　10. *C. solstitialis.*

1. Centaurea Cyanus L. Blue-bottle.
Corn Blue-bottle. Corn-flower.
Fig. 4656.

Cen'aurea Cyanus L. Sp. Pl. 911. 1753.

Annual, woolly, at least when young; stem leafy, slender, branched, $1°-2\frac{1}{2}°$ high, the branches ascending. Leaves linear or linear-lanceolate, mucronate, $3'-6'$ long, the basal and lower ones mostly remotely dentate, the upper, or sometimes all of them, entire; heads $1'-1\frac{1}{2}'$ broad, on long naked peduncles; involucre campanulate, its bracts greenish-yellow, or the inner with darker tips and margins, appressed, fimbriate with scarious teeth; flowers blue, purplish, pink or white, the marginal ones neutral with large radiant corolla-limbs; achenes slightly compressed, or 4-angled; pappus bristles unequal, nearly as long as the achene.

In waste places, escaped from gardens, and in ballast, Quebec to Ontario, western New York, Nebraska and Virginia. Witches'-bells or -thimbles. corn-centaury, corn-bottle or -binks. Brushes. Hurt-sickle. Blue-bonnets. Blaver. Blue poppy. Bachelor's-buttons. Blue caps. Barbeau. French pink. July–Sept.

2. Centaurea Jàcea L. Brown or Rayed Knapweed. Fig. 4657.

Centaurea Jacea L. Sp. Pl. 914. 1753.

Perennial, $2°$ high or less. Leaves entire or denticulate, rarely lobed; heads $1'-1\frac{1}{2}'$ broad; involucre globular to ovoid, its bracts closely imbricated, brown or with brown backs, the outer ones pale brown with fimbriate appendages, the middle ones lacerate, the inner entire or nearly so, dark brown; marginal flowers neutral with enlarged radiant corolla-limbs; achenes obscurely 4-sided; pappus none, or a minute crown.

In waste places, northern New York, Vermont and Massachusetts, and in ballast about eastern seaports. Also in British Columbia. Fugitive from Europe. June–Sept.

3. Centaurea nìgra L. Black Knapweed.
Horse-knops. Hardheads. Centaury.
Fig. 4658.

Cen aurea nigra L. Sp. Pl. 911. 1753.

Perennial, scabrous or pubescent; stem stiff, branched, $1°-2°$ high. Lower and basal leaves spatulate or oblong, acutish, entire, denticulate, dentate or lobed, not pinnatifid, $3'-6'$ long, narrowed into long petioles; upper leaves oblong or lanceolate, sessile, or partly clasping, entire or nearly so; heads rarely $1'$ broad, bracted by the small, uppermost leaves; involucre globose, its bracts lanceolate or ovate-lanceolate, closely imbricated, tipped with brown fimbriate appendages, or the uppermost merely lacerate; flowers rose-purple, all perfect, the marginal ones usually not at all enlarged, or sometimes radiant; achenes slightly 4-sided; pappus none, or a ring of minute scales.

In waste places and fields, Newfoundland to Ontario, New Jersey and Pennsylvania. Naturalized from Europe. Among many other English names are iron-head or -weed, club-weed, matfelon, hurt-sickle, tassel, horse-knobs; crop-, knob-, bole- or button-weed; loggerheads, lady's-cushion, blue-tops, hard-weed, bullweed, sweeps, bachelor's-buttons. July–Sept.

4. Centaurea Scabiosa L. Scabious Knapweed. Greater Centaury. Fig. 4659.

C. Scabiosa L. Sp. Pl. 913. 1753.

Slightly pubescent or villous, perennial; stem simple or branched, about 2° high. Leaves all pinnatifid, the lower and basal ones petioled, often 6' long, the upper sessile and much smaller; heads about 2' broad, on bracted peduncles 1'–4' long; involucre ovoid, its bracts all pectinate to or below the middle, and blackish-margined, the outer ovate, the inner oblong; corolla purple, the outer ones enlarged and neutral; pappus of stiff bristles, about as long as the achene.

Waste grounds and fields, Quebec and Ontario to Ohio. Naturalized from Europe. July–Sept.

5. Centaurea vochinensis Bernh. Tyrol Knapweed. Fig. 4660.

C. vochinensis Bernh.; Reichenb. Icon. Fl. Germ. **15**: 15. 1853.

Perennial, roughish, branched, 2° high or less. Leaves firm in texture, the basal and lower ones dentate or lyrate, 3'–5' long, the upper lanceolate to oblong-lanceolate, few-toothed or entire; heads about 1½' broad; involucre ovoid-cylindric, about 8" high, its faintly nerved bracts pectinate at the brownish tip, the lower short, ovate, the upper elongated; flowers rose-purple, the margined ones radiant.

Waste grounds, Ontario to Massachusetts and southern New York. Naturalized from Europe. Aug.–Oct.

6. Centaurea maculòsa Lam. Spotted Knapweed. Fig. 4661.

C. maculosa Lam. Encycl. **1**: 669. 1783.

Annual or biennial, loosely floccose-pubescent or glabrate, usually much branched, 2°–3° high, the stiff branches ascending. Leaves pinnatifid into linear segments, or the upper linear and entire, the lower up to 3' long; heads peduncled, about 10" broad; involucre ovoid, its ribbed bracts pectinate only at the black tip, the inner ones longer than the outer and entire or merely erose; corollas white to purple, the marginal ones radiant.

Waste grounds, Massachusetts to New Jersey and Pennsylvania. July–Aug.

7. Centaurea americàna Nutt.   American Star Thistle.   Fig. 4662.

C. americana Nutt. Journ. Phila. Acad. **2** : 117.   1821.

Plectocephalus americanus Don, in Sweet, Brit. Fl. Gard. (II) *pl. 51.*   1831.

Annual, roughish; stem stout, simple, or little branched, 2°–6° high.   Leaves entire or denticulate, the lower and basal ones spatulate or oblong, 2′–5′ long, narrowed into petioles, the upper oblong-lanceolate, sessile, mucronate; heads solitary at the much thickened ends of the leafy stem or branches, very showy, 2′–4′ broad; involucre nearly hemispheric, its bracts ovate or lanceolate with conspicuously pectinate appendages; flowers pink or purple, the marginal ones with enlarged and radiant corolla-limbs; achenes somewhat compressed, obliquely attached at the base; pappus of copious unequal bristles longer than the achene.

Dry plains, Missouri and Arkansas to Louisiana, Mexico and Arizona.   May–Aug.

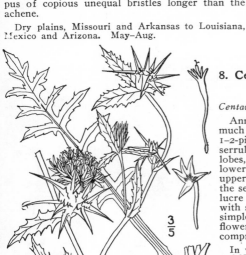

8. Centaurea Calcítrapa L.   Star Thistle.   Fig. 4663.

Centaurea Calcitrapa L. Sp. Pl. 917.   1753.

Annual, pubescent or glabrous, green; stem much branched, not winged, 1°–1½° high.   Leaves 1–2-pinnatifid into oblong-lanceolate to linear, serrulate-spinulose, dentate or entire mostly acute lobes, the upper sessile and slightly clasping, the lower and basal short-petioled, 4′–7′ long, the uppermost somewhat involucrate at the bases of the sessile heads which are about 1′ broad; involucre ovoid, its outer bracts ovate-oblong, tipped with stout, spreading, yellowish spines which are simple, or commonly with 2–6 bristles at the base; flowers purple, none of them radiant; achenes compressed or obscurely 4-sided; pappus none.

In waste places and ballast, southern New York and New Jersey to Virginia.   Also from British Columbia to California.   Adventive or naturalized from Europe.   Called also caltrops, maize- or mouse-thorn.   Knop-weed.   June–Oct.

9. Centaurea meliténsis L.   Rayless Winged Centaury.   Fig. 4664.

Centaurea melitensis L. Sp. Pl. 917.   1753.

Annual, 1°–4° high, grayish-pubescent, much branched, the stem and branches narrowly winged by the decurrent leaf-bases.   Basal leaves lyrate, their lobes obtuse; stem leaves few-lobed or entire, the upper ones 1′ long or less; heads sessile or nearly so; involucre about ½′ thick, its principal bracts tipped by a slender purplish divergent spine 5″ long or less, which is often branched below and with smaller spines at its base; flowers yellow, none of them radiant; pappus scales unequal.

Waste and cultivated grounds, Georgia to Missouri, Arizona, California and Oregon, and in ballast about the Atlantic seaports.   Naturalized or adventive from Europe.   Widely naturalized in South America.   April–Sept.

$\frac{1}{2}$

10. Centaurea solstitiàlis L. Barnaby's Thistle. Fig. 4665.

Centaurea solstitialis L. Sp. Pl. 917. 1753.

Annual, cottony-pubescent, branched, 1°-2° high, the stem and branches winged by the decurrent leaf-bases. Basal leaves pinnatifid, often 6′ long; stem leaves lanceolate to linear, mostly entire, the upper ½′-1′ long; involucre ovoid-globose, about ½′ thick, its principal bracts tipped by a stout, spreading or reflexed yellow spine, 6″-10″ long, with several much smaller ones at its base; flowers yellow, none radiant.

Waste and cultivated grounds, Massachusetts to New York, Pennsylvania, Ontario and Utah, and in California. Adventive from Europe. July–Sept.

109. CNICUS [Tourn.] L. Sp. Pl. 826. 1753.

An annual herb, with alternate pinnatifid or sinuate-dentate leaves, the lobes or teeth spiny, and large sessile heads of yellow tubular flowers, solitary at the ends of the branches, subtended by the upper leaves. Bracts of the involucre imbricated in several series, the outer ovate, the inner lanceolate, tipped by long pinnately branched spines. Receptacle flat, bristly. Achenes terete, striate, laterally attached, the horny margin 10-toothed at the summit; pappus of 2 series of awns, the inner fimbriate, the outer longer, naked; anther-appendages elongated, united to their tips. [Latin name of Safflower, early applied to thistles.]

A monotypic genus of the Old World.

1. Cnicus benedíctus L. Blessed Thistle. Our Lady's Thistle. Fig. 4666.

Cnicus benedictus L. Sp. Pl. 826. 1753.
Centaurea benedic a L. Sp. Pl. Ed. 2, 1296. 1763.

Hirsute or pubescent, much branched, seldom over 2½° high. Leaves oblong-lanceolate in outline, rather thin, reticulate-veined, 3′-6′ long, the upper clasping, the basal and lower ones narrowed at the base and petioled; heads about 2′ broad, subtended by several large lanceolate to ovate-lanceolate leaves; bristles of the receptacle soft, long; outer awns of the pappus alternating with the inner.

In waste places, Nova Scotia and New Brunswick to Maryland, Pennsylvania, Alabama and Michigan, and on the Pacific Coast. Adventive from southern Europe. Holy- or bitter thistle. Sweet-sultan. St. Benedict's-thistle. May–Aug.

Echinops sphaerocéphalus L., a tall spinose plant, with large clusters of 1-flowered involucres, the flowers white or bluish, occasionally escapes from cultivation into waste grounds. It is native of Europe.

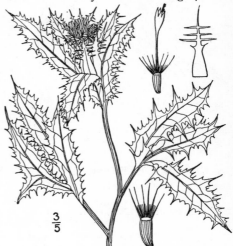

$\frac{3}{5}$

SUMMARY.

	Families.	Genera.	Species Figured.
PTERIDOPHYTA	11	31	130
SPERMATOPHYTA			
Gymnospermae	2	10	28
Angiospermae			
Monocotyledones	32	251	1261
Dicotyledones			
Choripetalae	103	526	1775
Gamopetalae	46	411	1472
TOTAL	194	1229	4666

General Index of Latin Genera and Species

[Classes, Families and Tribes in SMALL CAPITALS; genera in **heavy face**; synonyms in *italics*.
Heavy face figures indicate the volume; other figures, the page. Varietal synonyms are indented.

37

English Index, including Popular Plant Names.

[The heavy face figures **1, 2, 3** indicate the volume; those following them, the page. A few popular names, not printed in the text, are referred to the proper plant by the *number* of the *Illustration* or *Figure* (fig.) in the Index.]